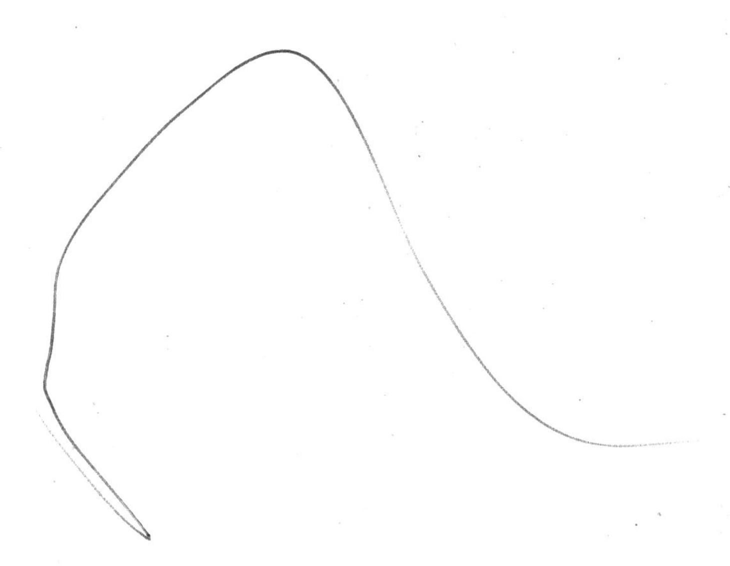

INTEGRATED MATHEMATICS

INTRODUCTORY COURSE

MARILYN OCCHIOGROSSO
Former Assistant Principal, Mathematics
Erasmus Hall High School • Brooklyn, New York

MICHAEL DAVID EPSTEIN
Mathematics Teacher
Elmont Memorial High School • Elmont, New York

STEVEN ADRIAN
Mathematics Teacher
Mahopac High School • Mahopac, New York

ANN ARMSTRONG
Mathematics Teacher, Retired
Schalmont High School • Schenectady, New York

PAUL O. ECKHARDT
Chairman, Department of Mathematics
Carmel High School • Carmel, New York

MARY C. GENIER
Coordinator of Mathematics, Retired
Rotterdam Mohonasen H. S. • Schenectady, New York

AMSCO SCHOOL PUBLICATIONS, INC.
315 Hudson Street / New York, N.Y. 10013

Consultants

CHARLES AMUNDSEN
Principal
High School of Telecommunication Arts and Technology
Brooklyn, New York

HOWARD BRENNER
Assistant Principal, Mathematics
Fort Hamilton High School • Brooklyn, New York

JOEL FRIEDBERG
Assistant Principal, Mathematics
Herbert Lehman High School • Bronx, New York

FRANK MERINGOLO
Assistant Principal, Mathematics
James Madison High School • Brooklyn, New York

Electronic Prepress Production by
A.W. Kingston Publishing, Inc., Chandler, AZ 85248

When ordering this book, please specify:
R 613 S *or* INTEGRATED MATHEMATICS: INTRODUCTORY COURSE, SOFTBOUND
or
R 613 H *or* INTEGRATED MATHEMATICS: INTRODUCTORY COURSE, HARDBOUND

ISBN 0-87720-293-1 (Softbound edition)
ISBN 0-87720-295-8 (Hardbound edition)

PRINTED IN THE UNITED STATES OF AMERICA

3 4 5 6 7 8 9 10 00 99 98 97 96 95

About This Book

"I understand." "I get it." "I see." "Wow, look at that."
"I think that ..." "I agree because ..." "I disagree because ..."

These are the kinds of responses that can be elicited in a mathematics classroom in which the student is an active participant.

AMSCO's *Integrated Mathematics: Introductory Course*

- **emphasizes exploration and discovery**

  The text is written in an interactive mode, using manipulatives to model situations, introducing the calculator as an investigative tool, and providing activities and experiments for individual and group experiences.

- **promotes reasoning and communication**

  Encouraged to explain their thinking, students are regularly asked to draw conclusions, judge validity, verify results, present arguments, and justify decisions.

- **builds skill in decision making and problem solving**

  Real-life situations connect concepts to daily life, involving students in relevant and meaningful mathematics.

This text

- **supports students at independent application**

  Try These problems, with solutions at the end of each chapter, are offered as an intermediary stage between detailed Examples and a variety of do-it-yourself Exercises.

- **encourages cooperative learning**

  A lead-in Activity for each chapter sets the stage for the mathematics of the chapter. Explorations at the ends of sets of Exercises are most suitable for group experiences.

- **provides opportunities for self-assessment**

  Self-Checks, Things You Should Know, Chapter Reviews, and Cumulative Chapter Reviews serve as cumulative checkpoints.

- **includes a Spiraled Review of Mathematical Skills**

  Incorporated as an Appendix, a series of 14 Spiraled Skill Reviews helps reaffirm mathematical fundamentals.

This contemporary text is offered in the spirit of the Standards of The National Council of Teachers of Mathematics, which promotes the belief that the study of mathematics can be made accessible to all students.

Contents

Communicating With Mathematics

1

ACTIVITY

Who was the better baseball player—Jackie Robinson or Roger Maris?

Is there only one correct answer to this question?

Is there a set of data we could agree to use to answer this question?

As you try to prepare an answer, you will be using some techniques that a mathematician uses to answer important questions for Science and Industry.

1.1 Good Definitions

Every day, we use words that have special meanings.

Let's look at the definition of a one-dollar bill to see what makes it a good definition.

Example 1 A *one-dollar bill* is a piece of paper money valued at 100 cents.

1. It names the term being defined.
   one-dollar bill
2. It uses only other terms that have been defined or whose meanings are understood.
   paper and cents
3. It places the term into a general set to which it belongs.
   money
4. It gives information that separates the term from all other members of the set.
   100 cents
5. It does not include unnecessary information.
   picture of Washington, colored green
6. It is reversible.
   A piece of paper money valued at 100 cents is a ***one-dollar bill.***

Here is a mathematical example of a good definition.

Example 2 A ***triangle*** is a geometric figure with three sides that are line segments that meet at their endpoints.

1. It names the term being defined.
   triangle
2. It uses only other terms that have been defined or whose meanings are understood.
   geometric figure, line segment, endpoint
3. It places the term into a general set.
   geometric figure
4. It gives information that separates the term from other geometric figures.
   three sides that meet at their endpoints
5. It does not include unnecessary information.
   It does not say that a triangle has three angles.
6. It can be turned around or reversed.
   A geometric figure with three sides that are line segments that meet at their endpoints is called a ***triangle.***

A few words in mathematics are not given definitions. Let's see why.

Look at the term *geometric figure*, used in the definition of a triangle.

A ***geometric figure*** is a set of points. But, what is a *set*? And what is a *point*?

A ***set*** is a collection of things, numbers, books, people, baseball cards, and so on. But, what is a *collection*? Trying to define the word *collection* will bring you back to the word *set*.

By trying to define *every* mathematical term, you are going around in a circle.

In mathematics, the meanings of certain basic words, such as *set, point,* and *line,* are left without definition. These *undefined* words are used to build other definitions.

What Makes a Good Definition?

1. It names the term being defined.
2. It uses only undefined terms or previously-defined terms.
3. It places the term into a general set.
4. It gives information, *distinguishing characteristics*, that separates the term from all other members of the set.
5. It does not include unnecessary information.
6. It can be turned around, or reversed.

Try These *(For solutions, see page 39.)*

Which of the following are good definitions? If the definition is not a good one, give at least one reason why it is not.

1. A *sophomore* is a student in the second year of high school who takes mathematics and art.
2. A *garage* is a building where automobiles are kept, repaired, or serviced.
3. *It* is a blue vehicle with four wheels.
4. An *acorn* is a nut that falls from a tree and grows into another tree.
5. A *shark* is an animal.

SELF-CHECK: SECTION 1.1

Use the characteristics of a good definition to decide if the following are good definitions. If the definition is not a good one, state at least one reason why it is not.

1. A *senior* is a fourth-year student.

   Answer: This is a good definition.

2. A *restaurant* is a place where people go to eat, and food is served.

   Answer: This definition has too much information.

EXERCISES: SECTION 1.1

In 1-10, tell if the sentence is a good definition. If it is not, state at least one reason why it is not.

1. A *dog* is an animal with four legs and a tail.
2. *It* is a yellow bird with a black beak.
3. A *line segment* is a part of a line between two endpoints.
4. A *barn* is a farm building.
5. A *product* is the result of the multiplication of two numbers.
6. A *blue spruce* is an evergreen tree.
7. A *bicycle* is a two-wheeled vehicle with a handle bar, seat, and pedals.
8. A *first baseman* is an infielder.
9. A *clock* is an instrument for measuring or indicating time.
10. A *pencil* is a long, thin instrument, usually made of wood, having a graphite center, and is used for writing.

In 11-15, a mathematical word is used in a nonmathematical way. Find the word and give it a mathematical definition.

11. My mother asked me to set the table for Christmas dinner.
12. Hans showed us an interesting product made in Bavaria.
13. My friend Carlos has a negative attitude about movies with subtitles.
14. Samantha is an excellent math student but in English she's just average.
15. The difference between beige and brown is that brown is darker.

EXPLORATIONS

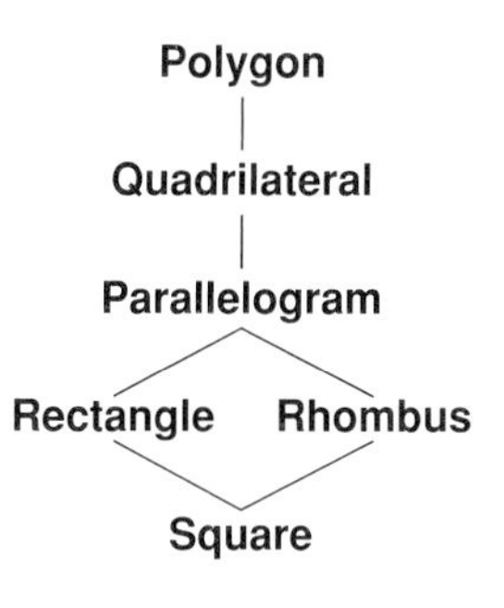

The diagram shows how some sets of special 4-sided polygons are related. A polygon named on a lower level of the diagram can also be called by a name above it. For example a *parallelogram* is also a *quadrilateral*, which, in turn, is also a *polygon*.

a. Give all the other names that apply to a *square*

Each special 4-sided polygon named on a lower level of the diagram is defined in terms of a figure above it. For example: A ***parallelogram*** is a quadrilateral in which both pairs of opposite sides are parallel.

A new figure is defined in terms of a previous figure to make it clear that the new figure shares all the characteristics (or *properties*) of the previous figure, and then has some new properties of its own.

b. Name some properties that a parallelogram has because a parallelogram is a quadrilateral.

You might know that the sum of the measures of the 4 angles of any quadrilateral is always 360 degrees.

There are other things that are true about the sides and angles of a parallelogram that are not mentioned in the definition. For example, in a parallelogram:

- both pairs of opposite sides are equal in length
- both pairs of opposite angles are equal in measure
- the sum of the measures of any pair of consecutive angles is 180 degrees

c. Keeping in mind that a good definition does not contain more information than is needed, write a definition of a rectangle in terms of:

(1) a quadrilateral **(2)** a parallelogram

A ***rhombus*** is a quadrilateral whose 4 sides are equal in measure.

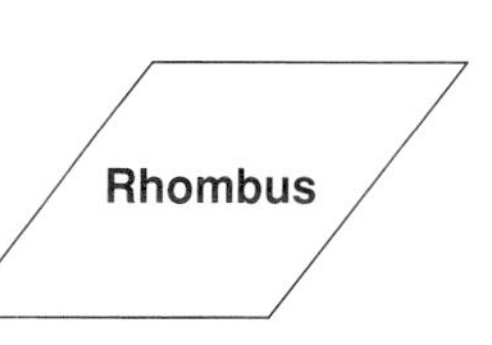

d. Use the set diagram above to tell all the other names that apply to a rhombus.

e. Write a definition of a rhombus in terms of a parallelogram.

f. Write a definition of a square in terms of:

(1) a quadrilateral **(2)** a parallelogram
(3) a rectangle **(4)** a rhombus

Square

1.2 Questions and Answers

Most people study mathematics because they can use it as a tool to solve practical problems at home or at work. Engineers, carpenters, business people, scientists, mechanics, chefs, doctors, teachers, and many others use mathematics to solve problems. Some problems in daily life and in mathematics do not even involve numbers.

Here's how the problem-solving process works:

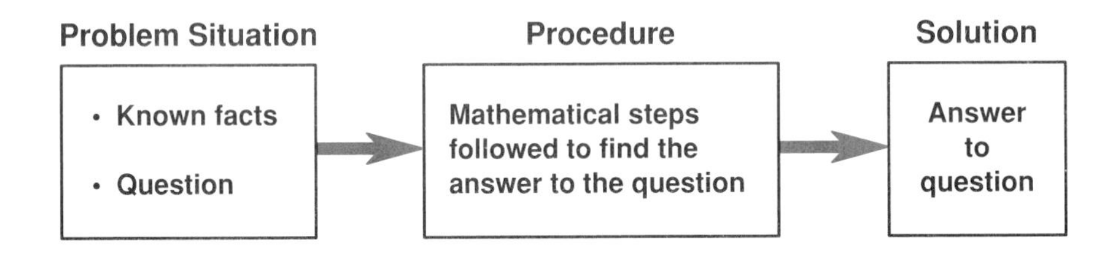

Every problem situation has facts that are known, and a question that must be answered. The answer to the question is called the ***solution*** to the problem.

For many problem situations, there are different questions that can be asked. Try to suggest other questions for each example that follows.

Example 1 There are five teams in a softball league. Each team plays every other team twice. No two teams have the same final record at the end of the season.

Some possible questions are:

How many games were played?
How many ways can the teams place first, second, and third?

Example 2 A toy manufacturer designs a puzzle with blocks to form the solid cube shown. Each block is either all black or all white.

Some possible questions are:

How many blocks can you see on each side of the cube?
How many blocks are there altogether?
What color is the middle block?
How many white blocks and how many black blocks are used?

Example 3 Roasting time for a 4-pound chicken is 20 minutes per pound. Potatoes take 30 minutes to cook. Dinner is to be served at 6:45 P.M.

Some possible questions are:

How long will it take for the chicken to cook?
What time should the chicken be put in the oven?
About what time should the potatoes be started?

Try These (For solutions, see page 39.)

For each problem situation, suggest at least one question using the known facts.

1. Liam puts a 90-minute video tape into his VCR at 7:45 in the evening.
2. Juan runs 3 miles on Mondays, Wednesdays, and Fridays. He runs 5 miles on Tuesdays and Saturdays.
3. A dog is tied by a rope to a ring that is attached to the corner of a garage. The rope is 25 feet long.

4. Soo Ling needs to buy a new hair dryer. The store has three models. The first dryer takes 1,000 watts and has two settings. The second dryer takes 1,200 watts and has three settings. The third dryer takes 1,500 watts and has five settings. They cost $9.95, $13.49, and $17.79 respectively.
5. The Beltz family is planning a trip by car to visit friends who live 1,200 miles away. They have 8 days for the trip and want to spend 5 days with their friends.
6. A total of 6 handshakes were exchanged at a party. Each person shook hands exactly once with each of the other people.

SELF-CHECK: SECTION 1.2

Tom bought a used car after graduating from high school. The car cost him $1,250. He needed two new tires and an engine tune-up, which cost $205. His insurance cost $785 for six months.

What are some possible questions?

Answer:

How much money did Tom spend for the car, tires, tune-up, and insurance for six months?
What will Tom's insurance cost for one year?
What other expenses are involved in owning a car?

EXERCISES: SECTION 1.2

What question(s) does each of the following bring to mind?

1. John reported to work at 10 A.M. on Saturday morning and worked for $3\frac{1}{2}$ hours.
2. Sara kept track of her grades for two weeks. She recorded the following in her notebook: 85, 92, 78, 98, 100, 88, and 84.
3. Conor's mom could buy 3 cans of beans for $2.37 or she could buy 5 cans of beans for $4.05.
4. Tom is 6'1" and weighs 185 pounds. Jerry is 5'11" and weighs 203 pounds.
5. Tom and his friends went on a trip and traveled 2,116 miles. They used 92 gallons of gasoline and paid an average of $1.20 per gallon.

6. In a flight of stairs just over 9 feet high, each step is 7 inches high. Each step is to be edged with a light-reflecting strip that is 2 feet long.

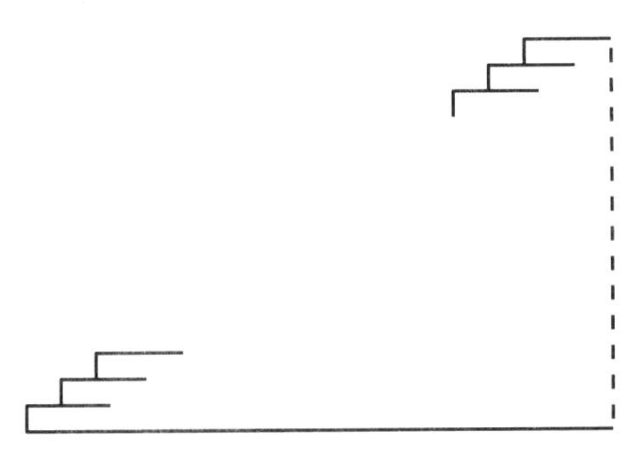

7. Kathy went to Thrift Mart and bought 3 pairs of stockings at $2 per pair and a hair clip for $1.75. She gave the clerk a $10 bill.
8. After two deposits, $38.50 and $120, Mr. Henry's balance in his savings account was $382.90.
9. Al, Scott, and Charlie competed in a race. Scott beat Al and Al beat Charlie.
10. Lucia is planning to construct a bulletin board 4' by 6'. She wants to cover it with 1' by 1' cork tiles that cost $.89 each.
11. Kimba works from 4 P.M. until 8 P.M. Monday through Friday. She earns $4.75 per hour.
12. A TV weather report said that there was a 50 percent chance of rain on Saturday and a 25 percent chance of rain on Sunday.
13. A game of skill calls for two players. The results of several rounds are shown in the table.

Player	Games Played	Points Won or Lost
Michael	4	won 12
Stephen	5	lost 25
Caryn	2	won 10
Roberta	7	won 28
Edie	3	lost 25

14. Bob runs 5 miles a day and uses 300 calories each day. Ann bicycles to work 5 days a week and uses 25 calories per mile. Sid sits in front of the TV and eats potato chips. He weighs 145 pounds.
15. Choose a number. Double it. Add 6. Multiply the answer by 10. Subtract 60.

EXPLORATIONS

In your textbook, you were asked to think up questions about the cube shown below. Now, in this activity, you will answer some questions about it. You might find it helpful to build a model with, for example, sugar cubes.

To begin, here is some information about the cube:

- The large cube has no empty space inside.
- The small cubes (or *blocks*) alternate in color, black-white-black.
- Each block has 6 sides, or *faces*.
- The faces of a block are joined by *edges*.

Find answers to these questions.

a. **(1)** How many blocks are there altogether in the cube?

(2) How many blocks are black? How many blocks are white?

(3) What color is the block in the very center of the cube?

b. **(1)** Sketch a diagram of the middle row of blocks.

(2) How many "middles" are there? Are they all the same?

c. **(1)** How many block edges lie along the edges of the large cube?

(2) How many block faces are on the outside of the cube? How many of them are black? How many are white?

d. **(1)** If you added more blocks to the outside of the cube, attaching a white block to every black one and a black to every white, how many blocks would there be altogether?

(2) The object in part (1) would not be a cube. How many more blocks would you need to complete the new larger cube?

1.3 Keys to Technology

A calculator is a tool you can use to do arithmetic calculations quickly. There are many different types of calculators.

In addition to the numerical keys on your calculator, the main operational keys are the following.

ON/OFF	turns the calculator on and off
C	clears the entire calculator
CE	clears the last entry
CE/C	clears the last number entered; if pressed twice, clears the entire problem
+ − × ÷	do addition, subtraction, multiplication, and division
=	displays the result of a calculation

Other keys you may use later are:

.	enters a decimal point
%	finds the percent of a number
+/−	changes the sign of a number
√	finds the square root of a number
M+ M− MR MC	work with numbers in the memory

Basic Calculator

Use your calculator to do the following examples. Enter the number from left to right just as you would write them. Commas are not used with a calculator.

Example 1 3,124 + 5,220 = ?

Enter: 3 1 2 4 + 5 2 2 0 =

Display: 8344.

From now on, calculator entries that are numbers are shown simply as numbers. Other entries, such as operational symbols, are shown as calculator keys.

Example 2 $427 \times 396 = ?$

Enter: 427 [×] 396 [=]

Display: 169092.

On many calculators, you can continue adding, subtracting, multiplying, or dividing the same number by pressing [=].

Example 3 $45 + 8 + 8 + 8 = ?$

Enter: 45 [+] 8 [=] *Display:* 53.

Now press [=] again. *Display:* 61.

Press [=] again. *Display:* 69.

Each time [=] is pressed, 8 is added.

45 [+] 8 [=] 53 [=] 61 [=] 69

Example 4 $3 \times 3 \times 3 \times 3 \times 3 \times 3 = ?$

Enter: 3 [×] 3 [=] 9 [=] 27 [=] 81 [=] 243 [=]

Display: 729.

Each time [=] is pressed, the number in the display is multiplied by 3.

Example 5 Subtract 119 from 289 and divide the difference by 10.

Enter: 289 [−] 119 [÷] 10 [=]

Display: 17.

If, on pressing [÷], your calculator did not display 170, enter instead:

289 [−] 119 [=] [÷] 10 [=]

The different ways calculators work will be discussed later.

Try These (For solutions, see page 39.)

In 1-10, use a calculator.

1. $3{,}574 + 2{,}111 + 5{,}432$

2. $345 - 168$

3. 23×147

4. $7{,}259 \div 17$

5. $1{,}476 - 37 - 37 - 37 - 37$

6. $15 \times 15 \times 15 \times 15 \times 15 \times 15$

7. $857 - 432 + 77$

8. $455 \times 23 \div 7$

9. $868 \div 124 \times 7$

10. $47{,}932 - 18{,}236 \div 4$

SELF-CHECK: SECTION 1.3

1. Use a calculator to add:

$$\begin{array}{r} 2{,}305 \\ 25 \\ 876 \\ \underline{+5{,}090} \end{array}$$

Answer: *Enter:* 2305 [+] 25 [+] 876 [+] 5090 [=]

Display: 8296.

2. Multiply: $6 \times 6 \times 6 \times 6$

Answer: *Enter:* 6 [×] 6 [=] 36 [=] 216 [=]

Display: 1296.

3. Subtract 27 from 789 and divide the difference by 3.

Answer: *Enter:* 789 [−] 27 [÷] 3 [=]

Display: 254.

EXERCISES: SECTION 1.3

1. Use a calculator to compute:

a. $732 + 1{,}003 + 128$

b. $1{,}276 - 843$

c. 146×58

d. $5{,}536 \div 32$

e. $876 + 428 - 917$

f. $763 - 410 + 136$

g. $195 \times 6 \div 13$

h. $642 \div 214 \times 5$

i. $15 \times 6 + 46 + 20$

j. Subtract 58 from the product of 32 and 17.

k. Multiply the sum of 758 and 106 by 8.

l. Divide the sum of 1,763 and 876 by 7.

2. Sometimes it is easier to do a problem in your head than to use a calculator. When you multiply a whole number by 100, simply add two zeros. For example: $7 \times 100 = 700$ $42 \times 100 = 4{,}200$

   If you know that $2 \times 50 = 100$ and $4 \times 25 = 100$, you can create 100 in your head to do quick multiplications.

   Try these problems without a calculator.

   a. $4 \times 7 \times 25$ **b.** $2 \times 27 \times 50$ **c.** $6 \times 4 \times 25 \times 3$

3. One way to check your answer to a problem is to do the *opposite* operation. If you added, then subtract. For example: $5 + 8 = 13$

   To check to see if 13 is the correct sum, enter 13 first and then subtract 5. If the answer is 8, then you did the problem correctly.

   Explain how you could check that 40 is the correct answer for 5×8.

4. John's father told him he would double John's allowance every day for a month. Do you think this will be a lot of money?

   Dad gave John a penny on January 1, and 2 cents on January 2.

   a. How much money did he get on January 10?

   b. What is the total amount of money he got through January 10?

5. The expression 2^5 means $2 \times 2 \times 2 \times 2 \times 2$ (the 5, called the ***exponent***, tells you to find the product of five 2's). Remember that you can find this product on your calculator by pressing

   2 [×] 2 [=] [=] [=] [=]. Thus, $2^5 = 32$.

   a. Use a calculator to determine which number of each of the following pairs is larger.

   (1) 3^4 or 4^3 **(2)** 2^5 or 5^2

   b. Using the pattern in part **a**, to what expression would you compare 7^3? Which do you think is larger? Try it and see.

6. **a.** Find each of the following values, and describe a pattern: 10^2, 10^3, 10^4, ...

   b. What name do we give to 10^9?

   Names have been given to two very large numbers.

   10^{100} is called a *googol*.

   $10^{1,000}$ is called a *googolplex*.

7. *Enter:* 4 [×] 3 [=] [=] [=]

   If the display reads 192, your calculator is repeatedly using 4 as a multiplier.

   What do you think would be the result of entering

   5 [×] 2 [=] [=] [=]?

   Guess, then check, using your calculator.

8. Write a key sequence to represent each computation, using the fewest possible number of key presses. Try your sequence on your calculator, and write the display. Is the result correct?

a. $8 + 7 + 7 + 7$ **b.** $80 - 6 - 6 - 6 - 6 - 6$ **c.** $2 \times 2 \times 2 \times 2 \times 2 \times 2 \times 2$

d. $72 \div 2 \div 2 \div 2$ **e.** $8 \times 3 \times 8 \times 8$

9. Brian is learning how his calculator does repeated operations when he keeps pressing [=]. Here are his entries, and the calculator displays:

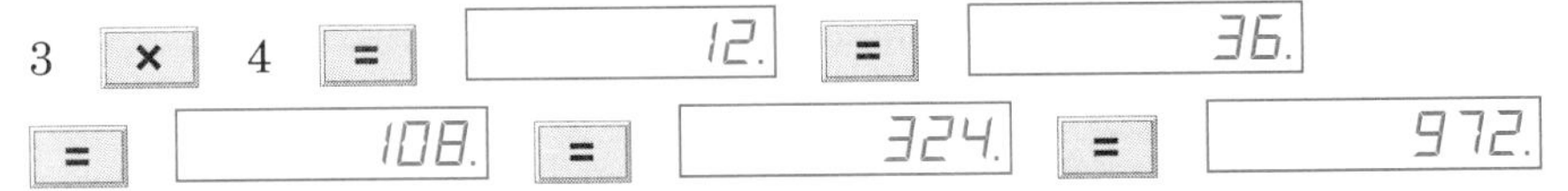

The next time Brian presses [=], what digit will appear in the ones place of the display?

10. In each of the following statements, the same number placed in both boxes makes the statement true.

☐	×	☐	= 4	The number is 2.
☐	×	☐	= 9	The number is 3.
☐	×	☐	= 5	The number is between 2 and 3.
				Guess and check.
2.5	×	2.5	= 6.25	Too big.
2.3	×	2.3	= 5.29	Too big.

Keep trying, and see how close you can come to the product 5.

11. Mr. King received his monthly checking account statement, but some of the entries were illegible.

Use the data shown in the statement to determine which of the following choices is his balance at the end of the month.

(a) $102.42

(b) $263.46

(c) $445.84

(d) $612.42

```
Account No.   00 - 421 - 87 - 5               Summary

      8/28    Opening balance                  354.65
              Total of 2 deposits              xxxxxx
              Total of 2 checks drawn          xxxxxx
              Service charge                   xxxxxx
      9/30    Closing balance                  xxxxxx
```

Transaction	Date	Amount	
Service charge	9/1	11.23	
Deposit	9/3	210.19	
Check drawn	9/8	215.27	Check No. 432
Deposit	9/18	150.00	
Check drawn	9/23	42.50	Check No. 433

EXPLORATIONS

You know that when you divide a number, you sometimes get a *remainder*. (When the remainder is 0, we say there is no remainder.)

When you divide by 9, you can get any one of 9 different remainders: 0, 1, 2, 3, 4, 5, 6, 7, or 8. On a calculator, these remainders appear as repeating digits: 0.111..., 0.222..., 0.333..., and so on.

You can amaze your friends by telling them the decimal part of a quotient without doing the division or using a calculator.

When dividing a number by 9, add the digits of the given number until the result is one digit. For example, for the number 2,357:

Add the digits: $2 + 3 + 5 + 7 = 17$

Add again: $1 + 7 = 8$ ⇑

This number repeats in the quotient.

Verify the result by looking at the actual long division, shown at the right.

```
     261.888. . .
 9)2357.000. . .
  -18
    55
   -54
     17
     -9
      8 0
     -7 2
        80
       -72
         80
          ↓
```

Then, use a calculator to divide 2,357 by 9.

Note that when you apply this speedy method to a problem and the answer is 9, there is no remainder. For example, using the number 2,367:

$2 + 3 + 6 + 7 = 18$

$1 + 8 = 9$

This means that 9 is a factor of 2,367.

Use the speedy method to find the decimal part of the quotient when each of the following numbers is divided by 9. Verify your result by doing the division and by using a calculator.

a. 197 **b.** 3,542 **c.** 147,238

1.4 Reading Tables

Tables are used to organize and study numbers or other information.

Example 1 Mr. Taylor sets up a table for the students in his class and records their scores on tests during a marking period.

From the table, he is able to make various decisions about each student.

Student	Test			
	1	2	3	4
Giordano, C.	82	85	98	88
Hernandez, R.	68	78	79	85
Meys, D.	90	81	75	65

Use the table to answer each question. *Answers*

a. Which student has been showing the most improvement? R. Hernandez

b. Which student seems to be having difficulty? D. Meys

c. Which student seems to be doing very well? C. Giordano

Example 2 This table shows the number of sides and angles in geometric figures called *polygons*.

A ***polygon*** is a geometric figure whose sides are line segments that meet at their endpoints.

Name	No. of Sides	No. of Angles
triangle	3	3
quadrilateral	4	4
pentagon	5	5
hexagon	6	6
heptagon	7	7
octagon	8	8
nonagon	9	9
decagon	10	10

Use the table to answer each question.

a. What is the number of sides of a decagon? a pentagon?

Answer: A decagon has 10 sides.
A pentagon has 5 sides.

b. How many angles in a decagon? in a pentagon?

Answer: A decagon has 10 angles. A pentagon has 5 angles.

c. What conclusion can you reach about the number of sides and angles of a polygon?

Answer: The number of sides and the number of angles in a polygon are the same.

d. What is a good definition of a decagon?

Answer: A decagon is a polygon with 10 sides.

Try These *(For solutions, see page 39.)*

Refer to the table to answer each question.

Times in Seconds			
Swimmer	**100-yard Freestyle**	**100-yard Backstroke**	**100-yard Butterfly**
Joe	61	69	65
Carlos	63	66	66
Brent	65	67	64
Franco	60	68	59

1. Which swimmer has the fastest time in the freestyle?

2. Which swimmer has the slowest time in the backstroke?

3. In which event did Brent swim faster than Carlos?

4. If a relay team is made up of three swimmers, each swimming a different event, which swimmers would make up the best team?

5. If each swimmer can be in no more than two events, which two swimmers would you put into each event?

SELF-CHECK: SECTION 1.4

1. Use the table in Example 1 to find the student with the lowest grade on any test.

Answer: D. Meys

2. Does the table in Example 2 tell you that there are only eight polygons?

Answer: No. It just lists eight polygons.

3. Use the table of polygons to write a good definition of an octagon.

Answer: An octagon is a polygon with 8 sides.

4. Draw a polygon of 3 sides, 4 sides, and 5 sides.

Answer:

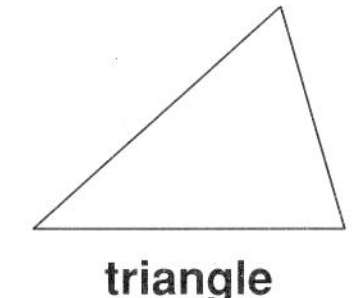

triangle

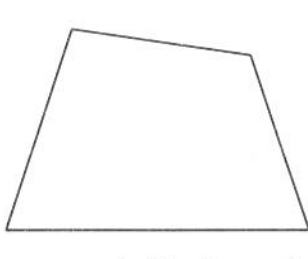

quadrilateral

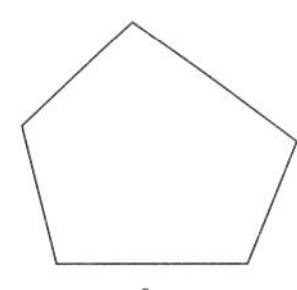

pentagon

EXERCISES: SECTION 1.4

1. This table shows the number of calories contained in popular lunch items.

Item	Calories
hot dog/roll	300
hamburger/roll	450
cheeseburger/roll	550
french fries	150
cola	140
milk	160
cheese pizza	150
apple juice	120

   a. Which lunch item has the most calories?

   b. Of the first three lunch items, which has the fewest calories?

   c. How many calories are in three pieces of cheese pizza and one can of cola?

   d. How many calories are in a lunch consisting of a cheeseburger, french fries, and milk?

   e. Select items for lunch totaling 820 calories.

   f. Select items for lunch totaling less than 600 calories.

   g. Name some foods that are not listed here, that are tasty and healthy and not high in calories.

2. In preparation for the upcoming sports season, two girls kept track of their exercise time over a 5-week period.

Exercise Time (in hours)

Name	Week				
	1	2	3	4	5
Mary Jo	$2\frac{1}{2}$	4	4	6	$6\frac{1}{2}$
Karma	3	4	$5\frac{1}{2}$	7	$7\frac{1}{2}$

   a. What is the total number of hours that Mary Jo exercised?

   b. What is the total number of hours that Karma exercised?

   c. During which week did they exercise the same number of hours?

   d. Which girl increased her exercise time the most over the 5-week period?

3. The table shows the number of sides and the number of diagonals of each polygon. A ***diagonal*** of a polygon connects two nonconsecutive ***vertices*** (corners).

Polygon	No. of Sides	No. of Diagonals
triangle	3	0
quadrilateral	4	2
pentagon	5	5
hexagon	6	9
heptagon	7	14
octagon	8	20

   a. How many diagonals can be drawn in a quadrilateral? Draw the quadrilateral and its diagonals.

   b. How many diagonals can be drawn in a pentagon? Illustrate your answer.

   c. How many more diagonals does a quadrilateral have than a triangle?

   d. How many more diagonals does a pentagon have than a quadrilateral?

   e. How many more diagonals does a hexagon have than a pentagon?

   f. Compare the number of diagonals in a hexagon, a heptagon, and an octagon.

4. The table shows the height and average points scored per game for eight players on North High's basketball team.

Player	Height	Avg. Points Scored
Jim	6'3"	14
Scott	6'5"	10
Paul	6'4"	8
Sid	6'2"	9
Mac	5'11"	8
Mike	6'1"	8
Tony	5'9"	12
Pete	5'10"	11

a. When playing against South High, Coach Brown wants his starting five players to have a height advantage. Which five players should he choose?

b. When playing against East High, Coach Brown wants to start the five highest scoring players. Name the starting players.

c. Find Coach Brown's ideal player by using the following grading method.

Start each player with 20 points.

Subtract 2 points for each inch the player is below 6 feet.

Add 2 points for each inch the player is above 6 feet.

Subtract 1 point for each point scored below 10.

Add 1 point for each point scored above 10.

For example, Jim's grade is 30, that is:
20 points to begin + 3 inches above 6 feet $\times$ (2) + 4 points scored above 10 points $\times$ (1) = 30

(1) Which person gets the highest grade?

(2) Which person gets the lowest grade?

(3) How many points would a 5'10" player have to score to get a higher grade than Jim?

(4) Show a possible height for Carlos, whose grade is exactly 20.

d. When playing against West High, Coach Brown wants to combine height and average for his starting five. Which five players would he start?

5. This table compares the lengths of a day on the planets in our solar system.

Planet	Length of One Day in Earth Hours
Mercury	1,406.4
Venus	5,832.0
Earth	23.9
Mars	24.6
Jupiter	9.8
Saturn	10.2
Uranus	23.2
Neptune	22.0
Pluto	153.0

a. Which planet has the longest day?

b. Which planet has the shortest day?

c. Which planet has a day closest to the length of an Earth day?

d. How much longer is one day on Saturn than one day on Jupiter?

e. How would you find out how many Earth days a day on Mercury lasts?

f. Write an exercise in which you would multiply to find the answer.

6. Carrie's diet allows her 2,000 calories a day. By dinner time yesterday, her calorie intake was just over 1,300 calories. Use the calorie table to write a dinner menu that will keep Carrie within her daily limit. Choose 1 main dish, 2 vegetables, 1 beverage, and 1 dessert.

Main Dishes		Vegetables	
beef stew	235	green beans	20
baked chicken	155	cauliflower	25
chili	240	spinach	25
hamburger	320	peas	75
fried fish	190	baked potato	145
Desserts		**Beverages**	
apple	80	orangeade	110
apple pie	300	cola	145
brownie	90	skim milk	90
peach pie	320	vegetable juice	35

7. Smith School has a program to encourage more reading. In a survey to see whether the program has helped, students were asked whether they had read a book in the past week. The administrators would like to find the greatest increase in reading among students that have been in the program longest (the higher grade levels). If the answers are organized in a table, which style of table would be best to show whether they got the desired results?

(a)

Student	Read a book?	
	Yes	No
Aarons	✓	
Alviero	✓	
Bennett		✓
Brody		✓

(b)

Grade Level	Read a book?	
	Yes	No
6	114	151
7	137	136
8	181	105

(c)

	Read a book?	
	Yes	No
Male	158	136
Female	173	157

(d)

Read a book?		Watched TV?	
Yes	No	Yes	No
432	392	584	240

EXPLORATIONS

This table has information about the distances and speeds of some animals during their migration season.

Animal	Distance (in miles)	Speed
Butterfly	1,870	32 miles per day
Hummingbird	500	50 miles per hour, or 70 wing beats per second
Tern (a bird)	22,000	100 miles per day
Shear (a bird)	20,000	
Seal	3,000	
Whale	4,000	

Complete the following story by giving answers for the lettered blanks. Some answers will not come from the table.

Many different kinds of animals travel long distances. Scientists are not certain about why the animals travel or how they know how to get where they are going. This type of travel, which scientists believe is instinctive, is called *migration.*

Monarch butterflies travel from the northern U. S. and Canada to as far south as Mexico, where they spend the winter in trees. A butterfly can travel as far as ____(a)____ miles and can move at speeds of ____(b)____. A mathematical question that can be asked is ____(c)____.

The 3-inch ruby-throated hummingbird, which is the smallest North American bird, weighs only $\frac{1}{8}$ of an ounce. It can fly at about ____(d)____ miles per hour, and its wings beat at ____(e)____ beats per second. The number of times its wings beat per hour is ____(f)____.

The shear and the Arctic tern are both long-distance travelers. The shear migrates in a loop over the Pacific that extends from Greenland to Australia. The tern follows the sun from the Arctic to the Antarctic. Of these two birds, the ____(g)____ travels ____(h)____ more miles than the ____(i)____.

Mothers and young Alaskan fur seals migrate from islands in the Bering Sea to islands off the coast of southern California, a distance of ____(j)____ miles. A reason birds can travel much greater distances than seals is ____(k)____.

Humpback whales swim long distances from the coast of Alaska to islands off the coast of Mexico. The distance from southern California (where the seals are) to the coast of Mexico (where the whales are) is about ____(l)____ miles.

1.5 Reading Graphs

In newspapers and magazines, numerical facts, called ***data***, are often shown as ***graphs***. Because they are visual, graphs are often easier to read and understand than tables. Some common types of graphs are *bar graphs, pictographs,* and *line graphs*.

Example 1 Cholesterol is a white, fatty substance. It can collect in blood vessels and block them. This blockage can lead to heart attacks. The higher the amount of cholesterol in the blood, the greater the risk of a heart attack. The bar graph shows the cholesterol levels for five adults. Levels from 200-240 indicate moderate risk, and above 240 are high risk.

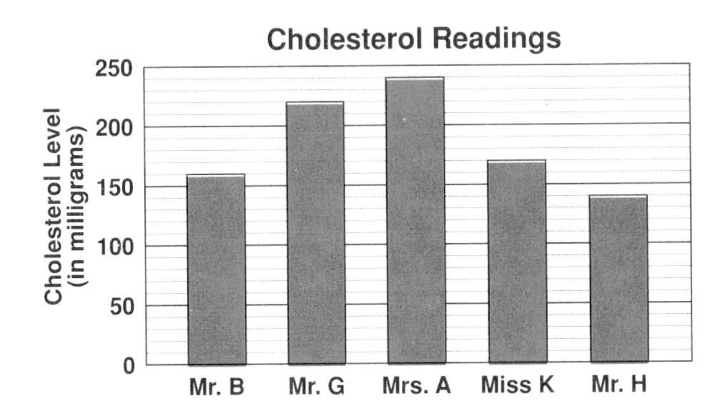

You can find the cholesterol level of each person by reading the heights of the bars. Mr. B. has a cholesterol level of 160. Mrs. A.'s level is 240.

Example 2 A pictograph shows data by using a symbol. Each symbol represents a given number of items.

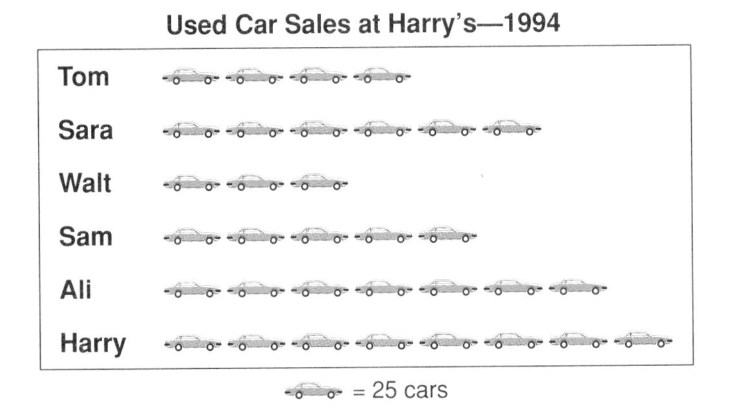

Tom sold 100 cars, and Harry sold 200. Harry sold twice as many cars as Tom. Walt, who sold only 75 cars, was the salesperson with the fewest sales.

Both a bar graph and a pictograph are used to compare *different* items. To show a *change* in a set of data, usually over a period of time, we use a line graph.

Example 3 This line graph shows the amount of money raised during each month by a small private college in a one-year fund drive.

In January, the college raised $50,000. The best month was June, when $150,000 was raised. During February and March, the same amount, $75,000, was raised.

The graph shows the greatest one-month increase in money raised was from May to June.

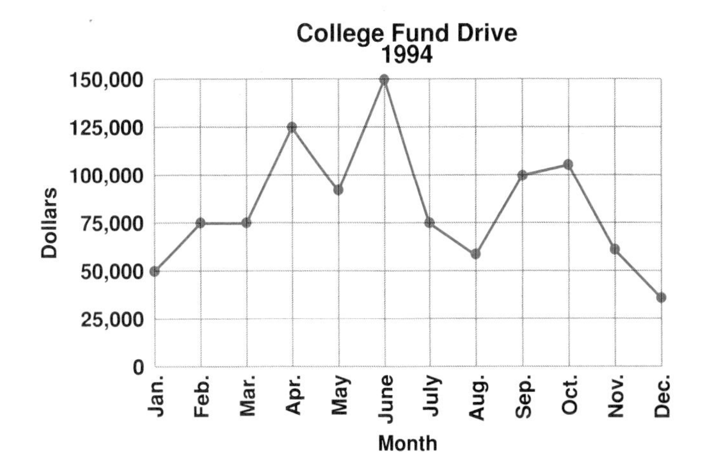

Try These *(For solutions, see page 39.)*

1. Some students were asked to select their favorite ice cream flavor from among cherry, vanilla, chocolate, and strawberry. The bar graph shows the results.
   a. What was the favorite flavor? the least favorite flavor?
   b. How many students chose each flavor?
   c. How many students were in the group altogether?
   d. How many students chose a pink ice cream?

2. This line graph gives information about the population in Tompkins County.
   a. What was the growth in population from 1985 to 1990?
   b. In which 5-year period did the population show the greatest increase? Explain.

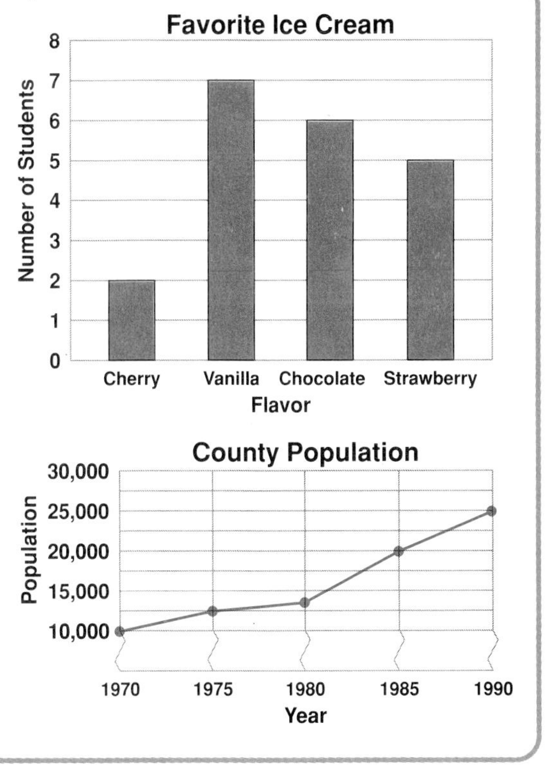

SELF-CHECK: SECTION 1.5

1. Use the bar graph in Example 1 to tell which person had the lowest cholesterol level.

Answer: Mr. H

2. Use the pictograph in Example 2 to tell which salesperson alone sold the same number of cars as did Tom and Walt together.

Answer: Ali

3. Use the line graph in Example 3 to tell how much money was raised in October.

Answer: About $110,000

EXERCISES: SECTION 1.5

1. Jake kept a record of time spent on his activities for one day, as shown on the bar graph.

a. At what activity did Jake spend the most time?

b. How many hours did Jake spend at work?

c. How many more hours did Jake spend at school than doing homework?

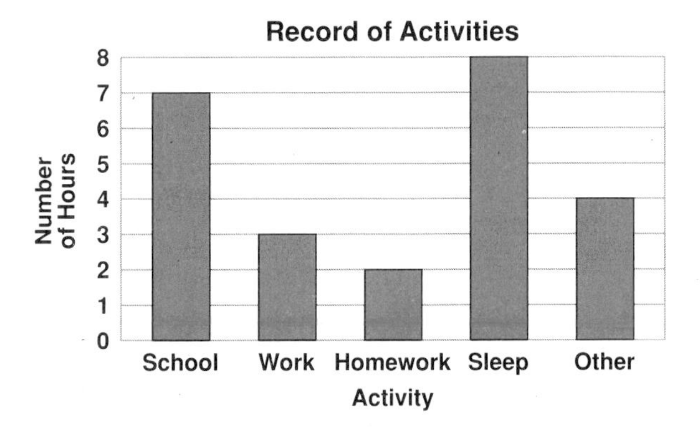

2. The bar graph shows the bowling scores for a five-person team.

a. What was the total team score?

b. What is the difference between the highest and the lowest scores?

c. Who scored between 100 and 150?

d. How could you use this graph to estimate a team average?

e. How would the graph be different if the scores were put in order?

f. Sometimes a table is a better way to display information than a graph. Do you think a table would be an improvement? Explain.

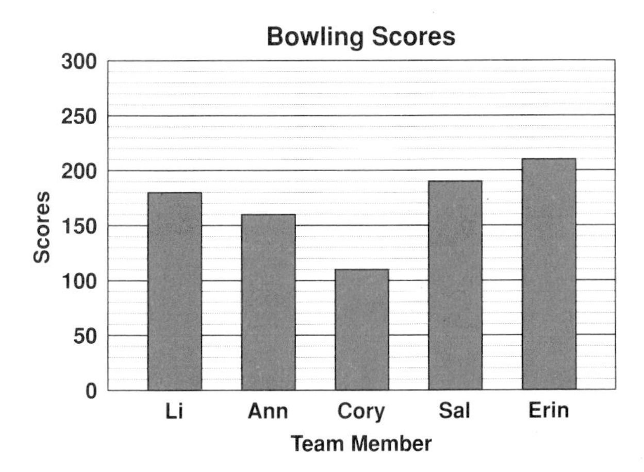

3. The heights of 5 basketball players are shown in the table.

Player	Sloan	Pike	Wood	Bean	Kors
Height (in inches)	75	71	81	72	77

Which of the following would be the best way to arrange a bar graph of the information? (The players are identified by their initials.)

(a)

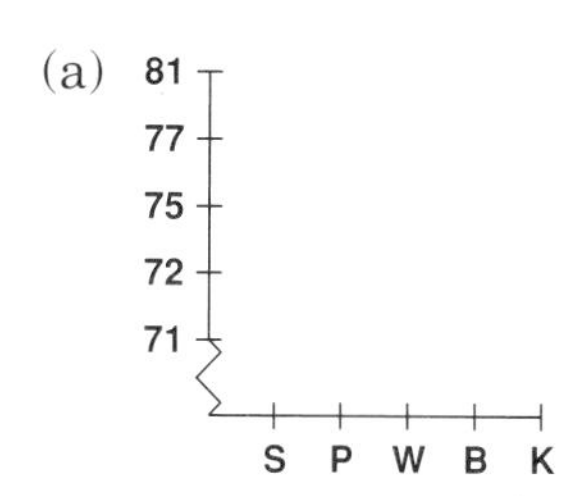

(b)

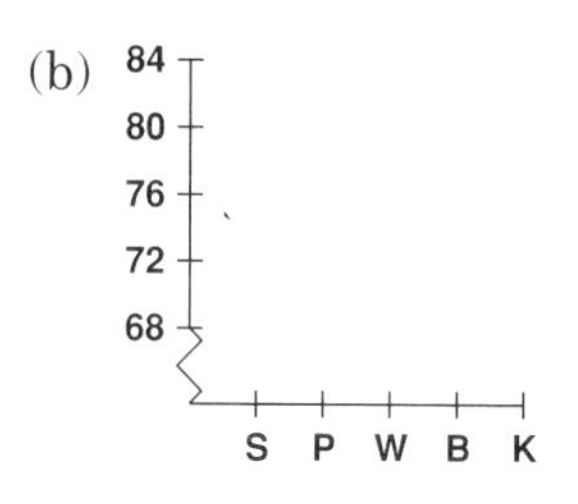

(c)

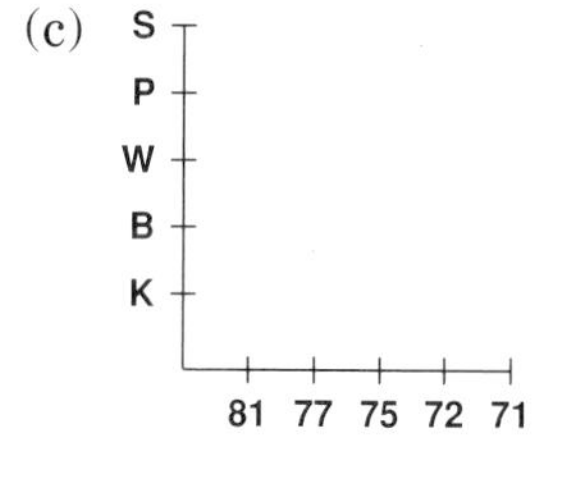

(d)

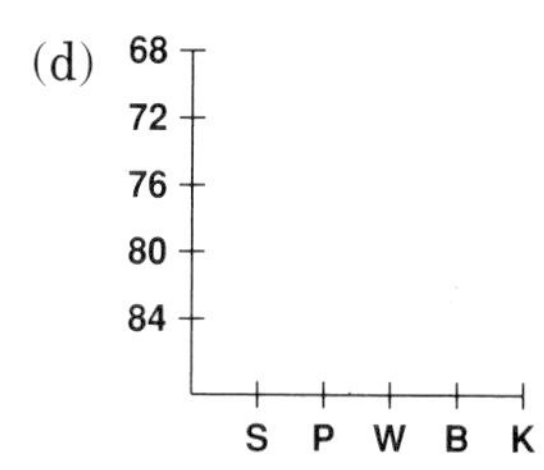

4. The Ecology Club and the Honor Society collected aluminum cans for recycling.

a. How many cans were collected by the Honor Society?

b. How many more cans were collected by the Ecology Club than by the Honor Society?

c. What was the total number of cans collected?

d. Which club do you think had more members? Explain.

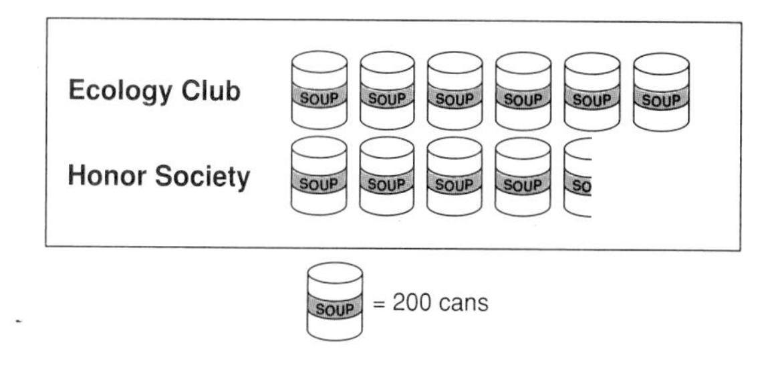

5.

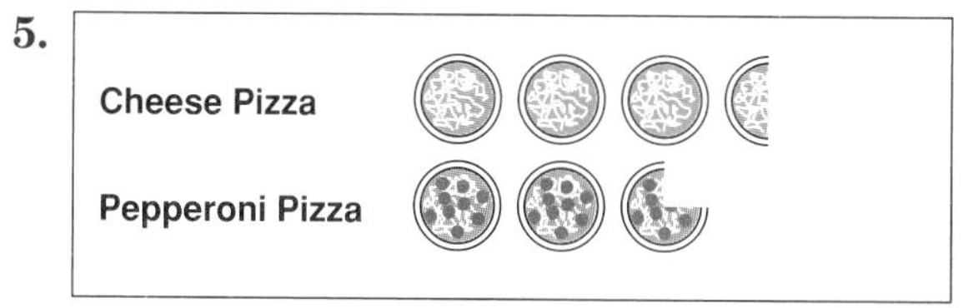

Each whole symbol represents 20 pizzas sold.

This pictograph shows sales at Peppe's Pizza.

a. How many cheese pizzas were sold?

b. How many pepperoni pizzas were sold?

c. How did you decide how many pepperoni pizzas were sold?

6. The pictograph shows the distribution of grades in 10th-grade English classes.

a. What grade was the most common?

b. How many students had a grade above C?

c. How many more students had a grade of B than had a grade of A?

d. List 4 strategies that students who earned grades of C, D, or F could use to improve their grades.

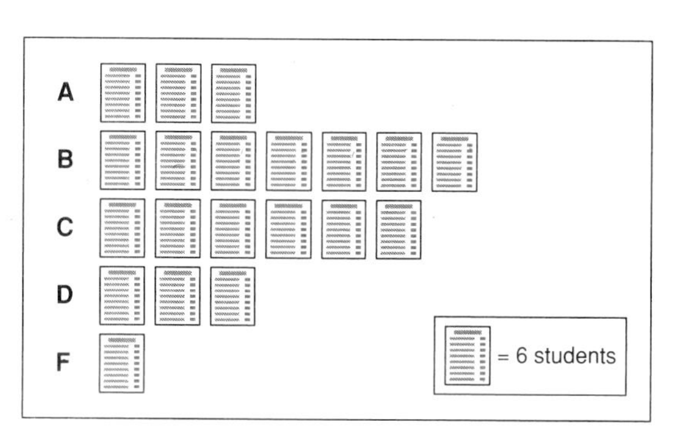

7. This graph shows temperature readings for a 6-hour period in February.

a. What was the highest temperature reading?

b. In which 2-hour period did the temperature increase the most?

c. During which time period did the temperature decrease?

d. What was the temperature at 11 A.M.?

e. Estimate the temperature at 8 A.M.

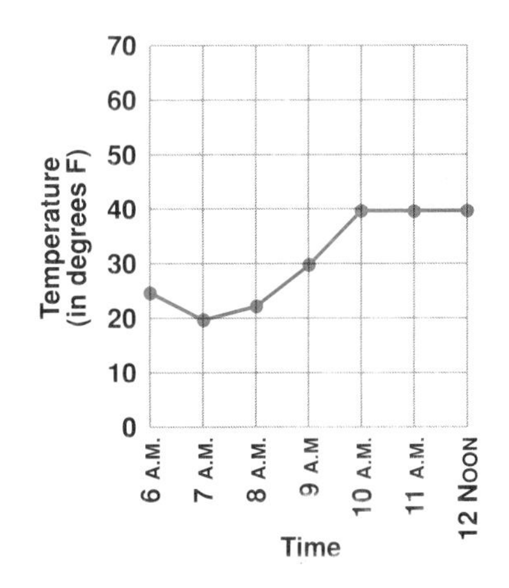

8. This graph relates distance traveled on a bicycle to the length of time the bicycle was ridden.

a. How far did the bicycle travel in 4 seconds?

b. After two seconds, how many feet did the bicycle travel?

c. What can you tell about the speed of the bicycle? (How fast is it traveling?)

9. This graph relates distance traveled to the length of time a bicycle was ridden.

a. Name the letters that show where the bicycle was stopped:

(1) for 1 second **(2)** for 2 seconds

b. Did the bike go faster from Start to A or from B to C?

c. At what point did the rider turn around to get something she dropped?

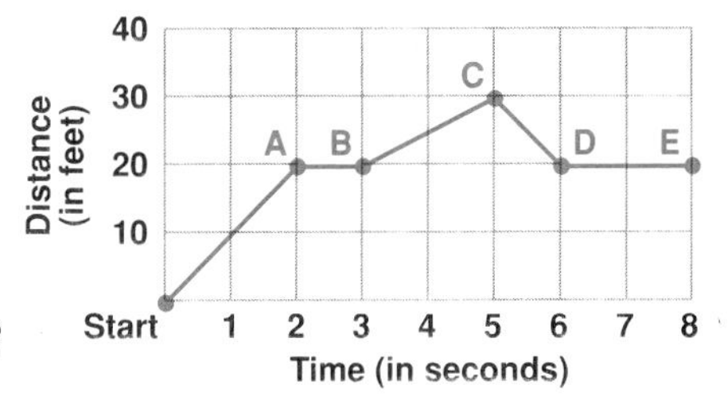

10. This graph represents blood pressure readings for two members of the wrestling team, monitored over a 12-hour period.

a. At what time were blood pressures the same?

b. What was Eli's blood pressure at noon?

c. What was Jake's blood pressure at 9 P.M.?

d. What is the difference in their blood pressures at noon?

e. Estimate Eli's blood pressure at 1:30 P.M.

f. What circumstances could have caused Jake's 3 P.M. blood pressure reading?

11. Sara goes to school in Overton, PA. For a science project, she recorded the noon outdoor temperatures over a period of 5 months, then found the average noon temperature for each of the months. Here is her graph of average noontime temperatures.

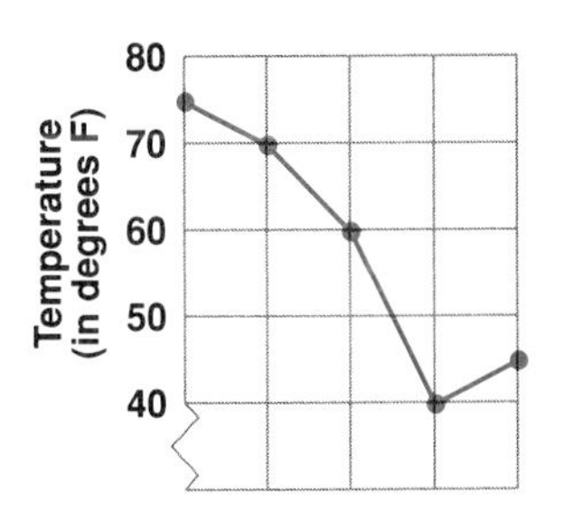

Sara forgot to write the months at the bottom of the graph. Which were probably the months in which she recorded the data?

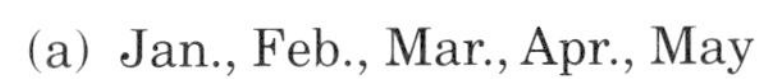

(a) Jan., Feb., Mar., Apr., May
(b) Mar., Apr., May, June, July
(c) June, July, Aug., Sept., Oct.
(d) Sept., Oct., Nov., Dec., Jan.

12. The backyard pool at the Browns' house is the same depth throughout. Because of restrictions on water use, the Browns could not add water during one summer month. The graph shows water levels for part of that time.

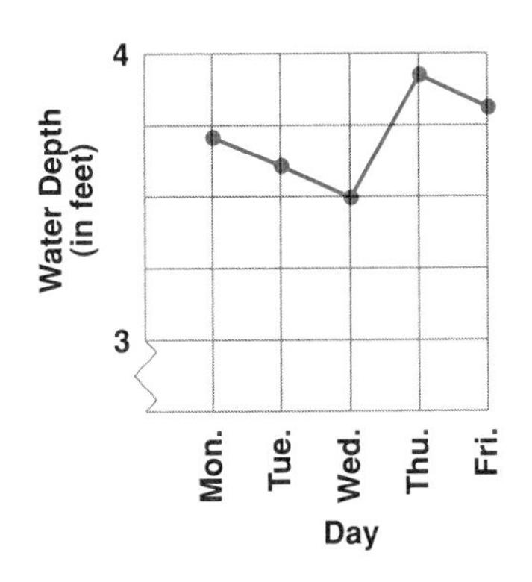

Which of the following might be a reason for the change from Wednesday to Thursday?

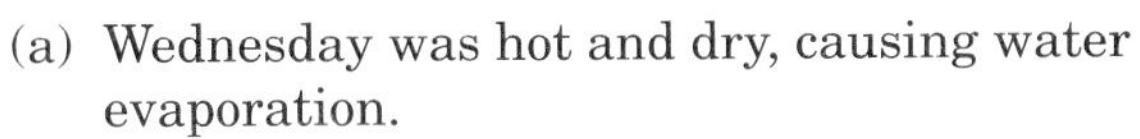

(a) Wednesday was hot and dry, causing water evaporation.
(b) The youngsters had a pool party, with a lot of water splashing out.
(c) There was a heavy rainstorm.
(d) Jeff Brown used pool water to wash his car.

13. The data in the line graph is also represented in which of the following pictographs?

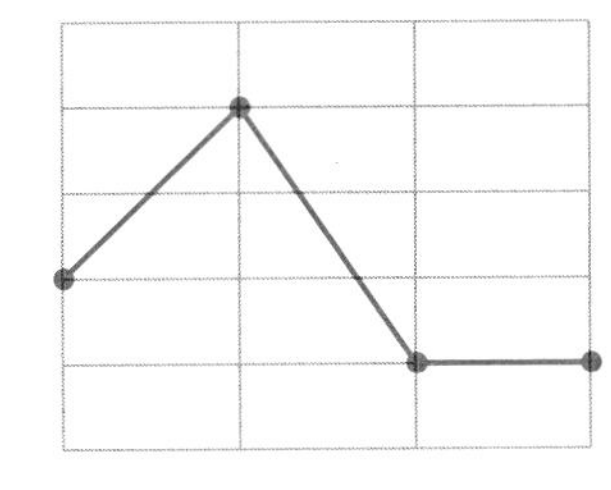

(a) 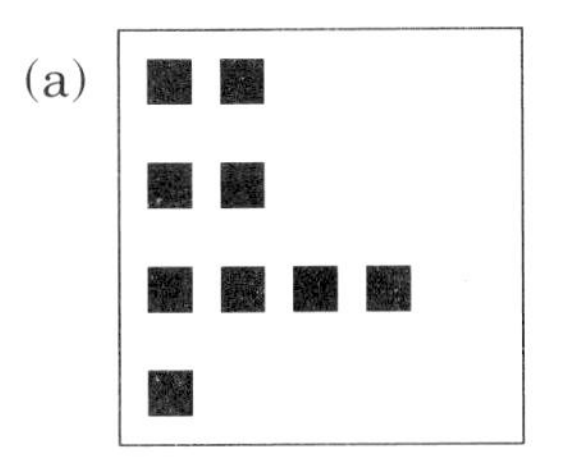

(b) 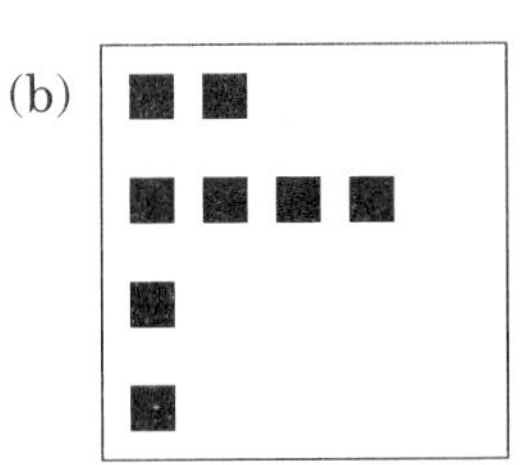

(c) 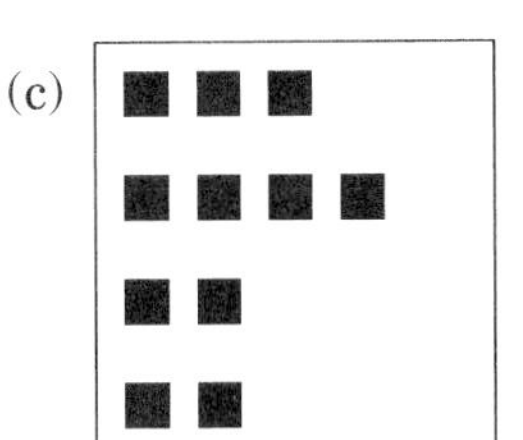

(d) 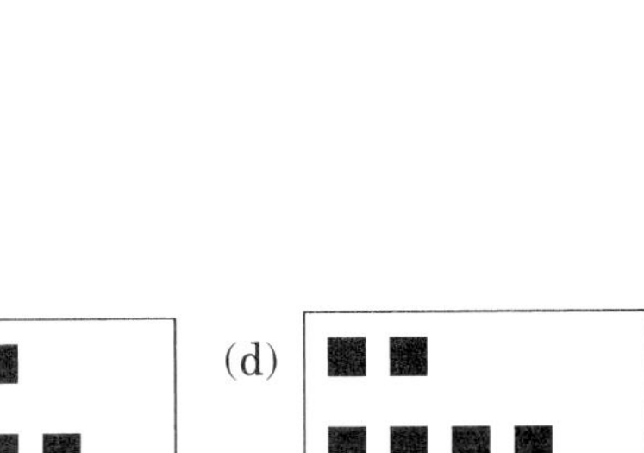

EXPLORATIONS

In Highland School's last basketball game, the team consisting of Cory, Brett, Liku, Jordan, and Lin beat Northside School by a score of 80 to 65.

Using this situation:

a. Make up data for the number of points scored by each player on Highland's team.

b. Create a symbol, telling the number of points it represents, and draw a pictograph for the data.

c. Write three questions that could be answered from your pictograph.

d. Draw a pictograph that compares the scores of the two schools. You may use the same symbol as before, but will you change the number of points it represents?

e. Write a question that could be answered from this pictograph.

1.6 Reaching Conclusions

Mario said that the figures shown are all polygons because they are made up of line segments. Is he right or wrong?

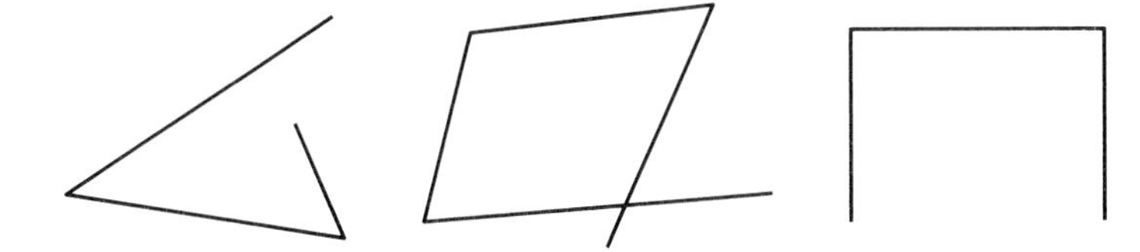

Mario's conclusion is wrong
because the definition of a polygon says the line segments *meet* at their endpoints.

The word ***conclusion*** means arriving at an answer based on other information. Mario did not understand the definition of a polygon, and this led him to draw a wrong conclusion.

Inductive Reasoning

An important goal of studying mathematics is to develop the ability to use information to arrive at correct conclusions. One way to obtain information is through simple observation.

Example 1 Karen planted 10 rose bushes. The first 8 flowers to bloom were yellow. She concluded that all her roses would be yellow.

This may or may not be a correct conclusion.
You must be careful in drawing conclusions from observations.

Example 2 In the figure, are line segments AB and CD straight?

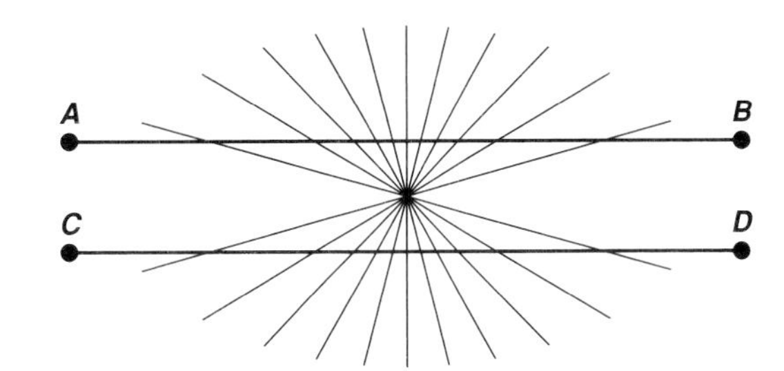

The segments appear to bend, but use a ruler to convince yourself that the line segments are straight.

Another type of observation is to find a pattern.

Example 3 Study this set of numbers. Can you predict the next three numbers?

123, 234, 345, 456, ?, ?, ?

The numbers form a pattern. Each group of three numbers begins with the second number in the group before it.

The next three numbers are 567, 678, and 789.

Be careful; sometimes patterns can lead to false conclusions.

A third type of observation is based on measurement.

Example 4 A ***square*** is a quadrilateral whose sides are equal in length. The diagonals are drawn in square *ABCD*.

Every square has two diagonals. Use a ruler to measure diagonals *AC* and *BD*. Are they equal in length? Do you think it is always true that the diagonals of a square are equal in length?

D C A B

You cannot draw a general conclusion from just one diagram. In this case, the general conclusion is correct, namely, the diagonals of a square are equal in length.

However, conclusions based upon measurements may or may not be correct because:

the measurements may not be accurate   you cannot measure all possible cases

Conclusions by Induction

Inductive reasoning **is a way of thinking that bases conclusions on observations, patterns, or measurements.**

Conclusions reached by inductive reasoning may or may not be correct because they are based on a limited number of examples.

Deductive Reasoning

Conclusions can also be drawn from known facts.

Example 5 A ***circle*** is a set of points that are all the same distance from a single point called the *center*.

The segment *PR*, which has one endpoint at the center of the circle and the other endpoint on the circle, is a ***radius*** of the circle.

Segment *PT* is also a radius of the circle. Is segment *PT* the same length as segment *PR*?

T R P

P is the center of the circle.

Conclusion: Segments *PT* and *PR* are equal in length.

This conclusion is correct because the definition of a circle tells us that points *T* and *R* are both the same distance from point *P*. This is an example of ***deductive reasoning***.

Example 6 This is a nonnumerical example of deductive reasoning.

1. All birds have wings.
2. Beni is a bird.

Conclusion: Beni has wings.

This conclusion is correct. Since Beni is a bird *and* all birds have wings, *then* Beni has wings.

Conclusions by Deduction

Deductive reasoning **is a way of thinking that bases conclusions on known facts.**
In mathematics, most conclusions are made by using deductive reasoning.

Try These *(For solutions, see page 39.)*

1. A boy sat outside the door of a supermarket in Cary, NC and counted the customers. He counted 238 females and 56 males. He concluded that in Cary, there are many more females than males. Is the boy correct?

2. A scientist discovered a medicine to cure a disease. She tries the medicine on 25 people and it cures 20 of them. She repeats the treatment on 25 more people and 22 are cured. She concludes that her medicine is successful in curing the disease. Is the scientist correct?

3. In order to be a member of the school band, a student must play a musical instrument. Sally is a band member. What conclusion can you make about Sally?

4. Frank measured the sides of six different polygons in a geometry book and found that, in each of the six, the sides were all the same length. He concluded that the sides of all polygons are always of equal length. Is Frank a good mathematician? Is this a correct conclusion?

5. Mr. Lapinski drew a line graph of his business sales for five years. In each year, his sales went up. Can he conclude from the graph that sales will go up again in the sixth year?

6. Mia observed that when multiplying two numbers on her calculator, it did not matter which number she entered first. The answer was always the same. She concluded that this was always true. What do you think?

7. Sato concluded that the next two numbers in the set are 64 and 81. Do you agree with her?

   2, 3, 4, 9, 4, 5, 16, 25, 6, 7, 36, 49, 8, 9, ?, ?

8. Avi told Jimmy that a circle is a *closed* figure because it has an inside and an outside. He drew this picture to show his idea. Avi also said that if Jimmy were standing inside the circle at point A, he would have to cross over the circle to get to point B.

   a. Avi asked Jimmy to name some other closed figures. Jimmy said a triangle is closed. Is Jimmy correct?

   b. Name another closed figure.

   c. Draw a figure that is not closed.

9. **a.** Enter the following multiplication problem on your calculator. Write your answer.

   b. Enter this: 123456 × 18 =
   Write down your answer.

   c. Enter this: 123456 × 27 =
   Write down your answer.

   d. Without using your calculator, write the answer to:

   $$123{,}456 \times 36$$

   Now check, using your calculator. Were you correct?

   e. Guess the missing multiplier, and check, using your calculator.

   $$123{,}456 \times ? = 6{,}666{,}624$$

SELF-CHECK: SECTION 1.6

1. While driving through Florida, Sue decides to count the number of white cars and the number of black cars that pass. After two hours, she has counted a total of 50 white cars, and 25 black cars. Sue concludes that the number of white cars is twice the number of black cars in Florida. Is her conclusion correct?

Answer: Sue's conclusion is probably not correct because she has counted a limited number of cars on one road for only two hours.

2. Jorge sees a pattern in this set of numbers: 2, 0, 22, 0, 222, 0, 2222, 0, 22222, 0

He concludes that the next number is 222222. Is he correct?

Answer: Jorge is probably correct. It is important to know, however, that a pattern can change.

3. If the sky is clear blue, then it is not raining. The sky is clear blue.

Conclusion: It is not raining. Is the conclusion correct?

Answer: The conclusion is correct by deductive reasoning.

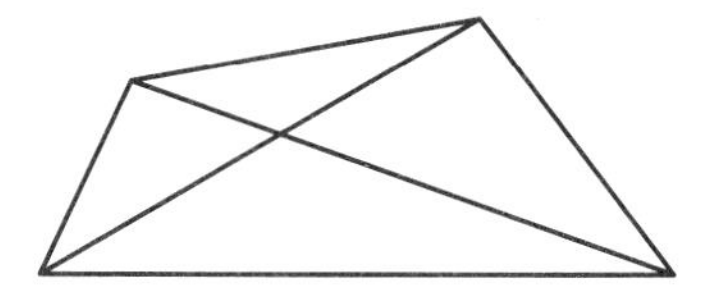

4. Louise draws a quadrilateral with its diagonals. She concludes that in every polygon the number of diagonals is two less than the number of sides. Is she correct?

Answer: No. A pentagon has 5 sides and 5 diagonals. Make your own drawing to show this.

EXERCISES: SECTION 1.6

1. Is each of the following conclusions necessarily correct? If not, tell why not.

a. On Monday, Marco counted the number of students who had chocolate milk with lunch. Out of 92 students that bought milk, he observed that 60 had chocolate milk. He concluded that the cafeteria manager should order twice as much chocolate milk as plain.

b. Lila rolled a 6-sided numbered cube (called a *die*) 200 times and recorded the frequency with which the numbers 1 through 6 appeared. After looking at the results, she concluded that it was a fair die.

Number	1	2	3	4	5	6
Frequency	30	28	28	31	30	53

c. As part of a project for his technology class, Al polled the students who ate in his lunch period. He counted the number who had part-time jobs versus those who did not. Out of the 375 polled, 190 said that they worked. Al concluded that about half of the students in the school had part-time jobs.

2. State a conclusion that can be correctly drawn from the given information.

   a. Monarch butterflies migrate in autumn to a warm climate.

      Milo saw thousands of Monarch butterflies hanging from a tree in November when he was on vacation.

   b. If a student is in homeroom 220, then he or she is in the 9th grade.

      Sue is in homeroom 220.

   c. Two angles are supplementary if the sum of their measures is 180 degrees.

      The sum of the measures of angle A and angle B is 180 degrees.

   d. Evergreen trees do not shed all their needles in one season.

      A pine is an evergreen.

   e. If Leah drives her car to work, she must get a parking permit.

      Leah drives to work.

3. Draw a conclusion, if possible.

   a. If a student is on the soccer team, his or her picture will be in the yearbook.

      Tom's picture is in the yearbook.

   b. To be in the World Series, a team must win its league pennant.

      In 1986, Boston was in the World Series.

   c. All elephants have large ears.

      Wendy is an elephant.

   d. When Lou misses the bus, he rides home with a friend.

      Lou rode home with his friend.

   e. At Wood High, all 9th graders take a mathematics course.

      Sam is in the 9th grade.

   f. If you bring your lunch to school, you must eat in the cafeteria.

      You ate in the cafeteria.

4. a. Enter each of the following on your calculator, and write the answer.

      (1) 101 [×] 11 [=]

      (2) 101 [×] 111 [=]

      (3) 101 [×] 1111 [=]

   b. Without using your calculator, guess the answer to: $101 \times 11{,}111$

      Check using your calculator.

   c. Draw a conclusion about what must be the missing multiplier:

      $101 \times\ ?\ = 2{,}244{,}422$

      Check, using your calculator.

5. Truman was doing his homework at the kitchen table and water dripped onto his notebook, blotting some exercises. Rewrite each problem, filling in the missing digits.

a.
$$\begin{array}{r} 89? \\ +\ \ ?7 \\ \hline ?60 \end{array}$$

b.
$$\begin{array}{r} ?713 \\ -\ \ ??8 \\ \hline 5785 \end{array}$$

c.
$$\begin{array}{r} 2?8? \\ \times\ \ \ 43 \\ \hline 1??198 \end{array}$$

d.
$$2?\overline{)?7207}\quad = ?283$$

6. A total of 103 social studies students signed up for the bus trip to the state capital. The buses accommodate no more than 32 riders, and there will be a teacher or a parent riding in each bus. Mr. Fenton ordered 3 buses. Was that a good decision?

(*a*) Yes, because $\frac{103}{32} = 3.21875$, which rounds to 3 as the nearest whole number.

(*b*) Yes, because even if he added an adult passenger in each bus, for a total of 103 + 3 = 106, dividing by 32 would still round to 3 as the nearest whole number.

(*c*) No, because he should have used exactly the number calculated, without rounding.

(*d*) No, because in this situation you must round up to the next larger whole number.

EXPLORATIONS

To find the number of degrees in the measure of an angle, we use an instrument called a *protractor*. For convenience in reading the measure of an angle regardless of its position, a protractor has two scales, each starting at 0° and ending at 180°.

Before using a protractor to read the measure of a particular angle, decide whether that angle contains more or less than 90°.

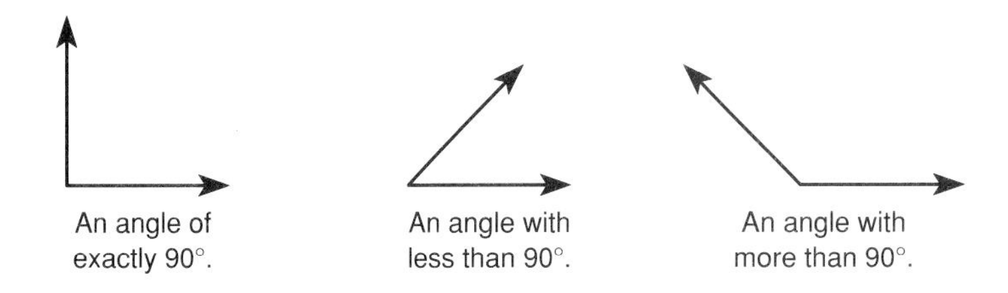

To get an accurate measure, it is important to properly place the protractor on the angle.

For example, to read the measure of angle ABC, which contains less than 90°:

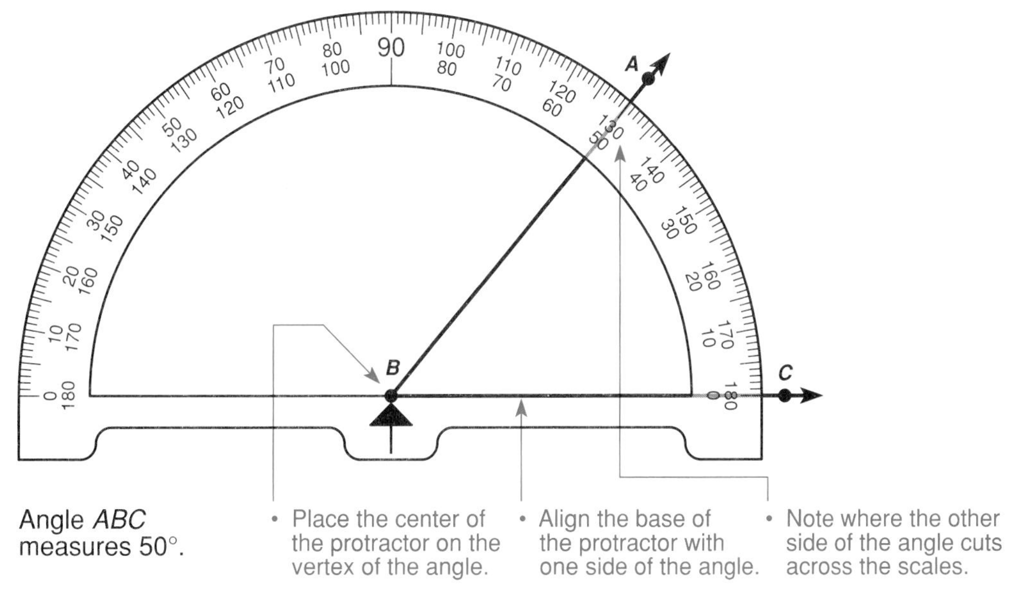

a. Use a protractor to find the measures of the angles numbered 1-5.

b. What is the sum of the measures of all 5 angles around point M?

c. Now you pick a point and draw angles around it, making sure that the point is the vertex of each angle. Use a protractor to find the measures of all these angles. What is their sum?

d. What general conclusion might you reach?

e. Draw any triangle. Use a protractor to find the sum of the measures of the 3 angles. Draw 2 more triangles, of different shapes and in different positions, and use a protractor to find the sum of the measures of the three angles. Use your observations to reach a conclusion about the sum of the measures of the angles of a triangle.

f. Conduct an experiment with a protractor to reach a conclusion about the sum of the measures of the angles of a quadrilateral.

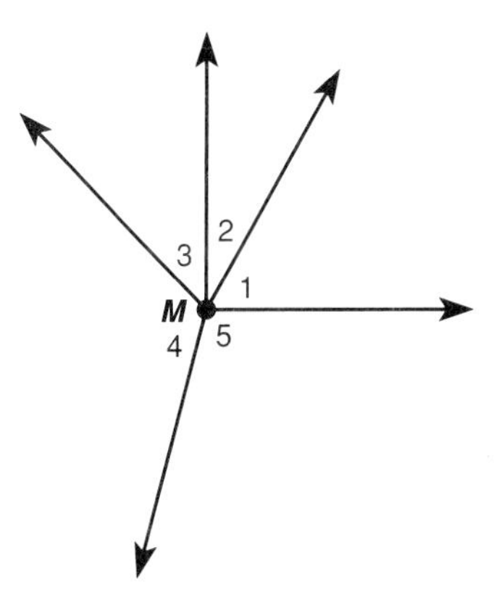

Things You Should Know After Studying This Chapter

KEY SKILLS

1.1 Recognize a good definition.

1.2 Identify the question in a problem situation.

1.3 Use a calculator.

1.4 Read a table.

1.5 Read a bar graph, pictograph, and line graph.

1.6 Reach a conclusion using inductive or deductive reasoning.

KEY TERMS

1.1 definition • point • line segment • triangle • geometric figure • set

1.2 problem situation • solution

1.3 calculator keys • exponent

1.4 table • polygon • diagonal

1.5 graphs • data

1.6 conclusion • pattern • inductive reasoning • deductive reasoning • square • circle • radius

CHAPTER REVIEW EXERCISES

1. Is each of the following a good definition? If not, give a reason for your decision.

a. A *senior* is a student in the last year of high school who graduates in June.

b. A *ruler* is an instrument used to measure length.

In 2 and 3, write a question that might be asked.

2. Doreen worked 32 hours last week. Her gross pay was \$236.80.

3. After a two-and-a-half-hour trip, the Taylor family arrived at Sea World at 1:15 P.M.

4. Use a calculator to compute.

a. $253 \times 69 \div 3 - 219$

b. Divide the sum of 1,038 and 743 by 13.

c. Find the sum $619 + 407 + 73 + 919 + 86$, then multiply by 32.

5. Use the table to answer the following questions about ball size.

Type	Diameter
baseball	7.6 cm
basketball	24.0 cm
golf ball	4.3 cm
soccer ball	22.0 cm
tennis ball	6.5 cm

a. Which ball has the largest diameter?

b. Which ball has a diameter that is closest to that of a golf ball?

c. Which ball has a diameter that is approximately three times that of a baseball?

6. This bar graph shows the money earned by four students during their summer vacation.

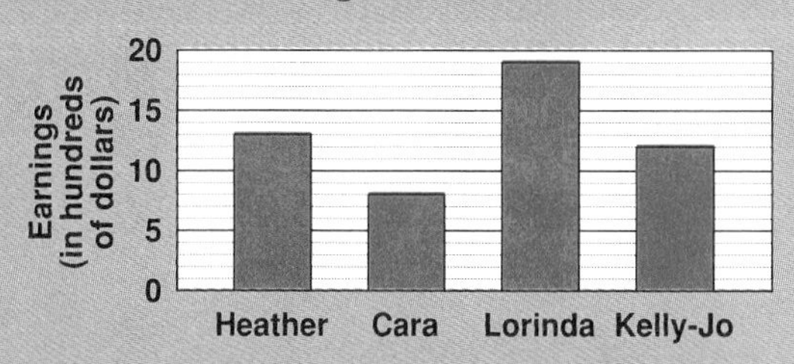

a. How much money did Heather earn?

b. How much more money did Lorinda earn than Cara?

c. What was the total earned by the four girls?

d. How much less than \$2,000 did Lorinda earn?

7. The pictograph shows how many bicycles were rented at a resort last summer.

a. How many more were rented in July than in June?

b. What was the total number rented last summer?

c. Why do you think July was the busiest month?

(Exercises continue)

CHAPTER REVIEW EXERCISES

8. This graph represents the speed of a race car during the last lap of a race, as clocked at 7 points around the track.

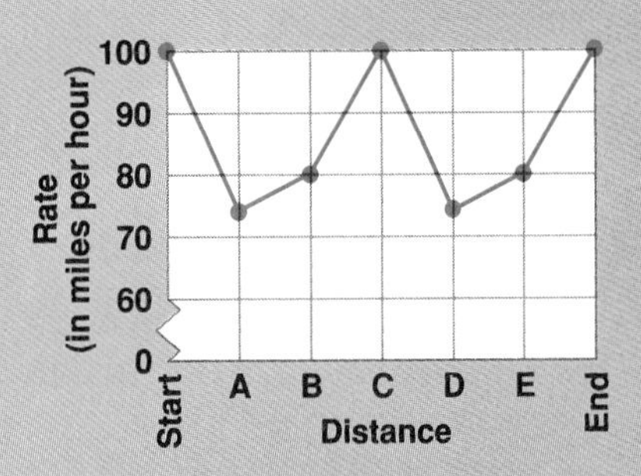

   a. What was the highest speed of the car during this lap?

   b. What was the difference between the highest and lowest speeds?

   c. Name the point(s) where the car is at its highest average speed.

   d. At what point has the car completed $\frac{2}{3}$ of a lap?

   e. Explain what the driver is doing between points A and C.

   f. Between what points would you say that the car is rounding a curve? Explain.

9. Caitlin observed that there are 18 boys and 9 girls in her math class. She concluded that twice as many boys as girls take high school math. Why is this not a correct conclusion? What conclusion could she draw from the given data?

10. The manager of a mall shoe store asked her clerk to sit in the mall and tally the type of sneaker worn by those passing by. The tally shows 100 pump type and 110 non-pump type. The manager decided that she should order an equal number of the two types. Do you agree? Justify your answer.

11. Where possible, draw a conclusion. If no conclusion is possible, explain why not.

    a. If John studies, he will pass.
       John studies.

    b. All sophomores must take a foreign language.
       Bill is taking French.

    c. All even numbers are divisible by two.
       X is divisible by 2.

    d. When Erica is sick, she doesn't go to school.
       Erica isn't in school.

    e. If it is raining, Mel carries an umbrella.
       Mel is not carrying an umbrella.

SOLUTIONS FOR TRY THESE

1.1 Good Definitions

Page 3

1. This is not a good definition. It contains the unnecessary information "who takes mathematics and art."
2. This is a good definition.
3. This is not a good definition. If does not name the term being defined.
4. This is not a good definition. It omits information (that it grows into an oak), and contains unnecessary information (that it falls from a tree).
5. This is not a good definition. It does not give distinguishing characteristics.

1.2 Questions and Answers

Page 7

1. At what time will the tape end?
2. What is the total distance for the 5 days?
3. How big is the dog's play area?
4. Are the extra power (more watts) and the extra settings worth the additional cost?
5. How many miles will they have to drive each day in traveling to their friends and back?
6. How many people were at the party?

1.3 Keys to Technology

Page 12

1. 11,117	**2.** 177
3. 3,381	**4.** 427
5. 1,328	**6.** 11,390,625
7. 502	**8.** 1,495
9. 49	**10.** 7,424

1.4 Reading Tables

Page 17

1. Franco **2.** Joe **3.** butterfly

4. freestyle, Joe; backstroke, Carlos; butterfly, Franco

5. freestyle, Joe and Franco; backstroke, Carlos and Brent; butterfly, Franco and Brent

1.5 Reading Graphs

Page 23

1. **a.** vanilla; cherry
   b. cherry, 2; vanilla, 7; chocolate, 6; strawberry, 5
   c. 20 **d.** 7
2. **a.** 5,000
   b. The graph shows the sharpest rise in the interval 1980-1985.

1.6 Reaching Conclusions

Page 31

1. Not necessarily. The females in Cary may be more likely to do the family marketing.
2. Probably. Though her results show that most people are cured, many more tests must be made.
3. Sally plays a musical instrument.
4. Frank is not a good mathematician; he should not base a conclusion on a few cases.
   The conclusion is not correct; example:

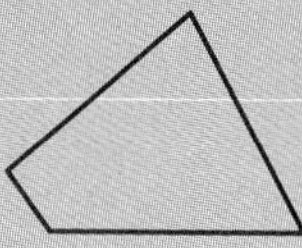

5. No; he cannot be sure that the pattern will continue.
6. Mia is right.
7. If the pattern continues, Sato is right: two numbers, then their squares, then the next two numbers and their squares, etc.
8. **a.** Jimmy is right.
   b. Since segments meet at their endpoints, any polygon is a closed figure.
   c. Some figures that are not closed:

9. **a.** 1111104 **b.** 2222208
   c. 3333312 **d.** 4444416
   e. 54

Problem Solving 2

Maria wanted to help her little brother learn about the numbers 1–9. She decided to make piles of pennies with 1 in the first pile, 2 in the second, 3 in the third, and so on, through 9 in the ninth pile. How many pennies did Maria need?

For success with problem solving, it is important to work in an organized way.

Key Problem-Solving Steps

1. ***Read the problem.***
2. ***Make a plan.***
3. ***Use the plan.***
4. ***Check the solution.***

In this chapter, you will see some approaches to solution that will help you plan. These approaches are called ***strategies***.

2.1 Finding a Pattern

Finding a pattern is a strategy that can be used to solve a problem.

Example 1 What are the next three numbers in this pattern?

1, 1, 2, 3, 5, 8, ?, ?, ?

Follow the four key steps to solve the problem.

1 ***Read the problem to understand it.***
You must find the next three numbers in the pattern.

2 ***Make a plan to solve the problem.***
Since the numbers are increasing, try addition or multiplication to find the pattern.

3 ***Use your plan to solve the problem.***
Try addition. Observe that the sum of the first two numbers equals the third.
$1 + 1 = 2$
See if this works for the next two numbers.
$1 + 2 = 3$
Yes, the sum of the second and third numbers is the fourth number.
Keep going. Is this pattern working?
$2 + 3 = 5$
$3 + 5 = 8$
Yes, the sum of the third and fourth numbers is the fifth number, and the sum of the fourth and fifth is the sixth.
Success. Then, if the pattern continues, the next three numbers are:
$5 + 8 = 13$
$8 + 13 = 21$
$13 + 21 = 34$

4 ***Check your solution.***
Write out the pattern. Redo your additions.
Answer: 1, 1, 2, 3, 5, 8, 13, 21, 34

Example 2 Find the sum of the first 25 odd numbers. The odd numbers are 1, 3, 5, 7, 9, 11, and so on.

Let's use the four key steps again.

1 ***Read the problem. What does it ask you to do?***
Add the first 25 odd numbers to find their sum.

2 ***Make a plan.***
Use a pattern to find the sum.

3 ***Solve the problem.***
Begin by finding the sum of the first 2 odd numbers.

$1 + 3 = 4$

Next, find the sum of the first 3 odd numbers.

$1 + 3 + 5 = 9$

Continue in this way.
The sum of the first 4 odd numbers is:

$1 + 3 + 5 + 7 = 16$

The sum of the first 5 odd numbers is:

$1 + 3 + 5 + 7 + 9 = 25$

Look at the sums. Do you see a pattern?

First 2: $1 + 3 = 4 \longrightarrow 4 = 2 \times 2$
First 3: $1 + 3 + 5 = 9 \longrightarrow 9 = 3 \times 3$
First 4: $1 + 3 + 5 + 7 = 16 \longrightarrow 16 = 4 \times 4$
First 5: $1 + 3 + 5 + 7 + 9 = 25 \longrightarrow 25 = 5 \times 5$

What do you predict as the sum of the first 6 odd numbers? Check to see that this sum is 6×6, or 36.
Using this pattern, you may conclude that the sum of the first 25 odd numbers is 25×25, or 625.

4 ***Check the solution by using a calculator.***
Answer: The sum of the first 25 odd numbers is 625.

Try These *(For solutions, see page 76.)*

1. Write the next three numbers in each pattern.
a. 0, 4, 8, 12, ?, ?, ?
b. 84, 78, 72, 66, ?, ?, ?

2. Use the following pattern to find $5 \times 9{,}999$ and $6 \times 9{,}999$.

$1 \times 9{,}999 = 9{,}999$
$2 \times 9{,}999 = 19{,}998$
$3 \times 9{,}999 = 29{,}997$
$4 \times 9{,}999 = 39{,}996$

3. a. Identify the following pattern. Write your answer in words.

$2 = 2 \longrightarrow 2 = 1 \times 2$
$2 + 4 = 6 \longrightarrow 6 = 2 \times 3$
$2 + 4 + 6 = 12 \longrightarrow 12 = 3 \times 4$
$2 + 4 + 6 + 8 = 20 \longrightarrow 20 = 4 \times 5$

b. Use the above pattern to find each sum.
(1) $2 + 4 + 6 + 8 + \ldots + 16$
(2) $2 + 4 + 6 + 8 + \ldots + 24$
(3) $2 + 4 + 6 + 8 + \ldots + 32$
Check your answers using a calculator.

Try These (continued)

4. Study the following pattern to find the missing value.

   $1 \times 1 = 1$
   $11 \times 11 = 121$
   $111 \times 111 = 12321$
   $1111 \times 1111 = ?$

   Use a calculator to check your answer.

5. John and Juan played a game called *Guess the Rule*. Juan told John to name any number and he would use the rule to name another number.

   John said 2; Juan replied 5.
   John said 3; Juan replied 7.
   John said 5; Juan replied 11.

   a. If John says 6, what would Juan reply?
   b. Write the rule that Juan is using.

SELF-CHECK: SECTION 2.1

1. What are the next three numbers in the pattern?

1, 1, 2, 3, 5, 8, 13, 21, 34, ?, ?, ?

This is the pattern in Example 1.

Answer: 55, 89, 144

2. Find the sum of the first 75 odd numbers.

Use the pattern in Example 2.

Answer: $75 \times 75 = 5{,}625$

EXERCISES: SECTION 2.1

1. Find the next three terms in each of the following:

a. 3, 7, 11, 15, ...	**b.** 4, 7, 10, 13, ...	**c.** 1, 7, 13, 19, ...
d. 5, 12, 19, 26, ...	**e.** 10, 25, 40, 55, ...	**f.** 37, 32, 27, 22, ...
g. 18, 15, 12, 9, ...	**h.** 113, 108, 103, 98, ...	**i.** 3, 6, 12, 24, ...
j. 2, 10, 50, 250, ...	**k.** 1, 4, 16, 64, ...	**l.** 2, 6, 18, 54, ...
m. $\frac{1}{4}, \frac{1}{9}, \frac{1}{16}, \frac{1}{25}, \ldots$	**n.** 1, 8, 27, 64, ...	**o.** 10; 100; 1,000; 10,000; ...
p. 2, 4, 8, 16, ...	**q.** 1, 3, 6, 10, 15, ...	**r.** 3, 5, 8, 12, 17, ...
s. 3, 5, 8, 13, 21, ...	**t.** 1, 3, 4, 7, 11, 18, ...	**u.** 128, 64, 32, 16, ...

2. a. Complete the next 2 terms of the sequence:

```
                + + +
        + +     + + +
  +     + +     + + +
```

b. What sequence of numbers does this pattern represent?

3. You have seen that to get the next number in the sequence 1, 1, 2, 3, 5, ..., you add the previous two numbers.

These special numbers, named after the mathematician who discovered them, are called the ***Fibonacci Numbers***.

The sums of the squares of these special numbers also form a pattern.

Complete the following statements.

$1^2 + 1^2 = 2 = 1 \times 2$

$1^2 + 1^2 + 2^2 = 6 = 2 \times 3$

$1^2 + 1^2 + 2^2 + 3^2 = 15 = 3 \times ?$

$1^2 + 1^2 + 2^2 + 3^2 + \boxed{?}^2 = ? = ? \times 8$

$1^2 + 1^2 + 2^2 + 3^2 + \boxed{?}^2 + \boxed{?}^2 = ? = 8 \times ?$

$1^2 + 1^2 + 2^2 + 3^2 + \boxed{?}^2 + \boxed{?}^2 + \boxed{?}^2 = ? = ? \times ?$

$1^2 + 1^2 + 2^2 + 3^2 + \boxed{?}^2 + \boxed{?}^2 + \boxed{?}^2 + \boxed{?}^2 = ? = ? \times ?$

4. a. Copy and complete the following statements.

$1 = 1 = 1 \times 2 \div 2$

$1 + 2 = 3 = 2 \times 3 \div 2$

$1 + 2 + 3 = 6 = 3 \times 4 \div 2$

$1 + 2 + 3 + 4 = ? = 4 \times ? \div 2$

$1 + 2 + 3 + 4 + 5 = ? = 5 \times ? \div ?$

$1 + 2 + 3 + 4 + 5 + 6 = ? = ? \times ? \div ?$

$1 + 2 + 3 + 4 + 5 + 6 + 7 = ? = ? \times ? \div ?$

b. Write a rule for adding a sequence of counting numbers beginning with 1.

c. Using the rule, add the following:

(1) $1 + 2 + 3 + ... + 20$

(2) $1 + 2 + 3 + ... + 56$

d. Liz was paid $1 for her first day of work, $2 the second, $3 the third, and so on.

(1) How much was she paid on the 100th day?

(2) What was the total amount paid for the 100 days of work?

e. How much money was Tony paid per day if he was paid exactly the same amount each day and earned the same amount as Liz for 100 days?

5. Parts of each of the following circles are shaded. The total amount shaded is the same in 3 of the 4 circles. In which one is a different amount shaded?

(a)

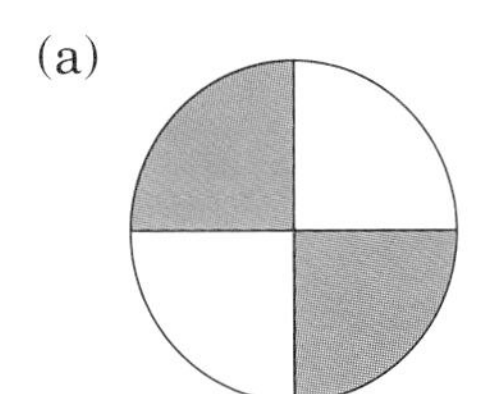

(b)

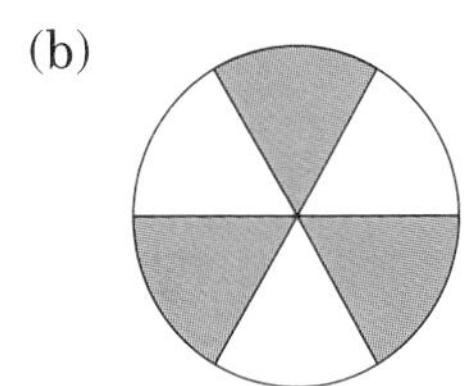

(c)

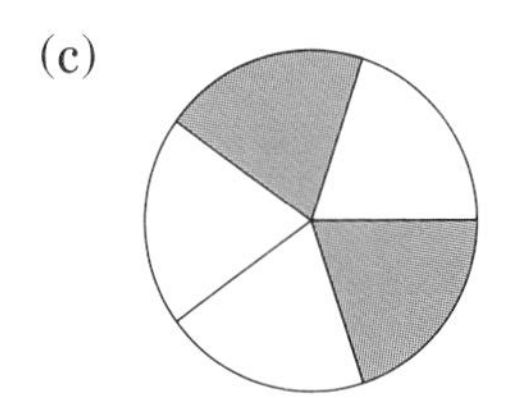

(d)

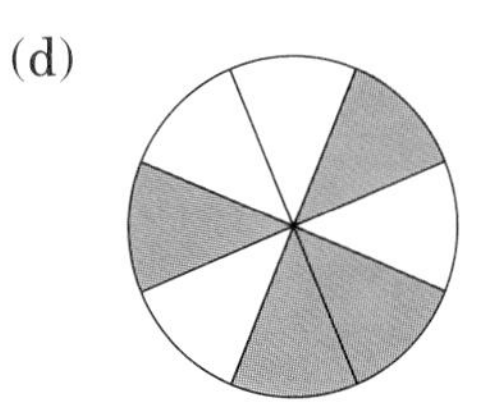

6. Starting with the innermost ○, the sequence ○, □, ☆, △ follows a pattern. If the pattern continues, what will be the next three figures?

△	△	☆	☆	☆	□	□	○	△
△								☆
△		△	△	△	△	☆		□
○		○				☆		○
□		□				☆		△
•		☆		○	□	□		△
•		△						△
•		○	□	□	☆	☆	☆	△

7. Fold a sheet of paper into two equal parts.
   a. If you fold the paper again, how many parts are there?
   b. How many parts are there after three folds?
   c. Use a pattern to find the number of parts after six folds. Check your answer by folding the paper.

8. In each successive diagram, the number of small triangles forming the base of the figure is increased by 1.
   If the pattern continues, what is the total number of small triangles there would be in a large triangle that has 8 small triangles forming its base?

△
1

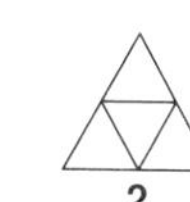

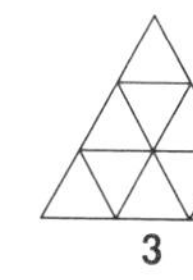

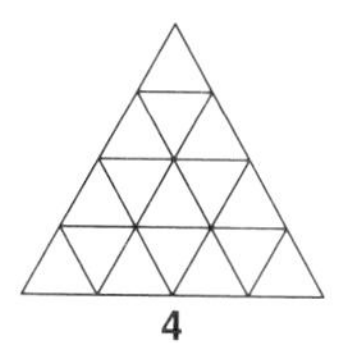

9. 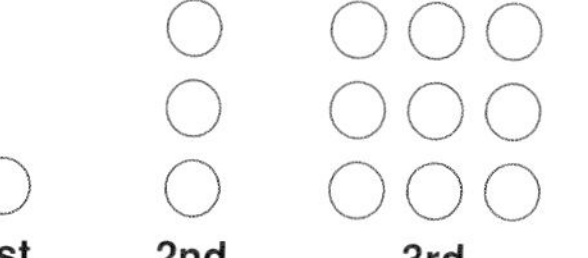

The figures show the growth pattern of an unusual cell.
To begin, it is a single cell and, in its second stage, it splits into 3. At each new stage each new cell splits into 3.
How many cells are there in the fifth stage?

10. **a.** Use a calculator to find the decimal equivalents of these fractions:
   $\frac{1}{11}, \frac{2}{11}, \frac{3}{11}, \frac{4}{11}, \frac{5}{11}$
   Write a rule to describe the pattern.
   b. Using the pattern, write the decimal equivalents of:
   (1) $\frac{9}{11}$ **(2)** $\frac{10}{11}$ **(3)** $\frac{11}{11}$

11. Some number patterns can be described by using the letter n to represent a given number. For example, $2n + 1$ says to double a given number and add 1.
   Here are some number games in which you will be asked to represent the rule using the letter n.
   a. In this game, Kay says a number and Paul replies.
   Kay says 8; Paul replies 24.
   Kay says 2; Paul replies 6.
   Kay says 7; Paul replies 21.
   (1) If Kay says 100, what number does Paul reply?
   (2) If Paul replies 33, what number has Kay said?
   (3) If n represents the number Kay chooses, write a rule using n that will represent Paul's response.

b. In this game, after Meg gives her Mom some beans, her Mom gives beans to Meg.

Meg gave Mom 1 bean; Mom gave 1.

Meg gave Mom 2 beans; Mom gave 4.

For 3 beans, Meg received 9 in return.

(1) How many beans did Meg get in return for 4 beans? for 6 beans?

(2) If Mom returns 25 beans, how many had Meg given?

(3) If n represents the number of beans Meg gives Mom, write a rule using n to represent the number Mom returns.

c. In this game, Bill says a number and Abe replies.

Bill says 5; Abe replies 9. Bill says 14; Abe replies 27. Bill says 40; Abe replies 79.

(1) If Bill says 8, what number does Abe reply?

(2) If Abe answers 11, what number has Bill said?

(3) If n represents the number Bill says, write a rule using n to represent the number Abe replies.

12. The figures show why the numbers 1, 3, 6 are called ***triangular numbers***.

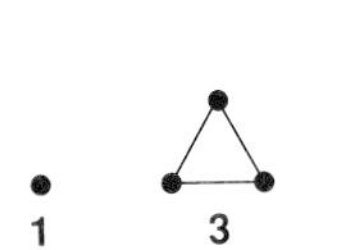

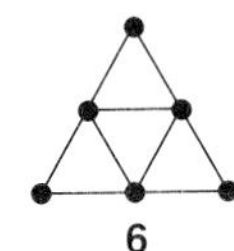

a. Draw the next two figures to find the next two triangular numbers.

b. Study the triangular numbers on the *right* side of the first three equations shown to determine a pattern, and find the next missing expressions.

Use a calculator to check your answers.

$1^3 = 1^2$

$1^3 + 2^3 = 3^2$

$1^3 + 2^3 + 3^3 = 6^2$

$1^3 + 2^3 + 3^3 + 4^3 = ?$

$1^3 + 2^3 + 3^3 + 4^3 + 5^3 = ?$

c. Using the pattern and a calculator, find the value of:

$1^3 + 2^3 + 3^3 + 4^3 + 5^3 + 6^3 + 7^3 + 8^3 + 9^3 + 10^3$

13. The figures show why the numbers 1, 5, 12, 22 are called ***pentagonal numbers***

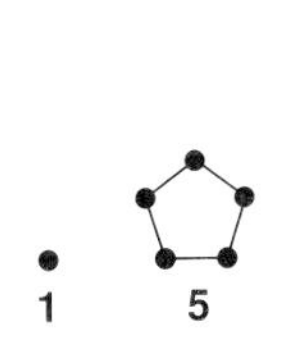

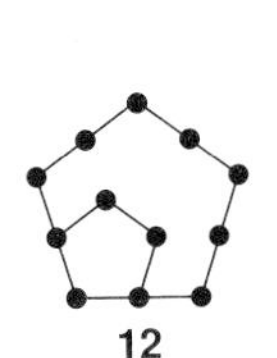

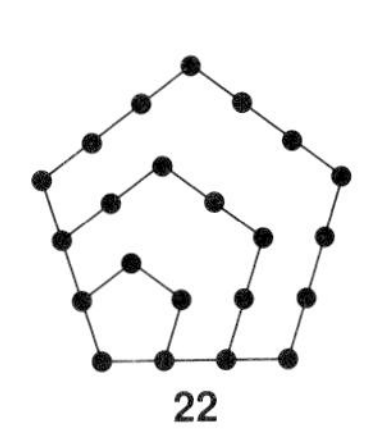

a. Draw the next figure to find the next pentagonal number.

b. Find the sixth pentagonal number by studying the first five to look for a pattern.

EXPLORATIONS

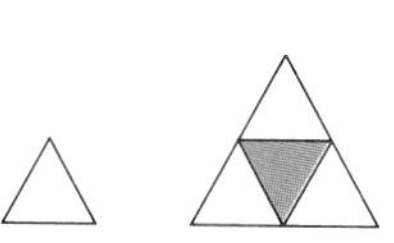
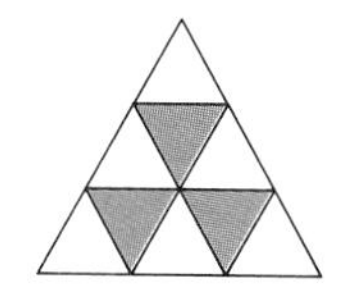

1. These drawings are related to triangular numbers. The individual triangles are placed together to form one large triangle.

Notice that when 3 triangles are placed together, a 4th (shaded) triangle is formed. When 6 triangles are placed together, 3 other (shaded) triangles are formed.

Artists and musicians call the original number of triangles the *figure* and the shaded triangles the *ground*.

a. Place 10 triangles together and tell the number of triangles in the ground. Do 15 triangles. Copy and complete this table.

Figure	Ground	Total
1	0	1
3	1	4
6	?	?
10	?	?
15	?	?

b. Study the table to look for a pattern in each column. Describe the patterns.

2. This pattern of numbers that visually forms a triangle, named after a mathematician, is called ***Pascal's Triangle***. Notice that each row begins and ends with 1. All other numbers in a row are found by adding the pairs of numbers to the right and left above.

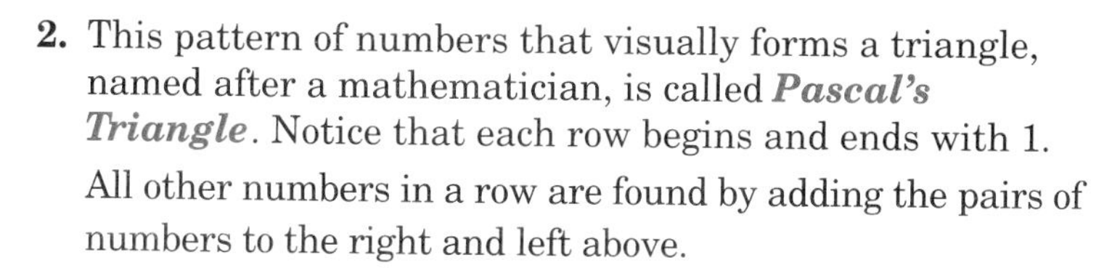

1	row 0
1 1	row 1
1 2 1	row 2
1 3 3 1	row 3
1 4 6 4 1	row 4

a. Write the numbers in rows 5–8.

b. Copy and complete this table.

Row Number	Sum of Numbers in a Row
1	2
2	4
3	?
4	?
5	?
6	?
7	?
8	?

c. Describe the pattern in the sum column.

d. **(1)** Use your calculator to find the value of 11^2, 11^3, and 11^4. Tell how these values are related to Pascal's Triangle.

(2) Now calculate 11^5. Does it fit your pattern?

e. Adding numbers from Pascal's Triangle, as circled in this diagram, produces a set of numbers you have seen before. What is this sequence of numbers called?

1
1 1
1 2 1
1 3 3 1
1 4 6 4 1
1

f. Many other patterns are hidden in this triangle. Find one more.

2.2 Organizing Data in a Table

Another problem-solving strategy is to organize your information in a table.

Example 1 Susan has 20 pennies, 4 nickels, and 2 dimes. In how many ways can she spend exactly 20 cents using these coins?

Use the four key problem-solving steps to solve this problem.

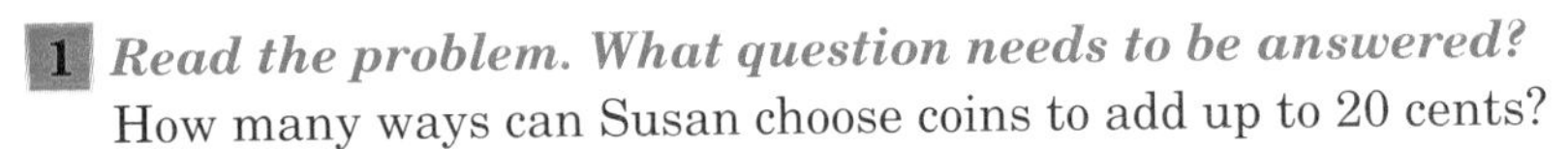

1 ***Read the problem. What question needs to be answered?***
How many ways can Susan choose coins to add up to 20 cents?

2 ***Make a plan.***
Use a table to show the different ways the coins can have a total value of 20 cents.

3 ***Use your plan.***
Solve the problem by making a table. Begin by using the highest value coin first.

Number of Different Ways	Number of dimes	Number of nickels	Number of pennies	Total cents
1	2	0	0	20
2	1	2	0	20
3	1	1	5	20
4	1	0	10	20
5	0	4	0	20
6	0	3	5	20
7	0	2	10	20
8	0	1	15	20
9	0	0	20	20

There appear to be 9 ways.

4 ***Check your answer.***
Reread the problem and check the table to make sure no arrangement of coins has been omitted.

Answer: There are 9 ways Susan can spend exactly 20 cents.

In addition to using a table to list all the possibilities for a situation, you can use a table to find a pattern.

Example 2 A bus leaves New York City with 27 passengers. At the first stop, 6 people get off. At the second stop, 3 people get on. At the third stop, 6 people get off. At the fourth stop, 3 people get on. If this pattern continues, at which stop will the last passenger get off?

1 ***Read the problem. What is the question?***

How many stops will it take for the bus to have no passengers?

2 ***Make a plan to solve the problem.***

Organize the information in a table.

3 ***Use your plan.***

Make a table and fill in what you know.

Stop	Change in Passengers	Total on Bus
NYC	—	27
1	6 off	21
2	3 on	24
3	6 off	18
4	3 on	21
5	6 off	15
6	3 on	18
7	6 off	12

Instead of continuing with the table, look for a pattern in how the total number of passengers changes at each stop.

One pattern is that after the first stop, the bus has 3 fewer passengers at every odd-numbered stop. Continue the pattern until there are no passengers.

Stop	Total on Bus
9	9
11	6
13	3
15	0

4 ***Check your solution.***

There are 8 odd-numbered stops.

At the first stop, 6 passengers get off: $27 - 6 = 21$

At each of 7 other odd-numbered stops, there are 3 fewer passengers: 7×3, or 21

After the 8th odd-numbered stop, no one is left: $21 - 21 = 0$

The 8th odd-numbered stop is the 15th stop.

Answer: The last passenger gets off at the 15th stop.

A third use for making a table is to assist in drawing a conclusion.

Example 3 Bob Teacher, Ben Butcher, and Bud Rancher are a teacher, a butcher, and a rancher. Their occupations do *not* match their last names. Bob Teacher is the rancher's uncle. Who is the teacher?

1 ***Read the problem. What is the question?***
Who is the teacher?

2 ***Make a plan to solve the problem.***
Use a table to organize the facts so that you will be able to draw a conclusion.

3 ***Use your plan.***

	teacher	butcher	rancher
Bob Teacher	*a.* no	*c.* yes	*b.* no
Ben Butcher		*a.* no	
Bud Rancher		*d.* no	*a.* no

The letters *a-d* match the steps below and show how the table was filled in.

a. The occupation of each person does *not* match his name.
b. Since Bob is the rancher's uncle, Bob cannot be the rancher.
c. Bob must be the *butcher* since he is not the teacher or rancher.
d. Since Bob is the butcher, Bud cannot be the butcher.

Since Bud Rancher is not the butcher or rancher, he must be the teacher.

4 ***Check the answer.***
Reread the problem and check the table to be sure each entry makes sense.

Answer: Bud Rancher is the teacher.

Try These *(For solutions, see page 76.)*

1. Sam has $100 to spend on balls for a sports program. Softballs are $9 each and baseballs are $7 each. Sam wants to use as much as he can of the $100. What are all possible orders he can place totaling at least $96, but not more than $100?

2. You have one penny, one nickel, and one dime.
How many different amounts of money can you make by taking 1, 2, or 3 of the coins?

3. Find the sum of the first fifteen odd numbers.

4. Art, Bill, and Cliff have the pets Allie, an alligator, Burt, a basset hound, and Cindy, a cat, not in that order.
Find each person's pet, if: *a.* No man has the same initial as his pet.
b. Bill is allergic to cats.

SELF-CHECK: SECTION 2.2

1. a. In how many ways can you give 65¢ in change using quarters, dimes, and nickels if you give no more than 5 of any one kind of coin?

Use a table to show the different ways.

	Number of quarters	Number of dimes	Number of nickels	Total cents
1	2	1	1	65
2	2	0	3	65
3	1	4	0	65
4	1	3	2	65
5	1	2	4	65
6	0	5	3	65
7	0	4	5	65

Answer: 7 ways

b. A quarter, a dime, and 6 nickels add up to 65¢. Why is this set of coins not included in the table?

Answer: More than 5 of one kind of coin is used.

2. Clyde's Carnival is giving away 500 bags of peanuts: 2 on day 1, 4 on day 2, 6 on day 3, and so on. On what day will all 500 bags have been given away?

Make a table to organize the information so that you can find a pattern.

Day	Number of Bags	Sum of Bags	Pattern for the Sums
1	2	2	2, or 1×2
2	4	$2 + 4$	6, or 2×3
3	6	$2 + 4 + 6$	12, or 3×4
4	8	$2 + 4 + 6 + 8$	20, or 4×5
5	10	$2 + 4 + 6 + 8 + 10$	30, or 5×6
6	12	$2 + 4 + 6 + 8 + 10 + 12$	42, or 6×7

Now, use a calculator to find the 2 consecutive numbers whose product is just over 500. $22 \times 23 = 506$

Answer: The 500th bag is given away on the 22nd day.

3. Jenny, Zoe, and Carrie are sisters who enjoy different sports. One likes diving, another bowls, and the third snorkels. Find the favorite sport of each sister, if:

a. Carrie does not like water sports.
b. Zoe is the youngest.
c. The oldest likes diving.

Make a table organizing the facts so that you can reason out the situation and draw a conclusion. The letters *a–f* show the order of the reasoning.

	diving	bowling	snorkel
Jenny	*f.* yes	*a.* no	*e.* no
Zoe	*c, b.* no	*a.* no	*d.* yes
Carrie	*a.* no	*a.* yes	*a.* no

Answer: Jenny: diving, Zoe: snorkeling, Carrie: bowling

EXERCISES: SECTION 2.2

1. In how many ways could you give $45 in change using $20's, $10's and $5's?

2. In how many ways could you give $65 in change using $20's, $10's and $5's, if you give no more than four of any kind of bill?

3. Shana has 6 coins in her purse, totaling 48¢. What are the coins?

4. **a.** The sophomore class took orders for candy at $5 and $3 per box. Lee sold 7 boxes and collected $29. How many boxes of each type of candy did he sell?

   b. Mark also took candy orders but lost his list of orders. He collected $27 and remembered that he sold more than one $5 box. How many boxes of each price candy should he order?

5. You are given $10 to spend on packages of baseball cards. Package A costs 80¢ and Package B costs $1.40. What possible combinations of Packages A and B can you buy, if you want to spend exactly $10?

6. In a rectangle:

   area = length × width
   perimeter = distance around

   The area of the rectangle shown is $3 \times 2 = 6$ square units.
   The perimeter is $2 + 3 + 2 + 3 = 10$ units.

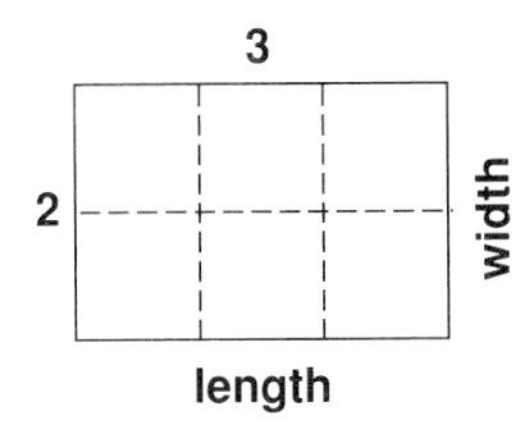

   In a given rectangle, the area is 30 square centimeters, and the lengths of the sides are whole numbers. Find the length and the width if the rectangle has:

   a. the least possible perimeter
   b. the greatest possible perimeter

7. Find the dimensions of the rectangle with area 270 sq. ft. and perimeter 66 ft.

8. A scientist trained three smart chimps.

Annie could take a number, multiply it by the number that comes *after*, divide by 2, and square the result.

Beth, starting with the same number as Annie, could multiply by the number that comes *before*, divide by 2, and square the result.

Carl could subtract Beth's answer from Annie's.

Annie and Beth started with 1, then 2, then 3, then 4.

By the time they worked through 4, Carl could see a pattern in the results of his subtractions. What pattern did Carl see?

9. Suppose you wrote an A, then two B's, then three C's and so on. What would you write as the 30th letter?

10. Ann, Dan, Jan, and Van all study Math, Biology, History and Spanish, but have different favorites. Find each student's favorite subject, if:

a. One of Ann's classmates in the group likes Biology.
b. Jan and Van do not like Spanish.
c. Dan's favorite is History.
d. Van liked Math but has changed his favorite.

11. Four friends, Regina, Roberto, Carlo, and Veronica, attend the same high school. Study halls are before lunch, period 2, and after lunch, period 5. Lunch is held two separate times, called Lunch A and Lunch B. Determine which lunch period each student has and what period each has study hall, if:

a. Veronica does not have Lunch B.
b. Carlo's study hall is before lunch.
c. Carlo and Veronica have the same lunch but not the same study hall.
d. Roberto and Carlo have the same study hall.
e. Regina has a lunch period different from all the others, but is with two of her friends in study hall.

12. Burger Barn is offering a free lunch to anyone who buys the proper combination of hamburgers, french fries, and soft drinks.

Here are the rules:

- The menu consists of three items:

Hamburgers	$1.89
French Fries	$0.59
Soft Drinks	$0.79

- You must buy at least one hamburger.
- You can't buy more than three of any item.
- The sales tax equals 8% of the subtotal for all the items ordered. (8% = 0.08)
- If the sales tax comes out as an *exact* number of cents, then the lunch is free!

Determine the way to get a free lunch.

EXPLORATIONS

Each of the following two statements can be true (T) or false (F) but, without further information, you cannot know the truth status of either.

Statement 1: I will go. Statement 2: She has a cat.

a. Copy and complete this table to show the possibilities for the truth status of the pair of statements.

Note that there are 4 possible combinations of truth status for the pair of statements.

1	2
T	T
T	?
F	T
F	?

b. Here are two ways that a pair of statements can be combined to make a single statement: using the word *and*, using the word *or*. Combine the previous Statement 1 with the previous Statement 2:

(1) using the word *and* **(2)** using the word *or*

c. The truth status of a combined statement depends upon the truth status of the individual statements.

Consider 2 individual statements in each of 4 cases.

CASE 1 Statement 1: Tuesday is after Monday.
Statement 2: 3 is an odd number.

CASE 2 Statement 1: Tuesday is after Monday.
Statement 2: 3 is an even number.

CASE 3 Statement 1: Tuesday is before Monday.
Statement 2: 3 is an odd number.

CASE 4 Statement 1: Tuesday is before Monday.
Statement 2: 3 is an even number.

(1) Tell the truth status of the individual statements.

(2) In Case 1-Case 4, combine Statement 1 with Statement 2 using the word *and*. Tell the truth status of each "*and*-statement."

(3) Use a table to summarize your results.

1	2	1 *and* 2

(4) Draw a conclusion about the truth status of an "*and*-statement."

(5) Combine Statement 1 with Statement 2 using the word *or*. Tell the truth status of each "*or*-statement."

(6) Use a table to summarize your results.

1	2	1 *or* 2

(7) Draw a conclusion about the truth status of an "*or*-statement."

2.3 Guessing and Checking

Some problems can be solved by making a series of guesses and checking the result of each guess to see if it is correct. As you check a guess, you will have a better idea of what the next guess should be.

Example 1 Angel and Ruby are playing a number game. Angel says, "I am thinking of a number. If I double it and then subtract 7, the result is 11. What is my original number?"

1 ***Read the problem. What question needs to be answered?***

What is Angel's original number?

2 ***Make a plan.***

Try to guess and check.

3 **4** ***Use the plan and check the answer.***

Ruby guesses 5.

$2 \times 5 = 10$ and $10 - 7 = 3$.

5 is not the answer.

Since the result 3 is smaller than the desired 11, Ruby tries a larger number. She guesses 10.

$2 \times 10 = 20$ and $20 - 7 = 13$.

10 is not the answer.

Ruby now knows the answer is between 5 and 10 because the first answer, 3, is smaller than 11, and the second answer, 13, is bigger than 11.

Ruby guesses 9.

$2 \times 9 = 18$ and $18 - 7 = 11$.

9 is the correct answer.

Answer: Angel's original number was 9.

Sometimes, two problem-solving strategies can be combined to solve a problem.

Example 2 Find two numbers whose product is 80 and whose sum is 21.

1 ***Read the problem. What is the question?***

What two numbers have 80 as a product and 21 as a sum?

2 ***Make a plan.***

Guess two numbers to start and organize the results in a table.

3 ***Carry out the plan.***

Record your guesses in a table.

4 ***Check the answer.***

The fourth guess is correct.

$5 \times 16 = 80$ and $5 + 16 = 21$

	Product	Sum
First guess →	$1 \times 80 = 80$	$1 + 80 = 81$
Second guess →	$2 \times 40 = 80$	$2 + 40 = 42$
Third guess →	$4 \times 20 = 80$	$4 + 20 = 24$
Fourth guess →	$5 \times 16 = 80$	$5 + 16 = 21$

Answer: The two numbers are 5 and 16.

Try These *(For solutions, see page 76.)*

1. Kay made up a number problem for you to solve. She said, "I'm thinking of a number. If I double it and add 3, the answer is 25. Find my number."
2. The product of two numbers is 100. Their sum is 25. Find the numbers.
3. Luigi went food shopping and spent $27.30 for meat. Hamburger was $2.10 per pound and sausage was $2.80 per pound. How many pounds of each did he buy?
4. Whopper World has three specials.

Special	Menu	Price
#1	Soup, Sandwich, Soda	$3.50
#2	Sandwich, Fries, Soda	$3.45
#3	Soup, Fries, Soda	$2.75

The bottom menu lists the price of each item.

Soup	$0.95
Sandwich	$1.75
Soda	$0.85
Fries	$1.10

When you compare the prices of the specials to the cost of buying the individual items, which special saves the most money?

SELF-CHECK: SECTION 2.3

Hector has $5.20 in quarters and dimes. He has 3 more dimes than quarters. How many quarters and dimes does Hector have?

	Quarters	Dimes	Total
First guess →	10	13	$3.80
Second guess →	13	16	$4.85
Third guess →	14	17	$5.20

The third guess gives the correct answer.

Answer: Hector has 14 quarters and 17 dimes.

EXERCISES: SECTION 2.3

1. **a.** Ian challenged his brother Andy with this puzzle: "Triple a number and subtract 5. The result is 7. Find the number." What should Andy answer?

   b. Andy made up a puzzle to try to stump his brother. He said to multiply a number by 4 and then add 2. The result was 30. What was the number?

2. The members of a folk dance club were divided into groups of 6, 8, or 12, depending on which dance they were doing. Whichever grouping was used, there was always 1 member who had to sit out the dance. Of the following numbers, which could be the number of members present?

   (a) 41 (b) 37 (c) 33 (d) 25

3. **a.** The product of two numbers is 36. Their difference is 9. Find the numbers.

   b. The product of two numbers is 96. Their sum is 28. Find the numbers.

   c. One number is 4 more than another. Their sum is 38. Find the numbers.

4. A new number can be formed when the digits of a number are reversed (unless the number reads the same forward and backward). For example, reversing the digits of 52 forms the number 25. Note that the number reversed, 25, is 27 less than the original number, 52.

   In each of the following, find the original number.

   a. The sum of the digits of a two-digit number is 9. If the digits are reversed, the new number formed is 45 less than the original number.

   b. The sum of the digits of a two-digit number is 6. If the digits are reversed, the new number is 18 less than the original number.

5. Mike bought compact discs and tapes and spent exactly $66. How many of each did he buy if CD's are $14 and tapes are $6, including tax?

6. Last week, Terri worked 6 hours more than Harry. Altogether, they worked 52 hours. How many hours did each work?

7. The length of a rectangle is 6 ft. more than its width. Find the dimensions of the rectangle if its perimeter is 28 feet.

8. **a.** Brett has $1.45 in nickels and dimes. If he has 5 more nickels than dimes, how many of each coin does he have?

   b. Suppose Brett's $1.45 consisted of nickels, dimes, and quarters. If he had 1 more nickel than quarters and 1 more dime than nickels, how many of each coin did he have?

9. Tatiana's little brother just learned to count. In the playground one day, he told her that he had counted the feet of all the pigeons and all the cats, and that there were 38 feet. If Tatiana saw that there were a dozen pigeons and cats altogether, how many of each were there?

10. Sandra bought a large Valentine card for a special friend. The card weighs 8 ounces. Postage costs 32¢ for the first ounce and 25¢ for each additional ounce. With the stamps she has on hand, Sandra cannot put the exact postage on the envelope. Which of the following sets of stamps would be enough to get the card mailed and would come closest to the required amount?

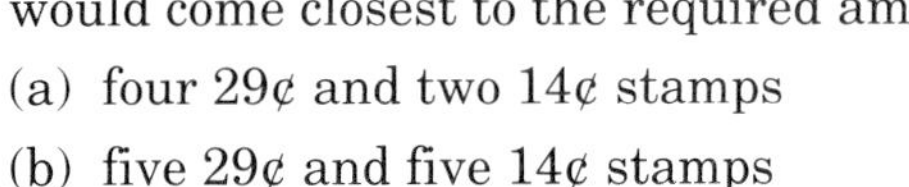

(a) four 29¢ and two 14¢ stamps

(b) five 29¢ and five 14¢ stamps

(c) six 29¢ stamps

(d) fourteen 14¢ cent stamps

EXPLORATIONS

1. **a.** Find the value of $2 + 3 \times 5$.

Explain how you could get 25 as a result.

Explain how you could get 17 as a result.

b. Show how you could use parentheses, ( ), as separators in the expression $2 + 3 \times 5$ so that the answer would have to be 17.

c. Rewrite each of the following sentences using parentheses to show the order in which the operations must be done to get the given answer.

$17 - 12 \div 2 = 11$ $\qquad$ $2 \times 3 + 4 \times 5 = 26$

d. Copy each of the following statements and replace ? with a number so that the sentence is true. When a sentence has more that one blank, use the same number in all.

$(? + 6) \times 3 = 51$ $\qquad$ $(? \times ?) + 3 = 67$ $\qquad$ $(? \times ?) - (2 \times ?) = 35$

e. Copy each of the following statements and fill in ? with one of the operation symbols +, –, ×, ÷. The operation symbols in each statement need not all be different. Insert parentheses as needed.

$4 ? 4 ? 4 ? 4 = 48$ $\qquad$ $4 ? 4 ? 4 ? 4 = 5$

$4 ? 4 ? 4 ? 4 = 28$ $\qquad$ $4 ? 4 ? 4 ? 4 = 6$

2. You use *memory* to remember information. So can a calculator. Here are some memory keys found on different calculators. See which you have on your calculator.

[M+] will store a number in the memory for future use

[MR] will recall a number that is stored in memory

[MC] will clear the memory (sets it to 0)

[MRC] pressed once recalls a number stored in memory
pressed twice clears the memory

For example, to calculate $(2 \times 3) + (4 \times 5)$ on a calculator that does not have parentheses, try this key sequence.

2 [×] 3 [=] [M+] 4 [×] 5 [+] [MR] [=]

The answer should be 26.

Write a key sequence for each of the following calculations to be done using memory on a calculator that does not have parentheses. Try your answers on your calculator.

a. $(10 \div 2) + (6 \times 3)$

b. $(4 \times 5 \times 3) + (18 \div 3)$

c. $(12 \times 2) - (8 \div 4)$

3. Christie is trapped in a maze of rooms at Great Escape Park. The number of rooms she may move at one time is the number written in the room in which she is standing. Combinations of horizontal (row) and vertical (column) moves are allowed, but no diagonal moves. Help her find her way out of the maze.

Begin 3	6	4	3	2	4	3
2	1	2	3	2	5	2
2	3	4	3	4	2	3
2	4	4	3	4	2	2
4	5	1	3	2	5	4
4	3	2	2	4	5	6
2	5	2	5	6	1	Escape

2.4 Drawing a Diagram

Sometimes, the circumstances of a problem can be modeled visually. Drawing a diagram is a useful strategy.

Example 1 Four people are introduced to one another at a meeting. Each person shakes hands with the three other people. Find the total number of handshakes.

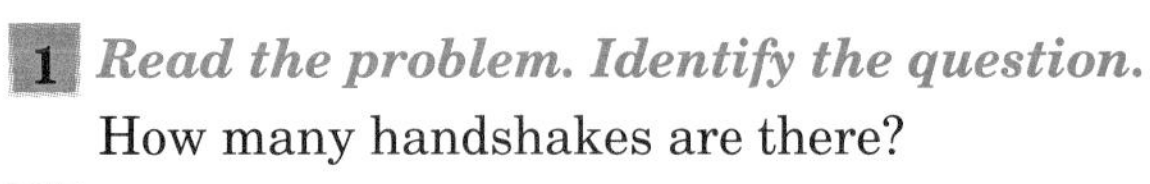

1 *Read the problem. Identify the question.*
How many handshakes are there?

2 *Make a plan.*
Draw a diagram.

3 *Solve the problem.*
Show each person as a point.

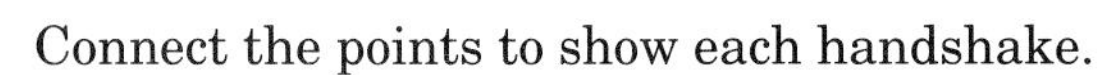

Connect the points to show each handshake.

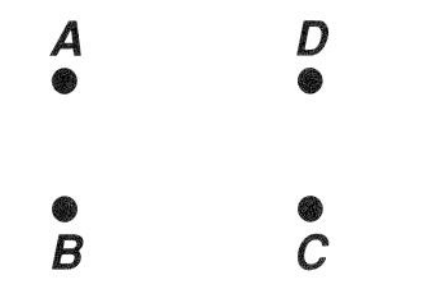

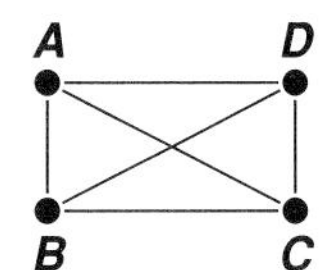

Handshakes

$A \longleftrightarrow D$	$B \longleftrightarrow A$	$C \longleftrightarrow D$	$D \longleftrightarrow A$
$A \longleftrightarrow B$	$B \longleftrightarrow C$	$C \longleftrightarrow B$	$D \longleftrightarrow C$
$A \longleftrightarrow C$	$B \longleftrightarrow D$	$C \longleftrightarrow A$	$D \longleftrightarrow B$

Eliminate repetitions: for example, the handshake $A \longleftrightarrow D$ is the same handshake as $D \longleftrightarrow A$. You are left with these 6 handshakes.

$A \longleftrightarrow D$	$B \longleftrightarrow D$	$C \longleftrightarrow D$
$A \longleftrightarrow B$	$B \longleftrightarrow C$	
$A \longleftrightarrow C$		

4 *Check your answer.*
Review the handshakes to check that each person shakes hands with three other people.

Answer: There are 6 handshakes.

Try These *(For solutions, see page 76.)*

1. A long railing is to be installed, with braces required every 5 feet. The first and last braces are 1 foot from the ends of the railing. If 10 braces are used, find the length of the railing.
2. On a street that is 1,200 feet long, how many houses can be constructed on one side of the street if each house is built in the center of a lot 200 feet wide?

Many different types of diagrams can be used to solve problems. Any diagram you can think of that leads to a solution is a good diagram. Some diagrams have special names and are often used for certain types of problems.

Tree Diagrams

Sometimes the possibilities of a solution can be shown by a series of branches called a ***tree diagram***.

This tree diagram shows the ways in which the 3 letters *ABC* can be arranged.

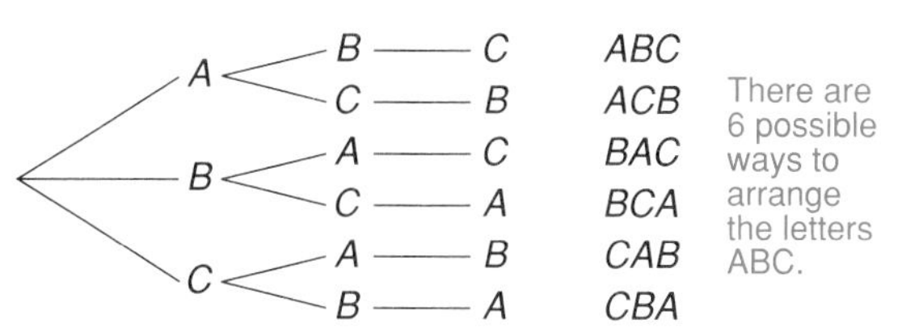

Example 2 Sally is making out her program for next term and wishes to take 2 of her 3 favorite majors — Math, Latin, Chemistry — in the first 2 morning periods. Name the ways she can arrange this part of her program.

1 ***Read the problem to identify the question.***

Name the ways in which any two of the subjects Math, Latin, and Chemistry can be arranged in periods 1 and 2.

2 ***Make a plan.***

Draw a tree diagram to show the information. Use *M* for Math, *L* for Latin, *C* for Chemistry.

Pd. 1	Pd. 2	Arrangements
M	L	ML
	C	MC
L	M	LM
	C	LC
C	M	CM
	L	CL

3 ***Solve the problem.***

Sally can take any one of the three subjects in period 1. After a subject has been placed in period 1, there are only two possibilities left for period 2.

4 ***Check the answer.***

Review the letters to make sure that if the letter is in period 1 it is not also in period 2.

Answer: There are 6 possible ways to arrange the subjects: Math, Latin; Math, Chem; Latin, Math; Latin, Chem; Chem, Math; Chem, Latin.

Try These *(For solutions, see page 77.)*

1. Two sisters, Amy and Ruth, are to share three sweaters, red, blue, and green. Draw a tree diagram to name the ways in which Amy and Ruth may wear the sweaters.
2. The table lists the activities available one morning at a conference. Use a tree diagram to write out all of the options available. Each person must go to breakfast (not necessarily first), and must finish an activity before starting the next one. Each person starts at 8 A.M. and finishes at 11 A.M.

Time			
8:00 A.M.–8:30 A.M.	Breakfast 1	Jogging	Weight room
8:30 A.M.–9:00 A.M.	Breakfast 2	Makeup Seminar	Free time
9:00 A.M.–9:30 A.M.	Breakfast 3	Tennis Instruction	Free time
9:30 A.M.–11:00 A.M.	Assembly	Assembly	Assembly

Venn Diagrams

The relationships between groups of data are often displayed on a ***Venn diagram.***

This Venn diagram shows that out of 600 seniors:

175 study Spanish only
100 study French only
50 study both Spanish and French

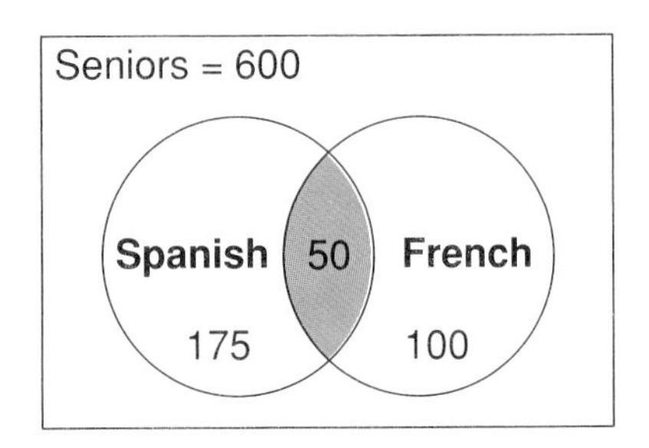

The diagram also shows that:

175 + 50 = 225 seniors study Spanish
100 + 50 = 150 seniors study French

Example 3 At a birthday party, eleven people ate pie and ten ate ice cream. Eight people ate both pie and ice cream. One person ate neither pie nor ice cream. How many people were at the party?

1 ***Read the problem. What is the question?***
How many people were at the party?

2 ***Make a plan.***
Draw a Venn diagram to show the information.

3 ***Solve the problem.***

a. Start with the 8 people that ate both pie and ice cream.
Write 8 where the circles overlap.

b. Since 11 people ate pie, then 11 – 8 = 3 people ate pie only.
Write 3 in the pie-only part of the circle.

c. Since 10 people ate ice cream, then 10 – 8 = 2 people ate ice cream only.
Write 2 in the ice-cream-only part of the circle.

d. One person ate neither pie nor ice cream.
Write 1 outside of the circles.

Add the numbers in the diagram: 8 + 3 + 2 + 1 = 14

4 ***Check the solution.***
Check the diagram with the given facts.

Answer: There were 14 people at the party.

Example 4 Two numbers that are multiplied together are ***factors*** in the multiplication. For example, in $4 \times 6 = 24$, the factors of 24 are 4 and 6.

a. A ***prime number***, such as 2 or 7, has only itself and 1 as factors. (A ***composite*** number has other factors as well.) Use tree diagrams to name all the prime factors of:

(1) 24

24 < 4 < 2, 2; 6 < 2, 3

(2) 30

30 < 5; 6 < 2, 3

The number 1 is not considered prime or composite.

Answer: **(1)** The prime factors of 24 are 2, 2, 2, and 3: $2 \times 2 \times 2 \times 3 = 24$
(2) The prime factors of 30 are 2, 3, and 5: $2 \times 3 \times 5 = 30$

b. When two numbers share the same factor, that factor is called a ***common factor***.

For example, 2 and 3 are common factors of 24 and 30, with 2×3, or 6, the ***greatest common factor***.

Use a Venn diagram to show the common factors of 24 and 30.

Answer:

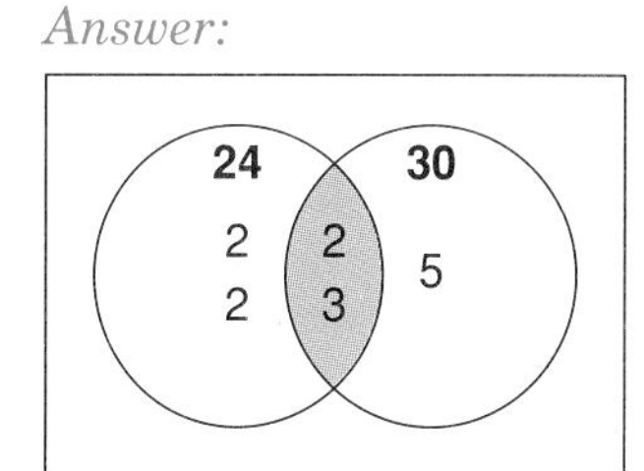

c. The ***least common multiple*** of two numbers is the smallest number the two numbers divide into evenly.

Find the least common multiple of 24 and 30.

To find the least common multiple, use the Venn diagram, multiply all the prime factors.

$$\textit{least common multiple} = 2 \times 2 \times 2 \times 3 \times 5 = 120$$

Or, find the product of the prime factors of the numbers, counting repetitions only once.

24: $2 \times 2 \times 2 \times 3$
30: $\qquad\quad 2 \times 3 \times 5$
least common multiple: $2 \times 2 \times 2 \times 3 \times 5 = 120$

Answer: The least common multiple of 24 and 30 is 120.

d. Two bells are rung at regular intervals, one every 24 minutes and the other every 30 minutes. After they sound together, how many minutes will it take before they sound together again?

The number of minutes after which the bells will again ring together is the least common multiple of the individual times. The least common multiple of 24 and 30 is 120.

Answer: The bells will ring together again after 120 minutes, or 2 hours.

Try These *(For solutions, see page 77.)*

1. In Homeroom 307, 12 students take only Spanish, 13 students take only French, 4 students take both French and Spanish, and 5 students take no foreign language. How many students are in the homeroom?

2. The circles in the Venn diagram contain the prime factors of two numbers.
   - **a.** What are the common prime factors of the two numbers?
   - **b.** What are the two numbers?

3. **a.** Use tree diagrams to name all the prime factors of:
   (1) 60 **(2)** 84
   Use a calculator to check the result.
   - **b.** Draw a Venn diagram to show the common factors of 60 and 84. What is the greatest common factor?
   - **c.** Find the least common multiple of 60 and 84.
   - **d.** Starting at the same time, two traffic lights are changing, one every 30 seconds and one every 36 seconds. How many seconds long are the intervals between the times they change together?

SELF-CHECK: SECTION 2.4

1. How many trees can be planted along one side of a 250-foot driveway if the centers of the trees are spaced 25 feet apart.

   Ten spacings of 25 feet, adding up to 250 feet, needs 11 trees.

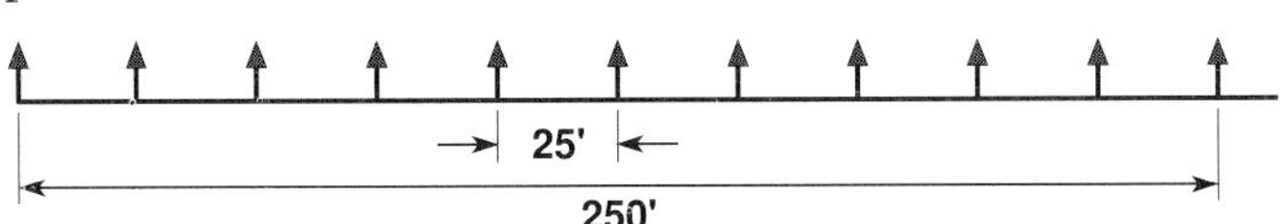

2. Draw a tree diagram for the handshake problem of Example 1.

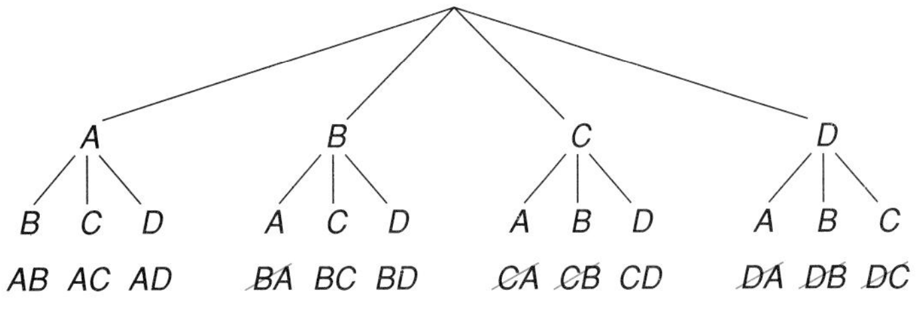

6 of the 12 possibilities read from the tree diagram are repeats, and need to be eliminated.

3. Use the Venn diagram showing the number of seniors studying Spanish or French (shown before Example 3) to find how many seniors are not studying either of these languages.

   From the 600 seniors, subtract those that do study Spanish or French, or both Spanish and French: $600 - 175 - 100 - 50 = 275$ not studying either

EXERCISES: SECTION 2.4

1. A highway is under construction. Traffic control cones are placed in a 1-mile (5,280- foot) section of the road to reduce traffic flow from 3 lanes to 2 lanes. If the cones are placed 55 feet apart, how many cones are needed?

2. Glow sticks that reflect light are to be installed on both sides of a straight 440-foot driveway that leads from a garage to the street. If the first stick is to be 30 feet from the garage, the last stick is to be 10 feet from the street, and the sticks are to be 40 feet apart, how many sticks are needed?

3. In a small rectangular patch, Sy is planning to grow three types of vegetables — carrots, radishes, and lettuce. He plans to give more room to the carrots than to the radishes, and the most room to the lettuce.

   Keeping each vegetable in its own rectangular block, draw a diagram to show how Sy could mark off this garden patch so that:

   a. the carrots are next to the lettuce, and the lettuce is next to the radishes, but the radishes are not next to the carrots

   b. each of the three blocks of vegetables touches each of the others

4. Find the greatest number of 3-inch-by-5-inch rectangles that can fit within a 10-inch-by-20-inch rectangle, if none of the rectangles overlap.

5. One of the boxes shown below is to hold a dozen pecan bars, each measuring 2 inches by 3 inches by 4 inches. Empty space in the box will be filled with packing material. Which of the boxes, inside measurements shown, would hold the dozen bars and need the least amount of packing material?

(a)

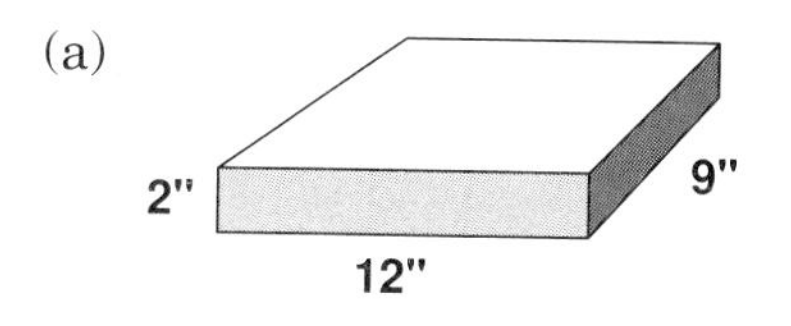

(b)

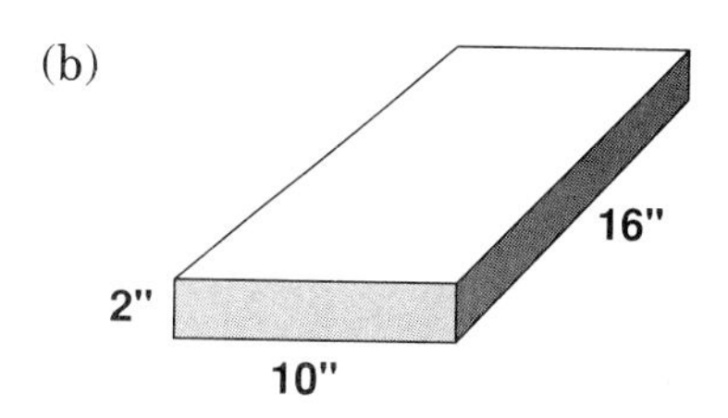

(c)

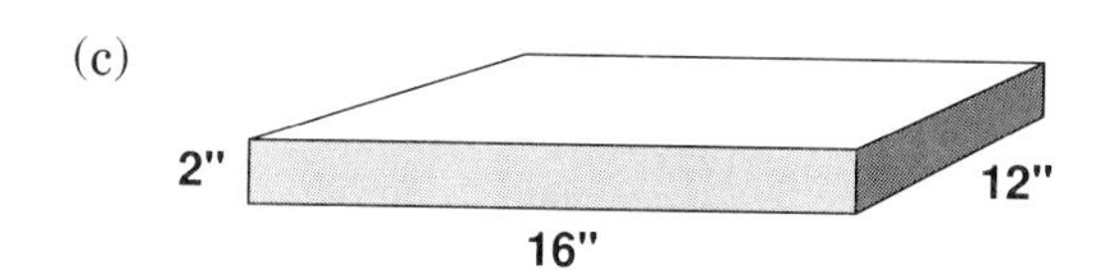

(d)

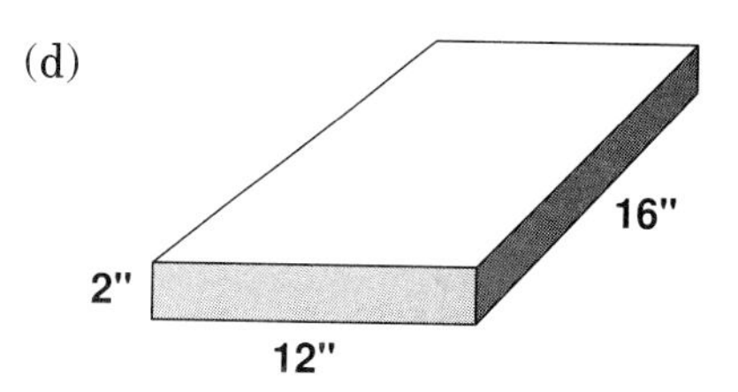

6. A *diagonal* of a polygon is a line segment that connects two nonconsecutive vertices. For example, a quadrilateral has 2 diagonals.

Complete the table to show the number of diagonals for each polygon listed.

Number of Sides	3	4	5	6
Number of Diagonals	0	2	?	?

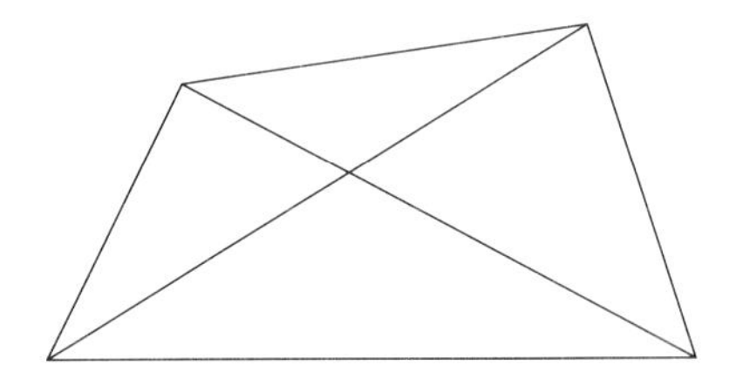

7. Draw a polygon (as in Exercise 6) to model the following situations.

a. Four teams — the reds, the yellows, the blues, and the greens— each played every other team once. How many games were played?

b. After that first round of play, the orange team joined in. If each of the five teams now played every other team once, how many games were played?

8. Three school buses come to pick up all the children at Lake Minnow. Among these children are the friends Liz, Tom, and Nan. For a reason known only to them, the 3 friends never ride in the same bus. Draw a tree diagram to name the ways that the 3 children can ride in the 3 buses.

9. At the school cafeteria, students have a lunch choice of a hot or a cold main dish, with milk, juice, or soda to drink, and fruit or ice cream for dessert. Draw a tree diagram to name all the possible ways in which a student may select one main dish, one drink, and one dessert.

10. A poll showed that 25 people read only the evening paper, 30 people read only the morning paper, 10 people read both, and 3 people read neither.
How many people were polled?

11. This Venn diagram shows all the prime factors of two numbers. Find :

a. the greatest common factor

b. the value of each number

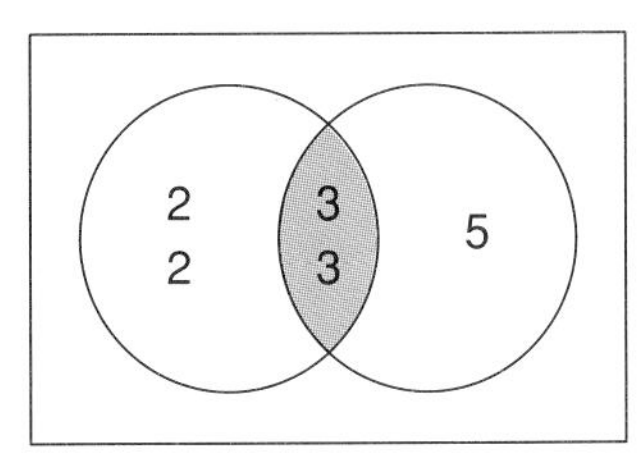

12. In a survey, 54 students said they owned a CD player, 86 students said they owned a tape player, 42 students said they owned both, and 8 students owned neither. How many students were surveyed?

13. Every student at Vernon High belongs to one of the following clubs: community service, computer, or chess. Some students belong to more than one. Which diagram best represents the club memberships?

(a) 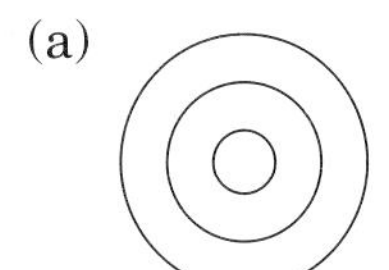

(b) 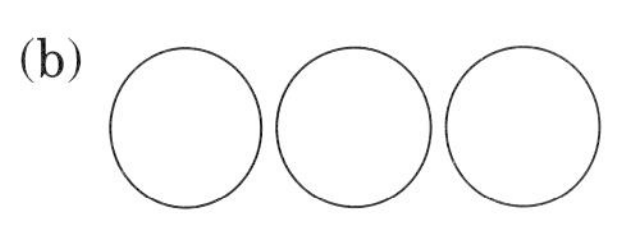

(c) 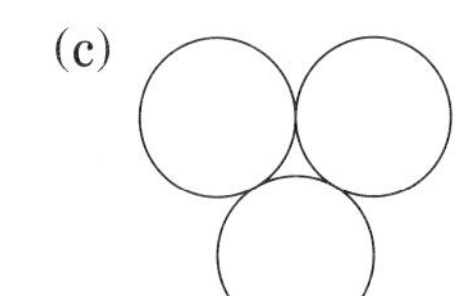

(d) 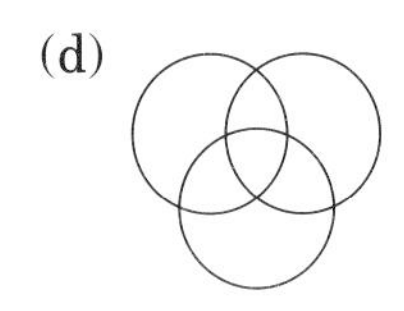

14. **a.** Use tree diagrams to name all the prime factors of:

(1) 72 **(2)** 120 **(3)** 144

Use a calculator to check the results.

b. Draw Venn diagrams to show the common prime factors of:

(1) 72 and 120 **(2)** 120 and 144

In each case, find the greatest common factor.

c. Find the least common multiple of:

(1) 72 and 120 **(2)** 120 and 144

Use a calculator to check the results.

d.

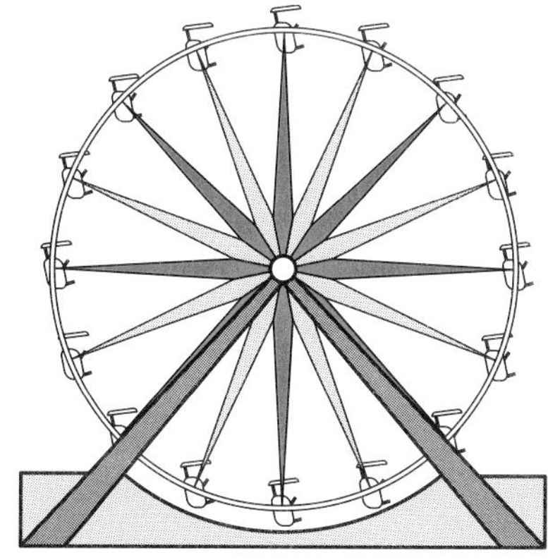

One Ferris wheel makes a complete rotation every 120 seconds, and another Ferris wheel makes a complete rotation every 144 seconds.

If they start at the same time, how many minutes will it take before the wheels begin a new rotation together?

15. Public School 53 has a field day each year. Students are allowed to plan their schedules according to the following offerings of the day. Each student must submit a planned schedule for the day. How many different schedules of complete activities can be made? (All must begin at 8 A.M. and stay until 1 P.M.)

<table>
<tr><td>8 A.M. – 9 A.M.</td><td>Relay Races</td><td rowspan="2">Hurdles</td><td>100-Meter Dash</td></tr>
<tr><td>9 A.M. – 10 A.M.</td><td rowspan="2">Pole Vaulting</td><td>Shot Put</td></tr>
<tr><td>10 A.M. – 11 A.M.</td><td>Discus Throw</td><td rowspan="3">Broad Jump</td></tr>
<tr><td>11 A.M. – 12 NOON</td><td rowspan="2">300-Meter Racing</td><td rowspan="2">High Jump</td></tr>
<tr><td>12 NOON – 1 P.M.</td></tr>
</table>

EXPLORATIONS

You have already seen that the number of diagonals in a polygon depends on the number of sides (or vertices) that the polygon has.

a. To summarize the information about the number of sides and diagonals in a polygon, copy and complete this table. (Look in your notebook. You have already done a pentagon and a hexagon.)

Number of Sides	3	4	5	6
Number of Diagonals	0	2	?	?

b. Now, let's try to find a general relation that will work for an n-sided polygon (where n is any whole number of 3 or more).

Here is a model of an n-sided polygon. Note that you cannot actually draw all the sides.

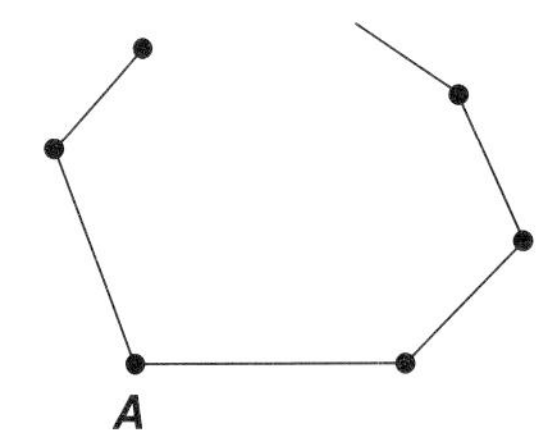

To use the model to find a general relationship between a polygon and the number of its diagonals, answer these questions:

(1) How many diagonals can you draw from any one vertex, say A? To answer in terms of n, think:

- One of the n vertices is A.
- How many of the n vertices cannot be connected to A to give a diagonal?
- How many of the n vertices are left?

(2) Now that you know how many diagonals you can draw from any one vertex, how may diagonals can you draw from n vertices?

(3) But, how many times has each diagonal been counted?

To avoid repetition, what must you do?

(4) Write your formula for the number of diagonals in terms of n, the number of sides.

Use the previous table to test your formula for $n = 4, 5, 6$.

c. Use your formula to tell how many diagonals there are in a polygon of 15 sides.

2.5 Simplifying the Problem

A difficult problem can sometimes be solved by first working with a series of simpler problems.

Example 1 Twin rock stars Carla and Clara are having a birthday bash. The first time the doorbell rings, one friend enters. Each time the doorbell rings again, a group enters that has two more friends than the group that just arrived. How many friends are at the party after the 10th ring?

1 *Read the problem. What is the question?*
How many friends are at the party after the 10th ring?

2 *Make a plan.*
First solve a simpler problem.

3 *Carry out the plan.*
Start with the first ring. How many friends arrived?

First ring: 1 friend arrives
Second ring: 1 friend + 2 more = 3
Third ring: 3 friends + 2 more = 5

Use a table to organize the information.

Doorbell Rings	Friends Entering	Total Arrived
1	1	1
2	3	4
3	5	9
4	7	16
5	9	25
6	11	36
7	13	49
8	15	64
9	17	81
10	19	100

4 *Check your answer.*
Observe the pattern that has emerged between the number of doorbell rings and the total arrived:
$1^2 = 1$, $2^2 = 4$, $3^2 = 9$, ..., $10^2 = 100$

Answer: After the 10th ring, 100 friends are at the party.

Example 2 P. J. and seven friends went to play basketball. They decided to pair up for some one-on-one games. If each friend played against each of the others exactly once, how many games were played?

1 *Read the problem. What is the question?*
How many games were played?

2 *Make a plan.*
Solve a simpler problem first.

3 *Carry out the plan.*
Draw a diagram to represent each of the simpler situations.

a. P. J. and *one* friend

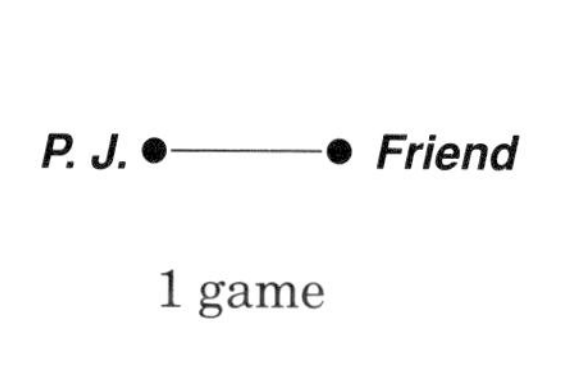

1 game

b. P. J. and *two* friends

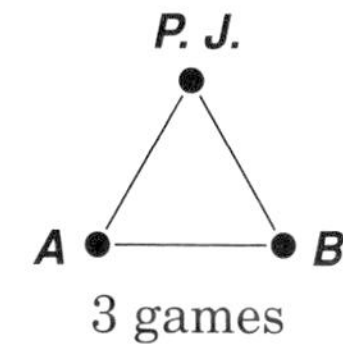

3 games

c. P. J. and *three* friends

P. J. C
A B

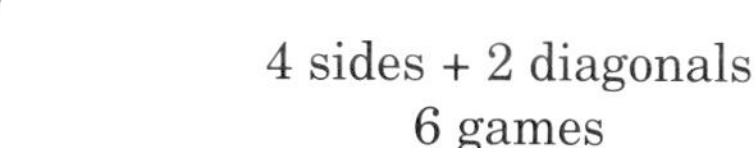

4 sides + 2 diagonals
6 games

d. The table shows the results so far. Note that a pattern is developing in the number of games played:

$1 + 2 = 3$ $\qquad$ $3 + 3 = 6$

Do you think P. J. and 4 friends played 10 games? Verify the prediction by diagram.

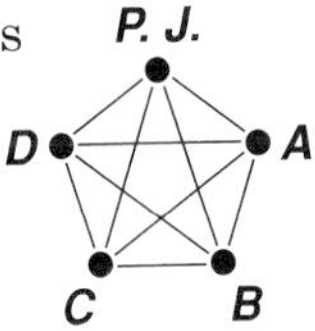

Number of Friends	1	2	3	4	5	6	7
Number of Games	1	3	6	?			

5 sides + 5 diagonals

10 games

e. Use the pattern to complete the table.

Number of Friends	1	2	3	4	5	6	7
Number of Games	1	3	6	10	15	21	28

4 ***Check your answer.***

Check the pattern to see that each entry in the table is correct.

Answer: P. J. and seven friends played a total of 28 games.

Try These *(For solutions, see page 77.)*

1. Find the total number of squares contained in the diagram. *Hint:* Simplify the problem by considering 3 separate problems:
(1) how many 1-by-1 squares
(2) how many 2-by-2 squares
(3) how many 3-by-3 squares

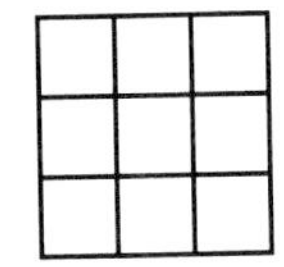

2. Find the number of line segments determined by 10 points on a line. For example, for 4 points, there are 6 possible segments.

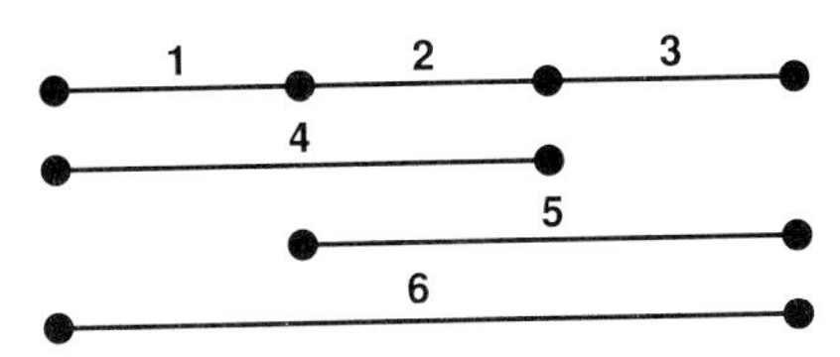

3. Cindy has 3 pairs of fashion slacks and 4 coordinating sweaters. How many different outfits can she wear?

4. Suppose you have one penny, one nickel, one dime, and one quarter. How many different amounts of money can you make by taking one, two, three, or four of the coins?

5. Twelve women all shake hands with each other before a panel discussion. How many handshakes are there altogether?

6. A ***palindrome*** reads the same backward as it does forward.
For example, the number 141 is a palindrome.
By definition, all one-digit numbers are palindromes.
To answer the question "How many whole numbers from 1 through 1,000 are palindromes?", simplify the problem by considering a series of separate problems.
How many whole-number palindromes are there:
a. from 1 through 9? **b.** from 10 through 99? **c.** from 100 through 1,000?
Now, answer the original question.

SELF-CHECK: SECTION 2.5

1. Use the pattern in the solution to Example 1 to find how many friends are at the party after the 12th ring.

Answer: $12 \times 12 = 144$ friends

2. Use the pattern in the solution to Example 2 to find how many games would be played by 11 players.

Answer: 55 games

3. What geometric figure can be used to show P. J. and five friends in Example 2?

Answer: A hexagon, which is a six-sided figure.

4. Name the different problem-solving strategies used in Example 2.

Answer: Simplifying the problem.
Drawing a diagram.
Making a table.
Finding a pattern.

EXERCISES: SECTION 2.5

1. A folk-dance caller calls one dancer onto the floor. With each succeeding call, two dancers more than on the last call join the group. After the 15th call, how many dancers are in the group?

2. **a.** Joshua and nine of his friends exchange cards at holiday time. If each sends one card to each of his friends, how many cards in all will be sent?

b. Suppose Joshua and his nine friends shake hands with each other at a holiday party.

(1) What polygon can be used to model the situation?

(2) How many diagonals does this polygon have?

(3) How many handshakes are exchanged?

3. Find the total number of squares contained in the diagram.

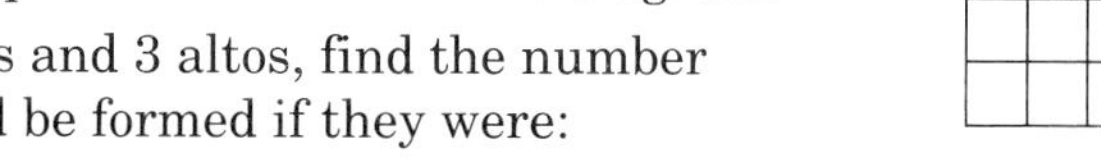

4. From a group of 5 sopranos and 3 altos, find the number of different duos that could be formed if they were:

a. both sopranos

b. both altos

c. one soprano and one alto

5. Eight points lie on a circle. Find the number of line segments needed to connect each point with each of the other points.

6. Erin has a new board game. The first time she lands on a square marked "treasure" she receives \$10 in play money. Each time that she again lands on the "treasure" square she receives \$5 more than she received the time before. How much money will Erin collect for landing on "treasure" 8 times?

7. Tony's Pizza Palace offers a luncheon special each day. The special consists of a thick or thin pizza with one topping from *Column B* and one topping from *Column C*. On the first day, Tony's special was a thick pizza with sausage and onions. Each day, Tony determined the special by replacing each item with the next item in that column. Thus, the 2nd-day special was a thin pizza with pepperoni and peppers. When the last item in each column was used, Tony went to the top of that column for the next day.

Column A	*Column B*	*Column C*
Thick	Sausage	Onions
Thin	Pepperoni	Peppers
	Meatballs	Mushrooms
	Anchovies	Black Olives
		Extra Cheese

a. What would the luncheon special be on:

(1) the 5th day? **(2)** the 13th day?

b. How many consecutive days will go by before Tony makes a thick pizza with sausage and onions again?

EXPLORATIONS

The diagram shows some smaller squares that can be found in a 3-by-3 square.

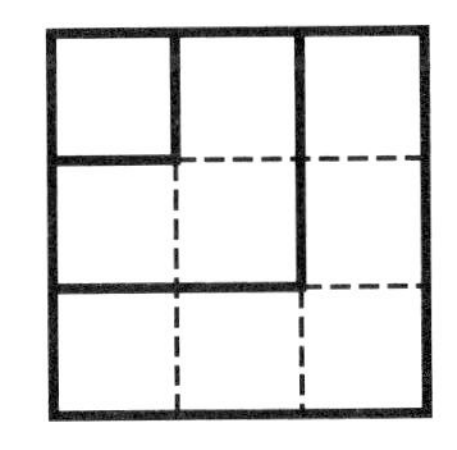

The table below is a record of squares within squares.

For each size shown in the left column of the table, draw a square on graph paper and count the number of squares inside the original square.

a. Copy the table and complete the last 3 rows. (You have already worked with a 3-by-3 and a 4-by-4 square. Look in your notebook.)

b. Use the information in the completed table to make an observation about:

(1) the numbers in the body of the table

(2) the numbers in each column of the table

(3) the numbers along diagonals in the table

(4) the number of 1's

(5) the number of 0's

(6) the numbers in the Total column

c. Describe a way to find the total number of squares in a 7-by-7 square.

Size of Outside Square	Number of Inside Squares						Total
	1 × 1	2 × 2	3 × 3	4 × 4	5 × 5	6 × 6	
1 × 1	1	0	0	0	0	0	1
2 × 2	4	1	0	0	0	0	5
3 × 3	9	4	1	0	0	0	14
4 × 4	?	?	?	?	?	?	?
5 × 5	?	?	?	?	?	?	?
6 × 6	?	?	?	?	?	?	?

Things You Should Know After Studying This Chapter

KEY SKILLS

Solve a problem by using these key steps:
1. Read 2. Plan 3. Solve 4. Check

Some problem-solving strategies, to be used alone or in combination, are:

2.1 Finding a Pattern

2.2 Organizing Data in a Table

2.3 Guessing and Checking

2.4 Drawing a Diagram

2.5 Simplifying the Problem

KEY TERMS

2.1 problem • plan • solve • check • strategy • pattern

2.2 organizing data in a table

2.3 guessing and checking

2.4 tree diagram • Venn diagram • prime number • composite number • factor • common factor • greatest common factor • least common multiple

2.5 simplifying the problem

CHAPTER REVIEW EXERCISES

1. Find the next three terms in each sequence.

a. 7,8,10,13,... **b.** 31,28,25,22,...

c. 1,4,5,9,14,... **d.** 1,3,9,27,...

2. Find the next row of numbers:

1
1 1
1 2 1
1 3 3 1
1 4 6 4 1
1 5 10 10 5 1

3. The large cube shown, which is made up of 27 small cubes, is painted on the outside only.

How many of the small cubes have exactly 2 of their sides painted?

4. The area of a rectangle is 60 sq. units. If the perimeter is 34 units, what are the dimensions of the rectangle?

5. Ava has 3 quarters, 4 dimes, and 5 nickels. Make a table to show all the possible ways she can select 90¢ worth of coins.

6. Meda ordered 30 tickets for an ice show. Adult tickets were $25 and student tickets were $12. If Meda paid $620, how many of each kind did she order?

7. Mei, Fred, Kurt and Sue are sitting around a circular table. Use the following clues to determine the seating order:

a. Fred never sits next to Mei.

b. Mei is sitting on Kurt's right.

c. Sue is sitting opposite Kurt.

8. Sharma told Kay, "I know a quick way to add the counting numbers, 1 + 2 + 3 + ..., up until whatever number you tell me. I take the number you tell me, multiply it by the next counting number, divide by 2, and that's the sum." When Kay gave Sharma a number, Sharma said the sum of the numbers from 1 through Kay's number was 66. What number did Kay give?

9. The table shows morning activities at a local youth center. How many different ways can a teen spend the morning? (All schedules begin at 9 A.M. and end at noon, and each activity must be completed before moving on.)

<table>
<tr><td>9:00 A.M.</td><td>watch cartoons</td><td>snack</td><td rowspan="3">magic show</td></tr>
<tr><td>9:30 A.M.</td><td rowspan="3">movie</td><td rowspan="2">talent show</td></tr>
<tr><td>10:00 A.M.</td></tr>
<tr><td>10:30 A.M.</td><td rowspan="3">band concert</td><td rowspan="3">science demonstration</td></tr>
<tr><td>11:00 A.M.</td><td rowspan="2">documentary film</td></tr>
<tr><td>11:30 A.M.</td></tr>
<tr><td>12:00 NOON</td><td></td><td></td><td></td></tr>
</table>

10. A chef decorated a birthday cake by making an 8"-by-12" rectangle of frosting on the top of the cake. Candles are placed around the perimeter of the rectangle every 2 inches, including a candle at each corner. If the number of candles represents Simon's age plus one to grow on, how old is Simon?

11. This Venn diagram shows all the prime factors of two numbers. Find:

a. the greatest common factor

b. the value of each number

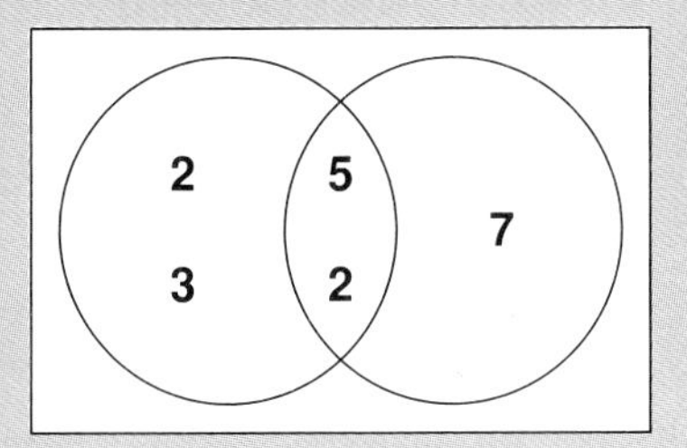

CHAPTER REVIEW EXERCISES

12. A poll taken in Mr. Lopino's math class showed that 15 students worked at part-time jobs, 20 students participated in school activities, 12 students did both, and 5 students did neither. How many students are in Mr. Lopino's class?

13. **a.** Use tree diagrams to name all the prime factors of

(1) 96 **(2)** 240

Use a calculator to check the results.

b. Draw Venn diagrams to show the common prime factors of 96 and 240. What is the greatest common factor?

c. Find the least common multiple of 96 and 240. Use a calculator to check the result.

d. Two tourist trams make their run at regular intervals, one every 16 minutes and the other every 40 minutes. If they start together at the beginning of the day, after how many minutes will they start together again?

14. Timothy drew lines inside a square to divide the square into smaller squares. To answer the question "How many small squares will he get if he draws 14 lines?", simplify the problem by considering a series of separate problems.

15. The sum of the digits of the number 550 is $5 + 5 + 0$, or 10.

To answer the question "How many of the whole numbers from 1 through 500 have 10 as the sum of their digits?", simplify the problem by considering a series of separate problems.

With 10 as the sum of their digits; find how many such whole numbers there are:

a. from 1 through 9

b. from 10 through 99

c. from 100 through 500

Now, answer the original question.

SOLUTIONS FOR TRY THESE

2.1 *Finding a Pattern*

Page 42

1. **a.** Multiples of 4: 16, 20, 24
   b. Subtract 6: 60, 54, 48
2. 49,995 and 59,994
3. **a.** To find the sum of the even numbers starting with 2, multiply half of the last number by the half plus 1.
   b. **(1)** $8 \times 9 = 72$
   (2) $12 \times 13 = 156$
   (3) $16 \times 17 = 272$
4. 1234321
5. **a.**
   $2 \rightarrow 5$
   $3 \rightarrow 7$
   insert: $4 \rightarrow 9$
   $5 \rightarrow 11$
   $6 \rightarrow 13$ Juan would reply 13.
   b. Juan doubles John's number and then adds 1.

2.2 *Organizing Data in a Table*

Page 50

1. Make a table to determine the different possibilities.

Number of $9 balls	11	10	8	7	6	4	3	1	0
Number of $7 balls	0	1	4	5	6	9	10	13	14
Total cost	99	97	100	98	96	99	97	100	98

There are 9 different orders he can place.

2. Make a table to determine the different possibilities.

Number of pennies	1	0	0	1	1	0	1
Number of nickels	0	1	0	1	0	1	1
Number of dimes	0	0	1	0	1	1	1
Total cents	1	5	10	6	11	15	16

There are 7 different amounts of money.

3. Make a table to look for a pattern.

Number of odd numbers	Odd number	Sum of the odd numbers
1	1	1, or 1^2
2	3	4, or 2^2
3	5	9, or 3^2
4	7	16, or 4^2
5	9	25, or 5^2
⋮	⋮	⋮

The pattern is that the sum is the square of the number of odd numbers. Thus, the sum of the first 15 odd numbers is 15^2, or 225.

4. Make a table to organize the information.

	Allie	Burt	Cindy
Art	*a.* no	*f.* no	*g.* yes
Bill	*c.* yes	*a.* no	*b.* no
Cliff	*d.* no	*e.* yes	*a.* no

Art has Cindy.
Bill has Allie.
Cliff has Burt.

2.3 *Guessing and Checking*

Page 56

1. 11
2. 5 and 20
3. 5 lb. of hamburger, 6 lb. of sausage.
4. Special #2 saves the most money.

2.4 *Drawing a Diagram*

Page 60

1. –|–|–|–|–|–|–|–|–|–|– 47 feet
   1 5 5 5 5 5 5 5 5 5 1

2. |□|□|□|□|□|□| 6 houses
   200' 200' 200' 200' 200' 200'
   |———— 1200' ————|

SOLUTIONS FOR TRY THESE

Page 61

1. Amy Ruth There are 6 possibilities:

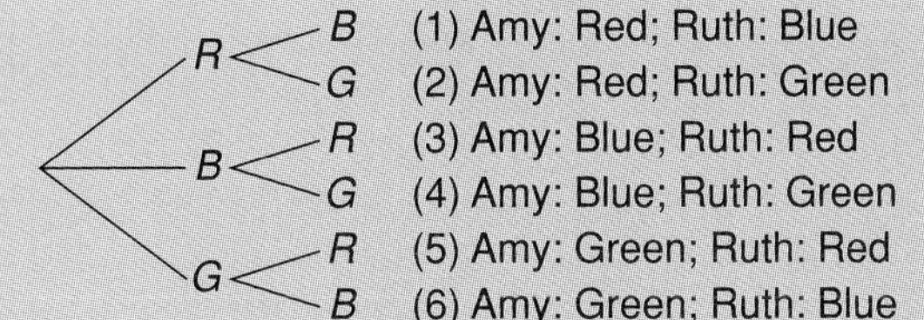

(1) Amy: Red; Ruth: Blue
(2) Amy: Red; Ruth: Green
(3) Amy: Blue; Ruth: Red
(4) Amy: Blue; Ruth: Green
(5) Amy: Green; Ruth: Red
(6) Amy: Green; Ruth: Blue

2.

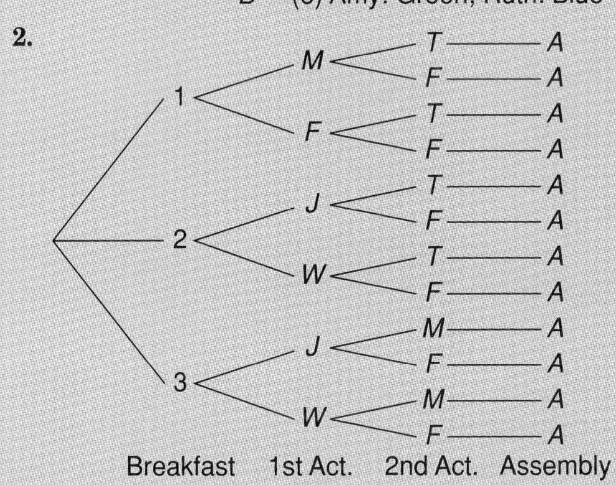

Page 64

1.

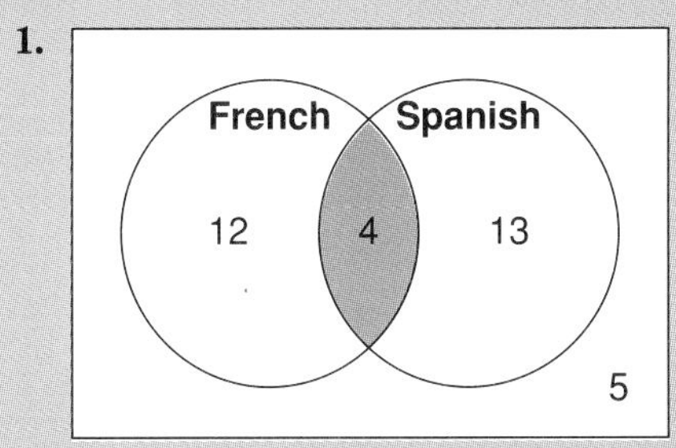

12 + 4 + 13 + 5 = 34 students in the homeroom

2. a. 2 and 3

The common factors are in the intersection of the circles.

b. $2 \times 3 \times 3 = 18$ $2 \times 3 \times 5 = 30$

The numbers are the products of all the prime factors.

3. a. (1) 60 → 5, 12; 12 → 4, 3; 4 → 2, 2

$60 = 5 \times 3 \times 2 \times 2$

(2)

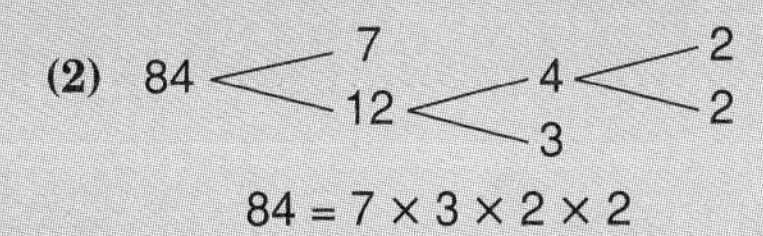

$84 = 7 \times 3 \times 2 \times 2$

b.

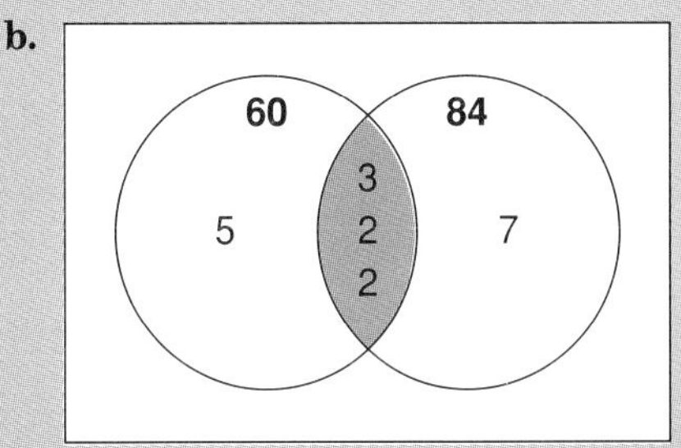

The greatest common factor is $3 \times 2 \times 2$, or 12.

c. From the Venn diagram, multiply all the prime factors.

least common multiple = $5 \times 3 \times 2 \times 2 \times 7 = 420$

d. 180 seconds

The time at which they flash together is the least common multiple of the individual flashing times.

2.5 *Simplifying the Problem*

Page 70

1. 1×1: 9 squares
2×2: 4 squares
3×3: 1 square
There are 14 squares altogether.

2. Think of the distance between 2 consecutive points as 1 unit, and count the number of line segments of each length.

Length	1	2	3	4	5	6	7	8	9	Total
Number of segments	9	8	7	6	5	4	3	2	1	45

There are 45 line segments altogether.

3. One pair of slacks can be worn with 4 different sweaters. This is true for each of the 3 pairs of slacks. There are 3×4, or 12, possibilities.

SOLUTIONS FOR TRY THESE

4. one coin: 4 amounts *p, n, d,* or *q*

two coins: 6 amounts *pn, pd, pq, nd, nq, dq*

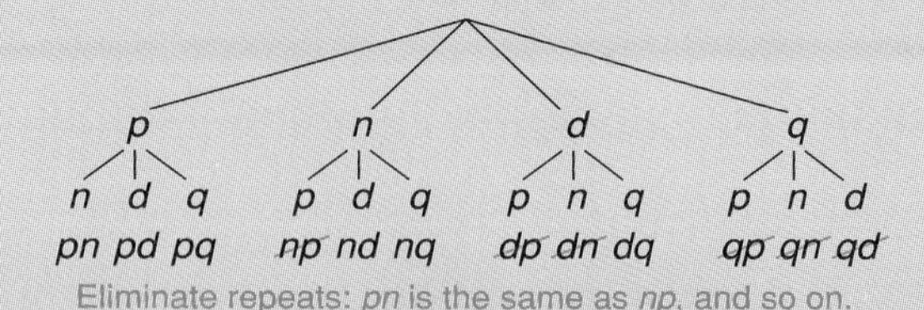

three coins: 4 amounts *pnd, pnq, pdq, ndq*

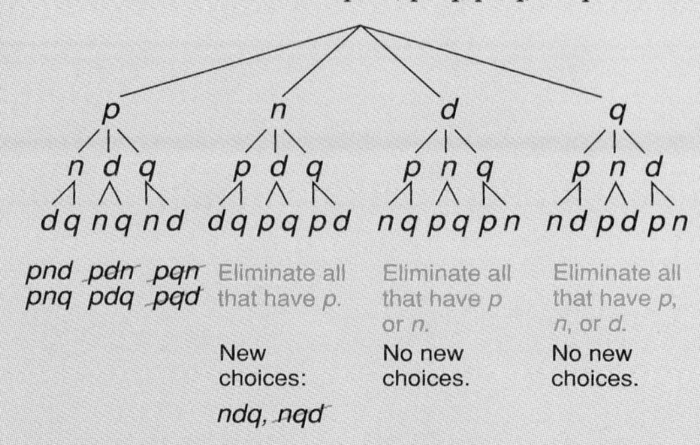

four coins: 1 amount *pndq*

Answer: There are 4 + 6 + 4 + 1, or 15, amounts.

---

5. Represent the situation with diagrams, using a point for a person, and a line segment for a handshake between 2 people.

When there are 3 or more people, the number of handshakes is the total of the number of sides of the polygon drawn and the number of diagonals in the polygon.

Look for a pattern in the number of handshakes.

	Number of: People	Handshakes	Pattern (Add)
	2	1	
			+2
	3	3	
			+3
	4	6	
			+4
	5	10	
			+5
	6	15	
			+6
	7	21	

Continue the pattern.

8 people	21 + 7 = 28
9 people	28 + 8 = 36
10 people	36 + 9 = 45
11 people	45 + 10 = 55
12 people	55 + 11 = 66 handshakes

Alternate Method

Think of each of the 12 women A, B, C, ... shaking hands with 11 others: $12 \times 11 = 132$. However since a handshake between A and B is the same as between B and A, divide by 2: $132 \div 2 = 66$

6. **a.** all: 9 palindromes

b. 11, 22, 33, 44, 55, 66, 77, 88, 99: 9 palindromes

c.

101	202	303	…	909
111	212	313		
121	222	323		
131	232	333		
141	242	343		
151	⋮	⋮		⋮
161				
171				
181				
<u>191</u>	<u>292</u>	<u>393</u>	…	<u>999</u>
10	10	10	…	10

9×10, or 90 palindromes

Thus, 9 + 9 + 90, or 108 whole-number palindromes from 1 through 1,000.

Operations With Numbers

3

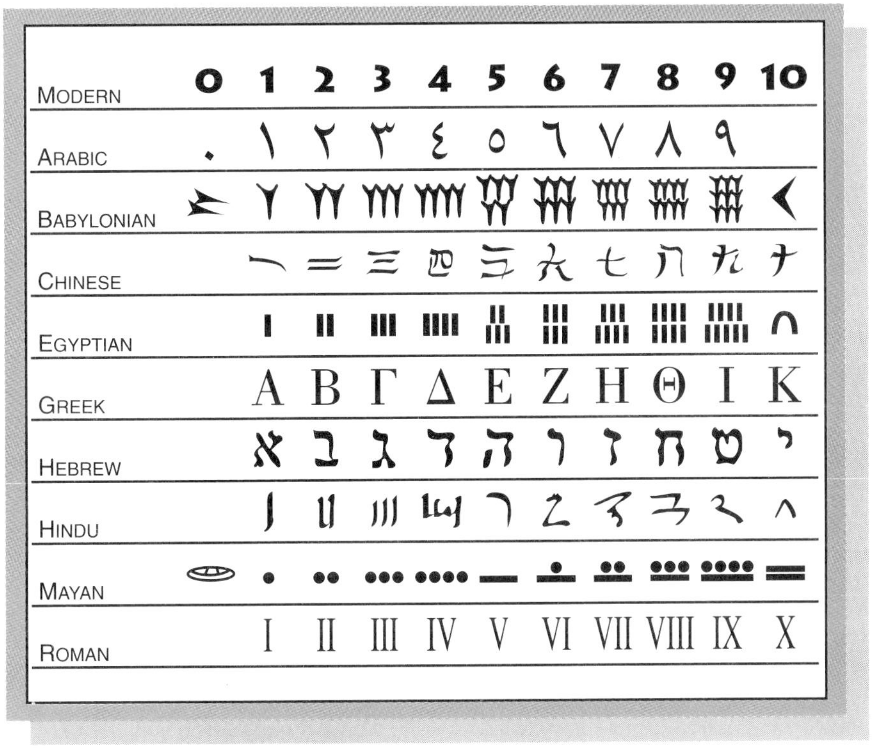

	0	1	2	3	4	5	6	7	8	9	10
MODERN	0	1	2	3	4	5	6	7	8	9	10
ARABIC	٠	١	٢	٣	٤	٥	٦	٧	٨	٩	
BABYLONIAN	[illegible]	𒐕	𒐖	𒐗	𒐘	𒐙	𒐚	𒐛	𒐜	𒐝	𒌋
CHINESE		一	二	三	四	五	六	七	八	九	十
EGYPTIAN		I	II	III	IIII	III/II	III/III	IIII/III	IIII/IIII	IIIII/IIII	∩
GREEK		Α	Β	Γ	Δ	Ε	Ζ	Η	Θ	Ι	Κ
HEBREW		א	ב	ג	ד	ה	ו	ז	ח	ט	י
HINDU		𑁒	𑁓	𑁔	𑁕	𑁖	𑁗	𑁘	𑁙	𑁚	𑁛
MAYAN	(shell)	•	••	•••	••••	—	•/—	••/—	•••/—	••••/—	=
ROMAN		I	II	III	IV	V	VI	VII	VIII	IX	X

ACTIVITY

Many kinds of numerals have been developed since ancient times.

You too can invent ways to write numbers.

Try to represent the numbers 0 through 9 using only the symbols △ and □.

3.1 Whole Numbers and Sets

Common Uses of Numbers

In daily life, we use numbers to:

① *Count and Compare*

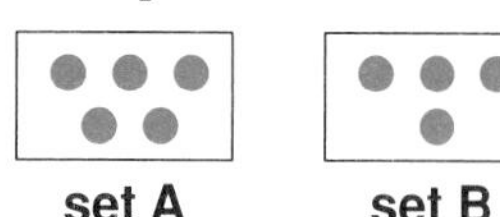

set A **set B**

There are 5 elements in set A and 4 in B.
5 is greater than 4, written $5 > 4$.
4 is less than 5, written $4 < 5$.

② *Order Members of a Set*

June is the 6th month of the year.

Numbers for counting or ordering are:

$\{1, 2, 3, \ldots\}$	$\{0, 1, 2, 3, \ldots\}$
Natural Numbers	***Whole Numbers***

③ *State Measurements*

Maseo, at 12 years old, is 5 feet tall and weighs 110 pounds.

④ *Solve problems*

Example 1 The Cliffords plan to drive from Philadelphia to Denver, and are deciding whether to take a route through Chicago or through Tulsa. How many miles would they save on the shorter route?

Use the 4 key problem-solving steps.

1 ***Read the problem to tell what questions must be answered.***

Which is the shorter route?
By how much?

2 ***Make a plan.***

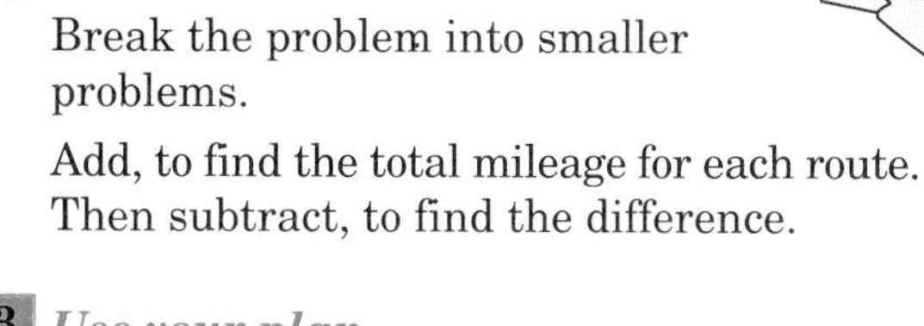

Break the problem into smaller problems.

Add, to find the total mileage for each route.
Then subtract, to find the difference.

3 ***Use your plan.***

For total distances, add.

<u>Via Chicago</u>	<u>Via Tulsa</u>	To compare, subtract.
$738 + 996 = 1{,}734$	$1{,}264 + 681 = 1{,}945$	$1{,}945 - 1{,}734 = 211$

4 ***Check your answer.***

Use inverse operations: check addition by subtracting and check subtraction by adding.
Estimate from the map whether your answer is reasonable.

Answer: The Chicago route is 211 miles shorter.

Try These *(For solutions, see page 118.)*

1. President Franklin Roosevelt said that the day Pearl Harbor was attacked, 12/7/41, was "a date which will live in infamy."

   In what year did this event take place? on what day of the month? in which month?

2. Use symbols to show that:
   a. 18 is greater than 11 **b.** 11 is less than 18 **c.** 15 is between 11 and 18

About Sets

A set can be described:

by listing its members	or	by stating a common property
{0, 2, 4, 6, ...}		{even numbers}

Example 2 Match each set in *Column A* with a set in *Column B* so that the two sets are equal.

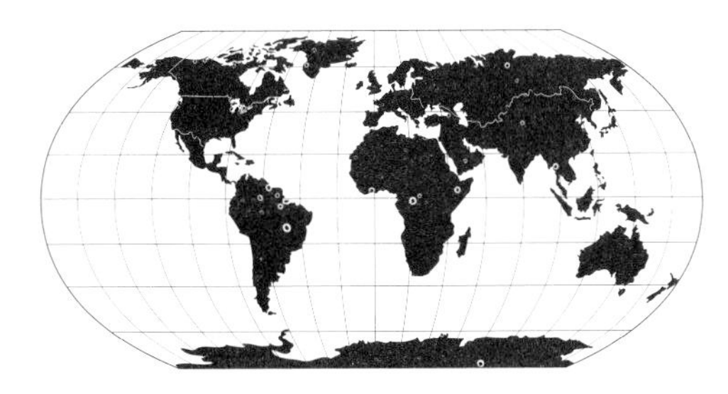

Column A	*Column B*
1. {the continents}	**a.** {Africa, Asia, Australia, Europe}
2. {the continents in the Western Hemisphere}	**b.** {Africa, Antarctica, Asia, Australia, Europe, North America, South America}
3. {the continents in the Eastern Hemisphere}	**c.** {North America, South America}

Answer: **1. b 2. c 3. a**

If the members of a set can be counted, the set is ***finite***. If the number of members continues without end, they cannot be counted, and the set is ***infinite***.

Example 3 Identify each set as finite or infinite.

K = {the people in China}

W = {whole numbers}

Z = {natural numbers less than 10,000,000}

$\varnothing$ = empty set (has no elements)

Answer: K finite, W infinite, Z finite, $\varnothing$ finite

When the elements of two sets can be matched so that each member has one and only one partner, the matching is called a ***one-to-one correspondence***.

For the sets {Luis, Joe, Phil} and {Jill, Rose, Vasha} the following pairs show a one-to-one correspondence:

Luis ↔ Jill   Joe ↔ Rose   Phil ↔ Vasha

Name a different one-to-one correspondence between these sets.

Try These *(For solutions, see page 118.)*

1. List the members in each set.
   a. {digits}
   b. {prime numbers between 1 and 20}
2. Describe each set by stating a property common to its members.
   a. {11, 13, 15, 17, 19}
   b. {90, 92, 94, 96, 98, 100}
3. List the subsets of $\{a, b\}$.
4. **a.** Write out a 1-to-1 correspondence between the two sets $\{e, g, w, z\}$ and {2, 4, 6, 8}.
   b. In how many different ways can a 1-to-1 correspondence between these sets be shown? Explain.

SELF-CHECK: SECTION 3.1

Consider these sets: $X = \{c, a, k, m, p, r\}$ and $Y = \{p, a, m, c, r\}$

a. State a property that the members of these sets have in common.

Answer: letters of the English alphabet

b. Are these sets equal?

Answer: No, X and Y do not have exactly the same members.

c. Can you set up a 1-to-1 correspondence between the members of X and Y?

Answer: No, the sets do not have the same number of elements.

d. Write a relation comparing the number of members of X and Y. There are 6 members in X and 5 in Y.

Answer: $6 > 5$ or $5 < 6$

e. Draw a Venn diagram to describe a relation between X and Y.

Answer:

Y is a subset of X.

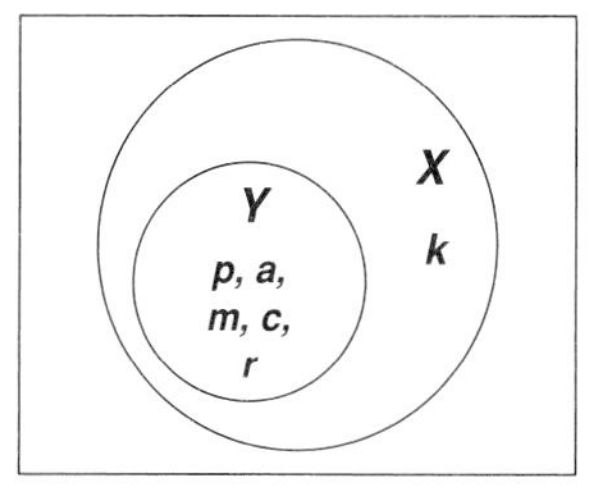

f. Are X and Y finite or infinite sets?

Answer: Finite, the number of members can be counted.

EXERCISES: SECTION 3.1

1. For each part **a-d**, choose the way of using numbers that the statement represents.

Statement	*Ways to Use Numbers*
a. Jan is 5'1" tall.	(1) comparing
b. Jan has Math 4th period.	(2) ordering
c. Jan has 3 fewer quarters than dimes.	(3) measuring
d. Jan has 4 quarters and 7 dimes.	(4) counting

2. For each group of numbers, write a relation using the symbol: **a.** > **b.** <

(1) 3, 9 **(2)** 4, 0 **(3)** 5, 2, 8

3. Parts *A* and *B* of a contest were given separate scores, which were then added to get the total score. This table shows the scores of 4 contestants. How much greater was the highest total than the lowest?

	A	B
Ross	221	407
Lainie	284	482
Eric	233	351
Ruta	181	446

4. Frank's Fruit Orchard has 22 rows of orange trees, with 18 trees in a row, and 16 rows of grapefruit trees, with 12 trees in a row. How many trees are in the orchard?

5. The table shows the inventory of sleeveless sweaters on hand a week ago at Mag's Rags. This week, Mag sold 47 of these sweaters: 9 style *A*, 5 style *B*, 14 style *C*, and the rest *D*.

Style	Number	Price
A	27	$25
B	16	38
C	31	26
D	46	18

a. What was the total retail value of the sweaters sold?

b. What is the value of the remaining sweaters?

6. In a *magic square*, every row, every column, and each diagonal add up to the same sum. Copy and complete these magic squares.

a.

45	17	?
?	33	?
29	?	21

b.

37	9	29
17	25	?
21	?	?

c.

50	8	?	?
?	35	?	26
29	23	20	38
14	?	?	?

7. List the members of these sets:

a. {the states in the USA whose names begin with *A*}

b. {the students in your math class over 5'6" tall}

c. {the even digits}

d. {the odd numbers between 700 and 708}

e. {the 7 dwarfs from Snow White}

8. Write a property common to the members in each of these sets:

a. {9, 11, 13, 15} **b.** {6, 9, 12, 15, 18} **c.** {23, 29, 31}

d. {1, 4, 9, 16} **e.** {Black, Red, Dead} **f.** {Taft, Truman, Tyler}

9. a. List all subsets of these four sets:

(1) {7} **(2)** {6, 7} **(3)** {0, 6, 7} **(4)** {0, 6, 7, 9}

b. Prepare a table to summarize your findings:

Number of Members in the Set	
Number of Subsets the Set Has	

c. Use the data in the table to find a pattern relating the number of members in a set and the number of subsets the set has.
Describe the pattern by telling the number of subsets for a set with n members.

d. Use your pattern to find the number of subsets for a set with 17 members.

10. Draw a Venn diagram to show the relation between each group of sets.

a. $A = \{m, a, q\}$
$B = \{q, r, m, n, a\}$

b. $P = \{7, 8, 9\}$
$Q = \{7, 8, 10\}$

c. $M = \{7, 8, 9\}$
$N = \{4, 5, 6\}$

d. $J = \{x, t, f, b\}$
$K = \{s, t, b, c, g, f, y, x\}$
$L = \{t, b, f, g, x, y\}$

11. Determine if the set is finite or infinite.

a. {people on Earth} **b.** {odd numbers} **c.** {digits > 6}

d. {digits > 9} **e.** {composite numbers} **f.** {whole numbers divisible by 4}

g. {whole numbers < 20} **h.** {grains of sand on a beach}

12. a. Write two different one-to-one correspondences between the members of $A = \{a, b\}$ and $B = \{c, d\}$. *Note:* $a \leftrightarrow c$ is the same as $c \leftrightarrow a$.

b. Write all possible one-to-one correspondences between the members of $C = \{1, 2, 3\}$ and $D = \{2, 4, 6\}$.

13. a. Here is a method of finding prime numbers. To find the prime numbers less than 100:

(1) Copy and complete this listing of natural numbers.

1 2 3 4 5 6 7 8 9 10
11 12 13 14 15 16 17 18 19 20
⋮ ⋮ ⋮ ⋮ ⋮ ⋮ ⋮ ⋮ ⋮ ⋮
91 92 93 94 95 96 97 98 99 100

(2) Circle 2, the first prime number, and cross out all multiples of 2, that is, 4, 6, 8, ...

(3) Circle 3, the next prime. Cross out any multiples of 3 that were not eliminated in step 2.

(4) Continue until all the prime numbers have been circled and the composite numbers have been crossed out. The first two rows will look like this:

1 (2) (3) ~~4~~ (5) ~~6~~ (7) ~~8~~ ~~9~~ ~~10~~

(11) ~~12~~ (13) ~~14~~ ~~15~~ ~~16~~ (17) ~~18~~ (19) ~~20~~

(5) List the prime numbers between 1 and 100.

b. In the process of circling prime numbers and crossing out their multiples, what was the smallest prime number you circled for which all its multiples had already been crossed out? Give a reason this was so.

EXPLORATIONS

1. Suppose you want to add 4 repeatedly, say five times, to the number 7.

a. Write out the repeated addition, and the answer.

b. Now see how your calculator handles the repeated addition.

Enter: 7 [+] 4 [=] [=] [=] [=] [=]

Explain how the calculator's method for repeated addition is different from what you wrote.

c. Under these conditions—repeatedly adding 4 to 7—the number 4 is called a ***constant***. Write a rule for using a calculator to repeatedly add a constant to a given number.

d. **(1)** Does your rule work for subtraction of a constant? Try it on your calculator by repeatedly subtracting the constant 4 from the given number 27.

(2) Does your rule work for multiplication by a constant? Try it on your calculator using 7 as the given number and 4 as the constant. Any changes needed?

(3) Which rule works for division by a constant? Try your prediction using 128 as the given number and 4 as the constant.

e. Summarize your findings about the way a calculator works with a constant under each of the four operations.

2. Suppose you want to add 4 to each of the series of numbers 7, 8, and 9.

a. Write out these three additions, and the answers.

b. Now see how your calculator handles this repeated addition.

(1) *Enter:* 7 [+] 4 [=] *Display:* 11.

What has the calculator done with 4?

(2) *Enter:* 8 [=] *Display:* 12.

What has the calculator done with 4?

(3) *Enter:* 9 [=] *Display:* 13.

What has the calculator done with 4?

c. Explain how the calculator's method is different from the three additions that you wrote.

(Explorations continue)

3. By using objects that you can move around, you can determine if a number is prime or composite, and you can find the factors of the number.

Remember: The number 1 is neither prime nor composite.

= 1 *polychip*

For a prime number, the polychips will form exactly two rectangles.

For example, for the number 2, use 2 polychips:

Since exactly two rectangles are possible, 2 is prime. Its factors are 1 and 2.

For a composite number, more than two rectangles (or squares) are possible.

Use 4 polychips for the number 4.

Since 4 polychips make more than two rectangles, the number 4 is composite. Its factors are 1, 2, and 4.

Observe that the number of rectangles is the same as the number of factors. The number 2 has two rectangles, and two factors, 1 and 2. The number 4 has three rectangles, and three factors, 1, 2, and 4.

Complete this table for the numbers 7 through 17.

Number	Dimensions of Rectangles	Number of Rectangles	Prime or Composite	Factors
2	1 × 2 2 × 1	2	prime	1, 2
3	1 × 3 3 × 1	2	prime	1, 3
4	1 × 4 2 × 2 4 × 1	3	composite	1, 2, 4
5	1 × 5 5 × 1	2	prime	1, 5
6	1 × 6 2 × 3 3 × 2 6 × 1	4	composite	1, 2, 3, 6

After completing the table, answer the question:
What kind of number has an odd number of factors?

4. Group Games: Arithmetic Tic-Tac-Toe

X	O	O
X	X	O
O	X	X

a. Like ordinary Tic-Tac-Toe, two players (or teams) alternate, placing **X**'s and **O**'s on a 3-by-3 gameboard.

The winner is the first one whose symbol fills a row, column, or diagonal.

1	2	3
4	5	6
7	8	9

Here's how to play: Number a gameboard 1 through 9, as shown.

Player X begins by choosing a problem from the 9 problems listed. The player finds the answer, and places an **X** in the numbered box that matches the answer.

Player O has a turn. The players then alternate, using each problem exactly once.

Problems:

(1) $12 - 5 - 5$ **(2)** $3 + 3 + 3$ **(3)** $1 + 1 + 1$

(4) $12 - 5 - 0$ **(5)** $6 + 2 + 0$ **(6)** $10 - 5 - 4$

(7) $6 - 1 - 1$ **(8)** $15 - 5 - 5$ **(9)** $12 + 3 - 9$

b. To play more games, the players create new sets of problems, whose answers are 1-9, but not in order.

c. Another Arithmetic Tic-Tac-Toe uses a pair of numbered cubes (dice) and a 5-by-5 gameboard numbered as shown.

16	4	12	15	9
8	18	6	10	12
1	11	0	7	6
5	14	13	3	10
15	4	20	8	5

Here's how to play: The box in which you place your **X** or **O** is determined by an operation of your choosing on the pair of numbers you get when you roll the dice.

For example, if you roll two 4's, you may add to get 8, subtract to get 0, multiply to get 16, or divide to get 1. Choose the operation that gives an answer that will help you fill a row or block your opponent.

5. Group Game: Guess the Set

A leader decides on a secret set, and the group tries numbers until the set is identified.

Here's how to play: Suppose the leader is thinking {multiples of 4}.

The leader draws a circle large enough for the group to see, and the group takes turns calling random numbers from 0 through 100. If the number called, say 8, is in the set, the leader tells the recorder to write the number in the circle. If the number called, say 5, is not in the set, the leader tells the recorder to write the number outside the circle.

When there are enough numbers in the circle to show a common property, the group can try to guess the set.

This game can be played in small groups or with the class as a whole.

3.2 Properties of Operations

From experience with addition, we notice some patterns.

Example 1

(1) When two whole numbers are added, the sum is always another whole number. We say that the set of whole numbers is ***closed*** under addition.

$5 + 2 = 7 \qquad 3 + 8 = 11 \qquad 14 + 36 = 50$

(2) The sum of two whole numbers is the same regardless of the order in which the numbers are added.

$$5 + 2 = 2 + 5$$
$$7 = 7$$

(3) When adding three whole numbers, begin by adding two of the numbers. The result is the same no matter which two numbers are added first.

$$(5 + 2) + 6 = 5 + (2 + 6)$$
$$7 + 6 = 5 + 8$$
$$13 = 13$$

(4) The sum of 0 and any number is that number.

$0 + 9 = 9 \qquad 9 + 0 = 9$

To summarize these facts about addition, we will not use specific whole numbers such as 5 or 8. Instead, we will use letters, called ***variables***, to stand for *any* of the various whole numbers.

Properties of Addition **(a, b, and c are whole numbers.)**

- **Closure Property** $a + b$ is a whole number.
- **Commutative Property** $a + b = b + a$
- **Associative Property** $(a + b) + c = a + (b + c)$
- **Addition Property of Zero** $a + 0 = a$

The operation of multiplication has some of the same properties as addition. Note that there are different symbols used to mean multiplication: $5 \times 2 \qquad 5 \bullet 2 \qquad 5(2)$

Example 2

(1) The product of two whole numbers is always another whole number. The set of whole numbers is closed under multiplication. $3 \cdot 5 = 15 \qquad 4 \cdot 7 = 28 \qquad 6 \cdot 9 = 54$

(2) The product of two whole numbers is the same regardless of the order in which the numbers are multiplied.

$$3 \cdot 5 = 5 \cdot 3$$
$$15 = 15$$

(3) When multiplying three whole numbers, it does not matter which two numbers are multiplied first.

$$(3 \cdot 5) \cdot 2 = 3 \cdot (5 \cdot 2)$$
$$15 \cdot 2 = 3 \cdot 10$$
$$30 = 30$$

(4) The product of 1 and any number is that number.

$1 \cdot 9 = 9 \qquad 9 \cdot 1 = 9$

(5) The product of 0 and any number is 0.

$0 \cdot 7 = 0 \qquad 7 \cdot 0 = 0$

Here is a summary of the facts about multiplication, using variables.

Properties of Multiplication
(*a*, *b*, and *c* are whole numbers.)

- Closure Property — $a \cdot b$ is a whole number.
- Commutative Property — $a \cdot b = b \cdot a$
- Associative Property — $(a \cdot b) \cdot c = a \cdot (b \cdot c)$
- Multiplication Property of One — $a \cdot 1 = a$
- Multiplication Property of Zero — $a \cdot 0 = 0$

The operations of addition and multiplication can be used together.

Example 3 What is the value of $3 \cdot (2 + 5)$?

Add first: $3 \cdot (2 + 5) = 3 \cdot 7 = 21$

Multiply first: $3 \cdot (2 + 5) = 3 \cdot 2 + 3 \cdot 5$
$= 6 + 15$
$= 21$

same answer

This shows that when multiplying a sum, you can either:

- *add first and then multiply, or*
- *multiply first and then add*

This computation can be modeled using polychips.

$3 \cdot (2 + 5) \quad = \quad 3 \cdot 2 \quad + \quad 3 \cdot 5$

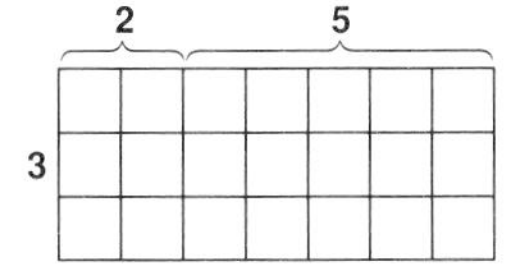

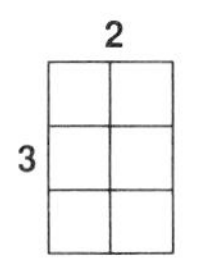

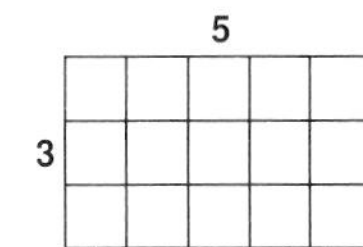

In general:

Distributive Property of Multiplication Over Addition

- If *a*, *b*, and *c* are whole numbers, then: $a \cdot (b + c) = a \cdot b + a \cdot c$
- You can also turn the sentence around: $a \cdot b + a \cdot c = a \cdot (b + c)$

Try These (For solutions, see page 118.)

1. Tell whether each statement is true or false.
   a. $0 \cdot 2 = 2 \cdot 0$
   b. $4 \cdot 0 = 0 + 4$
   c. $75 + 0 = 1 \cdot 75$
   d. $(2 + 3) + 4 = 2 + (3 + 4)$
2. Replace the letters in each sentence by a number to make the sentence true.
   a. $3 + a = 7 + 3$
   b. $f \cdot 1 = 23$
   c. $5 + (3 + c) = (5 + 3) + 9$
   d. $w \cdot 0 = 0$
3. Consider the set $D = \{0, 1, 2, 3, 4\}$. Note these sums for members of D.
   $0 + 1 = 1$ $\quad 0 + 2 = 2$ $\quad 1 + 2 = 3$
   $1 + 1 = 2$ $\quad 2 + 2 = 4$ $\quad 3 + 4 = 7$
   What do these sums tell you about set D? Explain.
4. Is $6 - 3$ a whole number?
   Is $3 - 6$ a whole number?
   Is the set of whole numbers closed under the operation of subtraction? Explain.

SELF-CHECK: SECTION 3.2

Fill in the missing numbers to make the sentences true.

1. **a.** $5 + 8 = ? + 5$
   b. $5 \times 8 = 8 \times ?$

   Answer: **a.** $5 + 8 = 8 + 5$
   b. $5 \times 8 = 8 \times 5$

2. **a.** $6 + (3 + 2) = (6 + ?) + 2$
   b. $6 \times (3 \times 2) = (6 \times 3) \times ?$

   Answer: **a.** $6 + (3 + 2) = (6 + 3) + 2$
   b. $6 \times (3 \times 2) = (6 \times 3) \times 2$

3. **a.** $0 + ? = 7$ **b.** $0 \times 7 = ?$ **c.** $? \times 7 = 7$

   Answer: **a.** $0 + 7 = 7$ **b.** $0 \times 7 = 0$
   c. $1 \times 7 = 7$

4. $2 \times (3 + 8) = 2 \times 3 + 2 \times ?$

   Answer: $2 \times (3 + 8) = 2 \times 3 + 2 \times 8$

EXERCISES: SECTION 3.2

1. Tell whether each given set is closed under the operation shown.
   a. {whole numbers}, $\times$ **b.** {even numbers}, + **c.** {5, 10, 15}, $\div$
   d. {0, 1}, $\times$ **e.** {0, 2, 4, 6, 8}, $\div$ by 2 **f.** {counting numbers}, +
2. Tell whether each statement is true or false.
   a. $23 + 8 = 8 + 23$ **b.** $11 \times 5 = 5 \times 11$ **c.** $7 - 3 = 3 - 7$
   d. $0 \cdot 3 = 4 \cdot 0$ **e.** $0 + 3 = 4 + 0$ **f.** $23 \cdot 0 = 23$
   g. $23 + 0 = 23$ **h.** $19 \times 1 = 19 + 0$
   i. $(6 + 8) + 3 = 6 + (8 + 3)$ **j.** $(4 \times 5) \times 8 = 4 \times (5 \times 8)$
   k. $3 \cdot (5 + 22) = 3 \cdot 5 + 3 \cdot 22$ **l.** $3 \cdot (7 + 4) = (3 + 7) + (3 + 4)$
   m. $0 \cdot (4 - 3) = (7 + 2) \cdot 0$

3. Replace each letter by a number to make the sentence true.

a. $3 + a = 9 + 3$ **b.** $2 \cdot (k + 8) = 2 \cdot 6 + 2 \cdot 8$ **c.** $f \cdot 1 = 16$ **d.** $w + 0 = w$

4. Consider these questions about properties.

a. If $F = \{0, 1\}$, is F closed under addition? Explain.

b. Does the commutative property hold for subtraction of whole numbers? Explain.

c. Does $(10 - 5) - 3 = 10 - (5 - 3)$?
Draw a conclusion about the operation of subtraction from this example.

d. Give an example using whole numbers to show that division is not a commutative operation.

e. If $a \cdot b = 0$, what must be true of a or b? What property justifies your answer?

f. Does $5 \cdot (7 - 1) = 5 \cdot 7 - 5 \cdot 1$? Draw a conclusion from this example.

g. Does $(12 \div 2) + (12 \div 3) = 12 \div (2 + 3)$? Draw a conclusion from this example.

5. Match a verbal statement from *Column B* with a numerical statement in *Column A*. (*Note:* ≠ means "is *not* equal to")

Column A	*Column B*
a. $9 + 3 = 3 + 9$	(1) The set of whole numbers is closed under multiplication.
b. $9 - 3 \neq 3 - 9$	(2) Addition is commutative.
c. $9 \times 3 = 3 \times 9$	(3) Multiplication is commutative.
d. $9 \div 3 \neq 3 \div 9$	(4) Subtraction is not commutative.
e. 9×3 is a whole number.	(5) The set of whole numbers is not closed under division.
	(6) Division is not commutative.

6. Use one of the words *addition*, *subtraction*, *multiplication*, or *division* to fill in each blank so that the resulting statement is true.

a. The set of whole numbers is closed under the operations of ___?___ and ___?___, but is not closed under ___?___ or ___?___.

b. The commutative property is true for the operations of ___?___ and ___?___, but is not true for ___?___ or ___?___.

c. The distributive property is true for ___?___ over ___?___.

7. For each part, **a – j**, choose the property of operations on the set of whole numbers that the statement illustrates.

	Properties
a. $2 \times 0 = 0$	(1) Closure Property of Addition
b. $2 \times (3 \times 4) = (2 \times 3) \times 4$	(2) Closure Property of Multiplication
c. $2 + 3 = 5$	(3) Commutative Property of Addition
d. $2 \times (3 \times 4) = 2 \times (4 \times 3)$	(4) Commutative Property of Multiplication
e. $2 + (3 + 4) = (2 + 3) + 4$	(5) Associative Property of Addition
f. $2 \times (3 + 4) = 2 \times 3 + 2 \times 4$	(6) Associative Property of Multiplication
g. $2 \times 1 = 2$	(7) Addition Property of 0
h. $2 \times 3 = 6$	(8) Multiplication Property of 0
i. $3 + 0 = 3$	(9) Multiplication Property of 1
j. $2 \times (3 + 4) = 2 \times (4 + 3)$	(10) Distributive Property of Multiplication Over Addition

8. Use polychips to show that: **a.** $2 \cdot (4 + 3) = 2 \cdot 4 + 2 \cdot 3$ **b.** $3 \cdot 5 + 3 \cdot 2 = 3(5 + 2)$

9. The distributive property can often help you to compute mentally.
For example, think of $5 \cdot 23$ as $5 \cdot (20 + 3)$.
Then $5 \cdot (20 + 3) = 5 \cdot 20 + 5 \cdot 3 = 100 + 15 = 115$.
Use the distributive property to show how each of the following could be computed mentally, and give the answer. **a.** $2 \cdot 59$ **b.** $3 \cdot 16$ **c.** $5 \cdot 46$

10. The commutative and associative properties, often used together, can also help you compute mentally. For example: $(2 \cdot 37) \cdot 5 = (37 \cdot 2) \cdot 5 = 37 \cdot (2 \cdot 5) = 37 \cdot 10 = 370$
Use the properties to show how each of the following could be computed mentally, and give the answer.
a. $24 + (6 + 47)$ **b.** $(2 \cdot 7) \cdot 5$ **c.** $(14 + 8) + (6 + 12)$ **d.** $4 \cdot (7 \cdot 5)$

EXPLORATIONS

1. You know how to do addition with the whole numbers. Copy and complete this addition table.
Here's how to tell if an operation is commutative by studying its operation table. Draw a diagonal through your table from upper left to lower right. Notice that each number on one side of the diagonal has a matching number on the other side of the diagonal.
If an operation "passes this diagonal test," the operation is commutative.

+	0	1	2	3	4
0	?	?	?	?	?
1	?	?	?	?	?
2	2	3	4	5	6
3	?	?	?	?	?
4	?	?	?	?	?

2.

+	△	◇	■
△	△	◇	■
◇	◇	■	△
■	■	△	◇

Here is a table that tells you how to "add" the members of {△, ◇, ■}. For example, to find the "sum" ■ + ◇:
at the left, find the row ■
at the top, find the column ◇.
The answer △ is where the ■ row meets the ◇ column.

a. Is this set closed under this addition? Explain.
b. Does ■ + ◇ = ◇ + ■? Draw a conclusion from this example. Test your conclusion on another example. Still true?
c. Use the diagonal test described in Exploration 1 to tell if this addition is commutative.

3. This table tells you how to do an operation ☺ on the members of $\{a, b, c, d, e\}$.
a. Suppose this operation ☺ is commutative. Copy the table and fill in the missing letters.
b. Study the effect that the letter e has under operation ☺. What whole number under addition does e remind you of? Explain. What whole number under multiplication? Explain.

☺	*a*	*b*	*c*	*d*	*e*
a	*a*	*a*	?	*b*	*a*
b	?	*b*	?	?	*b*
c	*b*	*d*	*c*	*c*	*c*
d	?	*e*	?	*d*	*d*
e	*a*	*b*	*c*	*d*	*e*

3.3 Order of Operations

The month of May has 3 days more than 4 weeks.

$$3 + 4 \times 7$$

MAY						
1	2	3	4	5	6	7
8	9	10	11	12	13	14
15	16	17	18	19	20	21
22	23	24	25	26	27	28
29	30	31				

Explain how you must do this calculation to get the number of days in May. If you said that you must multiply first, and then add, you are correct.

$$3 + 4 \times 7 = 3 + 28 = 31$$

Rules to avoid confusion in calculation tell us the *order* in which to carry out operations.

When a numerical expression contains two or more operations:

- **First, do all multiplications and divisions, from left to right.**
- **Then, do all additions and subtractions, from left to right.**

Example 1 Find the value of each expression.

a.
$9 - 6 \div 3$ There are 2 operations.
$= 9 - 2$ Divide first.
$= 7$ Then subtract.

b.
$4 \times 3 - 6 \div 2$ There are 3 operations. From the left, first multiply and divide.
$= 12 - 3$
$= 9$ Then subtract.

When an expression contains a ***power****, a repeated factor written with an* ***exponent****:*

- **Do the power before other operations.**

$3 + 7^2$ Power first.
$= 3 + 49$ $7^2 = 7 \times 7$
$= 52$ Then add.

When an expression has parentheses, the parentheses show how to group numbers.

- **Do the operation inside the parentheses first.**

$(8 + 3) \cdot 2$
$= 11 \cdot 2$
$= 22$

Order of Operations

1. Do all operations within parentheses.
2. Do all powers.
3. Working from left to right, do all multiplications and divisions.
4. Working from left to right, do all additions and subtractions.

Example 2 Find the value of:

$8 + (2 + 3)^2 \times 4$
$= 8 + (5)^2 \times 4$ Work within parentheses first.
$= 8 + 25 \times 4$ Do the power.
$= 8 + 100$ Do the multiplication.
$= 108$ Do the addition.

Try These (For solutions, see page 118.)

Find the value of each numerical expression.

1. $10 + 2 \times 3$ **2.** $30 \div 5 + 7$ **3.** 3×2^4

4. $9(6 - 4)$ **5.** $15 + 3 \cdot (8 - 5)$ **6.** $14 \div 2 + (3 + 4)^2$

Order on a Calculator

Some calculators automatically follow the order of operations; others do the operations in the order in which you enter them. To test your calculator, enter: 8 [+] 3 [×] 2 [=]

What display on the calculator will tell if it is following the order of operations?

A display of 14 shows that the calculator is following the order of operations.

A display of 22 shows that the calculator is working in the order you have entered the numbers.

How would you have to enter the numbers to get the correct answer 14?

On a calculator that does not follow the order of operations, you must enter:

3 [×] 2 [+] 8 [=]

Example 3 To find the value of the expression $12 - 8 \div 4$, write a key sequence for a calculator that does follow the order of operations (called *math order*).

For a math-order calculator, you can enter the operations in the order in which you see them and rely upon the calculator to follow the order of operations.

Answer: 12  8  4 [=]

After entering this key sequence, a math-order calculator displays 10.

Example 4 To find the value of the expression $12 - 8 \div 4$, write a key sequence for a calculator that does not follow the order of operations, but does the calculations in the order of entry (called *entry order*).

For an entry-order calculator, *you* must consider the order of operations. Use memory to do the division first.

① Do the division first.
(Your calculator may not need = here.) 8 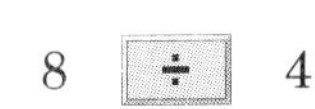

② Store the result of the division in memory. [M+]

③ Recall the result of the division from memory to subtract from 12.

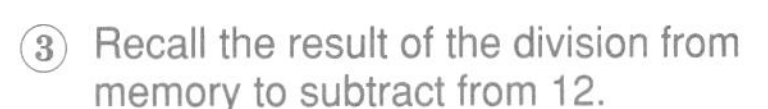

Answer:       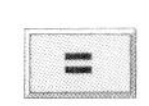

8 [÷] 4 [=] [M+] 12 [−] [MR] [=]

After entering this key sequence, an entry-order calculator will show 10.

Try These (For solutions, see page 118.)

For each of these numerical expressions

1. $15 + 9 \div 3$ **2.** $17 - 4^2$

write a key sequence for:

a. a math-order calculator

b. an entry-order calculator

SELF-CHECK: SECTION 3.3

Fill in the blanks to make the sentences true.

1. The first operation to do in $3 + 5 \times 4$ is _?_.

Answer: multiplication

2. In $(9 - 5) \times 3$, you must first _?_.

Answer: work within the parentheses

3. In $7 + (1 + 5)^2$, after first working within the parentheses, you must _?_.

Answer: do the power

4. To carry out the additions and subtractions in $7 + 8 - 4 - 2$, you must work from _?_.

Answer: left to right

5. A math-order calculator automatically follows _?_.

Answer: the order of operations

6. A calculator that does operations in the order entered is called an _?_.

Answer: entry-order calculator

EXERCISES: SECTION 3.3

1. Find the value of each numerical expression.

a. $3 + 7 \times 5$ **b.** $10 - 2 \cdot 3$ **c.** $12 + 9 \div 3$

d. $20 \div 2 + 3$ **e.** $7(12 - 8)$ **f.** $3(5 + 2)$

g. $7 + 4^2$ **h.** $(7 + 4)^2$ **i.** $15 - 8 + 9$

j. $19 - 7 - 5$ **k.** $5 \cdot 4 - 2 \cdot 3$ **l.** $6 \times 2 + 3 \times 5$

m. $5^2 - 2 \cdot 7$ **n.** $3^3 - 6 \div 2$ **o.** $8^2 - 5^2$

p. $(8 - 5)^2$ **q.** $5 \times 6 \div 2 + 8 \div 4$ **r.** $20 - 4 + 2 - 12 \div 3$

2. In each of the following, the beginning of the calculation is shown. Write the next step.

a. $3(7 - 2) + 5^2$
$= 3(5) + 5^2$

b. $3 \times (3 \times 2 - 2)$
$= 3 \times (6 - 2)$

c. $7^2 + 12 \div 4$
$= 49 + 12 \div 4$

d. $6 + 2(5 - 2) \times 3$
$= 6 + 2(3) \times 3$

3. In each of these calculations, a mistake has been made. Copy the problem, describe the error, and write the correct answer.

a. $3 + 2 \times 6$
$= 5 \times 6$
$= 30$

b. $20 - 5(2 + 1)$
$= 15(2 + 1)$
$= 15(3)$
$= 45$

c. $6 \times 5 - 4 \times 2$
$= 6 \times 1 \times 2$
$= 6 \times 2$
$= 12$

d. $6 \times 5^2 \times 5 - 2$
$= 6 \times 10 \times 5 - 2$
$= 60 \times 5 - 2$
$= 298$

Create some of these for your class. Make them tricky. Have more than one mistake per problem.

4. Use one of the symbols >, <, or = to replace ? in each of the following sentences so that the sentence is true.

a. $8 + 3 \times 6$? $8 \times 3 + 6$ **b.** $6 \times 4 - 3$? $6 \times (4 - 3)$ **c.** $4 + 2 \times 7$? $4 + (2 \times 7)$

d. 4^2 ? $(4 + 7)^2$ **e.** $6 \times 2 \times 0$? $6 + 2 + 0$ **f.** $6 \div 3 \times 1$? $6 \div 3 + 1$

5. Which of the following is a true statement?

(a) $12 \cdot 4 + 3 = 12(4 + 3)$ (b) $12 \cdot 4 + 12 \cdot 3 = 12(4 + 3)$

(c) $12 \div 4 \div 3 = 12 \div (4 \div 3)$ (d) $(12 - 4) - 3 = 12 - (4 - 3)$

6. For each of these numerical expressions

(1) $15 + 23 \times 8$ **(2)** $385 - 6^3$ **(3)** $14 \cdot 23 + 19 \cdot 35$ **(4)** $788 - (46 - 18)^2$

write a key sequence for: **a.** a math-order calculator **b.** an entry-order calculator

7. Choose the expression that means, "three times the difference between 9 and 5 is 12."

(a) $(3 \times 5) - (3 \times 9) = 12$ (b) $9 - 5 \times 3 = 12$ (c) $3 \times 9 - 5 = 12$ (d) $3 \times (9 - 5) = 12$

8. To prepare for a canoe race, Josh rowed 2 hours one morning and 3 hours on each of the next 5 mornings. Gregor rowed 2 hours in the morning and 3 hours in the afternoon every day for 5 days.

Which of the following expressions can be used to determine who rowed more?

(a) $2 + 3 \times 5 > (2 + 3) \times 5$

(b) $2 \times 3 + 5 > (2 \times 3) + 5$

(c) $2 + 3 \times 5 < (2 + 3) \times 5$

(d) $2 \times 3 + 5 < (2 \times 3) + 5$

9. At a party, 4 people shared 12 slices of pizza equally, and 6 people shared a dozen frankfurters equally. If Big George had both, which expression gives his total number of pizza slices and frankfurters?

(a) $12 \div 4 + 12 \div 6$

(b) $(12 - 4) + (12 - 6)$

(c) $(12 + 12) \div (4 + 6)$

(d) $(4 \div 12) + (6 \div 12)$

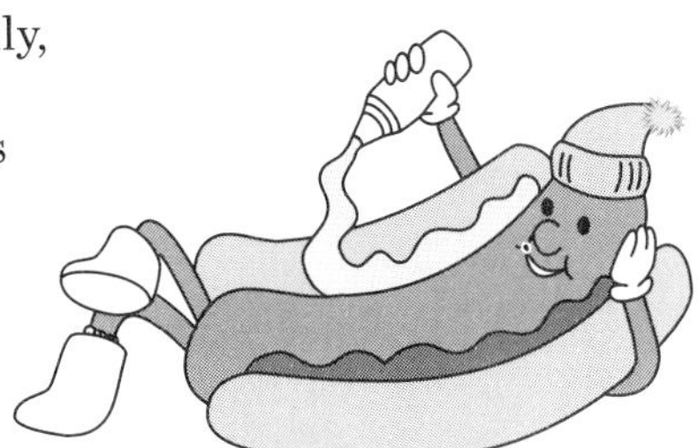

Courtesy of Delta Education, Inc.

10. At the beginning of their camping trip, Seth's backpack weighed 5 pounds less that Andrew's, and Ann's weighed 3 pounds less than Seth's. If Andrew's pack weighed 28 pounds, which expression best describes the weight of Ann's?

(a) $28 - 5 - 3$

(b) $28 - (5 - 3)$

(c) $5 - 3 < 28$

(d) $3 < 5 < 28$

11. Serena and Victor work in a flower shop. They had an order for 4 pots of daffodils at \$5 each and 3 pots of tulips at \$7 each.

Serena's calculation of the bill was:

$$4 \times 5 + 3 \times 7 = 161$$

Victor said it should be: $(4 + 3) \times (5 + 7) = 84$

What is the correct amount that should be charged? Explain what was done wrong.

EXPLORATIONS

To decode this message:

?	?	?		?	?	?	?	?	?	?	?	?	?
1	10	0		0	6	9	4	11	75	29	13	6	9

?	?	?	?	?		?	?
10	11	9	4	11		10	42

?	?	?	?	?	?	?	?	?	?
10	20	4	11	13	29	5	10	6	75

Find the value of each expression below. Locate your answer in the code. Each time the answer appears in the code, write the letter that is next to the expression in the blank above the answer. Use a separate sheet of paper onto which you have copied the code.

A $7 + 9 \times 2 \div 3$

F $2 \times 3^2 + 3 \times 2^3$

O $4^2 - 6(1)$

S $3(5)^2$

Y $48 \div 3 - 5 \times 3$

D $5^2 - 4^2$

I $17 - 4(3)$

P $(5 - 1)(20 \div 4)$

T $2 + 3^3$

E $(6 - 4)^2$

N $24 \div 6 + 2$

R $21 \div 3 + 20 \div 5$

U $0(7^{60})$

3.4 Fractions

How are all of these shaded representations alike? How are they different?

Although each represents one-fourth of a whole, they are not the same quarters since the wholes are in different units of time, money, and length.

When working in the same unit, there are many fractions that are equivalent to one-fourth.

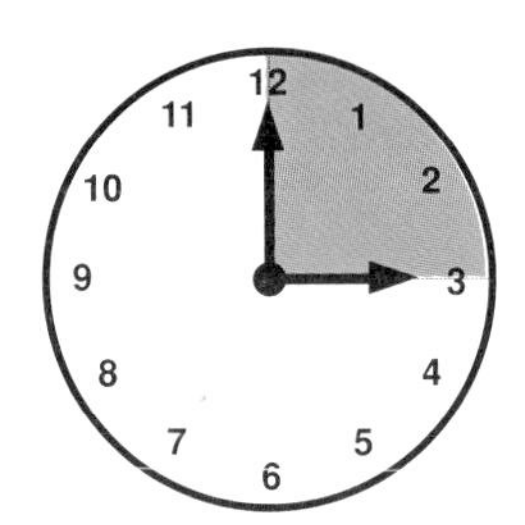

$$\frac{1}{4} = \frac{2}{8} = \frac{3}{12} = \frac{4}{16} = \ldots$$

Is $\frac{12}{48}$ equivalent to $\frac{1}{4}$?

$$\frac{12}{48} = \frac{12 \div 12}{48 \div 12} = \frac{1}{4}$$

To reduce a fraction, divide numerator and denominator by the greatest common factor.

How does $\frac{1}{4}$ compare in size to other fractions, say to $\frac{1}{3}$?

(1) Use each denominator as a factor to find a common denominator.

$$4 \times 3 = 12$$

⇑ common denominator

(2) Rewrite each fraction using the common denominator.

$$\frac{1}{4} = \frac{3}{12} \qquad \frac{1}{3} = \frac{4}{12}$$

(3) Compare numerators of the fractions with like denominators. The fraction with the larger numerator is the larger fraction.

Since $\frac{3}{12} < \frac{4}{12}$

Then $\frac{1}{4} < \frac{1}{3}$

To compare fractions, write them with the same denominator.

Try These *(For solutions, see page 118.)*

1. Write $\frac{36}{45}$ as an equivalent fraction in lowest terms.
2. Compare: $\frac{4}{5}$ and $\frac{7}{10}$

Using Fractions in Multiplication and Division

A diagram can model the operation of multiplication with fractions.

Example 1 To find $\frac{2}{5}$ of $\frac{5}{6}$, take a regular shape and divide it into 6 equal parts. Then:

① Lightly shade 5 of the 6 parts.

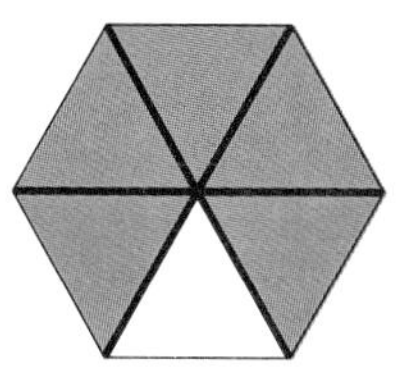

② Darken 2 of the 5 shaded parts.

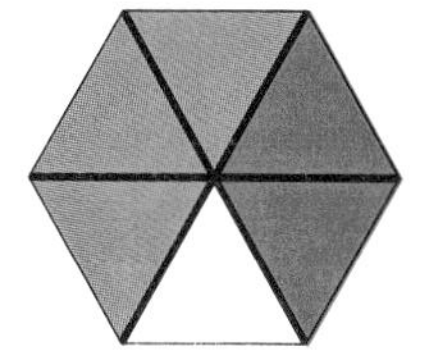

③ Compare the 2 darkly-shaded parts with the original 6.

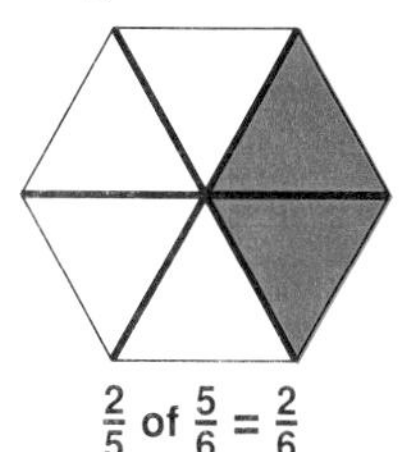

$\frac{2}{5}$ of $\frac{5}{6} = \frac{2}{6}$

④ The darkly-shaded part is also one-third of the original figure.

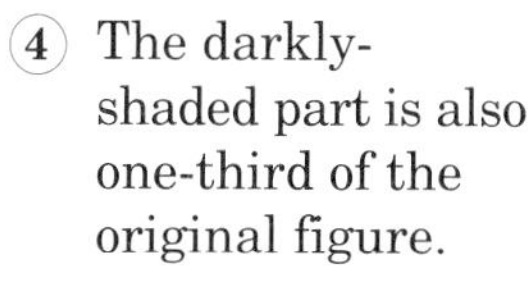

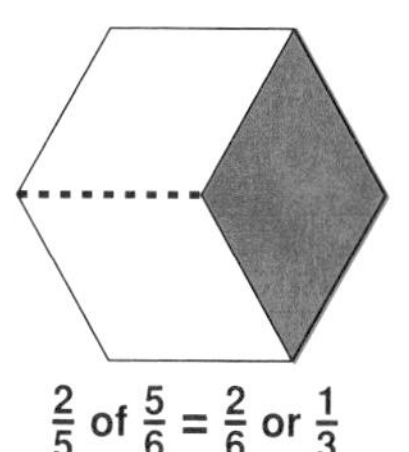

$\frac{2}{5}$ of $\frac{5}{6} = \frac{2}{6}$ or $\frac{1}{3}$

Examine the arithmetic result of this model to note a procedure for multiplying fractions:

$$\frac{\overset{1}{\cancel{2}}}{\underset{1}{\cancel{5}}} \times \frac{\overset{1}{\cancel{5}}}{\underset{3}{\cancel{6}}} = \frac{1 \times 1}{1 \times 3} = \frac{1}{3}$$

> **To multiply fractions:**
>
> **First, cancel any common factors, one from a numerator and the other from a denominator.**
>
> **Then, multiply the remaining factors in the numerator and in the denominator.**

Example 2 On his barbecue, Steve used half of the ground beef he found on a plate in the refrigerator to make a burger. If there was two-thirds of a pound of meat on the plate, how much was in the burger?

Take half of the two-thirds. $\quad \frac{1}{\underset{1}{\cancel{2}}} \times \frac{\overset{1}{\cancel{2}}}{3} = \frac{1}{3}$

Answer: The burger has $\frac{1}{3}$ of a pound of meat.

Since division is the inverse operation of multiplication:

> **To divide by a fraction, multiply by its inverse.**

Example 3 For the jewelry box she is making, Ora needs 6-inch sections of wood. How many such sections can she cut from a board that is 8 feet long?

Work in the same unit. $\quad 8 \div \frac{1}{2}$

6 inches = $\frac{1}{2}$ foot

Divide the 8-foot board into $\frac{1}{2}$-foot pieces. $\quad 8 \times \frac{2}{1}$ — Division is the same as multiplying by an inverse.

$8 \times \frac{2}{1} = \frac{16}{1}$ or 16 — Multiply numerators and multiply denominators.

Answer: From an 8-foot board, Ora will get 16 pieces that are each 6 inches long.

Try These *(For solutions, see page 119.)*

1. Multiply: $\frac{15}{4} \times \frac{2}{5}$ **2.** Divide: $16 \div \frac{4}{5}$

Using Fractions in Addition and Subtraction

To model addition of fractions that have the same denominator, work with a rectangle divided into the number of parts indicated by the denominator.

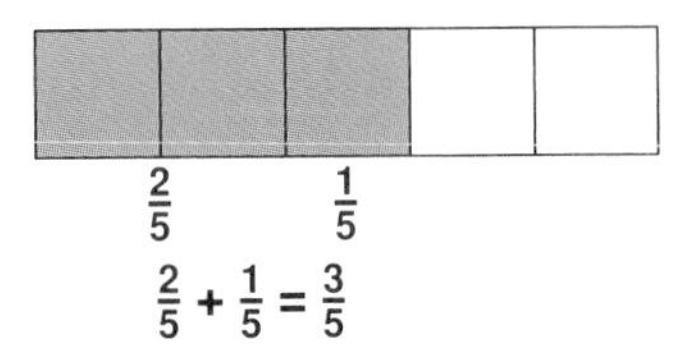

Examine the arithmetic result of the model to note a procedure:

To add (or subtract) fractions with like denominators, keep the denominator and add (or subtract) the numerators.

Example 4 Working with her chemistry set, Jasmine mixes 32 ounces of acid with one-quarter of a gallon of an acid solution and pours the mixture into an empty gallon container. To this, she adds one quarter of a gallon of water. After this mixture is in the gallon container, how much room is left for additional liquid?

Work in the same unit.

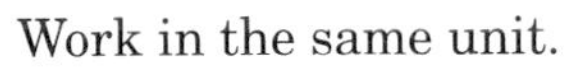

Since 1 gallon = 128 ounces,

then 32 ounces = $\frac{1}{4}$ gallon.

Add the amounts of acid, solution, and water.

$$\frac{1}{4} + \frac{1}{4} + \frac{1}{4} = \frac{3}{4}$$ With like denominators, just add the numerators.

The gallon container is now $\frac{3}{4}$ full.
How much space is left?

the whole ⇓ part used ⇓

$$1 - \frac{3}{4} = \frac{4}{4} - \frac{3}{4} = \frac{1}{4}$$ With like denominators, just subtract numerators.

⇑ rewrite the whole

Answer: The gallon is still one-quarter empty.

To model addition of fractions that have different denominators, work with two rectangles of the same size and shape.

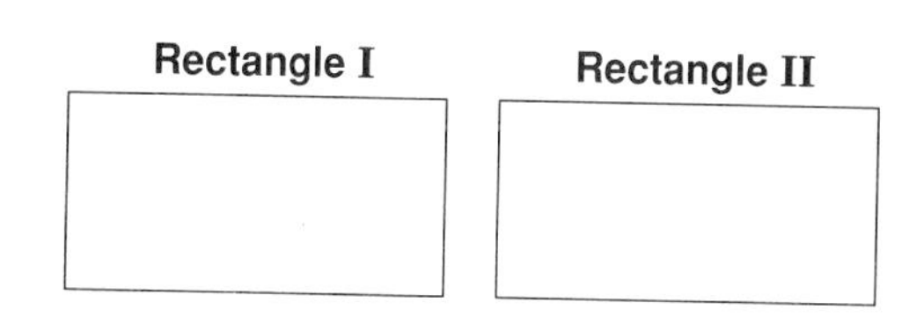

Example 5 Model: $\frac{1}{2}$ and $\frac{1}{3}$

① Shade $\frac{1}{2}$ of Rectangle I vertically.

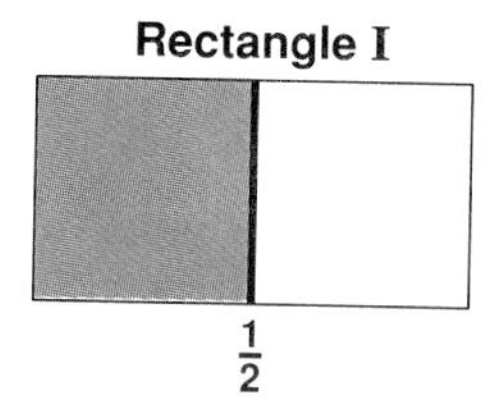

② Shade $\frac{1}{3}$ of Rectangle II horizontally.

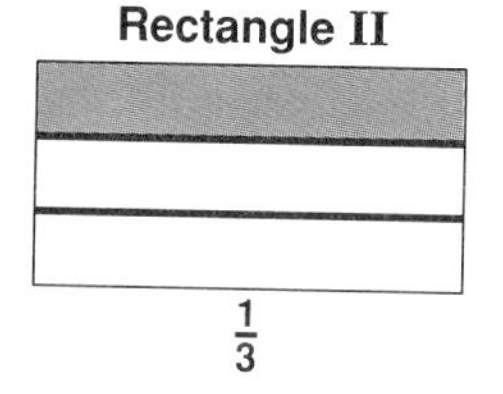

③ Separate the shaded portions into parts of equal size.

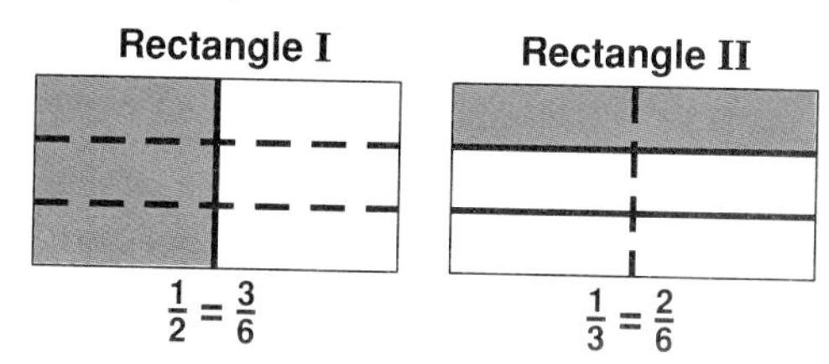

④ Use a clean rectangle to display their sum.

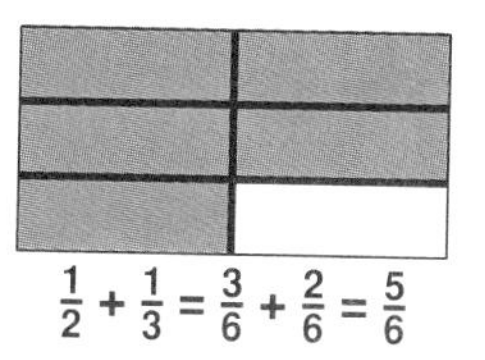

Examine the arithmetic result of the model to note a procedure:

To add (or subtract) fractions with unlike denominators, first rewrite each fraction using a common denominator.

Example 6 Cesar worked out $\frac{3}{4}$ hour this morning and $\frac{1}{2}$ hour this afternoon. Has he reached his goal of $1\frac{1}{2}$ hours of exercise for today?

Add the times:

$\frac{3}{4} + \frac{1}{2}$	These denominators are unlike. Get a common denominator: $4 \cdot 2 = 8$
$\frac{6}{8} + \frac{4}{8}$	Rewrite each fraction using the common denominator.
$\frac{10}{8}$	Now that the denominators are alike, add the numerators.
$1\frac{2}{8}$ or $1\frac{1}{4}$	Rewrite as a mixed number.

To compare $1\frac{1}{4}$ to $1\frac{1}{2}$, the denominators of the fractions must be alike.

$$1\frac{1}{4} \Downarrow \quad 1\frac{1}{2} \Downarrow$$

$$1\frac{1}{4} < 1\frac{2}{4}$$

From the comparison, you can see how far short of his goal he is.

Answer: Cesar has not yet reached his goal. He is $\frac{1}{4}$ hour short.

Try These *(For solutions, see page 119.)*

1. $\frac{7}{9} + \frac{2}{9}$ **2.** $\frac{2}{3} + \frac{4}{5}$ **3.** $\frac{3}{4} - \frac{5}{8}$

SELF-CHECK: SECTION 3.4

Fill in the blanks to best complete the sentences.

1. To reduce a fraction to lowest terms, divide by the __?__. *Answer:* greatest common factor
2. To compare fractions, they must have the __?__. *Answer:* same denominator
3. When multiplying fractions, first __?__ any common factors. *Answer:* cancel
4. To divide by a fraction, multiply by its __?__. *Answer:* inverse
5. To add fractions with like denominators, keep the denominator and __?__. *Ans.:* add numerators
6. To add fractions with unlike denominators, first rewrite the fractions using a __?__. *Answer:* common denominator

EXERCISES: SECTION 3.4

1. a. Write a fraction to represent the shaded portion of each of the following diagrams, and reduce the fraction to lowest terms.

(1)

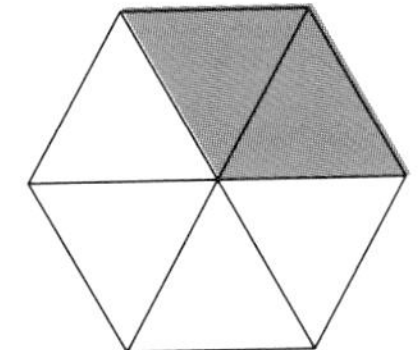

(2)

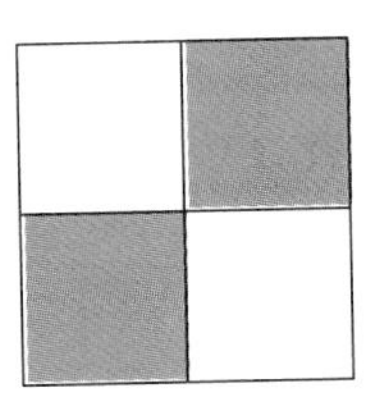

(3)

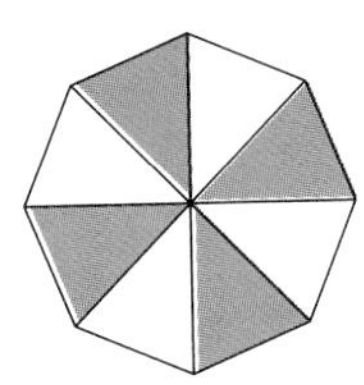

(4)

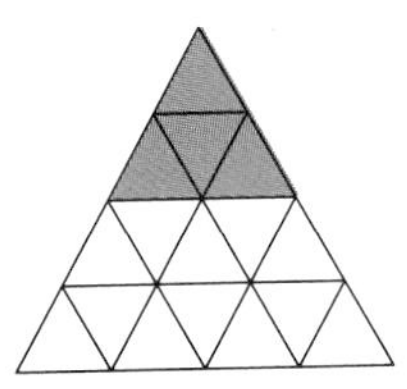

(5)

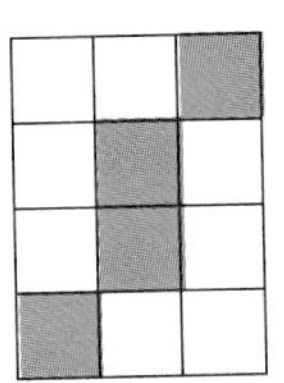

(6)

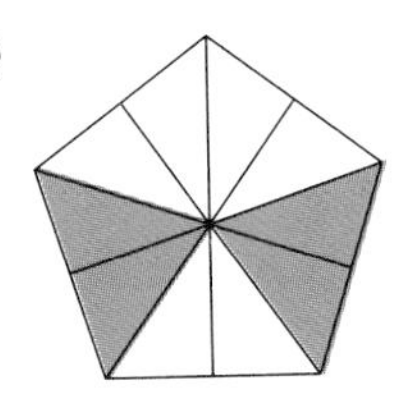

b. Replace ? to make each statement true.

(1) The diagrams that model $\frac{1}{2}$ have __?__ numbers of shaded and unshaded parts.

(2) The diagrams that model $\frac{1}{3}$ have __?__ as many __?__ parts as __?__ parts.

2. Find a number to complete each pair of equivalent fractions.

a. $\frac{1}{2} = \frac{?}{10}$ **b.** $\frac{?}{4} = \frac{6}{8}$ **c.** $\frac{2}{?} = \frac{8}{12}$ **d.** $\frac{4}{5} = \frac{8}{?}$

3. Find the greatest common factor of each pair of numbers.

a. 12, 18 **b.** 18, 27 **c.** 16, 48 **d.** 24, 36

4. Reduce each fraction to lowest terms.

a. $\frac{12}{18}$ **b.** $\frac{9}{12}$ **c.** $\frac{20}{36}$ **d.** $\frac{15}{45}$

5. Replace ? by >, <, or = to make a true comparison.

a. $\frac{5}{4}$? $\frac{3}{4}$ **b.** $\frac{1}{2}$? $\frac{3}{6}$ **c.** $\frac{1}{2}$? $\frac{1}{3}$ **d.** $\frac{3}{4}$? $\frac{5}{3}$

e. $\frac{5}{4}$? $\frac{4}{5}$ **f.** $\frac{5}{6}$? $\frac{11}{12}$ **g.** $\frac{8}{12}$? $\frac{2}{3}$ **h.** $\frac{7}{8}$? $\frac{13}{16}$

6. a. In each of the following models, use the lightly shaded and darkly shaded parts of the diagram to find the answer.

(1)

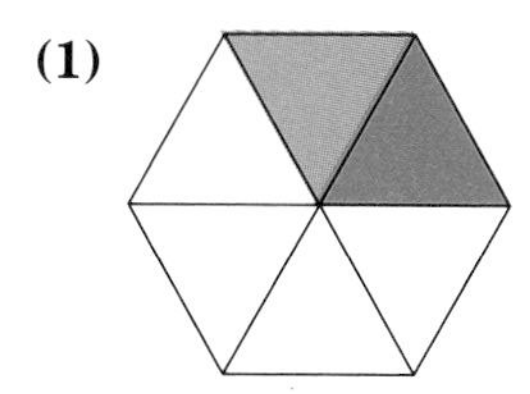

$\frac{1}{2}$ of $\frac{1}{3}$ = ?

(2)

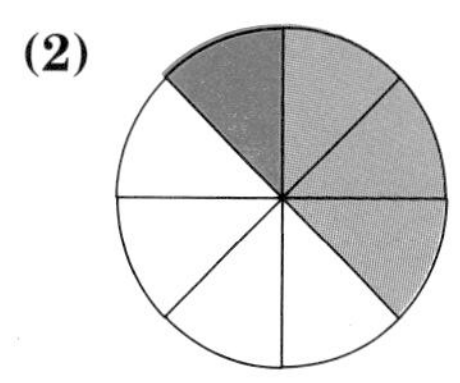

$\frac{1}{4}$ of $\frac{1}{2}$ = ?

(3)

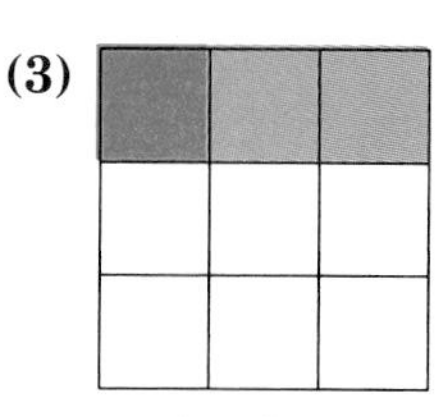

$\frac{1}{3}$ of $\frac{1}{3}$ = ?

(4)

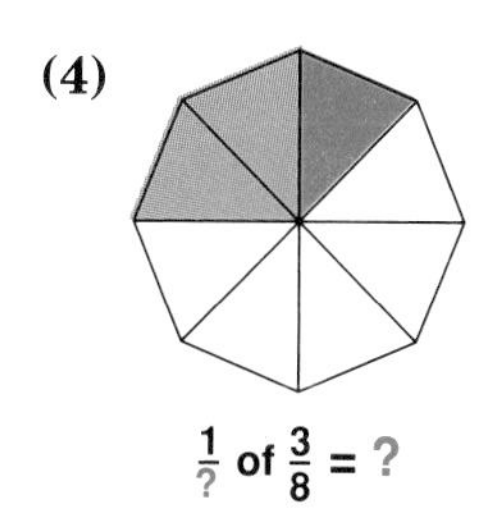

$\frac{1}{?}$ of $\frac{3}{8}$ = ?

(5)

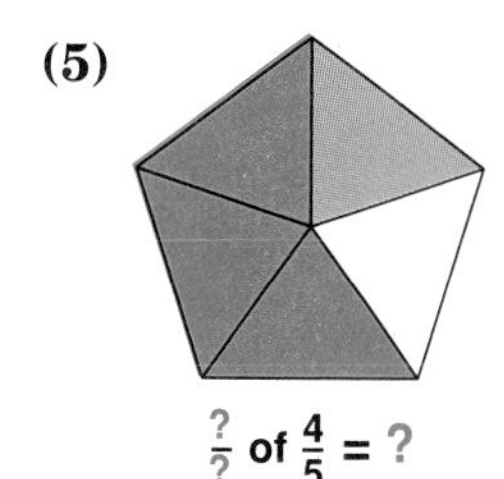

$\frac{?}{?}$ of $\frac{4}{5}$ = ?

(6)

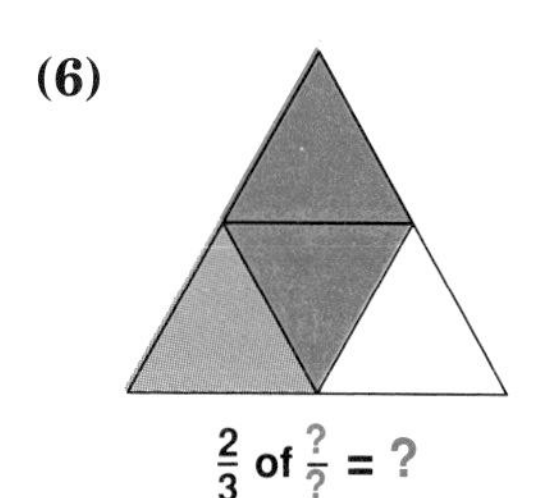

$\frac{2}{3}$ of $\frac{?}{?}$ = ?

b. Draw a model to find each product.

(1) $\frac{1}{3}$ of $\frac{3}{4}$ **(2)** $\frac{1}{2}$ of $\frac{1}{5}$ **(3)** $\frac{3}{4}$ of $\frac{1}{2}$ **(4)** $\frac{1}{2}$ of $\frac{1}{6}$

7. Multiply. Write answers in lowest terms.

a. $\frac{3}{4} \times \frac{1}{2}$ **b.** $\frac{5}{8} \times \frac{2}{3}$ **c.** $\frac{3}{8} \times \frac{16}{9}$ **d.** $\frac{5}{3} \times 12$

e. $\frac{3}{2} \times \frac{4}{9}$ **f.** $9 \times \frac{2}{3}$ **g.** $1\frac{1}{3} \times \frac{3}{8}$ **h.** $\frac{1}{10} \times 2\frac{1}{2}$

i. $\frac{3}{5} \times \frac{15}{18} \times \frac{12}{10}$ **j.** $\frac{4}{7} \times \frac{21}{8} \times \frac{16}{3}$ **k.** $5 \times \frac{3}{4} \times \frac{8}{15}$ **l.** $1\frac{1}{2} \times 1\frac{1}{9}$

8. Divide. Write answers in lowest terms.

a. $\frac{5}{8} \div \frac{15}{4}$ **b.** $\frac{2}{3} \div \frac{4}{9}$ **c.** $\frac{7}{6} \div 3$ **d.** $\frac{6}{8} \div \frac{10}{3}$

e. $8 \div \frac{1}{6}$ **f.** $\frac{3}{4} \div \frac{1}{2}$ **g.** $3\frac{3}{5} \div \frac{9}{10}$ **h.** $5 \div 2\frac{1}{2}$

9. Add or subtract, as shown. Write answers in simplest form.

a. $\frac{3}{4} + \frac{1}{4}$ **b.** $\frac{3}{5} - \frac{2}{5}$ **c.** $2\frac{2}{3} + \frac{2}{3}$ **d.** $2\frac{3}{5} + 3\frac{4}{5}$

e. $\frac{1}{4} + \frac{1}{2}$ **f.** $\frac{1}{2} - \frac{3}{6}$ **g.** $\frac{5}{6} + \frac{7}{12}$ **h.** $\frac{9}{10} - \frac{2}{5}$

i. $2\frac{1}{5} + \frac{3}{10}$ **j.** $\frac{3}{8} + 3\frac{7}{16}$ **k.** $\frac{3}{4} + \frac{2}{3}$ **l.** $\frac{1}{3} + \frac{4}{5}$

m. $3\frac{1}{5} + 1\frac{1}{2}$ **n.** $\frac{5}{8} + \frac{7}{10}$ **o.** $\frac{7}{12} - \frac{5}{18}$ **p.** $\frac{11}{15} + \frac{4}{9}$

10. Choose the best answer to each question. Test some values.

a. If the fraction $\frac{a}{b}$ reduces to $\frac{1}{2}$, how many pairs of whole numbers are possible for a and b?

(1) 1 (2) 2 (3) 3 (4) more than 3

b. If n represents a number greater than 1, which of the following is greatest?

(1) $n + \frac{1}{2}$ (2) $n - \frac{1}{2}$ (3) $n \cdot \frac{1}{2}$ (4) $n \div \frac{1}{2}$

c. If $\frac{a}{b}$ represents a fraction between $\frac{1}{2}$ and 1, dividing it by $\frac{1}{2}$ results in a number that

(1) > 1 (2) < 1 (3) ≥ 1 (4) ≤ 1

11. Like parentheses and brackets, a fraction bar is also a *grouping symbol.* When there are grouping symbols in an expression, the order of operations requires that the grouped calculations be done first.

Example: $\frac{4+1}{2 \times 5} = \frac{5}{10} = \frac{1}{2}$

Simplify:

a. $\frac{3(4+1)}{2+3}$ **b.** $\frac{2^3 + 4}{3^3 - 3}$ **c.** $\frac{2}{3} \cdot \frac{1+2}{2+2}$

d. $\frac{3}{4} + \left(\frac{2}{3} - \frac{1}{3}\right)$ **e.** $\frac{1}{5}\left(\frac{3}{4} + \frac{1}{2}\right)$ **f.** $\frac{6-1}{12} \div \frac{10}{3}$

12. Solve each of these problems.

a. A holiday pie crust recipe calls for $\frac{1}{3}$ cup butter, $\frac{3}{4}$ cup flour, and $\frac{1}{2}$ cup chopped walnuts. If Keith wants to bake 3 pies, how much of each ingredient does he need?

b. Nakia bought 3 packages of ground meat weighing $2\frac{1}{2}$ lb., $1\frac{3}{4}$ lb., and $2\frac{3}{4}$ lb. What was the total weight of this meat?

c. Rashida bought 8 packages of noodles, each with $14\frac{3}{4}$ ounces. Exactly how close to 7 pounds is this?

d. Emily bought $6\frac{7}{8}$ yards of material from a bolt that had $28\frac{1}{2}$ yards. How many yards are left on the bolt?

e. Sugar weighing $2\frac{1}{2}$ pounds was divided equally among 4 containers. How many *ounces* of sugar were in each?

f. Octavio bought a 5-foot-long carpet runner. Since this runner is too long for his hallway, he must trim it and, to maintain the pattern, he will trim the same amount from each end. How much must he trim from each end to wind up with a runner that is $56\frac{1}{2}$ inches long?

g. How many books $1\frac{1}{2}$ inches thick will fit on a rack that is 2 feet long?

h. For winter use in a parking lot, the attendant purchased $\frac{3}{4}$ of a ton of salt. If he used $\frac{2}{3}$ of the salt, how many pounds did he use?

i. A T-shirt regularly selling for $18 is on sale at $\frac{1}{3}$ off. What is the sale price?

EXPLORATIONS

1. Another model for multiplication of fractions uses graph paper.

Example: Model: $\frac{3}{4} \times \frac{2}{5}$

① Outline a vertical and a horizontal rectangle that meet at one corner. Using stripes, color 3 of the 4 squares of one rectangle and 2 of the 5 squares of the other rectangle.

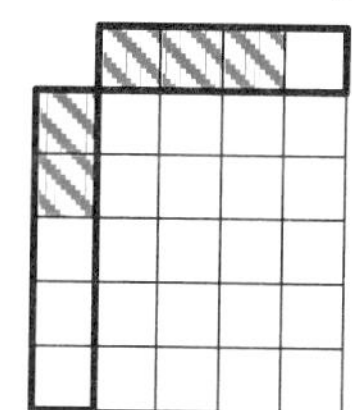

② Draw the large rectangle (shown lightly shaded), that is outlined by the vertical and horizontal rectangles.

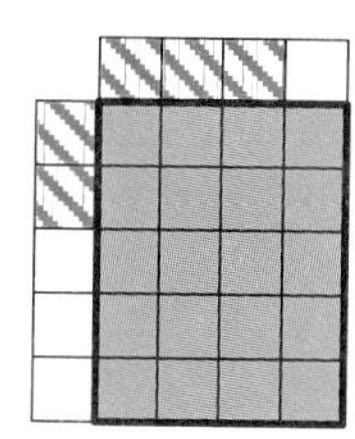

③ Use dark shading to color a region of the large rectangle to match the stripes in the smaller rectangles.

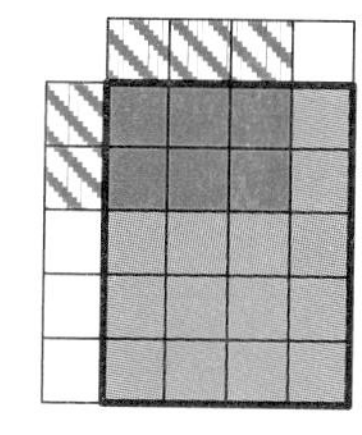

④ Of the total number of squares in the large rectangle, (20), how many are darkly shaded? (6)

What is the product $\frac{3}{4} \times \frac{2}{5}$? $\left(\frac{6}{20}\right)$

Use this method to find these products:

a. $\frac{2}{3} \times \frac{1}{4}$ **b.** $\frac{4}{5} \times \frac{5}{6}$ **c.** $\frac{1}{3} \times \frac{5}{8}$

2. You will need small paper or plastic squares to work with. Start with 16 squares of one color and 24 squares of another color, say red and white. Use all the squares to create a pattern that repeats.

a. **(1)** Here is one design: How many times can you make this design if you use all the squares?

= red

= white

(2) Write a fraction: $\frac{\text{number of red squares in the design}}{\text{number of white squares in the design}}$ Reduce the fraction to lowest terms.

(3) Compare the fraction in part **(2)** to this fraction: $\frac{\text{total number of red squares}}{\text{total number of white squares}}$ Are the fractions equivalent?

(4) Use all the squares to make another design 4 times. Draw your design.

(5) What is the greatest number of times you can make one design using all these squares? How is this number related to the total numbers of red and white squares?

(a) least common multiple (b) common prime factor

(c) greatest common factor (d) no relation

b. Create your own design to start with and follow part **a (1-5)** using:

(1) 12 red squares and 16 white squares **(2)** 12 red squares and 30 white squares

(3) 18 red squares and 36 white squares **(4)** 30 red squares and 45 white squares

3. It is often useful to get *estimates* as answers to computations that involve fractions.

a. **(1)** What whole number is a good estimate for these fractions? $\frac{11}{12}, \frac{99}{100}, \frac{942}{1{,}000}$

(2) When is the number 1 a good estimate for a fraction?

b. **(1)** What whole number is a good estimate for these fractions? $\frac{1}{20}, \frac{3}{50}, \frac{12}{1{,}000}$

(2) When is the number 0 a good estimate for a fraction?

c. **(1)** What "easy" fraction is a good estimate for these fractions? $\frac{51}{100}, \frac{32}{60}, \frac{4}{7}$

(2) When is the "easy" fraction $\frac{1}{2}$ a good estimate for a "hard" fraction?

d. **(1)** What "easy" fraction is a good estimate for these fractions? $\frac{4}{17}, \frac{5}{21}, \frac{6}{25}$

(2) When is the "easy" fraction $\frac{1}{4}$ a good estimate for a "hard" fraction?

e. **(1)** What "easy" fraction is a good estimate for these fractions? $\frac{13}{16}, \frac{15}{21}, \frac{19}{24}$

(2) When is the "easy" fraction $\frac{3}{4}$ a good estimate for a "hard" fraction?

f. Estimate answers to these calculations by rounding the given fractions to the numbers 0, 1, $\frac{1}{2}$, $\frac{1}{4}$, or $\frac{3}{4}$:

(1) $\frac{18}{19} + \frac{2}{19} + \frac{4}{19} + \frac{15}{19}$ Check your answer by finding the exact value.

(2) $\frac{28}{100} + \frac{75}{96} + \frac{8}{432} + \frac{112}{119}$ Check your answer on a calculator.

3.5 Decimals and Percents

In what special ways can the fractions in these sets be rewritten?

a. {fractions with numerators equal to denominators}

b. {fractions with denominators of 1}

c. {fractions whose denominators are powers of 10}

d. {fractions whose denominators are 100}

These special sets of fractions are familiar when written out.

a. $\left\{\frac{1}{1}, \frac{2}{2}, \frac{3}{3}, \frac{4}{4}, \ldots\right\}$ Each of these fractions can be rewritten as the number 1.

b. $\left\{\frac{1}{1}, \frac{2}{1}, \frac{3}{1}, \frac{4}{1}, \ldots\right\}$ is equivalent to the set of natural numbers {1, 2, 3, 4, ...}.

c. Fractions whose denominators are powers of 10 can be rewritten as ***decimals***.

$\left\{\frac{1}{10}, \frac{2}{10}, \frac{3}{10}, \ldots, \frac{1}{100}, \frac{2}{100}, \frac{3}{100}, \ldots, \frac{1}{1{,}000}, \frac{2}{1{,}000}, \ldots\right\}$

{0.1, 0.2, 0.3, ... , 0.01, 0.02, 0.03, ... , 0.001, 0.002, ...}

d. Fractions whose denominators are 100 can be rewritten as ***percents***.

$\left\{\frac{1}{100}, \frac{2}{100}, \frac{3}{100}, \ldots, \frac{99}{100}, \frac{100}{100}, \frac{101}{100}, \ldots\right\}$

{1%, 2%, 3%, ... , 99%, 100%, 101%, ...}

Different ways of writing fractions are convenient for computation. A calculator will show that *any* fraction can be rewritten as a decimal. Choose any "hard" fraction and find its decimal equivalent on a calculator.

For example, to find the decimal equivalent of $\frac{37}{158}$, enter:

37 [÷] 158 [=] *Display:* 0.2341772

From this key sequence, note a general method for converting a fraction to a decimal.

To change a fraction to a decimal, divide the numerator by the denominator.

Example 1 Use a calculator to find the decimal equivalents of the given fractions. Study the results and come to a conclusion.

	Fraction	Decimal		Fraction	Decimal
a.	$\frac{2}{9}$	0.2222222	**b.**	$\frac{4}{9}$	0.4444444
c.	$\frac{7}{9}$	0.7777777	**d.**	$\frac{13}{99}$	0.1313131

Some fractions are equivalent to decimals that repeat without end. Use your calculator to find other examples of fractions that repeat.

By converting different ways of expressing numerical information to a common form, you can make comparisons.

Converting a fraction to a decimal allows you to read the percent equivalent.

Also, a percent can be expressed as a decimal or as a fraction.

Example 2 Classy Clothes is selling stone-washed jeans for $\frac{1}{4}$ off its regular price of $24.99. For the same jeans, The Discount Den is taking 10% off its regular price of $20.99. Which is the better buy?

Classy Clothes		*The Discount Den*
$\frac{1}{4} = 0.25$ or 25%		10% = 0.10

To convert a decimal to a percent, move the decimal point 2 places to the right and add the percent symbol.

To convert a percent to a decimal, move the decimal point 2 places to the left and drop the percent symbol.

Classy Clothes		The Discount Den
24.99	← regular price →	20.99
× 0.25	← % of discount, as a decimal →	× 0.10
6.2475		2.099
≈ 6.25	← amount of discount →	≈ 2.10
24.99	← regular price →	20.99
− 6.25	← amount of discount →	− 2.10
the better buy → $18.74	← sale price →	$18.89

You may have used a calculator to find the amount of discount that each store offered. To do each entire calculation on a calculator, you will have to use memory. Try this key sequence to find the sale price $18.74 at Classy Clothes:

24.99 [×] .25 [M+] 24.99 [−] [MR] [=]

Write a similar key sequence to find the sale price at The Discount Den.

What common fraction is the equivalent of the decimal 0.5? Did you say $\frac{1}{2}$?

Note a general procedure so that you can convert any decimal to a fraction. A basic calculator does not perform this operation.

To convert a decimal to a fraction:

Read the decimal name to write a fraction whose denominator is a power of 10.

As possible, reduce the fraction.

0.5 is five tenths.

$$0.5 = \frac{5}{10}$$

$$0.5 = \frac{5}{10} = \frac{1}{2}$$

Example 3 To complete a survey, the school nurse needed to know the average weight of 5 children in a study group. The children's weights, in pounds, had been recorded as:

57.5 $62\frac{1}{4}$ $60\frac{3}{4}$ 73 59.8

To find the average, add the 5 weights and divide the sum by 5. To do these operations, all of the weights must first be expressed in a common form, say decimals.

57.50 + 62.25 + 60.75 + 73.00 + 59.80 = 313.30 ← Add the 5 weights.
Divide the sum by 5. → 313.30 ÷ 5 = 62.66
Round. → ≈ 62.7 ← the average weight

When values are expressed in decimal form, be sure they have the same number of decimal places before you make a comparison.

Which is greater, 61.8 or 61.27?
Since: 61.8 = 61.80
Then: 61.80 > 61.27

The most common everyday use you have for decimals is computation with money.

Example 4 On their way to the mall, two brothers counted their money. Roger had $7.45 and Mark had $11.79. They spent $16.48 on supplies for their fish tank, using up all of Roger's money. How much did Mark have left?

Roger →	7.45
Mark →	+ 11.79
total available →	19.24
amount spent →	– 16.48
amount remaining →	$2.76

In addition and subtraction of decimals, first line up the decimal points. Then add or subtract as usual, "carrying" numbers to the next column.

Example 5 Eliot types students' term papers at the rate of $1.75 per page. A page with a table counts as 1.5 pages. If Margo's paper turns out to be 45 pages of straight typing plus 7 pages of tables, how much does she owe Eliot?

First estimate an answer, using convenient numbers: 50 pages × $2 = $100

To get an exact answer, decide how many pages there are altogether. Since 7 pages of tables count as 7 × 1.5 or 10.5 pages, there are 45 + 10.5 or 55.5 pages in all.

number of pages × price per page = total cost
55.5 × $1.75 = $97.125
≈ $97.13 ← amount Margo owes

The number of decimal places in the product is the total number of decimal places in the factors.

Example 6 Zenia paid \$12.75 for gas. If gas costs \$1.179 per gallon, how many gallons did she get?

Note that gasoline prices are generally quoted in 3 decimal places even though the total is to the nearest cent.

The next time you are near a gas station, observe how the price is given.

For example: $1.17\frac{9}{10}$

To first estimate an answer, use convenient numbers.

$$\frac{\text{total } \$12}{\text{cost per gallon } \$1} = 12 \text{ number of gallons}$$

> **To divide by a decimal:**
>
> **In the divisor, move the decimal point to the right, after the last digit.**
>
> **In the dividend, move the decimal point the same number of places.**
>
> **Place the decimal point in the quotient above the decimal point in the dividend.**
> **If needed, add zeros to the dividend.**

divisor → $1.18.\overline{)12.75.}$ ← dividend (• ← quotient)

$$\begin{array}{r} 10.8 \\ 118\overline{)1275.0} \\ \underline{118} \\ 95\ 0 \\ \underline{94\ 4} \\ 6 \end{array}$$

10.8 ← gallons Zenia got

It is often useful to visually show information about how a whole quantity (100%) is broken down into percentages. A ***circle graph*** is used to do this.

Example 7 This circle graph shows the various ways in which the 1,450 employees of Comet Electronics commute to work. According to this graph:

a. What is the most popular way of commuting?

The largest percent of the employees use a train.

b. How many of these people come by car pool?

12% of 1,450 = 0.12 × 1,450
= 174 come by car pool

c. What percent of these people commute by methods other than those listed?

The listed methods account for 98% of the people:

34% + 30% + 18% + 12% + 4% = 98%

Thus, 100% – 98% or 2% of these people commute by other methods.

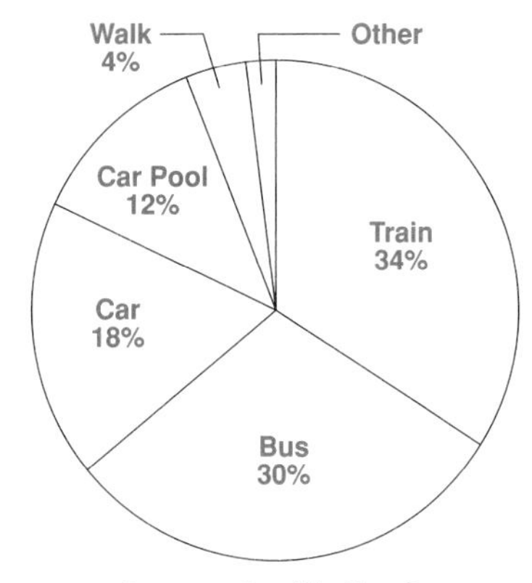

Commuter Methods

Try These (For solutions, see page 119.)

1. Copy and complete the chart to the right to show the decimal and percent equivalents of some frequently used fractions.
2. Of 8.7 and 8.293, which is the greater value?
3. Write a key sequence, using the memory of a calculator, to find the sale price of an item that is 20% less than its original price of $17.99.

Fraction	Decimal	Percent
$\frac{1}{2}$	?	50%
$\frac{3}{4}$	0.75	?
?	0.40	40%
?	0.375	?
?	?	$33.\overline{3}\%$
$\frac{2}{3}$	?	?

The bar over the 3 shows that this number repeats without end.

SELF-CHECK: SECTION 3.5

Fill in the blanks to best complete the sentences.

1. Every fraction can be rewritten as a _?_.

   Answer: decimal or percent
2. To change a fraction to a decimal, divide the _?_.

   Answer: numerator by the denominator
3. To round 7.84 to the nearest tenth, look at the number in the hundredths place. Since this number is less than 5, the rounded value is _?_.

   Answer: 7.8
4. Fractions whose denominators are 100 can be rewritten as _?_.

   Answer: percents
5. To convert a decimal to a percent, move the decimal point 2 places to the _?_ and insert the percent symbol.

   Answer: right
6. In a circle graph, the complete circle represents _?_.

   Answer: 100%

EXERCISES: SECTION 3.5

1. **a.** Use a calculator to find the decimal equivalents of these fractions: $\frac{17}{99}, \frac{48}{99}, \frac{63}{99}$

   b. Using the results of part **a**, what do you think is the decimal equivalent of $\frac{58}{99}$? of $\frac{435}{999}$? Check your predictions on a calculator.

   c. Use a calculator to find the decimal equivalents of these fractions: $\frac{17}{990}, \frac{48}{990}, \frac{63}{990}$

   d. Using the results of part **c**, what do you think is the decimal equivalent of $\frac{58}{990}$? of $\frac{62}{9{,}900}$? Check your predictions on a calculator.

2. Without using a calculator, find the decimal equivalent of each fraction. Check with a calculator.

a. $\frac{5}{8}$ **b.** $\frac{1}{6}$ **c.** $\frac{1}{25}$ **d.** $\frac{7}{20}$ **e.** $\frac{5}{12}$

3. Write each decimal as a fraction (or mixed number) in lowest terms.

a. 0.6 **b.** 2.75 **c.** 0.375 **d.** 0.15 **e.** 4.14

4. Round each number:

a. to the nearest tenth **(1)** 5.714 **(2)** 43.251

b. to the nearest hundredth **(1)** 2.548 **(2)** 27.023

5. Write each set of decimals in order, from least to greatest.

a. 0.02 0.20 0.022 **b.** 0.15 0.051 0.115 **c.** 7.3 0.773 7.37

6. Write each decimal as a percent.

a. 0.35 **b.** 0.01 **c.** 0.425 **d.** $0.66\overline{6}$

7. Write each percent as a decimal.

a. 10% **b.** 75.5% **c.** 1% **d.** $33.\overline{3}\%$

8. Write each percent as a fraction in lowest terms.

a. 80% **b.** 14% **c.** 150% **d.** $66.\overline{6}\%$

9. **a.** Convert each fraction to a percent by first rewriting the fraction as a decimal.

(1) $\frac{4}{5}$ **(2)** $\frac{3}{4}$ **(3)** $\frac{7}{8}$ **(4)** $\frac{5}{6}$

b. The % key on a calculator can be used to change a fraction to a percent.

Example: To change $\frac{4}{5}$ to a percent, use this key sequence: 4 [÷] 5 [%]

Display: 80.

Answer: 80% (Insert % symbol.)

Use a calculator to convert the following fractions to percents:

(1) $\frac{7}{8}$ **(2)** $\frac{1}{12}$ **(3)** $\frac{13}{25}$ **(4)** $\frac{15}{40}$

c. Without a calculator, convert the following fractions (or mixed numbers) to percents. Then check your work using a calculator.

(1) $\frac{1}{6}$ **(2)** $\frac{5}{9}$ **(3)** $2\frac{1}{2}$ **(4)** $3\frac{5}{8}$

10. Estimate, then compute.

a. $3.54 + 2.75 - 3.26$ **b.** $12.5 - 5.25 - 3.75$ **c.** $36.63 \div 0.3 - 22.1$

d. $6.48 + 1.13 \times 4$ **e.** $(1.5)^2 + 6.75$ **f.** $3.25 \times 6 \div 2.5$

11. Solve each problem. Where possible, first estimate an answer.

a. The diagram shows the lengths of the sides of a preschool play area. How many meters of fencing would be needed to enclose the area?

b. In his shopping cart, Gray Cloud has a can of tuna at $1.09, a head of lettuce at 99¢, a container of milk costing $1.37, and a loaf of bread for $1.29. At the checkout, what change should he get from a $10 bill?

c. The table shows the population, in thousands, of 5 towns in Greenfield County, in 2 census years. Did the total population of the 5 towns increase or decrease from 1960 to 1980? By how much?

Town	1960	1980
Kirby	23.5	22.9
Little Falls	36.7	38.2
Grafton	41.3	42.7
Strathmore	52.1	50.6
Clifton	47.6	49.3

d. Ms. Black placed a catalog order for 2 toddler jumpsuits at $12.45 each, 3 men's T-shirts at $14.95 each, and a sweater at $23.75. Tax and delivery charges come to $6.14. What is the total cost of the order?

e. On April 1st, John had a balance of $375.92 in his account. How much was in the account at the end of April after the transactions in this table?

April 3	Interest	9.02
10	Withdrawal	68.23
22	Deposit	124.75
27	Withdrawal	250.00
30	Withdrawal	18.68

f. At the hardware store, Naomi picked up a saucepan costing $6.19, a mop at $4.65, a package of light bulbs at $3.30, and a roll of mailing tape for $4.85. While waiting to check out, she mentally estimated the total cost by rounding each price to the nearest half-dollar.

(1) Her estimate was (a) $18 (b) $18.50 (c) $19 (d) $19.50

(2) What was the actual total?

g. The graph shows the heights, in centimeters, of members of a school track team. What is the average height of the 5 athletes?

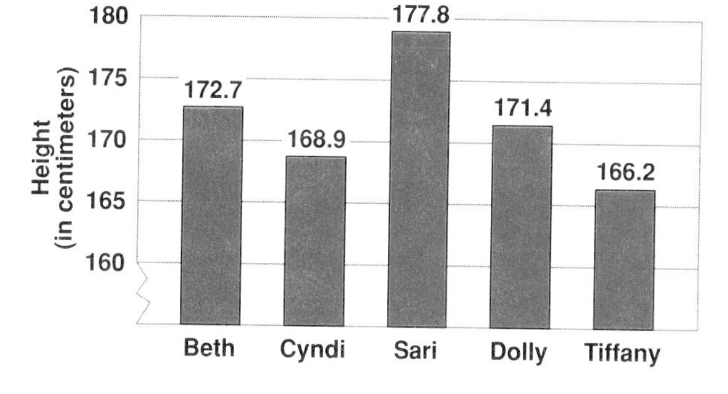

h. Mitch is paid $7.50 an hour for up to 35 hours in a week, and gets $11.25 an hour for hours worked over 35 hours. If he works 44 hours this week:

Describe how you would calculate his earnings. Do the calculations.

i. The A-1 Scientific Equipment Co. has 16 cartons ready for delivery: 3 cartons weighing 45.68 kilograms each, 4 cartons weighing 23.94 kilograms each, and 9 cartons weighing 35.378 kilograms each. The total weight of the cartons is over the maximum 540-kilogram capacity of their delivery van. What is the least number of cartons that must be left behind? Of what weights?

j. The *Newton News* is read by about 55% of the 2,404 households in Newton, and the *Pine Park Press* is read by three-fifths of the 2,335 households in Pine Park. If Lorna wants to place an ad to sell her guitar, in which newspaper will her ad likely reach more readers?

k. Jason has been offered two different selling jobs. One job pays a $150-a-week base salary and a 7% commission on sales, and the other job pays a $175-a-week base salary and a 6% commission on sales. If he expects to average $5,000 per week in sales, which is the better money deal?

l. Marla is buying a VCR, sale priced at $179.99. If she buys it on an installment plan, paying 35% down and $17 per month for 8 months, how much more than the sale price will she pay?

m. Of the 2,315 fans at a ball game, 32% bought hot dogs at $1.75 each, and three-quarters of those buying hot dogs bought soft drinks at 99 cents each. What was the total amount paid by all the fans who bought both of these items?

12. Read the circle graphs to answer the questions.

a. This graph shows how the Martin family budgets an income of $38,000.

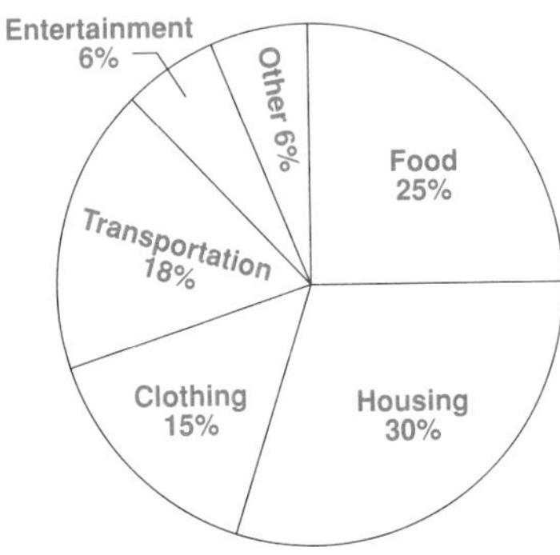

(1) How much money is spent on food?

(2) What two items total more than half the money spent?

(3) The largest expense is as much as which three other items combined?

(4) If Mr. Martin gets a raise of $3,000, which categories do you think will change?

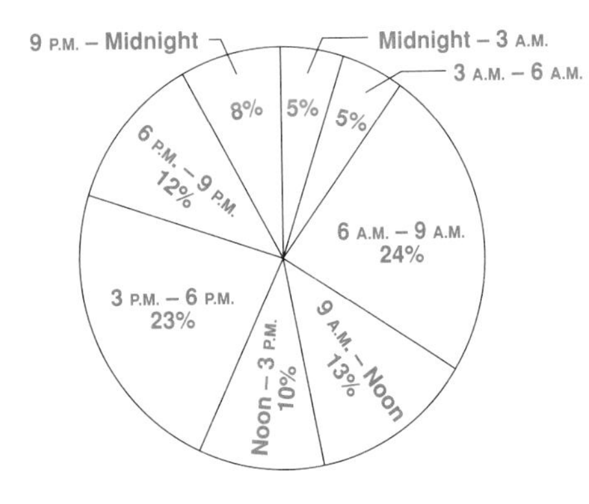

b. This graph shows the distribution of traffic in a 24-hour period, in which approximately 3,600 vehicles went over the bridge that connects Central City with its suburbs.

(1) In what 6-hour period was the bridge traffic lightest?

(2) How many vehicles crossed the bridge during the light 6-hour period?

(3) What was the combined percent of the two periods in which there was the most traffic?

(4) What would account for the heavier traffic in the two busiest periods?

c. This graph shows the results of Bill's 100 times at bat for the Bantams.

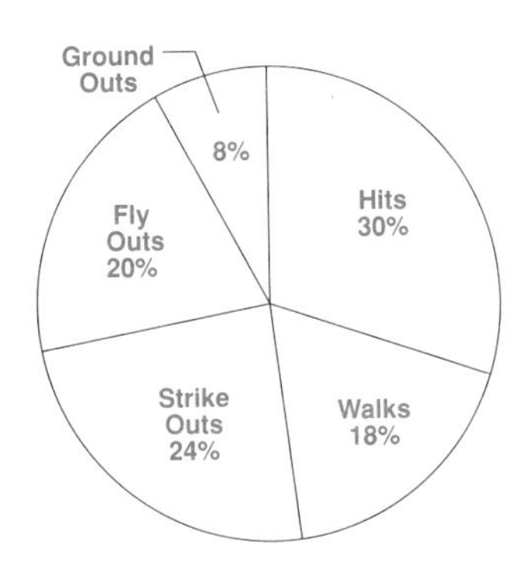

(1) How many hits did he get?

(2) What percent of the time was he out?

(3) The percent of Ground Outs was twice as much as the difference between which two other categories?

(4) After 100 times at bat, what was Bill's batting average?

EXPLORATIONS

1. Remarkable Repeats: Try these on a calculator.

a. Choose any one-digit number.

Multiply it by 0.9 and then multiply the result by 12,345,679.
What was the result? Try again with a different one-digit number.

b. Choose any one-digit number and divide it first by 9, then by 99, and then by 999. Predict what would happen if you divided your one-digit number by 9,999. Test your prediction on the calculator. Explain the display.

c. On a calculator, find the decimal equivalents of $\frac{1}{11}, \frac{2}{11}, \frac{3}{11}$.

Predict the decimal equivalent of $\frac{4}{11}$. Verify your prediction.

2. **a.** Compare each of the following decimals to 1. *Example:* $0.72 < 1$

(1) 0.25 **(2)** 3.08 **(3)** 0.984 **(4)** 1.000

Describe how you can tell whether a decimal is less than or greater than 1.

b. Use a calculator to multiply 25 by each of the following decimals. Record the results and tell whether the product is less than or greater than 25.

(1) 0.25 **(2)** 0.82 **(3)** 1.04 **(4)** 2.6

c. From the results of part **b**, write *two* conclusions in the form:

When the multiplier is _?_ than 1, the product is _?_ than the original number.

d. Use a calculator to divide 25 by each of the following decimals. Record the results and tell whether the quotient is less than or greater than 25.

(1) 0.25 **(2)** 0.82 **(3)** 1.04 **(4)** 2.6

e. From the results of part **d**, write *two* conclusions in the form:

When the divisor is _?_ than 1, the quotient is _?_ than the original number.

3. As with fractions, it is often useful to get ***estimates*** as answers to computations that involve decimals. In fact, it is most useful to estimate an "easy" fractional value or even a whole number for a "hard" decimal.

a. **(1)** What whole number is a good estimate for these decimals? 0.9875 0.89971

(2) What whole number is a good estimate for these decimals? 0.0045 0.10023

(3) What "easy" fraction is a good estimate for these decimals? 0.5123 0.4986

b. Estimate answers to these calculations by rounding the given numbers, using the values, 0, $\frac{1}{2}$, or 1. *Example:* $2.563 \approx 2\frac{1}{2}$

See how close your estimates are by using a calculator to find the actual answers.
Round the calculator answers to get closer to your estimate.

(1) $0.47 + 0.141 + 0.8997$ **(2)** $3.125 + 1.9761 + 6.015$

Things You Should Know After Studying This Chapter

KEY SKILLS

3.1 Use whole numbers in common ways, such as comparisons and problem solving.
Draw a Venn diagram to show set relations.
Write a one-to-one correspondence.
Identify a set as finite or infinite.

3.2 Identify the properties of addition and multiplication.

3.3 Apply the order of operations.
Write key sequences for calculators that do math order or entry order.

3.4 Carry out the basic operations with fractions.

3.5 Convert among the three forms: fraction, decimal, and percent.
Carry out the basic operations with decimals.
Solve problems involving percents.

KEY TERMS AND SYMBOLS

3.1 natural number • whole number • is greater than (>) • is less than (<) • finite vs. infinite sets • the empty set (∅) • one-to-one correspondence

3.2 variable • closure • commutative property for addition • commutative property for multiplication • associative property for addition • associative property for multiplication • distributive property for multiplication over addition • addition property of 0 • multiplication property of 0 • multiplication property of 1

3.3 order of operations • exponent • power • math-order calculator • entry-order calculator

3.4 fraction

3.5 decimal • percent (%)

CHAPTER REVIEW EXERCISES

1. To compare the number of members in $X = \{7, 9, 11, 13, 15\}$ and $Y = \{a, b, c, d\}$, write a relation using: **a.** $<$ **b.** $>$

2. Which of the following relations is not true?

(a) $2 < 6 < 9$ (b) $9 > 6 > 2$
(c) $2 < 9 > 6$ (d) $2 < 9$

3. For a club picnic, Jenny bought 2 packages of paper plates, with a dozen plates in each package. Now she learns that 42 people are coming to the picnic. What is the smallest number of additional packages of paper plates she should buy?

4. For each of the pairs of sets A and B:

a. Determine if one set is a subset of the other.

b. Draw a Venn diagram to show how sets A and B are related.

(1) $A = \{2, 3\}$ $B = \{2, 3, 4\}$
(2) $A = \{2, 3, 6\}$ $B = \{2, 3, 4\}$
(3) $A = \{0\}$ $B = \{\text{natural numbers}\}$

5. Determine whether each of the following sets is finite or infinite.

a. {telephone numbers in the United States}
b. {digits less than 5}
c. {odd numbers greater than 500}

6. Write a 1-to-1 correspondence between the members of $\{4, 3, 2, 1\}$ and $\{a, b, c, d\}$.

7. Is each of these sets closed under the operation shown? If a set is not closed, give an example showing why.

a. {even numbers} under multiplication
b. {0, 1, 2, 3} under addition
c. {whole numbers} under division

8. Tell whether each statement is true or false.

a. $0 \times 6 = 0 \times 5$ **b.** $1 \times 5 = 0 + 5$
c. $3 \times (5 + 2) = (3 \times 5) + 2$

9. Find the value of each expression.

a. $7 + 4 \times 3$ **b.** $12 - 8 \div 4$
c. $16 - 4 - 3$ **d.** $4 + 3^2$
e. $12 \div 4 + 16 \times 2$ **f.** $7 + 5^2 + (7 + 5)^2$

10. Do the indicated problems with fractions, and express the answers in lowest terms.

a. Reduce: $\frac{16}{96}$ **b.** Multiply: $\frac{3}{8} \times \frac{4}{15}$
c. Divide: $\frac{2}{3} \div \frac{12}{18}$ **d.** Subtract: $3\frac{3}{4} - 1\frac{1}{4}$
e. Add: $\frac{2}{7} + \frac{3}{14}$ **f.** Subtract: $7\frac{1}{8} - \frac{3}{4}$

11. Rice weighing $3\frac{3}{4}$ pounds was divided equally and placed in 4 containers. How many ounces of rice were in each?

12. Do these operations with fractions, decimals, and percents.

a. Rewrite $\frac{3}{8}$ as a decimal.
b. Rewrite 0.8 as a fraction in lowest terms.
c. Rewrite 0.09 as a percent.
d. Rewrite 75% as a fraction in lowest terms.
e. Arrange the following decimals in order from smallest to largest: 0.07 0.707 0.7

13. Bunker's August sale offered a 25% discount on home furnishings. The Cliffords bought 2 window shades regularly priced at \$13.50 each and 4 chair cushions regularly priced at 2 for \$25. If there was a sales tax of 8%, what was the total cost?

14. This graph shows how the Ping family spent their vacation budget of \$2,000.

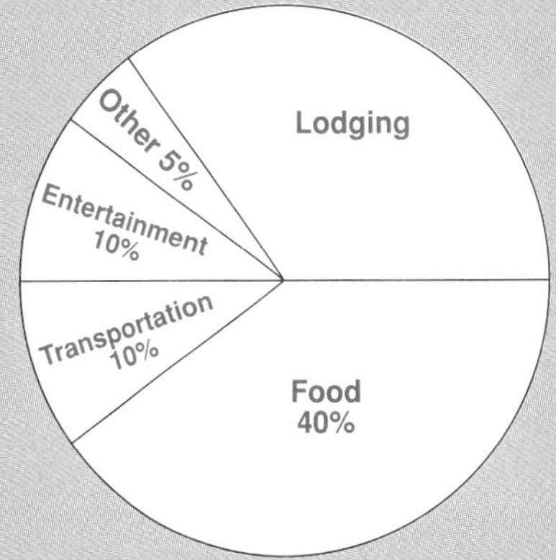

a. What percent of the money was spent on Lodging?
b. How much money was spent on Food?
c. What is a possible expense they might have included in Other?

SOLUTIONS FOR TRY THESE

3.1 Whole Numbers and Sets

Page 81

1. December 7, 1941
   the 7th day of the 12th month of the year 1941
2. **a.** $18 > 11$ **b.** $11 < 18$
   c. $11 < 15 < 18$ or $18 > 15 > 11$
   In one sentence, the symbols must be of the same kind, either both < or both >.

Page 82

1. **a.** {0, 1, 2, 3, 4, 5, 6, 7, 8, 9}
   b. {2, 3, 5, 7, 11, 13, 17, 19}
2. **a.** the odd numbers between 10 and 20
   b. the even numbers from 90 through 100
3. $\{a\}, \{b\}, \{a, b\}, \varnothing$. Every set is a subset of itself. The empty set is a subset of every set.
4. **a.** $e \leftrightarrow 2, g \leftrightarrow 4, w \leftrightarrow 6, z \leftrightarrow 8$
   b. 24 ways First pair e with any of the 4 numbers, then pair g with any of the remaining 3 numbers, then pair w with any of the remaining 2 numbers, then pair z with the 1 number left:
   $4 \times 3 \times 2 \times 1 = 24$

3.2 Properties of Operations

Page 90

1. **a.** True **b.** False **c.** True **d.** True
2. **a.** 7 **b.** 23 **c.** 9 **d.** any number
3. Set D is not closed under addition. There are sums, such as $3 + 4$, that do not have an answer in the set.
4. $6 - 3$ is a whole number but $3 - 6$ is not a whole number. The set of whole numbers is not closed under subtraction because when a larger number is subtracted from a smaller number, the difference is not a whole number.

3.3 Order of Operations

Page 94

1. $10 + 2 \times 3 = 10 + 6 = 16$
2. $30 \div 5 + 7 = 6 + 7 = 13$
3. $3 \times 2^4 = 3 \times 16 = 48$
4. $9(6 - 4) = 9(2) = 18$
5. $15 + 3 \cdot (8 - 5) = 15 + 3 \cdot 3 = 15 + 9 = 24$
6. $14 \div 2 + (3 + 4)^2 = 14 \div 2 + 7^2 = 14 \div 2 + 49 = 7 + 49 = 56$

Page 95

1. **a.** math-order:
   15 [+] 9 [÷] 3 [=]
   Result is 18.
   b. entry-order:
   9 [÷] 3 [+] 15 [=]
   Does not need memory.
2. **a.** math-order:
   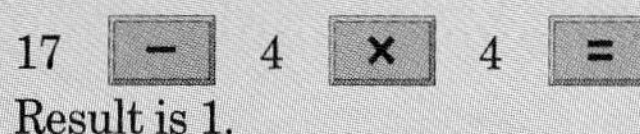

   Result is 1.
   b. entry-order:
   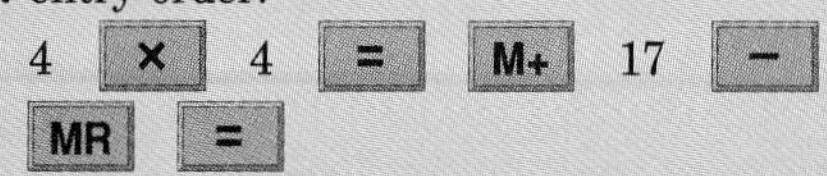

3.4 Fractions

Page 98

1. Divide numerator and denominator by their greatest common factor.
   $$\frac{36}{45} = \frac{36 \div 9}{45 \div 9} = \frac{4}{5}$$
2. Get a common denominator.
   $$\frac{4}{5} \ ? \ \frac{7}{10}$$
   $$\frac{8}{10} > \frac{7}{10}$$
   Answer: $\frac{4}{5} > \frac{7}{10}$

Page 100

1. $\dfrac{15}{4} \times \dfrac{2}{5}$

   $\dfrac{\overset{3}{\cancel{15}}}{\underset{2}{\cancel{4}}} \times \dfrac{\overset{1}{\cancel{2}}}{\underset{1}{\cancel{5}}}$ Cancel common factors.

   $\dfrac{3 \times 1}{2 \times 1}$ Multiply remaining factors.

   $\dfrac{3}{2}$ or $1\dfrac{1}{2}$ Improper fraction as a mixed number

2. $16 \div \dfrac{4}{5}$

   $\dfrac{16}{1} \div \dfrac{4}{5}$ Whole number as a fraction

   $\dfrac{16}{1} \times \dfrac{5}{4}$ $\div$ is the same as $\times$ by an inverse.

   $\dfrac{\overset{4}{\cancel{16}}}{1} \times \dfrac{5}{\underset{1}{\cancel{4}}}$ Cancel common factors.

   $\dfrac{4 \times 5}{1 \times 1}$ Multiply remaining factors.

   $\dfrac{20}{1}$ or 20

Page 101

1. $\dfrac{7}{9} + \dfrac{2}{9}$ Denominators are the same.

   $\dfrac{7 + 2}{9}$ Add numerators. Keep denominator.

   $\dfrac{9}{9} = 1$

2. $\dfrac{2}{3} + \dfrac{4}{5}$ Different denominators.

   $\dfrac{10}{15} + \dfrac{12}{15}$ LCD = 3×5

   $\dfrac{22}{15}$ or $1\dfrac{7}{15}$

3. $\dfrac{3}{4} - \dfrac{5}{8}$ Different denominators.

   $\dfrac{6}{8} - \dfrac{5}{8} = \dfrac{1}{8}$

3.5 Decimals and Percents

Page 111

1.

Fraction	Decimal	Percent
$\frac{1}{2}$	**0.50**	50%
$\frac{3}{4}$	0.75	**75%**
$\frac{2}{5}$	0.40	40%
$\frac{3}{8}$	0.375	**37.5%**
$\frac{1}{3}$	**$0.33\overline{3}$**	$33.\overline{3}\%$
$\frac{2}{3}$	**$0.66\overline{6}$**	**$66.\overline{6}\%$**

2. Use the same number of decimal places.

   8.7 = 8.700

   8.700 > 8.293

3. 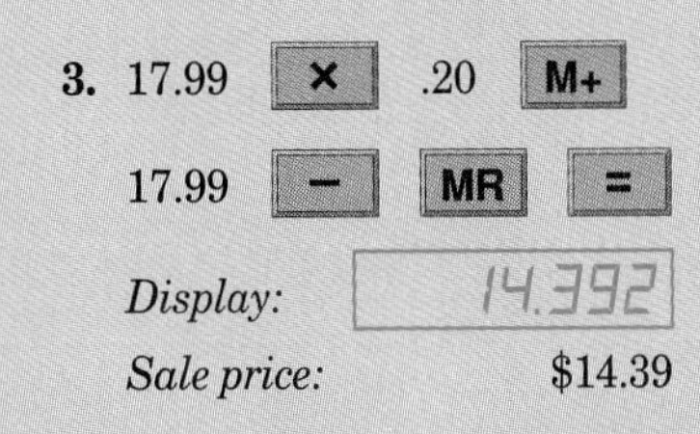

   Sale price: $14.39

Signed Numbers 4

ACTIVITY

Use a 6-sided numbered cube (a *die*) and a coin to play this game with a classmate. Each player begins with 15 chips and there is a pot of 30 chips in the center.

First flip the coin. *Heads* means *take from the pot* and *tails* means *put in the pot*. The number on the die tells how many to take or put in.

The game is over when the pot is empty. The winner is the player with the most chips.

4.1 The Number Line

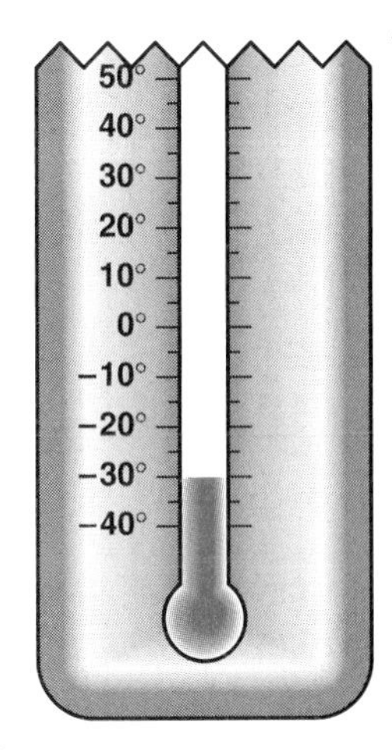

Your body would be quick to know the difference between an outdoor temperature of 30 degrees *below* zero as opposed to 30 degrees *above* zero!

To represent these opposite situations, we use ***signed numbers***.

Read –30 as negative 30

and +30 as positive 30.

Example 1 Express each situation as a signed number.

Express the opposite situation as a signed number.

	Situation	*Number*	*Opposite*	*Number*
a.	a loss of 20 pounds	–20	a gain of 20 pounds	+20
b.	50.5 meters above sea level	+50.5	50.5 meters below sea level	–50.5

As on a thermometer, signed numbers are represented by points on a number line.

−6 −5 −4 −3 −2 −1 0 +1 +2 +3 +4 +5 +6

Zero, the number that separates the positives from the negatives, has no sign. A positive number may be written without its sign. Thus, positive 10 is either +10 or just 10, but negative 10 must be written –10.

The set of signed numbers shown on the number line above is called the set of ***integers***. You can think of the integers as the whole numbers and their opposites.

Integers: $\{\ldots, -5, -4, -3, -2, -1, 0, 1, 2, 3, 4, 5, \ldots\}$

opposites

Think of the way a mirror reflects an image. The image point of an opposite is the ***reflection*** of the point of the number, with 0 as the point of reflection. For example, –1 is the opposite of 1 and 2 is the opposite of –2. The number 0 is its own opposite.

Between any two integers, there are infinitely many other numbers.

This diagram shows just some of the fractions and decimals that are between 0 and 1.

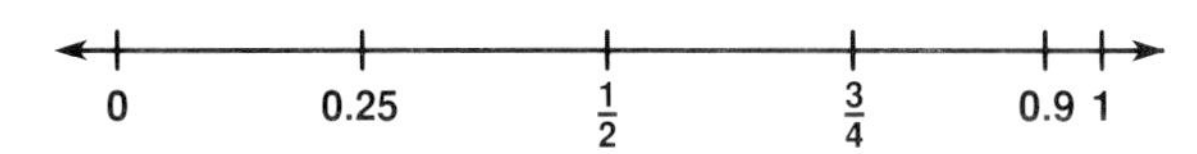

A number line can be used to compare two numbers.
The number to the right is always greater than the number to the left.

$5 > 3$ $\quad 2 > -1$ $\quad 3 > 0$ $\quad -1 > -4$

From the number line, observe that a positive number is always greater than a negative number. The positive numbers increase (are greater) as they get farther away from 0, while the negative numbers decrease (are less) as they get farther away from 0.

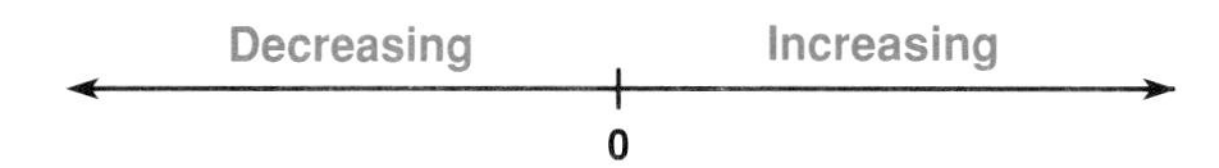

Example 2 Write a statement of inequality that compares the given numbers, first using the symbol > and then the symbol <.

	Numbers	Greater Than	Less Than
a.	–2, 0	$0 > -2$	$-2 < 0$
b.	–4, –7	$-4 > -7$	$-7 < -4$
c.	–3, –6, 0	$0 > -3 > -6$	$-6 < -3 < 0$

Try These *(For solutions, see page 160.)*

1. Express the given situation and its opposite as signed numbers.
  a. a gain of 6 yards in a football game
  b. 7 blocks to the South on a street map

2. List the set of integers from –4 through 7.

3. Write a statement of inequality that compares the given numbers, first using the symbol > and then the symbol <.
  a. 8, 10 **b.** 12, –12 **c.** 0, –5
  d. –9, –22 **e.** –14, –1, –10

4. Write these integers in order from least to greatest:
  a. 7, 9, 3, 0, 11, 5 **b.** –7, 9, –3, 0, –11, 5

Absolute Value

A measurement between two points is called a ***distance***. On a number line, the distance between any two numbers is the length of the segment between them, which can be found by counting the number of units.

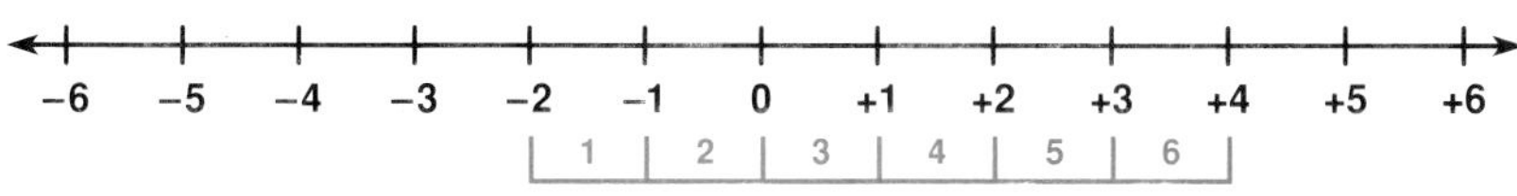

The distance between –2 and 4 is 6 units.

The number used to represent a distance is always a positive number.

The distance between a number and 0 is called the ***absolute value*** of that number.

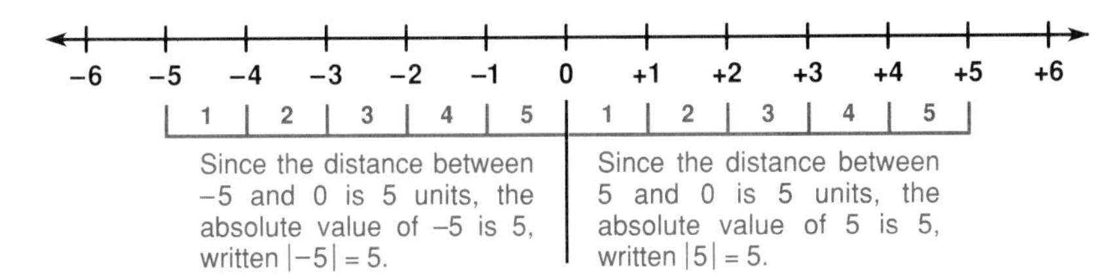

Example 3 Find the absolute value of each number.

	Number	Absolute Value
a.	10	$\lvert 10 \rvert$, or 10
b.	−13	$\lvert -13 \rvert$, or 13
c.	0	$\lvert 0 \rvert$, or 0

Try These *(For solutions, see page 160.)*

1. What is the distance on a number line between −17 and 0?
2. What is the absolute value of each number?
   a. 16 **b.** −16 **c.** 0 **d.** 5.3 **e.** $-7\frac{1}{2}$

SELF-CHECK: SECTION 4.1

1. What is the opposite situation of going up 6 floors in an elevator? Represent both situations by signed numbers.

   Answer: going up 6 floors: +6
   going down 6 floors: −6

2. The set of integers includes the whole numbers and __?__.

   Answer: their opposites

3. Is this inequality true or false? $4 > -13$

   Answer: True, a positive number is always greater than a negative number.

4. What is the distance between −3 and −9 on a number line?

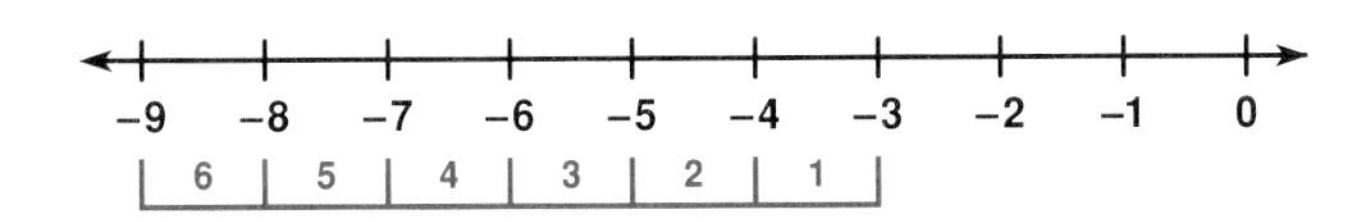

   Answer: 6 units

5. What is another name for the distance between 32 and 0?

   Answer: The absolute value of 32, written $\lvert 32 \rvert$.

EXERCISES: SECTION 4.1

1. Represent each situation by a signed number.

a. an archeologist digs 32 feet below ground

b. a library gets 47 new books

c. a stock clerk removes 48 cans of chili from a shelf

2. a. List the members of each set.
Tell whether each set is finite or infinite.

(1) {positive integers less than 5}

(2) {integers greater than –3 and less than 5}

(3) {positive integers that are not whole numbers}

(4) {even integers greater than –5}

b. Describe each set of numbers in words.

(1) {1, 2, 3, 4} **(2)** {–1, 0, 1, 2, 3, 4}

(3) {–8, –6, –4, –2, 0, 2, 4} **(4)** {–6, –3, 0, 3, 6, 9, 12}

3. Tell whether each statement is true or false.
If false, rewrite the statement so that it is true.

a. The set of whole numbers is a subset of the set of integers.

b. On a number line, a larger number is to the right of a smaller number.

c. On a number line, all positive numbers are to the left of zero.

4. Which is the greater number in each pair of integers?

a. –50, 5 **b.** 0, –12 **c.** –6, –2 **d.** –100, –1

5. Explain how you can help a friend remember that, although 5 is greater than 2:

a. –5 is less than +2 **b.** –5 is less than –2

6. Use the symbol > to order each set of three numbers.

a. 5, 2, 3 **b.** –11, 0, 12 **c.** –2, –5, –4

d. –5, –7, 0 **e.** 7.2, 9, 5 **f.** $-6\frac{1}{2}$, –6, –7

7. Use a number line to find the distance between each pair of numbers.

a. 1 and 8 **b.** 8 and 1 **c.** –3 and 0

d. 3 and 0 **e.** –5 and 2 **f.** 2 and –5

g. 6.5 and –1 **h.** –8 and –2 **i.** –4 and –10

8. Explain *absolute value* in your own words.

9. Find the absolute value of each number.

a. 5 **b.** –5 **c.** –6

d. 0 **e.** –7.2 **f.** $8\frac{1}{2}$

10. Tell whether each sentence is true or false.

a. $-2 < 1$ **b.** $-4 < -10$ **c.** $0 > -1$

d. $|-7| < 0$ **e.** $|-5| > -3$ **f.** $-3 < |-1|$

g. $|-5| = |5|$ **h.** $|-4| < |-10|$ **i.** $|-12| > |4|$

j. $|-2.1| = |2.1|$ **k.** $|-1\frac{1}{2}| = 1.5$ **l.** $|-2\frac{1}{2}| > 3\frac{1}{2}$

EXPLORATIONS

1. On separate strips of paper, prepare two number lines. For both lines, use the numbers 0 through 10, and the same equal spacing. Label one line *T* for Top Number Line and label the other line *B* for Bottom Number Line.
Place number line *B* below number line *T* with 0 on *B* at 3 on *T*.

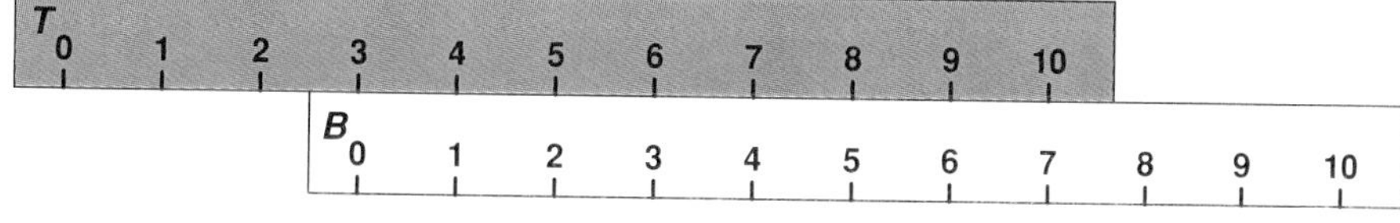

a. **(1)** 1 on *B* corresponds to what number on *T*?

(2) 5 on *B* corresponds to what number on *T*?

(3) Find other pairings between numbers on *B* and *T*.

(4) What pattern relates *T* numbers to *B* numbers when *T* and *B* are in this position?

(5) Move *B* so that 0 on *B* is at 2 on *T*, and answer questions **(1)**-**(4)** again.

(6) Describe a method for adding whole numbers using "top" and "bottom" number lines.

b. What number of *T* is at 0 on *B* when:

(1) 7 on *B* is at 10 on *T*?

(2) 4 on *B* is at 6 on *T*?

(3) 2 on *B* is at 9 on *T*?

c. Describe a method for subtracting whole numbers using "top" and "bottom" number lines.

2. Draw a Venn diagram to describe the relationship among these sets of numbers:
{integers}, {signed numbers}, {whole numbers}, {0}, {natural numbers}

4.2 Addition

Like Signs

What floor is an elevator on after starting at the ground level, going up 3 floors, and then going up 4 floors more?

The positive numbers act just like the natural numbers, the numbers you know from arithmetic.

$$(+3) + (+4) = 3 + 4 = 7$$

> The sum of two positive numbers is a positive number.

The elevator problem would be modeled vertically.

Addition of positive numbers can be modeled on a number line.

To show the sum (+3) + (+4):

Start at 0.
Move 3 units to the right.
Then move 4 more units to the right.
You are now at +7.

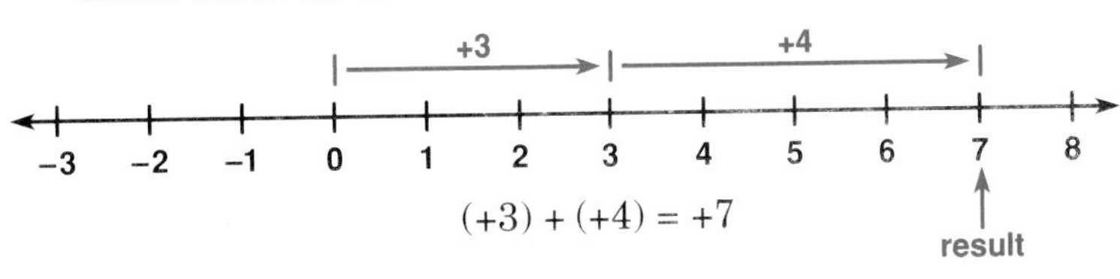

Another way to model addition of positive numbers is to use objects you can move around. We'll work with *integer chips*, where [+] represents +1.

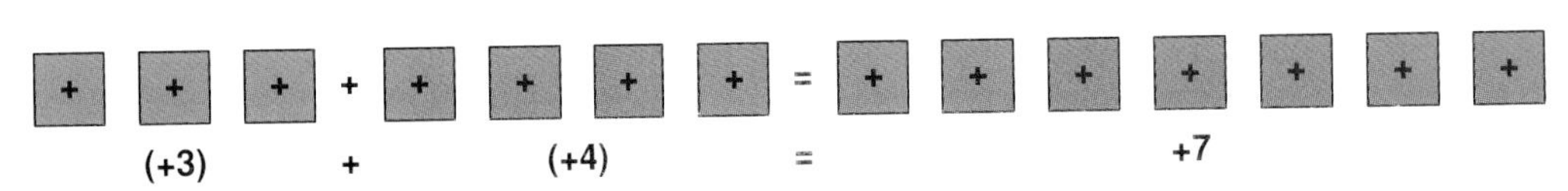

Reconsider the elevator. This time, it starts at the roof level, goes down 3 floors, and then goes down 4 floors more. Where is the elevator with respect to the roof level? Did you say 7 floors below the roof?

To find this sum (–3) + (–4) on a number line:

Start at 0.
Move 3 units to the left.
Then move 4 more units to the left.
You are now at –7.

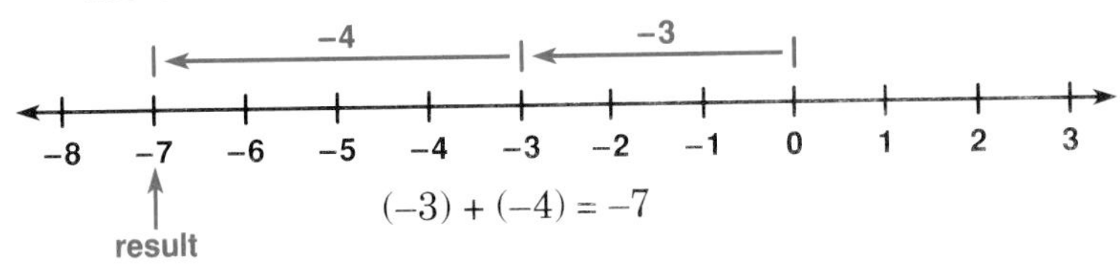

To find this sum using integer chips, let [–] represent –1.

[–] [–] [–] + [–] [–] [–] [–] = [–] [–] [–] [–] [–] [–] [–]

(–3) + (–4) = –7

From these models, we conclude:

The sum of two negative numbers is a negative number.

Look at the result of adding two positive numbers and the result of adding two negative numbers to observe a pattern.

$(+3) + (+4) = +7$ $\qquad$ $(-3) + (-4) = -7$

In both cases, the sum has:

- the sign of the kinds of numbers that were added
- the same number value disregarding the sign

We can use absolute value to describe a rule.

To Add Numbers With the Same Sign

① Add the absolute values of the two numbers.

② Give the result the same sign as the two numbers.

Example 1 Find each sum using absolute value.

a. $(+5) + (+12)$

Since the numbers have the same sign:

① Add absolute values.

② Use the same sign as the two numbers, +.

$$\begin{aligned} |+5| &= 5 \\ |+12| &= \underline{12} \\ & \quad 17 \end{aligned}$$

Answer: $(+5) + (+12) = +(5 + 12) = +17$

b. $(-5) + (-12)$

Since the numbers have the same sign:

① Add absolute values.

② Use the same sign as the two numbers, –.

$$\begin{aligned} |-5| &= 5 \\ |-12| &= \underline{12} \\ & \quad 17 \end{aligned}$$

Answer: $(-5) + (-12) = -(5 + 12) = -17$

Try These *(For solutions, see page 160.)*

1. Add, using a number line:
$(-4) + (-2)$

2. Add, using absolute value:
a. $(+7) + (+5)$ $\qquad$ **b.** $(-4) + (-2)$

Unlike Signs

After starting at the ground level, going up 3 floors, and then down 4 floors, where is the elevator with respect to the ground level? This time, the elevator lands one floor below the ground level.

The same models can be used to show how to add numbers with *different signs*.

Example 2

a. Use a number line to find the sum (+3) + (–4).

Start at 0.

Move 3 units to the right.

Then move 4 units to the left.

You are now at –1.

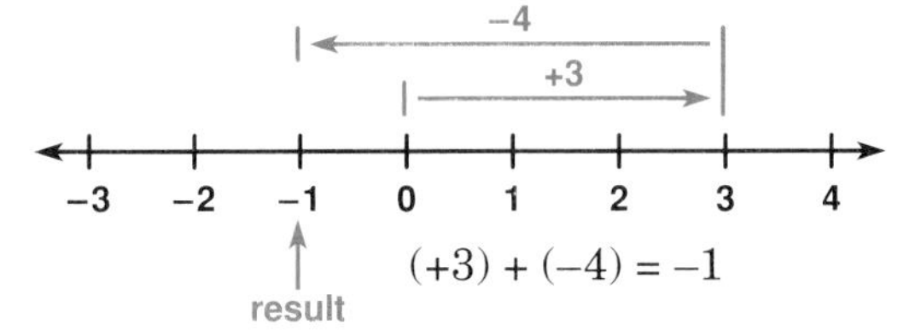

b. Use integer chips to find the sum (+3) + (–4).

Call the pair of integer chips [+]—[–] a *zero pair* because (+1) + (–1) = 0.

Represent the addition.

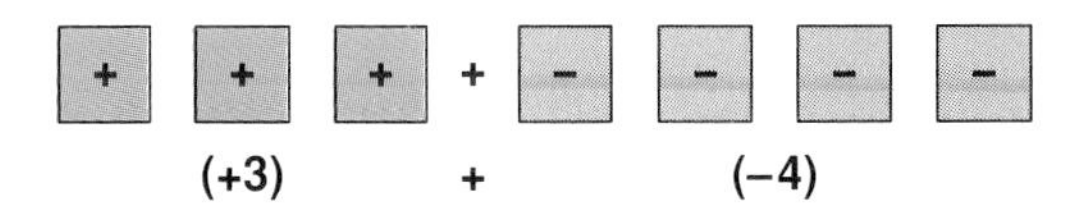

Group the chips to find all zero pairs.

([+]—[–]) + ([+]—[–]) + ([+]—[–]) + [–] = [–]

0 + 0 + 0 + (–1) = –1

Answer: (+3) + (–4) = –1

Example 3

a. Use a number line to find the sum (–3) + (+4).

Start at 0.

Move 3 units to the left.

Then move 4 units to the right.

You are now at +1.

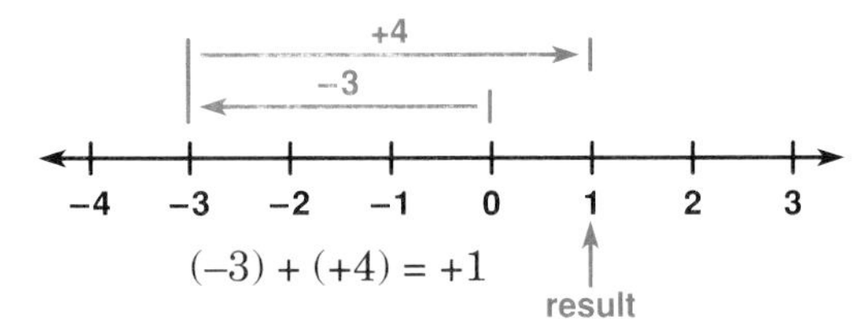

b. Use integer chips to find the sum $(-3) + (+4)$.

Represent the addition.

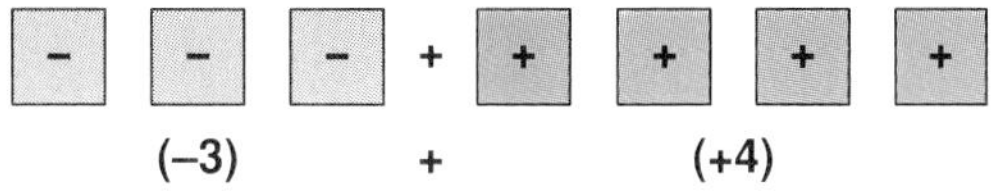

− − − + + + + +

(−3) + (+4)

Group the chips to find all zero pairs.

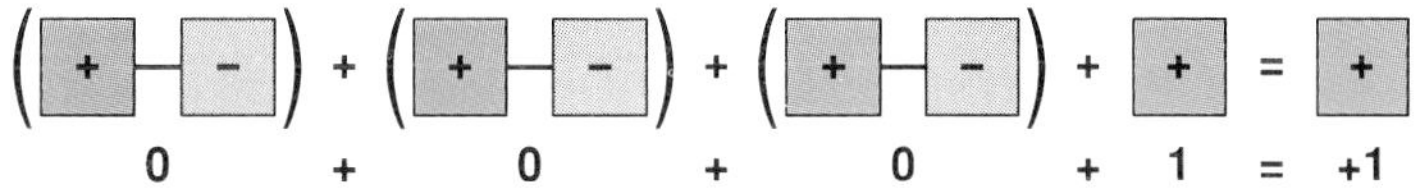

(+ −) + (+ −) + (+ −) + + = +

0 + 0 + 0 + 1 = +1

Answer: $(-3) + (+4) = +1$

Examples 2 and 3 show that when two numbers have different signs, the sum can be either a positive or a negative number.

$(+3) + (-4) = -1$ $\qquad$ $(-3) + (+4) = +1$

Note that in both cases, the number associated with the sum comes from a difference of the original values without signs.

Thus, again, absolute value can be used to describe a rule.

To Add Numbers With Different Signs

① **Subtract the absolute values of the two numbers.**

② **Give the result the sign of the number with the greater absolute value.**

Example 4 Find each sum using absolute value.

a. $(-12) + (+5)$

Since the numbers have different signs:

subtract absolute values

$(-12) + (+5) = -(12 - 5) = -7$

sign of number with greater absolute value

b. $(+12) + (-5)$

Since the numbers have different signs:

subtract absolute values

$(+12) + (-5) = +(12 - 5) = +7$

sign of number with greater absolute value

Try These *(For solutions, see page 160.)*

1. Add, using a number line: $(-5) + (+7)$
2. Add, using integer chips: $(+2) + (-6)$
3. Add, using absolute value:
   a. $(+7) + (-5)$ **b.** $(+3) + (-14)$

More about Addition

It is not always necessary to use parentheses when adding signed numbers or to use signs when writing positive numbers.

But, you must use parentheses when two signs follow one another.

Example 5 Write each addition in a simpler way. Find the value of the sum.

a. $(+12) + (+3) = 12 + 3 = 15$ **b.** $(-4) + (+7) = -4 + 7 = 3$

c. $(-11) + (+2) = -11 + 2 = -9$ **d.** $(+9) + (-6) = 9 + (-6) = 3$

Addition of signed numbers is useful in daily life.

Example 6 Leah has a savings account at a bank. Each month, she deposits some money into her account, and sometimes she withdraws money from her account. The table show her account's activity over a 4-month period.

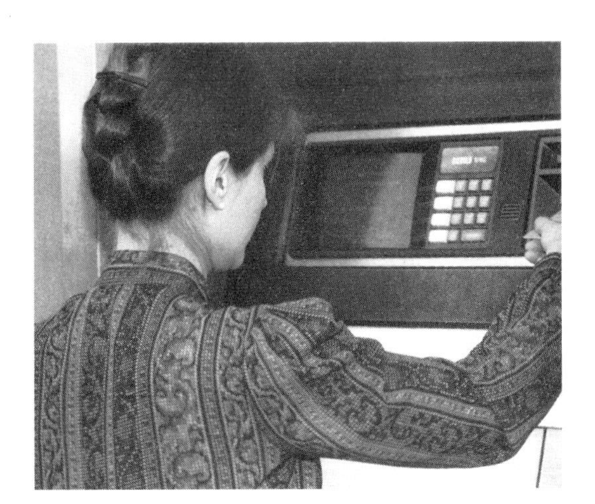

	Jan.	Feb.	Mar.	Apr.
Deposit	\$7	\$4	\$2	\$5
Withdrawal	\$3	\$4	\$2	\$8

If Leah had \$15 in her account at the beginning of January, how much does she have at the end of April?

Represent the *deposits* as *positive* numbers, and find their sum.

$7 + 4 + 2 + 5 = 18$

Represent the *withdrawals* as *negative* numbers, and find their sum.

$-3 + (-4) + (-2) + (-8) = -17$

Add the deposits to the beginning balance of \$15.

$15 + 18 = 33$

Add the negative value of the withdrawals.

$33 + (-17) = 16$

Answer: The balance at the end of April is \$16.

Try These *(For solutions, see page 160.)*

1. Write in a simpler way, and find the sum:

$(-6) + (+19) + (-3) + (+14)$

2. Bracken County has 4 school districts. The table shows changes in the enrollment in each district over a year. Write, as an integer, the change in the school enrollment for the entire county.

District	Increase	Decrease
North	147	
East		106
South		71
West	23	

SELF-CHECK: SECTION 4.2

1. The sum of two positive numbers is __?__.

Answer: positive

2. The sum of two negative numbers is __?__.

Answer: negative

3. The sum of numbers with different signs may be __?__.

Answer: either positive or negative

4. When adding numbers with like signs, what do you do with the absolute values? What sign does the result have?

Answer: Add absolute values.
Use the same sign as the original numbers.

5. When adding numbers with unlike signs, what do you do with the absolute values? What sign does the result have?

Answer: Subtract absolute values.
Use the sign of the number with the greater absolute value.

EXERCISES: SECTION 4.2

1. Use a number line to find each sum.

a. $(+4) + (+3)$ **b.** $(-7) + (-1)$
c. $(-2) + (+5)$ **d.** $(-4) + (-4)$
e. $(+3) + (-10)$ **f.** $(-4) + (+9)$

2. Use integer chips to find each sum.

a. $(+5) + (+2)$ **b.** $(-4) + (-3)$
c. $(+6) + (-2)$ **d.** $(-8) + (+3)$

3. Use absolute value to find each sum.

a. $(+3) + (+7)$ **b.** $(-8) + (-2)$
c. $(-10) + (+6)$ **d.** $(+7) + (-2)$
e. $(+8) + (-8)$ **f.** $(+7.3) + (-6)$
g. $(+2\frac{1}{2}) + (-3\frac{1}{2})$ **h.** $(-7\frac{1}{3}) + (+3\frac{2}{3})$
i. $(+6) + (+2) + (+5)$ **j.** $(-1) + (-3) + (-7)$
k. $(+5) + (+2) + (-4)$ **l.** $(-3) + (+7) + (+6)$
m. $(+7) + (-1) + (+3)$ **n.** $(+8.2) + (-7.4) + (0.1)$

4. Write each addition in a simpler way, and find the sum.

a. $(+4) + (+9)$ **b.** $(-5) + (-4)$
c. $(+6) + (-5)$ **d.** $(-8) + (0)$
e. $(+3) + (+6) + (+2)$ **f.** $(+7) + (-2) + (-9)$
g. $(-8) + (+3) + (+1)$ **h.** $(-5) + (+8) + (-3)$

5. Find the value of each expression.

a. $-3 + (-2) + 11$ b. $7 + (-4) + (-8)$
c. $8 + (-7) + |-2|$ d. $(-10) + |-6| + |4|$
e. $|-3| + |-2| + (-8)$ f. $|(+5) + (+2)|$
g. $|-6 + (-3)|$ h. $|2 + (-8)|$
i. $|-4 + 11|$ j. $|9 + (-12)|$

6. In each sentence, replace ? by an integer to make the sentence true.

a. $+6 + ? = 14$ b. $-8 + ? = -10$
c. $-6 + ? = -4$ d. $-5 + ? = 0$
e. $4 + ? = 0$ f. $2 + ? = -1$
g. $6 + ? = 2$ h. $5 + ? = -3$

7. a. Does the Closure Property hold for addition of integers? Explain.
 b. (1) Is this statement true or false? $(-5) + (+7) = (+7) + (-5)$
 (2) From your observation about the last statement, come to a conclusion about addition of integers.
 (3) Give another example to support your conclusion.
 c. (1) Is this statement true or false? $-5 + (7 + (-3)) = (-5 + 7) + (-3)$
 (2) From your observation about the last statement, come to a conclusion about addition of integers.
 (3) Give another example to support your conclusion.

8. Find the next 3 terms in each sequence.

a. $-6, -4, -2, \ldots$ b. $-7, -3, 1, \ldots$
c. $-10, -7, -4, \ldots$ d. $3, 1, -1, \ldots$
e. $9, 5, 1, \ldots$ f. $1, -2, -5, \ldots$

9. Solve each problem using signed numbers.

a. Ricki made pizza deliveries to an apartment building. His first delivery was on the 4th floor. He went up 3 floors for his 2nd delivery and then down 5 floors for his 3rd delivery. On what floor was his 3rd delivery?

b. An airplane is flying at an altitude of 20,000 feet over the ocean. A submarine is traveling at a depth of 1,500 feet below sea level. How many feet apart are the airplane and the submarine when the airplane is directly above the submarine?

c. Roberta gained 5 pounds, then went on a diet and lost 12 pounds. She gained back 3 pounds. If her starting weight was 123 pounds, how much does she weigh now?

d. Pablo deposited $50 into his savings account. For the next 3 months, he made the following transactions: a deposit of $23, a withdrawal of $15, and a withdrawal of $20. What was his balance at the end of the 3-month period?

10. The temperature on a winter day was taken every 2 hours.

a. Use the data in the table to answer the following questions:

(1) What was the lowest temperature recorded that day?

(2) At what time was the highest temperature recorded?

(3) How much higher was the temperature at 6 A.M. than at 2 A.M.?

(4) What was the average of the recorded temperatures?

b. Here is a line graph using the data from the table. What do you notice in the graph that was not obvious from the table?

Time	Temperature
2 A.M.	−3°
4 A.M.	−4°
6 A.M.	9°
8 A.M.	23°
10 A.M.	26°
12 NOON	29°
2 P.M.	34°
4 P.M.	31°
6 P.M.	23°
8 P.M.	10°
10 P.M.	−2°

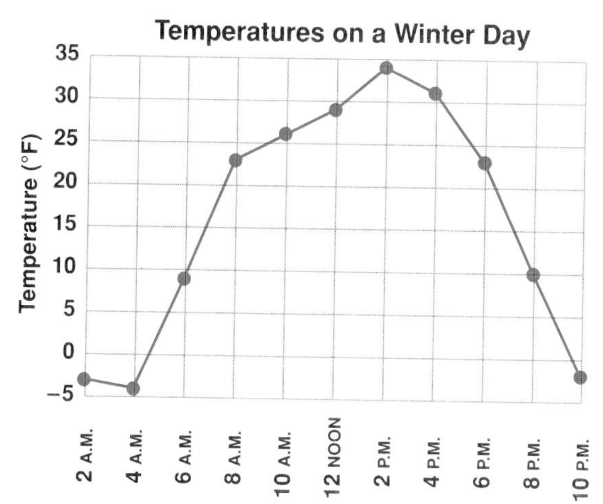

11. In each sum, find the value of A and the value of B.

a.
$$\begin{array}{r} 5A \\ +A1 \\ \hline 1B9 \end{array}$$

b.
$$\begin{array}{r} A49 \\ +23A \\ \hline B84 \end{array}$$

c.
$$\begin{array}{r} A7A \\ +B47 \\ \hline 1BB6 \end{array}$$

d.
$$\begin{array}{r} 89A \\ +B7B \\ \hline A466 \end{array}$$

EXPLORATIONS

1. On separate strips of paper, prepare two number lines. For both lines, use the numbers −5 through 5. Label one line T, for Top Number Line, and label the other line B, for Bottom Number Line.
Place number line B below number line T with 0 on B at 2 on T.

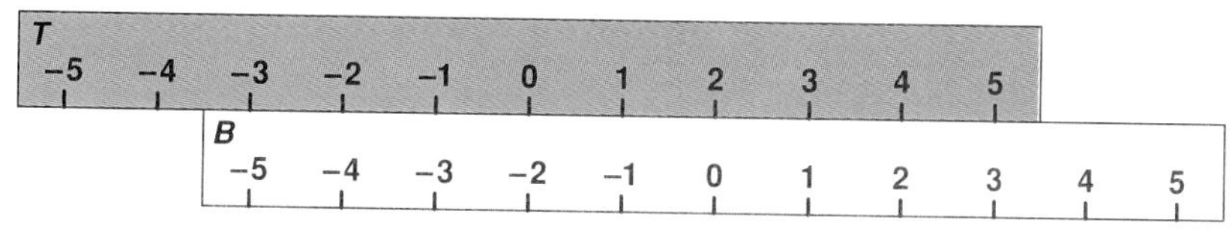

a. (1) 3 on B corresponds to what number on T?

(2) −2 on B corresponds to what number on T?

(3) Find other pairings between numbers on B and T.

(Continues on next page)

(4) What pattern relates T-numbers to B-numbers when T and B are in this position?

(5) Move B so that 0 on B is at –3 on T, and answer questions **(1)-(4)** again.

(6) Describe a method for adding integers using "top" and "bottom" number lines.

b. What number on T is at 0 on B when:

(1) –4 on B is at –1 on T?

(2) 3 on B is at –2 on T?

(3) –3 on B is at –1 on T?

c. Describe a method for subtracting integers using "top" and "bottom" number lines.

d. Compare the methods you have now described for addition and subtraction of integers using two number lines with the methods you previously described (in the Exploration of Section 4.1) for addition and subtraction of whole numbers.

2. A *magic triangle* is called "magic" because you get the same sum when you add the numbers along each side.

Verify that the magic sum of the magic triangle shown is 0.

Use different numbers from {–2, –1, 0, 1, 2, 3} to create a magic triangle:

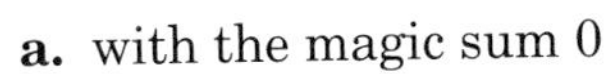

a. with the magic sum 0 **b.** with the magic sum 1

c. with the magic sum 2 **d.** with the magic sum 3

3. A *magic square* is called "magic" because you get the same sum when you add the numbers in each row, each column, and along each diagonal.

4	–3	2
–1	1	3
0	5	–2

Magic Square *A*

Verify that the magic square shown, Magic Square A, has a magic sum of 3.

a. By adding 2 to each integer in Magic Square A, create Magic Square B. What is the magic sum of B?

b. By adding 5 to each integer in Magic Square A, create Magic Square C. What is the magic sum of C?

c. By adding –3 to each integer in Magic Square A, create Magic Square D. What is the magic sum of D?

d. Explain the relationship between the magic sum of Magic Square A and the magic sum of a new magic square created by adding any number, say n, to each integer in Magic Square A.

e. (1) Use the relationship you just described to predict the magic sum of Magic Square E, the magic square you would get if you added –6 to each integer in Magic Square A.

(2) Verify your prediction by creating Magic Square E.

4.3 Subtraction

Integer chips can be used to illustrate the operation of subtraction with signed numbers.

Subtract: $(+7) - (-2)$

Start with 7 positive chips.

To be able to take away 2 negative chips, you must first place negative chips in the model without changing the value of the model. Use zero pairs.

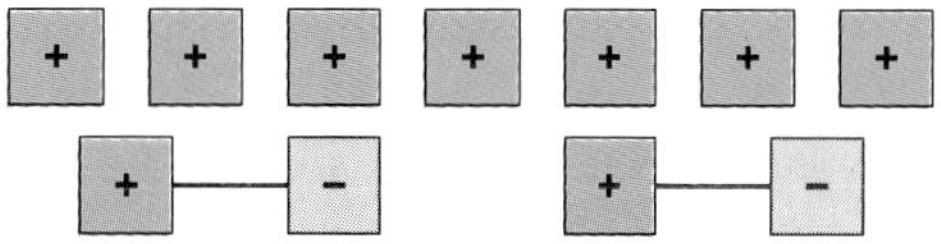

Now you can take away 2 negative chips.

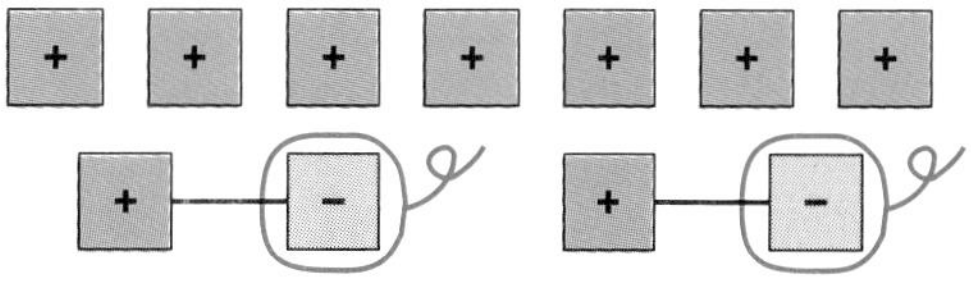

The result is 9 positive chips.

$(+7) - (-2) = +9$

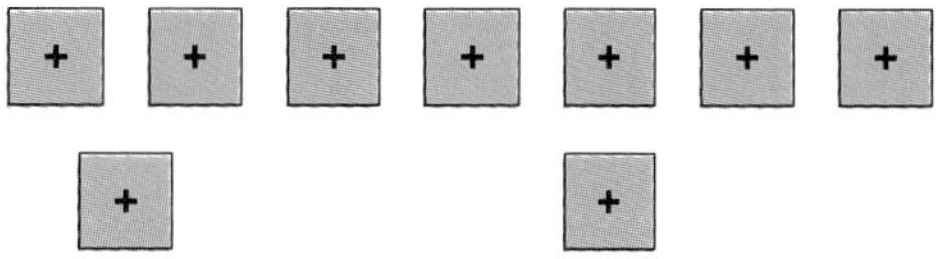

The model shows that when –2 is subtracted from +7, the result is +9.

$(+7) - (-2) = +9$

How else can we begin with +7 and have a result of +9?

$(+7) + (+2) = +9$

Note that subtracting –2 gives the same result as adding +2.

Use integer chips to model other subtractions to verify that:

Subtraction is equivalent to adding an opposite.

Example 1 Write each subtraction as an addition of an opposite and find the answer.

Subtraction	*Addition of an Opposite*
$6 - (+1)$	$6 + (-1) = 5$
$4 - (-7)$	$4 + (+7) = 11$
$-8 - (-2)$	$-8 + (+2) = -6$
$-5 - (-9)$	$-5 + (+9) = +4$
$-7 - (+11)$	$-7 + (-11) = -18$

Try These (For solutions, see page 160.)

1. Rewrite each subtraction as the addition of an opposite, and find the answer.

   a. $5 - (+9)$ **b.** $7 - 16$ **c.** $3 - (-6)$
   d. $0 - 8$ **e.** $-4 - (-12)$ **f.** $-1 - 1$

2. Find the value of each expression.

   a. $12 + 7 - 2$ **b.** $12 - 7 + 2$ **c.** $12 - 7 - 2$
   d. $-12 - 7 + 2$ **e.** $-12 - 7 - 2$ **f.** $-2 - 7 - 12$

Signed Numbers on a Calculator

Operations with signed numbers can be performed on a calculator that has the key [+/−], to obtain a negative number.

To write a key sequence to find the value of $(+7) - (-2)$,

Enter: 7 [−] 2 [+/−] [=] *Display:* 9.

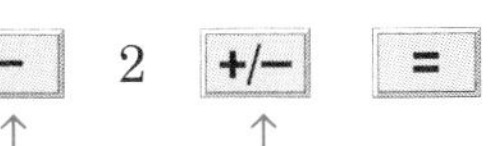

This is the operation subtraction. This changes a positive number to a negative number.

Example 2 Write a key sequence for each calculation. Then follow your sequence on a calculator.

a. $(+5) + (-2)$

Key sequence: 5 [+] 2 [+/−] [=] *Display:* 3.

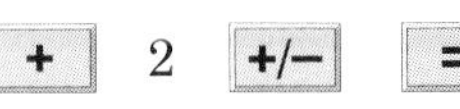

b. $(+5) - (-2)$

Key sequence: 5 [−] 2 [+/−] [=] *Display:* 7.

Try This (For solution, see page 161.)

Find the difference $5 - (-3)$ by:

a. using integer chips **b.** using a calculator (First write a key sequence.)

Order of Operations

The familiar order of operations applies to signed numbers.

Example 3 Find the value of $12 + (-3) - (-4)$.

$$12 + (-3) - (-4)$$ Begin with the sum at the left: $12 + (-3) = 9$

$$= 9 - (-4)$$ Now do the subtraction, which is equivalent to adding an opposite.

$$= 13$$ $9 - (-4) = 9 + 4 = 13$

Try This (For solution, see page 161.)

Find the value of $(-3) + (-2) - (-1)$.

SELF-CHECK: SECTION 4.3

1. Subtraction is equivalent to ___?___.

Answer: adding an opposite

2. How do you enter -3 on a calculator?

Answer: 3 [+/−]

3. Which calculator key do you use for subtraction?

Answer: [−]

EXERCISES: SECTION 4.3

1. Subtract, by adding the opposite.

a. $3 - (-7)$ **b.** $8 - (-2)$ **c.** $-10 - (-6)$ **d.** $-2 - (-5)$

e. $-8 - (-8)$ **f.** $-7 - (-2)$ **g.** $-7 - (+3)$ **h.** $-4 - (+9)$

i. $10 - (+9)$ **j.** $3 - (+11.2)$ **k.** $6\frac{2}{3} - (-5)$ **l.** $-6.4 - (-2.1)$

2. Find the value of each expression.

a. $7 + (-4) + (-12)$ **b.** $5 + (-2) + (+10)$ **c.** $8 - (+4) - (-8)$

d. $12 - (-15) + (-3)$ **e.** $7 - (-7) - 14$ **f.** $-8 + |-5| - (-3)$

g. $-6 - (-8) + |-2|$ **h.** $|-5| - |-1|$ **i.** $|-12 - (-8)|$ **j.** $|8 + (-4) - (-10)|$

3. Replace each ? by an integer that will make the sentence true.

a. $-2 - ? = 0$ **b.** $-4 - ? = 2$ **c.** $-7 - ? = -5$

d. $3 - ? = 5$ **e.** $1 - ? = 7$ **f.** $5 - ? = 5$

g. $3 - ? = -6$ **h.** $8 - ? = -8$ **i.** $-2 - ? = -7$

4. Find each of these differences:

a. $8 - (+2)$ **b.** $7 - (-3)$ **c.** $-4 - (+1)$ **d.** $-3 - (-2)$

by using

(1) integer chips

(2) a calculator (First write a key sequence.)

5. **a.** Does the Closure Property hold for subtraction of integers? Explain.

b. **(1)** Is this statement true or false? $(+7) - (+3) = (+3) - (+7)$

(2) From your observation about the last statement, come to a conclusion about subtraction of integers.

c. **(1)** Is this statement true or false? $7 - (5 - 3) = (7 - 5) - 3$

(2) From your observation about the last statement, come to a conclusion about subtraction of integers.

6. Find the next 3 terms in each sequence.

a. 8, 3, –2, ... **b.** –1, –4, –7, ... **c.** –5, –2, 1, ...

d. 4, 0, –4, ... **e.** –2.1, –2, –1.9 ... **f.** $0, -\frac{1}{2}, -1, \ldots$

7. Enter each key sequence on your calculator.

Then:

(1) write the result that is displayed.

(2) write out the problem that was calculated.

a. 4 [–] 3 [+/–] [=]

b. 4 [+/–] [–] 3 [=]

c. 4 [+] 3 [=]

d. 4 [+/–] [+] 3 [=]

e. 4 [+/–] [–] 3 [+/–] [=]

f. 4 [+/–] [+] 3 [+/–] [=]

g. Compare all the problems and their answers.
Tell which problems are equivalent, giving a reason.

8. Recall: Addition and subtraction are inverse operations.

Thus: If 2 + 3 = 5, then 5 – 3 = 2 and 5 – 2 = 3.

For each sentence in *Column A*, choose a matching sentence from *Column B*. Write the complete matching sentence by replacing ? with the number that makes the sentence true.

Column A	*Column B*
a. 2 + (–2) = 0	(1) –8 + ? = –10
b. –8 – (–6) = –2	(2) 8 + 2 = ?
c. 6 + 2 = 8	(3) 8 – (–2) = ?
d. 10 – 2 = 8	(4) 8 – ? = 2
e. –10 – (–2) = –8	(5) –8 + ? = –6
f. 10 + (–2) = 8	(6) –2 + (–6) = ?
g. –6 – 2 = –8	(7) 2 – ? = –8
h. –8 + 10 = 2	(8) –2 = 0 – ?

9. Solve each problem using signed numbers.

a. Luz earned $12 mowing lawns and $6 babysitting. If she paid $5 for admission to a movie, how much of her earnings remained?

b. At the end of the workday, Commuter Express Bus picked up 47 passengers downtown. At the first stop, 6 people got off and, at each of the next two stops, 9 people got off. At the next-to-last stop, 3 people got off.
How many passengers rode to the last stop?

c. Starting at noon, the temperature rose 3° in the first hour, rose 5° in the second hour, and dropped 4° in the next hour. If the 3 P.M. temperature was 58°, what was the temperature at noon?

d. To pass a fitness test, Lou lost weight at a steady rate. If his average weights in January and April were 211 pounds and 184 pounds respectively, what were his average weights in February and March?

e. There are 17 more boys than girls in a study hall. If there are 43 students altogether, how many are boys?

10. In a game, Jaime moves a marker *clockwise* around the board shown. He starts at 0.

To determine his moves, he rolls a die with sides numbered 1-6.

The number of spaces he can move is the absolute value of the difference between the number on the die and the number at his marker position.

For example, if his marker is at 4 and he rolls 5 on the die, he can move $|4 - 5|$ or 1 space clockwise on the board.

If his marker is now at 2, which of the following series of die rolls will land him on 6, the winning number?

(a) 5, 2, 4, 2 (b) 1, 5, 2, 4 (c) 3, 3, 5, 1 (d) 4, 5, 1, 3

11. The diagram shows the measurements, above and below ground level, of six buildings in the town of Crystal Falls.

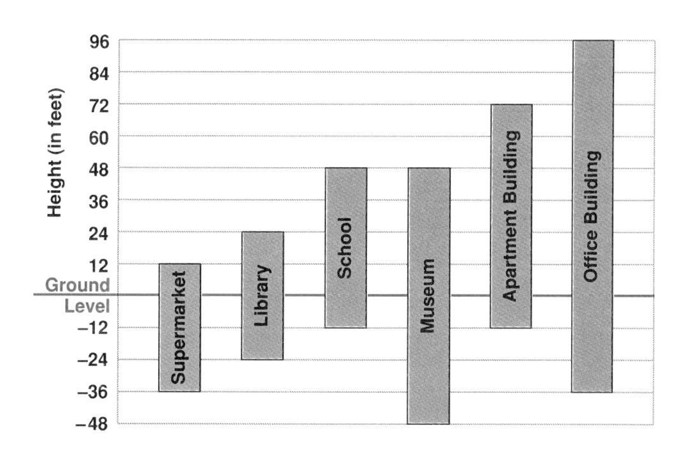

a. Which two buildings have the same overall height from basement to roof?

b. Which buildings measure the same distance above ground as below?

c. Which building has an overall height of 60 feet?

d. How much higher above ground is the office building than the supermarket?

e. How much greater is the overall height of the office building than that of the apartment building?

EXPLORATIONS

1. a. Look out the window to determine if the following statement is true or false.

It is raining.

b. From your observation as to the truth status of the previous statement, tell the truth status of the following statement.

It is not raining.

c. Describe a relationship between the two statements:

It is raining.

It is not raining.

d. Come to a conclusion about the truth status of pairs of statements that have the relationship you described.

2. Cory wanted his math average for 6 tests to be 85.

If his grades on the first 5 tests were 82, 91, 73, 83, and 88, what did he have to get on the 6th test?

To try to find the grade to aim for, he wrote the grades in order and subtracted the average from each grade.

Grade	73	82	83	88	91
Grade – Average	73 – 85 or –12	82 – 85 or –3	83 – 85 or –2	88 – 85 or +3	91 – 85 or +6

Then he added the positive and negative differences separately.

Negative Differences	*Positive Differences*
$-12 + (-3) + (-2) = -17$	$3 + 6 = 9$

Cory reasoned that, since the average is the "middle" grade, the differences below the average should balance the differences above.

a. If Cory's reasoning is correct, should the 6th grade be higher or lower than the average? By how many points?

b. What must the 6th grade be in order to achieve an average of 85?

Check your answer by using it as the 6th grade and finding the average by your usual method, that is, add the 6 grades and divide the sum by 6. Did your answer work?

c. Use Cory's method to find the necessary 6th grade for each of the following sets of grades if the given average is to be achieved.

(1) 80, 93, 75, 82, 85; average to be 81

(2) 85, 80, 90, 100, 80; average to be 88

4.4 Multiplication and Division

To experience the behavior of signs under the operation of multiplication, consider a nonmathematical situation with the following sets of opposites:

good = + bad = –

moving in = + moving out = –

Observe the results of the four possibilities:

Possibility			*Result*
When a good guy moves in to town, it is good for the town.			
+	+		+
When a bad guy moves out of town, it is good for the town.			
–	–		+

Operating with like signs results in +.

When a good guy moves out of town, it is bad for the town.		
+	–	–
When a bad guy moves in to town, it is bad for the town.		
–	+	–

Operating with unlike signs results in –.

To verify that these rules of signs apply to multiplication of signed numbers, recall that the operation of multiplication is the same as repeated addition.

Multiplication	*Repeated Addition*	*Result*
6×2	$6 + 6 = 12$	$(+6) \times (+2) = +12$

The product of two positive numbers is positive.

$(-6) \times 2$	$(-6) + (-6) = -12$	$(-6) \times (+2) = -12$

By the commutative property, $(-6) \times 2 = 2 \times (-6)$ and, thus:

$$(+2) \times (-6) = -12$$

The product of a positive number and a negative number is negative.

To establish the rule for the product of two negatives, study the following pattern.

$(-6) \times 3 = -18$

$(-6) \times 2 = -12$

$(-6) \times 1 = -6$

$(-6) \times 0 = 0$

$(-6) \times (-1) = +6$

$(-6) \times (-2) = +12$

$(-6) \times (-3) = +18$

Note that as the multiplier of -6 decreases

3, 2, 1, 0, –1, –2, –3

the product increases

–18, –12, –6, 0, +6, +12, +18

The product of two negative numbers is positive.

These observations may be summarized as follows:

Multiplying Signed Numbers

- **The product of two numbers with the same sign is positive.**
- **The product of two numbers with different signs is negative.**

Example 1 Find each product: **a.** $4 \cdot 9$ **b.** $(-4) \cdot (-9)$ **c.** $(-4) \cdot 9$ **d.** $4 \cdot (-9)$

The product of two numbers with the same sign is positive:

a. $4 \cdot 9 = 36$ **b.** $(-4) \cdot (-9) = 36$

The product of two numbers with different signs is negative:

c. $(-4) \cdot 9 = -36$ **d.** $4 \cdot (-9) = -36$

Since multiplication and division are inverse operations, we can use the rules of multiplication to observe the rules of signs for division.

Multiplication	*Related Division*
$(+6) \times (+4) = +24$	$(+24) \div (+6) = +4$

The quotient of two positive numbers is positive.

$(-6) \times (+4) = -24$	$(-24) \div (-6) = +4$

The quotient of two negative numbers is positive.

$(-6) \times (-4) = +24$	$(+24) \div (-6) = -4$
$(+6) \times (-4) = -24$	$(-24) \div (+6) = -4$

The quotient of a positive number and a negative number is negative.

These observations may be summarized as follows:

Dividing Signed Numbers

The rules of signs for division are the same as those for multiplication.

- **The quotient of two numbers with the same sign is positive.**
- **The quotient of two numbers with different signs is negative.**

Example 2 Find each quotient: **a.** $72 \div 8$ **b.** $(-72) \div (-8)$ **c.** $(-72) \div 8$ **d.** $72 \div (-8)$

The quotient of numbers with the same sign is positive:

a. $72 \div 8 = 9$ **b.** $(-72) \div (-8) = 9$

The quotient of numbers with different signs is negative:

c. $(-72) \div 8 = -9$ **d.** $72 \div (-8) = -9$

The familiar order of operations applies.

Example 3 Find the value of $12 \times (-3) \div (-4)$.

	$12 \times (-3) \div (-4)$	Begin with the product at the left.
$=$	$-36 \div (-4)$	The product of numbers with different signs is negative. Now do the next operation, division.
$=$	$+9$	The quotient of numbers with the same sign is positive.

Try These *(For solutions, see page 161.)*

1. Find the value of each product.

a. $3 \cdot 5$ **b.** $(-7)(-4)$ **c.** $6 \cdot (-8)$
d. $(-11) \times 9$ **e.** $(-3)(-6)(0)$ **f.** $(-4)(-3)(-6)$

2. Find the value of each quotient.

a. $16 \div (-4)$ **b.** $(-77) \div (-11)$ **c.** $(-60) \div 10$
d. $12 \div 1$ **e.** $0 \div -89$ **f.** $9 \div 0$

3. Use the order of operations to find the value of each expression.

a. $4(-6) \cdot 3$ **b.** $-84 \div (-2) \div 3$ **c.** $48 - 6 \div (-2)$

SELF-CHECK: SECTION 4.4

1. The product of numbers with like signs is $\underline{\ ?\ }$.

Answer: positive

2. The product of numbers with $\underline{\ ?\ }$ signs is negative.

Answer: unlike

3. How are the rules for division of signed numbers related to those for multiplication?

Answer: They are the same.

4. Describe the order of operations for $+, -, \times, \div$.

Answer: $\times$ and $\div$ before $+$ and $-$

EXERCISES: SECTION 4.4

1. Find the value of each product or quotient.

a. $5(4)$ **b.** $-4 \cdot 3$ **c.** $4(-3)$ **d.** $-5(-2)$
e. $0(-16)$ **f.** $6 \div (-3)$ **g.** $-4 \div (-4)$ **h.** $0 \div (-2)$
i. $3(-6.2)$ **j.** $(-2.1)(0.1)$ **k.** $-4.4 \div 0.1$ **l.** $5.2 \div (-0.4)$
m. $\frac{1}{2} \times \left(-\frac{3}{5}\right)$ **n.** $-\frac{3}{4} \times \left(-\frac{8}{9}\right)$ **o.** $6 \div \left(-\frac{1}{2}\right)$ **p.** $\frac{4}{5} \div \left(-\frac{8}{15}\right)$
q. $-20 \div (-5)$ **r.** $5 \cdot (-2) \cdot 2$ **s.** $-4 \cdot (-2) \cdot 6$ **t.** $12 \cdot 5 \cdot (-3)$
u. $-6 \cdot (-3) \cdot (-1)$ **v.** $-8 \cdot 2 \cdot (-5)$ **w.** $12 \div (-6) \div (-2)$ **x.** $-48 \div 8 \div (-3)$

2. Use the order of operations to find the value of each expression.

a. $-32 \div (-4) \cdot 3$ **b.** $9 \cdot (-8) \div 4$ **c.** $12 \div 2 \cdot (-5)$

d. $-28 \div (-4) \div (-7)$ **e.** $-16 + 24 \div (-2)$ **f.** $4 + 5 \cdot (-8) \div 2$

g. $-24 - 8 \div 2 + 2$ **h.** $6 + 3 \cdot (-4) - 8 \div (-2)$ **i.** $5(-6) - 30 \div (-6)$

j. $(-3 + 5)^2 \div (-1)$ **k.** $-100 \div (4 \cdot 5^2)$ **l.** $2(1 + 2)^2 \div (-6)$

3. a. Does the closure property hold for multiplication of integers? for division of integers? Explain.

b. (1) Is this statement true or false? $8 \times (-4) = (-4) \times 8$

(2) From your observation about the last statement, come to a conclusion about multiplication of integers.

(3) Is this statement true or false? $8 \div (-4) = (-4) \div 8$

(4) From your observation about the last statement, come to a conclusion about division of integers.

c. (1) Is this statement true or false? $-12 \times (2 \times 3) = (-12 \times 2) \times 3$

(2) From your observation about the last statement, come to a conclusion about multiplication of integers.

(3) Is this statement true or false? $-12 \div (2 \div 3) = (-12 \div 2) \div 3$

(4) From your observation about the last statement, come to a conclusion about division of integers.

4. a. Summarize your findings about which properties hold for operations with integers by copying and completing this table. (Enter *yes* or *no*.)

	Operation			
Property	+	−	×	÷
Closure	?	?	?	?
Commutative	?	?	?	?
Associative	?	?	?	?

b. (1) Which of these statements are true and which are false?

(a) $-12 \times (2 + 4) = (-12) \times 2 + (-12) \times 4$

(b) $-12 \times (2 - 4) = (-12) \times 2 - (-12) \times 4$

(c) $-12 \div (2 + 4) = (-12) \div 2 + (-12) \div 4$

(d) $-12 \div (2 - 4) = (-12) \div 2 - (-12) \div 4$

(2) From your observations about the previous statements, tell for which operations with integers the Distributive Property holds.

c. How do the properties of operations with integers compare to the properties of operations with whole numbers?

5. a. Find the value of each product.

(1) $(-1)(-1)$ **(2)** $(-1)(-1)(-1)$ **(3)** $(-1)(-1)(-1)(-1)$ **(4)** $(-1)(-1)(-1)(-1)(-1)$

b. Come to a conclusion relating the number of negative factors and the sign of the product.

6. Replace each ? by an integer that will make the sentence true. If no answer is possible, write *impossible*.

a. $-2 \times ? = 6$ **b.** $-4 \times ? = 0$ **c.** $5 \times ? = -5$

d. $-3 \times ? = 12$ **e.** $8 \div ? = -4$ **f.** $-27 \div ? = 3$

g. $10 \div ? = -1$ **h.** $0 \div ? = 0$ **i.** $4 \div ? = 0$

7. Find the next three terms in each sequence.

a. 1, 2, 4, ... **b.** 160, 80, 40, ... **c.** –1, –3, –9, ...

d. 96, 48, 24, ... **e.** 2, –6, 18, ... **f.** –32, 16, –8, ...

8. For each question, select the choice that is the best answer.

a. On the number line shown below, the point that best represents the quotient of the coordinates of points *A* and *F* is (1) *B* (2) *C* (3) *D* (4) *E*

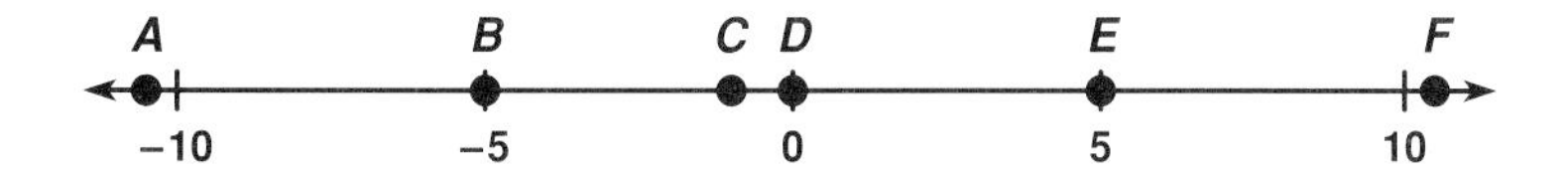

b. If you multiply three consecutive integers, like 7, 8, 9 or –3, –2, –1, when will the product be exactly divisible by 6?

(1) always (2) sometimes (3) never (4) cannot tell

c. If *a* and *b* are different positive or negative integers, which one operation symbol could replace ? to make both of these statements true? Try different numbers.

$a ? b = b ? a$ and $a ? 0 = 0$

(1) + (2) – (3) × (4) ÷

d. If *a* and *b* are integers, and $a \cdot b = 0$, which *must be* a value of one of the integers?

(1) 1 (2) –1 (3) 0 (4) cannot tell

e. If *a* and *b* are integers, and $a \cdot b = 18$, which *must be* a value of one of the integers?

(1) 18 (2) 9 (3) 0 (4) cannot tell

f. If *a* and *b* are integers, and $a \cdot b = 36$, which *could not be* a value of $a + b$?

(1) –12 (2) –13 (3) 35 (4) 37

9. Solve each problem.

a. On a winter night, the temperature at 7 P.M. was 15°*F*. If the temperature dropped 5° each hour until 11 P.M., what was the temperature at 11 P.M.?

b. Chloe and Cher do neighborhood jobs to earn money. Each of them has saved an amount that is equal to 6 times her average weekly earnings. On average, how much more does Cher earn in a week than Chloe?

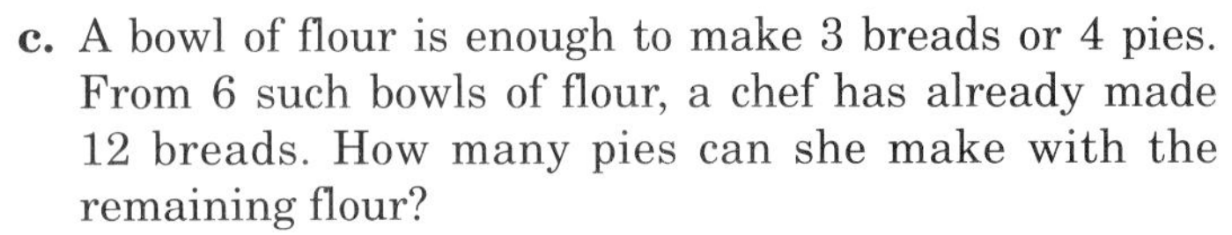
c. A bowl of flour is enough to make 3 breads or 4 pies. From 6 such bowls of flour, a chef has already made 12 breads. How many pies can she make with the remaining flour?

d. A group of students made a table to find the average of their combined math grades.

Al multiplied each grade by how many times it was earned (its *frequency*).

Bo added all Al's products.

Carrie divided Bo's sum by the total number of grades.

Grade	Frequency
90	2
85	4
80	3
70	4
Total	13

(1) What number did Carrie get as a result?

(2) Write out an equivalent calculation using only addition and division that will also give the average of these grades.

EXPLORATIONS

1. Use the following situation to explain the behavior of signs under the operation of multiplication.

Situation: The amount of water in a pool as the pool is filled or emptied.

Here is an explanation of the first possibility with signs:

When water flows in, then later, there is more water.

+ + +

The product of two positive numbers is positive.

a. Tell how you would assign + or – to the flow of water, the time, and the amount of water in the pool.

b. Write out the 3 remaining possibilities in words and in symbols, as shown in the given explanation.

c. If the pool can be filled or emptied at the rate of 4 gallons per minute, tell the change in the amount of water in the pool in 5 minutes, under the four possibilities.

2. You can perform a variety of operations on a magic square, and still obtain a magic square.

a. **(1)** Start with Magic Square A, and create Magic Square B by dividing each entry of A by 2.

What is the magic sum of B?

8	–6	4
–2	2	6
0	10	–4

Magic Square *A*

(2) Start with Magic Square B, and create Magic Square C by doing this to each entry: multiply by 3 and add 1.

What is the magic sum of C?

(3) Start with Magic Square C and create Magic Square D with a magic sum of -48.

What did you have to do to each entry of C?

(4) Start with Magic Square D, and create Magic Square E by doing this to each entry: divide by –2 and subtract 1.

What is the magic sum of E?

(5) Try other operations and see if the result is a magic square.

b. **(1)** Arrange the digits 1-9 to form Magic Square X, whose magic sum is 15.

(2) Using the arrangement you found for magic Square X, form square Y:

Replace the digits of X, in order, by nine consecutive odd numbers, beginning with 3.

Is Square Y a magic square?

(3) Use the arrangement you found for Magic Square X to arrange the nine consecutive even numbers that begin with –8, to form Square Z.

Is Square Z a magic square?

(4) Choose another sequence of nine numbers and use the pattern of X to arrange the numbers in a square.

Is your new square a magic square?

4.5 Coordinate Pairs

The number lines show dots at –3, 1, and 5. The dots are called the ***graphs*** of the numbers, and the numbers are the ***coordinates*** of the dots.

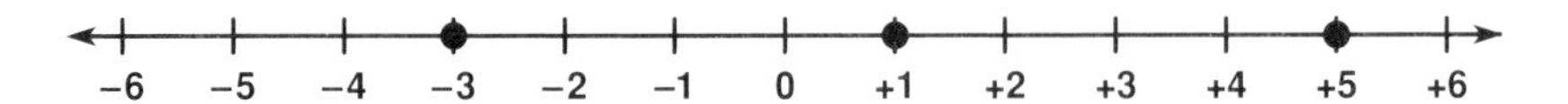

The coordinate of a point on a number line tells you where the point is located. You can think of the coordinate of a point as its *address*. For example, the coordinate of the point halfway between –3 and 5 is 1.

Example 1 Referring to points A and B on this number line

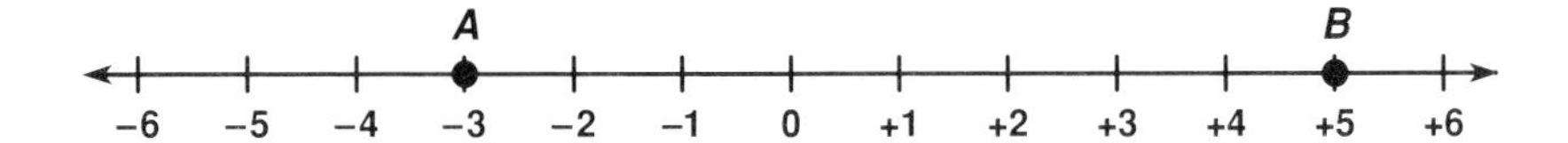

write the coordinate of a point that is:

a. 3 units from A

The coordinate of the point 3 units to the right of A is 0, and the coordinate of the point 3 units to the left of A is –6.

b. one-fourth of the way from A to B

Since there are 8 units from A to B, the point that is one-fourth of the way from A to B is

$$\frac{1}{4} \times 8, \text{ or } 2 \text{ units}$$

from A. The coordinate of that point is –1.

In cities and towns, people often use the intersections of streets and avenues, or different roads, as places to meet.

In New York City, for example, you might say to a friend, "I'll meet you on 5th Avenue at 34th Street, at the entrance to the Empire State Building."

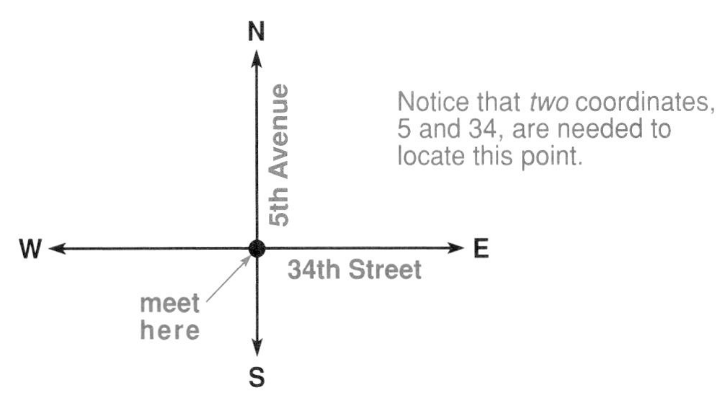

Notice that *two* coordinates, 5 and 34, are needed to locate this point.

The idea of intersecting streets and avenues can be extended by using two number lines.

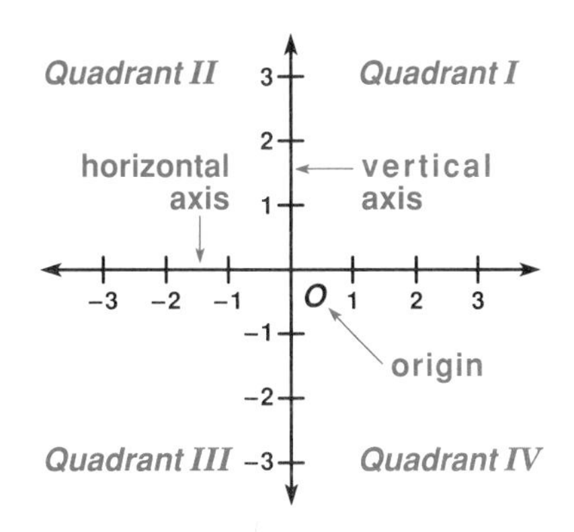

The number lines are made to intersect at their zero points. This point is called the ***origin***, and the letter O marks it.

Each number line is called an ***axis***.

Note that the two axes divide the area into 4 parts, which are called ***quadrants***. These quadrants are numbered as shown.

As usual, on the horizontal axis, the positive numbers are to the right of the origin, and the negative numbers are to the left. How are the numbers on the vertical axis arranged? If you said that the positive numbers are above the origin and the negative numbers are below, you are reading the diagram correctly.

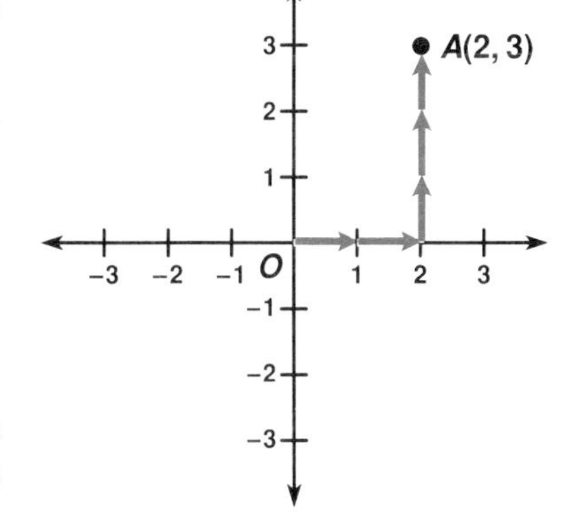

This model with two number lines can now be used to locate a point anywhere in a ***plane*** (a flat 2-dimensional area).

For example, the coordinate numbers for point A are 2 and 3. They are written as a pair within parentheses.

To locate point A:

Start at the origin.

First count 2 units to the right along the horizontal axis.

Then count 3 units up.

Since the *order* of counting the units is important, that is, you always count along the horizontal axis first, the coordinate numbers of a point are called an ***ordered pair***.

Example 2 Explain how the ordered pairs (2, 3) and (3, 2) are different.

(2, 3) means 2 right and 3 up.

(3, 2) means 3 right and 2 up.

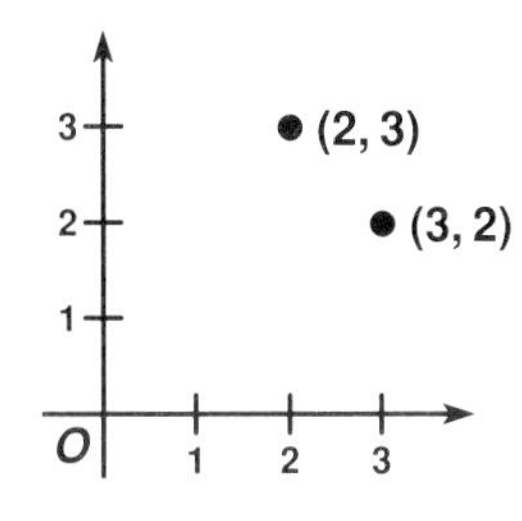

Coordinate Pairs

To write the coordinates of a point:

1. **Start at the origin.**
2. **First, count the units along the horizontal axis, positive to the right and negative to the left.**
3. **Next, count the units in the vertical direction, positive up and negative down.**
4. **The coordinates of a point are written as an ordered pair:**

(horizontal value, vertical value)

For convenience in working with coordinate pairs, you can use ruled graph paper.

Example 3 Write the ordered pair for each lettered point shown, and describe its location in the plane.

A, at (2, 1), is in Quadrant I.

B, at (–3, 4), is in Quadrant II.

C, at (–2, –1), is in Quadrant III.

D, at (4, –4), is in Quadrant IV.

E, at (3, 0), is on the horizontal axis.

F, at (0, –2), is on the vertical axis.

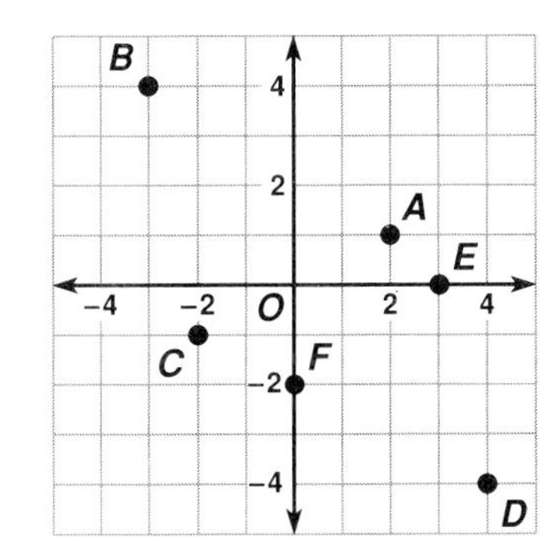

Example 4 Using one set of axes, graph and label these points:

$P(-4, 3)$ $Q(3, -1)$ $R(-3, 0)$ $S(0, 4)$

To begin, draw the axes. Show some numbers on each axis, and label the origin.

For point P, start at the origin and move 4 left, then 3 up.

For point Q, start at the origin and move 3 right, then 1 down.

For point R, start at the origin and move 3 left, then do not go up or down.

For point S, start at the origin and do not move right or left, then go up 4.

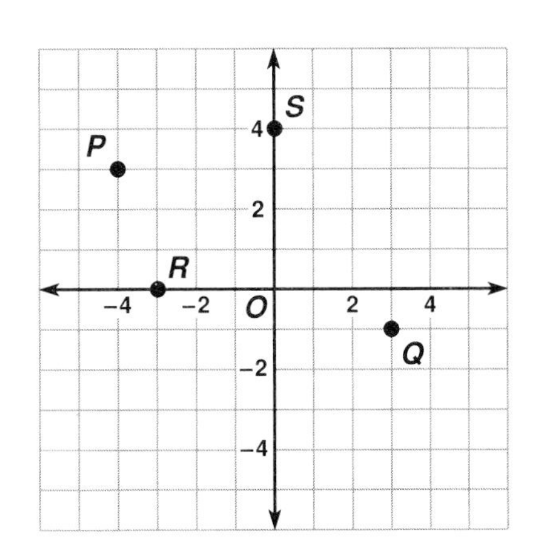

Try These *(For solutions, see page 161.)*

1. Write the coordinates of each lettered point shown.

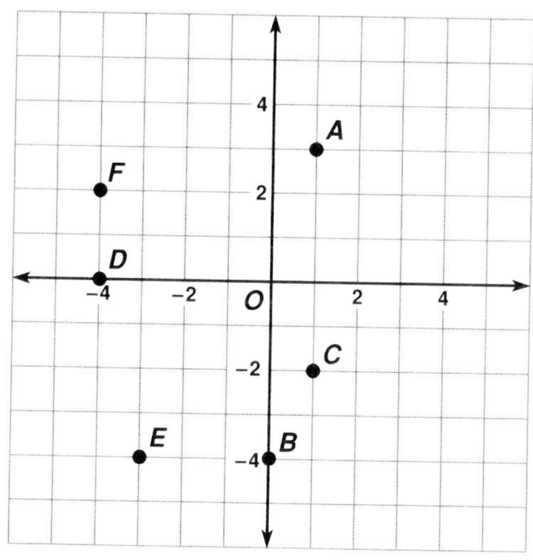

2. On one set of axes, graph and label these points:

   $P(3, 5)$ $Q(5, 3)$ $R(-4, -2)$
   $S(-2, -4)$ $T(-2, 6)$ $U(6, -2)$
   $V(4, 0)$ $W(0, 4)$ $X(-2, 0)$
   $Y(0, -2)$

3. **a.** On one set of axes, graph and label these points:

   $F(-2, -4)$ $C(-6, 2)$ $A(0, 8)$
   $E(-6, -2)$ $B(-2, 4)$ $K(6, 2)$
   $H(2, -4)$ $I(6, -2)$ $J(2, 0)$
   $D(-2, 0)$ $G(0, -8)$ $L(2, 4)$
   $M(0, 8)$

   b. Use straight line segments to connect these points in alphabetical order.

   c. The picture you get should be symmetric. Use the symmetry of the picture to correct any errors in graphing.

Reflection of a Point

Recall that when the point whose coordinate is 3 is reflected across 0 on the number line, the image point has the coordinate −3.

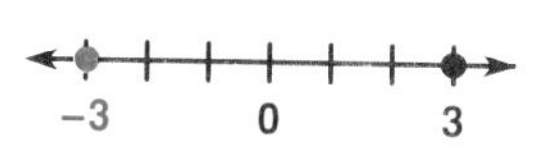

What are the coordinates of the image of $P(2, 3)$ when P is reflected over the horizontal axis? $P'(2, -3)$ is the image of $P(2, 3)$ when P is reflected over the horizontal axis.

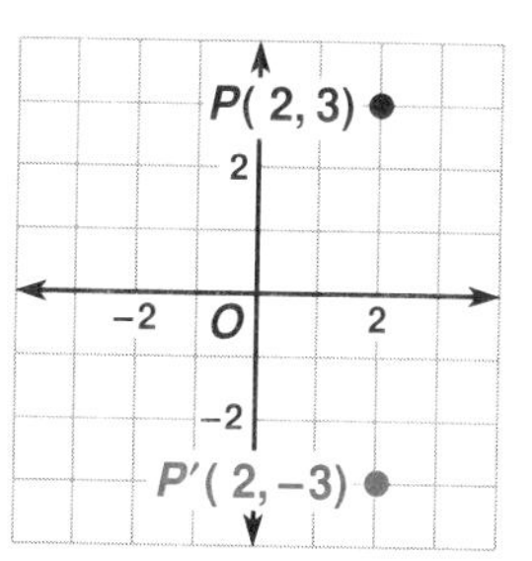

Example 5 Write the coordinates of the image point P' when $P(2, 3)$ is reflected over:

a. the vertical axis

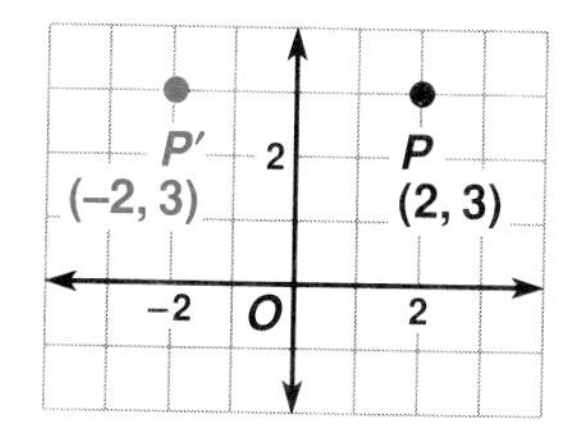

The image is $P'(-2, 3)$.

b. the line that evenly divides Quadrants I and III

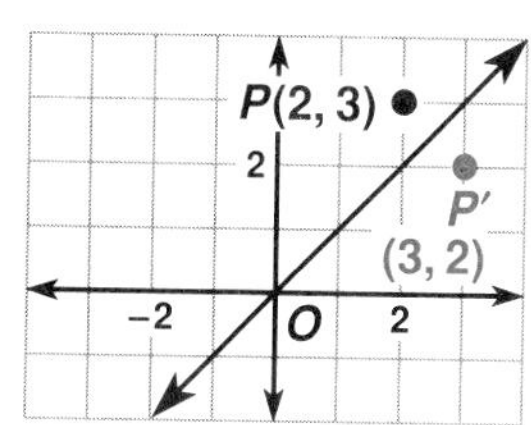

The image is $P'(3, 2)$.

Try This *(For solution, see page 161.)*

Write the coordinates of the image point P' when $P(4, 2)$ is reflected over:

a. the horizontal axis
b. the vertical axis
c. the line that evenly divides Quadrants I and III
d. the line that evenly divides Quadrants II and IV

SELF-CHECK: SECTION 4.5

1. How many numbers represent the address of a point on a number line?

Answer: one number

2. How many numbers represent the address of a point in a plane?

Answer: two numbers

3. To determine the location of a point in the plane, from which place do you begin your count?

Answer: from the origin

4. To locate point W at $(-6, 4)$, which direction should you move first from the origin?

Answer: to the left 6 units along the horizontal axis

5. Explain the difference between the ordered pairs $(1, -2)$ and $(-2, 1)$.

Answer: $(1, -2)$ means 1 right and 2 down
$(-2, 1)$ means 2 left and 1 up

6. What are the coordinates of the origin?

Answer: $(0, 0)$

7. If $P(4, 1)$ is reflected over the vertical axis, describe how the coordinates of the image are related to the coordinates of the original point.

Answer: The horizontal value is the opposite of the original value. The vertical value remains the same.

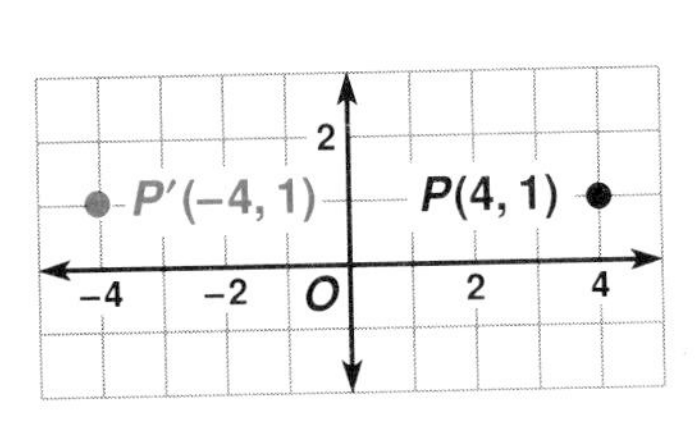

EXERCISES: SECTION 4.5

1. **a.** If O represents Ti-Hua's house and each square represents one block, give the directions that Ti-Hua must walk to get from his house to the:
   (1) Library
   (2) High School
   (3) Ice Cream Parlor
   (4) Recreation Center

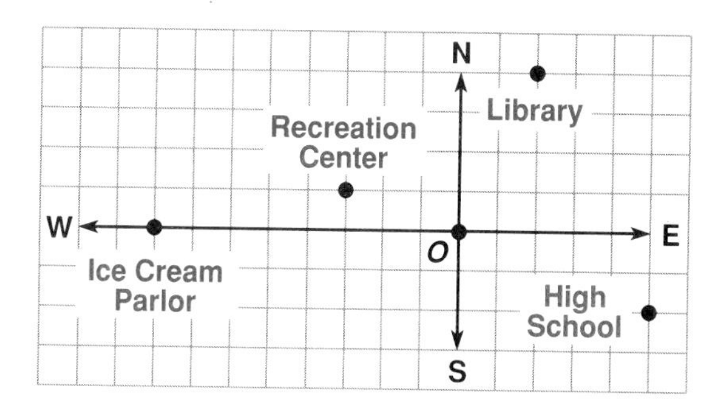

   b. Give directions to get from the:
   (1) Ice Cream Parlor to the Recreation Center
   (2) Library to the High School
   (3) Recreation Center to the Library
   (4) High School to the Recreation Center

2. Write the coordinates of each lettered point shown on the graph, and describe its location in the plane.

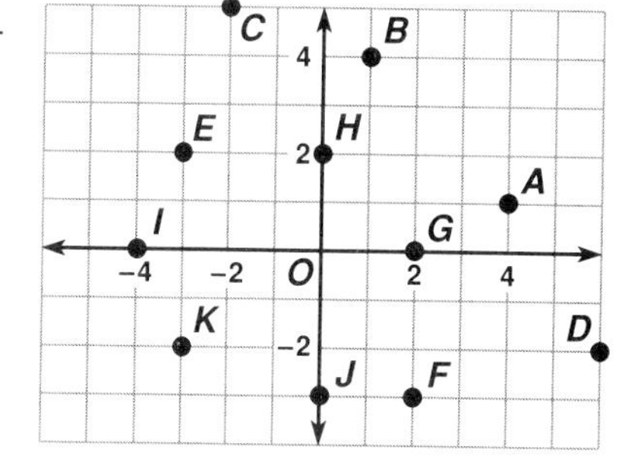

3. Using this graph

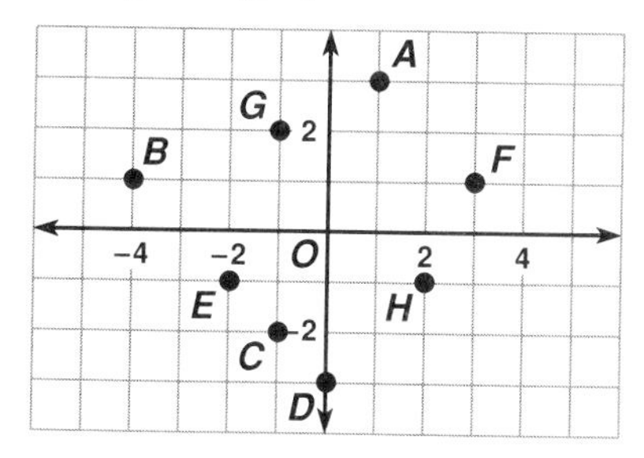

copy and complete this chart

Point	Coord.	Point	Coord.
A	?	?	(3, 1)
B	?	?	(–1, 2)
C	?	?	(2, –1)
D	?	?	(–2, –1)

4. Use the letters of the points given in the graph to read the following message, which is spelled out by the coordinates of the points.

$\frac{?}{(2, 1)}$ $\frac{?}{(-1, 3)}$ $\frac{?}{(2, -1)}$ $\frac{?}{(-3, -1)}$ $\frac{?}{(3, 3)}$ $\frac{?}{(-1, 2)}$

$\frac{?}{(2, -1)}$ $\frac{?}{(-1, 3)}$ $\frac{?}{(-2, -3)}$

$\frac{?}{(5, 0)}$ $\frac{?}{(2, -3)}$ $\frac{?}{(-1, 0)}$

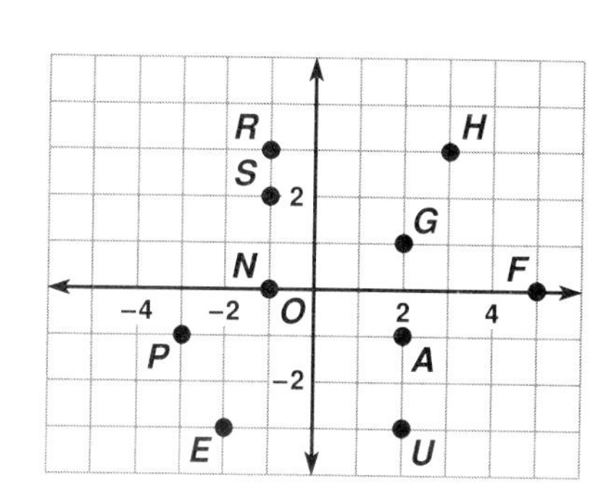

5. On graph paper, plot the following points. Read down the columns and, in the order given, connect the points with line segments. Start a new line after each "stop."

When you are done, you will read a message from Dr. Martin Luther King, Jr.

(–12, 16)
(–13, 15)
STOP

(–13, 16)
(–14, 15)
STOP

(–12, 11)
(–11, 11)
(–11, 15)
(–12, 15)
STOP

(–11, 15)
(–10, 15)
STOP

(–11, 11)
(–10, 11)
STOP

(–4, 15)
(–4, 13)
(–7, 13)
(–7, 15)
STOP

(–7, 13)
(–7, 11)
STOP

(–4, 11)
(–4, 13)
STOP

(–3, 11)
(–1, 15)
(1, 11)
STOP

(–2, 13)
(0, 13)
STOP

(1, 15)
(3, 11)
(5, 15)
STOP

(8, 11)
(6, 11)
(6, 15)
(8, 15)
STOP

(6, 13)
(7, 13)
STOP

(11, 11)
(13, 15)
(15, 11)
STOP

(12, 13)
(14, 13)
STOP

(–10, –10)
(–10, –6)
(–7, –6)
(–6, –7)
(–6, –9)
(–7, –10)
(–10, –10)
STOP

(–5, –10)
(–5, –6)
(–2, –6)
(–1, –7)
(–1, –8)
(–3, –8)
(–1, –10)
STOP

(2, –6)
(0, –6)
(0, –10)
(2, –10)
STOP

(0, –8)
(1, –8)
STOP

(3, –10)
(5, –6)
(7, –10)
STOP

(4, –8)
(6, –8)
STOP

(8, –10)
(8, –6)
(10, –10)
(12, –6)
(12, –10)
STOP

(12, –5)
(13, –6)
STOP

(13, –5)
(14, –6)
STOP

6. On different graphs, plot each of the following sets of points.
Use line segments to connect the points, *in alphabetical order*, to see a "picture" emerge.
Note that the coordinates of the last point are the same as those of the first point so that the figure closes.

a. $A(2, 22)$ $S(-4, 4)$ $P(-14, -16)$ $F(17, -6)$ $T(-11, 3)$ $K(4, -24)$
$Q(-5, -6)$ $B(12, 12)$ $I(12, -20)$ $N(-8, -20)$ $J(4, -17)$ $M(0, -17)$
$L(0, -24)$ $V(-8, 12)$ $G(9, -5)$ $R(-12, -7)$ $H(18, -16)$ $D(14, 2)$
$U(-3, 13)$ $C(7, 13)$ $E(7, 4)$ $W(2, 22)$

b. $A(-7, 7)$ $P(2, -9)$ $F(5, -3)$ $D(9, 1)$ $S(-1, -2)$ $H(5, -9)$
$N(-4, -11)$ $T(-2, 1)$ $C(12, 1)$ $G(6, -7)$ $M(-5, -12)$ $J(12, -12)$
$R(2, -3)$ $Q(1, -7)$ $E(8, -2)$ $L(0, -12)$ $K(5, -12)$ $U(-5, 1)$
$I(11, -11)$ $B(14, 7)$ $V(-7, 7)$

c. $A(-8, 5)$ $I(7, 1)$ $S(-12, -7)$ $U(-13, -1)$ $B(-8, 10)$ $N(-1, -12)$
$R(-8, -9)$ $T(-15, -1)$ $L(3, -9)$ $C(-6, 9)$ $V(-14, 4)$ $G(3, 4)$
$Q(-4, -8)$ $F(1, 10)$ $D(-3, 14)$ $J(9, 1)$ $E(-1, 9)$ $K(7, -6)$
$P(-2, -13)$ $H(9, 4)$ $W(-8, 5)$ $M(-3, -7)$

7. Write the coordinates of the image point when each of the following points is reflected over:

(1) the horizontal axis
(2) the vertical axis
(3) the line that evenly divides Quadrants I and III
(4) the line that evenly divides Quadrants II and IV

a. $(2, 5)$ **b.** $(5, 2)$ **c.** $(-2, 5)$ **d.** $(-5, 2)$ **e.** $(-2, -5)$ **f.** $(-5, -2)$ **g.** $(2, -5)$ **h.** $(5, -2)$

8. You have seen how to find the coordinates of an image point under *reflection*.
Think of reflection as *flipping* the original point over the line of reflection.

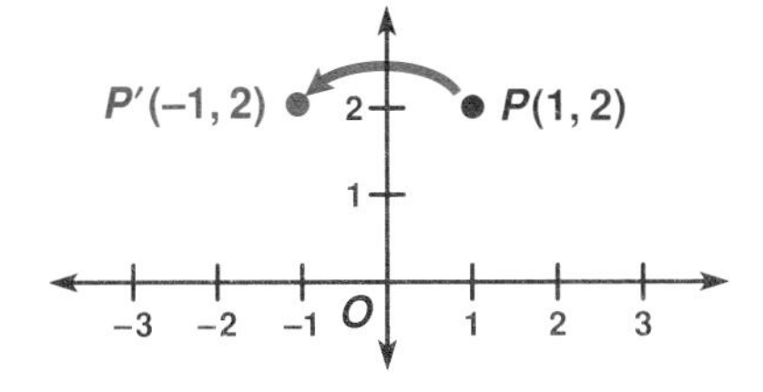

What happens if we *slide* the point $P(1, 2)$ 2 units to the right and 1 unit down?
The image point is $P'(3, 1)$.
This kind of change is called a ***translation***.

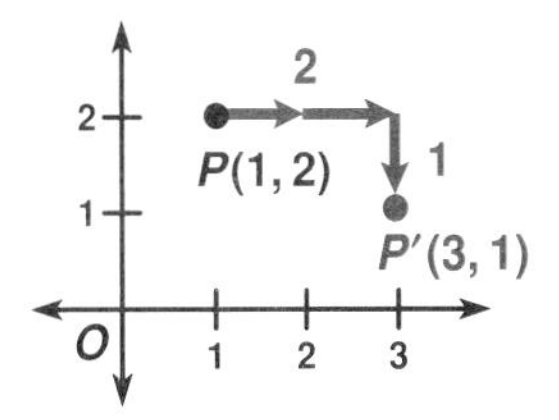

For each of the following points:

- Graph the image point under a translation of 1 unit to the right and 3 units up.
- Write the coordinates of the image point.

a. $A(4, 2)$ **b.** $B(-4, 2)$ **c.** $C(-4, -2)$ **d.** $D(4, -2)$

9. Each of these sets of coordinates gives the locations of three of the four vertices of a rectangle. Graph the three vertices on a set of axes so that you can complete the rectangle and write the coordinates of the fourth vertex.

a. (–6, 3) (6, 3) (6, –2)

b. (3, 5) (8, 5) (8, –3)

c. (–8, –1) (–1, –1) (–1, –4)

d. (–2, –2) (–2, 5) (0, 5)

10. **a.** What do the coordinates of these points have in common?

$A(7, 2)$ $B(4, 2)$ $C(0, 2)$ $D(-3, 2)$ $E(-5, 2)$

b. Graph these points on one set of axes and describe a way to connect all five of them. (Disregard order of connection.)

c. Consider the point $P(5, -1)$, but do not graph it.

Write the coordinates of four more points (call them Q, R, S, T) that will share with P a characteristic similar to that shared by points A-E.

d. Without graphing points P-T, describe a way to connect all five of them. (Disregard order of connection.) Then graph points P-T to verify your answer.

e. From your observations about these two sets of points, come to a general conclusion.

11. **a.** What do the coordinates of these points have in common?

$A(3, 7)$ $B(3, 2)$ $C(3, 0)$ $D(3, -2)$ $E(3, -5)$

b. Graph these points on one set of axes and describe a way to connect all five of them.

c. Consider the point $P(-4, -1)$, but do not graph it.

Write the coordinates of four more points (call them Q, R, S, T) that will share with P a characteristic similar to that shared by points A-E.

d. Without graphing points P-T, describe a way to connect all five of them. Then graph points P-T to verify your answer.

e. From your observations about these two sets of points, come to a general conclusion.

12. **a.** **(1)** What do the coordinates of these points have in common?

$A(8, 0)$ $B(5, 0)$ $C(0, 0)$ $D(-2, 0)$ $E(-6, 0)$

(2) Graph these points on one set of axes and describe a characteristic that all five locations have in common.

(3) From your observations about this set of points, come to a general conclusion.

b. **(1)** What do the coordinates of these points have in common?

$P(0, 9)$ $Q(0, 4)$ $R(0, 0)$ $S(0, -3)$ $T(0, -7)$

(2) Graph these points on one set of axes and describe a characteristic that all five locations have in common.

(3) From your observations about this set of points, come to a general conclusion.

c. Tell some things that you know about the origin.

EXPLORATIONS

Game: Battleships

Here's what each of two players needs:
Two graph sheets, both labeled *A-J* across and 1-10 down. The players have one of these graph sheets for their own ships, called *Ships Graph*, and the other for keeping track of strikes against the opponent's ships, called *Strikes Graph*.

Here's how to position ships on the *Ships Graph*:
There are 4 types of ships, with each taking up a particular number of squares on the graph.

battleship = 4 squares   destroyer = 2 squares

cruiser = 3 squares   submarine = 1 square

A ship may be positioned across or down, but not diagonally.
Ships may not touch, even at corners.

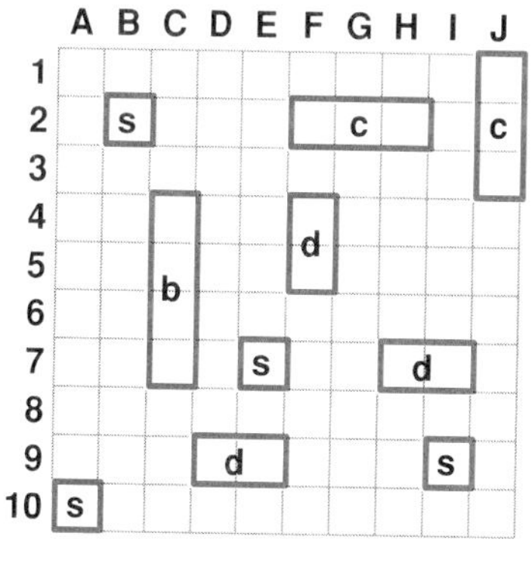

This sample *Ships Graph* shows the positions of:
1 battleship, labeled **b**   2 cruisers, each labeled **c**

3 destroyers, each labeled **d**   4 submarines, each labeled **s**

These labels are important for the *Strikes Graph*.
They need not be included on the *Ships Graph*.

Here's how to play:
Each player sets up a *Ships Graph*, which the opponent never sees, using 1 battleship, 2 cruisers, 3 destroyers, and 4 submarines.

The players take turns guessing at the location of the opponent's ships by calling out coordinates such as *E* 3.

Sam, guessing *E* 3, marks a dot in that square of his *Strikes Graph*.

Mary answers the guess by telling whether a ship has been hit and, if yes, what kind it is.

Say a cruiser has been hit. Sam marks c along with the dot on his *Strikes Graph*.

If no ship is hit, the dot on the *Strikes Graph* stands alone.

This sample *Strikes Graph* shows that Sam:
guesses *E* 3 and is told that a cruiser was hit

guesses *D* 3 and is told "no hit"

guesses *E* 2 and is told that a cruiser was hit

knows that either *E* 1 or *E* 4 will "finish off" the cruiser

guesses *E* 1, but is wrong

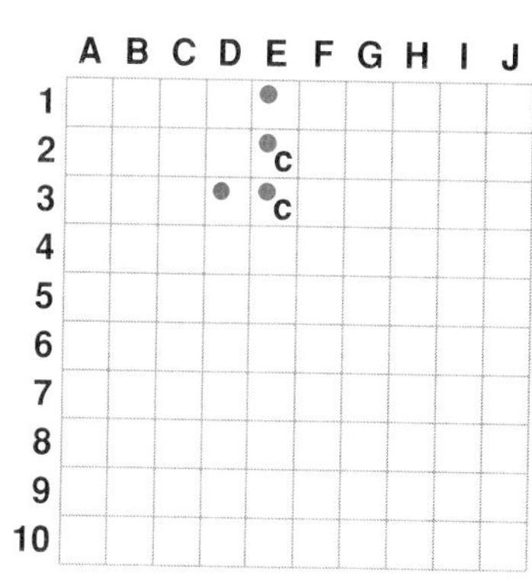

As Sam is keeping track of his guesses about Mary's fleet, she is marking Sam's guesses on her *Ships Graph*, with ✔.

Sam and Mary take turns guessing at hits on the other's ships. Both keep track of their own guesses, on the *Strikes Graph*, and their opponent's guesses, on the *Ships Graph*. When all parts of a ship have been hit, the ship sinks.

The winner is the player who first sinks the opponent's entire fleet.

Things You Should Know After Studying This Chapter

KEY SKILLS

4.1 Represent opposite situations by signed numbers.
Use inequality symbols to compare signed numbers.
Find the absolute value of a signed number.

4.2 Add signed numbers.

4.3 Subtract signed numbers.

4.4 Multiply and divide signed numbers.

4.5 Locate a point in a plane using an ordered pair of signed numbers.

KEY TERMS AND SYMBOLS

4.1 opposites • positive • negative • signed number • integer • inequality (> or <) • distance • absolute value (| |) • reflection • image

4.2 like signs • unlike signs • integer chips • zero pair

4.3 adding an opposite

4.4 product • quotient • inverse operations

4.5 graph • coordinate • plane • origin • axis (axes) • quadrants • ordered pair • translation

CHAPTER REVIEW EXERCISES

1. Represent each situation by a positive or a negative integer.

a. The water level in the reservoir rose 2 inches.

b. Jody lost a quarter.

c. Erin lost 3 pounds and then gained back $\frac{1}{2}$ pound.

2. Tell whether each statement is true or false.

a. A positive number is to the right of every negative number on a number line.

b. The set of positive integers is a finite set.

c. $|-3| = 3$

3. Replace each ? by <, >, or = to make a true statement.

a. $-4 \; ? \; -20$ **b.** $|-6| \; ? \; |+6|$

c. $|-1| \; ? \; -5$ **d.** $|-2| \; ? \; 0$

4. Arrange each set of numbers in descending order (greatest to least).

a. 1, 4, –10, –8, –1

b. 0, –1, 1, 2, –2

5. Find the value of each expression.

a. $(-10) + (-1) + 6$ **b.** $(-4) + (-5) + (-1)$

c. $8 - (-1)$ **d.** $|-12| + (-2) + (-4)$

e. $-4 - (-6)$ **f.** $12 - 20$

g. $|-6| - |-2|$ **h.** $2(-8)$

i. $(-1)(-9)(-2)$ **j.** $(-5)(3)(-1)$

k. $15 \div (-3)$ **l.** $-18 - 3(-6)$

m. $4 \cdot 8 - 10 \div 2$ **n.** $6 + 2(-4) - (-10)$

o. $10 \div 2 + 3$ **p.** $(-5)(-8 + 5)^2 - (12)(-4)$

6. Replace each ? by an integer that will make the sentence true.

a. $-5 + ? = 2$

b. $6 - ? = 1$

c. $3 \times ? = -3$

7. Find the next 3 terms in each sequence.

a. 7, 4, 1, ...

b. –1, –3, –9, ...

c. 4, 1, $\frac{1}{4}$, ...

8. If the temperature at 6 A.M. was –2° *F* and at 2 P.M. it was 28° *F*, what was the change in temperature from 6 A.M. to 2 P.M.?

9. Lonny's cat climbed 5 rungs up from the bottom of a ladder, then slipped back 2 rungs, then climbed 7 rungs to reach the top. How many rungs were on the ladder?

10. Match each property in the list that follows with a statement in **a** through **d** below.

Properties	*Statements*
(1) Closure Property	**a.** $7 + (-2) = (-2) + 7$
(2) Commutative Property for Addition	**b.** The sum of 7 and (–2) is an integer.
(3) Associative Property for Multiplication	**c.** $(-7) \times (2 + 3) = (-7) \times 2 + (-7) \times 3$
(4) Distributive Property	**d.** $(-7 \times 2) \times 3 = -7 \times (2 \times 3)$

11. Write the coordinates of each lettered point shown, and describe its location in the plane.

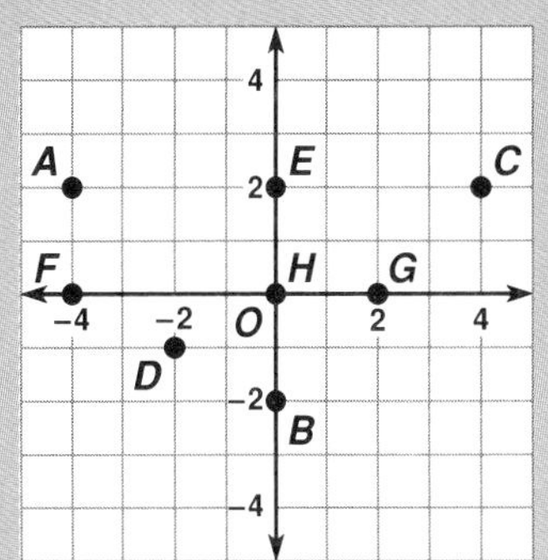

12. What are the coordinates of the image point P' when the point $P(3, 1)$ is reflected over the horizontal axis?

13. What are the coordinates of the image point Q' when the point $Q(3, 4)$ is translated 5 units to the left and 1 unit up?

SOLUTIONS FOR TRY THESE

4.1 *The Number Line*

Page 122

1. **a.** gain: +6 loss: −6
   b. South: −7 North: +7
2. {−4, −3, −2, −1, 0, 1, 2, 3, 4, 5, 6, 7}
3. **a.** $10 > 8$ $8 < 10$
   b. $12 > -12$ $-12 < 12$
   c. $0 > -5$ $-5 < 0$
   d. $-9 > -22$ $-22 < -9$
   e. $-1 > -10 > -14$ $-14 < -10 < -1$
4. **a.** 0, 3, 5, 7, 9, 11 **b.** −11, −7, −3, 0, 5, 9

Page 123

1. 17 units
2. **a.** $|16| = 16$ **b.** $|-16| = 16$ **c.** $|0| = 0$
   d. $|5.3| = 5.3$ **e.** $\left|-7\frac{1}{2}\right| = 7\frac{1}{2}$

4.2 *Addition*

Page 127

1.
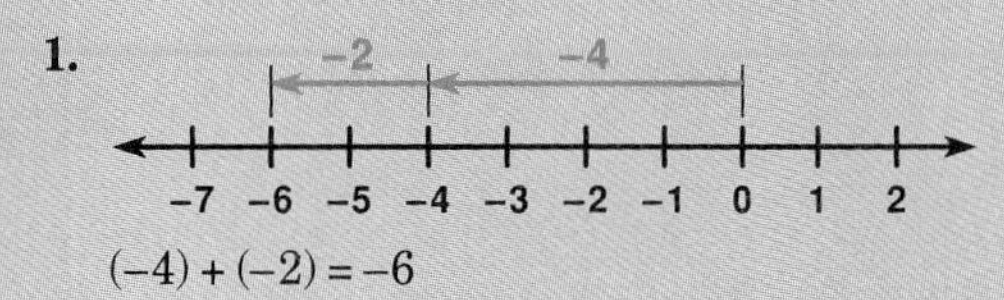

(−4) + (−2) = −6

2. **a.** (+7) + (+5) = +(7 + 5) = +12 (add absolute values)
   b. (−4) + (−2) = −(4 + 2) = −6 (add absolute values)

Page 129

1.
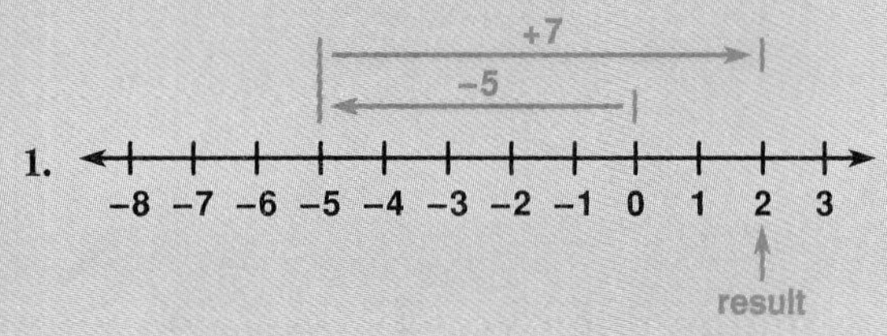

Answer: (−5) + (+7) = +2

2. [+] [+] + [−] [−] [−] [−] [−] [−]
   +2 + −6
   ([+][−]) + ([+][−]) + [−] [−] [−] [−] = [−] [−] [−] [−]
   0 + 0 + (−4) = −4

   Answer: (+2) + (−6) = −4
3. **a.** Since the integers have different signs:

   subtract absolute values

   (+7) + (−5) = +(7 − 5) = +2

   sign of integer with greater absolute value

   b. Since the integers have different signs:

   subtract absolute values

   (+3) + (−14) = −(14 − 3) = −11

   sign of integer with greater absolute value

Page 130

1. (−6) + (+19) + (−3) + (+14) = −6 + 19 + (−3) + 14
   = 13 + (−3) + 14
   = 10 + 14
   = 24
2. Increases are positive. 147 + 23 = 170
   Decreases are negative. (−106) + (−71) = −177
   Sum the results. 170 + (−177) = −7
   Answer: The total change was a decrease of 7 students.

4.3 *Subtraction*

Page 136

1. **a.** 5 − (+9)
   = 5 + (−9) Rewrite the subtraction as the addition of the opposite.
   = −4

   b. 7 − 16
   = 7 − (+16) This subtracts positive 16.
   = 7 + (−16) Rewrite the subtraction as addition of the opposite.
   = −9

c. $3-(-6)=3+(+6)=9$

d. $0-8=0-(+8)=0+(-8)=-8$

e. $-4-(-12)=-4+(+12)=8$

f. $-1-1=-1-(+1)=-1+(-1)=-2$

2. a. $12+7-2=19-2=17$

b. $12-7+2=5+2=7$

c. $12-7-2=5-2=3$

d. $-12-7+2=-19+2=-17$

e. $-12-7-2=-19-2=-21$

f. $-2-7-12=-9-12=-21$

Page 136

a.

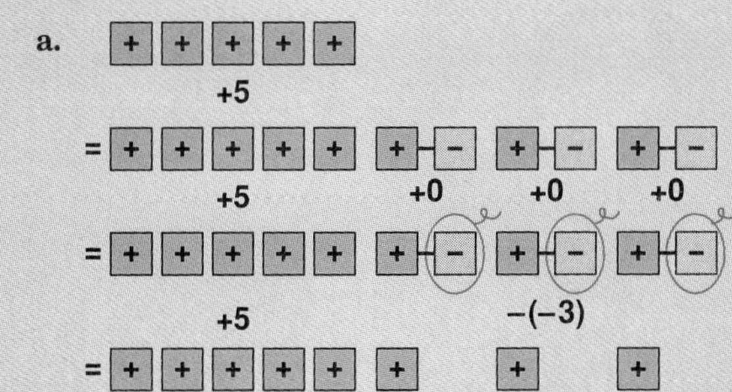

b. 5 [−] 3 [+/−] [=] Display: 8.

Page 136

	$(-3)+(-2)-(-1)$	Begin with the sum at the left:
=	$-5 \quad -(-1)$	$(-3)+(-2)=-5$
=	-4	Now do the subtraction, which is equivalent to adding an opposite. $-5-(-1)=-5+1=-4$

4.4 Multiplication and Division

Page 143

1. a. 15 b. 28
c. −48 d. −99
e. 0 f. −72

2. a. −4 b. 7
c. −6 d. 12
e. 0 f. cannot divide by 0

3. a. $4(-6)\cdot 3=-24\cdot 3=-72$

b. $-84\div(-2)\div 3=42\div 3=14$

c. $48-6\div(-2)=48-(-3)=48+(+3)=51$

4.5 Coordinate Pairs

Page 151

1. $A(1,3)$ $B(0,-4)$ $C(1,-2)$
$D(-4,0)$ $E(-3,-4)$ $F(-4,2)$

2.

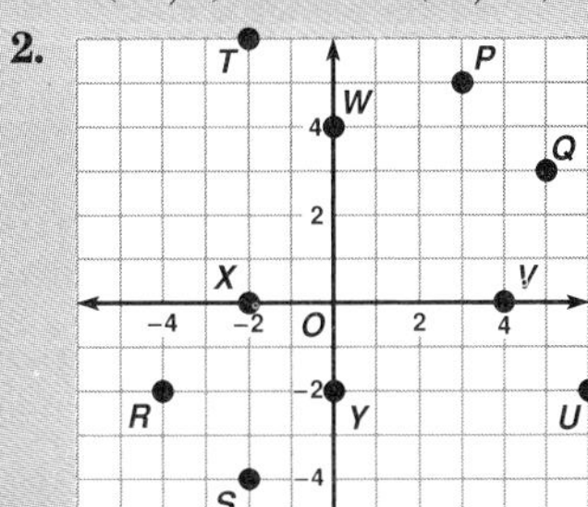

3.

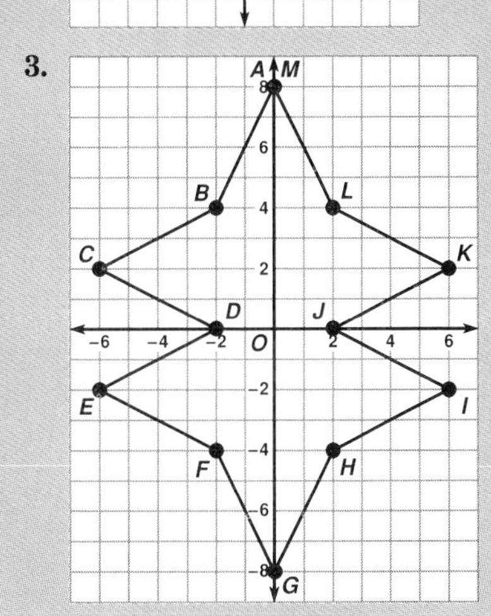

Page 152

a.

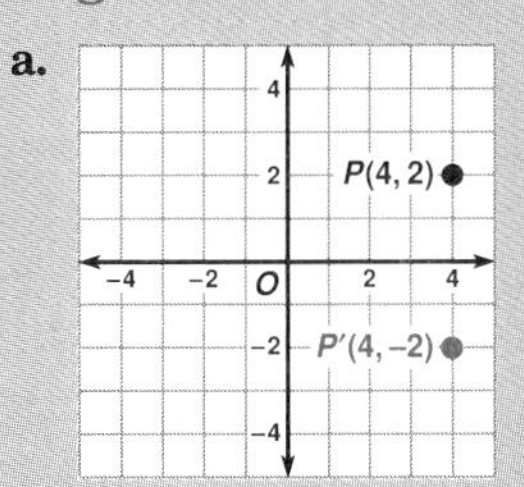

The image is $P'(4,-2)$.

b.

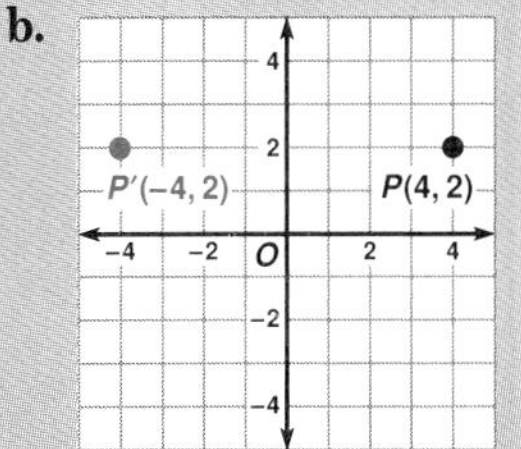

The image is $P'(-4,2)$.

c.

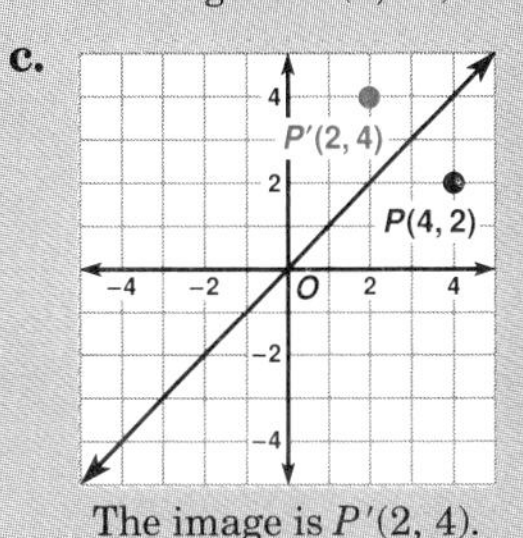

The image is $P'(2,4)$.

d.

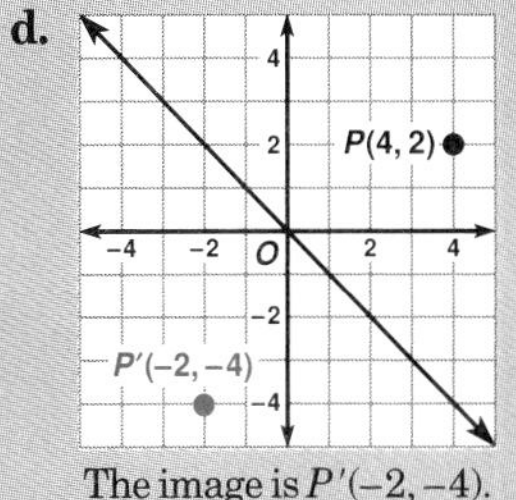

The image is $P'(-2,-4)$.

Using Symbols 5

ACTIVITY

Why is it often more convenient to use symbols instead of words?

Tell the meaning of the road symbols pictured.

Give examples of other symbols used in daily life.

Create some symbols to represent familiar situations.

5.1 Evaluating a Variable Expression

Luz has a job after school at which she earns \$6 per hour. How would you calculate the amount she earned on a day that she worked 3 hours? She earned 6×3, or \$18.

The number of hours Luz works *varies* from day to day. To write a rule that shows the calculation for the amount of money she earns in a day, it is convenient to use a symbol, called a ***variable***, to represent the number of hours worked.

Let h = the number of hours worked.

Then $6 \cdot h$ = the amount of money earned.

How could you use the expression $6 \cdot h$ to find how much money Luz earned on Saturday when she worked 7 hours?

Substitute the number 7 for the variable h in the expression.

When $h = 7$, the value of $6 \cdot h$ is $6 \cdot 7$, or 42.

Example 1 Find the value of each variable expression for the given value of the variable.

Variable Expression	*Value of Variable*	*Calculation*
$7t$	$t = 3$	$7 \cdot 3 = 21$
$x - 6$	$x = 11$	$11 - 6 = 5$
$y + 3$	$y = -5$	$-5 + 3 = -2$
$b \div 7$	$b = -63$	$-63 \div 7 = -9$
$3(m - 6)$	$m = 4$	$3(4 - 6) = 3(-2) = -6$

Luz deposits her earnings into a savings account. In this account, she also deposits all the money she receives as gifts. If $6h$ represents the money Luz earns and g represents the money she gets as gifts, what expression represents the amount of a deposit to the account?

$6h + g$ = the amount in a deposit.

If Luz worked 14 hours this week and received \$10 as a gift from her grandmother, how much does she have to deposit?

When $h = 14$ and $g = 10$, the value of $6h + g$ is $6 \times 14 + 10$, or $84 + 10$, or 94.
Luz has \$94 to deposit.

Evaluating a Variable Expression

1. Replace each variable by its value.
2. Find the numerical value of the expression by following the order of operations.

Example 2 Find the value of each variable expression for the given values of the variables.

Variable Expression	*Value of Variable*	*Calculation*
$3c + 4d$	$c = 3, d = 4$	$3(3) + 4(4)$
		$= 9 + 16$
		$= 25$
$2x^2 - y$	$x = 3, y = 5$	$2(3)^2 - 5$
		$= 2(9) - 5$
		$= 18 - 5$
		$= 13$

Try These *(For solutions, see page 187.)*

1. $2x + 3$ when $x = 5$
2. $10 - 3n$ when $n = 2$
3. $t^2 - 5$ when $t = -3$
4. $3p^2 + p - 1$ when $p = -1$
5. $2(x + w)$ when $x = 7$ and $w = 4$
6. $9y - 5z$ when $y = \frac{2}{3}$ and $z = -2$
7. $3x + y$ when $x = 0.5$ and $y = 2$
8. $a(b - c)$ when $a = -1$, $b = -2$, and $c = 3$

SELF-CHECK: SECTION 5.1

1. Since $5x$ is a rule for a calculation that can be used for a variety of specific values of x, then $5x$ is called a _?_.

   Answer: variable expression

2. How many variables are in the expression $5x$?

   Answer: one variable, x
   (5 is called a ***constant*** because its value never changes.)

3. How would you find the value of $5x$ when x is 8?

   Answer: Substitute 8 for x and multiply 5(8).

4. Name the variables in the expression $w + 5x$.

   Answer: There are two variables, w and x.

5. How would you find the value of $w + 5x$ when w is 3 and x is 2?

   Answer: Substitute 3 for w and 2 for x, and follow the order of operations.
   $3 + 5(2) = 3 + 10 = 13$

EXERCISES: 5.1

1. Evaluate each expression.

a. $3h + 4$ when $h = 9$

b. $20 - 3t$ when $t = 2$

c. $-p$ when $p = -7$

d. $4 + 8q$ when $q = -1$

e. $a^2 + 4$ when $a = 3$

f. $\frac{1}{2}x + 8$ when $x = 12$

g. $x^2 + 3x$ when $x = -2$

h. $3t^2 + 5t + 10$ when $t = 2$

2. Evaluate each expression when $k = 4$.

a. $3 + k^2$

b. $(3 + k)^2$

c. $3k^2$

d. $(3k)^2$

e. $(k - 3)^2$

f. $(3 - k)^2$

g. $-3k^2$

h. $(-3k)^2$

3. Evaluate each expression.

a. $6x + 4y$ when $x = 3$ and $y = 2$

b. $5mn$ when $m = -2$ and $n = -3$

c. $(r + s)^2$ when $r = 1$ and $s = 7$

d. $(x - y)^2$ when $x = 8$ and $y = 5$

e. $4rs$ when $r = \frac{3}{4}$ and $s = 2$

f. $10a + 20b$ when $a = 0.2$ and $b = 0.1$

g. $ab + ac$ when $a = \frac{2}{5}$, $b = 5$, and $c = \frac{5}{2}$

4. Find the area of the figure shown by using the rule given below the diagram and the numbers in the diagram.

a.

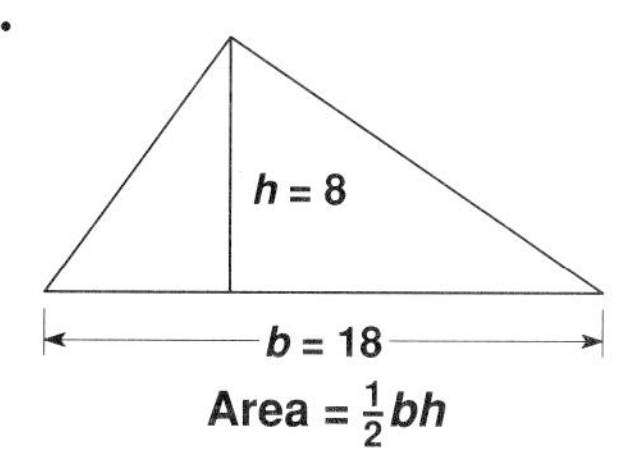

Area = $\frac{1}{2}bh$

b.

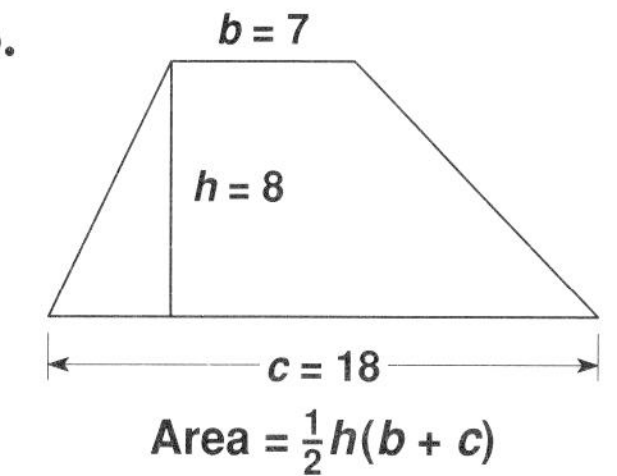

Area = $\frac{1}{2}h(b + c)$

5. To evaluate each of the expressions that follow:

(1) Estimate a value by rounding.

(2) Write a key sequence for an entry-order calculator.

(3) Try your key sequence on a calculator and write the display.
Is your estimate close to the calculator result?

a. ab when $a = 3.12$ and $b = 2.9$

b. $18p - q$ when $p = 7.95$ and $q = 35$

c. $2x + 4y$ when $x = -3.8$ and $y = 1.9$

d. $x^2 - 4x + 4.75$ when $x = 2.9$

6. Read each description to determine if the calculation was done correctly. If you do not agree with the method, write how you would do the calculation.

a. To evaluate $y(3 + 5)$ for $y = 4$, Norma substituted 4 for y. Then she added 3 and 5, and multiplied the sum by 4.

b. To evaluate $3 + 4n$ for $n = 2$, Carlos substituted 2 for n. Then he added 3 and 4, and multiplied the sum by 2.

c. To evaluate $4x^2$ for $x = 3$, Yolanda substituted 3 for x. Then she multiplied 4 by 3, and squared the result.

7. Use the strategy of guessing and checking to answer these questions.

a. Find the value of x that will make each sentence true.

(1) $\frac{1}{2}x = 12$ **(2)** $2 + 3x = 17$ **(3)** $4x + 1 = 3x - 2$

b. Find the two values of x that will make each sentence true.

(1) $(x - 5)^2 = 4$ **(2)** $x^2 - 10x + 21 = 0$

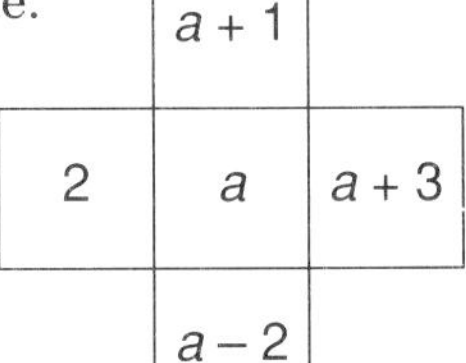

c. Find the value of a for which the sum in the row will equal the sum in the column.

8. Select the choice that best answers the question.

a. Which of the following always gives an even number when any natural number is substituted for n?

(1) $2 + n$ (2) $2 - n$ (3) $2 \div n$ (4) $2n$

b. Which of the following always gives an odd number when any natural number is substituted for n?

(1) $n - 1$ (2) $n^2 + 1$ (3) $2n + 1$ (4) $3n + 1$

c. If x and y are integers greater than 1, which symbol can replace ? to make the following sentence true?

$xy + 1 \ ? \ x(y + 1)$

(1) $>$ (2) $<$ (3) $=$ (4) $+$

9. To decode this message:

$\frac{?}{11}\ \frac{?}{5.5}\ \frac{?}{100}\ \frac{?}{11}\ \frac{?}{-14}\ \frac{?}{100}\ \frac{?}{5.5}\ \frac{?}{16}\ \frac{?}{4.5}\ \frac{?}{20}\ \frac{?}{16.69}$

$\frac{?}{5.5}\ \frac{?}{20}\ \frac{?}{45}\quad \frac{?}{-90}\ \frac{?}{5.5}\ \frac{?}{4}\ \frac{?}{0.7}\ \frac{?}{-15.6}$

$\frac{?}{16}\ \frac{?}{4.5}\ \frac{?}{4.5}\ \frac{?}{100}\ \frac{?}{16.69}$

Use a calculator to evaluate each expression below. Locate your answer in the code. Each time the answer appears in the code, write the letter that is next to the expression in the blank above the answer. Use a separate sheet of paper onto which you have copied the code.

To evaluate the following expressions, use:

$x = 2.5$, $y = -3$, and $z = 1.2$

A $x - y$ **O** $xy + 10z$

C $2x + 5z$ **R** $8x + 2y + 5z$

D $x + y + z$ **S** $x^2 + y^2 + z^2$

E $-6xy$ **T** $10x + 5(y + z)$

H $10xyz$ **U** $x + xy + xyz$

L $16x^2$ **Y** $x(y + z) + y(x + z)$

N $y^2 - 2x$

EXPLORATIONS

a. You can form the set of whole numbers by substituting values for a and b into the expression $3a + 8b$. Any integers may be used to replace a, but only 0, 1, or 2 may be used for b. For example, the following table shows how to form the whole numbers 0 through 3 from the expression $3a + 8b$.

Number to Form	Values for a and b		$3a + 8b$	Calculation
0	$a = 0$	$b = 0$	$3(0) + 8(0)$	$0 + 0 = 0$
1	$a = -5$	$b = 2$	$3(-5) + 8(2)$	$-15 + 16 = 1$
2	$a = -2$	$b = 1$	$3(-2) + 8(1)$	$-6 + 8 = 2$
3	$a = 1$	$b = 0$	$3(1) + 8(0)$	$3 + 0 = 3$

(1) Continue the table by finding values for a and b to form the numbers 4 through 20.
(2) Consider all the successive a-values used to form the numbers 0 through 20. Describe a pattern.
(3) Consider all the successive b-values used to form the numbers 0 through 20. Describe a pattern.

b. Any three successive whole numbers can also be written in the forms $3n$, $3n + 1$, and $3n + 2$. Verify that every whole number is of one of these forms by writing the n-values needed to obtain the whole numbers 0 through 20.

For example, if $n = 4$: $12 = 3n$, or $3(4)$
$13 = 3n + 1$, or $3(4) + 1$
$14 = 3n + 2$, or $3(4) + 2$

c. Since every whole number can be formed by substituting values for the variables

$$\text{in } 3a + 8b \qquad \text{or} \qquad \text{in one of } \begin{cases} 3n \\ 3n + 1 \\ 3n + 2 \end{cases}$$

it seems reasonable that there should be a relationship between the variables a, b, and n. The following table shows such a relationship for numbers of the form $3n + 1$.

$3n + 1$	n	$a = n - 5$	$b = 2$	$3a + 8b$
$10 = 3(3) + 1$	3	$3 - 5 = -2$	2	$3(-2) + 8(2) = 10$
$25 = 3(8) + 1$	8	$8 - 5 = 3$	2	$3(3) + 8(2) = 25$

Prepare tables for $3n$ and for $3n + 2$, with the headings shown. Remember that only 0, 1, or 2 may be used for b. Find a relationship between a and n.

(1)

$3n$	n	$a = ?$	$b = ?$	$3a + 8b$

(2)

$3n + 2$	n	$a = ?$	$b = ?$	$3a + 8b$

5.2 Simplifying a Variable Expression

Combining Like Terms

Carmen bought 6 apples, 8 bananas, and 4 pears. If she already had 2 apples and 3 bananas at home, how many pieces of each fruit does she now have for her family?

Let's represent this situation using symbols, with a = apples, b = bananas, and p = pears.

In symbols, what did Carmen buy?	$6a + 8b + 4p$
What did Carmen already have?	$2a + 3b$
What was the fruit total?	$8a + 11b + 4p$

To get the total, why was $6a$ added to $2a$ and not, say, to $3b$? Just as the fruits that are alike can be combined into a simpler total, terms that are alike in the variable expression can be combined.

It is easy to recognize that bananas are alike, and are different from pears. When working with symbols, you must be sure to recognize terms that are like, and know when they are different (unlike).

Like Terms	*Unlike Terms*
$6a$, $2a$	$6a$, $2b$
$6ab$, $2ab$	$6ab$, $2a$, $2b$
$3x^2$, $5x^2$	$3x^2$, $5x$
5, 28	$5a$, 28

Like terms have exactly the *same letters.*
Like terms have the *same powers.*
Terms that are just numbers are like.

Example 1 In each variable expression, tell which are like terms.

	Variable Expression	*Like Terms*	
a.	$6a + 8b + 4p + 2a + 3b$	$6a$, $2a$	$8b$, $3b$
b.	$5x^2 + 3x - 7 + 2x^2 + x$	$5x^2$, $2x^2$	$3x$, x
c.	$-3ad + 5 + 8ad + 7$	$-3ad$, $8ad$	5, 7
d.	$2y + y^2 - xy$	none	

When you added 6 apples to 2 apples, not only did you recognize that these are like terms, you also focused on the number of each:

6 apples + 2 apples = (6 + 2) apples or 8 apples

In symbols: $6a + 2a = (6 + 2)a$ or $8a$

To add like terms, add the numbers in front of the variable.

The number in front of a variable is called the ***coefficient***.

coefficient → $7x^2$ ← exponent; ↖ variable

If no number is written in front of a variable, the coefficient is 1. For example: x means $1x$

To combine like terms by addition or subtraction, combine the coefficients.

Example 2 Simplify each variable expression by combining like terms.

a. $10c - 7c = (10 - 7)c$ or $3c$

b. $8x^2 + 3x^2 = (8 + 3)x^2$ or $11x^2$

c. $6y + 5y - y = (6 + 5 - 1)y$ or $10y$

d. $2x + 3y - 8x + 4y = (2 - 8)x + (3 + 4)y = -6x + 7y$

e. $-6x + 2 - 8 + x = (-6 + 1)x + (2 - 8) = -5x - 6$

Try These *(For solutions, see page 187.)*

Write each expression in simplest form.

1. $2n + 3n$	**2.** $4x - x$	**3.** $7t^2 + 3t^2$
4. $p - 2p$	**5.** $7x + y - 3x$	**6.** $-2ab + b + 4ab + a$

Removing Parentheses

Would you rather write 8 apples or (6 + 2) apples? You would probably prefer the simpler way, especially if you had to write the term several times.

Similarly, in ***Algebra***, the mathematical language that uses symbols, it is more convenient to work with simpler written expressions. You have already seen that combining like terms makes a variable expression simpler. $6a + 8b + 4p + 2a + 3b = 8a + 11b + 4p$

What can be done to an expression like $3(n + 1) + n$ to make it simpler?
Why can't we immediately combine like terms?

To visualize the meaning of the expression, let's use chips that can be moved around. Let [long chip] represent n and [small chip] represent 1.

Then $3(n + 1)$ is 3 groups of $n + 1$.

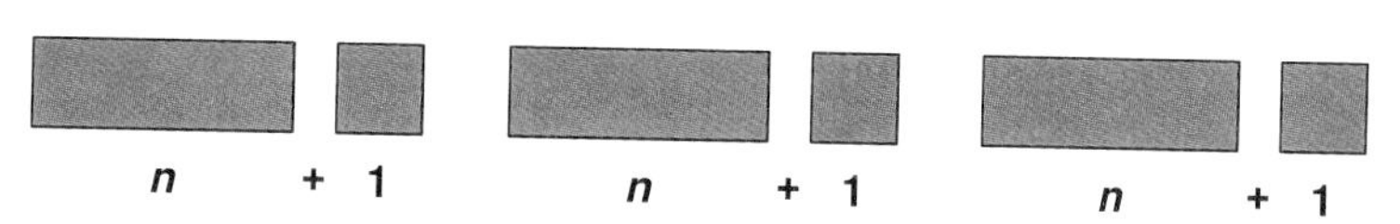

Regroup the chips to see that $3(n + 1) = 3n + 3$.

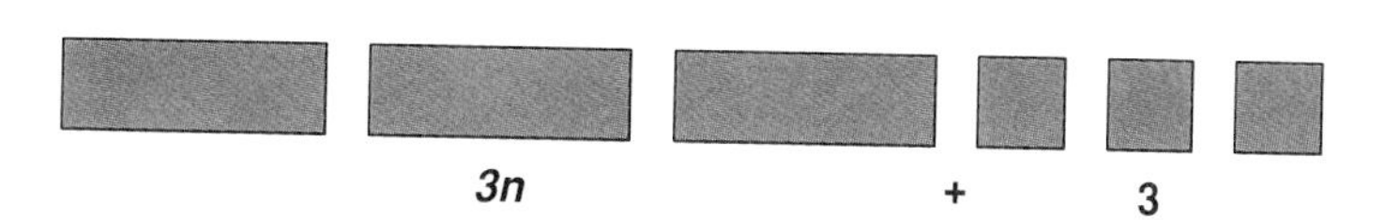

Add n to complete the given expression, $3(n + 1) + n$.

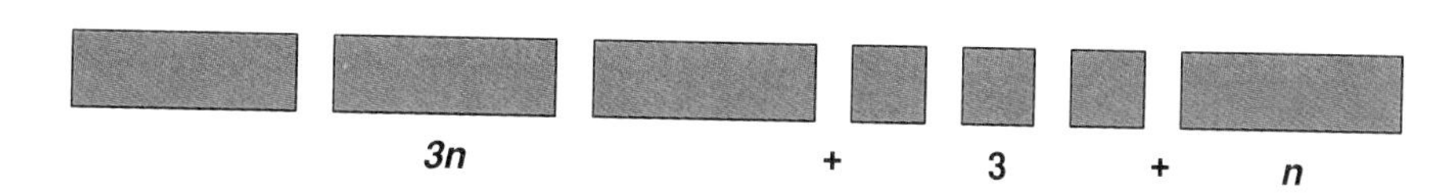

Combine like chips.

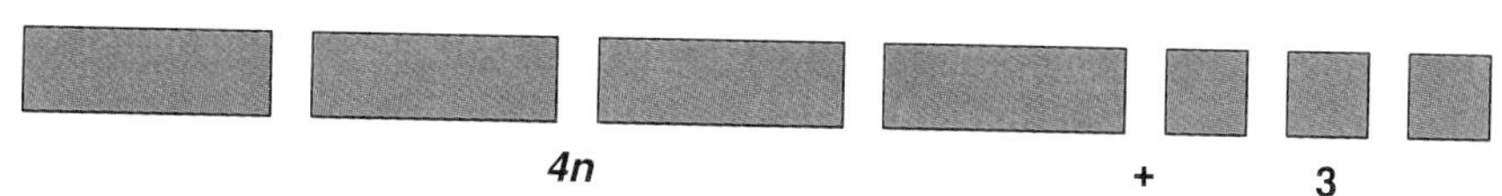

Summarizing the result of the model, we note that like terms cannot be combined until the parentheses are removed.

$3(n + 1) + n$	3 multiplies n and 1.
$= 3n + 3 + n$	Distribute the 3 to remove parentheses.
$= 4n + 3$	Combine like terms.

To Write an Expression in Simplest Form

1. Multiply to remove parentheses.

2. Combine like terms.

Example 3 Simplify the expression.

$3x + 4(x + d) + 6d$
$= 3x + 4x + 4d + 6d$ — Multiply to remove parentheses.
$= (3 + 4)x + (4 + 6)d$ — Combine like terms by adding coefficients.
$= 7x + 10d$

Try These *(For solutions, see page 187.)*

Write each expression in simplest form.

1. $2(x + y) - x$

2. $3y + 4(2y + 1)$

SELF-CHECK: SECTION 5.2

1. Terms that have exactly the same variable are called __?__.
Answer: like terms

2. In the expression $5x + 3y + 6x$, which are the like terms?
Answer: $5x$ and $6x$

3. Are $5x^2$ and $6x$ like terms? Explain.
Answer: No, like terms have the same power of the variable.

4. In the expression $5 + 9a + 7 + 3b$, which are the like terms?
Answer: 5 and 7 are like since they are both just numbers.

5. What is a *coefficient*?
Answer: the number multiplying the variable

6. In the expression $3x + y$, what is the coefficient of y?
Answer: 1

7. How would you combine the like terms $7x$ and $-3x$?
Answer: Combine the coefficients $(7 - 3)x$ or $4x$.

8. How would you combine the like terms $7x^2$ and $-3x^2$?
Answer: Combine the coefficients $(7 - 3)x^2$ or $4x^2$

9. Before you can combine the like terms in $x + 3(x - 2)$, what must you do? How?
Answer: Remove parentheses by multiplication

10. In the expression $5 + 4(x + 2)$, which terms are to be multiplied by 4?
Answer: x and 2

EXERCISES: SECTION 5.2

1. Simplify each variable expression by combining like terms, as possible.

a. $6x + 2x$ **b.** $5y - 3y$ **c.** $3m - m$ **d.** $-6xy + 10xy$

e. $5y^2 - 3y^2$ **f.** $3m^2 - m^2$ **g.** $3.5x - 2x$ **h.** $4\frac{1}{2}y - \frac{1}{2}y$

i. $6.1x^2 - x^2$ **j.** $4x + 7x + 2x$ **k.** $4x + 7x + 2y$ **l.** $4x + 7y + 2z$

m. $4x^2 + 7x^2 + 2x^2$ **n.** $4x^2 + 7x^2 + 2y^2$ **o.** $4x^2 + 7x + 2y^2$

p. $16g + 5 - 11g + 8$ **q.** $13x + 12 + 12x - 9$ **r.** $x^2 + 7x + 12 - x + 3x^2$

s. $12de + 6 - 4de - de - 2$ **t.** $-2a + b - 3a - b$ **u.** $7xy + 4x - 2y$

2. Write each representation in simplest form.

a. Max has $4x$ compact discs and Rosanne has $7x - 1$. Represent the total number of discs.

b. John spent n hours on his homework and Reba spent $(4 - 2n)$ hours on hers. Represent the total number of hours.

c. Bojan had $(8y - 2)$ books in his collection. When he moved, he donated $4y$ books to the library. Represent the remaining number of books.

3. Express the perimeter of each polygon in simplest form. (Recall that the perimeter is the sum of the lengths of the sides of the figure.)

a.

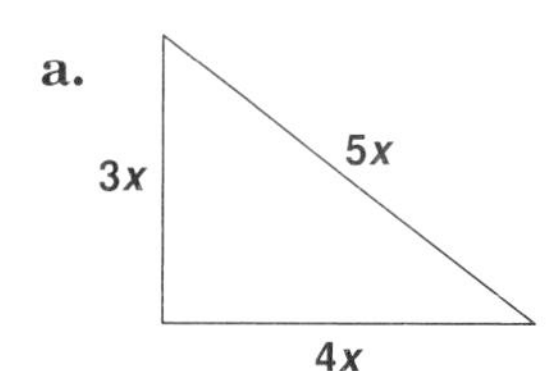

b.

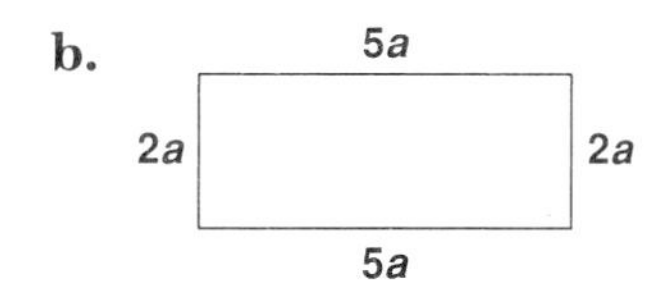

c.

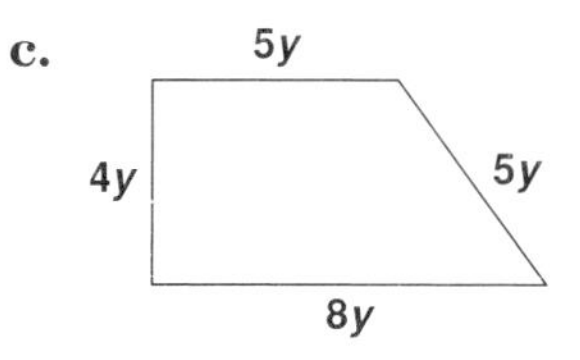

4. Solve each of these problems.

a. Find the missing length of the side of the polygon, given that the perimeter is $68x$.

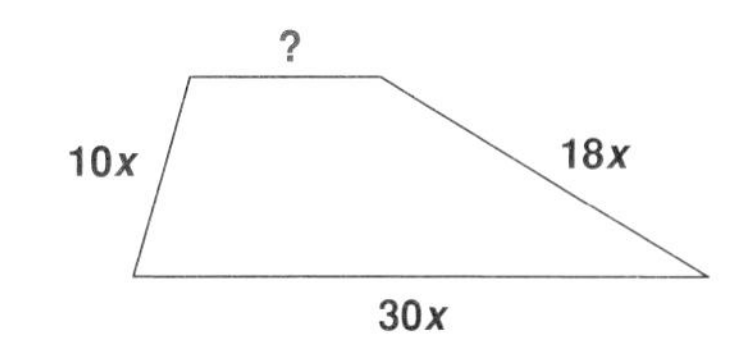

b.

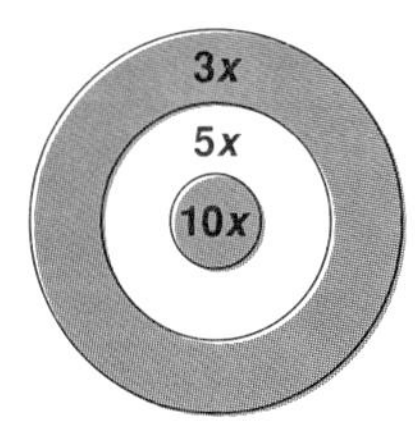

When thrown at this target, a dart that lands in the circled areas receives the scores shown. If a dart lands on a line of the target, $2x$ points are deducted.

(1) Gino threw 6 darts with these results:

in the center, on a line, on a line,
in the outer ring, in the outer ring, on a line

What is his final score?

(2) If Ann's score for 6 darts thrown was $36x$, and one of her darts landed on a line, describe how her other 5 darts could have landed.

(3) Using 6 darts is it possible to get a score of 0? Explain.

5. Do you agree or disagree with each result? Explain.

a. Mike got $12x^4$ as a result when he simplified the expression $4x^2 + 8x^2$.

b. Sara's first step when simplifying $1 + 2(3x + 4)$ resulted in $1 + 6x + 4$.

c. When simplifying $4 - 2(3x - 1)$, Darryl got $4 - 6x - 2$ or $2 - 6x$ but Lila got $4 - 6x + 2$ or $6 - 6x$.

6. Using [long tile] to represent n and [small square tile] to represent 1, show how to simplify each expression.

a. $2(3n + 1) + 1$ **b.** $3(n + 3) + 3$

7. Simplify each expression.

a. $4(2k + 3) + 5$ **b.** $2(3m + 1) - 2$ **c.** $3(2t - 1) + 6$

d. $4(p - 2) - 8$ **e.** $5x + 7(x + 3)$ **f.** $5x + 7(x - 3)$

g. $7x + 3(5 + x)$ **h.** $5x + 7(3 - x)$ **i.** $\frac{3}{4}(8a - 12) + 1$

j. $7t + \frac{1}{2}(10t + 8)$ **k.** $6(0.5r - 1) + 2r$ **l.** $\frac{1}{2}(4x + 6y) + \frac{1}{3}(6y - 9x)$

8. In each expression, combine like terms, then evaluate the resulting expression when $x = 2$, $y = -5$, and $z = 0$.

a. $3x - y - x + 6y$ **b.** $12xy - 16z^2 - 10xy + z$

c. $5(x^2 - y) + 8y - 2x^2$ **d.** $4xy - 2yz + 14xz - 6xy$

9. To decode this message:

$\frac{?}{5x}\ \frac{?}{2x}\ \frac{?}{-5x^2}\ \frac{?}{5x^2}\ \frac{?}{4x + 3}\ \frac{?}{2x + 3}\ \frac{?}{5x}\quad \frac{?}{-x}\ \frac{?}{2x + 2}\quad \frac{?}{6x}\ \frac{?}{0}\ \frac{?}{2 - 2x}$

Simplify each expression below. Locate your answer in the code. Each time the answer appears in the code, write the letter that is next to the expression in the blank above the answer. Use a separate sheet of paper onto which you have copied the code.

A $2x + 3x$

B $x + 3(x + 1)$

E $7x^2 - 2x^2$

F $3(2x + 1) - 3$

G $2x^2 - 7x^2$

I $x - 2x$

L $x + 1 + x - 1$

N $1 - x - x + 1$

R $3(x + 1) - x$

S $x + 1 + x + 1$

U $3x - 4x + x$

EXPLORATIONS

More models are possible using chips with signs. Let's call them *algebra chips*.

Let 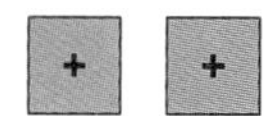 represent n and  represent 1, Then zero pairs are: 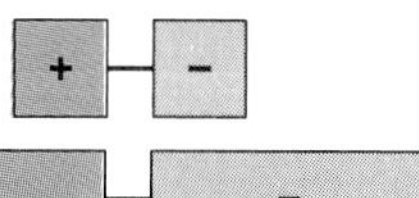

 represent $-n$ and  represent -1. and 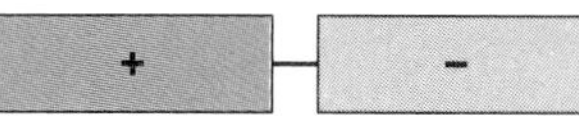 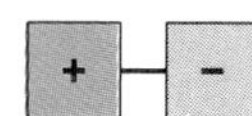

Example: Model $3 - 2(n + 1)$.

Examining the expression shows that we are to subtract 2 groups of $n + 1$ from 3 chips. In order to subtract, zero pairs must be included in the original layout.

Lay out 3 chips. Lay out 2 sets of zero pairs for n and 2 sets for 1.

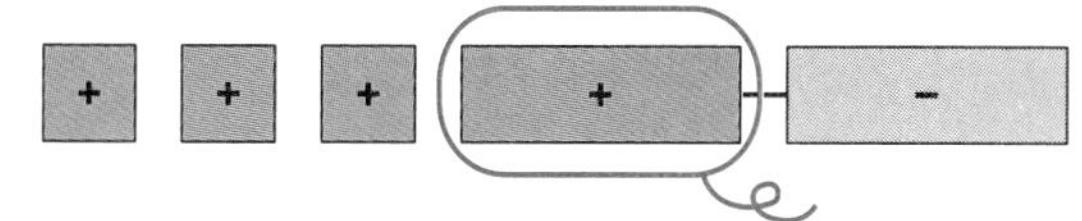
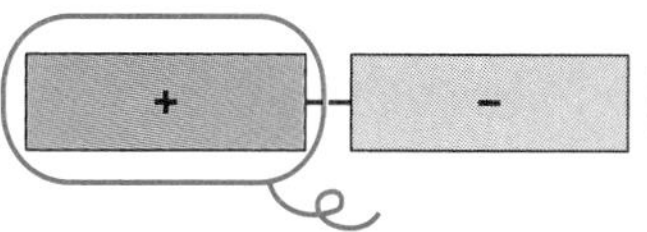
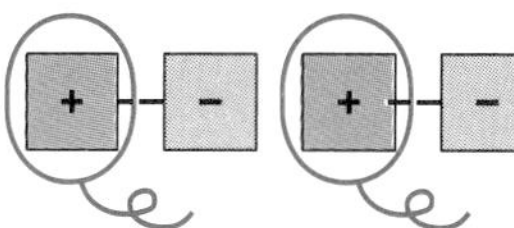

From this arrangement, subtract 2 sets of $n + 1$, or $2n + 2$.

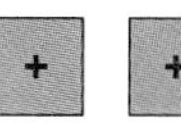

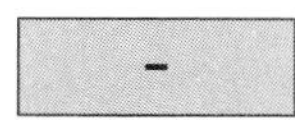

These are the remaining chips.

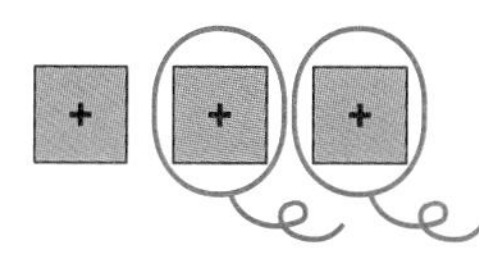
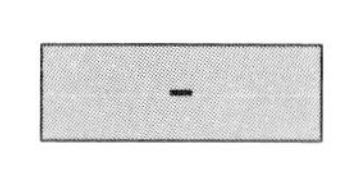
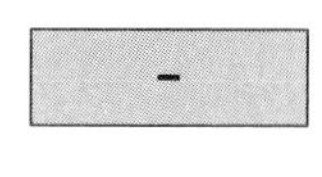
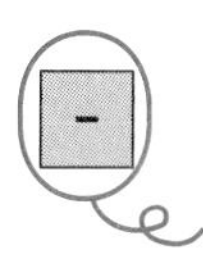
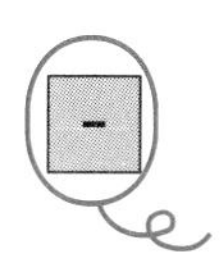

Remove the equivalent of zero pairs.

These are the remaining chips: 1 and $2(-n)$, or $1 - 2n$. The model shows: $3 - 2(n + 1) = 1 - 2n$

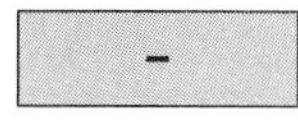
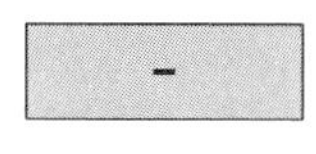

Let's simplify the original expression according to the operations shown, and match the result against the model.

$3 - 2(n + 1)$	
$= 3 + (-2)(n + 1)$	Subtraction is the same as addition of an opposite.
$= 3 + (-2)n + (-2)1$	Distribute the multiplier, -2.
$= 3 + (-2n) + (-2)$	Multiply.
$= 1 + (-2n)$	Combine like terms: $3 + (-2) = 1$
$= 1 - 2n$	Addition of an opposite is the same as subtraction.
	This result does match the result of the model.

Simplify each expression by: **(1)** using algebra chips **(2)** using the operations of algebra

a. $3 - 2(n + 2)$ **b.** $5 - 3(2n - 1)$ **c.** $3n - 4(n - 1)$

5.3 Translating Between Words and Symbols

Ensign Robinson is a signal corpsman in the U.S. Navy. His job is to translate between words and radio signals.

In this section, you will learn how to translate back and forth between words and the symbols of Algebra.

You already know many of these translations. For example, what symbol would you use for the word *add*? for the word *six*? for the word *equals*?

There are words that suggest the use of a particular symbol. This table shows some words that can be translated into operation symbols.

+	−	×	÷
add	subtract	multiply	divide
plus	minus	times	divided by
sum	difference	product	quotient
more	less	twice	half
increase	decrease	triple	into
total	fewer	of	share equally
			part of

Example 1 Translate each word phrase into a numerical expression.

	Word Phrase	*Numerical Expression*
a.	the sum of two and six	$2 + 6$
b.	twelve decreased by seven	$12 - 7$
c.	twice fifty	2×50
d.	ten divided by negative two	$10 \div (-2)$ or $\frac{10}{-2}$
e.	seven more than 10	$10 + 7$
f.	seven less than 10	$10 - 7$

When a word phrase refers to an *unknown* number, a variable is used to represent that number. You can use any letter for the variable. The letters n and x are often used.

Example 2 Translate each word phrase into an algebraic expression.

	Word Phrase	*Algebraic Expression*
a.	5 more than three times a number	$3n + 5$
b.	2 less than four times a number	$4x - 2$
c.	twice the sum of a number and 6	$2(y + 6)$
d.	half the difference of 20 and a number	$\frac{1}{2}(20 - n)$

In the English language, a sentence contains a *verb* that joins phrases.

Ian's father is three years older than his mother.

(Ian's father = phrase; is = verb; three years older than his mother = phrase)

In the language of Algebra, expressions can be joined by the equal sign to form a sentence. When used in this way, the symbol = represents the verb *is*, and a sentence called an ***equation*** is formed.

Five more than a number is ten.

$$x + 5 = 10$$

Example 3 Translate each sentence into an equation.

Let n = the unknown number.

a. Four times a number is 24.

$$4n = 24$$

b. Seven less than a number is 13.

$$n - 7 = 13$$

c. Three times the sum of a number and 18 is 60.

$$3(n + 18) = 60$$

The symbol = is used to represent other forms of the verb *is*, such as *are*, *were*, *will be*.
The symbol = is also used to represent different forms of verbs like *have*.

If Sue bought 4 more CD's, she would have 50 CD's.

$c + 4 = 50$ (where c = the number of CD's Sue has now)

Translating a Word Sentence Into an Equation

1. **Representing the unknown number by a variable, write an algebraic expression using the variable.**
2. **Using the symbol = to represent forms of verbs such as *is* and *have*, write an equation that relates the algebraic expression to the given numbers.**

Example 4 Select the equation that best represents the situation.

a. In 3 hours, the time will be 6 P.M. What time is it now?

Let t = the time now.

(1) $3t = 6$ (2) $t + 3 = 6$ (3) $6 + 3 = t$ (4) $6(3) = t$

Answer: (2)

b. If this wedge of cheese is shared equally by 3 people, each will have 4 ounces. How much does the wedge weigh?

Let w = the weight of the wedge of cheese.

(1) $\frac{w}{3} = 4$ (2) $\frac{3}{w} = 4$ (3) $w - 3 = 4$ (4) $3w = 4$

Answer: (1)

Example 5 Translate each situation into an equation. Tell what the variable represents.

a. If Maia decreased the weight of her backpack by 5 pounds, it would weigh 25 pounds. What is the weight now?

Let w = the weight now. $w - 5 = 25$

b. Seven years from now, Roy will be 35 years old. How old is he now?

Let a = Roy's age now. $a + 7 = 35$

Try These *(For solutions, see page 187.)*

1. Write each phrase as an algebraic expression.
Let n represent the unknown number.

a. a number increased by seven
b. fifty less than a number
c. seventeen divided by the sum of nine and a number
d. twice the product of a number and twelve
e. one-third the sum of fifteen and a number

2. Write each sentence as an equation. Tell what the variable represents.

a. A number increased by 7 is 21.
b. A number decreased by 50 is 31.
c. Twenty four divided by the sum of 4 and a number is 3.
d. One-fourth the product of 15 and a number is 11.
e. If Abbie had 16 more baseball cards, she would have 150.
f. If Ron lost 6 pounds, he would weigh 175.
g. Ten years ago, Alex was 15 years old.

SELF-CHECK: SECTION 5.3

1. What symbol represents the word *product*?
Answer: $\times$ or $\cdot$

2. Name some words that the symbol – could represent.
Answer: subtract, minus, difference, less, decrease, fewer

3. How can you write "4 divided by n" in symbols?
Answer: $\frac{4}{n}$ or $4 \div n$

4. Which of these expressions can be used to represent 2 *more than* x: $x + 2$ or $2 + x$? Explain.
Answer: Both $x + 2$ and $2 + x$ are appropriate. Addition is commutative (can be done backwards or forwards).

5. Which of these expressions can be used to represent 2 *less than* x: $2 - x$ or $x - 2$? Explain.
Answer: $x - 2$ For this phrase, you must subtract 2 from x. Subtraction is not commutative.

6. Which symbol represents the verb *is*?
Answer: =

7. An algebraic sentence with the symbol = is called an _?_.
Answer: equation

EXERCISES: SECTION 5.3

1. Write each word phrase as an algebraic expression. Let n represent the unknown number.
   - **a.** 3 more than a number
   - **b.** 3 less than a number
   - **c.** a number decreased by 7
   - **d.** 7 decreased by a number
   - **e.** 4 increased by a number
   - **f.** a number increased by 4
   - **g.** one-half a number
   - **h.** the product of 8 and a number
   - **i.** two-thirds of a number
   - **j.** 4 divided by a number
   - **k.** 3 more than twice a number
   - **l.** 5 less than half a number
   - **m.** 3 times a number decreased by 7
   - **n.** 4 times a number increased by 5
   - **o.** 6 times the sum of 3 and a number
   - **p.** the sum of a number and 6, divided by 3
   - **q.** the difference of a number and 2, divided by the number
   - **r.** 7 more than the square of a number
   - **s.** 7 less than the square of a number
   - **t.** the square of a number increased by 7
   - **u.** the square of a number increased by the number

2. Using the given variable(s), write an algebraic expression for each word phrase.
   - **a.** the total number of books, if there are p paperbacks and h hardcovers
   - **b.** the number of tennis balls remaining, if t of the 8 balls are lost
   - **c.** Wendy's weight, if she weighed p pounds before she gained 3 pounds
   - **d.** the total cost of s stamps that cost 32 cents each
   - **e.** the cost of one light bulb, if 5 bulbs cost d dollars
   - **f.** Dave's age now, if he was y years old 4 years ago

3. If the given variable represents *a number*, write each algebraic expression as a word phrase.
   - **a.** $t + 3$
   - **b.** $m - 4$
   - **c.** $10x$
   - **d.** $\frac{t}{4}$
   - **e.** $\frac{1}{3}y$
   - **f.** $3(x + 2)$
   - **g.** $2p - 9$
   - **h.** $x^2 + 4$

4. Using n to represent *a number*, write each sentence as an equation.
   - **a.** Five more than a number is 3.
   - **b.** The difference between a number and 6 is 8.
   - **c.** The sum of 8 and a number is 12.
   - **d.** A number decreased by 1 is 7.
   - **e.** Four decreased by a number is 10.
   - **f.** Three more than twice a number is 15.
   - **g.** Five less than four times a number is 11.
   - **h.** Twice the sum of 5 and a number is 8.
   - **i.** Three times a number, decreased by 4, is 7.
   - **j.** One-third of a number, increased by 7, is 24.

5. Select the equation that best represents each situation.

a. After Gary gave Peg 3 of his Mexican jumping beans, he had 8 left. How many did he start with?

Let b = the original number of beans.

(1) $3b = 8$ (2) $b = 8 - 3$ (3) $b + 3 = 8$ (4) $b - 3 = 8$

b. If Sue increases the weight load she now works out with by 1.5 pounds, she will be lifting a total of 10 pounds. How many pounds does Sue lift now?

Let p = the number of pounds now.

(1) $1.5p = 10$ (2) $p + 1.5 = 10$ (3) $10 + 1.5 = p$ (4) $p + 10 = 1.5$

c. Two hours ago, the time was 3 P.M. What time is it now?

Let t = the time now.

(1) $2 - t = 3$ (2) $t - 2 = 3$ (3) $t + 2 = 3$ (4) $3 - 2 = t$

d. Half the number of seats in an auditorium are occupied. If there are 50 people sitting, how many seats are there?

Let s = the number of seats.

(1) $s = \frac{1}{2} \cdot 50$ (2) $2s = 50$ (3) $\frac{1}{2}s = 50$ (4) $\frac{1}{2}s = 100$

e. On the last test in Mrs. O's math class, the class was so well prepared that all of the 27 students got either A or B. If there were twice as many B's as A's, how many A's were there?

Let A = the number of A grades.

(1) $A + 2A = 27$ (2) $2A = 27$ (3) $A = 2(27)$ (4) $A(2A) = 27$

6. Translate each situation into an equation. Tell what the variable represents.

a. Four years ago, Samantha was 3 years old. How old is she now?

b. In 5 hours, it will be 11 A.M. What time is it now?

c. After spending half his savings for the month, Ari had $25 left. How much had he saved for the month?

d. With 240 toy soldiers in his collection, Ernest has three times the number that Amy has in her collection. How many toy soldiers does Amy have?

e. At his barbecue, Marco made some beef burgers and twice as many veggie burgers. If there were 15 burgers in all, how many were beef?

f. At North Kennels, two poodles had pups on the same day, for a total of 8 new pups. If there were 2 more pups in Cleo's litter than in Sheba's, how many pups were in Sheba's litter?

g. Stan and Sandy pooled their quarters for the laundromat. Sandy had 3 times as many quarters as Stan. If, together, they had 24 quarters, how many did Stan have?

EXPLORATIONS

Game: Concentration

Here's what 2 players need:

12 pairs of cards containing algebraic expressions. One card of a pair has the expression written in words, and the other card has the same expression written in symbols.

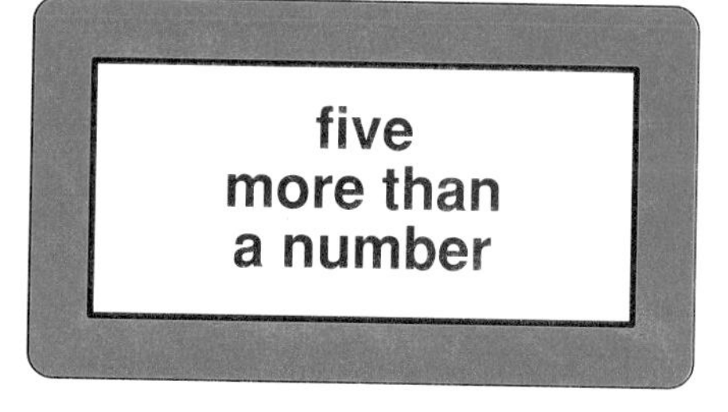

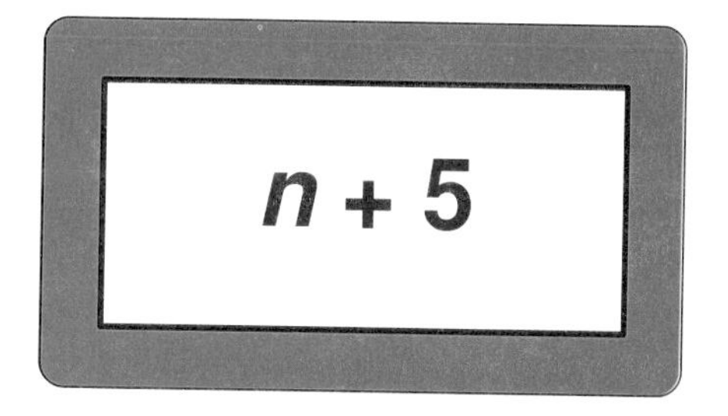

Here's how to play:

Shuffle all 24 cards together, and lay them face down in 4 neat rows.

Player 1 turns any 2 cards face up, for both players to see.

If the cards are a matched pair, Player 1 keeps them, leaving all the other cards in position, and goes again, turning up another 2 cards.

If the cards are not a matched pair, Player 1 replaces the cards face down, and Player 2 goes, turning up any 2 cards.

The players concentrate on the positions of the cards whose faces they have seen, so that when they turn up a card, they can remember where to find its match.

When all the cards have been removed, the winner is the player who has taken the most pairs.

5.4 Open Sentences

Is this sentence true or false? *He was the first president of the United States.*

You cannot tell if the sentence is true or false because it does not say who *he* is. Such a sentence is called an ***open sentence***.

Replace the word *he* by a name to make the open sentence into a true statement. Replace the word *he* to make the open sentence into a false statement.

True Statement: George Washington was the first president of the U. S.

False Statement: Abraham Lincoln was the first president of the U. S.

Example 1 What kinds of sentences are the following equations?

$2 + 4 = 6$	true statement
$2 + 4 = 5$	false statement
$x + 4 = 6$	open sentence

In the open sentence *He was the first president of the U. S.*, only one name, George Washington, could replace the word *he* to make a true statement. How is that open sentence different from the open sentence *He was a president of the U. S.*? Here, you can use any of nearly 40 names to replace the word *he* to make a true statement.

Which of these numbers {1, 2, 3} could you use to replace the variable x in the equation $x + 4 = 6$ to make a true statement? From the ***replacement set*** {1, 2, 3}, only the number 2 will make the open sentence $x + 4 = 6$ a true statement. The numbers that make an open sentence true are called the ***solution set***.

Example 2 What is the solution set of the open sentence $t + 4 = 9$ if the replacement set for t is {3, 4, 5}?

Replace t by each value from the replacement set. Do the calculation. Determine if the result is true or false.

Try $t = 3$ in $t + 4 = 9$
$3 + 4 \stackrel{?}{=} 9$
$7 \stackrel{?}{=} 9$ false

Try $t = 4$ in $t + 4 = 9$
$4 + 4 \stackrel{?}{=} 9$
$8 \stackrel{?}{=} 9$ false

Try $t = 5$ in $t + 4 = 9$
$5 + 4 \stackrel{?}{=} 9$
$9 \stackrel{?}{=} 9$ true

Answer: The solution set is {5}.

Some sentences contain the symbol $<$ or $>$ instead of $=$. Such a sentence is called an ***inequality***, and may also be true, false, or open.

Example 3 What is the solution set for the open sentence $2x > 10$ if the replacement set for x is $\{4, 6, 8\}$?

Replace x by each value from the replacement set. Do the calculation. Determine if the result is true or false.

Try $x = 4$ in $2x > 10$	Try $x = 6$ in $2x > 10$	Try $x = 8$ in $2x > 10$
$2(4) \overset{?}{>} 10$	$2(6) \overset{?}{>} 102$	$(8) \overset{?}{>} 10$
$8 \overset{?}{>} 10$ false	$12 \overset{?}{>} 10$ true	$16 \overset{?}{>} 10$ true

Answer: The solution set is $\{6, 8\}$.

Try These *(For solutions, see page 187.)*

1. Tell whether each sentence is true, false, or open.

a. October is the tenth month of the year.
b. April is the fifth month of the year.
c. It is the third month of the year.
d. $7 + 13 = 20$ **e.** $7 + t = 20$ **f.** $7 + 3 \cdot 2 = 20$
g. $7 + 3 > 20$ **h.** $7 + t < 20$ **i.** $3(7 + 3) > 20$

2. Using $\{-3, 1, 5\}$ as the replacement set, find the solution set for each open sentence.

a. $t - 3 = 2$ **b.** $2x = -6$ **c.** $y > -1$ **d.** $z < 6$

3. Use the given replacement set to find the solution set for each open sentence.

	Open Sentence	*Replacement Set*
a.	It is a day of the week.	{Mon., Jan., Thurs., Fri.}
b.	$a + 9 = 16$	$\{5, 7, 8\}$
c.	$w - 4 = -11$	$\{-7, 7, 15\}$
d.	$2x > 10$	$\{3, 6, 9\}$
e.	$y + 2 < 8$	$\{4, 5, 6, 7\}$
f.	$t^2 = 4$	$\{2, -2\}$
g.	$y^2 = -9$	$\{3, -3\}$

SELF-CHECK: SECTION 5.4

1. Why is this sentence called an *open sentence?*

It is a state of the United States.

Answer: You cannot tell whether the sentence is true or false because of the word *it*.

2. For the open sentence *It is a state of the United States.*:

   a. Using {Alabama, Arkansas, Arizona, Atlanta} as the replacement set, what is the solution set?

      Answer: {Alabama, Arkansas, Arizona}
      Atlanta is a city in the state of Georgia.

   b. How many solutions do you know there are?

      Answer: There are 50 states in the U. S.

3. Give an example of an open sentence that has:

   a. exactly one solution

      Answer: It is the first month of the year.

   b. more than one solution

      Answer: It is a month of the year.

   c. no solutions

      Answer: It is the thirteenth month of the year.

EXERCISES: SECTION 5.4

1. Tell whether each sentence is true, false, or open.

   a. Humans have two feet. **b.** Birds have four feet. **c.** They have two feet.
   d. He has two feet. **e.** $x + 3 = 7$ **f.** $3^2 = 6$
   g. $8 + 2(3 + 1) = 40$ **h.** $(-2)^2 = 4$ **i.** $7 + (2 + 3) = (7 + 2) + 3$
   j. $7 - 2 = 2 - 7$ **k.** $3a = 12$ **l.** $(0.1)^2 = 0.01$
   m. $12 + 0 = 12(1)$ **n.** $\frac{1}{2} + \frac{1}{3} = \frac{2}{5}$ **o.** $5 > 2$
   p. $-5 > -2$ **q.** $x > 2$ **r.** $y < 0$

2. Using {–6, 0, 6} as the replacement set, find the solution set for each open sentence.

   a. $6 + a = 0$ **b.** $m + 6 = 6$ **c.** $x - 2 = -8$ **d.** $2r = 0$
   e. $3t = 18$ **f.** $2(w + 2) = -8$ **g.** $x^2 = 36$ **h.** $2k = 6$
   i. $x > 1$ **j.** $y < 4$ **k.** $z > -10$ **l.** $q < -10$

3. Use the given replacement set to find the solution set for each open sentence.

	Open Sentence	*Replacement Set*		*Open Sentence*	*Replacement Set*
a.	$5 + x = 12$	{7, 0, –7}	**b.**	$a + 4 = 3$	{–1, 1, 7}
c.	$2y = 0$	{–2, 0, 2}	**d.**	$1 = 8 - p$	{–7, 7, 9}
e.	$4d + 2 = 10$	{2, 4, 6}	**f.**	$12 - m = 15$	{–5, –3, 3}
g.	$-3y = -18$	{–15, –6, 6}	**h.**	$3w = 2$	$\{-1, \frac{2}{3}, -\frac{1}{3}\}$
i.	$\frac{1}{2}k = 10$	{5, 10, 20}	**j.**	$0.9 + x = 1$	{0.01, 0.1, 1}
k.	$0.1h = 0.01$	{0.01, 0.1, 1}	**l.**	$2(x + 2) = 10$	{–3, 0, 3}
m.	$3z + 7 = 10$	{–1, 0, 1}	**n.**	$x^2 = 16$	{–8, –4, 4, 8}

4. Use the given replacement sets to find the solution set for each open sentence.

	Open Sentence	*Replacement Set*		*Open Sentence*	*Replacement Set*
a.	$x + 2 > 7$	{3, 7, 9}	**b.**	$y - 3 < 9$	{10, 14, 18}
c.	$p - 7 > 2$	{10, 11, 12}	**d.**	$r + 6 < 10$	{ 2, 3, 4}
e.	$r + 5 < 4$	{–4, –3, –2}	**f.**	$9 + x > 10$	{–2, –1, 0, 1}
g.	$7 - k > 2$	{–10, –6, 4, 6, 10}	**h.**	$10 - g < 12$	{–3, –2, 2, 3}
i.	$2k > 8$	{5, 6, 7}	**j.**	$3m < 9$	{3, 4, 5}
k.	$-2a > 10$	{–6, –5, 5, 6}	**l.**	$-3w > 12$	{–9, –5, 5, 9}
m.	$-t > 7$	{–6, –5, –4}	**n.**	$-x > 0$	{–2, –1, 0, 1, 2}

5. Use the strategy of guessing and checking to find the solution set for each equation.

a. $r + 5 = 7$ **b.** $3t = 12$ **c.** $t - 1 = 8$ **d.** $4y = -20$

e. $8 - w = 10$ **f.** $9 - q = -2$ **g.** $k - 1 = -8$ **h.** $2 - y = -3$

i. $\frac{1}{4}x = 8$ **j.** $\frac{y}{3} = 9$ **k.** $5y + 1 = 16$ **l.** $4z - 2 = 10$

m. $2x + 3 = 11$ **n.** $5b - 1 = 4$ **o.** $2(x + 1) = 10$ **p.** $3(y - 1) = 9$

EXPLORATIONS

1. A variable that appears more than once in an equation must be given the same value throughout the equation.

Example: Using {1, 2} as the replacement set, find the solution set for $3x + 4x = 14$.

Try $x = 1$ in $3x + 4x = 14$

$3(1) + 4(1) \stackrel{?}{=} 14$

$3 + 4 \stackrel{?}{=} 14$

$7 \stackrel{?}{=} 14$ false

Try $x = 2$ in $3x + 4x = 14$

$3(2) + 4(2) \stackrel{?}{=} 14$

$6 + 8 \stackrel{?}{=} 14$

$14 \stackrel{?}{=} 14$ true

Answer: The solution set is {2}.

Use the given replacement set to find the solution set for each equation.

a. $2m + 4 = m - 3$ {–7, 0, 1}

b. $7 - m = m - 7$ {–2, 0, 7}

c. $4g + 1 + g = 2g - 2$ {–1, 0, 1}

d. $x^2 + x - 2 = 0$ {–2, 1, 0}

2. a. Using {0, 1, 2, 3} as the replacement set, find the solution set for each equation.

(1) $2a = 3a - 3$ **(2)** $3b = 2(b + 1)$ **(3)** $x^2 - 5x + 6 = 0$

(4) $p^2 - p = 0$ **(5)** $y^3 - 3y^2 + 2y = 0$ **(6)** $z^3 - 6z^2 + 11z - 6 = 0$

b. Use the results of part *a* to draw a conclusion relating the nature of the equation to the number of members in its solution set.

Things You Should Know After Studying This Chapter

KEY SKILLS

5.1 Evaluate a variable expression.

5.2 Combine like terms.
Remove parentheses.
Write an expression in simplest form.

5.3 Translate from words to algebraic symbols.

5.4 Find the solution set for an open sentence by testing the members of a replacement set.

KEY TERMS

5.1 variable • constant • variable expression • evaluate • substitute

5.2 like terms • unlike terms • coefficient • Algebra

5.3 translate • word phrase • numerical expression • algebraic expression • equation

5.4 open sentence • replacement set • solution set • solution • inequality

CHAPTER REVIEW EXERCISES

1. Evaluate each expression.

a. $5 + 7a$ when $a = 2$

b. $y^2 + 3y$ when $y = 4$

c. $4x + 2y$ when $x = 3$ and $y = -3$

d. $12cd - 8c$ when $c = \frac{1}{2}$ and $d = 5$

2. Find the value of the indicated variable.

a. If $p = 4s$, find p when $s = 3.5$

b. If $S = a + (n - 1)d$, find S when $a = 10$, $n = 8$, and $d = 2$.

c. If $A = \frac{1}{2}h(b + c)$, find A when $h = 6$, $b = 8$, and $c = 5$.

3. Select the like terms in each group of terms.

a. $2x, 2y, 3x, 2xy$

b. $2, 2x, 9$

c. $3x^2, 4y^2, 5x^2, y^2$

4. Simplify each expression by combining like terms.

a. $3p + 5p$

b. $5r + 7 - r$

c. $3x^2 + 2x^2$

d. $5a + 3b - 2a - 3b$

5. Use the diagrams to answer the questions.

a. Express the perimeter in simplest form.

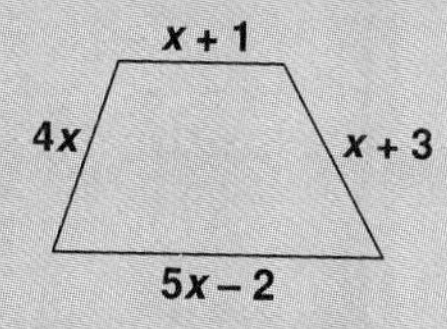

b. If the perimeter is $40m$, find the missing length.

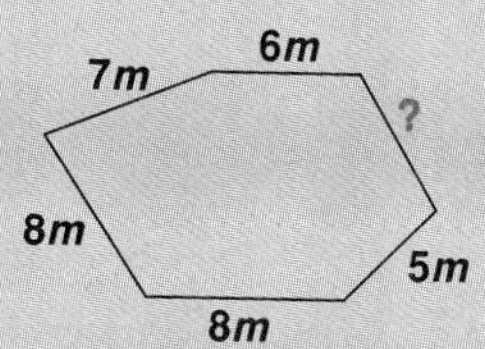

6. Simplify each expression by removing parentheses, and combining like terms.

a. $a + 3(a + 2)$

b. $3y + 5(2y - 2)$

c. $4(x + y) + 8x - 6y$

d. $2(p + m) + 4(p - m)$

7. Write each word phrase as an algebraic expression.

Let n represent the unknown number.

a. ten more than a number

b. three less than a number

c. twice a number, increased by ten

d. one-half the sum of a number and five

8. Using the given variable(s), write an algebraic expression for each word phrase.

a. the total number of students in a class that has b boys and g girls

b. the number of days in w weeks

c. the distance covered in h hours when traveling at 50 miles per hour

9. If the given variable represents *a number*, write each algebraic expression as a word phrase.

a. $n + 5$

b. $y - 4$

c. $2(x + 8)$

(Exercises continue)

CHAPTER REVIEW EXERCISES

10. Using n to represent *a number*, write each sentence as an equation.

a. A number decreased by 3 is 11.

b. The sum of 2.5 and a number is 16.

c. Seventeen increased by 3 times a number is 50.

d. Eight less than twice a number is 10.

11. Select the equation that best represents the situation.

a. After Arnon lost 5 of his marbles to Loni, he had 37 marbles left. How many did he start with?

Let m = the original number of marbles.

(1) $5 - m = 37$

(2) $m - 5 = 37$

(3) $\frac{m}{5} = 37$

(4) $5m = 37$

b. One fourth of the passenger seats in a plane were reserved for economy class. If there were 48 economy seats, how many passenger seats were in the plane?

Let p = the number of passenger seats.

(1) $\frac{1}{4}(48) = p$

(2) $4p = 48$

(3) $\frac{1}{4}p = 48$

(4) $p = \frac{48}{4}$

12. Translate each situation into an equation. Tell what the variable represents.

a. Before she lost 5 pounds, Rose weighed 125. How much does she weigh now?

b. Three hours ago, it was 5 P.M. What time is it now?

c. Raoul has some hardcover books and some softcover, with twice as many soft as hard. If he has 9 books in all, how many are hardcover?

13. Tell whether each sentence is *true*, *false*, or *open*.

a. He is in my class.

b. $d + 8 = 11$

c. $2 + 3(4) = 20$

d. $(-4)(-2) = -8$

e. $x > 5$

f. $|3 - 5| = 2$

14. Using $\{-2, -1,. 0, 1, 2\}$ as the replacement set, find the solution set for each open sentence.

a. $a + 7 = 9$

b. $3x + 5 = 11$

c. $6y + 2 = -4$

d. $q > -1$

e. $5d = 10d$

f. $2r < 6$

g. $2t > 10$

h. $x^2 = 1$

SOLUTIONS FOR TRY THESE

5.1 Evaluating a Variable Expression

Page 164

1. $2x + 3$
$= 2(5) + 3$
$= 10 + 3 = 13$

2. $10 - 3n$
$= 10 - 3(2)$
$= 10 - 6 = 4$

3. $t^2 - 5$
$= (-3)^2 - 5$
$= 9 - 5$
$= 4$

4. $3p^2 + p - 1$
$= 3(-1)^2 + (-1) - 1$
$= 3(1) - 1 - 1$
$= 3 - 1 - 1 = 1$

5. $2(x + w)$
$= 2(7 + 4)$
$= 2(11) = 22$

6. $9y - 5z$
$= 9\left(\frac{2}{3}\right) - 5(-2)$
$= 6 + 10 = 16$

7. $3x + y$
$= 3(0.5) + 2$
$= 1.5 + 2 = 3.5$

8. $a(b - c)$
$= -1(-2 - 3)$
$= -1(-5) = 5$

5.2 Simplifying a Variable Expression

Page 169

1. $2n + 3n = (2 + 3)n = 5n$

2. $4x - x = (4 - 1)x = 3x$

3. $7t^2 + 3t^2 = (7 + 3)t^2$
$= 10t^2$

4. $p - 2p = (1 - 2)p$
$= -1p = -p$

5. $7x + y - 3x = (7 - 3)x + y = 4x + y$

6. $-2ab + b + 4ab + a = (-2 + 4)ab + b + a$
$= 2ab + b + a$

Page 170

1. $2(x + y) - x = 2x + 2y - x$
$= (2 - 1)x + 2y$
$= x + 2y$

2. $3y + 4(2y + 1) = 3y + 8y + 4$
$= (3 + 8)y + 4$
$= 11y + 4$

5.3 Translating Between Words and Symbols

Page 176

1. **a.** $n + 7$ **b.** $n - 50$ **c.** $\frac{17}{9 + n}$
d. $2(12n)$ **e.** $\frac{1}{3}(15 + n)$

2. **a.** Let n = the number.
$n + 7 = 21$
b. Let n = the number.
$n - 50 = 31$
c. Let n = the number.
$\frac{24}{4 + n} = 3$
d. Let n = the number.
$\frac{1}{4}(15n) = 11$
e. Let c = the number of cards now.
$c + 16 = 150$
f. Let p = Ron's weight now.
$p - 6 = 175$
g. Let a = Alex's age now.
$a - 10 = 15$

5.4 Open Sentences

Page 181

1. **a.** true **b.** false **c.** open
d. true **e.** open **f.** false
g. false **h.** open **i.** true

2. **a.** Try -3 in $t - 3 = 2$
$-3 - 3 \stackrel{?}{=} 2$
$-6 \stackrel{?}{=} 2$ false

Try 1 in $t - 3 = 2$
$1 - 3 \stackrel{?}{=} 2$
$-2 \stackrel{?}{=} 2$ false

Try 5 in $t - 3 = 2$
$5 - 3 \stackrel{?}{=} 2$
$2 \stackrel{?}{=} 2$ true

Answer: The solution set is {5}.

b. Try -3 in $2x = -6$
$2(-3) \stackrel{?}{=} -6$
$-6 \stackrel{?}{=} -6$ true

Try 1 in $2x = -6$
$2(1) \stackrel{?}{=} -6$
$2 \stackrel{?}{=} -6$ false

Try 5 in $2x = -6$
$2(5) \stackrel{?}{=} -6$
$10 \stackrel{?}{=} -6$ false

Answer: The solution set is {–3}.

c. Try -3 in $y > -1$
$-3 \stackrel{?}{>} -1$ false

Try 1 in $y > -1$
$1 \stackrel{?}{>} -1$ true

Try 5 in $y > -1$
$5 \stackrel{?}{>} -1$ true

Answer: The solution set is {1, 5}.

d. Try -3 in $x < 6$
$-3 \stackrel{?}{<} 6$ true

Try 1 in $x < 6$
$1 \stackrel{?}{<} 6$ true

Try 5 in $x < 6$
$5 \stackrel{?}{<} 6$ true

Answer: The solution set is {–3, 1, 5}

3. **a.** {Mon., Thurs., Fri.} **b.** {7}
c. {–7} **d.** {6, 9}
e. {4, 5} 6 is not a solution: $6 + 2 \stackrel{?}{<} 8$
$8 \stackrel{?}{<} 8$ false
f. {2, –2} $(2)^2 = 2 \cdot 2 = 4$ and $(-2)^2 = (-2)(-2) = 4$
g. ∅ Both 3^2 and $(-3)^2$ equal +9.

CUMULATIVE REVIEW • CHAPTERS 1-5

1. Give a reason why each of the following is not a good definition.
   a. A carrot is a vegetable.
   b. A chair is furniture made of wood, to sit on.

2. Write a question that is suggested by the following: Super Thrifty Store is advertising 3 notebooks for $5. The Penny Saver has 4 of the same notebooks for $6.

3. The table shows the heights of female basketball players picked for the all-star team. All on the team must be over 5'7" or average at least 15 points per game.

Heights of Players			
Crystal Karen Kayla	5'8" 5'11" 5'6"	Eliza Missy Kris	5'5" 5'10" 6'1"

   a. Which two are the tallest?
   b. What is the difference in height between the tallest and the shortest?
   c. Write a conclusion about Eliza.
   d. Does Karen score more than 15 points per game?

4. The bar graph shows the weights of the 5 Taino children.

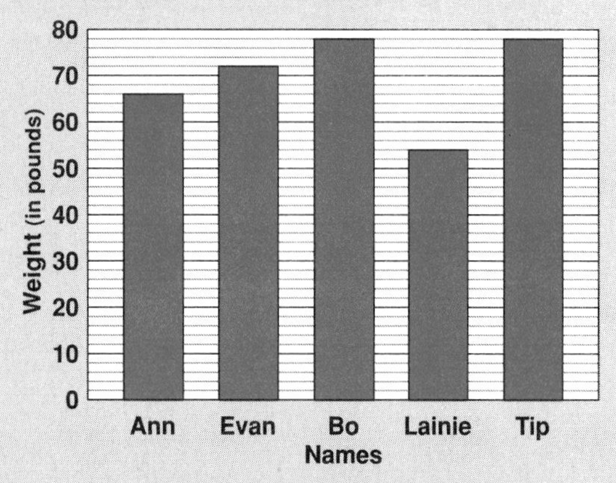

   a. Who weighs between 60 and 70 pounds?
   b. Which 2 children weigh the same?
   c. About how much heavier is Evan than Lainie?

5. Tricia and Jamie compared the number of calories they ate over a 5-day period. The graph shows the number of calories recorded by each girl.

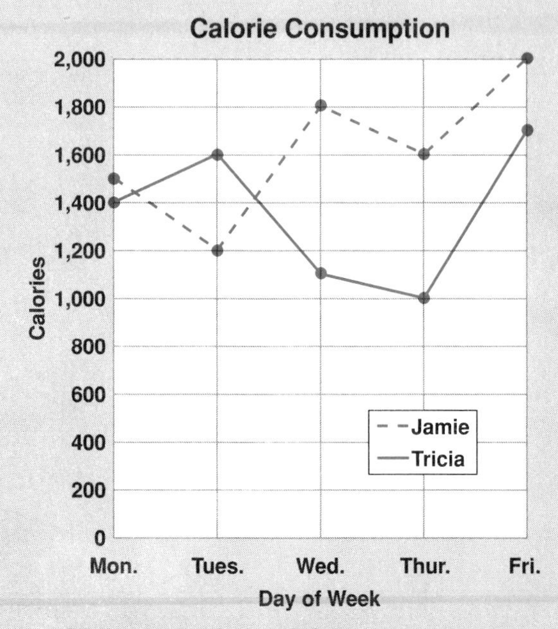

   a. How many calories did Jamie eat on Friday?
   b. How many more calories did Tricia eat than Jamie did on Tuesday?
   c. Andrea was dieting, and was careful about calories. What might a graph of her calorie consumption look like?

(Continued on next page)

6. The pictograph shows the number of cars rented by Hav-a-Car Rentals in one week.

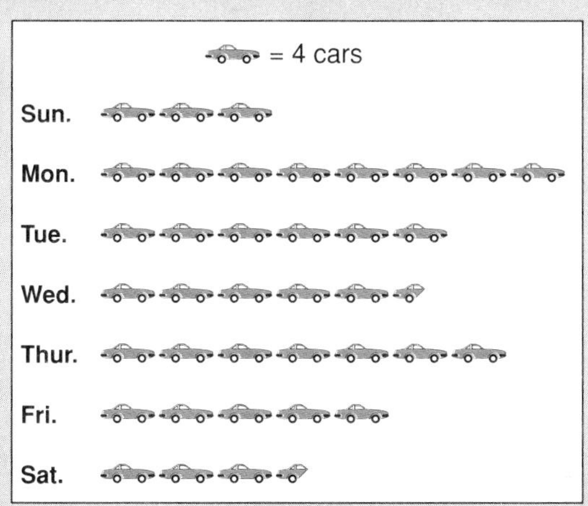

a. What was the greatest number of cars rented in one day?

b. On what day were 22 cars rented?

c. How many more cars were rented on Monday than on Friday?

7. Draw a conclusion, if possible, for each situation. If no conclusion is possible, explain why.

a. At Bryant School, all seniors study history. Hallie is studying history.

b. When Shula has a fever, she drinks lots of juice. Shula has a fever.

8. Find the next term in each sequence:

a. 2, 4, 7, 11, . . .

b. △, □, ⬠,...

9. Make a table to show how many ways 70¢ can be selected from 2 quarters, 4 dimes, and 7 nickels, if at least 1 of each kind of coin must be included.

10. Joan, Jean, Jill and Jessie all volunteered to sing a duet in the fall concert. How many different choices for a duo does the music teacher have?

11. The product of two numbers is 176. Their sum is 27. Find the numbers.

12. Find each of the following in the set $S = \{3, 4, 18, 48, 54, 60\}$.

a. the GCF of 12 and 16

b. the LCM of 3 and 18

c. the LCM of 12 and 15

13. The Venn diagram shows the prime factors of R and S.

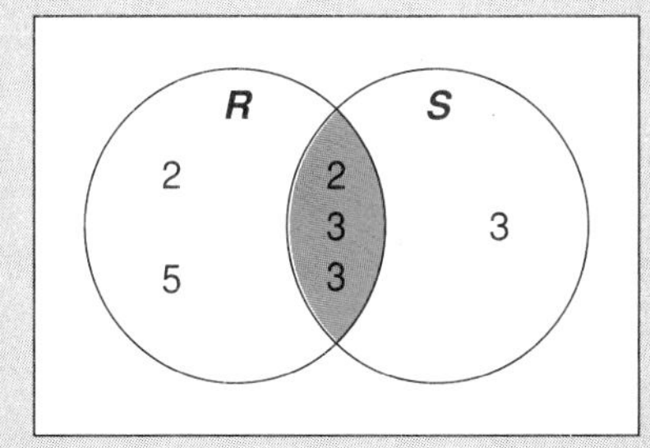

a. Find the greatest common factor of R and S.

b. Find the value of R.

c. Find the value of S.

14. Tell whether each statment is true or false.

a. The set of even numbers is closed under addition.

b. $3 \times 0 + 4 = 4$

c. $5 + (3 \times 7) = (5 + 3) \times 7$

d. $4 + 3^2 + (1 + 5)^2 = 7^2$

e. $12 \div 2 \times 3 + 7 = 9$

15. Write in order, from least to greatest:

a. $\frac{1}{2}, \frac{5}{12}, \frac{2}{3}$

b. 1.07, 1.3, 1.037

16. Write answers in simplest form.

a. Reduce: **(1)** $\frac{5}{15}$ **(2)** $\frac{9}{18}$

b. Add: **(1)** $\frac{1}{2} + \frac{2}{3}$ **(2)** $\frac{3}{4} + \frac{5}{8}$

c. Multiply: **(1)** $\frac{3}{5} \times \frac{7}{9}$ **(2)** $\frac{2}{5} \times \frac{15}{18}$

d. Divide: **(1)** $\frac{1}{2} \div \frac{3}{4}$ **(2)** $\frac{3}{5} \div \frac{9}{10}$

17. Copy and complete the table, to show equivalent values.

Fraction	Decimal	Percent
$\frac{1}{4}$	0.25	?
?	0.5	?
?	0.6	?
?	?	20%
?	?	75%
$\frac{2}{3}$	?	?

18. The gross pay of 5 employees of the Acme Supply Company is listed in the table.

Employee	Gross Pay
Miguel	\$175.60
Will	\$278.50
Lou	\$157.25
Mia	\$314.85
Ana	\$139.25

a. What is the total gross pay earned by the 5 employees?

b. Who earns twice as much as Ana?

c. Mia's pay is the same as the total pay of which 2 other employees?

19. The graph shows how Mr. Hill spent his time for one 24-hour period.

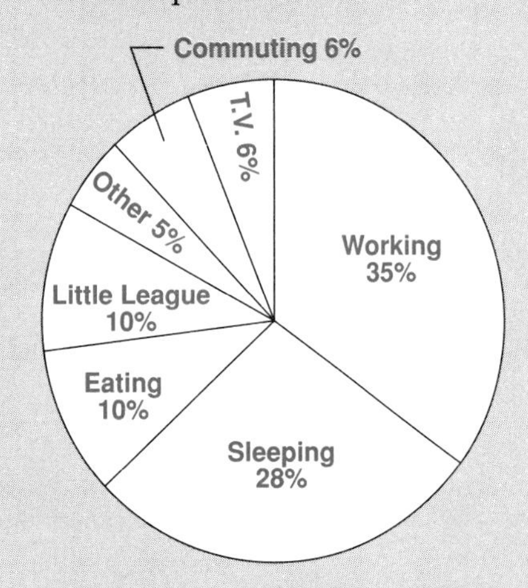

a. What 2 activities together took more than half his time?

b. What percent of the time was for Other?

c. What activity took the same amount of time as commuting?

d. In hours and minutes, how much time was spent on Little League?

20. Use the order of operations to find the value of each expression.

a. $3(-4) + 14$ **b.** $18 + 10 \div (-2)$

c. $-8(3) - 4(-3)$ **d.** $36 \div 6 \div 3$

e. $3 + 4(8-11)^2$ **f.** $24 - 12 \div (-4)$

21.

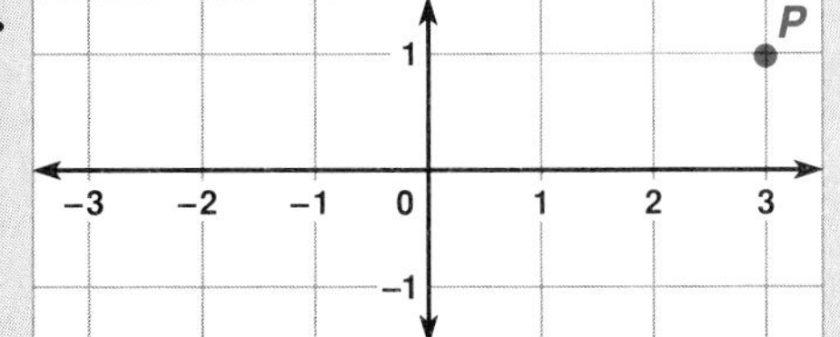

a. What are the coordinates of point P?

b. If point P is reflected over the horizontal axis, what are the coordinates of the image point P'?

c. In which quadrant is P' located?

22. Find the value of each expression when $x = 4$, $y = 3$, and $z = -2$.

a. xyz **b.** $x + y^2$ **c.** $2x - z$

d. $x(y + z)$ **e.** $x + 2y + 3z$ **f.** $\frac{x - 2y}{z}$

23. Simplifying each expression by removing parentheses and combining like terms.

a. $2d + 3(d + 5)$ **b.** $4(a + b) + 2(a - b)$

24. Using the given variables, write an algebraic expression for each word phrase.

a. five more than the number n

b. the number of months in y years

c. the total amount of precipitation if there were r inches of rain and s inches of snow

d. Ron's weight if he weighs three pounds less than twice Warren's weight w

Solving Equations 6

ACTIVITY

Draw a picture to show how the seesaw would look if these children sat on it so that:

- one twin is on either end
- a twin is on one end and the big boy is on the other end

Describe other ways the children could sit on the seesaw, and draw a picture to show the position of the seesaw.

6.1 Using Addition or Subtraction

The pans on this scale are in perfect balance. What does this mean about the weights in the two pans?

If you removed a 3-ounce weight from the right pan, what would you have to do to keep the two pans in balance?

Think of the two sides of the balanced scale as the two sides of an equation.

If you subtracted a number from one side of the equation, what would you have to do to keep the balance of the equation?

If you added a number to one side of the equation, what would you have to do to keep the balance?

> **To keep the balance of an equation:**
>
> **If you subtract a number from one side, you must subtract the same number from the other side.**
>
> **If you add a number to one side, you must add the same number to the other side.**

$$4 + 1 + 1 + 1 = 3 + 3 + 1$$
$$4 + \cancel{1} + \cancel{1} + \cancel{1} = \cancel{3} + 3 + 1$$
$$4 = 3 + 1$$
$$4 + 2 = 3 + 1 + 2$$

Consider this model of an equation.

How many circles would you need to replace the square on the left side of the model to see that the two sides are in balance?

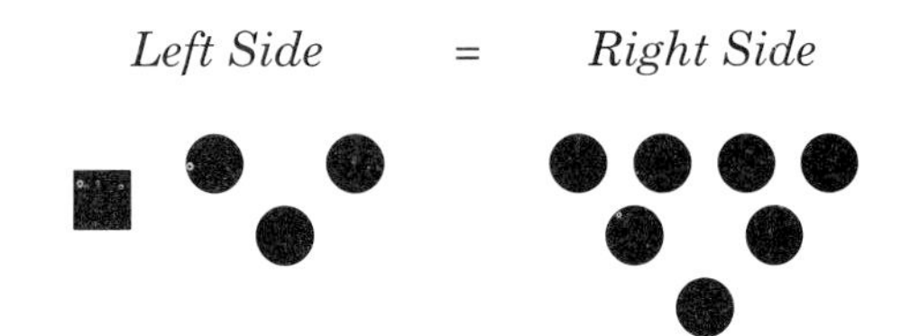

Use words and symbols to write the equation that is modeled.

1 square + 3 circles = 7 circles
s + 3 circles = 7 circles
4 circles + 3 circles = 7 circles

If s is replaced by 4 circles, the equation is balanced.
In the language of Algebra, $s = 4$ is the ***solution*** to the equation.

Summarize the result of the model using symbols.

$$s + 3 = 7$$
$$s = 4$$

What operation must you perform on the equation to get the solution?

To *solve* this equation, to get s alone on one side, you have to subtract 3 from each side.

$$s + 3 = 7$$
$$s + 3 - 3 = 7 - 3$$
$$s = 4$$

Example 1 Solve each equation.

a. $x + 39 = 211$

Since this equation contains an addition, think of it as an ***addition equation***. To solve an addition equation, subtract the same number from both sides.

$x + 39 = 211$	An addition equation
$x + 39 - 39 = 211 - 39$	Subtract 39 from both sides.
$x + 0 = 172$	$39 - 39 = 0$
$x = 172$	$x + 0 = x$

b. $n - 17 = 353$

Since this equation contains a subtraction, think of it as a ***subtraction equation***. To solve a subtraction equation, add the same number to both sides.

$n - 17 = 353$	A subtraction equation
$n - 17 + 17 = 353 + 17$	Add 17 to both sides.
$n + 0 = 370$	$-17 + 17 = 0$
$n = 370$	$n + 0 = n$

Solving an Equation by Addition or Subtraction

- **To solve an equation means to get the variable alone on one side of the equation.**
- **If a number has been added to a variable, get the variable alone by subtracting that number from both sides of the equation.**
- **If a number has been subtracted from the variable, get the variable alone by adding that number to both sides of the equation.**

Example 2 Solve each equation and check your answer.

a. $x + 7 = 5$

To solve an addition equation, subtract the same number from both sides.

$x + 7 = 5$

$x + 7 - 7 = 5 - 7$

$x = -2$

To check, substitute your answer for the variable in the original equation, and do the calculation.

$x + 7 = 5$

$-2 + 7 \stackrel{?}{=} 5$

$5 = 5$ ✓

b. $y - 5 = -1$

To solve a subtraction equation, add the same number to both sides.

$y - 5 = -1$

$y - 5 + 5 = -1 + 5$

$y = 4$

To check, substitute your answer for the variable in the original equation, and do the calculation.

$y - 5 = -1$

$4 - 5 \stackrel{?}{=} -1$

$-1 = -1$ ✓

An important problem-solving strategy is to translate the words of a problem into an equation, and solve the equation. It's a good idea to first "guesstimate" an answer.

Example 3 Polly and Esther spent \$237 while shopping. If they had \$71 left, how much money did they start with?

Follow the four key steps of problem solving.

1 ***Read the problem to understand it.***
You know the amount spent and the amount left. You must find the original amount.

2 ***Make a plan to solve the problem.***
Translate the given information into an equation.

Let x = the original amount.
Then $x - 237$ = the amount left after spending \$237.

$$x - 237 = 71$$

3 ***Use your plan to solve the problem.***
Solve the equation that models the problem situation.

$x - 237 = 71$ — A subtraction equation.
$x - 237 + 237 = 71 + 237$ — Add 237 to both sides.
$x = 308$

4 ***Check your solution.***
Is your answer reasonable for the original problem?

$$x - 237 = 71$$
$$308 - 237 \stackrel{?}{=} 71$$
$$71 = 71 \checkmark$$

Answer: Polly and Esther started with \$308.

Try These *(For solutions, see page 232.)*

1. Solve each equation and check your answer.

a. $x + 4 = 12$ **b.** $y - 6 = 18$
c. $t + 7 = 2$ **d.** $r - 8 = -4$
e. $b + 0.7 = 2.9$ **f.** $a - \frac{1}{2} = 7\frac{1}{2}$
g. $20 = x + 2$ **h.** $14 = y - 3$

2. Use an equation to solve this problem. First, guesstimate an answer.

Between midnight and noon, the temperature rose 36 degrees. If the temperature at noon was 84 degrees, what was it at midnight?

SELF-CHECK: SECTION 6.1

1. What symbol shows that two quantities are balanced, or have the same value?

   Answer: =

2. If you subtract a number from one side of an equation, what must you do to keep the balance of the equation?

   Answer: Subtract the same number from the other side.

3. In the equation $x + 3 = 7$, why would you want to subtract 3 from both sides?

   Answer: To get the variable x alone, which solves the equation.

4. In the equation $y - 4 = 9$, what would you do to solve the equation?

   Answer: Add 4 to both sides.

5. How would you know if $t = -1$ is the solution to the equation $t - 7 = -8$?

   Answer: Substitute -1 for t in the equation and do the calculation.

6. What is an important use for equations?

   Answer: solving problems

7. When you use an equation to model a problem, why must your answer be checked in the original problem and not just in the equation?

   Answer: Your equation may not correctly model the problem.

EXERCISES: SECTION 6.1

1. Solve by subtraction, and check.

   a. $x + 7 = 23$ **b.** $n + 9 = 2$ **c.** $y + 8 = -3$

   d. $27 = w + 14$ **e.** $2 = z + 5$ **f.** $-7 = p + 1$

   g. $a + 0.7 = 7.1$ **h.** $q + 1.4 = 7.5$ **i.** $9.5 = x + 6.23$

   j. $d + \frac{1}{3} = 1\frac{2}{3}$ **k.** $k + \frac{1}{3} = \frac{5}{6}$ **l.** $3 = b + 2\frac{1}{3}$

2. Solve by addition, and check.

   a. $x - 5 = 13$ **b.** $n - 3 = 2$ **c.** $y - 7 = -3$

   d. $9 = w - 11$ **e.** $-2 = z - 5$ **f.** $-7 = p - 1$

   g. $a - 1.7 = 7.1$ **h.** $q - 3.6 = 4.2$ **i.** $8.51 = x - 2.3$

   j. $d - \frac{3}{4} = 1\frac{1}{4}$ **k.** $k - \frac{1}{4} = \frac{5}{8}$ **l.** $6\frac{2}{3} = b - 1\frac{2}{3}$

3. Solve and check.

   a. $x - 8 = 19$ **b.** $n + 2 = 2$ **c.** $y - 2 = -10$

   d. $13 = w + 14$ **e.** $-12 = z - 5$ **f.** $-7 = p - 2$

   g. $a + 0.1 = 8.2$ **h.** $q - 2.4 = 5.12$ **i.** $7.5 = x - 4.2$

   j. $d - \frac{1}{5} = 1\frac{3}{5}$ **k.** $k + \frac{1}{5} = \frac{3}{10}$ **l.** $6 = b - 1\frac{1}{5}$

4. Use an equation to find each unknown number. First, guesstimate an answer.
   a. A number increased by 12 is 103.
   b. The sum of 9 and a number is –3.
   c. Seventeen more than a number is 84.
   d. A number decreased by 27 is 98.
   e. The difference of a number and 6 is –10.
   f. Fourteen less than a number is 23.

5. Use an equation to solve each problem. First guesstimate an answer.
   a. Fifteen students from Ms. Stanton's Social Studies class have gone to do research in the library today. If 8 students remain in the classroom, and everyone is present today, how many students are in the class?

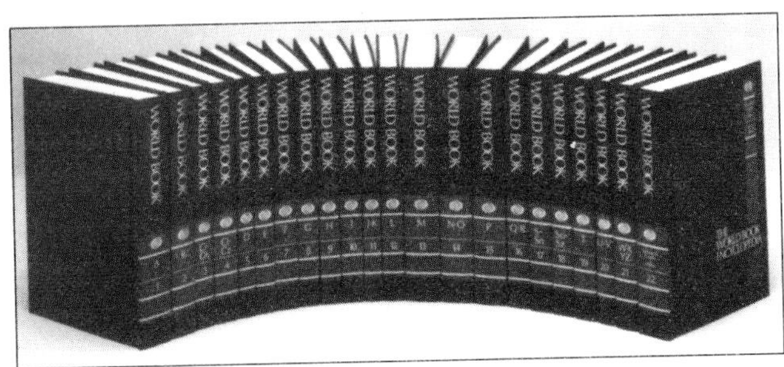

   b. Josh bought 6 guppies for his fish tank, bringing the number of guppies in the tank to 17. How many were there before the new guppies were added?
   c. After Felipe took $15 from his wallet to pay a bill, he had $23 left. How much was in the wallet to start?
   d. At the end of a summer day, Sally's Sea Shop had sold 19 snorkel tubes, leaving a dozen tubes in stock. How many snorkel tubes had there been in stock at the beginning of the day?
   e. The club treasurer reported that $32.75 was still owed in dues. If $83.25 had already been paid, what was the total amount of dues to be collected?
   f. Dom mixed some water with crystals to make a plant food solution. David added $1\frac{1}{4}$ cups of water to Dom's mixture, so that there were 4 cups of water in all. How much water did Dom use?
   g. Luz bought some fabric and a pattern to make an outfit. After reading the pattern carefully, she found that she needed 7 yards of fabric, which was a yard and a half more than she had bought. How much fabric had she bought originally?

6. Make up a problem that could be solved using each equation.
   a. $x + 6 = 25$ b. $x - 28 = 60$ c. $10 = 2 + x$

7. If $y - 3 = 7$, find the value of:
   a. $y + 3$ b. $2y$ c. $y - 6$

EXPLORATIONS

You have seen that in an equation like $n - 2 = 3$, the symbol – can represent the operation of subtraction. To solve this equation, you undo the operation of subtraction by using addition.

You can also think of the left side of this equation as containing two terms: n and -2. Then, algebra chips can be used to model the equation and its solution.

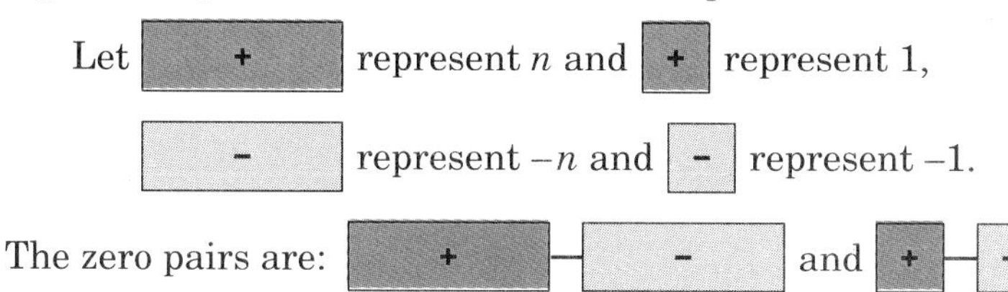

To model the equation $n - 2 = 3$ and its solution, lay out a left side and a right side.

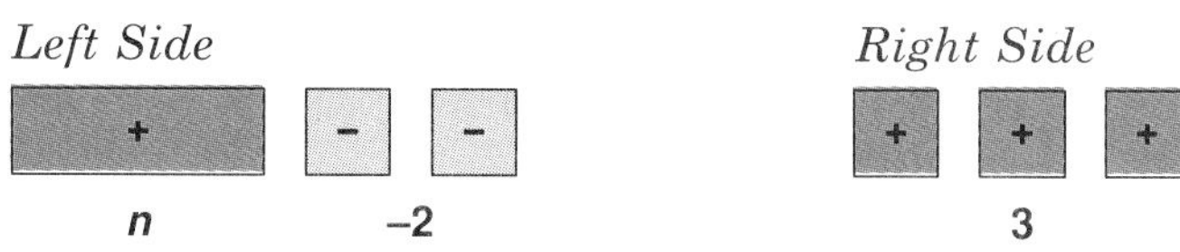

To get n alone on the left side, we will want to remove – – from the left.

To keep the equation in balance, we will also need to remove – – from the right side.

Since there are no such chips on the right, we need to insert zero pairs on the right side.

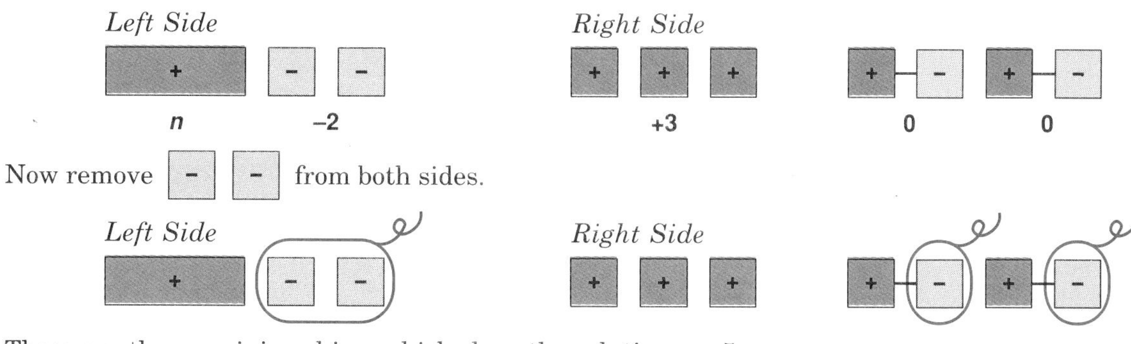

These are the remaining chips, which show the solution $n = 5$.

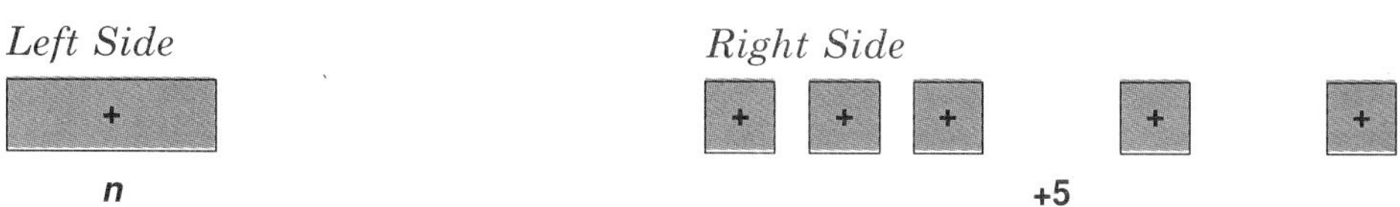

Use algebra chips to model each equation and its solution.

a. $n - 1 = 4$ **b.** $n - 2 = -3$ **c.** $n + 3 = -1$ **d.** $5 = n + 3$ **e.** $3 = n - 1$

f. $-2 = n - 1$ **g.** $2 + n = 4$ **h.** $2 - n = 2$ **i.** $3 - n = -1$ **j.** $4 - n = -2$

6.2 Using Multiplication or Division

Micki can win an arcade game in 6 minutes, which is twice as long as it takes Kim to win. How long does it take Kim to win?

You probably guessed that Kim's time is 3 minutes.

Now consider the equation that can be used to model this situation.

Let n = Kim's time.
Then $2n$ = Micki's time.

$$2n = 6$$

Use chips to model the equation so that you can see the solution.

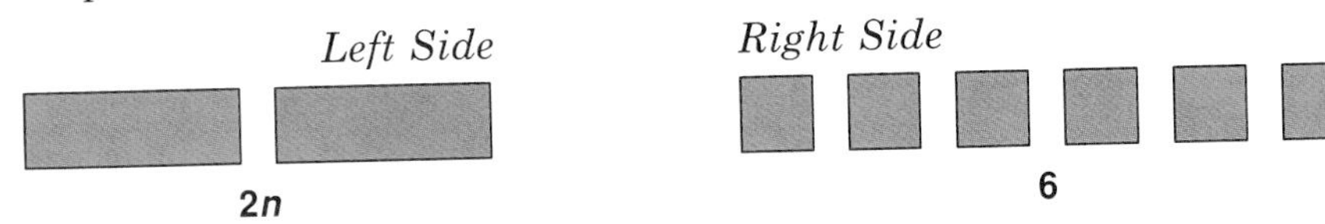

Separate the chips into two identical groups.

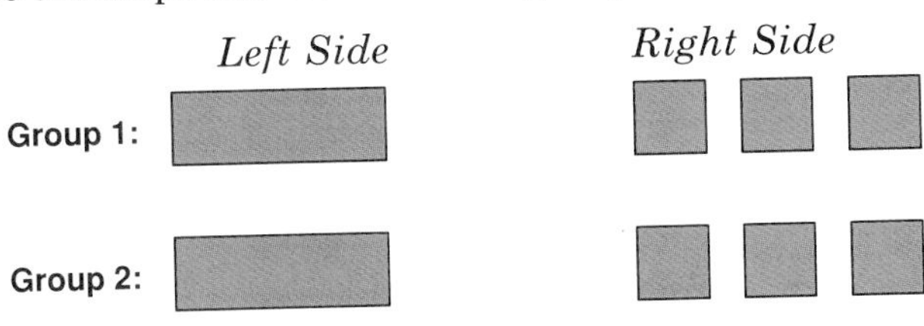

Each group shows than $n = 3$.

Summarize the result of the model.

$$2n = 6$$
$$n = 3$$

What operation must you perform on the equation to get the solution? To solve this equation, you have to divide both sides by 2.

$$2n = 6$$
$$\frac{2n}{2} = \frac{6}{2}$$
$$n = 3$$

Suppose the information in the arcade problem were reversed:

It takes Kim 3 minutes to win the game, which is half of Micki's time. What is Micki's time?

Let t = Micki's time.

Then $\frac{t}{2}$ = Kim's time.

Write a new equation and note that to solve, you have to multiply both sides by 2.

$$\frac{t}{2} = 3$$
$$2 \cdot \frac{t}{2} = 2 \cdot 3$$
$$t = 6$$

To keep the balance of an equation:

If you divide one side by a number, you must divide the other side by the same number.

If you multiply one side by a number, you must multiply the other side by the same number.

Example 1 Solve each equation.

a. $3y = 12$

Since this equation contains a multiplication, think of it as a ***multiplication equation***. To solve a multiplication equation, divide both sides by the same number, which is the multiplier of the variable.

$3y = 12$	A multiplication equation
$\frac{3y}{3} = \frac{12}{3}$	Divide both sides by 3, the multiplier of the variable.
$1y = 4$	$3 \div 3 = 1$
$y = 4$	$1y = y$

b. $\frac{w}{4} = 8$

Since this equation contains a division, think of it as a ***division equation***. To solve a division equation, multiply both sides by the same number, which is the divisor of the variable.

$\frac{w}{4} = 8$	A division equation
$4 \cdot \frac{w}{4} = 4 \cdot 8$	Multiply both sides by 4, the divisor of the variable.
$1w = 32$	$4 \div 4 = 1$
$w = 32$	$1w = w$

Solving an Equation by Multiplication or Division

- To solve an equation means to get the variable alone on one side of the equation.
- If the variable has been multiplied by a number, get the variable alone by dividing both sides of the equation by that number.
- If the variable has been divided by a number, get the variable alone by multiplying both sides of the equation by that number.

Example 2 Solve each equation and check your answer.

a. $-5y = 20$

To solve a multiplication equation, divide both sides by the same number, which is the multiplier of the variable.

$$-5y = 20$$
$$\frac{-5y}{-5} = \frac{20}{-5}$$
$$y = -4$$

To check, substitute your answer for the variable in the original equation, and do the calculation.

$$-5y = 20$$
$$-5(-4) \stackrel{?}{=} 20$$
$$20 = 20 \checkmark$$

b. $\frac{x}{2} = -10$

To solve a division equation, multiply both sides by the same number, which is the divisor of the variable.

$$\frac{x}{2} = -10$$
$$2 \cdot \frac{x}{2} = 2(-10)$$
$$n = -20$$

To check, substitute your answer for the variable in the original equation, and do the calculation.

$$\frac{x}{2} = -10$$
$$\frac{-20}{2} \stackrel{?}{=} -10$$
$$-10 = -10 \checkmark$$

Try These *(For solutions, see page 232.)*

1. Solve each equation and check your answer.

a. $5n = 75$ **b.** $-4y = 44$ **c.** $\frac{1}{4}t = 8$ **d.** $\frac{n}{8} = -6$

e. $\frac{c}{-5} = 15$ **f.** $0.2x = 1.6$ **g.** $10 = 2x$ **h.** $14 = \frac{y}{7}$

2. Use an equation to solve each problem. First, guesstimate an answer.
a. If twice a number is 126, what is the number?
b. Joe's pay rate is one-third of Henry's. If Joe's rate is $5 per hour, what is Henry's?

SELF-CHECK: SECTION 6.2

1. If you multiply one side of an equation by a number, what must you do to keep the balance of the equation?

Answer: Multiply the other side by the same number.

2. To solve the equation $6q = 12$, how would you get the variable q alone?

Answer: Divide both sides of the equation by 6, the multiplier of the variable q.

3. Can you also use division to get the variable q alone in the equation $q + 6 = 12$? Explain.

Answer: No, since this equation contains an addition, you must use subtraction to get q alone.

4. Which of the following equations is equivalent to $\frac{1}{2}x = 10$?

(a) $2x = 10$ (b) $\frac{x}{2} = 10$

Answer: (b), $\frac{1}{2}x$ means half of x, or x divided by 2, or $\frac{x}{2}$.

5. In which of the following equations would you divide by 3 to get the variable k alone?

(a) $\frac{1}{3}k = 9$ (b) $\frac{k}{3} = 9$ (c) $3k = 9$ (d) $k + 3 = 9$

Answer: (c), division undoes multiplication. In $3k$, the variable is multiplied by 3.

EXERCISES: SECTION 6.2

1. Solve by division, and check.

a. $6y = 30$ **b.** $7c = -42$ **c.** $5b = 5$

d. $-3z = 15$ **e.** $-6k = -18$ **f.** $24 = 6d$

g. $8m = 4$ **h.** $6p = 0$ **i.** $5b = 4$

j. $2a = 0.6$ **k.** $0.7y = 84$ **l.** $0.4x = 3.2$

m. $\frac{1}{2}k = 10$ **n.** $\frac{2}{3}x = 12$ **o.** $\frac{3}{4}y = 24$

2. Solve by multiplication, and check.

a. $\frac{b}{4} = 8$ **b.** $\frac{r}{3} = 12$ **c.** $\frac{n}{5} = 10$

d. $\frac{x}{7} = -3$ **e.** $\frac{m}{-3} = 9$ **f.** $\frac{n}{-2} = -14$

g. $8 = \frac{a}{9}$ **h.** $-2 = \frac{b}{2}$ **i.** $-6 = \frac{c}{-3}$

j. $\frac{z}{3} = 3.4$ **k.** $\frac{x}{0.2} = 6$ **l.** $\frac{y}{1.2} = 1.2$

m. $\frac{1}{2}k = 10$ **n.** $\frac{2}{3}x = 12$ **o.** $\frac{3}{4}y = 24$

3. Solve and check.

a. $5w = 35$ **b.** $\frac{r}{6} = 4$ **c.** $t + 4 = 18$

d. $q - 7 = 17$ **e.** $36 = 9a$ **f.** $\frac{x}{7} = -6$

g. $16 = y + 6$ **h.** $12 = k - 2$ **i.** $4d = 7$

j. $3.2 = \frac{x}{5}$ **k.** $\frac{1}{3}y = 12$ **l.** $-4b = -6$

m. $-6 = 0.2q$ **n.** $-6 = k + 0.2$ **o.** $\frac{2}{3}z = 24$

4. Use an equation to find each unknown number. First, guesstimate an answer.
   - **a.** Twice a number is 68.
   - **b.** The product of 9 and a number is 108.
   - **c.** The quotient of a number and 3 is –4.
   - **d.** One-fifth of a number is 13.
   - **e.** When 18 is multiplied by a number, the result is 54.
   - **f.** Sixteen more than a number is 24.
   - **g.** Thirteen less than a number is 72.
5. Use an equation to solve each problem. First, guesstimate an answer.
   - **a.** If 5 identical backpacks weigh 135 pounds, what is the weight of one backpack?
   - **b.** An auditorium has 605 seats in 55 identical rows. How many seats are in a row?
   - **c.** Todd has one-fourth as much money as his brother Scott. If Todd has $24, how much does Scott have?
   - **d.** One-fifth of the jellybeans in a jar are yellow. If there are 22 yellow ones, how many jellybeans are in the jar?
   - **e.** The Simpsons took a second mortgage of $15,000 on their home, bringing their mortgage debt to $80,000. What was the amount of the first mortgage?
   - **f.** In a science experiment, Marla evaporated 3.5 ounces of water from a mixture, leaving 17 ounces of water in the mixture. How many ounces of water were there in the original mixture?

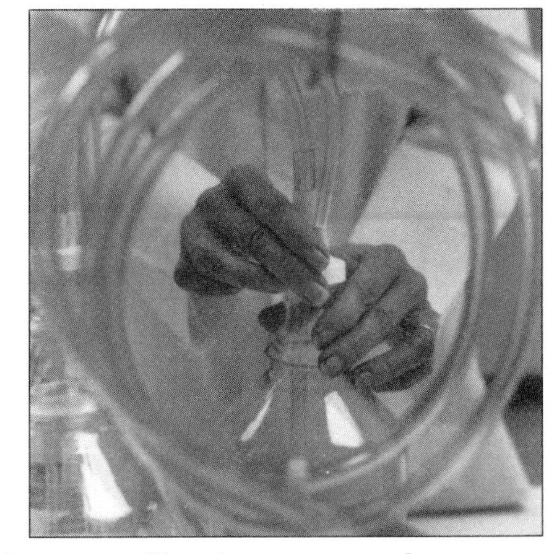

   - **g.** Mia worked for 38 hours and earned $171. How much does she earn per hour?
   - **h.** Two-thirds of the students in the sophomore class have part-time jobs. If 100 sophomores have part-time jobs, how many sophomores are in the class?
   - **i.** A magazine page has 3,220 letters with an average of 70 letters per line. How many lines are on the page?
   - **j.** A rectangular crossword-puzzle grid contains 323 squares. If each row across has 17 squares, how many squares are in each column going down?
   - **k.** If each bus can take no more than 38 passengers, how many buses are needed for 188 people going on a school trip?
   - **l.** What is the largest number of 32-cent stamps that Perry can get for $10?
6. Make up a problem that could be solved using each equation.

   a. $5x = 75$ **b.** $\frac{n}{3} = 8$ **c.** $y + 4 = 25$ **d.** $t - 18 = 60$

7. If $2a = 20$ and $6b = 12$, find the value of:

   a. ab **b.** $\frac{a}{b}$ **c.** $\frac{b}{a}$ **d.** $a + ab$

EXPLORATIONS

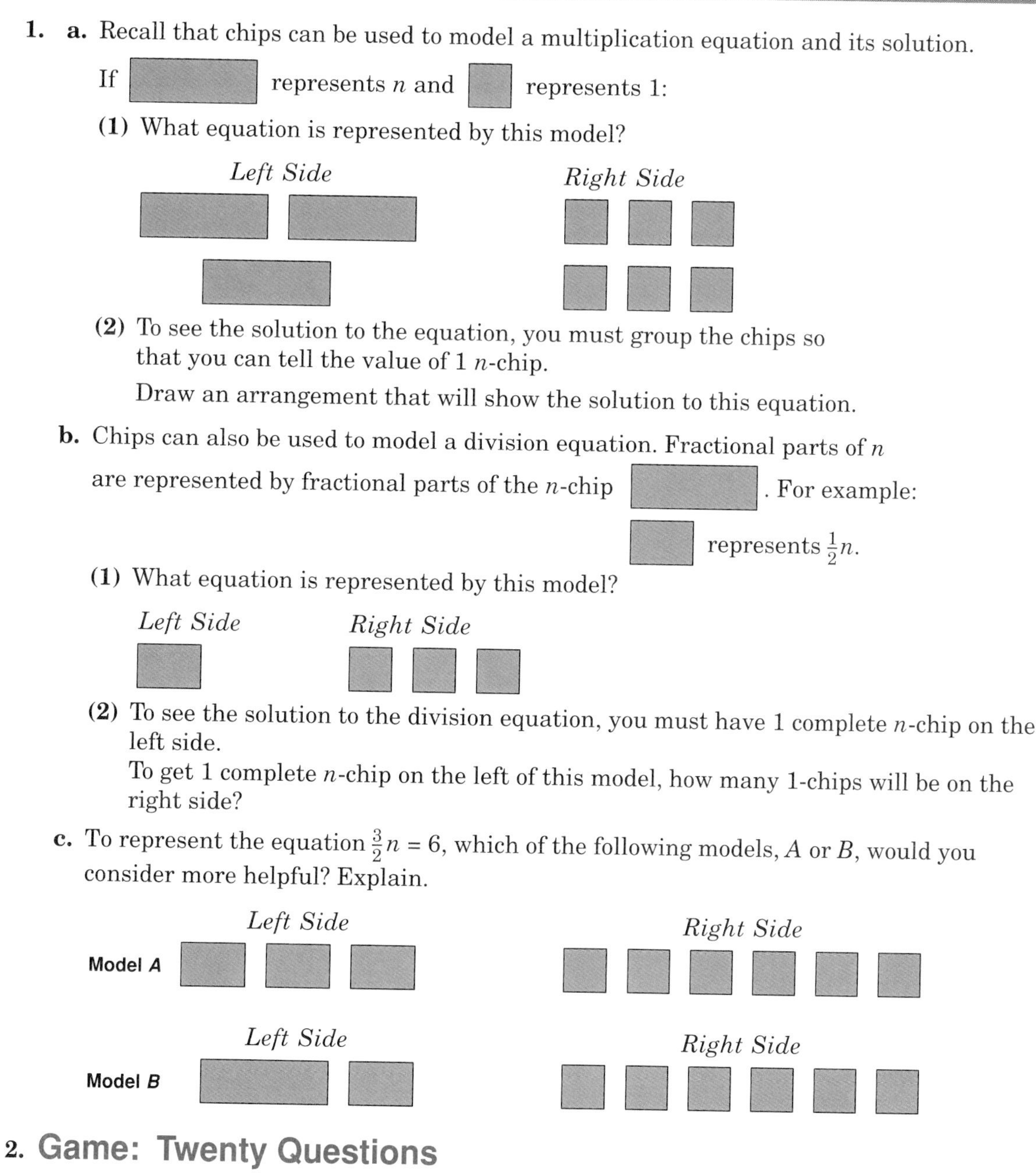

1. **a.** Recall that chips can be used to model a multiplication equation and its solution.

   If [chip] represents n and [chip] represents 1:

   (1) What equation is represented by this model?

   Left Side *Right Side*

   (2) To see the solution to the equation, you must group the chips so that you can tell the value of 1 n-chip.

   Draw an arrangement that will show the solution to this equation.

   b. Chips can also be used to model a division equation. Fractional parts of n are represented by fractional parts of the n-chip [chip]. For example:

   [chip] represents $\frac{1}{2}n$.

   (1) What equation is represented by this model?

   Left Side *Right Side*

   (2) To see the solution to the division equation, you must have 1 complete n-chip on the left side.
   To get 1 complete n-chip on the left of this model, how many 1-chips will be on the right side?

   c. To represent the equation $\frac{3}{2}n = 6$, which of the following models, A or B, would you consider more helpful? Explain.

   Model *A* *Left Side* *Right Side*

   Model *B* *Left Side* *Right Side*

2. ### Game: Twenty Questions

Team A secretly picks the name of a famous person, living or in the past.
Team B must ask questions that can be answered yes or no.
By using at most 20 questions, Team B must guess the secret.
The game can be expanded to more than one category, for example, people and places.

6.3 Using Two Operations

One more than twice a number is 7. What is the number? Explain your reasoning.

Since 1 more than twice the number is 7, twice the number must be 6. Since twice the number is 6, the number must be 3.

Consider an equation to model this problem.

Let n = the number.

$2n + 1 = 7$

Use chips to model the equation.

If ▭ represents n and ▫ represents 1:

Left Side | *Right Side*

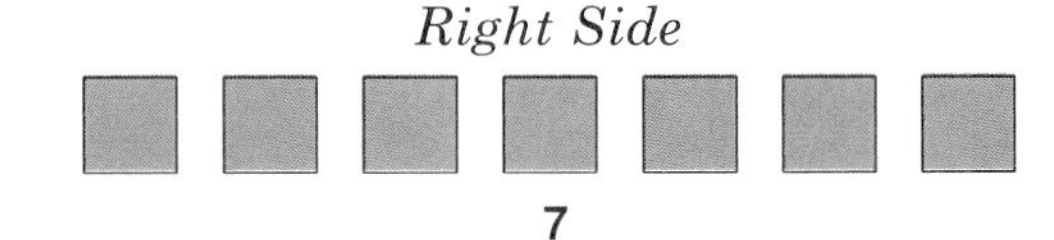

$2n$ + 1 | **7**

Use the model of the equation to see the solution. The object is to get n alone on the left side.

To get only n-chips on the left, remove the 1-chip.

To keep the equation in balance, remove a 1-chip from the right.

Left Side | *Right Side*

Since the left side has 2 n-chips, separate the chips into 2 identical groups.

Left Side | *Right Side*

Each group shows that $n = 3$.

Summarize the result of the model.

$2n + 1 = 7$	An equation with two operations: + and ×
$2n + 1 - 1 = 7 - 1$	First, subtract 1 from both sides.
$2n = 6$	
$\frac{2n}{2} = \frac{6}{2}$	Next, divide both sides by 2.
$n = 3$	

To solve an equation that contains the 2 operations addition and multiplication:
First, undo the addition by using subtraction.
Next, undo the multiplication by using division.

Example 1 Solve $2y + 4 = 12$.

$2y + 4 = 12$ Two operations: + and ×

$2y + 4 - 4 = 12 - 4$ First, undo + by using −.

$2y = 8$

$\frac{2y}{2} = \frac{8}{2}$ Next, undo × by using ÷.

$y = 4$

An equation with 2 operations may contain combinations of operations besides addition and multiplication.

Example 2 Solve each equation.

a. $2y - 4 = 12$ Two operations: − and ×

$2y - 4 + 4 = 12 + 4$ First, undo − by using +.

$2y = 16$

$\frac{2y}{2} = \frac{16}{2}$ Next, undo × by using ÷.

$y = 8$

b. $\frac{y}{2} + 4 = 12$ Two operations: + and ÷

$\frac{y}{2} + 4 - 4 = 12 - 4$ First, undo + by using −.

$\frac{y}{2} = 8$

$2 \cdot \frac{y}{2} = 2 \cdot 8$ Next, undo ÷ by using ×.

$y = 16$

Solving an Equation with Two Operations

- To solve an equation, you must get the variable alone on one side of the equation.
- First, undo the addition or subtraction by using the inverse operation.
- Next, undo the multiplication or division by using the inverse operation.

Example 3 Shilo paid $27 for earphones for her portable audio player. If the earphones were on sale for $5 less than half the regular price, what was the regular price?

Write an equation to model the problem.

Let p = the regular price.

Then $\frac{p}{2}$ = half the regular price

and $\frac{p}{2} - 5$ = the amount that Shilo paid.

$$\frac{p}{2} - 5 = 27 \qquad \text{Two operations: } - \text{ and } \div$$

$$\frac{p}{2} - 5 + 5 = 27 + 5 \qquad \text{First, undo } - \text{ by using } +.$$

$$\frac{p}{2} = 32$$

$$2 \cdot \frac{p}{2} = 2 \cdot 32 \qquad \text{Next, undo } \div \text{ by using } \times.$$

$$p = 64$$

Check: Does 64 work in the original problem?
If $64 is the regular price, half is $32. Does 5 less than $32 equal $27? Yes.

Answer: The regular price is $64.

Try These *(For solutions, see page 233.)*

1. Solve each equation, and check.

   a. $3x + 2 = 8$ **b.** $2y - 5 = 15$ **c.** $\frac{t}{3} + 2 = 11$ **d.** $\frac{z}{2} - 1 = 5$

2. Use an equation to solve this problem. First, guesstimate an answer.

   The sweater Mary wants to buy costs $8 less than three times the amount she has saved. If the price of the sweater is $52, how much has Mary saved?

When Subtraction Involves the Variable

In which ways are the following two equations alike? How are they different?

$$n - 3 = 2 \qquad 3 - n = 2$$

A way in which the equations are alike is that they both contain subtraction. A way in which the equations differ is that in the second equation, it is the variable term that is subtracted. So far, you have seen solutions where it is a constant term that is subtracted.

Before looking at the model for the equation $3 - n = 2$, do you think you know the value of n that makes this open sentence true?

To lay out chips, think of the left side as containing two terms: 3 and $-n$.

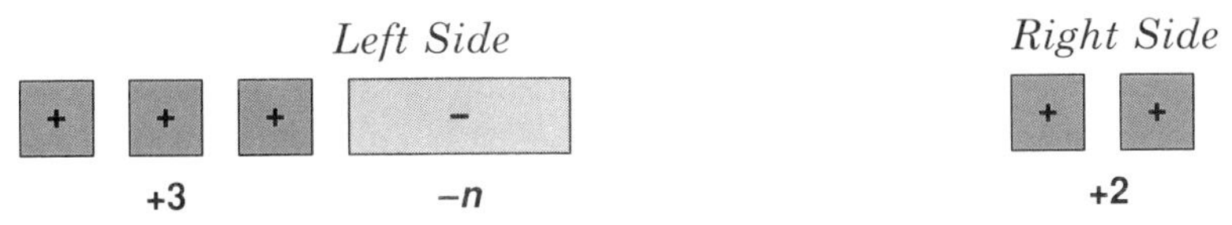

To see the solution, we will need to get n, not $-n$, alone on one side. We can put n in the model by adding an n-chip to each side.

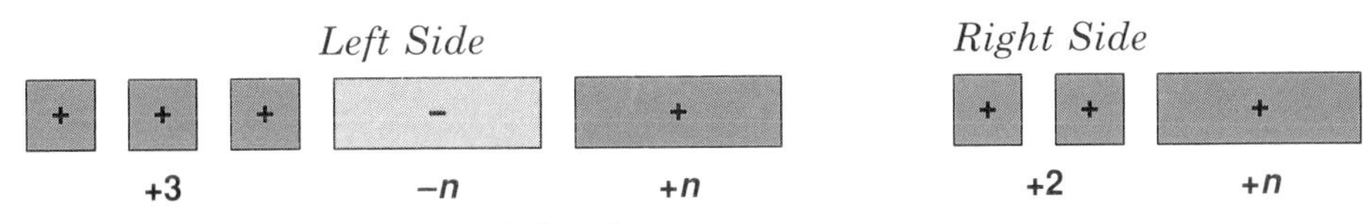

Remove the zero-pair from the left side.

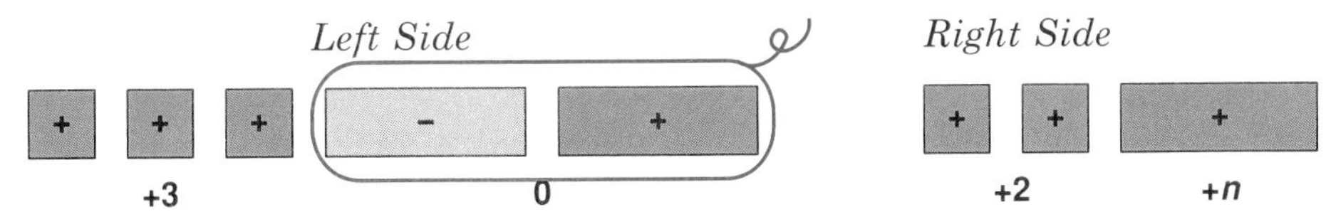

An n-chip is now on the right side of the equation.

To get the n-chip alone, remove + + from each side.

The remaining chips show the solution, $1 = n$ or $n = 1$.

Summarize the result of the model.

$3 - n = 2$	The subtraction involves the variable.
$3 - n + n = 2 + n$	Undo – by +.
$3 = 2 + n$	The variable is now on the right.
$3 - 2 = 2 - 2 + n$	To get n alone, subtract 2 from both sides.
$1 = n$	

When subtraction involves the variable, undo the subtraction by addition. The variable term will then be on the opposite side of the equation.

Example 4 Solve $5 - n = 1$, and check.

Before solving algebraically, do you think you know the value of n that makes this open sentence true?

$5 - n = 1$	Subtraction involves the variable.
$5 - n + n = 1 + n$	Undo – by +.
$5 = 1 + n$	The variable is on the opposite side.
$5 - 1 = 1 - 1 + n$	To get n alone, subtract 1 from both sides.
$4 = n$	

Check: $5 - n = 1$

$5 - 4 \stackrel{?}{=} 1$

$1 = 1$ ✓

Answer: $n = 4$

Try These *(For solutions, see page 234.)*

Solve each equation, and check: **1.** $7 - r = 2$ **2.** $2 - w = -3$

SELF-CHECK: SECTION 6.3

1. What is the aim in order to solve an equation?

Answer: Get the variable alone on one side of the equation.

2. When the variable is alone on one side of the equation, what is on the other side?

Answer: The solution, that is, the value of the variable that makes the open sentence true.

3. How many operations are in the equation $x + 4 = 10$?

Answer: one operation: addition

4. How many operations are in the equation $2x + 4 = 10$?

Answer: two operations: addition and multiplication

5. What operation must you perform to undo addition?

Answer: subtraction, the inverse operation of addition

6. What operation must you perform to undo multiplication?

Answer: division, the inverse operation of multiplication

7. To solve the equation $2x + 4 = 10$, which operation do you work on first? Next?

Answer: First, undo the addition by using subtraction. Next, undo the multiplication by using division.

EXERCISES: SECTION 6.3

1. Do you agree or disagree with each procedure? If you disagree, tell what you would do to solve the equation.

a. To solve $x + 3 = 16$, Kevin added 3 to both sides.

b. To solve $y - 3 = 16$, Laura added 3 to both sides.

c. To solve $3m + 2 = 11$, Dana first divided both sides by 3.

d. To solve $\frac{t}{2} + 3 = 15$, Luis first subtracted 3 from both sides.

2. Solve by first undoing addition or subtraction, and then undoing multiplication. Check.

a. $4x + 3 = 15$ **b.** $5y - 2 = 13$ **c.** $17 = 4r + 1$

d. $12 = 5w - 3$ **e.** $3m + 1.1 = 7.7$ **f.** $0.2x + 1 = 15$

g. $5p + 1\frac{1}{2} = 11\frac{1}{2}$ **h.** $2k - \frac{1}{4} = 3\frac{3}{4}$ **i.** $\frac{1}{2}q - 3 = 4$

3. Solve by first undoing addition or subtraction, and then undoing division. Check.

a. $\frac{x}{4} + 3 = 11$ **b.** $\frac{y}{5} - 2 = 8$ **c.** $15 = \frac{r}{4} + 3$

d. $18 = \frac{w}{7} - 3$ **e.** $\frac{m}{3} + 1.1 = 7.7$ **f.** $\frac{w}{6} - 0.2 = 4$

g. $\frac{c}{2} + 2\frac{1}{2} = 6\frac{1}{2}$ **h.** $\frac{a}{3} + \frac{1}{3} = 4$ **i.** $\frac{1}{4}x - 2 = 6$

4. Solve and check.

a. $x + 8 = 20$ **b.** $y - 2 = 17$ **c.** $3m = 6$

d. $\frac{r}{2} = 14$ **e.** $16 = x + 4$ **f.** $12 = \frac{w}{3}$

g. $3x + 6 = 21$ **h.** $10y + 17 = 7$ **i.** $-8 = 2 + 5a$

j. $\frac{x}{4} + 2 = 14$ **k.** $\frac{z}{3} - 6 = 12$ **l.** $20 = 5 + \frac{w}{3}$

m. $-2b + 11 = 3$ **n.** $2c - 9 = -11$ **o.** $-8x - 23 = 1$

p. $2s - 8 = 0$ **q.** $3r - 8 = -8$ **r.** $5x + 10 = 0$

s. $3z + 3.2 = 5$ **t.** $0.2y - 3 = 0.6$ **u.** $\frac{r}{2} + 1.6 = 10$

v. $\frac{1}{5}k - 7 = 3$ **w.** $\frac{4}{5}y + 1 = 25$ **x.** $\frac{3}{8}t - 2 = -23$

5. Solve each equation: **(1)** using algebra chips **(2)** algebraically, and check

a. $n - 4 = 1$ **b.** $4 - n = 1$ **c.** $n - 2 = 5$

d. $2 - n = 5$ **e.** $n - 3 = -2$ **f.** $3 - n = -2$

g. $2n - 1 = 7$ **h.** $1 - 2n = 7$ **i.** $5 = 4 - n$

6. A group of students came up with several different ways of solving the equation $2y - 12 = 8$. Try each of their methods, which follow, and tell if the method works.

a. Add 12 to both sides, than multiply both sides by $\frac{1}{2}$.

b. Add 12 to both sides, then divide both sides by 2.

c. Add 12 to both sides, then divide both sides by $\frac{1}{2}$, then divide both sides by 4.

d. Divide both sides by 2, then add 6 to both sides.

7. Use an equation to find each unknown number. First, guesstimate an answer.

a. If 5 is subtracted from 3 times a number, the result is 13.

b. Five more than twice a number is 29.

c. Six less than 3 times a number is 18.

d. Four times a number, decreased by 11, is 37.

e. If 6 is added to one-half a number, the sum is 67.

f. If twice a number is increased by 12, the result is 6.

8. Use an equation to solve each problem. First guesstimate an answer. Use the answer you get from the equation to answer the second question about the problem.

a. Phaedra would like to buy a blouse costing \$27, which is \$3 more than 4 times the amount of money she has saved so far.

(1) How much money has she saved so far?

(2) How much more does she need to buy the blouse?

b. In a second game, Jack's bowling score was 47 points less than twice the score of his first game. If his second score was 165:

(1) What was his first score?

(2) What was the difference in scores of the two games?

c. Mark bought 3 student tickets and a \$25 adult ticket for a concert. If the total cost of the tickets was \$76:

(1) What was the cost of 1 student ticket?

(2) How much less than an adult ticket is a student ticket?

d. After deductions of \$45 were taken from Kate's weekly earnings, her take-home pay was \$165.

(1) If she worked 30 hours, how much did she earn per hour?

(2) At the same hourly rate, what would be Kate's pay before taxes for a 40-hour week?

e. An auditorium has 50 identical rows of seats and 35 additional seats. If there are 685 seats in all:

(1) How many seats are there in each of the rows?

(2) If 2 seats were added to each of the rows, how many seats would be in the auditorium?

f. For a group of friends, Tammy filled a candy dish by taking pieces of candy from a large package. After she had used one-third of the number of pieces in the package, her brother took 4 pieces from the bowl.

(1) If 18 pieces of candy remained in the bowl, how many pieces were in the original package?

(2) If Tammy's mother took half of the remaining pieces from the original package for her club meeting, how many pieces did she have?

g. Two hundred and forty people are going on a school trip in buses and cars. The available cars will hold a total of 40 people. If each bus can take no more than 35 people:

(1) How many buses are needed?

(2) Describe how you would assign people to be seated on the buses.

h. The cost of sending a package is \$5 plus \$2 for each pound or part of a pound.

(1) If Colin paid \$17 to send a package, how much did it weigh?

(2) Could Colin be charged \$20 for sending a package? Explain.

9. Make up a problem that could be solved using each equation.

a. $n + 6 = 20$ **b.** $3x = 24$ **c.** $\frac{n}{3} = 15$

d. $2y - 7 = 17$ **e.** $3x + 5 = 20$ **f.** $\frac{n}{3} + 5 = 11$

10. If $\frac{x}{3} - 8 = -2$ and $4y + 6 = 12$, find the value of:

a. $x + y$ **b.** $\frac{x}{y}$ **c.** $x + xy$

EXPLORATIONS

1. For each of the equations below:

(1) Use algebra chips to model a solution. Check your result in the original equation.

(2) Following the solution suggested by the model, write out an algebraic solution.

a. $1 - 3n = 7$ **b.** $3 - 2n = 1$ **c.** $5 - 3n = -1$

2. Consider the equation $2x + 1 = 12$.

a. Describe the way you would solve this equation. Follow your method and get an answer.

b. Now try this solution.

Step 1: Divide each of the 3 terms of the equation by 2.

Step 2: Get x alone on the left by undoing the addition.
Did you get the same answer as before? Explain.

c. Which of the two methods do you prefer? Why?

6.4 Combining Like Terms

In Martin's apartment house, all the apartments are numbered. On Martin's floor, the numbers are 5, 6, 7.

Describe a way in which the numbers 5, 6, 7 are related.

Since each of these integers follows directly after the one before it, they are called ***consecutive*** integers.

Name three other consecutive integers.

If 8 is the first of two consecutive integers, what is the next?

If 12 is the first of two consecutive integers, what is the next?

If n is the first of two consecutive integers, what is the next?

Just as $8 + 1$, or 9, is the next integer after 8, and $12 + 1$, or 13, is the next integer after 12, the next integer after n is $n + 1$.

Name two consecutive integers whose sum is 7.

Did you say 3 and 4?

Consider an equation to model this problem.

Let n = the first integer.
Then $n + 1$ = the next integer.

$n + n + 1 = 7$ The sum of the two integers is 7.

How else could you write the left side of this equation?

Look at a model to help decide:

$2n \quad + \quad 1$

How is the simplified equation, $2n + 1 = 7$, familiar?

The simplified equation contains two operations.
Complete the solution to see that the first integer, n, is 3.

If there are like terms on one side of an equation, combine them to simplify the equation.

Example 1 Solve $x + 3x - 8 = 20$, and check.

$x + 3x - 8 = 20$	Like terms on the same side.
$4x - 8 = 20$	Combine like terms.
$4x - 8 + 8 = 20 + 8$	Undo − by +.
$4x = 28$	
$\frac{4x}{4} = \frac{28}{4}$	Undo × by ÷.
$x = 7$	

Check: $x + 3x - 8 = 20$

$$7 + 3(7) - 8 \stackrel{?}{=} 20$$
$$7 + 21 - 8 \stackrel{?}{=} 20$$
$$28 - 8 \stackrel{?}{=} 20$$
$$20 = 20 \checkmark$$

Answer: $x = 7$

Example 2 Find two consecutive odd integers whose sum is 24.

Before writing an equation to model the problem, think about *consecutive odd* integers. Name some. Tell how they are related.

Consecutive odd integers differ by 2, as in 7, 9, 11.

To write a model equation:

Let n = the first odd integer.
Then $n + 2$ = the next odd integer.

$n + n + 2 = 24$	The sum of the integers is 24.
$2n + 2 = 24$	Combine like terms on the same side.
$2n + 2 - 2 = 24 - 2$	Undo + by −.
$2n = 22$	
$\frac{2n}{2} = \frac{22}{2}$	Undo × by ÷.
$n = 11$	This is to be the first integer.
$n + 2 = 13$	Use n to find the second integer.

Check: Are 11 and 13 consecutive odd integers? Yes.
Does the sum of 11 and 13 equal 24? Yes.

Answer: 11 and 13 are the two consecutive odd integers whose sum is 24.

Try These *(For solutions, see page 234.)*

1. Solve each equation and check.
   a. $4y - 3y + 5 = 8$ **b.** $8 = 7z - 6z - 2$ **c.** $3x + 2x + 7 = 27$
2. Use an equation to solve this problem. First, guesstimate an answer.
   Find two consecutive even integers whose sum is 50.

When the Variable is on Both Sides

What equation does the following model represent?

The model represents the equation $2n = n + 1$. To get 1 n-chip alone on one side, what must you do? Removing an n-chip from each side of the model results in 1 n-chip alone on the left.

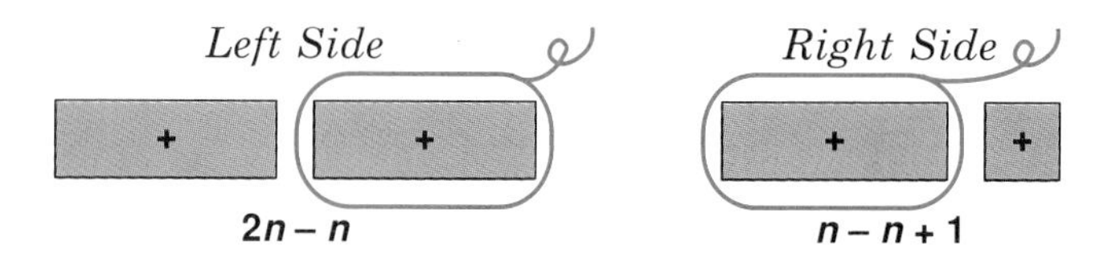

Now that a single n-chip is alone on one side, you can see the solution, $n = 1$.

Summarize the result of the model.

$2n = n + 1$	There are variable terms on both sides.
$2n - n = n - n + 1$	Collect the variable terms on the left.
$n = 1$	Combine like terms.

When there are variable terms on both sides of an equation, collect them on one side. Then, combine like terms.

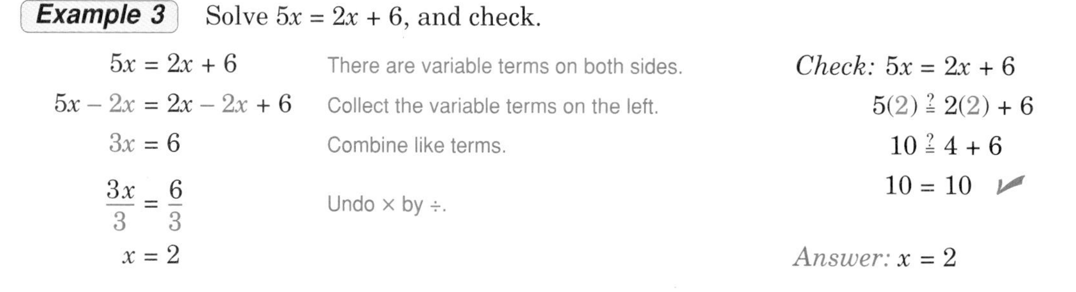

Example 3 Solve $5x = 2x + 6$, and check.

$5x = 2x + 6$	There are variable terms on both sides.	*Check:* $5x = 2x + 6$
$5x - 2x = 2x - 2x + 6$	Collect the variable terms on the left.	$5(2) \stackrel{?}{=} 2(2) + 6$
$3x = 6$	Combine like terms.	$10 \stackrel{?}{=} 4 + 6$
$\frac{3x}{3} = \frac{6}{3}$	Undo × by ÷.	$10 = 10$ ✓
$x = 2$		*Answer:* $x = 2$

What equation is represented by the following model?

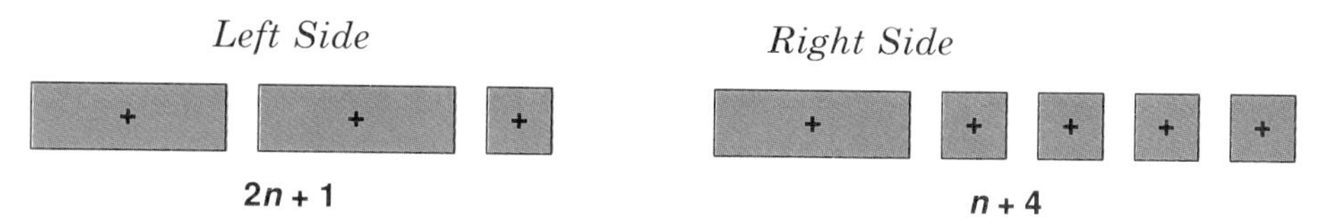

The model represents the equation $2n + 1 = n + 4$. To deal first with the variable terms, what would you do? Subtract 1 n-chip from each side of the equation.

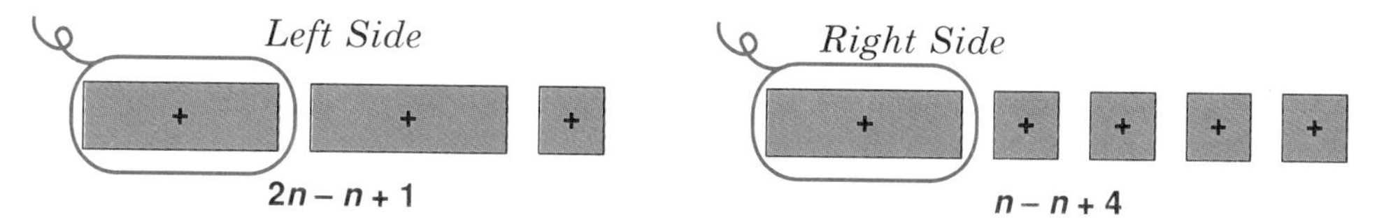

These are the remaining chips:

To get the n-chip alone on the left, what would you do? Subtract a 1-chip from each side.

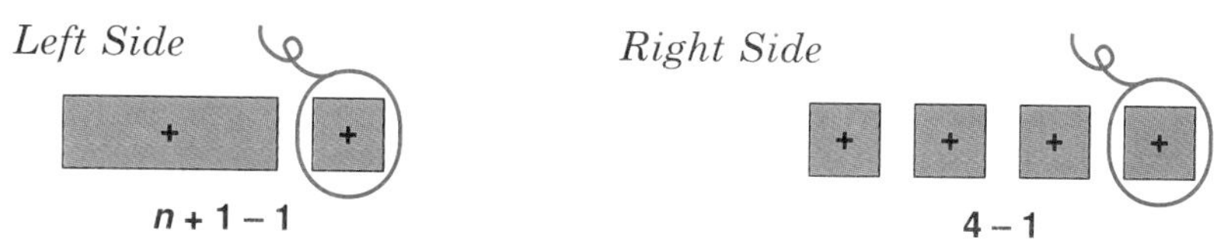

These are the remaining chips:

The solution is $n = 3$.

Summarize the result of the model.

$2n + 1 = n + 4$ The variable is on both sides.

$2n - n + 1 = n - n + 4$ Collect the variables on the left.

$n + 1 = 4$ Combine like terms.

$n + 1 - 1 = 4 - 1$ Undo + by −.

$n = 3$

Example 4 Solve $4x + 2 = 2x + 4$, and check.

$$4x + 2 = 2x + 4 \quad \text{The variable is on both sides.}$$
$$4x - 2x + 2 = 2x - 2x + 4 \quad \text{Collect the variables on the left.}$$
$$2x + 2 = 4 \quad \text{Combine like terms.}$$
$$2x + 2 - 2 = 4 - 2 \quad \text{Undo + by −.}$$
$$2x = 2$$
$$\frac{2x}{2} = \frac{2}{2} \quad \text{Undo × by ÷.}$$
$$x = 1$$

Check: $4x + 2 = 2x + 4$

$$4(1) + 2 \stackrel{?}{=} 2(1) + 4$$
$$4 + 2 \stackrel{?}{=} 2 + 4$$
$$6 = 6 \ \checkmark$$

Answer: $x = 1$

Many different types of problems can be modeled by equations where the variable is on both sides.

Example 5 Six times a number is the same as 5 more than the number. What is the number?

Try some guesses before we use an equation to model the problem.

Let x = the number.
Then $6x$ = 6 times the number
and $x + 5$ = 5 more than the number.

$$6x = x + 5 \quad \text{6 times the number equals 5 more than the number.}$$
$$6x - x = x - x + 5 \quad \text{Collect the variables on the left.}$$
$$5x = 5 \quad \text{Combine like terms.}$$
$$\frac{5x}{5} = \frac{5}{5} \quad \text{Undo × by ÷.}$$
$$x = 1$$

Check: Does 1 work in the problem?

If the number is 1, then 6 times the number is 6.
If the number is 1, then 5 more than the number is 6.
Are the results equal? Yes.

Answer: The number is 1.

Try These *(For solutions, see page 234.)*

Solve each equation and check.

1. $3t = 2t + 27$ **2.** $9m = m + 16$ **3.** $5z + 1 = 2z + 10$

SELF-CHECK: SECTION 6.4

1. What first step would simplify the equation $x + 2x + 4 = 10$?

Answer: Combining the like terms x and $2x$ on the left side.

2. What is the position of the like variable terms $2x$ and x in the equation $2x + x + 4 = 10$? in the equation $2x = x + 4$?

Answer: In $2x + x + 4 = 10$, the variable terms $2x$ and x are on the same side. In $2x = x + 4$, the variable terms are on opposite sides of the equation.

3. Before you can combine the like variable terms in the equation $2x = x + 4$, what must you do? How?

Answer: Collect them on one side, by subtracting x from both sides.

4. Name a set of 3 consecutive integers — *Answer:* 7, 8, 9
3 consecutive odd integers — *Answer:* 7, 9, 11
3 consecutive even integers — *Answer:* 6, 8, 10

5. Describe: consecutive integers; consecutive odd integers; consecutive even integers

Answer: Each consecutive integer follows directly after the integer before it, and is 1 more than the integer before it.

Each consecutive odd integer follows directly after the odd integer before it, but there is an even integer between them. The next odd integer is 2 more than the odd integer before it.

Each consecutive even integer follows directly after the even integer before it, but there is an odd integer between them. The next even integer is 2 more than the even integer before it.

6. If x is an odd integer, represent:

the next consecutive integer after x — *Answer:* $x + 1$
the next consecutive odd integer after x — *Answer:* $x + 2$
the next consecutive even integer after x — *Answer:* $x + 1$

EXERCISES: SECTION 6.4

1. Solve by first combining like terms to simplify. Check.

a. $x + x + 9 = 16$ **b.** $5y + y - 8 = 28$ **c.** $2z + 3z + 6 = -29$

d. $18 = 7y - y$ **e.** $-11 = y + 3y - 31$ **f.** $x - 5x + 2 = 10$

g. $2t - 7t + 10 = 5$ **h.** $8m - 12m + 12 = 4$ **i.** $w + 3w + 0.2 = 8.6$

j. $0.2r + 3.8r - 5 = 11$ **k.** $\frac{1}{3}x + 2\frac{2}{3}x - 7 = 5$ **l.** $q + 5q + \frac{1}{2} = 3\frac{1}{2}$

2. Solve by first collecting the variable terms on the same side. Check.

a. $2y = y + 9$ **b.** $4w = 10 + 3w$ **c.** $6x = 2x + 8$

d. $3z = 15 - 2z$ **e.** $9 - 2c = c$ **f.** $14 - 3h = 4h$

g. $8q + 1 = 7q + 5$ **h.** $9k + 7 = 8k + 3$ **i.** $5x - 2 = 2x + 4$

j. $4x + 8 = x - 1$ **k.** $6t + 8 = 10t + 4$ **l.** $4p + 7 = 6p - 9$

m. $26 - 5n = 2 - 2n$ **n.** $8 - 5\mathrm{r} = 26 - 2r$ **o.** $12 - 3t = 36 + 5t$

3. Solve and check.

a. $x + 4 = 1$ **b.** $y - 2 = -8$ **c.** $6m = -42$

d. $-8q = -48$ **e.** $\frac{w}{6} = -12$ **f.** $\frac{a}{-3} = -5$

g. $2x + 3 = 1$ **h.** $-6y + 4 = -8$ **i.** $7q - 6q = 9$

j. $5 = 10s - 9s$ **k.** $8r + 2r = 20$ **l.** $5p + p = 3$

m. $0.8c - 0.2c = 24$ **n.** $5\frac{1}{2}x - \frac{1}{2}x = 6$ **o.** $9b = 8b + 4$

p. $8a = 3a + 10$ **q.** $7d = 44 - 4d$ **r.** $2b + 24 = 6b$

s. $1.1y = 30 + 0.5y$ **t.** $0.7a + 36 = 2.7a$ **u.** $7d - 4 = 4d + 35$

v. $5p + 7 = 2p + 19$ **w.** $8k + 5 = 2k - 7$ **x.** $6y + 5 = y$

y. $3x + 9 = 7x + 1$ **z.** $10p - 1 = 2p + 1$ **z′.** $2m + 8 = 6m + 4$

4. Use an equation to find each unknown number. First, guesstimate an answer.

a. If a number is added to twice itself, the result is 36.

b. If 6 more than a number is added to the number, the result is 20.

c. If 4 less than a number is added to the number, the result is 18.

d. Twelve more than a number equals 7 times the number.

e. A number is equal to 8 more than one-third of the same number.

f. If 9 is subtracted from 4 times a number, the result is the same as if 1 is added to 3 times the number.

g. If twice a number is increased by 50, the result is the same as if 3 times the number is decreased by 31.

5. Use an equation to find:

a. two consecutive integers whose sum is 61

b. two consecutive integers whose sum is 83

c. two consecutive odd integers whose sum is 36

d. two consecutive even integers whose sum is 46

e. two consecutive odd integers whose sum is 204

f. two consecutive even integers whose sum is 406

6. Do you agree or disagree? Explain.
   - **a.** Jose says it is impossible to find two consecutive integers whose sum is 20.
   - **b.** Jim says it is possible to find two consecutive odd integers whose sum is 26.
   - **c.** Joy says it is impossible to find two consecutive even integers whose sum is 31.
   - **d.** Jane says it is impossible to find two consecutive integers whose difference is 2.
   - **e.** Jorge says it is possible to find two consecutive even integers whose difference is 4.
   - **f.** Juan says it is impossible to find two consecutive odd integers whose difference is 3.

7. Use an equation to solve each problem. First, guesstimate an answer.
   - **a.** You and your friend are paid $24 to do a job. Your friend works twice as long as you do. To be fair, how much of the $24 should each of you get?
   - **b.** The Sears Tower in Chicago is 104 feet taller than one tower of the World Trade Center in New York City. If the combined height of these two buildings is 2,804 feet, how tall is one tower of the World Trade Center?
   - **c.** While shopping, Juan spent $15 less than Manuel. If, together, they spent $63, how much did Manuel spend?
   - **d.** Li and Mei weigh the same and Bette weighs 101 pounds. If the total weight of the three friends is 313 pounds, what does Li weigh?
   - **e.** Jack, Al, and Mike had an equation-solving contest. In the five-minute time period, Al solved 3 fewer equations than Jack, and Mike solved 4 equations. If, together, they solved 15 equations, how many equations did Jack solve?
   - **f.** Jody has twice as many mints as Serena. If Jody gave Serena 7 mints, they would both have the same number. How many mints does Serena have now?

g. One morning, Cory's Pet Shop had twice the number of dogs as Pat's. During the day, Cory sold 4 pups and Pat took in 2 pups, so that at the end of the day they both had the same number of dogs in their places. How many dogs did Pat have in the morning?

h. At Bart's Bakery, the number of apple pies was 5 more than twice the number of cherry pies. When 7 more cherry pies were baked, there were just as many cherry pies as apple. How many cherry pies were there originally?

i. In 8 years, Ron will be exactly 3 times as old as he is now. How old is he now?

j. Sam and Vic are to meet at their favorite diner. Sam's driving distance to the diner is usually 3 times Vic's distance. However, today, Vic had to take an 8 mile detour, so that both of them drove the same distance. What is Vic's usual distance from the diner?

8. Make up a problem that could be solved using each equation.

a. $x + 2x = 10$ **b.** $n + n + 5 = 25$ **c.** $2z = z + 3$

9. If $x + 4x + 7 = 27$ and $4y = 3y + 2$, find the value of:

a. x^2 **b.** $(y + 1)^2$ **c.** $(xy)^2$

EXPLORATIONS

1. To decode this message:

$$\frac{?}{12}\ \frac{?}{7}\ \frac{?}{15}\ \frac{?}{-22}\ \frac{?}{2}\quad \frac{?}{2}\ \frac{?}{11}\ \frac{?}{1}\ \frac{?}{3}\ \frac{?}{-16}\ \frac{?}{4}\ \frac{?}{7}\ \frac{?}{8}\ \frac{?}{12}\ ?$$

$$\frac{?}{-1}\ \frac{?}{2}\ \frac{?}{12}\quad \frac{?}{4}\quad \frac{?}{5}\ \frac{?}{3}\ \frac{?}{8}\ !$$

Solve each equation below. Each time the solution to an equation appears in the code, write the letter that is the variable of that equation.

Use a separate sheet of paper onto which you have copied the code.

$2A + 1 = 7$

$2C + 4 = 14$

$7E - 2 = 12$

$3I - 4 = 8$

$5L - 28 = 2L + 17$

$4N + 3 = N + 27$

$2O + 5O + 2 = 51$

$\frac{3}{5}Q = \frac{33}{5}$

$S - 16 = 8 - S$

$\frac{1}{4}T = -4$

$5U + 53 = U + 57$

$3V - 2 = 4V + 20$

$2Y + Y + 1 = 4Y + 2$

2. To solve this problem, first guess and check, then use an equation.

Find three consecutive integers such that 3 times the first is equal to 18 more than the last.

6.5 Removing Parentheses

The doorbell rang for Trick or Treat. To each of the little children there, Mrs. Wilson gave a caramel apple and a number of pennies. As the children were leaving, Mrs. Wilson found enough pennies in her apron pocket to give each child one more penny. If she had given out 15 pennies in all, and there were 3 children, how many pennies had each child first gotten, before the extra penny?

Did you say that each child first got 4 pennies?
Explain your reasoning.

If there were 15 pennies in all, each of the 3 children wound up with 5 pennies. Before the extra penny, each had 4 pennies.

Consider an equation to model this problem.

Let n = the original number of pennies for each child.
Then $n + 1$ = the final number of pennies for each child
and $3(n + 1)$ = the total number of pennies distributed.

$3(n + 1) = 15$

Use chips to model the equation.

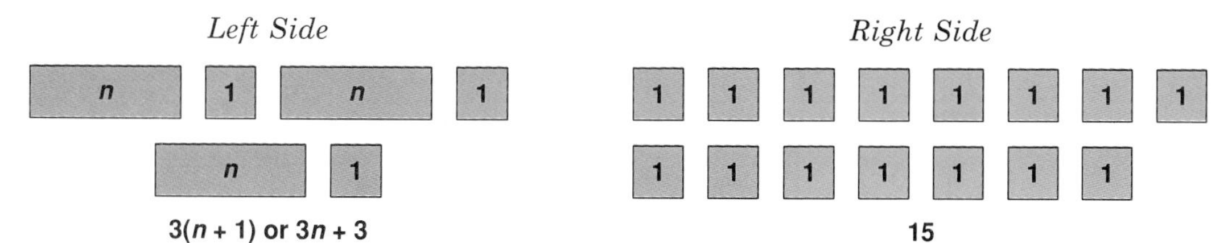

Use the model of the equation to see the solution. The object is to get n alone on the left side. To get only n-chips on the left, remove three 1-chips. To keep the balance, remove three 1-chips from the right.

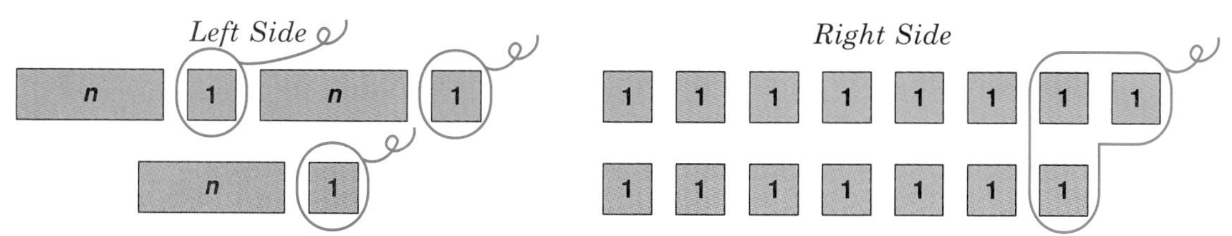

Since the left side has 3 n-chips, separate the chips into 3 identical groups.

Each group shows that $n = 4$.

Summarize the result of the model.

$$3(n + 1) = 15 \quad \text{Equation contains parentheses.}$$
$$3n + 3 = 15 \quad \text{Multiply to remove parentheses.}$$
$$3n + 3 - 3 = 15 - 3 \quad \text{Undo + by −.}$$
$$3n = 12$$
$$\frac{3n}{3} = \frac{12}{3} \quad \text{Undo × by ÷.}$$
$$n = 4$$

If there are parentheses in an equation, remove them by multiplication to simplify the equation.

Example 1 Solve $2(t - 5) = 52$, and check.

$$2(t - 5) = 52 \quad \text{Equation contains parentheses.}$$
$$2t - 10 = 52 \quad \text{Multiply to remove parentheses.}$$
$$2t - 10 + 10 = 52 + 10 \quad \text{Undo − by +.}$$
$$2t = 62$$
$$\frac{2t}{2} = \frac{2}{2} \quad \text{Undo × by ÷.}$$
$$t = 31$$

Check: $2(t - 5) = 52$

$$2(31 - 5) \stackrel{?}{=} 52$$
$$2(26) \stackrel{?}{=} 52$$
$$52 = 52 \checkmark$$

Answer: $t = 31$

Example 2 Twice the sum of a number and 8 is 64. Find the number.

Before writing an equation to model the problem, try some guesses.

Let n = the number.

Then $n + 8$ = the sum of the number and 8

and $2(n + 8)$ = twice the sum of the number and 8.

$$2(n + 8) = 64 \quad \text{Equation contains parentheses.}$$
$$2n + 16 = 64 \quad \text{Multiply to remove parentheses.}$$
$$2n + 16 - 16 = 64 - 16 \quad \text{Undo + by −.}$$
$$2n = 48$$
$$\frac{2n}{2} = \frac{48}{2} \quad \text{Undo × by ÷.}$$
$$n = 24$$

Check: Does 24 work in the original problem?

If 24 is the number, then the sum of 24 + 8 is 32.
Does twice that sum equal 64? Yes.

Answer: The number is 24.

Try These *(For solutions, see page 235.)*

1. Solve and check.
   a. $3(2x + 1) = 15$ **b.** $2(t - 2) = 14$
2. Use an equation to solve this problem. First, guesstimate an answer.
   Twice the sum of a number and 3 is 26. Find the number.

More About Equations With Parentheses

What equation is represented by this model?

Left Side *Right Side*

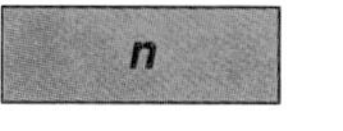

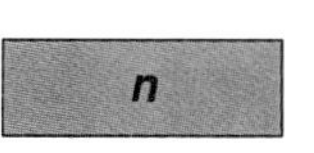

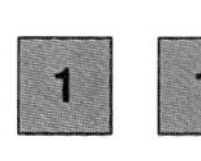

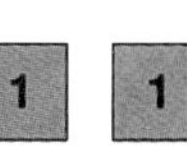

$n + 2(n + 1)$ 8

or $n + 2n + 2$ or $3n + 2$

From the model, you can read the left side of this equation in three different ways. Summarize these steps, and complete the solution.

$n + 2(n + 1) = 8$ Equation contains parentheses.

$n + 2n + 2 = 8$ Multiply to remove parentheses.

$3n + 2 = 8$ Combine like terms.

$3n + 2 - 2 = 8 - 2$ Undo + by –.

$3n = 6$

$\frac{3n}{3} = \frac{6}{3}$ Undo × by ÷.

$n = 2$

After removing parentheses in an equation, you may need to combine like terms.

Example 3 Solve $3x + 4(x + 2) = 1$, and check.

$3x + 4(x + 2) = 1$ Equation contains parentheses.

$3x + 4x + 8 = 1$ Multiply to remove parentheses.

$7x + 8 = 1$ Combine like terms.

$7x + 8 - 8 = 1 - 8$ Undo + by –.

$7x = -7$

$\frac{7x}{7} = \frac{-7}{7}$ Undo × by ÷.

$x = -1$

Check: $3x + 4(x + 2) = 1$

$3(-1) + 4(-1 + 2) \stackrel{?}{=} 1$

$-3 + 4(1) \stackrel{?}{=} 1$

$-3 + 4 = 1$

$1 = 1$ ✔

Answer: $x = -1$

Example 4 Before going to the mall, Gali added $3 to the money she had in her purse. At the last minute, she decided to take twice as much.

In his pocket, Alex had the same amount of money that Gali originally had in her purse.

If, together, Gali and Alex now had $72, how much was originally in Gali's purse?

Before writing an equation to model the problem, try some guesses.

Let p = the original amount in Gali's purse.
Then $p + 3$ = the original amount plus $3
and $2(p + 3)$ = twice the new amount.

$$p + 2(p + 3) = 72 \qquad \text{Alex's amount + Gali's amount = \$72.}$$

$$p + 2p + 6 = 72 \qquad \text{Multiply to remove parentheses.}$$

$$3p + 6 = 72 \qquad \text{Combine like terms.}$$

$$3p + 6 - 6 = 72 - 6 \qquad \text{Undo + by −.}$$

$$3p = 66$$

$$\frac{3p}{3} = \frac{66}{3} \qquad \text{Undo × by ÷.}$$

$$p = 22$$

Check: Does $22 work in the problem?

If Gali's original amount was $22, when she added $3, she had $25.
Twice this sum is $50. So, Gali took $50 to the mall.
If Alex had $22, do they together have $72?
$50 + $22 = $72 ✓

Answer: The original amount in Gali's purse was $22.

Try These *(For solutions, see page 235.)*

Solve and check.

1. $3y + 2(y - 1) = 8$ **2.** $4 + 3(x + 4) = 19$

SELF-CHECK: SECTION 6.5

1. Consider the equation $3(2x + 4) = 30$.

a. What would you do to remove the parentheses?

Answer: Distribute the 3 by multiplication.

b. When you distribute the 3 on the left side of the equation, do you get $6x + 4$? Explain.

Answer: No, you get $6x + 12$. Both terms in the parentheses must be multiplied by 3.

c. When you distribute the 3 on the left side of the equation, do you also multiply the right side by 3? Explain.

Answer: No, you are not inserting the multiplier 3 on the left side. (If you were, you would have to also multiply the right side by 3.) You are just rewriting the left side.

2. Consider the equation $x + 2(3x - 4) = 6$.

a. What is the first step in the solution?

Answer: Remove the parentheses by distributing the 2.

b. What will the result be after distributing the 2?

Answer: $x + 6x - 8 = 6$

c. What is the next step?

Answer: Combine like terms, $7x - 8 = 6$.

EXERCISES: SECTION 6.5

1. Solve by first removing parentheses. Check.

a. $3(y + 4) = 21$ **b.** $5(x - 2) = 15$ **c.** $2(3z + 1) = 14$

d. $7(2c - 1) = 14$ **e.** $3(n - 4) = -60$ **f.** $-2(b + 3) = 8$

g. $9 = 3(x - 2)$ **h.** $4 = 2(2t - 6)$ **i.** $-5(n - 2) = -20$

j. $-(a - 4) = 6$ **k.** $-2(3 - y) = 12$ **l.** $\frac{1}{2}(2x - 4) = 7$

m. $\frac{3}{5}(5x + 10) = 18$ **n.** $2(3k + 1.1) = 3.4$ **o.** $0.25(d - 8) = 2.5$

2. Solve by first removing parentheses and then combining like terms. Check.

a. $b + 2(b + 6) = 3$ **b.** $3x + 2(x - 6) = 8$ **c.** $y + 3(2y + 4) = 26$

d. $4z + 2(3z - 1) = 18$ **e.** $2n + (3n - 8) = 28$ **f.** $4y + 5(5 + y) = -2$

g. $3t + 2(6 + 2t) = 33$ **h.** $8 + 3(k - 2) = 14$ **i.** $3t + 4(1 - t) = 3$

j. $5 + 2(z + 3) = -11$ **k.** $9 + 0.5(2g - 4) = 19$ **l.** $2(3 + x) + x = 6$

3. Solve and check.

a. $k - 3 = -5$
b. $r + 5 = -9$
c. $-3t = -9$
d. $\frac{w}{-2} = -8$
e. $6y + 2 = -4$
f. $7z - 3 = -17$
g. $2x + 6x + 8 = 36$
h. $3p - 9p + 2 = -10$
i. $10c = 2c + 24$
j. $8d = 36 - 4d$
k. $2p = 4p + 10$
l. $3q = 6q - 12$
m. $6k + 3 = 2k + 23$
n. $8m + 2 = 3m - 3$
o. $4(y + 3) = 20$
p. $5(2q - 1) = 15$
q. $\frac{1}{2}(4n - 6) = 11$
r. $3(5k + 0.2) = 15.6$
s. $7s + 2(s - 3) = 12$
t. $8w + 2(4 - w) = 44$
u. $4 + 2(3y - 1) = 26$
v. $3 + 5(2z - 1) = -12$
w. $4(t + 6) - 7 = -15$
x. $a + 7 + 2(a - 1) = 36$

4. Do you agree or disagree with the first step shown as the solution of each of the following equations? If you disagree, tell what you would do.

a. $3x - 1 = 6$
Add 1 to both sides.

b. $\frac{w}{2} = 18$
Divide both sides by 2.

c. $3(d + 2) = 9$
Multiply both sides by 3.

d. $6 + 4(x - 1) = 14$
Add 6 and 4.

e. $4y + 2y + 10 = 34$
Add $4y$ and $2y$.

f. $4y + 10 = 2y$
Add $4y$ and $2y$.

5. Use an equation to find each unknown number. First guesstimate an answer.

a. Twice the sum of a number and 6 is 40.
b. One-half the sum of a number and 8 is 11.
c. The product of 4 and 2 more than a number is 48.
d. The product of 3 and 4 less than a number is 9.
e. Three times a number is increased by 4. When that sum is doubled, the product is 20.
f. Twice a number is decreased by 10. When that difference is multiplied by 6, the product is 24.

6. Use an equation to find:

a. two consecutive integers such that twice the second is 40
b. two consecutive integers such that three times the second is 90
c. two consecutive even integers such that 3 times the second is 144
d. two consecutive odd integers such that twice the second is 42
e. two consecutive integers such that if the first is increased by 3 times the second, the result is 103
f. two consecutive odd integers such that if the first is increased by twice the second, the result is 127

7. Use an equation to solve each problem. First, guesstimate an answer.

   a. At Sam's Sounds, a CD costs \$6 more than a tape. If 3 CD's cost \$45, what is the cost of a tape?

   b. Renee switched to a lighter yogurt because it has 25 fewer calories than her regular kind. She calculated that if she ate a portion of the light yogurt for each of 5 days, she would be getting 630 calories. How many calories are in a serving of her regular yogurt?

   c. John runs an hour and a half longer each day than Ryan. If, at the end of a week, John has run a total of 21 hours, how many hours does Ryan run each day?

   d. 

   For a cake he is baking, a chef added 1 cup of flour to the amount that was already in a bowl. Deciding on a larger cake, he doubled this amount and then added still another cup of flour. If, in all, the cake contained 9 cups of flour, how many cups of flour were originally in the bowl?

   e. At Soo's Produce, an average honeydew melon weighs 2 pounds more than a cantaloupe. If, together, 4 cantaloupes and 2 honeydews weigh 13 pounds, what does a cantaloupe weigh?

   f. Joan lives next door to Guy and their house numbers are consecutive even integers. If the sum of their house numbers is 714, what are their house numbers?

   g. The price of Cindy's favorite brand of sweatsocks increased by 50 cents per pair. When she was preparing for camp, Cindy was able to buy 3 pairs of the socks at the old price, and she bought 3 pairs at the new price. If the total price of the socks (before tax) was \$19.50, what was the old price for a pair of socks?

   h. Cole College is offering financial assistance to 8 students from Hunter High. Two students will get loans and 6 students will get scholarships. The amount of a scholarship is \$1,500 less than the amount of a loan. If the total money package offered to the 8 students is \$47,000, what is the value of a scholarship?

8. Make up a problem that could be solved using each equation.

   a. $2(n + 1) = 50$ b. $x + 2(x + 1) = 32$

9. If $4(y - 2) = 20$ and $x + 2(x + 1) = 11$, find the value of:

   a. $x + y$ b. $(x + y)^2$ c. $x + y^2$

EXPLORATIONS

1. To complete the cross-number puzzle, solve each equation.
Each decimal point gets its own square.
Trace the diagram onto a separate sheet of paper.

1 ?	?	2 ?		3 ?	4 ?	
?		5 ?	6 ?		7 ?	8 ?
			?		9 ?	?
10 ?	11 ?	?		12 ?	?	?
13 ?	?		14 ?			
15 ?	?		16 ?	17 ?		18 ?
	19 ?	?		20 ?	?	?

ACROSS

1. $x + 0.7 = 2$
3. $3x - 1 = 35$
5. $-2x + 8 = -74$
7. $4x + 6 = 8$
9. $\frac{x}{5} = 6$
10. $\frac{x}{3} - 40 = 30$
12. $12x - 2x - 0.5 = 0.2$
13. $2x - 1.2 = 0.4$
15. $120 - x - x = 0$
16. $2(x + 15) = 100$
19. $\frac{x}{4} + 8 = 13$
20. $9 = 5(x - 20) - 4x$

DOWN

1. $5x + 20 = 80$
2. $x - 20 = 14$
4. $2x - 1.7 = 2.9$
6. $-3x - 2 = -38$
8. $\frac{x}{13} = 39$
10. $5(2x - 3) = 11$
11. $\frac{x}{2} - 400 = 501$
14. $36 - 2x - x = -3$
17. $0.3x = 15.3$
18. $7x - 5x - 32 = 6$

2. For each of the equations below:
   (1) Use algebra chips to model a solution. Check your result in the original equation.
   (2) Following the solution suggested by the model, write out an algebraic solution.

   a. $2n - (n + 3) = 8$ **b.** $3n - (n - 1) = 5$ **c.** $6n - 2(n + 1) = 6$
   d. $4 - 3(n - 1) = 13$ **e.** $4n - 2(n - 2) = 8$ **f.** $5n - 3(n - 2) = 0$

Things You Should Know After Studying This Chapter

KEY SKILLS

6.1 Solve an equation using addition or subtraction.
Use an equation to solve a problem.

6.2 Solve an equation using multiplication or division.
Use an equation to solve a problem.

6.3 Solve an equation using 2 operations.
Use an equation to solve a problem.

6.4 Solve an equation by first combining like terms.
Use an equation to solve a problem.

6.5 Solve an equation by first removing parentheses.
Use an equation to solve a problem.

KEY TERMS

6.1 balance • solution • solve •
addition equation • subtraction equation

6.2 multiplication equation •
division equation

6.3 inverse operation

6.4 like terms

6.5 parentheses • distribute

CHAPTER REVIEW EXERCISES

1. Solve by addition or subtraction, and check.

a. $h - 4 = 7$ **b.** $y + 3 = -5$

c. $16 = q + 5$ **d.** $6 + x = 4$

e. $a - 3.2 = 5.7$ **f.** $p + \frac{3}{4} = \frac{5}{6}$

2. Solve by multiplication or division, and check.

a. $\frac{x}{5} = 15$ **b.** $18 = -6y$

c. $0.3p = 1.5$ **d.** $2.8d = 56$

e. $\frac{1}{5}b = 30$ **f.** $-12 = \frac{3}{4}w$

3. Solve by using two operations, and check.

a. $3d - 7 = 17$ **b.** $2x + 3 = 11$

c. $\frac{y}{2} + 4 = 10$ **d.** $\frac{z}{3} - 2 = 7$

e. $28 = 2a + 16$ **f.** $-3t + 2 = 1$

4. Solve by first combining like terms. Check.

a. $5h + 2h = 35$ **b.** $6y - 2y = 20$

c. $5r - r + 3 = 15$ **d.** $9q + 2q - 2 = -13$

e. $25 = 4k - 2 + 5k$ **f.** $4 = 3p + 8 + p$

5. Solve by first collecting variables on one side. Check.

a. $3y = 2y + 9$ **b.** $9 + w = 4w$

c. $6b = 3b - 4$ **d.** $6x = 10x - 4$

e. $5x + 4 = x + 10$ **f.** $y + 16 = 3y - 2$

6. Solve by first removing parentheses. Check.

a. $2(a - 3) = 18$ **b.** $5(z + 8) = -45$

c. $\frac{1}{3}(3x - 6) = 7$ **d.** $\frac{1}{2}(y + 4) = 8$

e. $3h + 2(h - 7) = 1$ **f.** $6 + 2(p - 1) = 24$

7. Solve and check.

a. $\frac{x}{3} = 18$

b. $15 = 5 + y$

c. $2.5a = 7.5$

d. $z - 2 = -7$

e. $4d - 6 = 26$

f. $\frac{1}{2}t + 8 = 18$

g. $8w + 2 - 3w = 12$

h. $5 + y - 3y = 19$

i. $4(x - 7) = 24$

j. $28 = 2(b + 3)$

k. $7q + 2(q - 1) = 7$

l. $8(3y + 1) + 4 = 24$

m. $8m - 1 = 6m + 13$

n. $3x - 12 = 16 + x$

8. Use an equation to find each unknown number. First guesstimate an answer.

a. One-half of a number is 18.

b. The sum of a number and 8 is 35.

c. Four less than a number is 28.

d. If 3 times a number is decreased by 8, the result is 19.

e. The difference between 6 times a number and 4 times the number is 32.

f. Three times a number is equal to 6 more than the number.

g. Twice the sum of a number and 4 is equal to 42.

CHAPTER REVIEW EXERCISES

9. Use an equation to find:

a. two consecutive integers whose sum is 199

b. two consecutive odd integers whose sum is 44

c. two consecutive even integers whose sum is 98

d. two consecutive integers such that the sum of the first and twice the second is 182

10. Use an equation to solve each problem. First guesstimate an answer.

a. A dealer sold a radio for $29.99, which was $13 more than he paid for it. What had he paid?

b. At a garage sale, Cindy sold a sweater for $12 less than she had paid for it last year. If she sold the sweater for $17, how much had she paid for it?

c. In a city parking lot, the number of unrestricted parking spaces is 12 more than 7 times the number of spaces reserved for the handicapped. If there are 52 parking spaces in all, how many are set aside for the handicapped?

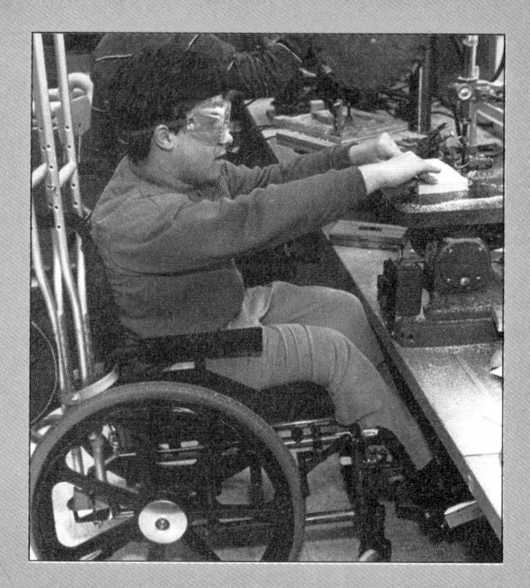

d. George needs $90 to buy a camera. He has already saved $12. If he saves $13 each week, how many weeks will it take him to reach his goal?

e. Nick noticed that all the boats on the lake this morning had either 2 sails or 3 sails. If there were the same number of sailboats of each type, and he counted 40 sails in all, how many sailboats were on the lake?

f. Eric started with twice as many comic books as Ali. After Eric bought 4 more comics and Ali bought 9 more, they each had the same number. How many comics did Ali start with?

g. Cherie bought a bag of plums. She gave 2 plums to her friend, and then shared the remaining plums equally with her sister. If Cherie wound up with 6 plums for herself, how many were in the bag she bought?

SOLUTIONS FOR TRY THESE

6.1 Using Addition or Subtraction

Page 194

1. a. $x + 4 = 12$; $x + 4 - 4 = 12 - 4$; $x = 8$
   Check: $x + 4 = 12$; $8 + 4 \stackrel{?}{=} 12$; $12 = 12$ ✓
   Answer: $x = 8$

   b. $y - 6 = 18$; $y - 6 + 6 = 18 + 6$; $y = 24$
   Check: $y - 6 = 18$; $24 - 6 \stackrel{?}{=} 18$; $18 = 18$ ✓
   Answer: $y = 24$

   c. $t + 7 = 2$; $t + 7 - 7 = 2 - 7$; $t = -5$
   Check: $t + 7 = 2$; $-5 + 7 \stackrel{?}{=} 2$; $2 = 2$ ✓
   Answer: $t = -5$

   d. $r - 8 = -4$; $r - 8 + 8 = -4 + 8$; $r = 4$
   Check: $r - 8 = -4$; $4 - 8 \stackrel{?}{=} -4$; $-4 = -4$ ✓
   Answer: $r = 4$

   e. $b + 0.7 = 2.9$; $b + 0.7 - 0.7 = 2.9 - 0.7$; $b = 2.2$
   Check: $b + 0.7 = 2.9$; $2.2 + 0.7 \stackrel{?}{=} 2.9$; $2.9 = 2.9$ ✓
   Answer: $b = 2.2$

   f. $a - \frac{1}{2} = 7\frac{1}{2}$; $a - \frac{1}{2} + \frac{1}{2} = 7\frac{1}{2} + \frac{1}{2}$; $a = 8$
   Check: $a - \frac{1}{2} = 7\frac{1}{2}$; $8 - \frac{1}{2} \stackrel{?}{=} 7\frac{1}{2}$; $7\frac{1}{2} = 7\frac{1}{2}$ ✓
   Answer: $a = 8$

   g. $20 = x + 2$; $20 - 2 = x + 2 - 2$; $18 = x$
   Check: $20 = x + 2$; $20 \stackrel{?}{=} 18 + 2$; $20 = 20$ ✓
   Answer: $x = 18$

   h. $14 = y - 3$; $14 + 3 = y - 3 + 3$; $17 = y$
   Check: $14 = y - 3$; $14 \stackrel{?}{=} 17 - 3$; $14 = 14$ ✓
   Answer: $y = 17$

2. **1** *Read for information.*

   You know that the midnight temperature increased by 36° to get to the noon temperature of 84°. You must find the midnight temperature.

   2 *Translate to an equation.*

   Let m = the midnight temperature.
   Then $m + 36$ = the noon temperature.
   $m + 36 = 84$

   3 *Solve the equation.*

   $m + 36 = 84$
   $m + 36 - 36 = 84 - 36$
   $m = 48$

   4 *Check in the problem.*

   $48 + 36 \stackrel{?}{=} 84$
   $84 = 84$ ✓

   Answer: The midnight temperature was 48°.

6.2 Using Multiplication or Division

Page 200

1. a. $5n = 75$; $\frac{5n}{5} = \frac{75}{5}$; $n = 15$
   Check: $5n = 75$; $5(15) \stackrel{?}{=} 75$; $75 = 75$ ✓
   Answer: $n = 15$

   b. $-4y = 44$; $\frac{-4y}{-4} = \frac{44}{-4}$; $y = -11$
   Check: $-4y = 44$; $-4(-11) \stackrel{?}{=} 44$; $44 = 44$ ✓
   Answer: $y = -11$

   c. $\frac{1}{4}t = 8$; $4 \cdot \frac{1}{4}t = 4 \cdot 8$; $t = 32$
   Check: $\frac{1}{4}t = 8$; $\frac{1}{4}(32) \stackrel{?}{=} 8$; $8 = 8$ ✓
   Answer: $t = 32$

   d. $\frac{n}{8} = -6$; $8 \cdot \frac{n}{8} = 8(-6)$; $n = -48$
   Check: $\frac{n}{8} = -6$; $\frac{-48}{8} \stackrel{?}{=} -6$; $-6 = -6$ ✓
   Answer: $n = -48$

e. $\frac{c}{-5} = 15$ *Check:* $\frac{c}{-5} = 15$

$-5 \cdot \frac{c}{-5} = -5 \cdot 15$ $\frac{-75}{5} \stackrel{?}{=} 15$

$c = -75$ $15 = 15$ ✓

Answer: $c = -75$

f. $0.2x = 1.6$ *Check:* $0.2x = 1.6$

$\frac{0.2x}{0.2} = \frac{1.6}{0.2}$ $0.2(8) \stackrel{?}{=} 1.6$

$x = 8$ $1.6 = 1.6$ ✓

Answer: $x = 8$

g. $10 = 2x$ *Check:* $10 = 2x$

$\frac{10}{2} = \frac{2x}{2}$ $10 \stackrel{?}{=} 2(5)$

$5 = x$ $10 = 10$ ✓

Answer: $x = 5$

h. $14 = \frac{y}{7}$ *Check:* $14 = \frac{y}{7}$

$7 \cdot 14 = \frac{y}{7} \cdot 7$ $14 \stackrel{?}{=} \frac{98}{7}$

$98 = y$ $14 = 14$ ✓

Answer: $y = 98$

2. a. Let n = the number.

$2n = 126$ *Check:* $2(63) \stackrel{?}{=} 126$

$\frac{2n}{2} = \frac{126}{2}$ $126 = 126$ ✓

$n = 63$ *Answer:* The number is 63.

b. Let r = Henry's rate.
Then $\frac{1}{3}r$ = Joe's rate.

$\frac{1}{3}r = 5$ *Check:* $\frac{1}{3}(15) \stackrel{?}{=} 5$

$3 \cdot \frac{1}{3}r = 3 \cdot 5$ $5 = 5$ ✓

$r = 15$ *Answer:* Henry's rate is $15 per hour.

6.3 Using Two Operations

Page 206

1. a. $3x + 2 = 8$ *Check:* $3x + 2 = 8$

$3x + 2 - 2 = 8 - 2$ $3(2) + 2 \stackrel{?}{=} 8$

$3x = 6$ $6 + 2 \stackrel{?}{=} 8$

$\frac{3x}{3} = \frac{6}{3}$ $8 = 8$ ✓

$x = 2$ *Answer:* $x = 2$

b. $2y - 5 = 15$ *Check:* $2y - 5 = 15$

$2y - 5 + 5 = 15 + 5$ $2(10) - 5 \stackrel{?}{=} 15$

$2y = 20$ $20 - 5 \stackrel{?}{=} 15$

$\frac{2y}{2} = \frac{20}{2}$ $15 = 15$ ✓

$y = 10$ *Answer:* $y = 10$

c. $\frac{t}{3} + 2 = 11$ *Check:* $\frac{t}{3} + 2 = 11$

$\frac{t}{3} + 2 - 2 = 11 - 2$ $\frac{27}{3} + 2 \stackrel{?}{=} 11$

$\frac{t}{3} = 9$ $9 + 2 \stackrel{?}{=} 11$

$3 \cdot \frac{t}{3} = 3 \cdot 9$ $11 = 11$ ✓

$t = 27$ *Answer:* $t = 27$

d. $\frac{z}{2} - 1 = 5$ *Check:* $\frac{z}{2} - 1 = 5$

$\frac{z}{2} - 1 + 1 = 5 + 1$ $\frac{12}{2} - 1 \stackrel{?}{=} 5$

$\frac{z}{2} = 6$ $6 - 1 \stackrel{?}{=} 5$

$2 \cdot \frac{z}{2} = 6 \cdot 2$ $5 = 5$ ✓

$z = 12$ *Answer:* $z = 12$

2. Let s = the amount Mary saved.
Then $3s - 8$ = the cost of the sweater.

$3s - 8 = 52$

$3s - 8 + 8 = 52 + 8$

$3s = 60$

$\frac{3s}{3} = \frac{60}{3}$

$s = 20$

Check: Does $20 work in the problem?
If Mary saved $20, then $3(20) - 8$, or $52, is the cost of the sweater. ✓

Answer: Mary saved $20.

SOLUTIONS FOR TRY THESE

Page 208

1.

$$
\begin{aligned}
7 - r &= 2 \\
7 - r + r &= 2 + r \\
7 &= 2 + r \\
7 - 2 &= 2 - 2 + r \\
5 &= r
\end{aligned}
$$

Check:

$$
\begin{aligned}
7 - r &= 2 \\
7 - 5 &\stackrel{?}{=} 2 \\
2 &= 2 \checkmark
\end{aligned}
$$

Answer: $r = 5$

2.

$$
\begin{aligned}
2 - w &= -3 \\
2 - w + w &= -3 + w \\
2 &= -3 + w \\
2 + 3 &= -3 + 3 + w \\
5 &= w
\end{aligned}
$$

Check:

$$
\begin{aligned}
2 - w &= -3 \\
2 - 5 &\stackrel{?}{=} -3 \\
-3 &= -3 \checkmark
\end{aligned}
$$

Answer: $w = 5$

6.4 Combining Like Terms

Page 213

1. a.

$$
\begin{aligned}
4y - 3y + 5 &= 8 \\
y + 5 &= 8 \\
y + 5 - 5 &= 8 - 5 \\
y &= 3
\end{aligned}
$$

Check:

$$
\begin{aligned}
4y - 3y + 5 &= 8 \\
4(3) - 3(3) + 5 &\stackrel{?}{=} 8 \\
12 - 9 + 5 &\stackrel{?}{=} 8 \\
3 + 5 &\stackrel{?}{=} 8 \\
8 &= 8 \checkmark
\end{aligned}
$$

Answer: $y = 3$

b.

$$
\begin{aligned}
8 &= 7z - 6z - 2 \\
8 &= z - 2 \\
8 + 2 &= z - 2 + 2 \\
10 &= z
\end{aligned}
$$

Check:

$$
\begin{aligned}
8 &= 7z - 6z - 2 \\
8 &\stackrel{?}{=} 7(10) - 6(10) - 2 \\
8 &\stackrel{?}{=} 70 - 60 - 2 \\
8 &\stackrel{?}{=} 10 - 2 \\
8 &= 8 \checkmark
\end{aligned}
$$

Answer: $z = 10$

c.

$$
\begin{aligned}
3x + 2x + 7 &= 27 \\
5x + 7 &= 27 \\
5x + 7 - 7 &= 27 - 7 \\
5x &= 20 \\
\frac{5x}{5} &= \frac{20}{5} \\
x &= 4
\end{aligned}
$$

Check:

$$
\begin{aligned}
3x + 2x + 7 &= 27 \\
3(4) + 2(4) + 7 &\stackrel{?}{=} 27 \\
12 + 8 + 7 &\stackrel{?}{=} 27 \\
20 + 7 &\stackrel{?}{=} 27 \\
27 &= 27 \checkmark
\end{aligned}
$$

Answer: $x = 4$

2. Let x = the first even integer.
Then $x + 2$ = the next even integer.

$$
\begin{aligned}
x + x + 2 &= 50 \\
2x + 2 &= 50 \\
2x + 2 - 2 &= 50 - 2 \\
2x &= 48 \\
\frac{2x}{2} &= \frac{48}{2} \\
x &= 24 \\
x + 2 &= 26
\end{aligned}
$$

Check: Are 24 and 26 consecutive even integers? Yes. Is the sum of 24 and 26 equal to 50? Yes.

Answer: 24 and 26 are the two consecutive even integers whose sum is 50.

Page 216

1.

$$
\begin{aligned}
3t &= 2t + 27 \\
3t - 2t &= 2t - 2t + 27 \\
t &= 27
\end{aligned}
$$

Check:

$$
\begin{aligned}
3t &= 2t + 27 \\
3(27) &\stackrel{?}{=} 2(27) + 27 \\
81 &\stackrel{?}{=} 54 + 27 \\
81 &= 81 \checkmark
\end{aligned}
$$

Answer: $t = 27$

2. $9m = m + 16$

$9m - m = m - m + 16$

$8m = 16$

$\frac{18}{2} = \frac{16}{8}$

$m = 2$

Check: $9m = m + 16$

$9(2) \stackrel{?}{=} 2 + 16$

$18 = 18$ ✓

Answer: $m = 2$

3. $5z + 1 = 2z + 10$

$5z - 2z + 1 = 2z - 2z + 10$

$3z + 1 = 10$

$3z + 1 - 1 = 10 - 1$

$3z = 9$

$\frac{3z}{3} = \frac{9}{3}$

$z = 3$

Check: $5z + 1 = 2z + 10$

$5(3) + 1 \stackrel{?}{=} 2(3) + 10$

$15 + 1 = 6 + 10$

$16 = 16$ ✓

Answer: $z = 3$

6.5 Removing Parentheses

Page 223

1. a. $3(2x + 1) = 15$

$6x + 3 = 15$

$6x + 3 - 3 = 15 - 3$

$6x = 12$

$\frac{6x}{6} = \frac{12}{6}$

$x = 2$

Check: $3(2x + 1) = 15$

$3(2 \cdot 2 + 1) \stackrel{?}{=} 15$

$3(4 + 1) \stackrel{?}{=} 15$

$3(5) \stackrel{?}{=} 15$

$15 = 15$ ✓

Answer: $x = 2$

b. $2(t - 2) = 14$

$2t - 4 = 14$

$2t - 4 + 4 = 14 + 4$

$2t = 18$

$\frac{2t}{2} = \frac{18}{2}$

$t = 9$

Check: $2(t - 2) = 14$

$2(9 - 2) \stackrel{?}{=} 14$

$2(7) \stackrel{?}{=} 14$

$14 = 14$ ✓

Answer: $t = 9$

2. Let n = the number.
Then $n + 3$ = the sum of the number and 3
and $2(n + 3)$ = twice that sum.

$2(n + 3) = 26$

$2n + 6 = 26$

$2n + 6 - 6 = 26 - 6$

$2n = 20$

$\frac{2n}{2} = \frac{20}{2}$

$n = 10$

Check: Does 10 work in the problem?
If the number is 10, then the sum of the number and 3 is 13. Does twice that sum equal 26? Yes.

Answer: The number is 10.

Page 224

1. $3y + 2(y - 1) = 8$

$3y + 2y - 2 = 8$

$5y - 2 = 8$

$5y - 2 + 2 = 8 + 2$

$5y = 10$

$\frac{5y}{5} = \frac{10}{5}$

$y = 2$

Check: $3y + 2(y - 1) = 8$

$3 \cdot 2 + 2(2 - 1) \stackrel{?}{=} 8$

$6 + 2(1) \stackrel{?}{=} 8$

$6 + 2 \stackrel{?}{=} 8$

$8 = 8$ ✓

Answer: $y = 2$

2. $4 + 3(x + 4) = 19$

$4 + 3x + 12 = 19$

$3x + 16 = 19$

$3x + 16 - 16 = 19 - 16$

$3x = 3$

$\frac{3x}{3} = \frac{3}{3}$

$x = 1$

Check: $4 + 3(x + 4) = 19$

$4 + 3(1 + 4) \stackrel{?}{=} 19$

$4 + 3(5) \stackrel{?}{=} 19$

$4 + 15 \stackrel{?}{=} 19$

$19 = 19$ ✓

Answer: $x = 1$

Inequalities 7

A sign on a playground says:

CHILDREN
8 AND UNDER
MAY PLAY HERE

A sign on another playground says:

CHILDREN
UNDER THE AGE OF 8
MAY PLAY HERE

Do the signs have the same meaning? Explain.

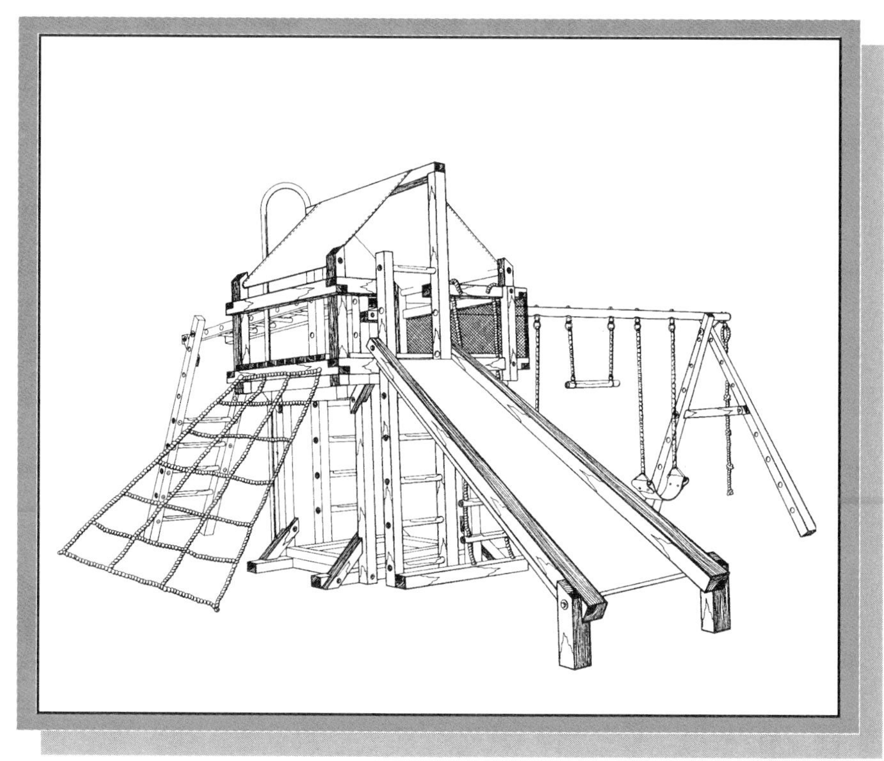

7.1 Symbols and Graphing

As required by law, the Promenade Diner displays this sign. According to this rule, is it legal for 139 people to be in the restaurant?

> **OCCUPANCY BY MORE THAN 139 PERSONS IS DANGEROUS AND UNLAWFUL.**
>
> **Commissioner**
> **Dept. of Buildings**

This rule forbids more than 139, but allows up to and including 139.

What is another way of expressing this rule?

The number of people can be less than or equal to 139. In symbols: ≤ 139

You are familiar with some symbols that show two quantities are *not equal*. Still other symbols may be new to you.

Example 1 Translate from words to symbols.

In Words	*Using Symbols*
2 is less than 5	$2 < 5$
5 is greater than 2	$5 > 2$
2 is not equal to 5	$2 \neq 5$
n is less than or equal to 2	$n \leq 2$
x is greater than or equal to 5	$x \geq 5$

Is the sentence $5 < 2$ true or false?

Since 5 is greater than 2, the sentence is false.

Is the sentence $n \leq 2$ true or false?

Since n has not been given a specific value, this is an open sentence.

Name a value to replace n in the open sentence $n \leq 2$ that will make the sentence true. How many values will make this open sentence true?

Some of the infinite number of values that make this open sentence true are 2, 1.9, 1, $\frac{1}{3}$, 0, –1, –1,000.

Here is a way of showing the solution set of this open sentence on a number line.

Notice that the points corresponding to the number 2 and all the signed numbers to the left of 2 are shaded in this picture, or ***graph***, of the solution set.

Example 2 Using n as the variable, write the inequality that is shown in each graph.

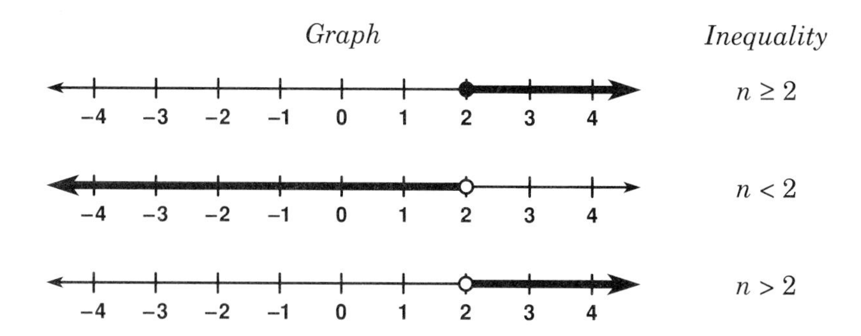

Observe that when the number that starts the graph is included in the graph, there is a full dot (called a ***closed hole***) at the number. When the starting number is not included in the graph, there is an ***open hole*** at the number.

Example 3 Graph the solution set of each inequality.

a. $x \leq 4$

The solution set of this open sentence is the number 4 and all signed numbers less than 4.

① Draw a number line that shows integers less than 4, and include a few integers greater than 4.

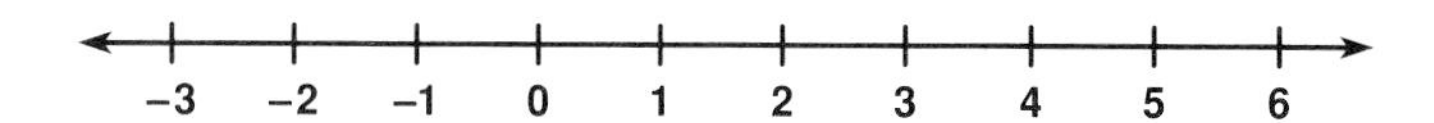

② Since 4 is in the solution set, draw a closed hole at 4.

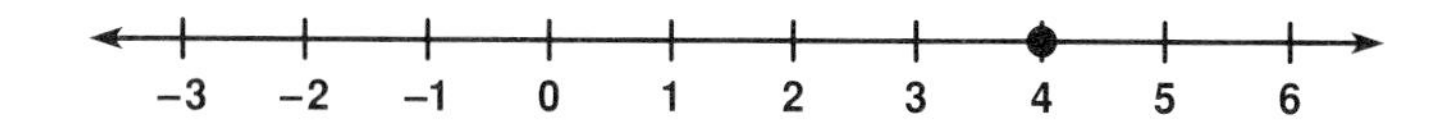

③ Since all signed numbers less than 4 are in the solution set, shade in the number line to the left of 4. The arrow at the end of the shading means that the graph continues without end.

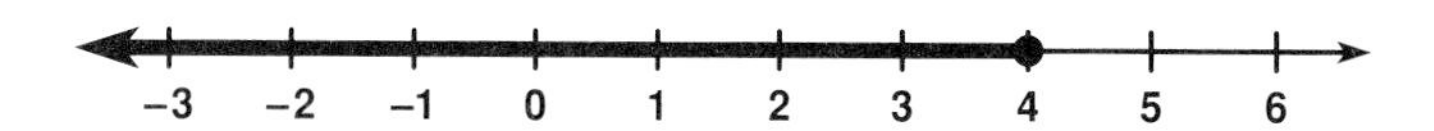

b. $n > -1$

The solution set of this open sentence is all signed numbers that are greater than -1. The number -1 is not included.

Draw a number line that shows integers greater than -1, and include a few integers less than -1.

Since -1 is not in the solution set, draw an open hole at -1.

Since all signed numbers greater than -1 are in the solution set, shade in the number line to the right of -1. Include an arrow to show that the graph continues without end.

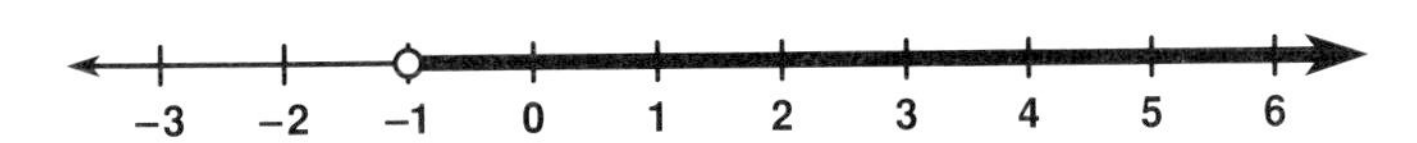

Try These *(For solutions, see page 276.)*

1. Write each of these sentences of inequality in symbols.
   a. 8 is greater than 7.2
   b. -1 is less than 0
   c. x is less than or equal to 9
   d. 9 is not equal to $10 - 2$
2. Write each of these sentences of inequality in words.
   a. $9 < 17$
   b. $-4 > -9$
   c. $x \geq 5$
   d. $7 - 1 \neq \frac{12}{3}$
3. Using n as the variable, write the inequality that is shown in each graph.
   a. −4 −3 −2 −1 0 1 2 3 4
   b. −4 −3 −2 −1 0 1 2 3 4
4. Draw the graph of each inequality.
   a. $x > 3$
   b. $n \leq -2$

Between Two Numbers

For a conference day, Ms. Norris had scheduled workshops from 11 A.M. through 1 P.M. and from 3 P.M. through 5 P.M.

During which hours could participants have lunch and relax?

Did you say that the lunch time was *between* 1 and 3?

Here is a graph of the lunch time, using a number line to represent the conference day.

Lunch time is between 1 and 3.
$1 < \text{lunch} < 3$

The numbers that are between 1 and 3 are greater than 1 *and* less than 3. In this inequality, neither end number is included.

Example 4 Graph the solution set of $-1 \leq n < 2$.

The solution set of this open sentence is between the numbers -1 and 2, including the number -1 but not including the number 2.

Here is a summary of some inequality graphs, where a and b represent any two signed numbers, with a less than b.

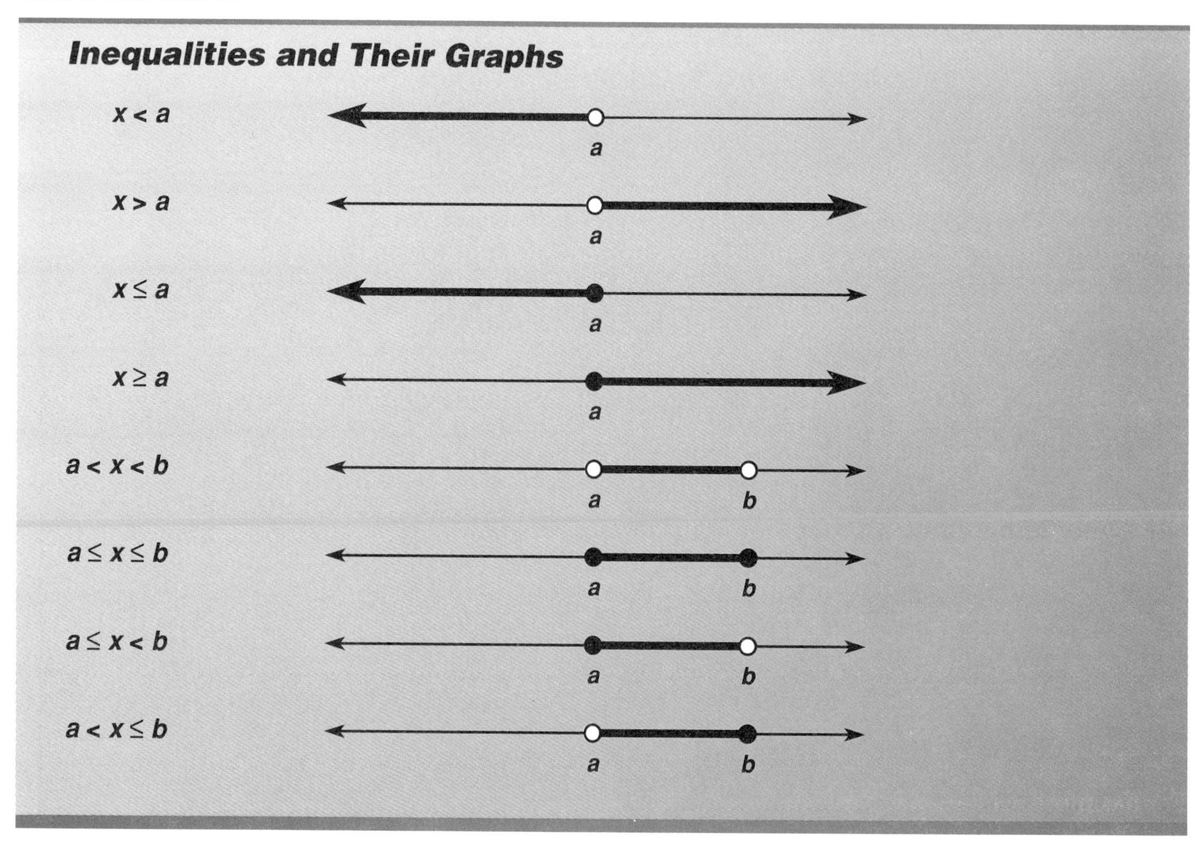

Try These (For solutions, see page 276.)

1. Write this sentence in symbols.
   x is between 5 and 9
2. Write this sentence in words.
   $-1 < y < 4$
3. Using n as the variable, write the inequality that is shown on the graph.

4. Draw the graph of each inequality.
   a. $-2 < n < 3$ **b.** $0 \leq n < 4$

SELF-CHECK: SECTION 7.1

1. What symbol represents *is greater than*?

Answer: $>$

2. What words does the symbol $\leq$ represent?

Answer: is less than or equal to

3. What symbol represents *is not equal to*?

Answer: $\neq$

4. What words does the symbol $\geq$ represent?

Answer: is greater than or equal to

5. Consider the inequality $x > 1$.

a. Is 3 in the solution set? Explain.

Answer: Yes, the solution set contains all signed numbers that are greater than 1.

b. When graphing the solution set, how would you represent the number 1? Explain.

Answer: By an open hole, since 1 is the starting number of the graph but is not included in the graph.

c. On which side of 1 would you draw the graph to represent the solution set?

Answer: To the right of 1.

d. Why must you include an arrow at the end of the shaded portion of the graph?

Answer: To show that the solution set continues without end.

e. Since the number –2 is not included in the solution set, would you represent –2 by an open hole on this graph? Explain.

Answer: No, an open or closed hole is used only at the start of the solution set.

f. How does the graph show that a number like 3.5 is included in the solution set?

Answer: When you draw the shaded bar between two integers, such as between 3 and 4, you are showing that all the numbers between these integers, such as 3.5 are included in the graph.

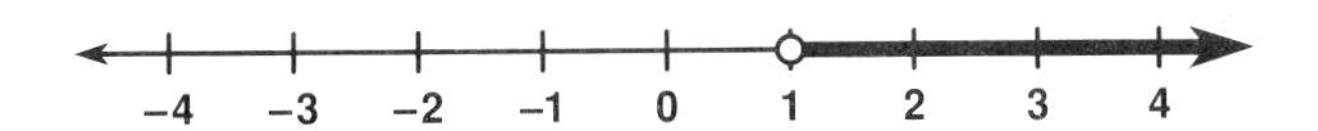

This graph represents the solution set of the inequality $x > 1$. We also say that this graph represents the inequality $x > 1$.

EXERCISES: SECTION 7.1

1. Match each word phrase in *Column A* to a symbol in *Column B*.

Column A	*Column B*	
a. is greater than	**(1)** $\neq$	**(5)** $<$
b. is not equal to	**(2)** $=$	**(6)** $\not<$
c. is less than or equal to	**(3)** $\not>$	**(7)** $>$
d. is less than	**(4)** $\geq$	**(8)** $\leq$
e. is greater than or equal to		

2. Translate from words to symbols.

a. 2 is greater than 0
b. –1 is less than 1
c. –4 is less than –3
d. 6 is not equal to 7 – 3
e. 7 is between 3 and 9
f. –8 is between –9 and –7
g. 6 is between 8 and 5
h. –3 is between –2 and –4
i. 7.3 is between 7 and 8
j. 6.2 is between 7 and 6
k. $\frac{3}{5}$ is between $\frac{4}{5}$ and $\frac{1}{5}$
l. $\frac{1}{2}$ is between $\frac{5}{6}$ and $\frac{1}{6}$
m. n is greater than or equal to 7
n. n is less than or equal to –2

3. Match each *sentence* with a *symbolic statement* from the list below.

Sentences	*Symbolic Statement*
a. n is greater than 4 and n is less than 7	**(1)** $-2 < n < 5$
b. n is less than 9 and n is greater than 2	**(2)** $2 \leq n < 10$
c. n is greater than –2 and n is less than 5	**(3)** $0 \leq n \leq 5$
d. n is less than –1 and n is greater than –4	**(4)** $1 < n \leq 3$
e. n is less than or equal to 3 and n is greater than 1	**(5)** $2 < n < 9$
f. n is greater than or equal to 2 and n is less than 10	**(6)** $4 < n < 7$
g. n is greater than or equal to –3 and n is less than or equal to 1	**(7)** $-3 \leq n \leq 1$
h. n is less than or equal to 5 and n is greater than or equal to 0	**(8)** $-4 < n < -1$

4. Translate from symbols to words.

a. $3 < 9$
b. $9 > 3$
c. $3 \neq 9$
d. $n \geq 8$
e. $n \leq 2$
f. $n \neq 7$
g. $1 < 6 < 9$
h. $8 > 5 > 3$
i. $-3 < -2 < -1$
j. $3 < n < 5$
k. $9 > n > 6$
l. $2 \leq n < 5$
m. $-1 \leq n \leq 0$
n. $8 \geq n \geq 1$
o. $3 < n \leq 9$

5. In each set of sentences, select the sentence that is always true.

a. (1) $8 < 2$ (2) $2 < 8$ (3) $n > 8$ (4) $n < 8$

b. (1) $-2 > -1$ (2) $-2 > 0$ (3) $-2 > -3$ (4) $-2 > n$

c. (1) $\frac{1}{2} < \frac{1}{3}$ (2) $\frac{1}{2} < 0$ (3) $-\frac{1}{2} < 0$ (4) $\frac{1}{2} < n$

d. (1) $0.01 > 0.1$ (2) $0.1 > 0.01$ (3) $0.1 > \frac{1}{10}$ (4) $n > 0.1$

e. (1) $10\% < \frac{1}{10}$ (2) $10\% > \frac{1}{10}$ (3) $10\% = \frac{1}{10}$ (4) $10\% \neq 0.1$

f. (1) $\frac{1}{5} = 1.5$ (2) $\frac{1}{5} \neq 0.15$ (3) $\frac{1}{5} < 10\%$ (4) $\frac{1}{5} = 15\%$

g. (1) $0 > 1 > 2$ (2) $2 > 1 < 0$ (3) $0 < 1 < 2$ (4) $0 \neq 1 < 2$

h. (1) $-3 < -2 < -1$ (2) $-3 > -2 > -1$ (3) $-3 > -2 < -1$

i. (1) $0.1 < 0.2 < 0.03$ (2) $0.03 > 0.2 > 0.1$ (3) $0.03 < 0.1 < 0.2$

6. In each set of sentences, select the sentence that is false.

a. (1) $2 < 6$ (2) $6 < 7$ (3) $7 > 6$ (4) $7 < 2$

b. (1) $-9 > -8$ (2) $-8 < -7$ (3) $-8 \neq -9$ (4) $-9 \neq -8$

c. (1) $0.4 < 0.5$ (2) $0.5 > 0.4$ (3) $0.5 \neq 0.4$ (4) $0.5 = 0.6 - 1$

d. (1) $0.5 = \frac{1}{2}$ (2) $0.5 \neq 0.6$ (3) $0.6 > 0.5$ (4) $0.5 > 0.6$

e. (1) $\frac{1}{3} < \frac{2}{3}$ (2) $\frac{2}{3} > 1$ (3) $-\frac{2}{3} < \frac{1}{3}$ (4) $-1 < -\frac{1}{3}$

f. (1) $10\% = \frac{1}{10}$ (2) $10\% = 0.10$ (3) $10\% = 0.1$ (4) $10\% = 0.01$

g. (1) $\frac{1}{5} = 20\%$ (2) $\frac{1}{5} \neq 0.02$ (3) $\frac{1}{5} = 0.02$ (4) $\frac{1}{5} = 0.20$

h. (1) $-1 < -2 < -3$ (2) $-3 < -2 < -1$ (3) $-3 > -4 > -5$

i. (1) $0.1 < 0.2 < 0.3$ (2) $0.1 < 0.20 < 0.3$ (3) $0.1 < 0.02 < 0.03$

7. Select the graph that best represents the given inequality.

a. $x > 2$

(1) 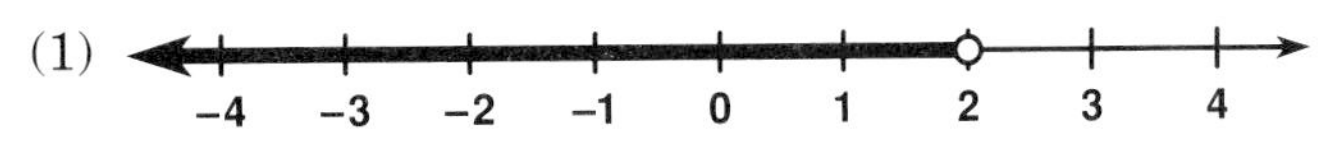

(2) 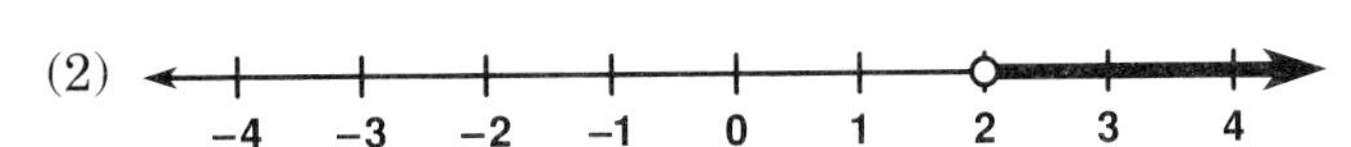

(3) 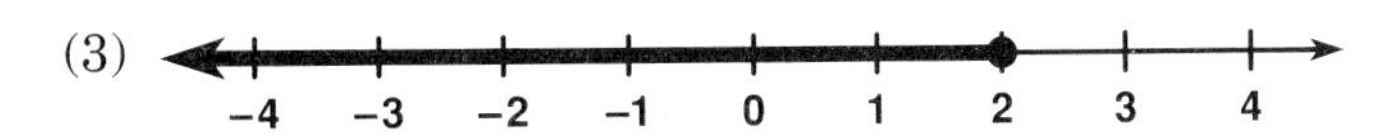

(4) 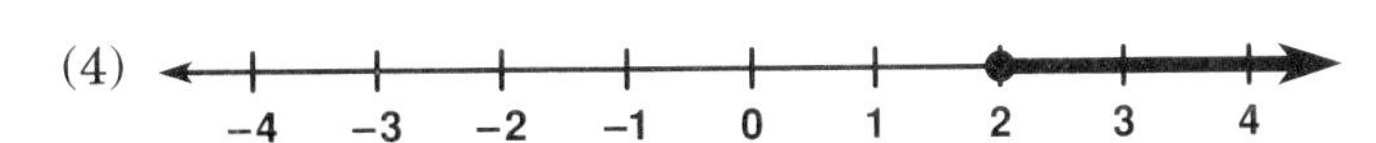

b. $x \leq -1$

(1) 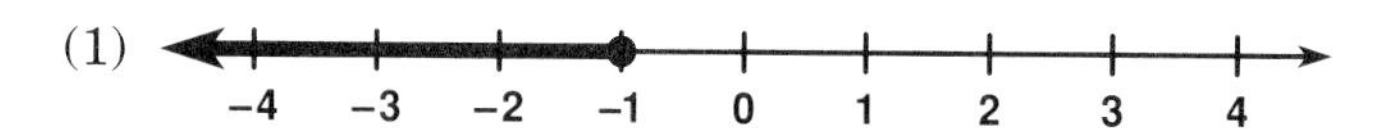

(2) 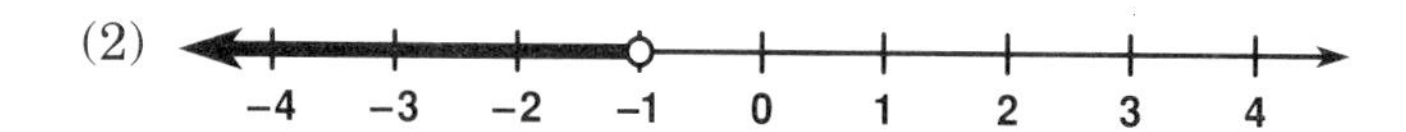

(3) 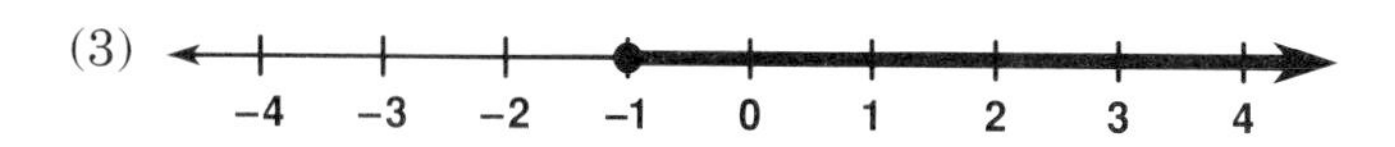

(4) 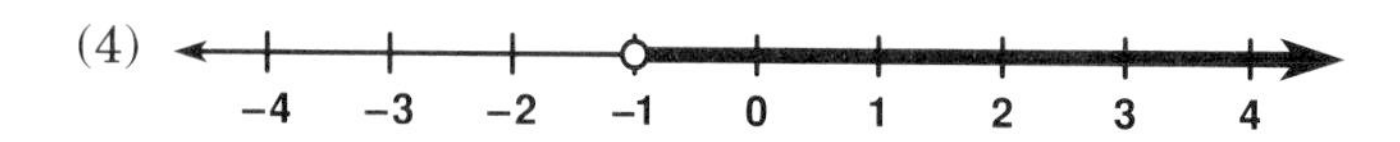

c. $-1 \leq n \leq 2$

(1) 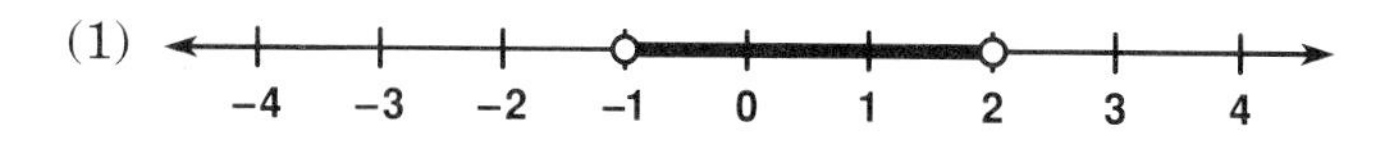

(2) 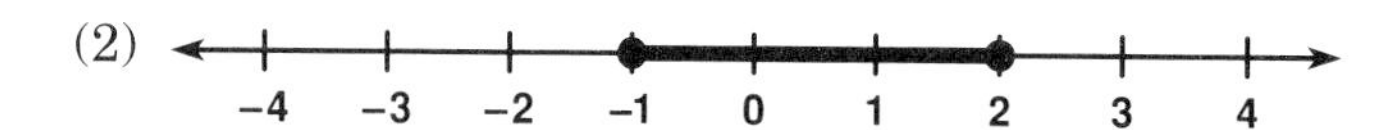

(3) 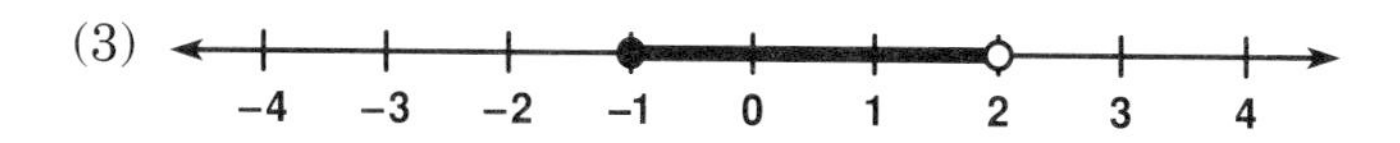

(4) 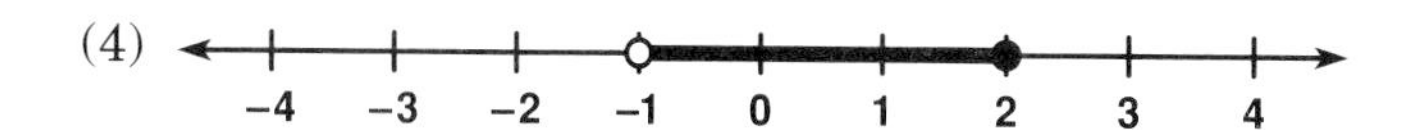

d. $3 \geq n > 1$

(1) 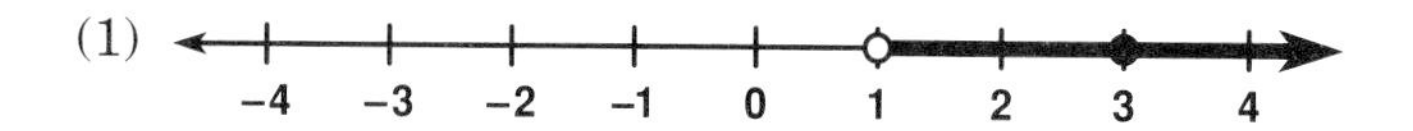

(2) 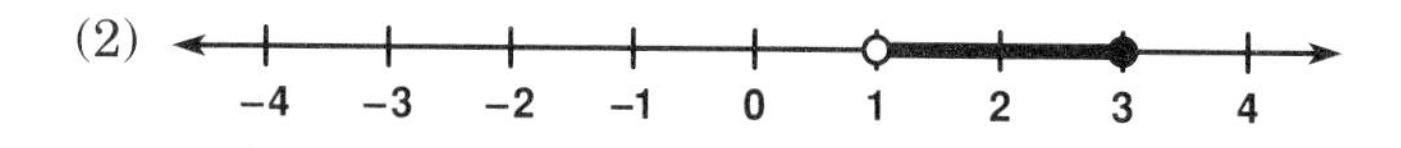

(3) 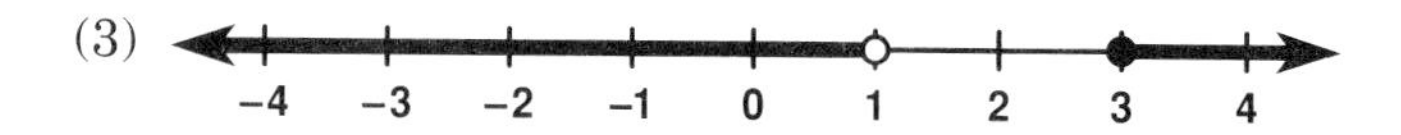

(4) 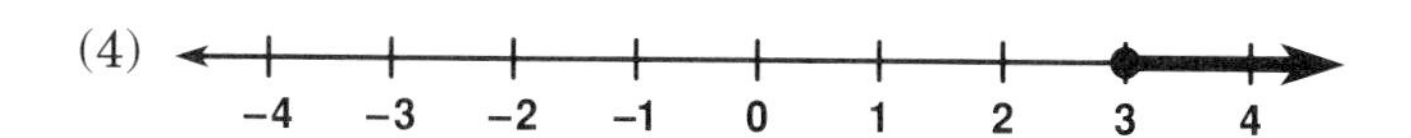

e. $1 < n$

(1) 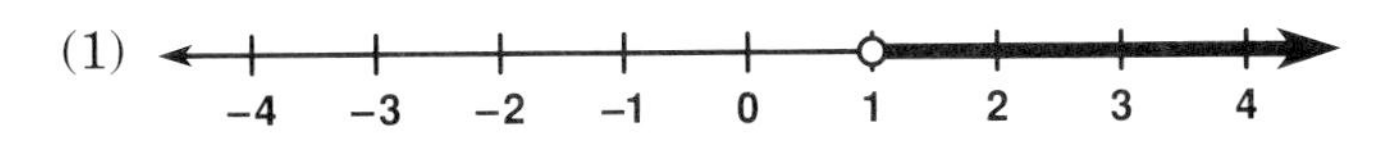

(2) 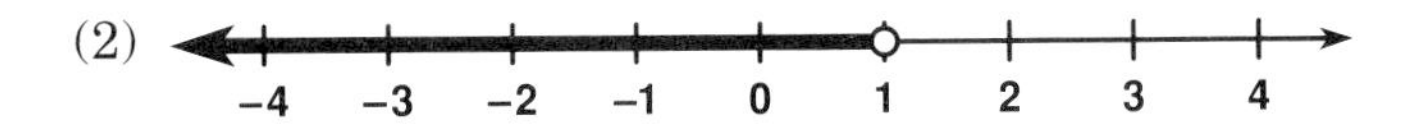

(3) 

(4) 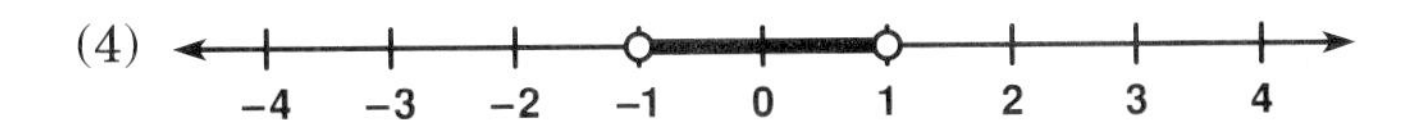

f. $-1 \geq n$

(1) 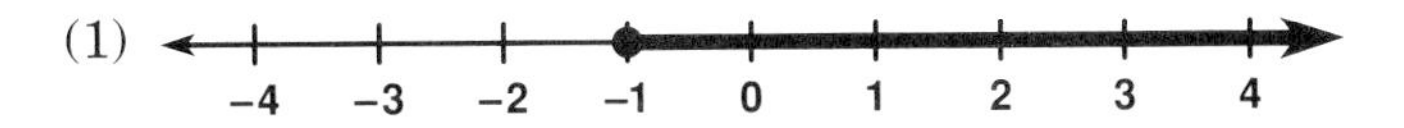

(2) 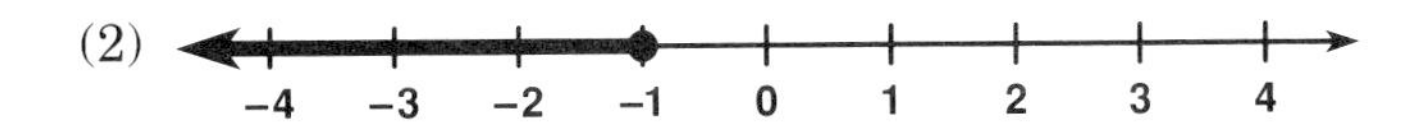

(3) 

(4) 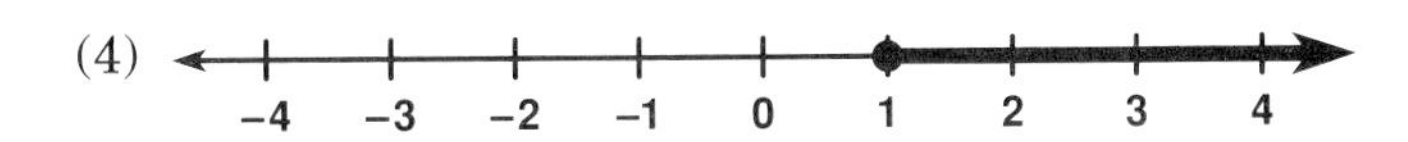

g. $-2 \leq n$

(1) 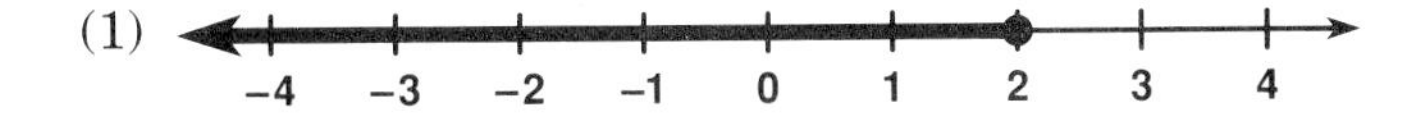

(2) 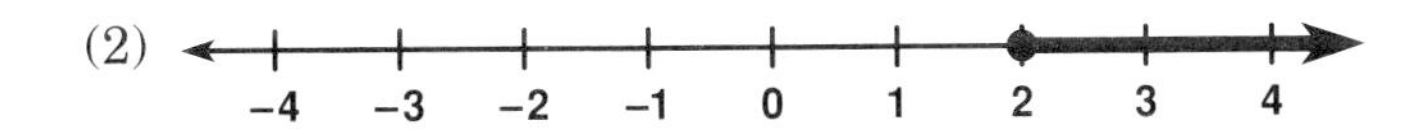

(3) 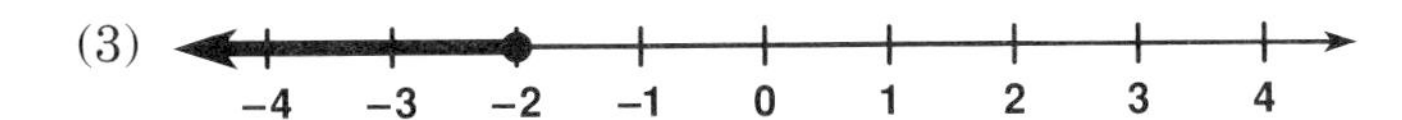

(4) 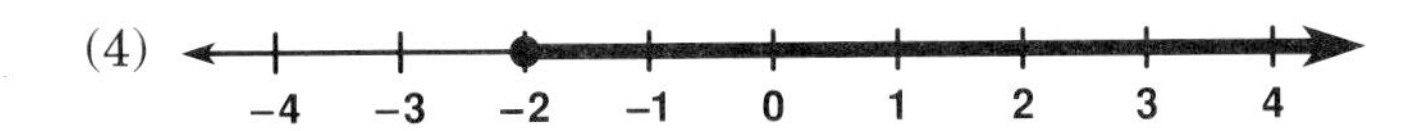

8. Using n as the variable, write the inequality that is shown in each graph.

a. −3 −2 −1 0 1 2 3

b. −3 −2 −1 0 1 2 3

c. −3 −2 −1 0 1 2 3

d. −3 −2 −1 0 1 2 3

e. −3 −2 −1 0 1 2 3

f. −3 −2 −1 0 1 2 3

g. −3 −2 −1 0 1 2 3

h. −3 −2 −1 0 1 2 3

9. Draw the graph of each inequality.

a. $x \geq 5$

b. $x < 7$

c. $x > -3$

d. $x \leq 0$

e. $8 > x$

f. $4 < x$

g. $-4 > x$

h. $2 \geq x$

i. $-6 \leq x$

j. $-2 < x < 4$

k. $5 > x > 1$

l. $0 \leq x < 3$

m. $-1 \leq x \leq 8$

n. $-5 < x \leq 0$

o. $-1 \geq x \geq -4$

EXPLORATIONS

1. Game:Concentration

Here's what 2 players need:

Make 12 pairs of matching cards containing inequalities. One card of a pair has the inequality written in words, and the other card has the same inequality written in symbols.

Sample Pair of Cards

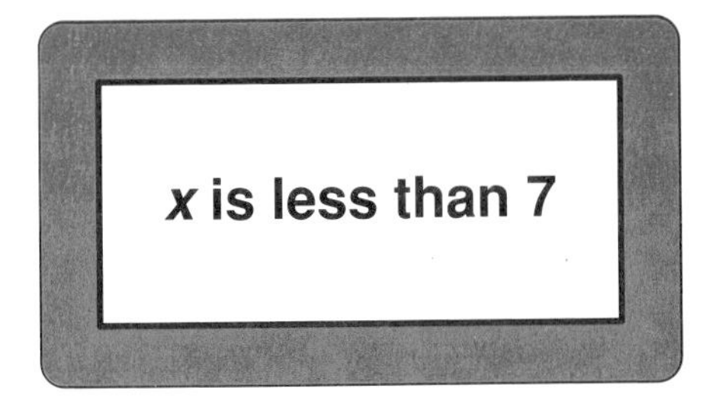

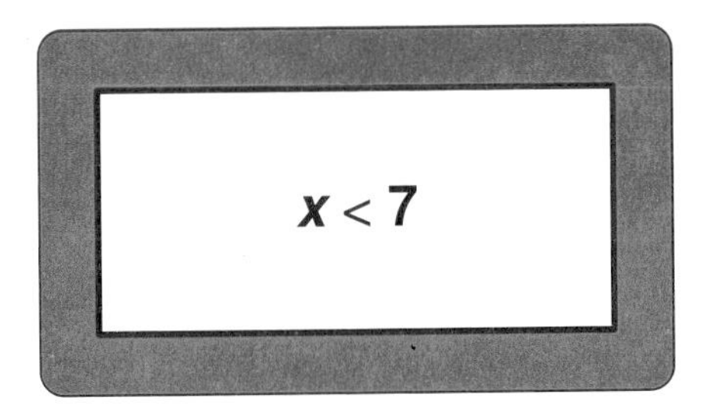

Here's how to play:

Shuffle all 24 cards together, and lay them face down in 4 neat rows.

Player 1 turns up any 2 of the cards face up, for both players to see.

If the cards are a matched pair, Player 1 keeps them, leaving all the other cards in position, and goes again, turning up another 2 cards.

If the cards are not a matched pair, Player 1 replaces the cards face down and Player 2 goes, turning up any 2 cards.

The players concentrate on the positions of the cards whose faces they have seen, so that when they turn up a card, they can remember where to find its match.

When all the cards have been removed, the winner is the player who has taken the most pairs.

2. Draw a graph for each inequality. Explain your graph.

a. $x \neq 2$ **b.** $x > 5$ or $x < -1$

3. This tree diagram shows the possibilities for the relationship between any two signed numbers, a and b.

Describe the possibilities, and give examples.

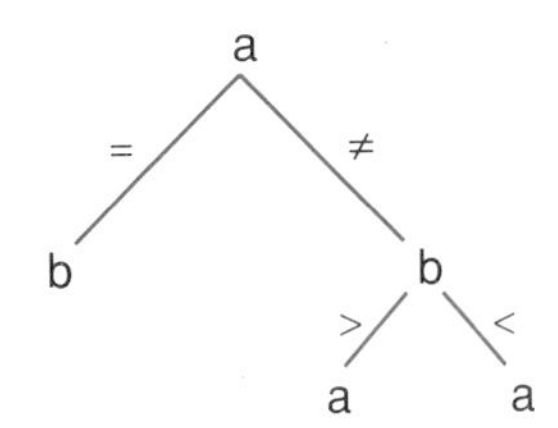

7.2 Properties of Inequality

Addition and Subtraction

Roy earns $13 an hour and Stan earns $9.

If they each get a raise of $1 per hour, who earns more per hour?

Roy Stan

$13 > 9$

$+1 = +1$

$13 + 1 \ ? \ 9 + 1$

$14 > 10$

Roy still earns more per hour than Stan.

Adding the same number to both sides of an inequality does not change the order of the inequality. (Use the same symbol.)

If they each have $1 per hour deducted for savings, who will have more to take home?

Roy Stan

$13 > 9$

$-1 = -1$

$13 - 1 \ ? \ 9 - 1$

$12 > 8$

Roy will still have more per hour than Stan.

Subtracting the same number from both sides of an inequality does not change the order of the inequality. (Use the same symbol.)

Example 1 Tell whether each sentence is true or false. Explain.

a. If $3 < 8$, then $3 + 2 < 8 + 2$.

Answer: True, $5 < 10$.

Adding the same number to both sides of an inequality does not change the order of the inequality.

b. If $2 < 7$, then $2 + 4 > 7 + 4$.

Answer: False, 6 is not greater than 11 (that is, $6 \not> 11$). $6 < 11$

Adding the same number to both sides of an inequality does not change the order of the inequality.

c. If $14 > 8$, then $14 - 9 > 8 - 9$.

Answer: True, $5 > -1$.

Subtracting the same number from both sides of an inequality does not change the order of the inequality.

d. If $9 > 5$, then $9 - 3 < 5 - 3$.

Answer: False, $6 \not< 2$. $6 > 2$

Subtracting the same number from both sides of an inequality does not change the order of the inequality.

Addition and Subtraction Properties of Inequality

The order of an inequality does not change if the same number

- **is added to both sides.**
- **is subtracted from both sides.**

Example 2 Consider the numbers -4 and 8.

a. Write a relation of inequality using : **(1)** $<$ **(2)** $>$

Answer: **(1)** $-4 < 8$ **(2)** $8 > -4$

b. Explain what happens if 3 is added to both sides of the inequality that has the symbol: **(1)** $<$ **(2)** $>$

Answer: **(1)** If $-4 < 8$, then

$-4 + 3 \ ? \ 8 + 3$ Adding 3 to both sides

$-1 < 11$ does not change the order.

(2) If $8 > -4$, then

$8 + 3 \ ? \ -4 + 3$ Adding 3 to both sides

$11 > -1$ does not change the order.

c. Explain what happens if 1 is subtracted from both sides of the inequality that has the symbol: **(1)** $<$ **(2)** $>$

Answer: **(1)** If $-4 < 8$, then

$-4 - 1 \ ? \ 8 - 1$ Subtracting 1 from both sides

$-5 < 7$ does not change the order.

(2) If $8 > -4$, then

$8 - 1 \ ? \ -4 - 1$ Subtracting 1 from both sides

$7 > -5$ does not change the order.

Try These *(For solutions, see page 276.)*

1. Replace ? by $>$ or $<$, and give a reason.

a. If $6 > 1$, then $6 + 2 \ ? \ 1 + 2$.

b. If $3 < 7$, then $3 - 1 \ ? \ 7 - 1$.

2. Write the sentence that results when the operation stated is performed on the given inequality.

a. 6 is added to both sides of $-6 < 0$

b. 2 is subtracted from both sides of $2 > 1$

Multiplication and Division

In October, Alice saved \$12 and Lenore saved \$8.

If, in November, each girl was able to double the amount she had saved in October, who had saved more in November?

Alice Lenore

$12 > 8$

$\times 2 = \times 2$

$2 \times 12 \ ? \ 2 \times 8$

$24 > 16$

Alice still saved more money than Lenore.

Multiplying both sides of an inequality by the same positive number does not change the order of the inequality. (Use the same symbol.)

If, in December, each girl was only able to save half as much as she had saved in October, who saved more in December?

Alice Lenore

$12 > 8$

$\div 2 = \div 2$

$12 \div 2 \ ? \ 8 \div 2$

$6 > 4$

Alice still saved more money than Lenore.

Dividing both sides of an inequality by the same positive number does not change the order of the inequality. (Use the same symbol.)

Example 3 Tell whether each sentence is true or false. Explain.

a. If $5 < 11$, then $3 \times 5 < 3 \times 11$.

Answer: True, $15 < 33$.

Multiplying both sides of an inequality by the same positive number does not change the order of the inequality.

b. If $8 > 1$, then $2 \times 8 < 2 \times 1$.

Answer: False, $16 \nless 2$. $16 > 2$

Multiplying both sides of an inequality by the same positive number does not change the order of the inequality.

c. If $3 < 12$, then $3 \div 3 < 12 \div 3$.

Answer: True, $1 < 4$.

Dividing both sides of an inequality by the same positive number does not change the order of the inequality.

d. If $15 > 5$, then $15 \div 5 < 5 \div 5$.

Answer: False, $3 \nless 1$. $3 > 1$

Dividing both sides of an inequality by the same positive number does not change the order of the inequality.

Multiplying or Dividing An Inequality by a Positive Number

The order of an inequality does not change if the same positive number

- **multiplies both sides.**
- **divides both sides.**

Example 4 Consider the numbers -4 and 8.

a. Write a relation of inequality using: **(1)** $<$ **(2)** $>$

Answer: **(1)** $-4 < 8$ **(2)** $8 > -4$

b. Explain what happens if 3 multiplies both sides of the inequality that has the symbol: **(1)** $<$ **(2)** $>$

Answer: **(1)** If $-4 < 8$, then

$3(-4) \;?\; 3(8)$ — Multiplying both sides by 3 does not change the order.

$-12 < 24$

(2) If $8 > -4$, then

$3(8) \;?\; 3(-4)$ — Multiplying both sides by 3 does not change the order.

$24 > -12$

c. Explain what happens if 4 divides both sides of the inequality that has the symbol: **(1)** $<$ **(2)** $>$

Answer: **(1)** If $-4 < 8$, then

$-4 \div 4 \;?\; 8 \div 4$ — Dividing both sides by 4 does not change the order.

$-1 < 2$

(2) If $8 > -4$, then

$8 \div 4 \;?\; -4 \div 4$ — Dividing both sides by 4 does not change the order.

$2 \quad -1$

Try These *(For solutions, see page 276.)*

1. Replace ? by $>$ or $<$, and give a reason.

a. If $5 > 3$, then $2 \times 5 \;?\; 2 \times 3$.

b. If $4 < 8$, then $4 \div 2 \;?\; 8 \div 2$.

2. Write the sentence that results when the operation stated is performed on the given inequality.

a. 4 multiplies both sides of $2 < 7$

b. 2 divides both sides of $10 > 8$

The temperature in Nome, Alaska is 20° at the same time that the temperature in Fairbanks, Alaska is 10°.

If the temperature in each city drops by twice its number, which temperature drops farther?

Nome Fairbanks

$20 > 10$

$\times(-2) = \times(-2)$

$-2(20) \ ? \ -2(10)$

$-40 < -20$

Nome's higher temperature results in a deeper drop.

> **Multiplying both sides of an inequality by the same negative number *changes* the order of the inequality.**
> **(Use the opposite symbol.)**

If the temperature in each city drops by half its number, which temperature drops farther?

Nome Fairbanks

$20 > 10$

$\div(-2) = \div(-2)$

$20 \div (-2) \ ? \ 10 \div (-2)$

$-10 < -5$

Nome's higher temperature results in a deeper drop.

> **Dividing both sides of an inequality by the same negative number *changes* the order of the inequality.**
> **(Use the opposite symbol.)**

Example 5 Tell whether each sentence is true or false. Explain.

a. If $5 < 11$, then $-3 \times 5 > -3 \times 11$.

Answer: True, $-15 > -33$.
Multiplying both sides of an inequality by the same negative number changes the order of the inequality.

b. If $8 > 1$, then $-4 \times 8 > -4 \times 1$.

Answer: False, $-32 \not> -4$. $-32 < -4$
Multiplying both sides of an inequality by the same negative number changes the order of the inequality.

c. If $3 < 12$, then $3 \div (-3) > 12 \div (-3)$.

Answer: True, $-1 > -4$.
Dividing both sides of an inequality by the same negative number changes the order of the inequality.

d. If $15 > 5$, then $15 \div (-5) > 5 \div (-5)$.

Answer: False, $-3 \not> -1$. $-3 < -1$
Dividing both sides of an inequality by the same negative number changes the order of the inequality.

Multiplying or Dividing An Inequality by a Negative Number

The order of an inequality changes if the same negative number

- **multiplies both sides.**
- **divides both sides.**

Example 6 Consider the numbers -4 and 8.

a. Write a relation of inequality using: **(1)** $<$ **(2)** $>$

Answer: **(1)** $-4 < 8$ **(2)** $8 > -4$

b. Explain what happens if -3 multiplies both sides of the inequality that has the symbol: **(1)** $<$ **(2)** $>$

Answer: **(1)** If $-4 < 8$, then

$$-3(-4) \ ? \ -3(8)$$

$$12 > -24$$

Multiplying both sides by -3 changes the order.

(2) If $8 > -4$, then

$$-3(8) \ ? \ -3(-4)$$

$$-24 < 12$$

Multiplying both sides by -3 changes the order.

c. Explain what happens if -4 divides both sides of the inequality that has the symbol: **(1)** $<$ **(2)** $>$

Answer: **(1)** If $-4 < 8$, then

$$-4 \div (-4) \ ? \ 8 \div (-4)$$

$$1 > -2$$

Dividing both sides by -4 changes the order.

(2) If $8 > -4$, then

$$8 \div (-4) \ ? \ -4 \div (-4)$$

$$-2 < 1$$

Dividing both sides by -4 changes the order.

Try These *(For solutions, see page 276.)*

1. Replace ? by $>$ or $<$, and give a reason.

a. If $4 > 1$, then $-3 \times 4 \ ? \ -3 \times 1$.

b. If $16 > 12$, then $16 \div (-4) \ ? \ 12 \div (-4)$.

2. Write the sentence that results when the operation stated is performed on the given inequality.

a. -2 multiplies $7 > 2$

b. -3 divides $6 < 9$

Multiplication and Division Properties of Inequality

- **The order of an inequality does not change if the same positive number is used as a multiplier or divisor.**
- **The order of an inequality changes if the same negative number is used as a multiplier or divisor.**

Example 7 Consider the inequality $8 > -4$.

a. Explain what happens if 2 is a multiplier of both sides.

Answer: If $8 > -4$, then

$2(8) \; ? \; 2(-4)$ — Multiplying both sides by 2

$16 > -8$ — does not change the order.

b. Explain what happens if 2 is a divisor of both sides.

Answer: If $8 > -4$, then

$8 \div 2 \; ? \; -4 \div 2$ — Dividing both sides by 2

$4 > -2$ — does not change the order.

c. Explain what happens if -2 is a multiplier of both sides.

Answer: If $8 > -4$, then

$-2(8) \; ? \; -2(-4)$ — Multiplying both sides by -2

$-16 < 8$ — changes the order.

d. Explain what happens if -2 is a divisor of both sides.

Answer: If $8 > -4$, then

$8 \div (-2) \; ? \; -4 \div (-2)$ — Dividing both sides by -2

$-4 < 2$ — changes the order.

Try These (For solutions, see page 276.)

1. Replace ? by > or <, and give a reason.

a. If $4 < 5$, then $4 \times 4 \; ? \; 4 \times 5$.

b. If $10 > 5$, then $10 \div 5 \; ? \; 5 \div 5$.

c. If $8 > 4$, then $-2 \times 8 \; ? \; -2 \times 4$.

d. If $8 > 4$, then $8 \div (-2) \; ? \; 4 \div (-2)$.

2. Write the sentence that results when the operation stated is performed on the given inequality.

a. 3 multiplies both sides of $-4 < -3$

b. 4 divides both sides of $0 > -4$

c. -3 multiplies both sides of $1 < 2$

d. -2 divides both sides of $-4 > -6$

SELF-CHECK: SECTION 7.2

1. **a.** Describe the order of the inequality $3 < 5$.

   Answer: The left side is less than the right side.

   b. Write a different order in which the numbers 3 and 5 can be compared. Describe this order.

   Answer: $5 > 3$ The left side is greater than the right side.

2. Consider the inequality $9 < 12$. Explain what happens to the order of the inequality when:

   a. 3 is added to both sides or subtracted from both sides

   Answer: When 3 is added to both sides, the result is $12 < 15$.
   When 3 is subtracted from both sides, the result is $6 < 9$.
   In both cases, the order remains unchanged.

   b. 3 is a multiplier of both sides or a divisor of both sides

   Answer: When 3 is a multiplier, the result is $27 < 36$.
   When 3 is a divisor, the result is $3 < 4$.
   In both cases, the order remains unchanged.

   c. -3 is a multiplier or a divisor or both sides

   Answer: When -3 is a multiplier, the result is $-27 > -36$.
   When -3 is a divisor, the result is $-3 > -4$.
   In both cases, the order changes.

3. **a.** Under what operations does the order of an inequality not change?

   Answer: The order remains unchanged under addition or subtraction by any number, and under multiplication or division by a positive number.

   b. Under what operations does the order of an inequality change?

   Answer: The order changes under multiplication or division by a negative number.

EXERCISES: SECTION 7.2

1. Match each word phrase in *Column A* to a symbol in *Column B*.

Column A

a. is equal to
b. is not equal to
c. is greater than
d. is not greater than
e. is less than
f. is not less than

Column B

(1) $\neq$ **(5)** $<$
(2) $=$ **(6)** $\not<$
(3) $\not>$ **(7)** $>$
(4) $\geq$ **(8)** $\leq$

2. Replace ? by $>$ or $<$, and give a reason.

a. If $8 > 1$, then $8 + 3$? $1 + 3$.

b. If $2 < 5$, then $2 - 1$? $5 - 1$.

c. If $1 < 7$, then 5×1 ? 5×7.

d. If $16 > 8$, then $16 \div 8$? $8 \div 8$.

e. If $9 > 3$, then -3×9 ? -3×3.

f. If $9 > 3$, then $9 \div (-3)$? $3 \div (-3)$.

g. If $5 < 10$, then -5×5 ? -5×10.

h. If $5 < 10$, then $5 \div (-5)$? $10 \div (-5)$.

3. Write the sentence that results when the operation stated is performed on the given inequality.

a. 7 is added to both sides of $5 < 8$

b. 2 is added to both sides of $-3 > -4$

c. 4 is subtracted from both sides of $9 < 13$

d. 5 is subtracted from both sides of $-4 < -2$

e. 6 multiplies both sides of $-2 < 1$

f. -3 multiplies both sides of $8 < 14$

g. 5 divides both sides of $25 > 15$

h. -5 divides both sides of $25 > 15$

4. Each of the following statements is false. Rewrite the statement to make it true.

a. $-8 > -1$

b. $\frac{1}{2} \neq 50\%$

c. $7 \ngtr 1$

d. $9 \nless 10$

e. If $5 > -2$, then $5 + 3 < -2 + 3$.

f. If $7 < 9$, then $-2 \times 7 < -2 \times 9$.

g. If $6 > 1$, then $6 + 3 > 1 + 9$.

h. If $9 < 10$, then $9 + 1 < 10 - 1$.

5. The properties of equality and inequality can be written in symbols as well as words. Use a, b, and c to represent any three signed numbers.

Example 1: The addition property of equality can be written

in words: Adding the same number to both sides of an equation maintains the balance of the equation.

in symbols: If $a = b$, then $a + c = b + c$.

Example 2: The addition property of inequality can be written

in words: Adding the same number to both sides of an inequality does not change the order of the inequality.

in symbols: If $a > b$, then $a + c > b + c$.

Using Examples 1 and 2 as models, write the following properties of equality and inequality in words and in symbols.

a. subtraction property of equality

b. subtraction property of inequality

c. multiplication property of equality

d. multiplication property of inequality

(Remember: For inequality, this property must be considered in two parts, when the multiplier c is positive and when the multiplier c is negative.)

e. division property of equality

f. division property of inequality

(Remember: For inequality, this property must be considered in two parts, when the divisor c is positive and when the divisor c is negative.)

EXPLORATIONS

1. Write the sentence that results when the operation stated is performed on all members of the inequality $9 < 12 < 15$.

a. 1 is added

b. 2 is subtracted

c. 4 is a multiplier

d. 3 is a divisor

e. -4 is a multiplier

f. -3 is a divisor

2. **a.** When two quantities are equal, such as $x - 1 = 3$, what happens when the same number is added to both sides of the equation?

b. When two quantities are unequal, such as $x - 1 > 3$, what happens when the same number is added to both sides of the inequality?

c. What do you think is the meaning of the sentence $x - 1 \geq 3$?

d. Do you think that the addition property that is true for = and the addition property that is true for > is also true for ≥? Explain.

3. **a.** Name a value that makes the open sentence $x - 1 = 3$ true. How many values make this open sentence true?

b. Name a value that makes the open sentence $x - 1 > 3$ true. How many values make this open sentence true?

c. Does the value that satisfies the equation $x - 1 = 3$ (makes the open sentence true) also satisfies the inequality $x - 1 > 3$?

d. How would you rewrite the inequality $x - 1 > 3$ if you wanted the value that satisfies the equation $x - 1 = 3$ to also satisfy the inequality?

7.3 Solving a Simple Inequality

Jill's fish tank can hold no more than 30 fish. If she has just put 8 fish in the tank, how many were already there?

Explain why you will not be able to give just one number as the answer.

> If Jill has now put 8 fish in the tank, then there is room for at most 22 more. But you don't know if there were 22 fish already there or 21 fish or 20 fish or 3 fish.

All you can say is that the number of fish already in the tank could be at most 22.

Let's look at an open sentence to model the situation.

If n = the number of fish already in the tank,
then $n + 8$ = the number of fish after 8 were added.

$n + 8 \leq 30$ The number of fish is at most 30.

If this open sentence were the equation $n + 8 = 30$, how would you solve it?

> To get n alone and to maintain the balance of the equation, subtract 8 from both sides.

Treat the inequality in the same fashion as the equation.

$n + 8 \leq 30$ To undo the addition, subtract 8 from both sides of the inequality.

$n + 8 - 8 \leq 30 - 8$ Subtracting the same number from both sides of an inequality does not change the order of the inequality.

$n \leq 22$

This solution shows that there were at most 22 fish in the tank.

As with a simple equation, to solve a simple inequality, apply the inverse operation to the operation shown.

Example 1 Solve each inequality by applying the inverse operation.

a. $x - 9 > 4$ — This is a subtraction inequality.

$x - 9 + 9 > 4 + 9$ — Undo − by +. Adding the same number to both sides of an inequality does not change the order of the inequality.

$x > 13$

b. $y + 6 < 8$ — This is an addition inequality.

$y + 6 - 6 < 8 - 6$ — Undo + by −. Subtracting the same number from both sides of an inequality does not change the order of the inequality.

$y < 2$

c. $2n \geq 16$ — This is a multiplication inequality.

$\frac{2n}{2} \geq \frac{16}{2}$ — Undo × by ÷. Dividing both sides of an inequality by the same positive number does not change the order of the inequality.

$n \geq 8$

d. $\frac{t}{3} < 9$ — This is a division inequality.

$3 \cdot \frac{t}{3} < 3 \cdot 9$ — Undo ÷ by ×. Multiplying both sides of an inequality by the same positive number does not change the order of the inequality.

$t < 27$

Remember that when you multiply or divide an inequality by a negative number, the order of the inequality changes.

Example 2 Solve $-3x < 18$, and check.

$-3x < 18$

$\frac{-3x}{-3} > \frac{18}{-3}$ — Undo × by ÷. Dividing both sides of an inequality by the same negative number changes the order of the inequality.

$x > -6$

Since you cannot check all of the numbers in the solution set, choose one number that is greater than −6, say −5, and substitute it in the inequality.

Check: Let $x = -5$.

$-3x < 18$

$-3(-5) \overset{?}{<} 18$

$15 < 18$ ✓

Example 3 Solve $\frac{w}{-2} \geq 2$, and graph the solution set.

$\frac{w}{-2} \geq 2$

$-2 \cdot \frac{w}{-2} \leq -2 \cdot 2$ — Undo ÷ by ×. Multiplying both sides of an inequality by the same negative number changes the order of the inequality.

$w \leq -4$

Check: Let $w = -6$.

$\frac{w}{-2} \geq 2$

$\frac{-6}{-2} \overset{?}{\geq} 2$

$3 \geq 2$ ✓

Answer: The solution set is all signed numbers less than or equal to −4.

−6 −5 −4 −3 −2 −1 0

When using an inequality to model a problem, follow the four key steps of problem solving.

Example 4 At the end of one year, Mandy wants to have saved at least \$1,300. What is the least she should save each week to reach her goal?

1 ***Read the problem to understand it.***

You know the minimum amount she wants to have at the end of a year. You want to find the minimum weekly amount she should save. You should try to first estimate an answer. How would you round the number of weeks in the year? What operation would you perform on that number and \$1,300?

2 ***Make a plan to solve the problem.***

Translate the given information into an inequality.

Let w = the weekly amount.

Then $52w$ = the yearly amount.

$52w \geq 1{,}300$ The yearly amount is to be at least \$1,300.

3 ***Use your plan to solve the problem.***

Solve the inequality that models the problem.

$52w \geq 1{,}300$ This is a multiplication inequality.

$\frac{52w}{52} \geq \frac{1{,}300}{52}$ Undo × by ÷. Dividing by a positive number does not change the order of the inequality.

$w \geq 25$

4 ***Check your solution.***

Choose a number greater than 25.

$52 \times 26 \overset{?}{>} 1{,}300$

$1{,}352 > 1{,}300$ ✓

Answer: If Mandy saves at least \$25 a week, she will have at least \$1,300 at the end of one year.

Key Words That Translate to the Symbols of Inequality

$<$	$>$	$\leq$	$\geq$
is less than	is greater than	is less than or equal to	is greater than or equal to
is below	is above	is at most	is at least
is under	is over	is no more than	is no less than
		is a maximum of	is a minimum of

Try These (For solutions, see page 276.)

1. Solve each inequality, check your answer, and graph the solution set.

   a. $x + 4 < 7$ **b.** $y - 6 > 0$ **c.** $2n \leq 10$ **d.** $-4x > 8$ **e.** $\frac{z}{3} > 2$ **f.** $\frac{n}{-2} \leq 2$

2. Use an inequality to solve this problem. First estimate an answer.

   Kevin would like to have a job that pays at least $280 for a 35-hour week. To reach this goal, what is the lowest hourly rate Kevin should aim for?

SELF-CHECK: SECTION 7.3

1. Consider the inequality $x + 6 > 9$.

a. What would you do to solve?

Answer: Subtract 6 from both sides.

b. When you subtract 6 from both sides, does the order of the inequality change? Explain.

Answer: No, subtracting the same number from both sides of an inequality does not change the order.

2. Consider the inequality $y - 2 < 7$.

a. What would you do to solve?

Answer: Add 2 to both sides.

b. When you add 2 to both sides, does the order of the inequality change? Explain.

Answer: No, adding the same number to both sides of an inequality does not change the order.

3. Consider the inequality $3z \geq 21$.

a. What would you do to solve?

Answer: Divide both sides by 3.

b. When you divide both sides by 3, does the order of the inequality change? Explain.

Answer: No, dividing both sides by the same positive number does not change the order.

4. Consider the inequality $\frac{n}{-2} < 8$.

a. What would you do to solve?

Answer: Multiply both sides by –2.

b. When you multiply both sides by –2, does the order of the inequality change? Explain.

Answer: Yes, multiplying both sides by the same negative number does change the order.

5. How is solving a simple inequality the same as solving a simple equation? How is it different?

Answer: As long as you do the same thing to both sides of an equation or an inequality, you do not upset the balance. However, in an inequality, you must remember to reverse the order of the inequality (use the opposite symbol) when you multiply or divide by a negative number.

EXERCISES: SECTION 7.3

1. Solve each inequality by subtraction. Check your answer. Graph the solution set.

a. $x + 4 < 9$ **b.** $y + 2 \geq 5$ **c.** $n + 3 \leq 1$

d. $z + 1 > -2$ **e.** $t + 1.1 \geq 0.1$ **f.** $p + \frac{1}{2} \leq -2\frac{1}{2}$

2. Solve each inequality by addition. Check your answer. Graph the solution set.

a. $x - 4 < 1$ **b.** $y - 2 \geq 3$ **c.** $n - 3 \leq -5$

d. $z - 1 > -2$ **e.** $t - 1.1 \geq 0.9$ **f.** $p - \frac{1}{2} \leq -2\frac{1}{2}$

3. Solve each inequality by division. Check your answer. Graph the solution set.

a. $5k > 10$ **b.** $3p \leq 9$ **c.** $2x > -8$

d. $-3x < 6$ **e.** $-2y \geq 10$ **f.** $-z < 4$

4. Solve each inequality by multiplication. Check your answer. Graph the solution set.

a. $\frac{x}{2} < 1$ **b.** $\frac{y}{3} > 3$ **c.** $\frac{z}{4} \leq -1$

d. $\frac{n}{-2} \geq 3$ **e.** $\frac{z}{-3} < 1$ **f.** $\frac{w}{-2} > -1$

5. Solve each inequality by applying the inverse operation, and check.

a. $x + 9 < 10$ **b.** $y - 3 > 4$ **c.** $2n < 10$

d. $-3y \geq 6$ **e.** $\frac{w}{2} < 4$ **f.** $\frac{t}{-3} > 1$

g. $3 + y > 5$ **h.** $-2 + z \leq 1$ **i.** $4 < 4w$

j. $6k < 3$ **k.** $-3 > -3t$ **l.** $\frac{k}{-2} > 0$

6. Do you agree or disagree with each solution? If you disagree, explain why.

a. To solve $y + 3 \leq 5$, Lenny added 3 to both sides, with the result $y \leq 8$.

b. To solve $w - 3 > 8$, Loni added 3 to both sides, with the result $w < 11$.

c. To solve $z - 3 \geq 1$, Larry added 3 to both sides, with the result $z \geq 4$.

d. To solve $-3x < 9$, Laura divided both sides by -3, with the result $x < -3$.

e. To solve $-3x > 6$, Lilly divided both sides by -3, with the result $x < -2$.

7. Use an inequality to find the set of values for each unknown number.

a. A number increased by 5 is more than 7.

b. A number decreased by 3 is less than 5.

c. Twice a number is at most 12.

d. One-half a number is at least 4.

e. Six more than a number is at least 10.

f. Four less than a number is at most 20.

8. Select the inequality that can best be used to model each problem.

a. A bookshelf can hold no more than 20 volumes of an encyclopedia. If 8 volumes are already on the shelf, what is the maximum number of volumes that can still be put there?

If n = the number of volumes to be put on the shelf, then:

(1) $n + 8 < 20$ (2) $n + 8 \leq 20$ (3) $n - 8 < 20$ (4) $n - 8 \leq 20$

b. Harvey can type an average of 40 words per minute. If he has to type his term paper of at least 3,000 words, what is the amount of time he must plan on?

If m = the number of minutes he must plan on typing, then:

(1) $\frac{m}{40} < 3{,}000$ (2) $\frac{m}{40} \geq 3{,}000$ (3) $40m < 3{,}000$ (4) $40m \geq 3{,}000$

c. Counting overtime, Wilma's maximum weekly earnings are \$375. If she sets aside 10% of her budget for clothing, what is the most she can spend on clothing?

If b = the amount of Wilma's weekly budget, then:

(1) $\frac{b}{10} \leq 375$ (2) $\frac{b}{10} < 375$ (3) $10b \leq 375$ (4) $10b < 375$

d. To reduce their roster to no more than 47 players, the Football Flooders had to trade off 13 players. How many players did they have on their roster before the reduction?

If p = the number of players before the reduction, then:

(1) $47 - 13 \leq p$ (2) $47 + 13 \leq p$ (3) $p - 13 \leq 47$ (4) $p - 13 < 47$

9. Use an inequality to solve each problem. First, estimate an answer.

a. The number of horses regularly boarded at Greenvale Stables is under 50. If, in May, 14 stalls are kept empty for repairs, what is the maximum number of horses that can be boarded?

b. A textbook storeroom has a capacity of no more than 1,500 books. After Mr. Schneider returned 35 books, the storeroom was full. How many books were already there?

c. Julia has to read at least 380 pages of a book in the next 4 days. What is the least number of pages she should read each day?

d. Ms. Wilson has at most \$500 available in her checking account. After writing a check for \$175, how much money will she have left?

e. The Jones family plans to pay no more than \$150,000 for a house. If their down payment will be \$40,000, what is the maximum mortgage they will have to apply for?

f. Two-thirds of the Freshman class at Cole College will get some form of financial aid. If there is to be a maximum of 1,100 Freshmen on financial aid, what could be the size of the Freshman class?

g. On his trip, John plans to drive at least 350 miles the first day. If he averages 50 miles per hour, what is the least number of hours he should plan on driving?

EXPLORATIONS

Algebra chips can be used to model the solution of an inequality.
Be sure to include the symbol of inequality.

Example: Model the solution of $3 - n < 1$.

Lay out chips to represent each side of the inequality.

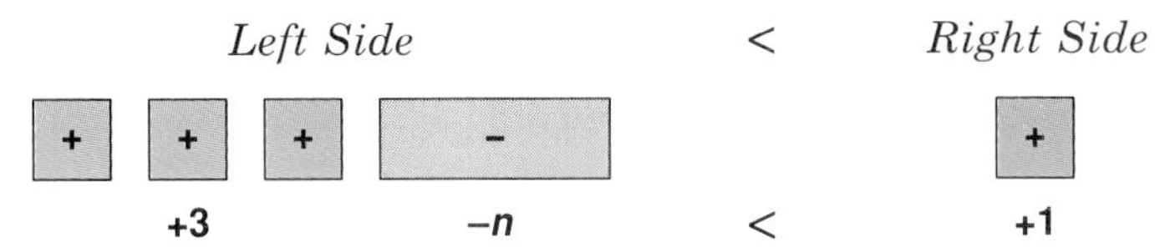

As with an equation, the object is to get $+n$ alone on one side. Since there is no $+n$-chip in the model, we can add such a chip to each side without upsetting the balance of the inequality.

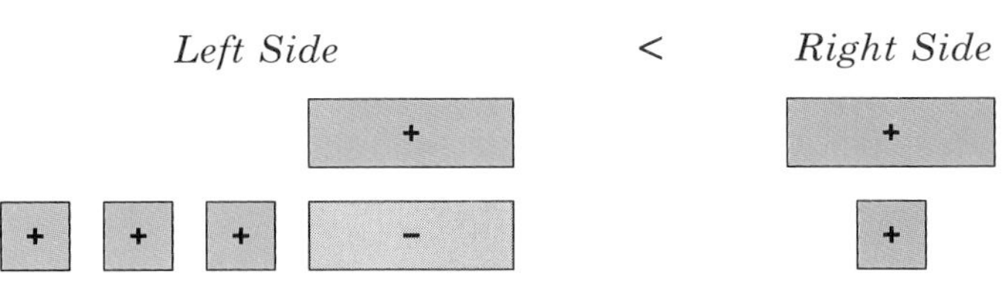

Remove the zero pair from the left side.

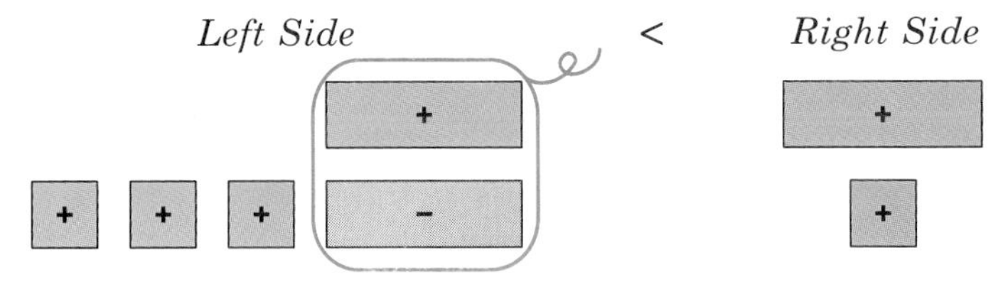

To get the n-chip alone on the right side, remove + from each side.

These are the remaining chips, which shows the solution $2 < n$.

How else can you write the inequality $2 < n$?

Use algebra chips to model the solution of each inequality.

a. $5 - n > 2$ **b.** $4 - n < -2$ **c.** $5 > 2 - n$ **d.** $3 < 1 - n$

7.4 Using Two Operations to Solve

The value of 1 more than twice a positive integer must be under 7. What are the possibilities for the integer?

Did you say that the integer could be 2 or 1? Explain your reasoning.

> Since 1 more than twice the integer is less than 7, twice the integer is less than 6.
> Since twice the integer is less than 6, the integer is less than 3.
> Since the integer is positive, the only possibilities are 2 or 1.

Consider an inequality to model this problem.

Let n = the number.

$2n + 1 < 7$ — The value must be under 7.

If this open sentence were the equation $2n + 1 = 7$, how would you solve it?

> Carry out the solution in two steps, first subtracting 1 from both sides, and then dividing both sides by 2.

Treat the inequality in the same fashion as the equation.

$2n + 1 < 7$ — There are 2 operations: + and ×.

$2n + 1 - 1 < 7 - 1$ — Undo + by –. Subtraction does not change the order of the inequality.

$2n < 6$

$\frac{2n}{2} < \frac{6}{-2}$ — Undo × by ÷. Division by a positive number does not change the order of the inequality.

$n < 3$

Since the integer is positive and less than 3, the integer may be 1 or 2.

> **As with an equation that requires two steps to solve, if an inequality contains the operation addition and multiplication:**
> **First, undo the addition by using subtraction.**
> **Next, undo the multiplication by using division.**

Example 1 Solve $-2x + 3 > 11$, and check. Graph the solution set.

$-2x + 3 > 11$ — There are 2 operations: + and ×.

$-2x + 3 - 3 > 11 - 3$ — First, undo + by –. Subtraction does not change the order of the inequality.

$-2x > 8$

$\frac{-2x}{-2} < \frac{6}{-2}$ — Undo × by ÷. Dividing by a negative number changes the order of the inequality.

$x < -4$

Check: Let $x = -5$.

$-2x + 3 > 11$

$-2(-5) + 3 \overset{?}{>} 11$

$10 + 3 \overset{?}{>} 11$

$13 > 11$ ✓

Answer: The solution set is all signed numbers less than –4.

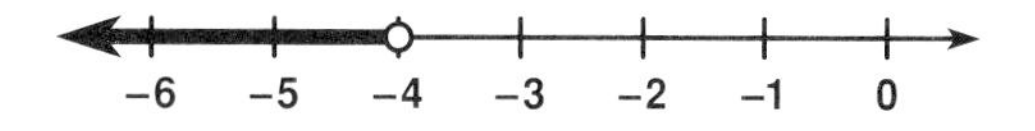

> **Remember that multiplying or dividing both sides of an inequality by a negative number changes the order of the inequality.**

When using an inequality to model a problem, follow the four key steps of problem solving.

Example 2 Cheapie Car Rental charges \$39 per day, plus 35 cents per mile driven. If Joe rents a car for a day, what distance can he drive and have his total rental charge at most \$60?

1 ***Read the problem to understand it.***
You know the basic charge for the car and the charge per mile.
You want to find the number of miles Joe can drive and have a total charge of \$60 or less.

You should try to first estimate an answer. Consider the charge for the car first. How much is left for the mileage charges? Use convenient numbers to make the arithmetic easier.

2 ***Make a plan to solve the problem.***
Translate the given information into an inequality.

Let m = the number of miles he can drive.
Then $0.35m$ = the mileage cost
and $0.35m + 39$ = the mileage cost plus the car cost.

$0.35 + 39 \le 60$ The total cost is to be no more than \$60.

3 ***Use your plan to solve the problem.***
Solve the inequality that models the problem.

$0.35m + 39 \le 60$ Two operations: + and ×
$0.35 + 39 - 39 \le 60 - 39$ Undo + by −.
$0.35m \le 21$ No change in order.
$\frac{0.35m}{0.35} = \frac{21}{0.35}$ Undo × by ÷. Dividing by a positive, no change in order.
$m \le 60$

4 ***Check your solution.***
Choose a number less than 60.
Suppose he drove 50 miles.

Then $0.35(50) + 39 \overset{?}{\le} 60$
$17.50 + 39 \overset{?}{\le} 60$
$56.50 < 60$ ✔

Answer: To keep his total cost to no more than \$60, Joe should drive no more than 60 miles.

Try These ***(For solutions, see page 277.)***

1. Solve each inequality, check your answer, and graph the solution set.

a. $2x + 1 < 7$ **b.** $3y - 2 \ge 4$ **c.** $-2z + 3 \le 5$ **d.** $\frac{n}{2} - 1 > 1$

2. Use an inequality to solve this problem. First estimate the answer.
Devon can spend no more than \$180 for two pairs of boots. If one pair costs \$40 more than the other, what is the maximum he can spend on each pair?

Combining Like Terms

Name two consecutive integers whose sum is at least 100.

To be safe, you could name any two consecutive integers that are each more than 100, say 1,000 and 1,001. What is the smallest pair of integers that satisfy the condition? Did you say 50 and 51?

Consider an inequality to model this problem.

Let x = the first of the integers.
Then $x + 1$ = the next integer.
$x + x + 1 \geq 100$ — The sum is greater than or equal to 100.

If this open sentence were the equation $x + x + 1 = 100$, how would you solve it?

First, simplify the equation by combining like terms.

Treat the inequality in the same fashion as the equation.

$x + x + 1 \geq 100$

$2x + 1 \geq 100$ — Combine like terms.

$2x + 1 - 1 \geq 100 - 1$ — Undo + by −. Subtraction does not change the order of the inequality.

$2x \geq 99$

$\frac{2x}{2} \geq \frac{99}{2}$ — Undo × by ÷. Division by a positive does not change the order.

$x \geq 49.5$

50 is the first integer that is greater than 49.5.

Answer: 50 and 51 are the two smallest consecutive integers whose sum is at least 100.

Example 3 Charles and Joel need at least \$50 in cash to pay greens fees for their golf game. If Joel is to pay \$10 more than Charles, what is the least amount of cash Charles must have?

You know the relation between the amounts the men pay, and you know the minimum total. You are to find the minimum amount Charles must have.

First try to estimate an answer. What would be the minimum each needed if they both had the same amount?

Let c = Charles's cash.
Then $c + 10$ = Joel's cash.

$c + c + 10 \geq 50$ — The total cost is at least \$50.

$2c + 10 \geq 50$ — Combine like terms.

$2c + 10 - 10 \geq 50 - 10$ — Undo + by −. No change in order.

$2c \geq 40$

$\frac{2c}{2} \geq \frac{40}{2}$ — Undo × by ÷. Division by a positive does not change the order.

$c \geq 20$

Check: Suppose Charles has \$20. Then Joel has \$30.
Is that enough for the \$50 minimum? ✔

Answer: To meet the minimum fee, Charles must have at least \$20.

Try These (For solutions, see page 278.)

Solve each inequality, check your answer, and graph the solution set.

1. $x + 2x < 9$ **2.** $n + n - 1 > 3$

SELF-CHECK: SECTION 7.4

1. Consider the inequality $3x - 1 > 8$.

a. To solve, what would you do first? Will this change the order of the inequality?

Answer: First, undo the subtraction by adding 1 to both sides. Addition does not change the order of the inequality.

b. After adding 1 to both sides, the inequality reads $3x > 9$. What is the next step in the solution? Will this change the order of the inequality?

Answer: Divide both sides by 3. The order of the inequality does not change when the divisor is positive.

c. How is the method of solution of the inequality $3x - 1 > 8$ the same as that of the equation $3x - 1 = 8$? How is it different?

Answer: The steps of the solution are the same, first undoing subtraction and then undoing multiplication.

For an inequality, you must be aware of the order, remembering that the order will change if multiplying or dividing by a negative number.

2. What would you do first to simplify the inequality $n + 2n < 6$?

Answer: Combine like terms, $3n < 6$.

EXERCISES: SECTION 7.4

1. Solve by first undoing addition or subtraction, and then undoing multiplication. Check.

a. $4x + 3 < 11$ **b.** $5y - 2 > 3$ **c.** $2n + 5 \leq 9$

d. $3g - 1 \geq -4$ **e.** $2m + 6 < 2$ **f.** $17 > 4r + 1$

g. $-2g + 3 < 7$ **h.** $-3k - 1 \geq 2$ **i.** $-y + 4 < 5$

2. Solve by first undoing addition or subtraction, and then undoing division. Check.

a. $\frac{x}{2} + 3 > 7$ **b.** $\frac{y}{3} - 1 < 8$ **c.** $\frac{z}{2} + 4 \geq 2$

d. $\frac{m}{3} + 2 \leq -1$ **e.** $\frac{w}{4} - 2 > -10$ **f.** $\frac{r}{2} + 2 \leq 0$

g. $\frac{w}{-2} + 3 > 5$ **h.** $\frac{t}{-3} - 4 \leq -1$ **i.** $\frac{n}{-1} - 1 < -1$

3. Solve by first combining like terms. Check.

a. $x + 3x > 8$ **b.** $4n - n < 9$ **c.** $y - 2y > 7$

d. $z + 2z + 1 \leq 10$ **e.** $5t - t - 2 \geq 6$ **f.** $6m + m + 3 < -4$

g. $x - 2x + 2 > 7$ **h.** $-3y + 4 - 1 < 3$ **i.** $m - 2m + 7 - 1 > 8$

4. Solve and check. Graph each solution set.

a. $m + 8 > 7$ **b.** $x - 2 < 3$ **c.** $4x > -8$

d. $\frac{y}{4} \leq -2$ **e.** $21 < 3m$ **f.** $-6g \geq 24$

g. $6m + 8 \leq 20$ **h.** $5t - 2 > 3$ **i.** $-2k + 1 \leq 2$

j. $\frac{t}{2} - 2 > 2$ **k.** $\frac{r}{-1} + 3 \leq 8$ **l.** $3 - 2k > 11$

m. $7 - 3n < -2$ **n.** $4 - m > 5$ **o.** $8 - 3x \leq -1$

p. $2x + x > -3$ **q.** $4p - 2p + 1 \leq 4$ **r.** $2x - 4x + 3 > 1$

5. **a.** To solve the inequality $3x + 5 < 14$, students in a group offered the following suggestions. Do you agree or disagree with each suggestion? If you disagree, explain why.

(1) Marvin advised that both sides first be divided by 3.

(2) Mel agreed to first divide both sides by 3 and he said, "Don't forget to change the order of the inequality."

(3) Marla said that 5 should first be subtracted from both sides, with the result $3x > 9$.

(4) Maryann said that 5 should first be subtracted from both sides, but the result would be $3x < 9$.

b. Describe the steps you would follow to solve the inequality $-2n - 1 > 3$.

6. Select the inequality that can best be used to find the set of values for each unknown number, n.

a. Five more than twice a number is at most 25.

(1) $2n + 5 > 25$ (2) $2n + 5 \geq 25$ (3) $2n + 5 < 25$ (4) $2n + 5 \leq 25$

b. Three less than twice a number is no more than 5.

(1) $3 - 2n < 5$ (2) $3 - 2n \leq 5$ (3) $2n - 3 < 5$ (4) $2n - 3 \leq 5$

c. One-half a number, increased by 4 is greater than 16.

(1) $\frac{1}{2}(n + 4) > 16$ (2) $\frac{1}{2}n + 4 > 16$ (3) $\frac{1}{2} + 4n > 16$ (4) $\frac{1}{2}(4 + n) > 16$

7. Use an inequality to find the set of values for each unknown number.

a. Twice a number increased by 4 is under 14.

b. If three times a number is added to 6, the result is at most 12.

c. Four more than twice a number is greater than 14.

d. One-half a number decreased by 2 is below 10.

8. **a.** **(1)** Use an inequality to find the two largest consecutive integers whose sum is less than 50.

   (2) Name 3 other sets of two consecutive integers whose sum is less than 50.

   b. **(1)** Use an inequality to find the two smallest consecutive odd integers whose sum is at least 204.

   (2) Name 3 other sets of two consecutive odd integers whose sum is at least 204.

   c. **(1)** Use an inequality to find the two largest consecutive even integers whose sum is under 400.

   (2) Name 3 other sets of two consecutive even integers whose sum is under 400.

9. Select the inequality that can best be used to model each problem.

   a. In Emelda's shoe closet, the top 4 shelves each contain the same number of pairs of shoes, and there are 6 pairs on the bottom shelf. If there are at least 50 pairs of shoes in all, what is the least number there could be on the top shelf?

   If p = the number of pairs of shoes on the top shelf, then:

   (1) $4p + 6 \geq 50$ (2) $4(p + 6) > 50$ (3) $4p + 6 \leq 50$ (4) $4(p + 6) \leq 50$

   b. To listen to your horoscope on 1-900-DESTINY, the charge is \$2.95 for the first minute and \$1.50 for each additional minute. For how many minutes can Jon listen if he wishes to keep the charge under \$15?

   If m = the number of minutes Jon listens, then:

   (1) $2.95 + 1.50m < 15$ (2) $(2.95 + 1.50)m < 15$

   (3) $4.45m < 15$ (4) $4.45(1 + m) < 15$

   c. Joan is 6 years older than Mike. Together they are no older than half a century.

   If m = Mike's age, then:

   (1) $m + m + 6 \leq 100$ (2) $m + m + 6 \leq 500$

   (3) $m + m + 6 \leq 50$ (4) $m + m + 6 \leq 1{,}000$

   d. When packing her books to move, Carla put the same number of books in each of 10 cartons. Thinking the cartons would be too heavy, she removed 2 books from each carton. If there were at least 150 books left in the cartons, what was the least number of books Carla had originally put in a carton?

   If b = the number of books originally in each carton, then:

   (1) $10b - 2 > 150$ (2) $10(b - 2) > 150$

   (3) $10b - 2 \geq 150$ (4) $10(b - 2) \geq 150$

10. Use an inequality to solve each problem. First estimate an answer.

a. In Sandra's bookcase, the top 3 shelves each contain the same number of books, and there are 12 books on the bottom shelf. If there are more than 36 books in all, what is the least number there could be on the top shelf?

b. At Harvey High, each of 5 clubs baked the same number of pies for a bake sale, where 68 pies were sold. If fewer than a dozen pies remained, what is the greatest number of pies that each club could have baked?

c. To prepare for her piano recital, Lucy plans to practice for at least 15 hours this week. If she practices the same length of time on Monday through Saturday, and then spends 3 hours on Sunday, what is the least time she should practice on Monday?

d. Phil's Photocopy charges 15 cents for the first copy and 8 cents for each additional copy. What is the greatest number of copies Orrie can get for no more than $1?

e. Each of 3 Math quizzes had the same number of questions, and the next quiz had 8 questions. If there were fewer than 25 questions in all, what is the greatest number there could have been on the first quiz?

f. Tim, Tom, and Ted all had the same number of marbles until Tim lost 2. If the 3 children together then had no more than 16 marbles, what is the most each could have started with?

g. On a school trip, there are the same number of students in each of 6 buses, and 8 students in a van. If there are at least 200 students on the trip, what is the least number of students in each bus?

h. Anna tossed a flying saucer twice as far as Seth did. If their combined distances totaled at most 42 feet, what was Seth's greatest possible distance?

i. At the Holt Animal Shelter, a collie's litter had 2 pups more than a poodle's litter. If, together, both litters contain no more than 10 pups, what is the greatest number of poodle pups there can be?

j. A bowl of fruit has 4 pears and twice as many plums as peaches. If there are at least 13 pieces of fruit in all, what is the fewest possible number of peaches?

k. Chairs are arranged in 5 equal rows in a class-room. By adding a chair to each row, the room can seat at most 35 students. What is the maximum number of students that can be seated if no chairs are added?

l. If Marla increased her weekly payments to her Christmas Club Account by $5, she would save at least $1,400 over the 40 week period. Without the increase, what is the most she is now paying in weekly?

m. The Grimes triplets each weighed the same, and each lost 4 pounds during the first week of their diet. If their total combined weight at the end of the week was no more than 360 pounds, what was the most each could have weighed before starting the diet?

EXPLORATIONS

1. Use chips to model the solution of each inequality.

a. $3 - 2n > 1$ **b.** $5 - 3n < -1$ **c.** $7 < 1 - 2n$ **d.** $5 > 2 - 3n$

2. In a barnyard with chickens and horses, there are at most 60 legs. What is the greatest possible number of chickens?

Explain your reasoning.

Things You Should Know After Studying This Chapter

KEY SKILLS

7.1 Use symbols of inequality to translate between words and symbols.

Graph the solution set of an inequality on a number line.

7.2 Use the properties of inequality to tell whether a sentence involving inequality is true or false.

7.3 Solve an inequality by applying the inverse operation.

Use an inequality to solve a problem.

7.4 Solve an inequality by using two operations.

Use an inequality to solve a problem.

KEY TERMS AND SYMBOLS

7.1 inequality ($>$, $<$, $\geq$, $\leq$, $\neq$) • graph • closed hole • open hole • between

7.2 order of an inequality • properties of inequality

7.3 inverse operation • at most • no more than • at least • no less than

7.4 using two operations • combining like terms

CHAPTER REVIEW EXERCISES

1. Write a word phrase for each symbol.

a. $>$ **b.** $<$ **c.** $\leq$ **d.** $\geq$ **e.** $\neq$

2. Translate from words to symbols.

a. 5 is less than 12
b. –4 is greater than –5
c. 6 is between 5 and 7
d. 9 – 1 is not equal to 7
e. *n* is greater than or equal to 6
f. *y* is at most 17
g. *x* is less than or equal to –4
h. *z* is at least 10
i. *k* is greater than 7 and *k* is less than 10
j. *t* is greater than or equal to 0 and *t* is less than 6

3. Translate from symbols to words.

a. $8 < 17$ **b.** $-2 < -1$
c. $7 \neq 9$ **d.** $1 < 3 < 7$
e. $x \geq 7$ **f.** $y \leq 4$
g. $2 < n < 7$ **h.** $0 \leq x < 9$

4. Using *n* as the variable, write the inequality that is shown in each graph.

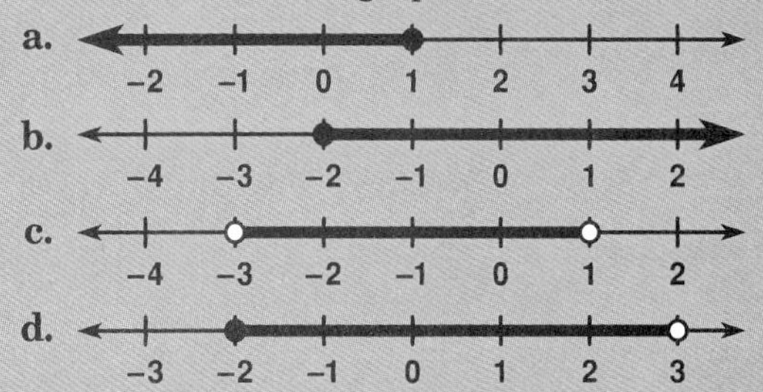

5. Select the graph that best represents the given inequality.

a. $x \geq -1$

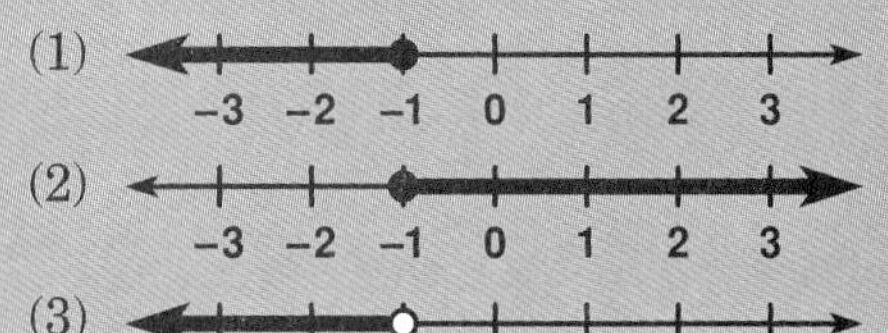

b. $-2 < n \leq 2$

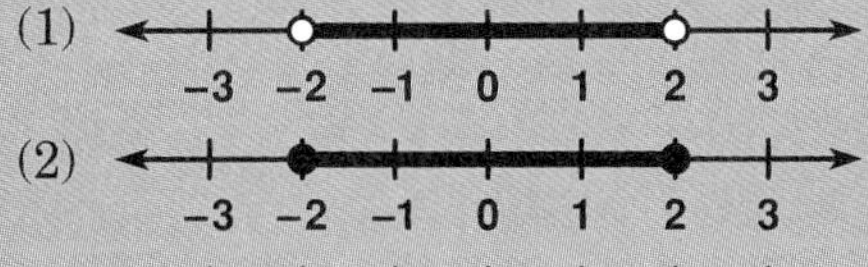

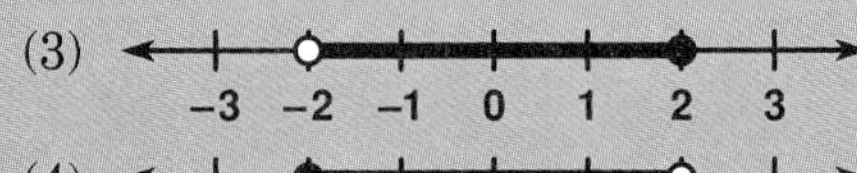

6. Draw the graph of each inequality.

a. $x > 0$ **b.** $x < 3$
c. $x \geq -1$ **d.** $x \leq 2$
e. $3 > x$ **f.** $-2 \leq x$
g. $-1 < x < 3$ **h.** $-2 \leq x < 4$

7. Replace ? by > or <, and give a reason.

a. If $5 > 2$, then $5 + 2$? $2 + 2$.
b. If $-3 < 0$, then $-3 - 1$? $0 - 1$.
c. If $4 > 1$, then 3×4 ? 3×1.
d. If $6 < 7$, then -1×6 ? -1×7.
e. If $9 > 6$, then $9 \div (-3)$? $6 \div (-3)$.

8. Write the statement that results when the operation stated is performed on the given inequality.

a. –1 is added to both sides of $7 > 5$
b. 3 is subtracted from both sides of $1 < 2$
c. 5 is a multiplier of both sides of $2 > 0$
d. –2 is a divisor of both sides of $-2 < 4$
e. –4 is a multiplier of both sides of $-2 > -3$

9. Solve each inequality by applying the inverse operation. Check your answer. Graph the solution set.

a. $y - 7 > -1$ **b.** $x + 5 \leq 7$
c. $4z < 12$ **d.** $\frac{w}{2} \geq 2$
e. $-3t > -6$ **f.** $\frac{a}{-1} \leq -3$

CHAPTER REVIEW EXERCISES

10. Solve each inequality by first undoing addition or subtraction, and then undoing multiplication or division. Check.

a. $2x + 1 < 9$ **b.** $3y - 2 > 1$

c. $4z + 2 \leq -10$ **d.** $\frac{w}{2} + 2 > 10$

e. $-3t + 3 \leq 9$ **f.** $\frac{a}{-2} - 1 \geq 3$

11. Solve each inequality by first combining like terms. Check.

a. $2x + x < 7$ **b.** $3y - y > 4$

c. $4z + z + 2 \leq -13$ **d.** $5x + 1 + x - 3 > 4$

12. Solve each inequality and check.

a. $-2m < 12$ **b.** $x - 3 \geq -6$

c. $\frac{a}{4} > 4$ **d.** $y - 1 \leq -1$

e. $2z + 1 > 11$ **f.** $x + 3x + 4 > 12$

13. Use an inequality to find the set of values for each unknown number.

a. A number increased by 7 is no more than 10.

b. The maximum value of a number is 3.

c. Four less than a number is greater than 14.

d. Six more than twice a number is at least 30.

e. Five times a number decreased by 10 is at most 45.

14. Select the inequality that can best be used to model each problem.

a. Janet would like to lose a minimum of 10 pounds. If her diet will allow her to lose about 1.5 pounds a week, what is the least number of weeks it will take her to reach her goal?

If w = the number of weeks on the diet, then:

(1) $w - 1.5 < 10$ (2) $w - 1.5 \geq 10$

(3) $1.5w < 10$ (4) $1.5w \geq 10$

b. After Frank had prepared 15 burger patties for his barbeque, some additional guests were invited. Frank bought more meat and used one-quarter of each pound to make a patty. If he wanted at least 25 patties in all,what was the least additional amount of meat he should have bought?

If p = the number of pounds of additional meat, then:

(1) $4p + 15 > 25$ (2) $4p + 15 \geq 25$

(3) $4(p + 15) > 25$ (4) $4(p + 15) \geq 25$

c. What are the smallest two consecutive odd integers whose sum is at least 300?

If x = the first of the odd integers, then:

(1) $x + x + 1 \geq 300$ (2) $x + x + 1 \leq 300$

(3) $x + x + 2 \geq 300$ (4) $x + x + 2 \leq 300$

15. Use an inequality to solve each problem. First estimate an answer.

a. For her first apartment rental, Lila plans to spend no more than one-quarter of her monthly income. If her monthly income is $1,400, what is the maximum rent she has budgeted?

b. The seating capacity for the Greatful To Be Alive concert is at most 900 people. If 420 people are already seated, what is the greatest number of seats still available?

c. Mike has twice as many video tapes as Talia. If, together, they have more than 30 tapes, what is the least number of tapes Talia might have?

d. For her trip to Europe, Nancy's luggage can weigh no more than 40 pounds. If her larger suitcase weighs about 10 pounds more than her smaller case, what is the maximum weight of each case?

e. The record for the 5-mile race walk at Truman High is 37.5 minutes. If Norma is planning to break that record, what is, on the average, the greatest number of minutes she can use per mile?

SOLUTIONS FOR TRY THESE

7.1 *Symbols and Graphing*

Page 239

1. **a.** $8 > 7.2$ **b.** $-1 < 0$
 c. $x \leq 9$ **d.** $9 \neq 10 - 2$
2. **a.** 9 is less than 17
 b. -4 is greater than -9
 c. x is greater than or equal to 5
 d. $7 - 1$ is not equal to $\frac{12}{3}$
3. **a.** $n < 1$ **b.** $n \geq 0$
4. **a.**
 b.

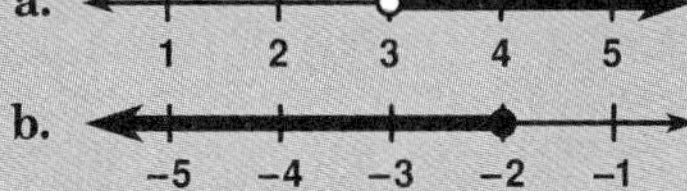

Page 240

1. $5 < x < 9$
2. y is between -1 and 4.
3. $-1 < n < 2$
4. **a.**
 b.

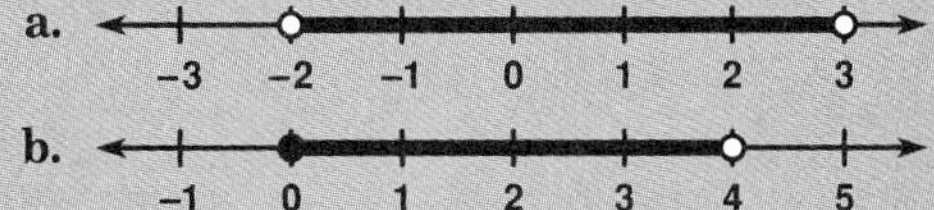

7.2 *Properties of Inequality*

Page 249

1. **a.** If $6 > 1$, then $6 + 2 > 1 + 2$.
 The order of an inequality does not change when the same number is added to both sides.
 b. If $3 < 7$, then $3 - 1 < 7 - 1$.
 The order of an inequality does not change when the same number is subtracted from both sides.
2. **a.** $0 < 6$ **b.** $0 > -1$

Page 251

1. **a.** If $5 > 3$, then $2 \times 5 > 2 \times 3$.
 The order of an inequality does not change when the same positive number multiplies both sides.
 b. If $4 < 8$, then $4 \div 2 < 8 \div 2$.
 The order of an inequality does not change when the same positive number divides both sides.
2. **a.** $8 < 28$ **b.** $5 > 4$

Page 253

1. **a.** If $4 > 1$, then $-3 \times 4 < -3 \times 1$.
 The order of an inequality changes when the same negative number multiplies both sides.
 b. If $16 > 12$, then $16 \div (-4) < 12 \div (-4)$.
 The order of an inequality changes when the same negative number divides both sides.
2. **a.** $-14 < -4$ **b.** $-2 > -3$

Page 254

1. **a.** If $4 < 5$, then $4 \times 4 < 4 \times 5$.
 The order of an inequality does not change when the same positive number multiplies both sides.
 b. If $10 > 5$, then $10 \div 5 > 5 \div 5$.
 The order of an inequality does not change when the same positive number divides both sides.
 c. If $8 > 4$, then $-2 \times 8 < -2 \times 4$.
 The order of an inequality changes when the same negative number multiplies both sides.
 d. If $8 > 4$, then $8 \div (-2) < 4 \div (-2)$.
 The order of an inequality changes when the same negative number divides both sides.
2. **a.** $-12 < -9$ **b.** $0 > -1$
 c. $-3 > -6$ **d.** $2 < 3$

7.3 *Solving a Simple Inequality*

Page 261

1. **a.**

$$x + 4 < 7$$
$$x + 4 - 4 < 7 - 4 \quad \text{Undo + by –. Order does not change.}$$
$$x < 3$$

Check: Choose a number less than 3, say 2.

$$x + 4 < 7$$
$$2 + 4 \overset{?}{<} 7$$
$$6 < 7 \checkmark$$

Answer: $x < 3$

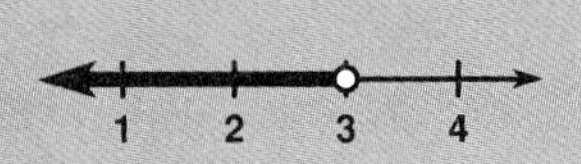

SOLUTIONS FOR TRY THESE

b. $y - 6 > 0$

$y - 6 + 6 > 0 + 6$ Undo – by +. Order does not change.

$y > 6$

Check: Choose a number greater than 6, say 7.

$y - 6 > 0$

$7 - 6 \overset{?}{>} 0$

$1 > 0$ ✓

Answer: $y > 6$

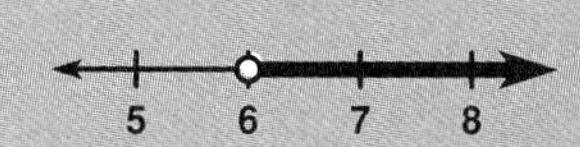

c. $2n \leq 10$

$\frac{2n}{2} \leq \frac{10}{2}$ Undo × by ÷. Dividing by a positive does not change the order of the inequality.

$n \leq 5$

Check: Choose a number less than 5, say 4.

$2n \leq 10$

$2(4) \overset{?}{\leq} 10$

$8 \leq 10$ ✓

Answer: $n \leq 5$

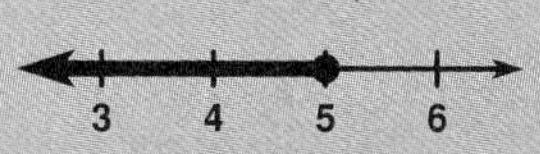

d. $-4x > 8$

$\frac{-4x}{-4} < \frac{8}{-4}$ Undo × by ÷. Dividing by a negative changes the order of the inequality.

$x < -2$

Check: Choose a number less than –2, say –3.

$-4x > 8$

$-4(-3) \overset{?}{>} 8$

$12 > 8$ ✓

Answer: $x < -2$

–4 –3 –2 –1

e. $\frac{z}{3} > 2$

$3 \cdot \frac{z}{3} > 3 \cdot 2$ Undo ÷ by ×. Multiplying by a positive does not change the order of the inequality.

$z > 6$

Check: Choose a number greater than 6, say 7.

$\frac{z}{3} > 2$

$\frac{7}{3} \overset{?}{>} 2$

$2\frac{1}{3} > 2$ ✓

Answer: $z > 6$

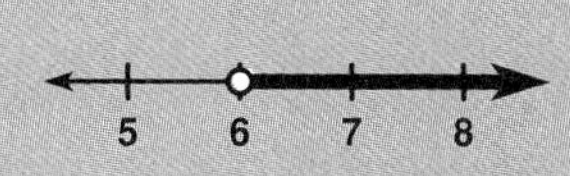

f. $\frac{n}{-2} \leq 2$

$-2 \cdot \frac{n}{-2} \geq -2 \cdot 2$ Undo ÷ by ×. Multiplying by a negative changes the order of the inequality.

$n \geq -4$

Check: Choose a number greater than –4, say –3.

$\frac{n}{-2} \leq 2$

$\frac{-3}{-2} \overset{?}{\leq} 2$

$1\frac{1}{2} \leq 2$ ✓

Answer: $n \geq -4$

–5 –4 –3 –2

2. Let h = the hourly rate

$35h \geq 280$

$\frac{35h}{35} \geq \frac{280}{35}$

$h \geq 8$

Check: Choose a number greater than 8, say 9.

$35(9) = 280$

$315 \geq 280$ ✓

Answer: To make at least \$280 in a 35-hour week, Kevin must earn at least \$8 per hour.

7.4 Using Two Operations to Solve

Page 266

1. a. $2x + 1 < 7$

$2x + 1 - 1 < 7 - 1$ Undo + by –. Order does not change.

$2x < 6$

$\frac{2x}{2} < \frac{6}{2}$ Undo × by ÷. Dividing by a positive does not change the order .

$x < 3$

Check: Choose a number less than 3, say 2.

$2x + 1 < 7$

$2(2) + 1 \overset{?}{<} 7$

$4 + 1 \overset{?}{<} 7$

$5 < 7$ ✓

Answer: $x < 3$

1 2 3 4

b. $3y - 2 \geq 4$

$3y - 2 + 2 \geq 4 + 2$ Undo – by +. Order does not change.

$3y \geq 6$

$\frac{3y}{3} \geq \frac{6}{3}$ Undo × by ÷. Dividing by a positive does not change the order .

$y \geq 2$

Check: Choose a number greater than 2, say 4.

$3y - 2 \geq 4$

$3(4) - 2 \overset{?}{\geq} 4$

$12 - 2 \overset{?}{\geq} 4$

$10 \geq 4$ ✔

Answer: $y \geq 2$

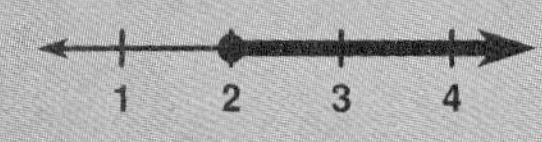

c. $-2z + 3 \leq 5$

$-2z + 3 - 3 \leq 5 - 3$ Undo + by –. Order does not change.

$-2z \leq 2$

$\frac{-2z}{-2} \geq \frac{2}{-2}$ Undo × by ÷. Dividing by a negative changes the order .

$z \geq -1$

Check: Choose a number greater than –1, say 0.

$-2z + 3 \leq 5$

$-2(0) + 3 \overset{?}{\leq} 5$

$0 + 3 \overset{?}{\leq} 5$

$3 \leq 5$ ✔

Answer: $z \geq -1$

–2 –1 0 1

d. $\frac{n}{2} - 1 > 1$

$\frac{n}{2} - 1 + 1 > 1 + 1$ Undo – by +. Order does not change.

$\frac{n}{2} > 2$

$2 \cdot \frac{n}{2} > 2 \cdot 2$ Undo ÷ by ×. Multiplying by a positive does not change the order.

$n > 4$

Check: Choose a number greater than 4, say 5.

$\frac{n}{2} - 1 > 1$

$\frac{5}{2} - 1 \overset{?}{>} 1$

$2\frac{1}{2} - 1 \overset{?}{>} 1$

$1\frac{1}{2} > 1$ ✔

Answer: $n > 4$

3 4 5 6

2. Let p = the price of the cheaper pair.

Then $p + 40$ = the price of the more expensive pair.

$p + p + 40 \leq 180$ The total price is no more than 180.

$2p + 40 \leq 180$ Combine like terms.

$2p + 40 - 40 \leq 180 - 40$ Undo + by –. No change in order.

$2p \leq 140$

$\frac{2p}{2} \leq \frac{140}{2}$ Undo × by ÷. No change the order.

$p \leq 70$

Check: Choose a number less than 70, say 60.

If the cheaper pair costs 60, then the more expensive pair costs 60 + 40, or 100. Is the total 60 + 100, or 160, less than 180? ✔

Answer: To keep the total cost under $180, Devon can spend up to $70 on one pair of boots and up to $110 on the other pair.

Page 268

1. $x + 2x < 9$

$3x < 9$ Combine like terms.

$\frac{3x}{3} < \frac{9}{3}$ Undo × by ÷. Dividing by a positive does not change the order .

$x < 3$

Check: Choose a number less than 3, say 2.

$x + 2x < 9$

$2 + 2(2) \overset{?}{<} 9$

$2 + 4 \overset{?}{<} 9$

$6 < 9$ ✔

Answer: $x < 3$

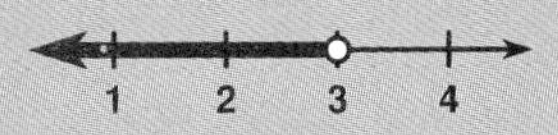

2. $n + n - 1 > 3$

$2n - 1 > 3$

$2n - 1 + 1 > 3 + 1$ Undo – by +. Order does not change.

$2n > 4$

$\frac{2n}{2} > \frac{4}{2}$ Undo × by ÷. Dividing by a positive does not change the order .

$n > 2$

Check: Choose a number greater than 2, say 3.

$n + n - 1 > 3$

$3 + 3 - 1 \overset{?}{>} 3$

$6 - 1 \overset{?}{>} 3$

$5 > 3$ ✔

Answer: $n > 2$

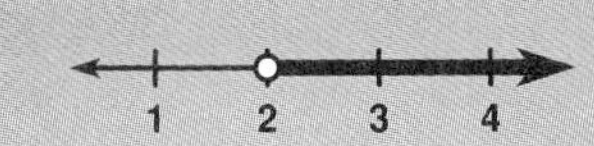

Using Algebra in Geometry

8

Lines and shapes are everywhere around you.

Study the dark pair of lines in each drawing and describe how they are related to each other.

What happens if each line of the pair is extended?

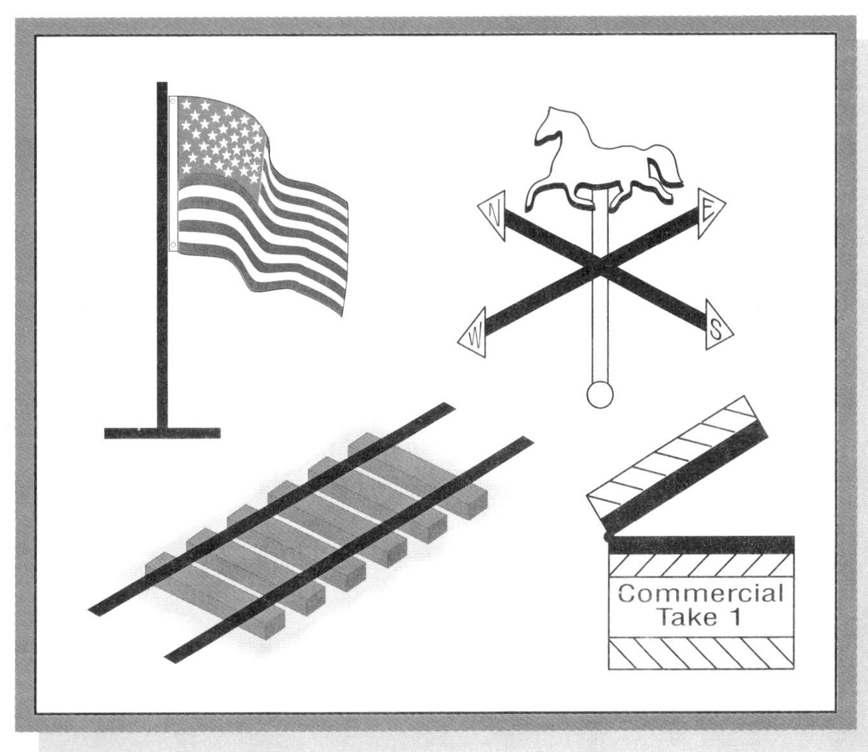

8.1 Basic Geometric Figures

Point, Line, Line Segment

Think of yourself in relation to a globe of our planet Earth.

How would you represent your location on Earth?

How would you represent the distance between yourself and another student who is in Brazil?

To make drawings that represent location and distance, we use different geometric figures.

The most basic geometric figure is a ***point***, which is used to represent a location.

• *P*
point *P*

A collection of points can form a ***straight line***.

Since a line extends without end in opposite directions, shown by arrowheads, a line cannot be measured.

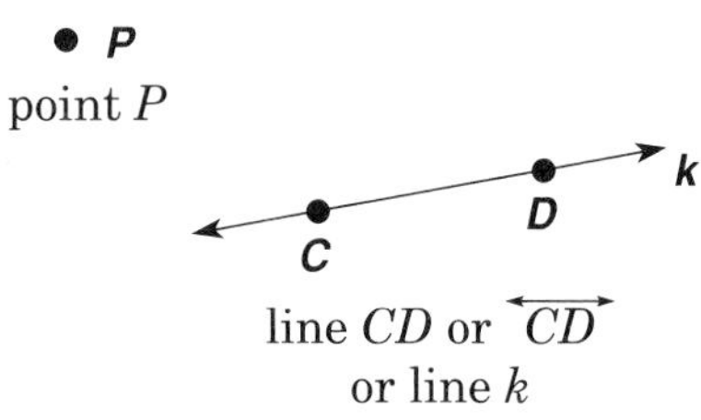

line CD or $\overleftrightarrow{CD}$ or line k

A line is named by any two points on it, or by using a lowercase letter.

Line CD is also written $\overleftrightarrow{CD}$, where the symbol $\leftrightarrow$ uses the arrowheads showing that a line never ends.

A ***line segment*** is part of a line. Since a line segment is between two points, called ***endpoints***, the length of a line segment can be measured.

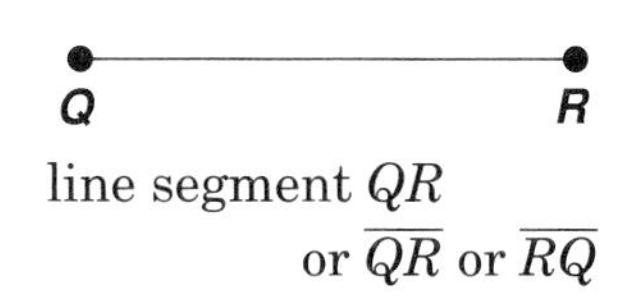

line segment QR or $\overline{QR}$ or $\overline{RQ}$

Try These (For solutions, see page 347.)

Write the word name of each figure whose symbol is given. Draw a diagram.

1. $\overline{AB}$ **2.** $\overleftrightarrow{CD}$

Length, Midpoint

On the number line shown, X has the coordinate 1 and Y has the coordinate 4. How would you find the length of $\overline{XY}$?

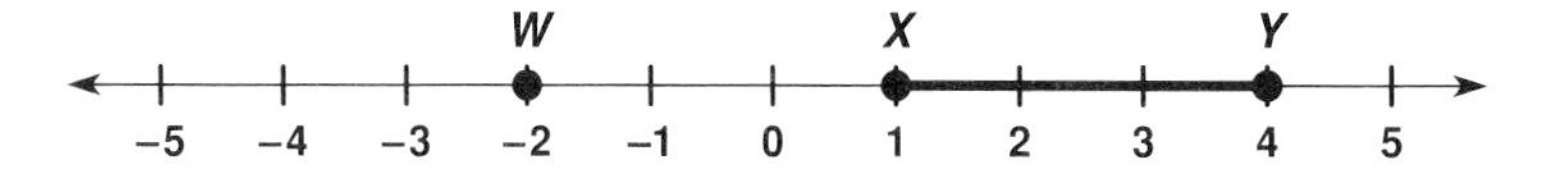

To find the length of $\overline{XY}$, you could count units on the number line. From 1 through 4, there are 3 units.

$\overline{XY}$ has a length of 3 units, or $XY = 3$.

Note that we represent the length XY without writing the line segment symbol above it.

To find the length of $\overline{XY}$, you could also subtract the coordinates of the endpoints.

$$XY = 4 - 1 = 3$$

Using the previous number line, describe the location of point X with respect to points W and Y. Did you say that X is in the *middle*? X is called the ***midpoint*** of $\overline{WY}$.

Example 1 Point T is the midpoint of $\overline{RS}$.
The length of $\overline{TS}$ is 9 units.
Draw some conclusions using this figure.

Since T is the midpoint: $RT = TS = 9$

$$RS = RT + TS = 9 + 9 = 18$$

A midpoint divides a line segment into two segments that are equal in length. Each of the smaller segments is half the length of the original segment.

The length of the original segment is the sum of the lengths of the two smaller segments, or twice the length of one of them.

Example 2 Point G is the midpoint of $\overline{EF}$.
If $EG = 2x + 2$ and $GF = x + 7$, find the length of $\overline{EF}$.

1 ***Read the problem for information.***
Since G is the midpoint, you know
$EG = GF$ and $EG + GF = EF$.

2 ***Make a plan.***
First, find the value of x from the equation

$$EG = GF$$
$$2x + 2 = x + 7$$

Then, use the value of x to find EG, GF, and EF.

3 ***Carry out your plan.***

$$EG = GF$$
$$2x + 2 = x + 7$$
$$2x - x + 2 = x - x + 7 \quad \text{Subtract } x \text{ from both sides.}$$
$$x + 2 = 7$$
$$x + 2 - 2 = 7 - 2 \quad \text{Subtract 2 from both sides.}$$
$$x = 5$$

$$EG = 2x + 2 = 2(5) + 2 = 10 + 2 = 12 \quad \text{Substitute } x = 5.$$
$$GF = x + 7 = 5 + 7 = 12$$
$$EF = EG + GF = 12 + 12 = 24$$

4 ***Check.***
Does $EG = GF$? $12 = 12$ ✔

Answer: $EF = 24$

Try These (For solutions, see page 347.)

1. Draw a diagram for each line segment.
   - **a.** Find the length of $\overline{AB}$ if the coordinate of A is 4 and the coordinate of B is 9.
   - **b.** Find the length of $\overline{RS}$ if the coordinate of R is -3 and the coordinate of S is 4.
2. M is the midpoint of $\overline{AB}$.
   If $AM = 4x - 2$ and
   $MB = 3x + 1$:

   - **a.** Write an equation in x that tells the relationship between AM and MB.
   - **b.** Solve the equation.
   - **c.** Use the value for x to find the number of units in AM, MB, and AB.

Ray, Angle

Suppose you are at a point on Earth, in a rocketship that takes off into space. How would you represent the takeoff?

A ***ray*** is a part of a line that has one endpoint and extends without end in one direction.

The name of a ray always mentions the endpoint first.

ray AB or $\overrightarrow{AB}$

When two rays have the same endpoint, an ***angle*** is formed. The endpoint of the rays is the ***vertex*** of the angle, and the rays are the ***sides*** of the angle.

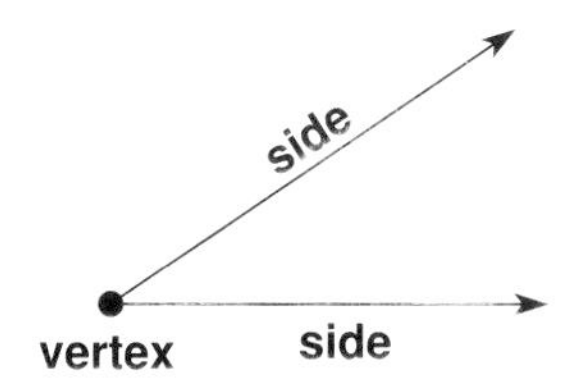

Using the symbol $\angle$ for angle, the name of an angle can be written:

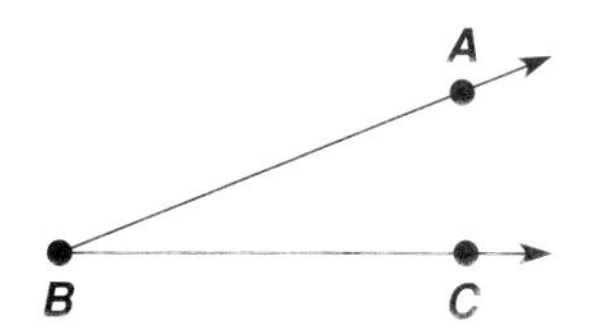

- by the vertex alone
- by the vertex and a point on each side
  The vertex is always the middle letter.

This angle may be called $\angle B$, $\angle ABC$, or $\angle CBA$.

Try These (For solutions, see page 347.)

1. Use letters to name 6 angles in this diagram.
2. Explain why $\angle Y$ is not a good way to name an angle in this diagram.
3. Use letters to name 6 angles with vertex at point Y.

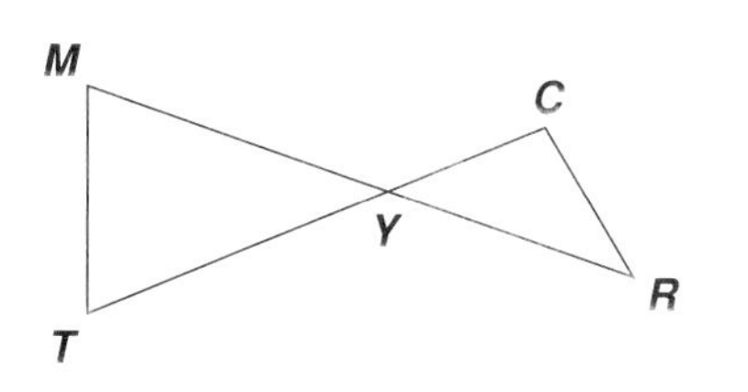

Measuring an Angle

Consider three angles whose sides are line segments of the same length.

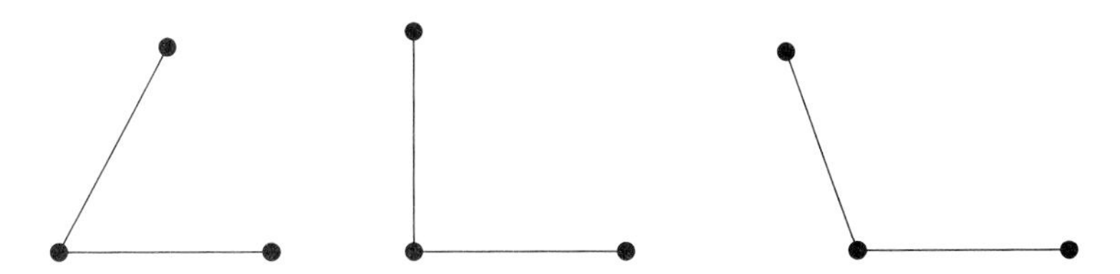

Would you say that the size of these angles is the same? What do you mean by the *size of an angle*?

The size of an angle is the amount of openness between the sides.

The size of an angle is measured in units called ***degrees***. An angle of 1 degree, written as 1°, is $\frac{1}{360}$ of a complete circle.

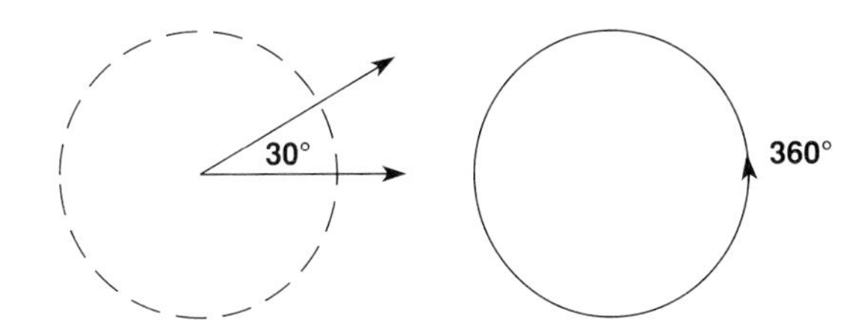

A ***protractor*** is an instrument used to measure an angle. In symbols, the measure of angle ABC is written $m\angle ABC$.

Example 3 Read the measures of some of the angles shown on this protractor.

For convenience in reading the measure of an angle regardless of its position, a protractor has two scales, each starting at 0° and ending at 180°.

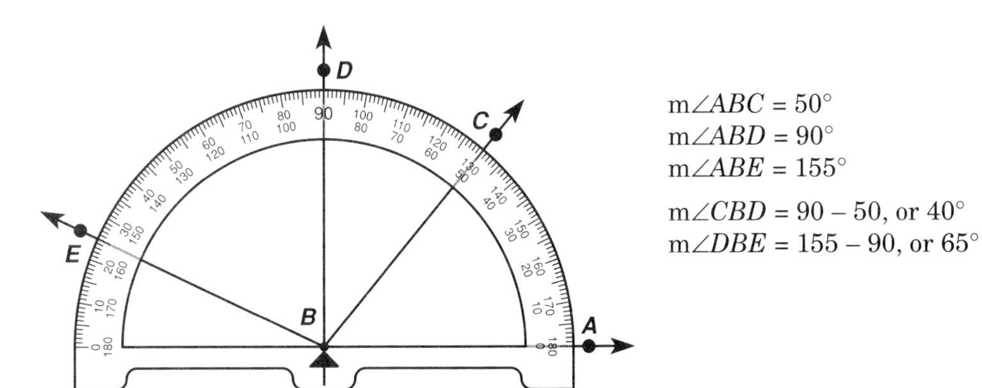

$m\angle ABC = 50^\circ$
$m\angle ABD = 90^\circ$
$m\angle ABE = 155^\circ$

$m\angle CBD = 90 - 50$, or 40°
$m\angle DBE = 155 - 90$, or 65°

To use a protractor to measure an angle:

Place the center of the protractor at the vertex of the angle.

Align the 0° mark on the protractor with one side of the angle.

Try These (For solutions, see page 347.)

Use a protractor to measure each angle.

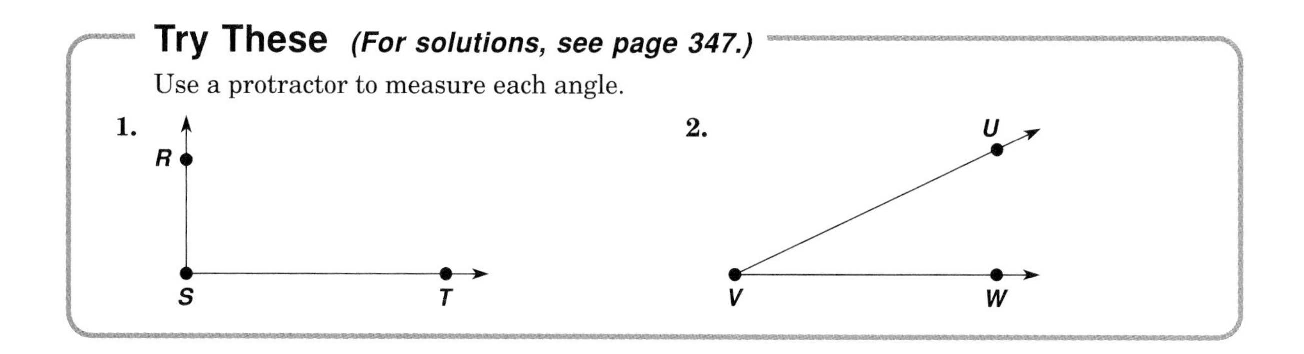

Classifying an Angle

Angles are named according to their degree measure.

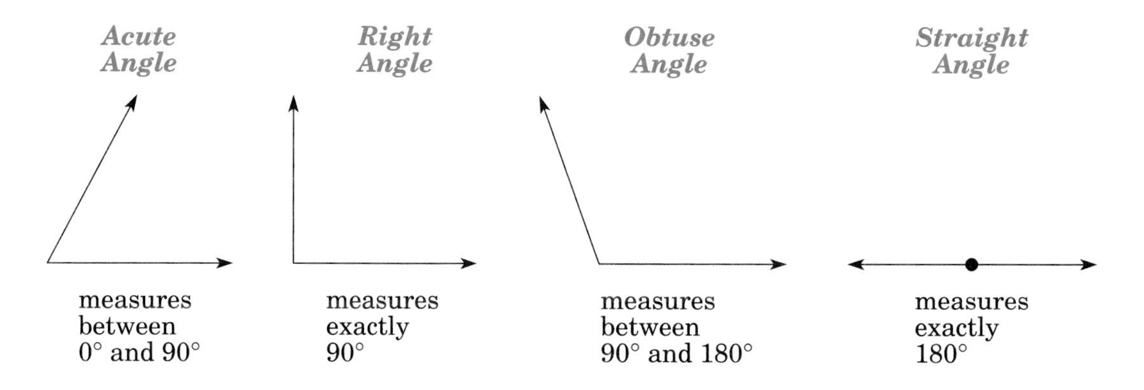

Example 4 Classify each angle and estimate its degree measure. Use a protractor to check your estimate.

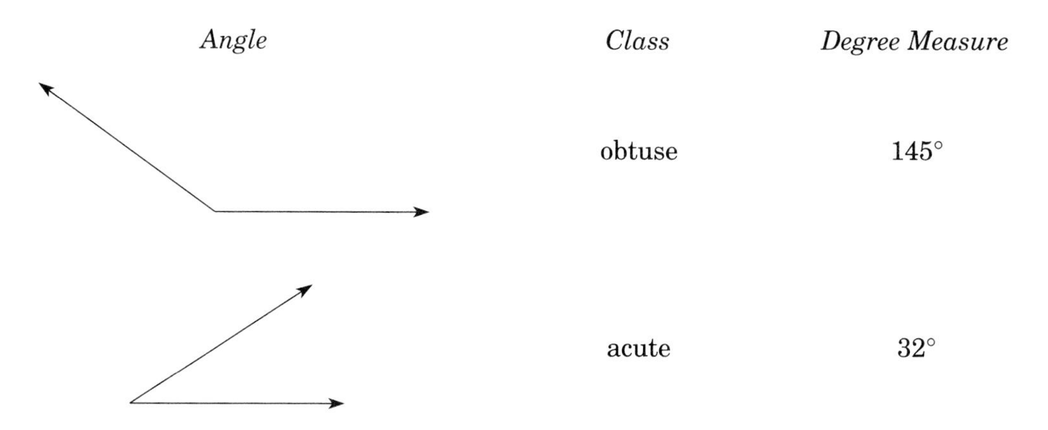

Try These (For solutions, see page 347.)

Classify each angle.

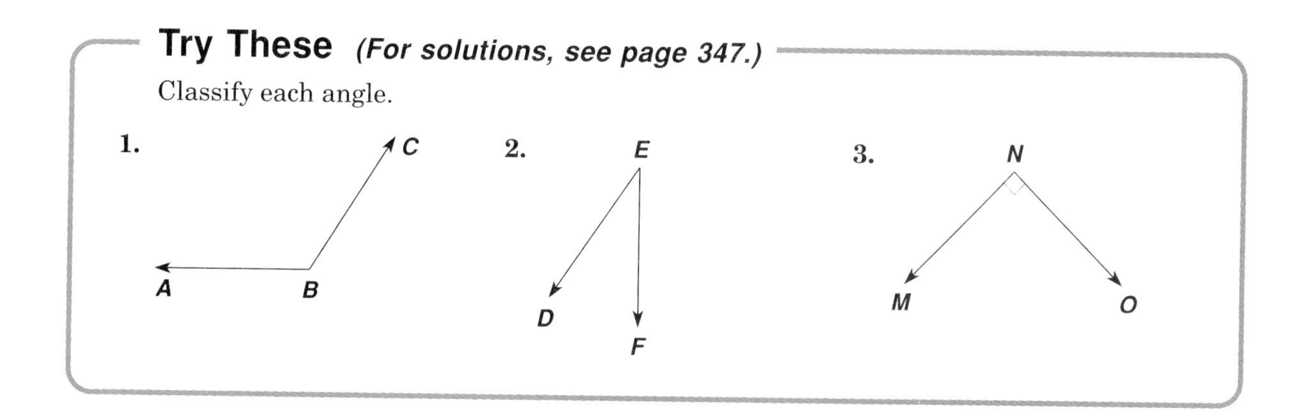

Congruence

These three figures all have the same length. How else are they the same? How are they different?

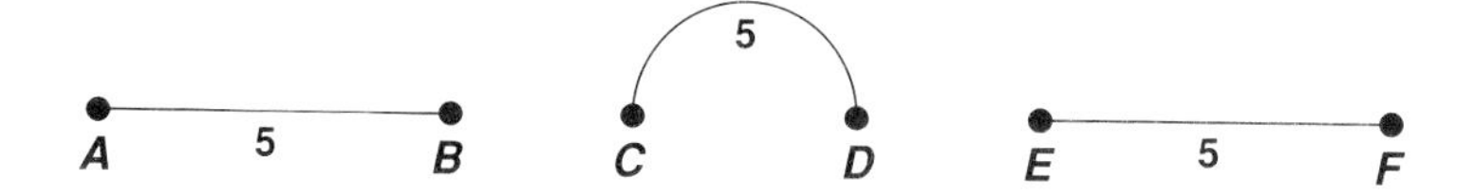

Segment AB is the same as segment EF in that they are both straight. Segment CD is a different shape; it is not straight.

Figures that are the same size and the same shape are called *congruent*.

$\cong$ is the symbol for the words *is congruent to*.

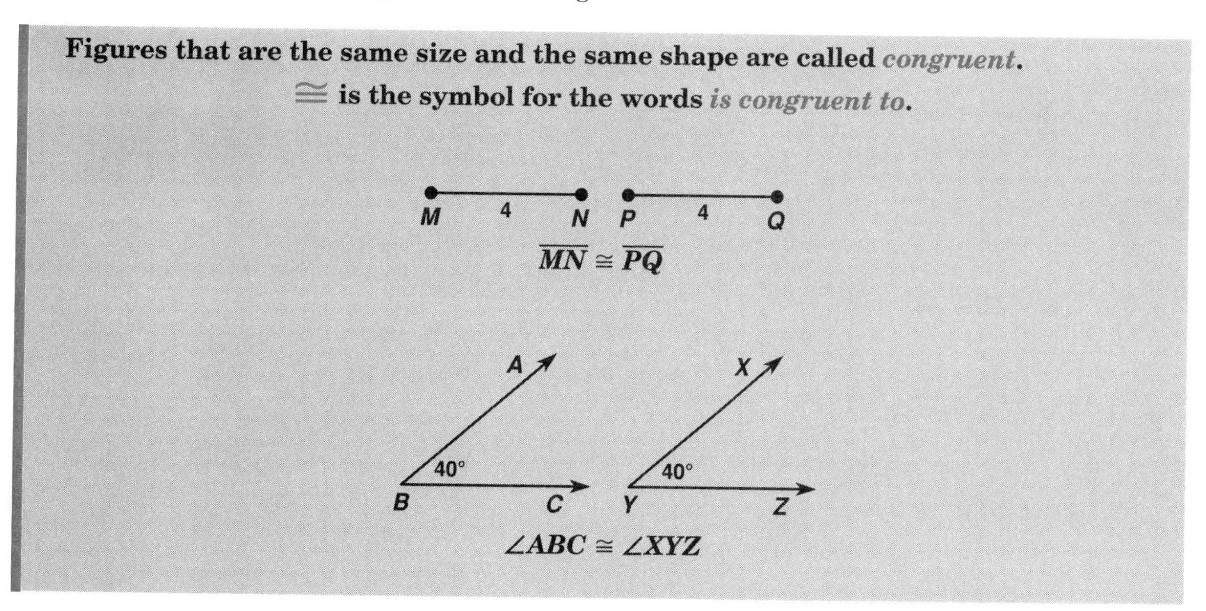

Example 5 These two angles are congruent. Without using a protractor, find their degree measure.

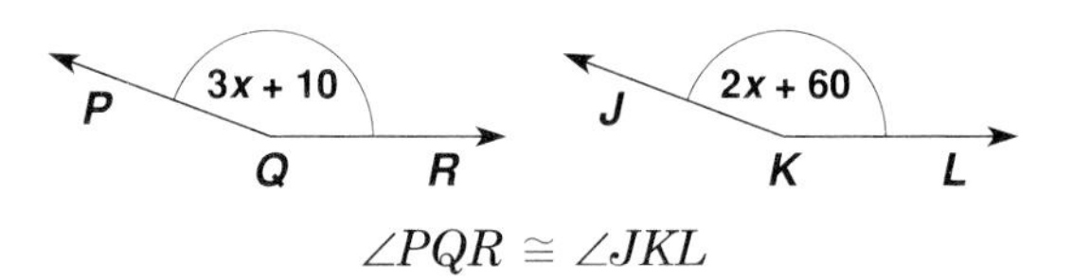

$\angle PQR \cong \angle JKL$

1 ***Read the problem for information.***
Since the angles are congruent, you know they have the same degree measure.

2 ***Make a plan.***
First, find the value of x from the equation

$$m\angle PQR = m\angle JKL$$
$$3x + 10 = 2x + 60$$

Then, use the value of x to find $m\angle PQR$ and $m\angle JKL$.

3 ***Carry out your plan.***

$$m\angle PQR = m\angle JKL$$
$$3x + 10 = 2x + 60$$
$$3x - 2x + 10 = 2x - 2x + 60 \quad \text{Subtract } 2x \text{ from both sides.}$$
$$x + 10 = 60$$
$$x + 10 - 10 = 60 - 10 \quad \text{Subtract 10 from both sides.}$$
$$x = 50$$
$$m\angle PQR = 3x + 10 \quad \text{Substitute } x = 50.$$
$$= 3(50) + 10$$
$$= 150 + 10 = 160$$
$$m\angle JKL = 2x + 60$$
$$= 2(50) + 60$$
$$= 100 + 60 = 160$$

4 ***Check.***
Does $m\angle PQR = m\angle JKL$? $160 = 160$ ✔

Answer: Each of the two angles measures 160°.

Try This *(For solution, see page 347.)*

These two angles are congruent. Without using a protractor, find their degree measure.

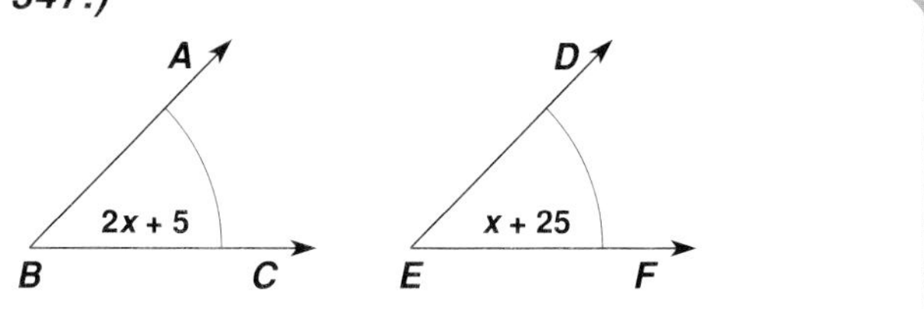

SELF-CHECK: SECTION 8.1

1. What geometric figure is used to represent a location? a distance between two points?

Answer: A point represents a location and a line segment represents a distance between two points.

2. What is the meaning of the symbol $\overleftrightarrow{AB}$? What does the symbol tell about the figure?

Answer: $\leftrightarrow$ is the symbol for line, telling that line AB continues without end in opposite directions.

3. What is the difference between a line and a line segment?

Answer: A line continues without end in opposite directions, while a line segment has two endpoints.

A line cannot be measured. A line segment has a specific length.

4. **a.** What is another name for $\overline{AB}$?

Answer: $\overline{BA}$, a line segment may be named mentioning either endpoint first.

b. If the measure of $\overline{AB}$ is 4 units, write this information in symbols.

Answer: $AB = 4$ Note that for the length of a line segment, the line segment symbol is not used.

5. How is a ray different from a line segment?

Answer: A ray has only one endpoint, and continues without end in one direction.

6. **a.** Explain why $\overrightarrow{AB}$ is different from $\overrightarrow{BA}$. Draw the 2 figures.

A B A B

Answer: $\overrightarrow{AB}$ has the endpoint A, while $\overrightarrow{BA}$ has the endpoint B.

b. Can you find the length of $\overrightarrow{AB}$? Explain.

Answer: No, a ray cannot be measured since it continues without end in one direction.

7. What instrument do you use to measure a line segment? Can you use the same instrument to measure an angle? Explain.

Answer: A ruler is used to measure a line segment. Since a ruler is straight, it cannot be used to measure the amount of openness in an angle, which is measured on a circle. A protractor is used to measure an angle.

8. What are some units that are used to measure a "short" line segment? a "long" line segment?

Answer: For a short line segment: inch, centimeter
For a longer line segment: foot, meter, yard
For still a longer line segment: mile, kilometer

9. If M is the midpoint of $\overline{PQ}$ and $PQ = 8$ units, how long is $\overline{PM}$?

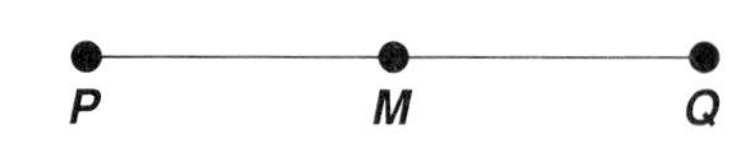

Answer: $PM = 4$ units, a midpoint divides a line segment into two exact halves.

10. Write the name of this angle in symbols.

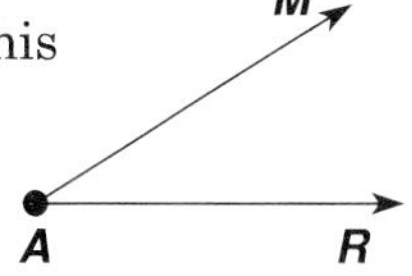

Answer: $\angle A$, $\angle MAR$, or $\angle RAM$

11. What do we call an angle whose measure is:

	Answer
a. exactly 90°	a right angle
b. exactly 180°	a straight angle
c. between 0° and 90°	an acute angle
d. between 90° and 180°	an obtuse angle

12. For two figures to be called *congruent,* what must be true?

Answer: The figures must have the same size and the same shape.

EXERCISES: SECTION 8.1

1. Refer to the diagram to answer:

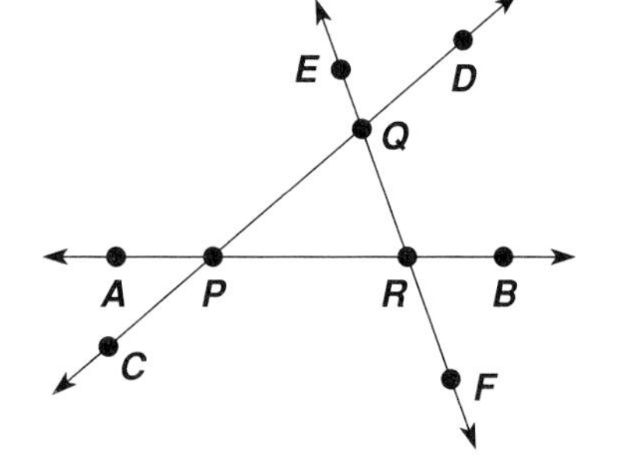

a. (1) How many lines are shown? Name them.

(2) Which lines *intersect* (cross) at point R?

(3) What is the point of intersection of $\overleftrightarrow{AB}$ and $\overleftrightarrow{CD}$?

b. How many line segments are shown? Name them.

c. (1) Explain why $\overrightarrow{CP}$, $\overrightarrow{CQ}$, and $\overrightarrow{CD}$ are all the same ray.

(2) How many different rays have the endpoint P? Name them.

d. (1) Explain why $\angle DPB$ and $\angle QPR$ are the same angle.

(2) How many different angles do you see that have point P as the vertex? Name them.

2. a. Using this number line, find the lengths of these line segments:

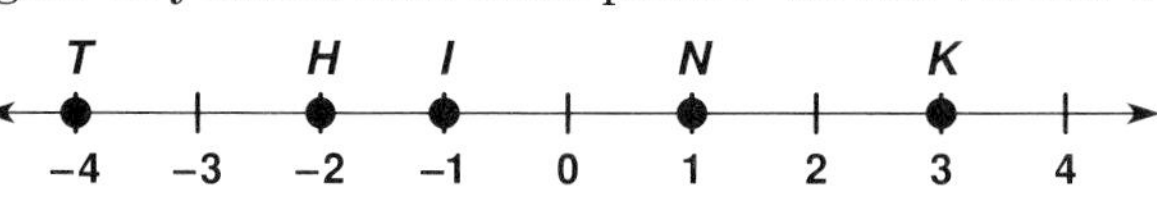

(1) $\overline{NK}$ **(2)** $\overline{IK}$ **(3)** $\overline{HK}$ **(4)** $\overline{TK}$ **(5)** $\overline{IN}$ **(6)** $\overline{TI}$

b. Explain why you can use either $|3 - 1|$ or $|1 - 3|$ to find the length of $\overline{NK}$.

c. Write an absolute-value expression to find the length of $\overline{HN}$.

d. Name a line segment that is congruent to $\overline{TH}$.

e. Copy and complete this table.

Points	Names of All Line Segments Formed by These Points	Number of Points	Number of Line Segments
T, *H*	$\overline{TH}$	2	1
T, *H*, *I*	$\overline{TH}, \overline{TI}, \overline{HI}$	3	3
T, *H*, *I*, *N*	?	4	?
T, *H*, *I*, *N*, *K*	?	5	?

How many line segments do you think could be formed by 6 points?

3. In the diagram, M is the midpoint of $\overline{RS}$.

a. Explain why each of the following is true.

(1) $RM = \frac{1}{2}RS$ and $MS = \frac{1}{2}RS$ **(2)** $RM = MS$ **(3)** $RS = 2RM$

b. If $RM = 8$, find: **(1)** MS **(2)** RS

c. If $RM = 5$, and the length of $\overline{MS}$ is represented by $3x - 7$, find the value of x.

d. If $MS = 12$ and $RS = 4y + 8$, find the value of y.

e. If $RM = 2a + 3$ and $MS = a + 7$, find : **(1)** RM **(2)** RS

4. a. What time is it on each of these clocks?

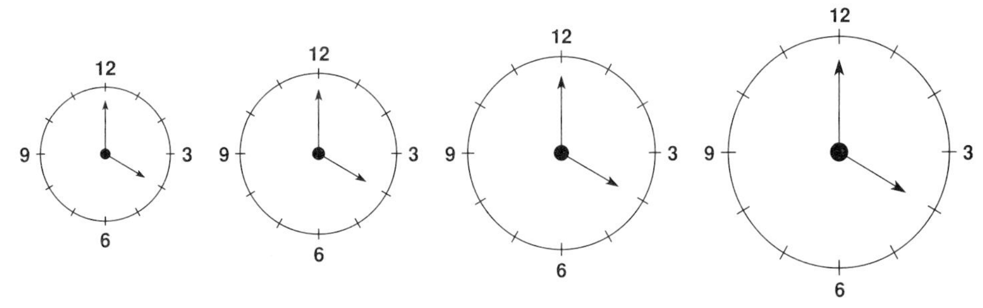

b. Use a protractor to find the number of degrees in the angle formed by the hands on each clock.

Make an observation about the size of the clock and the number of degrees in the angle formed by the hands when the time is 4 o'clock.

5. What kind of angle is formed by the hands of a clock when the time is:

a. 3 o'clock **b.** 2 o'clock **c.** 5 o'clock **d.** 6 o'clock

6. How many degrees are there in the angle formed by the hands of a clock when the time is:

a. 3 o'clock **b.** 6 o'clock **c.** 1 o'clock **d.** 4 o'clock

e. 2:30 **f.** 5:15 **g.** 3:45 **h.** 4:40

7. Use a coaster or other round object to trace a circle on paper. Cut out the circle and fold it in half to find a diameter.

From each end of the diameter, fold the circle down so that the two folds meet at a point on the circle.

a. Unfold your circle, and look at the angle whose vertex is at the point where the folds met. What kind of angle is this? Use a protractor to verify your conclusion.

b. From each end of the diameter, fold the circle down again so that the two new folds meet at a point different from the previous point. Unfold your circle. What kind of angle has its vertex at the point where these new folds met?

c. Describe the line segments that form the sides of the angles you have been forming. Describe the location of the vertex of these angles. Make an observation about the angles formed in this way.

8. In the diagram, angles are formed by 3 intersecting lines.

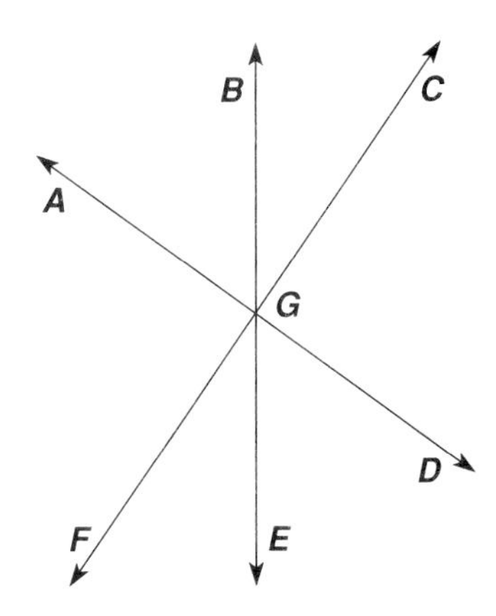

a. Name the angles that look as though they are right angles. Use a protractor to check.

b. What kind of an angle is $\angle BGC$?

c. Name 3 acute angles.

d. What kind of an angle is $\angle BGD$? Name another such angle.

e. What kind of angle is $\angle AGD$? Name another such angle.

f. Name an angle that measures about:

(1) 30° **(2)** 60°

(3) 120° **(4)** 150°

Use a protractor to check.

g. Name an angle that appears to be congruent to $\angle DGE$.

9. In this diagram, each angle is congruent to another angle.

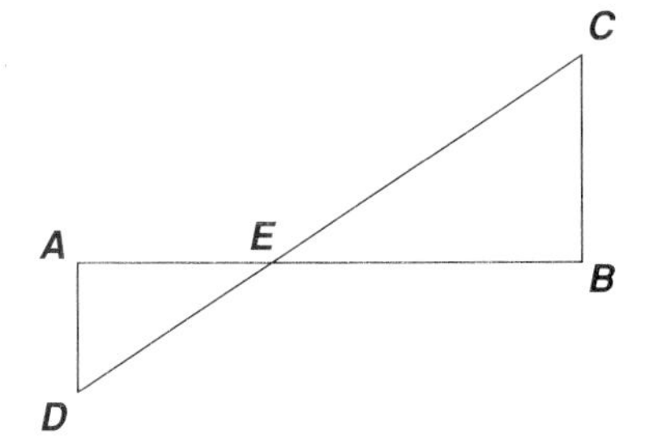

a. By observation, complete each statement.

(1) $\angle A \cong$? **(2)** $\angle ADE \cong$?

(3) $\angle BEC \cong$? **(4)** $\angle DEB \cong$?

b. If $\angle C \cong \angle D$, and $m\angle C = 4x + 18$ and $m\angle D = 6x - 2$, find:

(1) the value of x **(2)** the measure of $\angle C$

EXPLORATIONS

1. **a.** Refer to the graph.

   (1) Write the coordinates of point P.

   (2) Write the coordinates of its image P' if P is reflected over the horizontal axis.

   (3) Write the coordinates of point Q.

   (4) Write the coordinates of its image Q' if Q is reflected over the horizontal axis.

   (5) What is the length of $\overline{PQ}$? of $\overline{P'Q'}$?

   (6) Use an appropriate symbol to tell the relationship between the lengths of $\overline{PQ}$ and its image $\overline{P'Q'}$. $\overline{PQ}$? $\overline{P'Q'}$

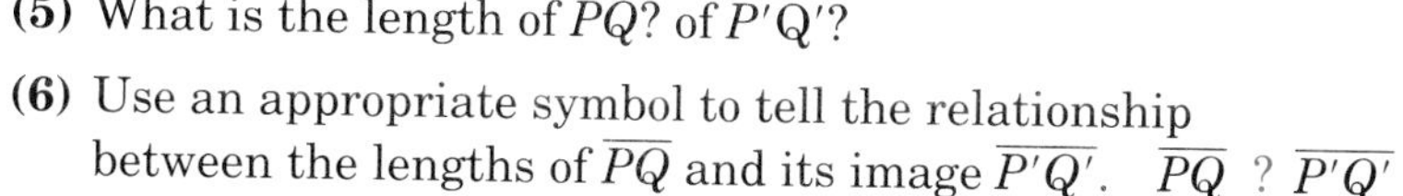

   b. Graph $\overline{AB}$ with endpoints $A(1, 3)$ and $B(1, -2)$. Write the coordinates and draw the graph of:

   (1) the image $\overline{A'B'}$ when $\overline{AB}$ is reflected over the vertical axis

   (2) the image $\overline{A''B''}$ (read *A double prime B double prime*) when $\overline{AB}$ is reflected over the line that evenly divides Quadrants I and III

   (3) the image $\overline{A'''B'''}$ (read *A triple prime B triple prime*) when $\overline{AB}$ is reflected over the line that evenly divides Quadrants II and IV

   c. Write the coordinates and draw the graph of:

   (1) the image $\overline{C'D'}$ after $\overline{CD}$ with endpoints $C(-3, 1)$ and $D(2, 1)$ is translated 2 units right and 3 units up

   (2) the image $\overline{E'F'}$ after $\overline{EF}$ with endpoints $E(2, 4)$ and $F(2, -1)$ is translated 3 units left and 1 unit down

   (3) the image $\overline{G'H'}$ after $\overline{GH}$ with endpoints $G(-2, -1)$ and $H(1, 3)$ is translated 3 units right and 2 units down

(Explorations continue)

2. In Geometry, the most simple tools are:

- an unmarked ruler, called a ***straightedge***
- an unmarked instrument, called a ***compass***, that will draw a complete circle or part of a circle (an ***arc***)

The drawings made using a straightedge and a compass are called ***constructions***.

Here's how to copy the measure of a line segment using a straightedge and compass.

This is the line segment to be copied.

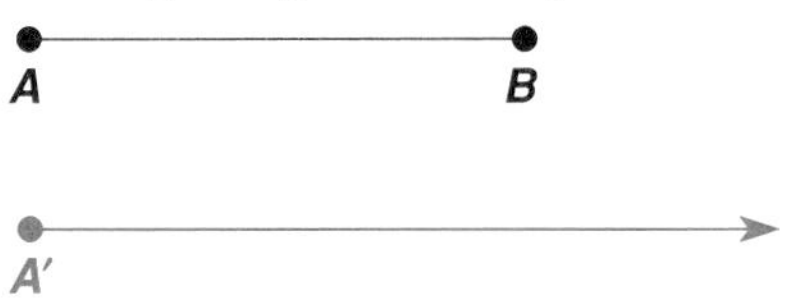

(1) Use a straight edge to draw a ray on which $\overline{AB}$ will be copied. Label the endpoint of the ray A'. (Read A' as A *prime*.)

(2) Measure $\overline{AB}$ with the compass: With the point at A, open the compass so that the pencil is at B.

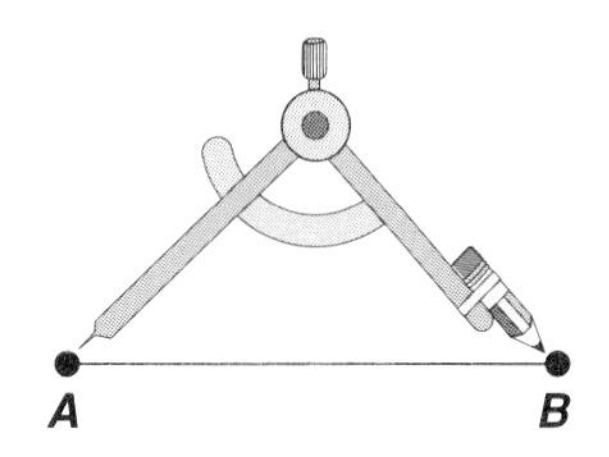

(3) Using this setting on the compass, place the point of the compass at A' and let the pencil draw an arc that intersects the ray at B'.

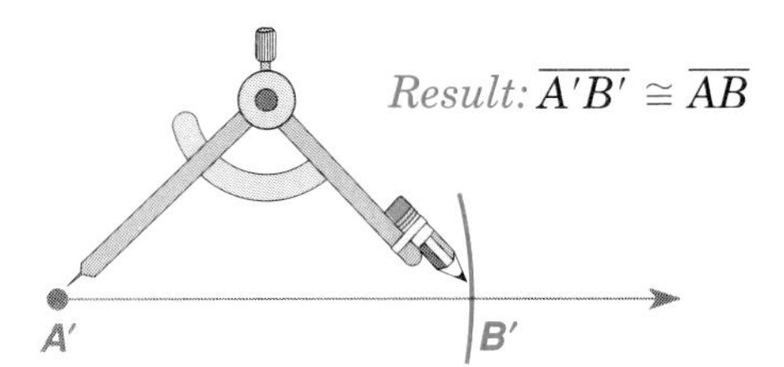

a. In the construction, if you had let the pencil draw a complete circle instead of just an arc, you would be able to see the location of point A' with respect to that circle. What is it?

b. Practice the construction by copying this line segment onto separate paper.

P ——— **Q**

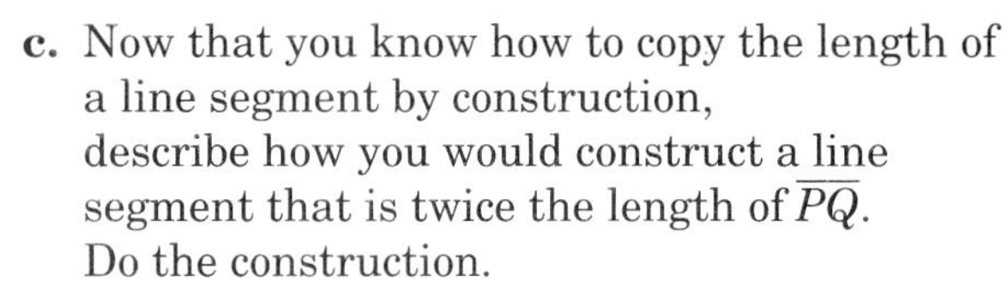

c. Now that you know how to copy the length of a line segment by construction, describe how you would construct a line segment that is twice the length of $\overline{PQ}$. Do the construction.

d. Suppose you had two different line segments. Describe how you would construct a line segment with measure equal to the sum of the measures of the two line segments.

e. Using line segments of lengths a and b, construct a line segment of length:

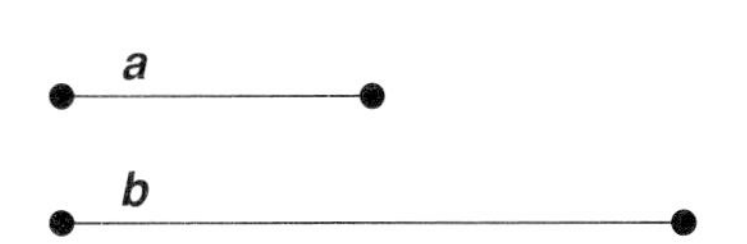

(1) $a + b$

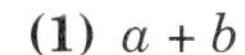

(2) $b - a$

3. Here's how to copy an angle. This is the angle to be copied.

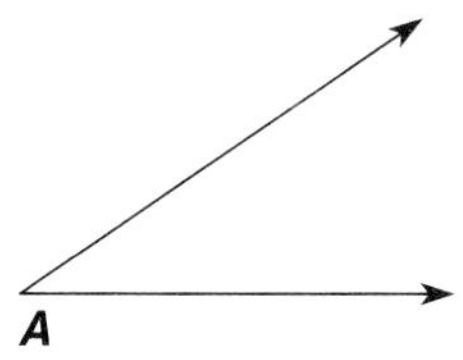

① Use a straightedge to draw a ray that will be one side of the new angle. The endpoint of the ray, A', will be the vertex of the new angle.

② With any setting on the compass, use A as the center and draw an arc that intersects the sides of the given angle. Label the points of intersection B and C.

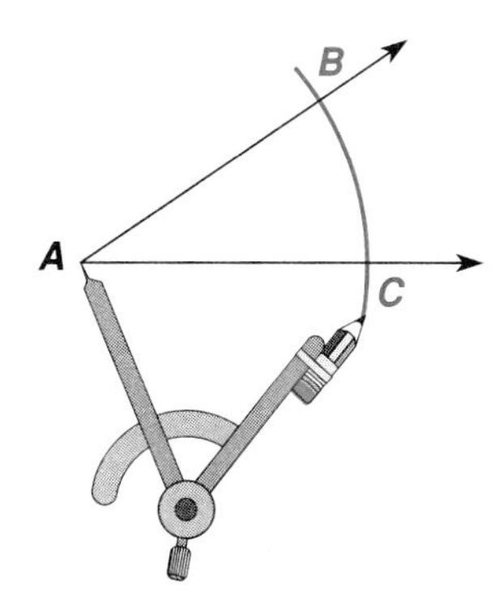

③ With the same setting on the compass, use A' as the center and draw an arc onto which the given angle will be copied. Label the point of intersection of the arc and the ray C'.

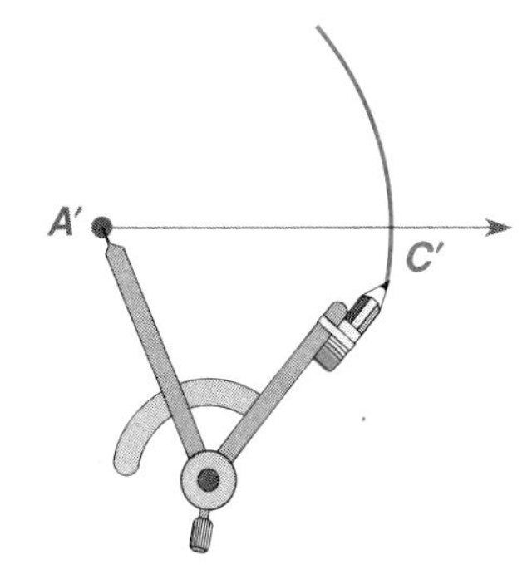

④ Measure the arc constructed on $\angle A$:

With the point at C, open the compass so that the pencil is at B.

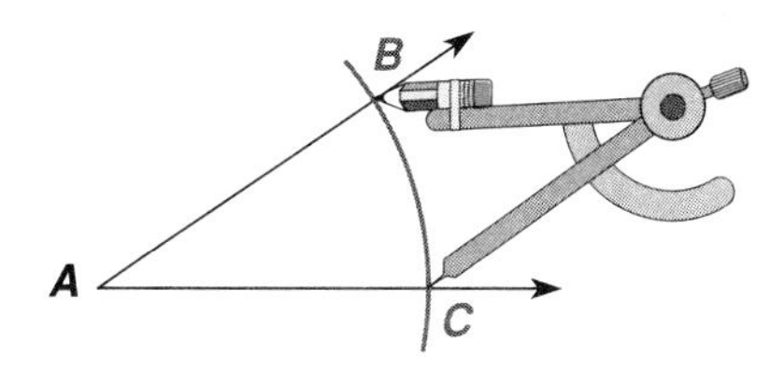

⑤ Copy the measure of this arc onto the arc with the ray:

Keeping the radius of step 4, use C' as the center and draw an arc that intersects the existing arc. Label this point of intersection B'.

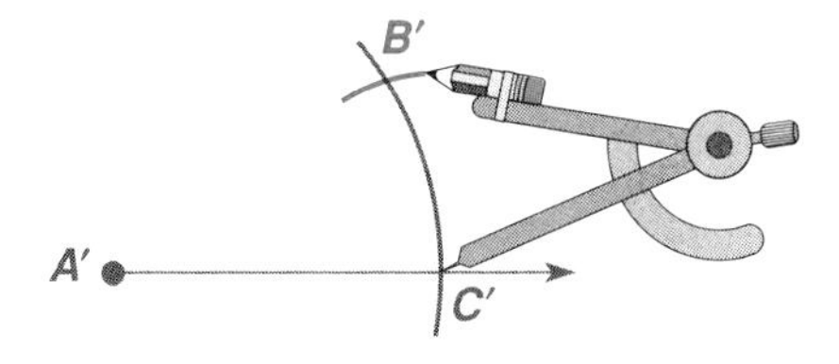

⑥ With a straightedge, draw a ray from A' through B'.

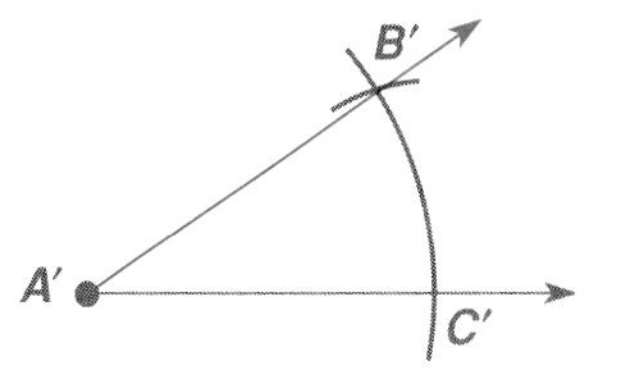

Result: $\angle A' \cong \angle A$

a. Practice this construction by copying each of these angles onto separate paper.

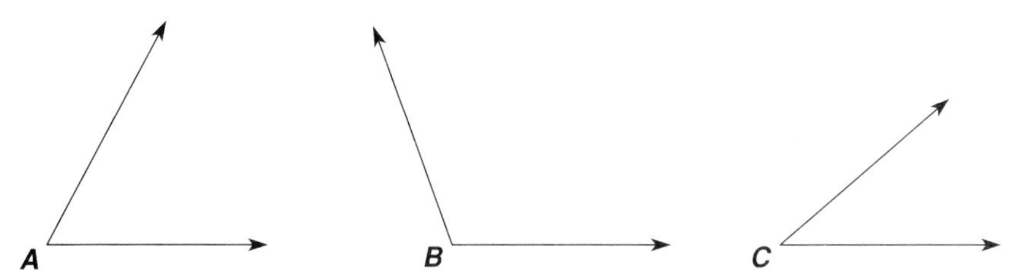

b. Now that you know how to copy an angle by construction, describe how you would construct an angle that is twice the measure of a given angle. Do the construction using $\angle A$ above.

c. Suppose you had two different angles. Describe how you would construct an angle with measure equal to the sum of the measures of the two angles.

d. Using angles of measures a and b, construct an angle of measure:

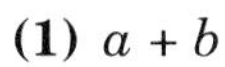

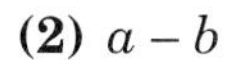

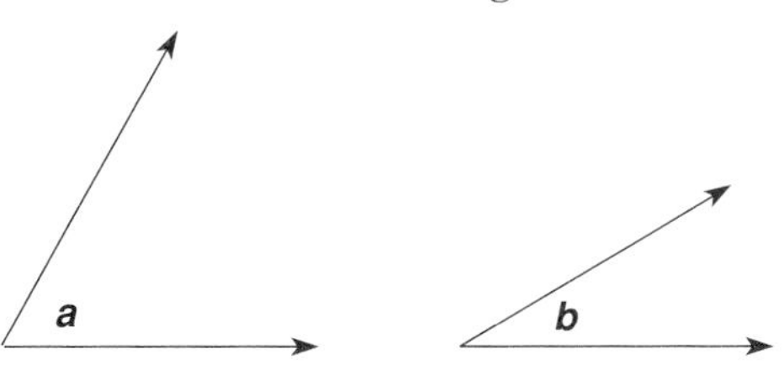

4. Here's how to find the midpoint of a line segment.

(1) Open the compass to a setting that is more than half the length of $\overline{CD}$.

Using this setting and C as the center, draw an arc above and below $\overline{CD}$.

(2) With the same setting and D as the center, draw arcs above and below $\overline{CD}$.

(3) With a straightedge, draw a line through the points where the arcs cross.

This line crosses $\overline{CD}$ at its midpoint, M.

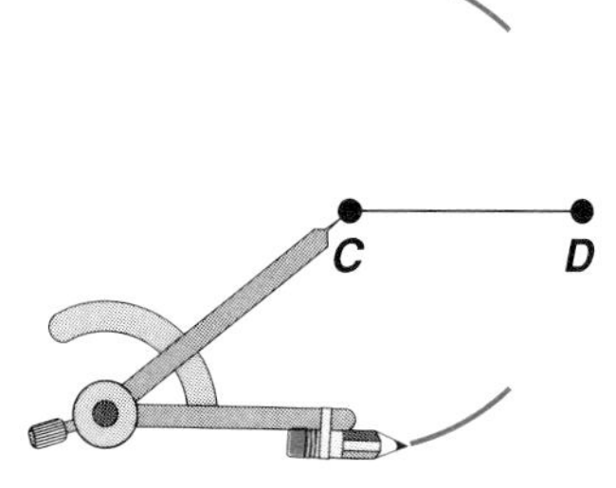

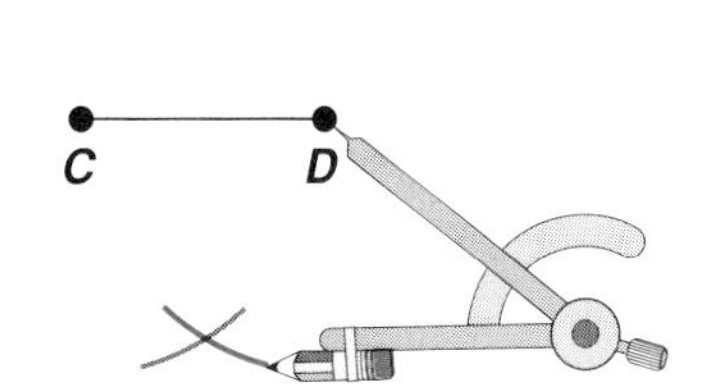

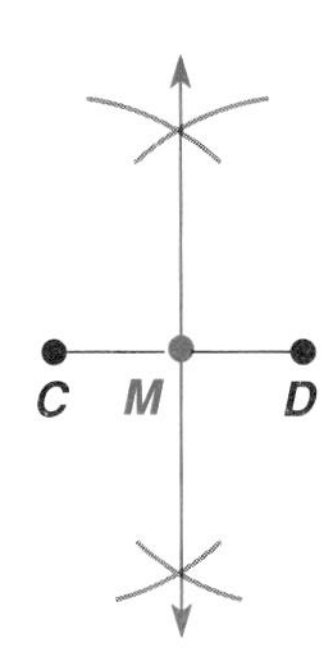

Result: M is the midpoint of $\overline{CD}$.

a. **(1)** Use a ruler to draw $\overline{AB}$ of length 1 inch.

(2) Use a compass and straightedge to find the midpoint of $\overline{AB}$.

(3) Use a ruler to check the accuracy of your construction.

b. Now that you know how to copy the length of a line segment, say $\overline{PQ}$, and how to find the midpoint of $\overline{PQ}$, describe how you would construct a line segment whose length is $1\frac{1}{2}$ times the length of $\overline{PQ}$. Do the construction, using this length for $\overline{PQ}$:

P ●————————● Q

3. Here's how to copy an angle. This is the angle to be copied.

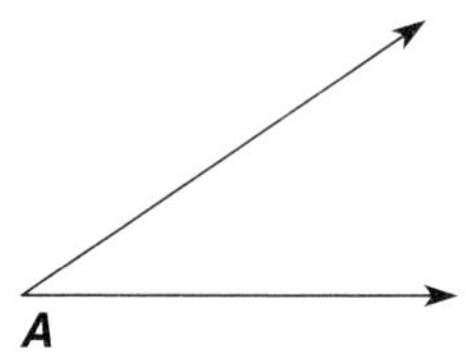

① Use a straightedge to draw a ray that will be one side of the new angle. The endpoint of the ray, A', will be the vertex of the new angle.

② With any setting on the compass, use A as the center and draw an arc that intersects the sides of the given angle. Label the points of intersection B and C.

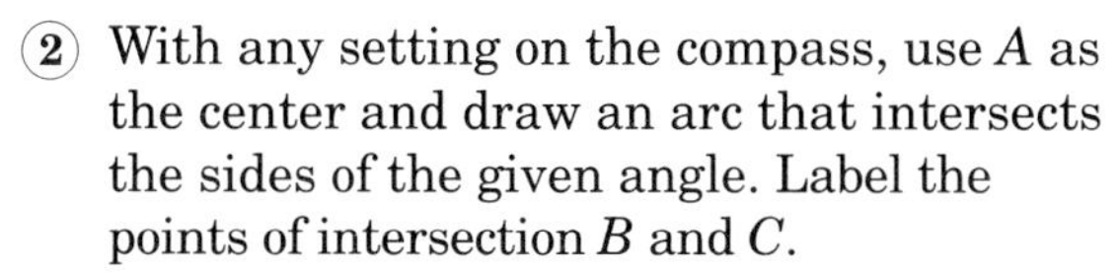

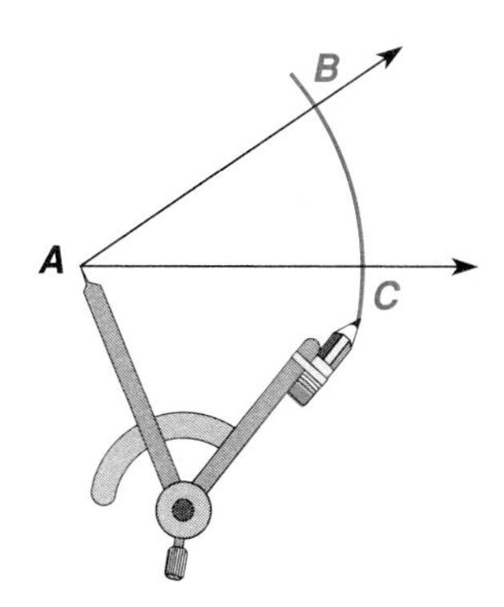

③ With the same setting on the compass, use A' as the center and draw an arc onto which the given angle will be copied. Label the point of intersection of the arc and the ray C'.

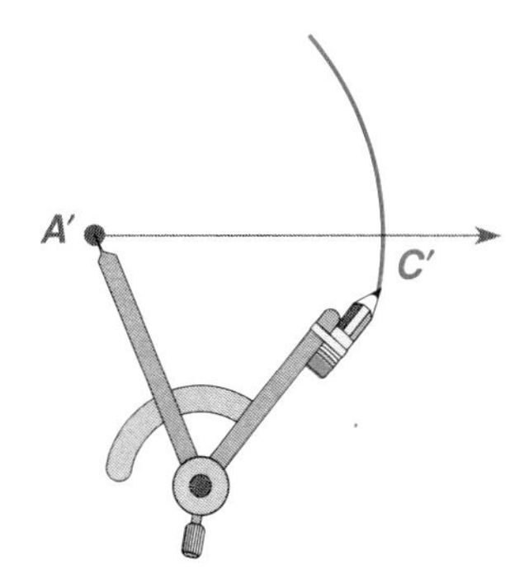

④ Measure the arc constructed on $\angle A$:

With the point at C, open the compass so that the pencil is at B.

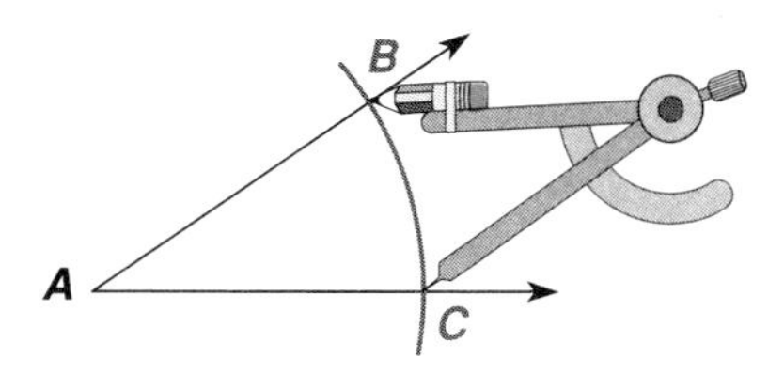

⑤ Copy the measure of this arc onto the arc with the ray:

Keeping the radius of step 4, use C' as the center and draw an arc that intersects the existing arc. Label this point of intersection B'.

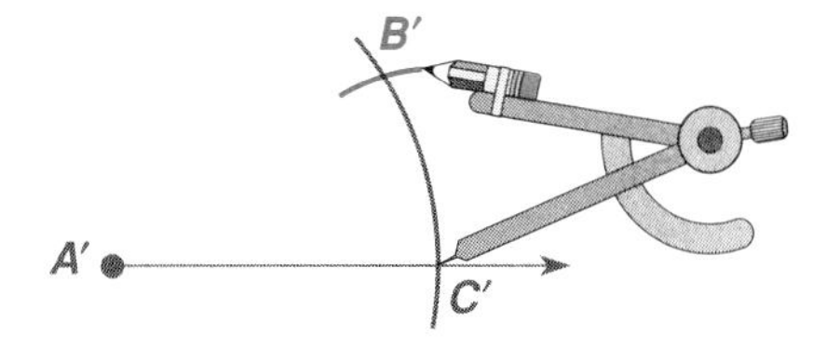

⑥ With a straightedge, draw a ray from A' through B'.

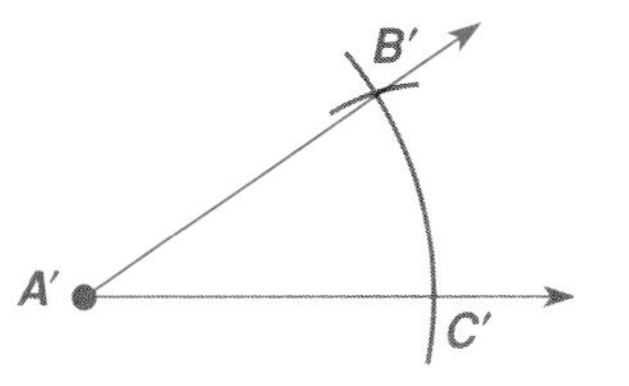

Result: $\angle A' \cong \angle A$

a. Practice this construction by copying each of these angles onto separate paper.

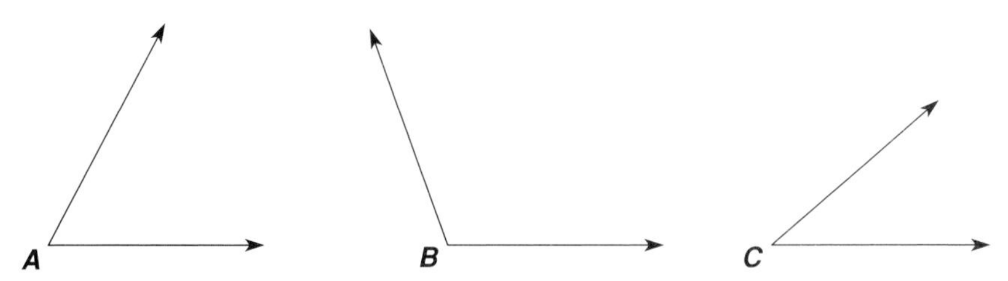

b. Now that you know how to copy an angle by construction, describe how you would construct an angle that is twice the measure of a given angle. Do the construction using $\angle A$ above.

c. Suppose you had two different angles. Describe how you would construct an angle with measure equal to the sum of the measures of the two angles.

d. Using angles of measures a and b, construct an angle of measure:

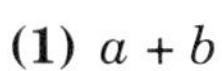
(1) $a + b$

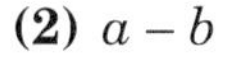
(2) $a - b$

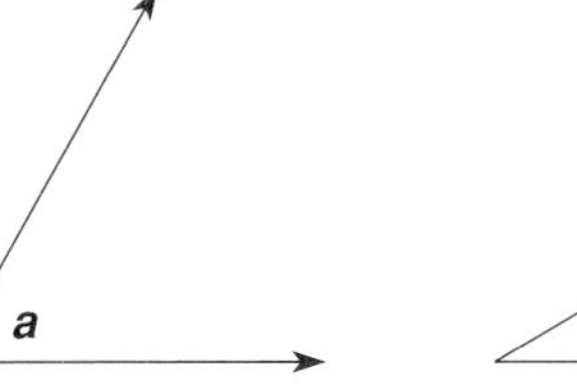

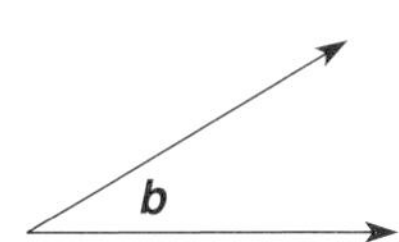

4. Here's how to find the midpoint of a line segment.

① Open the compass to a setting that is more than half the length of $\overline{CD}$.

Using this setting and C as the center, draw an arc above and below $\overline{CD}$.

② With the same setting and D as the center, draw arcs above and below $\overline{CD}$.

③ With a straightedge, draw a line through the points where the arcs cross.

This line crosses $\overline{CD}$ at its midpoint, M.

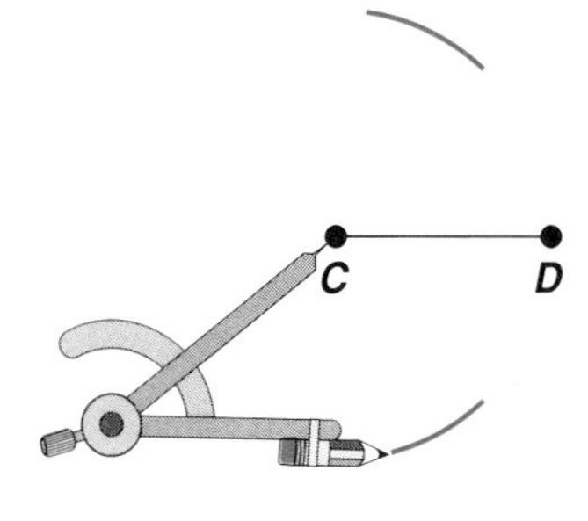

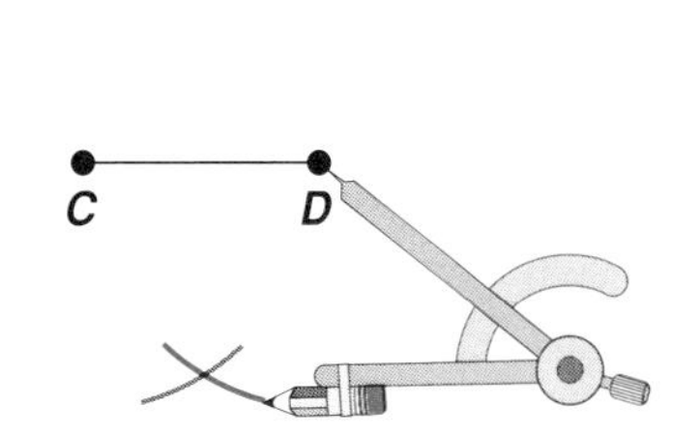

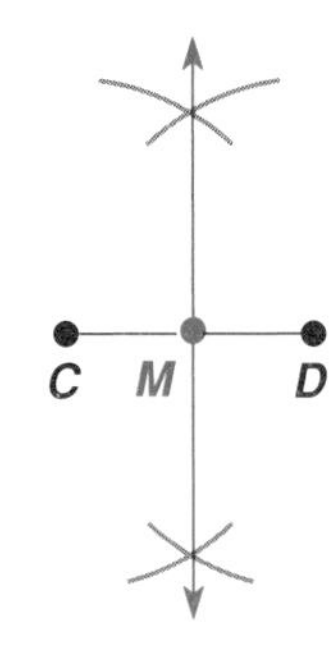

Result: M is the midpoint of $\overline{CD}$.

a. **(1)** Use a ruler to draw $\overline{AB}$ of length 1 inch.
(2) Use a compass and straightedge to find the midpoint of $\overline{AB}$.
(3) Use a ruler to check the accuracy of your construction.

b. Now that you know how to copy the length of a line segment, say $\overline{PQ}$, and how to find the midpoint of $\overline{PQ}$, describe how you would construct a line segment whose length is $1\frac{1}{2}$ times the length of $\overline{PQ}$. Do the construction, using this length for $\overline{PQ}$:

P ●————————● Q

8.2 Pairs of Lines

Perpendicular Lines

This building is called the Leaning Tower of Pisa.

Why do you think it is called *leaning*? How do we expect a building to stand?

> **Two lines that meet at right angles are called *perpendicular*.**

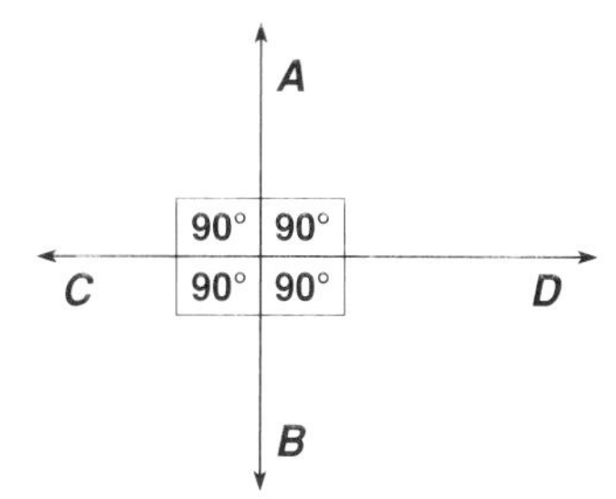

$\overleftrightarrow{AB}$ is perpendicular to $\overleftrightarrow{CD}$

$\overleftrightarrow{AB} \perp \overleftrightarrow{CD}$

Example 1

If $\overleftrightarrow{MQ} \perp \overleftrightarrow{RS}$, what can you say about the angles at P?

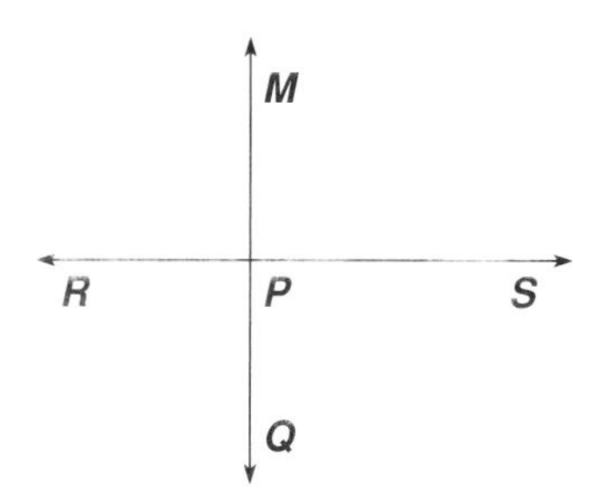

Since $\overleftrightarrow{MQ} \perp \overleftrightarrow{RS}$, there are 4 right angles at P.

If the angles at K are right angles, what can you say about the lines XY and AB?

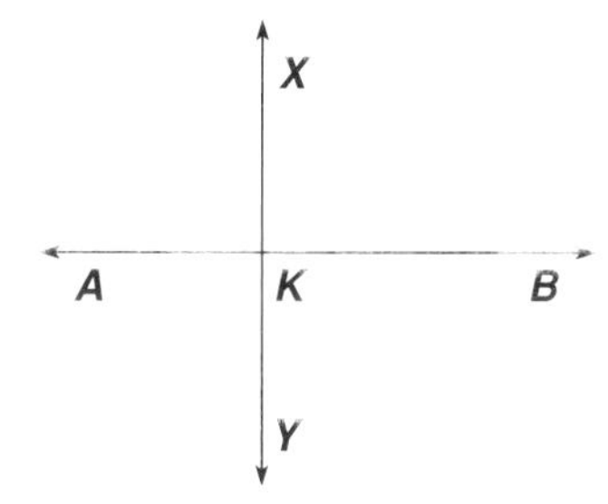

Since the angles at K are right angles, $\overleftrightarrow{XY} \perp \overleftrightarrow{AB}$.

> **If two lines are perpendicular, then they meet at right angles.**
> **If two lines meet at right angles, then they are perpendicular.**

Example 2 In the figure, $\overrightarrow{SR} \perp \overrightarrow{ST}$. If m$\angle RST$ is represented by $4x + 30$, find the value of x.

The corner marking at $\angle S$ shows a right angle.

1 ***Read the problem for information.***
Since the rays are perpendicular, you know $\angle S$ is a right angle. You know the measure of a right angle is 90°.

2 ***Make a plan.***
Write an equation to find the value of x.

$$m\angle S = 90°$$
$$4x + 30 = 90°$$

3 ***Solve the equation.***

$$4x + 30 = 90$$
$$4x + 30 - 30 = 90 - 30 \quad \text{Subtract 30 from both sides.}$$
$$4x = 60$$
$$\frac{4x}{4} = \frac{60}{4} \quad \text{Divide both sides by 4.}$$
$$x = 15$$

4 ***Check.***
If $x = 15$, does $4x + 30$ equal 90?

$$4(15) + 30 \stackrel{?}{=} 90$$
$$60 + 30 \stackrel{?}{=} 90$$
$$90 = 90 \ ✓$$

Answer: $x = 15$

While building a tool shed, Brian and Irene each nailed two strips of wood together. How are the sets of strips the same? How are they different?

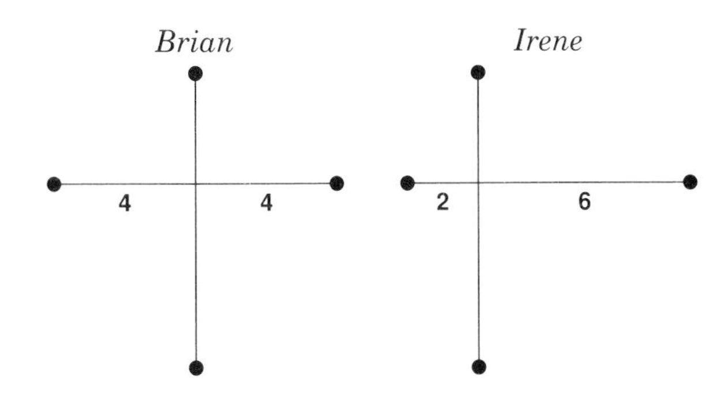

Both Brian and Irene nailed the strips so that they are perpendicular to each other. One of Brian's strips cuts the other strip in half.

A perpendicular that goes through the midpoint of a line segment is called the *perpendicular bisector*.

In the diagram, the square corner shows a right angle, meaning that the lines are perpendicular.

The double "tick" marks show that $\overline{CM} \cong \overline{MD}$, meaning that M is the midpoint of $\overline{CD}$.

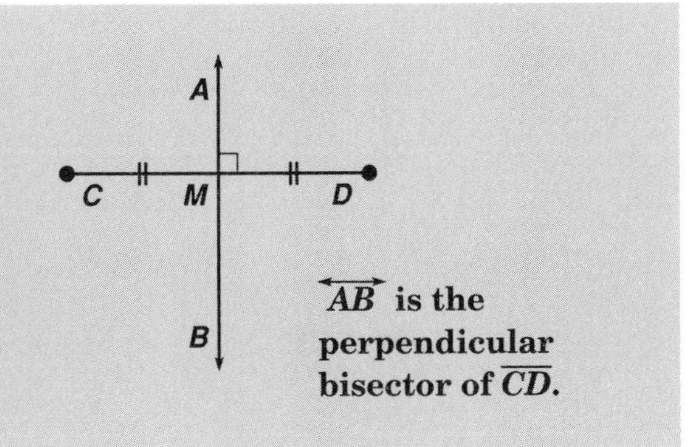

$\overleftrightarrow{AB}$ is the perpendicular bisector of $\overline{CD}$.

Example 3 In each diagram, could $\overleftrightarrow{AB}$ be the perpendicular bisector? Explain.

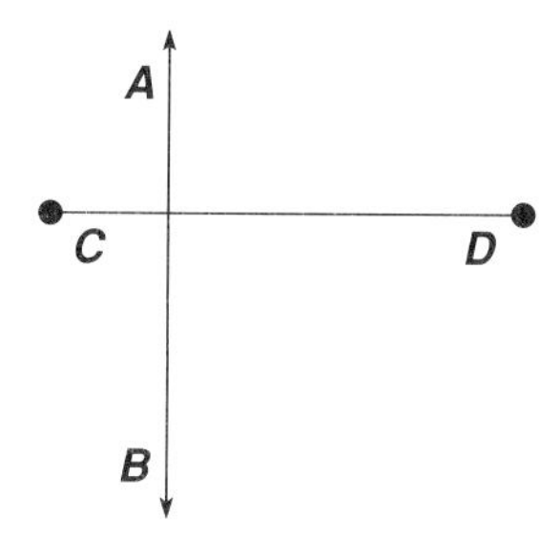

No, $\overleftrightarrow{AB}$ does not go through the midpoint of $\overline{CD}$.

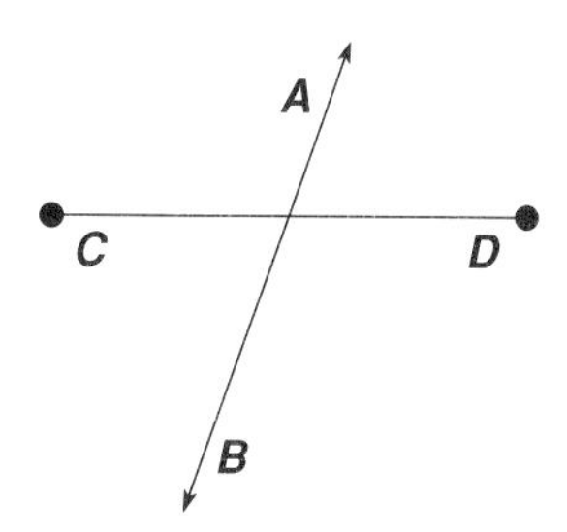

No, $\overleftrightarrow{AB}$ is not perpendicular to $\overline{CD}$.

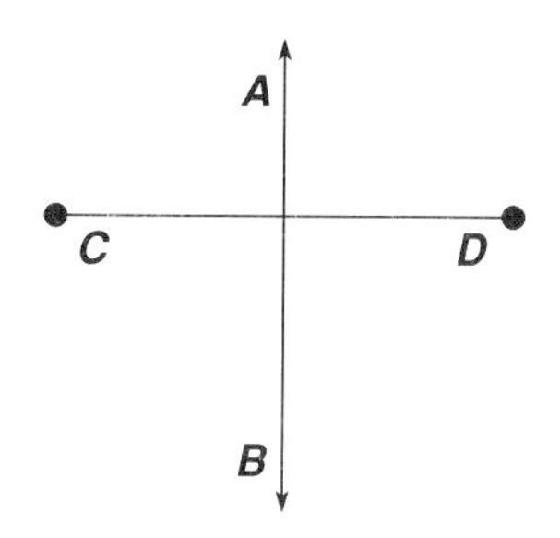

Yes, it looks as though the lines are perpendicular and that $\overleftrightarrow{AB}$ goes through the midpoint of $\overline{CD}$.

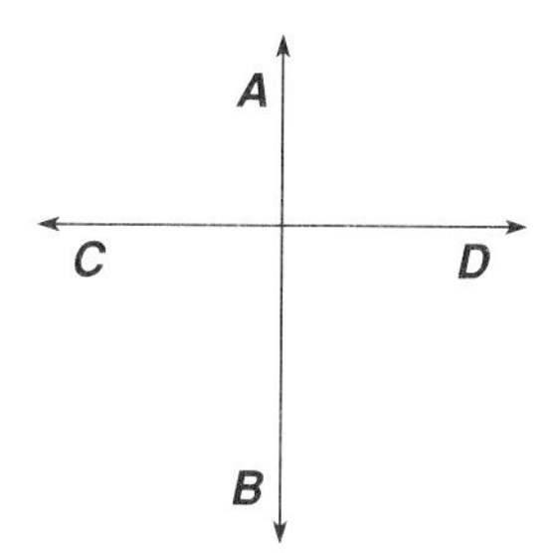

No, since $\overleftrightarrow{CD}$, a line, does not have a midpoint.

Try These (For solutions, see page 347.)

1. In both of these diagrams, two line segments are perpendicular. What is different about the two situations?

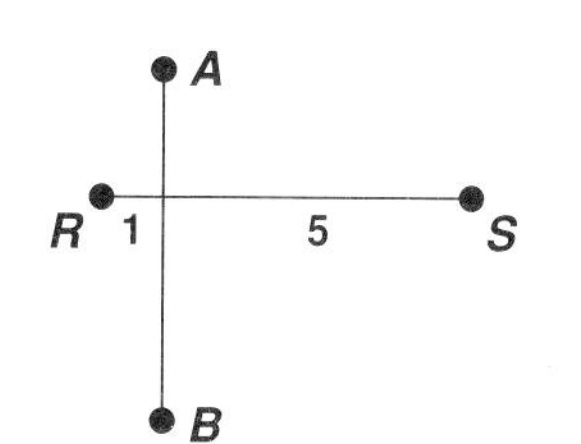

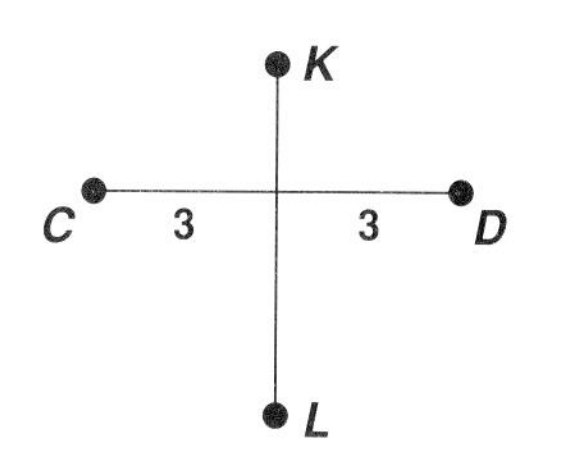

2.

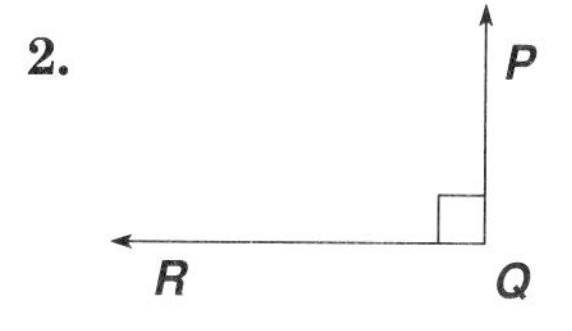

In the diagram, $\overrightarrow{QP} \perp \overrightarrow{QR}$. If m$\angle PQR$ is represented by $5x - 10$, find the value of x.

Parallel Lines

An AMTRAK train runs between Boston and New York City. If the train tracks in Boston are standard 4 feet 8.5 inches apart, how far apart do you think the tracks are 50 miles out of Boston? 100 miles out? in New York?

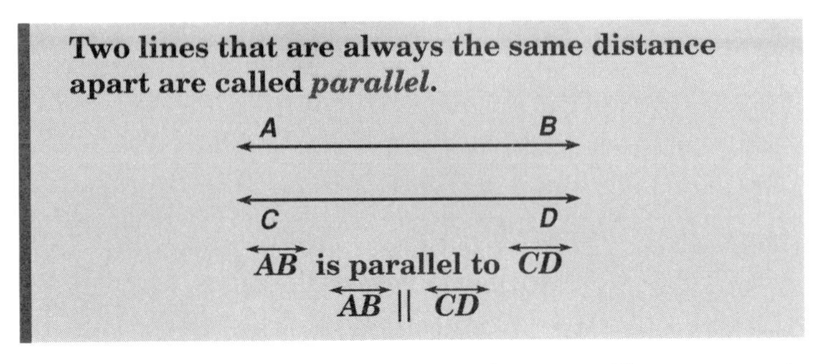

Two lines that are always the same distance apart are called *parallel*.

$\overleftrightarrow{AB}$ is parallel to $\overleftrightarrow{CD}$

$\overleftrightarrow{AB} \parallel \overleftrightarrow{CD}$

Since parallel lines are always the same distance apart, they will never meet no matter how far extended.

Example 4 Name the sides that are parallel and the sides that are perpendicular in this rectangle. Which sides are congruent?

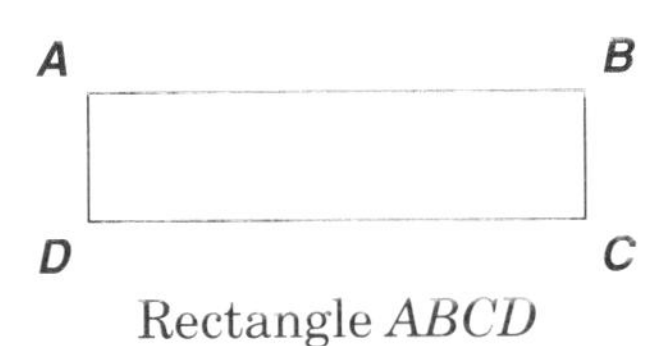

Rectangle *ABCD*

In a rectangle:

Opposite sides are parallel.	$\overline{AB} \parallel \overline{DC}$ and $\overline{AD} \parallel \overline{BC}$
Consecutive sides are perpendicular.	$\overline{AD} \perp \overline{DC}$, $\overline{DC} \perp \overline{CB}$, $\overline{CB} \perp \overline{BA}$, $\overline{BA} \perp \overline{AD}$
Opposite sides are congruent.	$\overline{AB} \cong \overline{DC}$ and $\overline{AD} \cong \overline{BC}$

Parallel Lines and Angle Measure

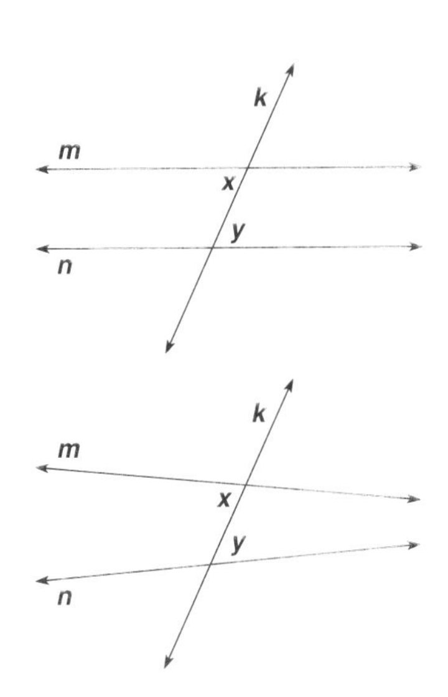

Consider the pair of parallel lines m and n, and a third line k that intersects the pair. Line k is called a ***transversal***, meaning *crossing over*.

Together with the parallel lines, the transversal forms different angles. Let's study the angles labeled $\angle x$ and $\angle y$.

What do you think is true about the measures of $\angle x$ and $\angle y$? Use a protractor to verify your conclusion.

Suppose m and n are not parallel. What can you say now about the measures of $\angle x$ and $\angle y$? Use a protractor to verify your conclusion.

Try again, using the pair of angles labeled $\angle r$ and $\angle s$.

What is true about the measures of these angles:
when the lines are parallel? when the lines are not parallel?

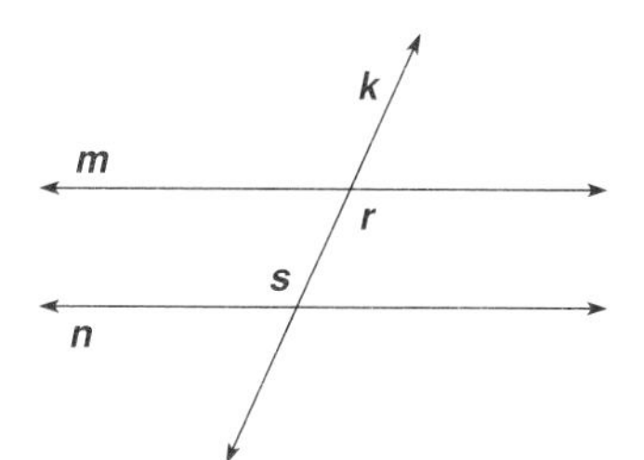

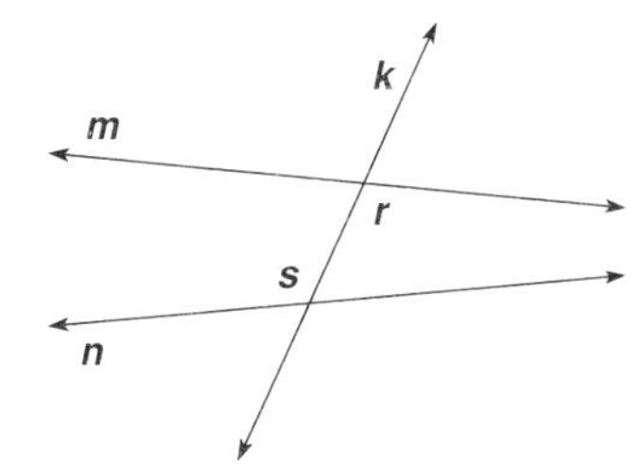

See if you can pick out a different pair of angles whose measures are the same when the lines are parallel, and are not the same when the lines are not parallel.

When parallel lines are crossed by a transversal, certain pairs of angles are equal in measure.

To classify the pairs of angles formed by two lines and a transversal, notice that 8 angles are formed.

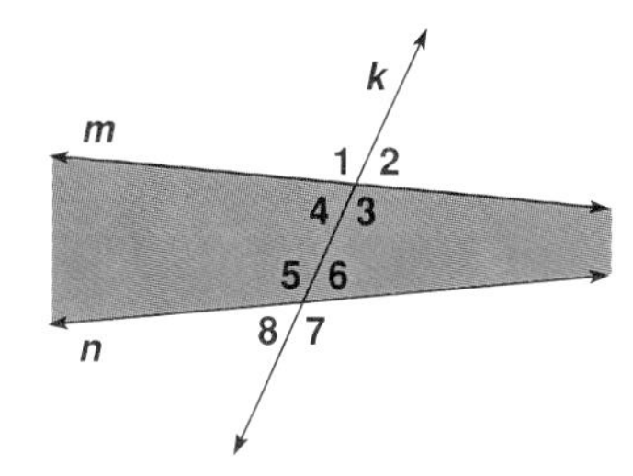

Four of these angles are inside the pair of lines, and four are outside.

interior angles: $\angle 3$, $\angle 4$, $\angle 5$, $\angle 6$

exterior angles: $\angle 1$, $\angle 2$, $\angle 7$, $\angle 8$

A further classification of the angles is made according to their position on either side of the transversal.

Alternate interior angles are interior angles on opposite sides of the transversal.

pairs of alternate interior angles: $\angle 4$ and $\angle 6$ $\angle 3$ and $\angle 5$

What would you call the pair $\angle 1$ and $\angle 7$? (*alternate exterior*)

Corresponding angles have the same position with respect to each line and the transversal.

pairs of corresponding angles:

$\angle 1$ and $\angle 5$ above each line, left of transversal

$\angle 2$ and $\angle 6$ above each line, right of transversal

$\angle 4$ and $\angle 8$ below each line, left of transversal

$\angle 3$ and $\angle 7$ below each line, right of transversal

Which pairs of angles are congruent when the lines are parallel?

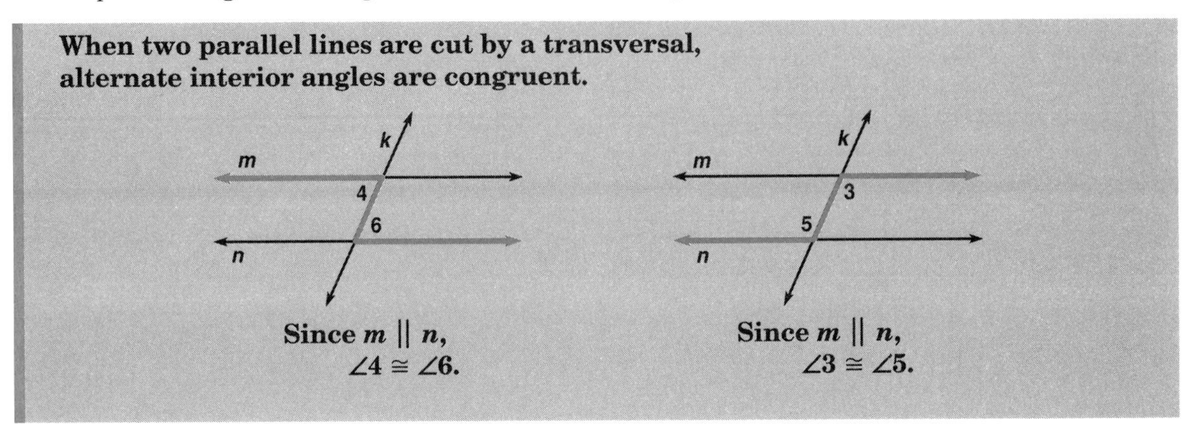

When two parallel lines are cut by a transversal, alternate interior angles are congruent.

Since $m \parallel n$, $\angle 4 \cong \angle 6$.

Since $m \parallel n$, $\angle 3 \cong \angle 5$.

To find alternate interior angles of parallel lines, look for a "zee shape."

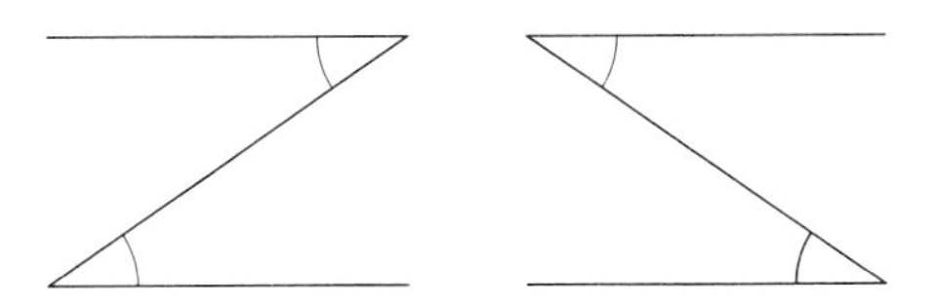

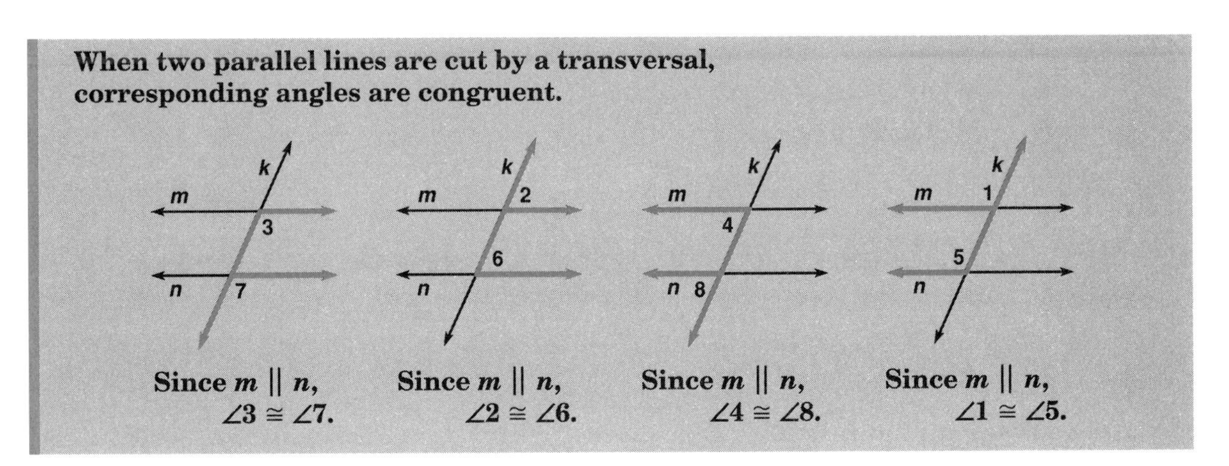

When two parallel lines are cut by a transversal, corresponding angles are congruent.

Since $m \parallel n$, $\angle 3 \cong \angle 7$.

Since $m \parallel n$, $\angle 2 \cong \angle 6$.

Since $m \parallel n$, $\angle 4 \cong \angle 8$.

Since $m \parallel n$, $\angle 1 \cong \angle 5$.

To find corresponding angles of parallel lines, look for an "eff shape".

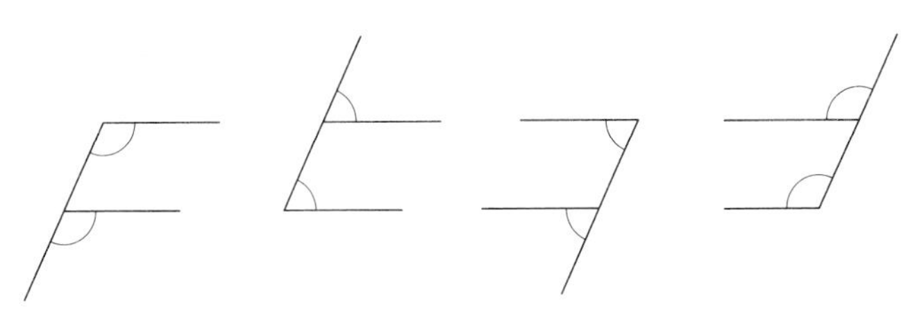

Example 5 Parallel lines p and q are cut by transversal r. If $m\angle a = 120°$, find:

a. $m\angle b$ **b.** $m\angle c$

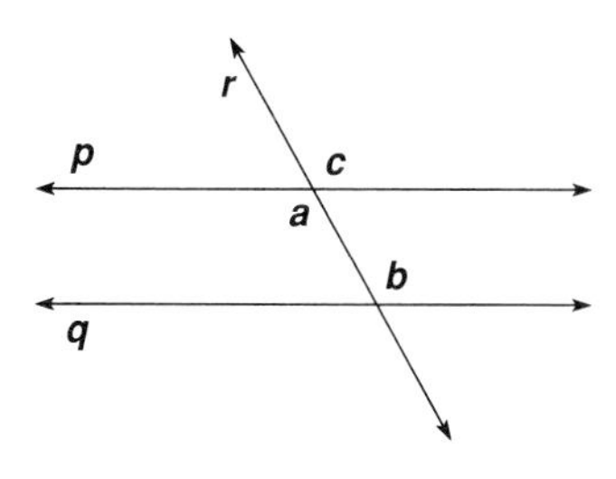

a. Angles a and b are alternate interior angles. Since the lines are parallel, these angles are equal in measure.

$$m\angle b = m\angle a = 120°$$

b. Angles b and c are corresponding angles. Since the lines are parallel, these angles are equal in measure.

$$m\angle c = m\angle b = 120°$$

Example 6 In the diagram, parallel lines are cut by a transversal. If $m\angle DAB = 7x + 4$ and $m\angle FBE = 2x + 54$, find the measure of each angle.

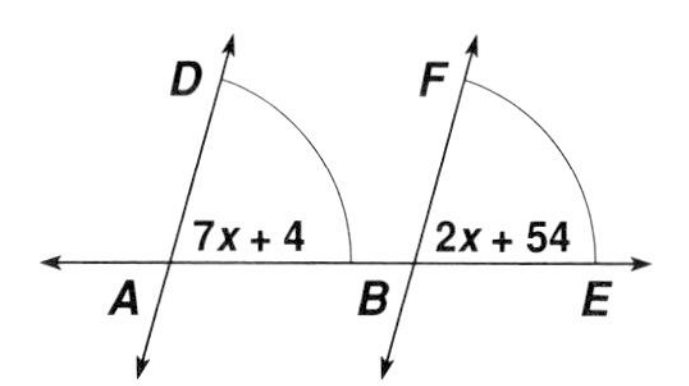

1 ***Read the problem for information.***

Classify the angles as corresponding angles.
Since the lines are parallel, these angles are equal in measure.

2 ***Make a plan.***

Write an equation to find the value of x.

$$m\angle DAB = m\angle FBE$$

$$7x + 4 = 2x + 54$$

3 ***Solve the equation.***

$$7x + 4 = 2x + 54$$

$$7x - 2x + 4 = 2x - 2x + 54 \quad \text{Subtract } 2x \text{ from both sides.}$$

$$5x + 4 = 54$$

$$5x + 4 - 4 = 54 - 4 \quad \text{Subtract 4 from both sides.}$$

$$5x = 50$$

$$\frac{5x}{5} = \frac{50}{5} \quad \text{Divide both sides by 5.}$$

$$x = 10$$

$$m\angle DAB = 7x + 4 = 7(10) + 4 = 70 + 4 = 74 \quad \text{Substitute } x = 10.$$

$$m\angle FBE = 2x + 54 = 2(10) + 54 = 20 + 54 = 74$$

4 ***Check.***

If $x = 10$, are the two angles equal in measure? $74 = 74$ ✔

Answer: $m\angle DAB = m\angle FBE = 74°$

Try These (For solutions, see page 348.)

(For solutions, see page 348.)

1. Parallel lines a and b are cut by transversal c. If $m\angle x = 110°$, find $m\angle y$ and $m\angle z$.

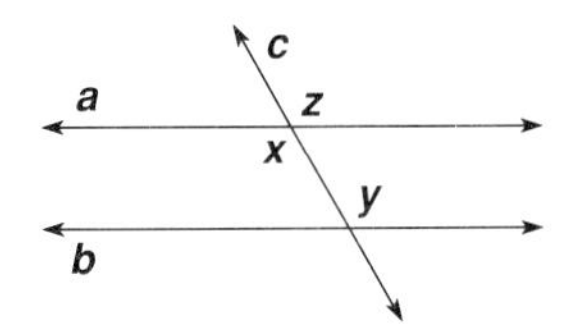

2. In the diagram, $\overleftrightarrow{AB} \parallel \overleftrightarrow{CD}$. Transversal $\overleftrightarrow{EF}$ cuts $\overleftrightarrow{AB}$ at G and $\overleftrightarrow{CD}$ at H. If $m\angle AGH = 8x - 20$ and $m\angle DHG = 4x + 44$, find the measure of each angle.

SELF-CHECK: SECTION 8.2

1. **a.** Describe perpendicular lines.

   Answer: Perpendicular lines meet at right angles.

   b. Describe parallel lines.

   Answer: Parallel lines are always the same distance apart.

2. **a.** Using symbols, write: line AB is perpendicular to line CD.

   Answer: $\overleftrightarrow{AB} \perp \overleftrightarrow{CD}$

   b. Using symbols, write: line MN is parallel to line PQ.

   Answer: $\overleftrightarrow{MN} \parallel \overleftrightarrow{PQ}$

3. In the diagram, $\overleftrightarrow{AB} \perp \overline{CD}$ at M, where M is the midpoint of $\overline{CD}$. What is $\overleftrightarrow{AB}$ called?

   Answer: $\overleftrightarrow{AB}$ is the perpendicular bisector of $\overline{CD}$.

4. In the diagram, lines m and n are cut by transversal p.

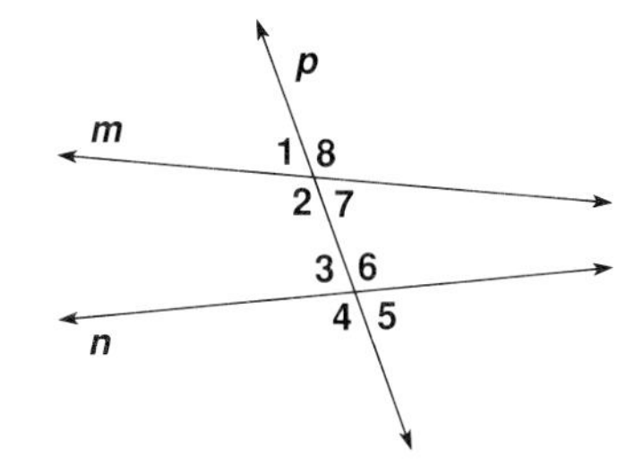

   a. Name the pairs of alternate interior angles.

   Answer: $\angle 2$ and $\angle 6$
   $\angle 7$ and $\angle 3$

   b. Name the pairs of corresponding angles.

   Answer: $\angle 1$ and $\angle 3$, $\angle 2$ and $\angle 4$, $\angle 8$ and $\angle 6$, $\angle 7$ and $\angle 5$

5. **a.** When two lines are cut by a transversal, are alternate interior angles always congruent? Explain.

   Answer: No, alternate interior angles are congruent only when the lines are parallel.

   b. Name another pair of angles that are congruent when the lines are parallel.

   Answer: Corresponding angles of parallel lines are congruent.

EXERCISES: SECTION 8.2

1. Label the corners of a sheet of paper A, B, C, D, as shown.

   A B
   D C

   a. Place edge $\overline{AB}$ on edge $\overline{DC}$, and fold carefully. Unfold.
   - **(1)** How is this fold related to edges $\overline{AB}$ and $\overline{DC}$?
   - **(2)** How is this fold related to edges $\overline{BC}$ and $\overline{AD}$?

   b. Now place edge $\overline{BC}$ on edge $\overline{AD}$, and fold again. Unfold.
   - **(1)** How is this second fold related to the first fold?
   - **(2)** How is this second fold related to edges $\overline{AB}$ and $\overline{DC}$?

2. **a.** For each quadrilateral, name the sides that are parallel and any sides that are perpendicular.

Rectangle *ABCD*

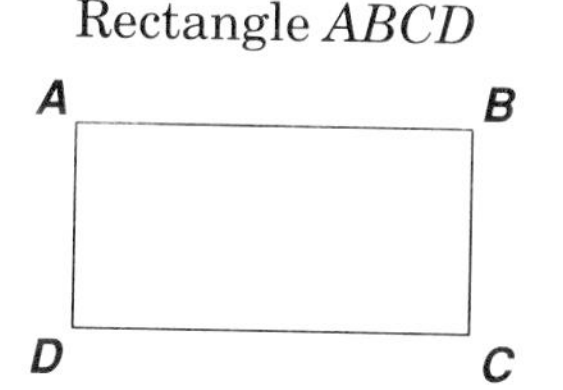

Parallelogram *MNOP*

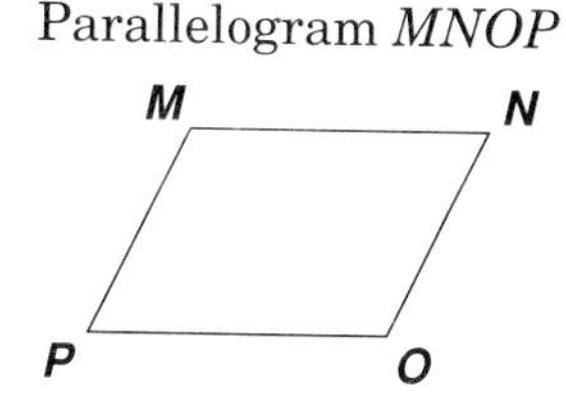

Trapezoid *QRST*

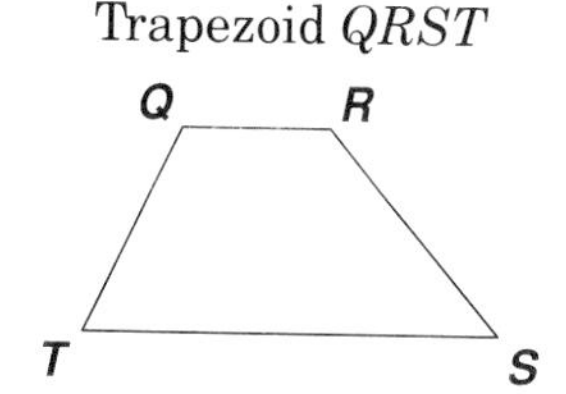

   b. Draw these quadrilaterals (make them larger) on separate paper and cut them out.

Square *ABCD*

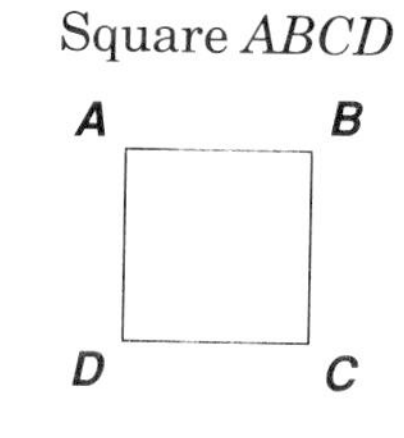

Rectangle *MNOP*

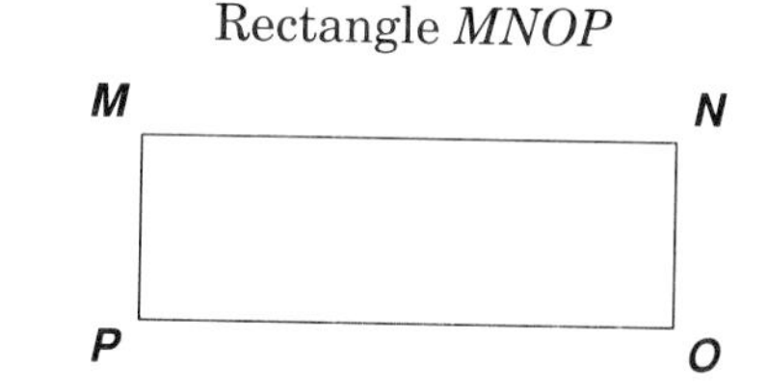

Parallelogram *PQRS*

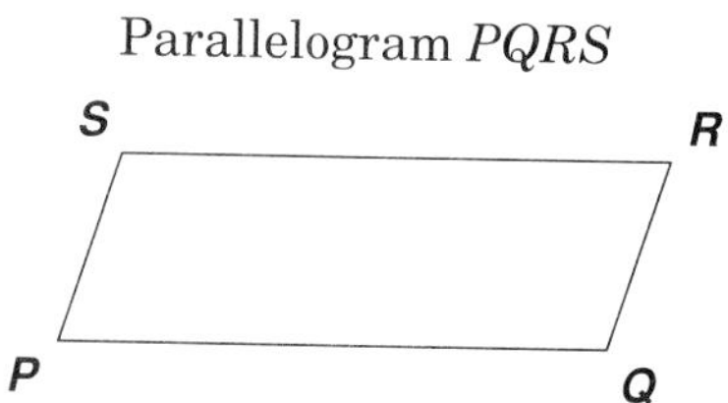

Rhombus *EFGH*

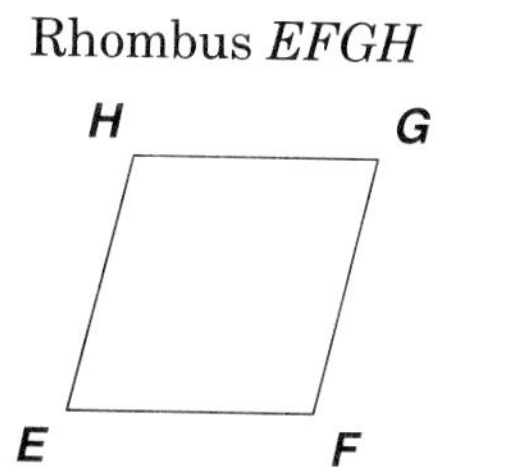

Fold each quadrilateral to create a diagonal. Unfold.
Fold again to create the other diagonal. Unfold.
Study the diagonal folds. In which kinds of quadrilaterals are the diagonals perpendicular?

c. Draw the capital letters of the alphabet that have:

(1) parallel line segments

(2) perpendicular line segments

(3) both parallel and perpendicular line segments

3. In the diagram, $\overleftrightarrow{AB}$ and $\overline{CD}$ intersect at P.

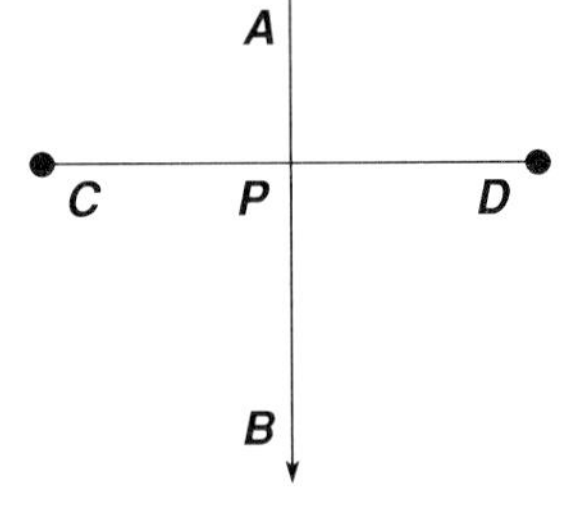

a. If m$\angle APC = 5x + 15$, find the number of degrees in m$\angle APC$ when $x = 15$.

b. If m$\angle APC = 90°$, what is m$\angle APD$? Explain.

c. Draw a conclusion about $\overleftrightarrow{AB}$ and $\overline{CD}$.

d. If $CP = PD = 5$ units, what else is true about $\overleftrightarrow{AB}$?

4. In the diagram, $\overrightarrow{SR} \perp \overrightarrow{ST}$.

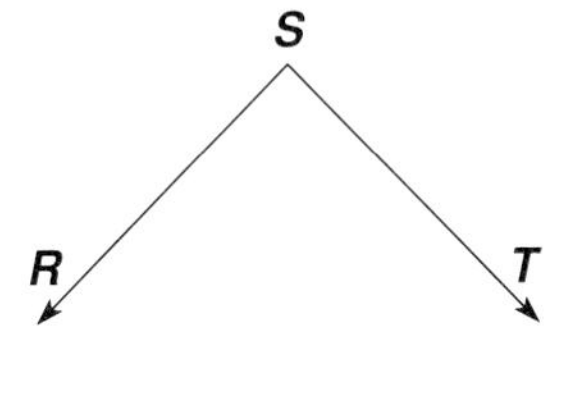

a. What must be true about $\angle S$?

b. If m$\angle S$ is represented by $5x - 10$, find the value of x.

c. If x has the value that you found in part b, could $4x + 10$ also represent m$\angle S$? Explain.

d. If x has the value that you found in part b, make up another way in terms of x that could represent m$\angle S$.

5. In square $ABCD$, $\overline{NM}$ is the perpendicular bisector of $\overline{AD}$.

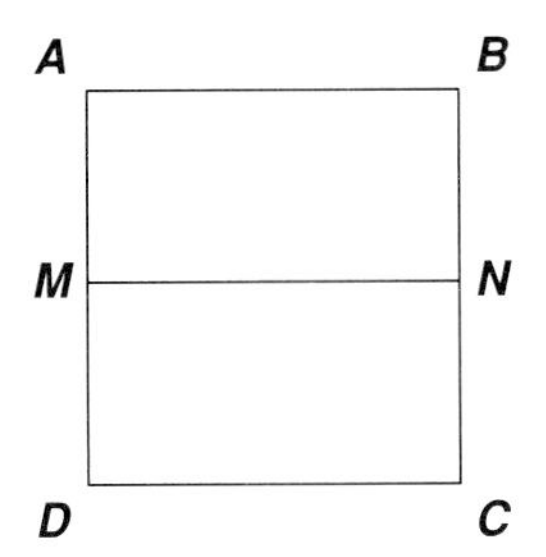

a. If $AM = 7x + 1$ and $MD = 2x + 6$, find the value of x.

b. Using the value of x that you found, make some number statements about this square.

6. Refer to the diagram to complete each statement.

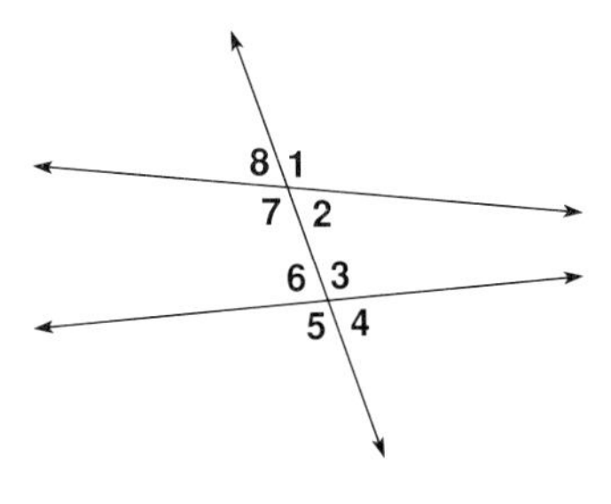

a. The interior angles are __?__.

b. The exterior angles are __?__.

c. $\angle 2$ and __?__ are alternate interior angles.

d. $\angle 7$ and $\angle 3$ are __?__ angles.

e. $\angle 1$ and __?__ are corresponding angles.

f. $\angle 8$ and $\angle 6$ are __?__ angles.

g. $\angle 5$ and $\angle 7$ are __?__ angles.

7. In each of these diagrams, parallel lines are cut by a transversal. Find the measure of the angle marked x. Give a reason for your answer.

a.

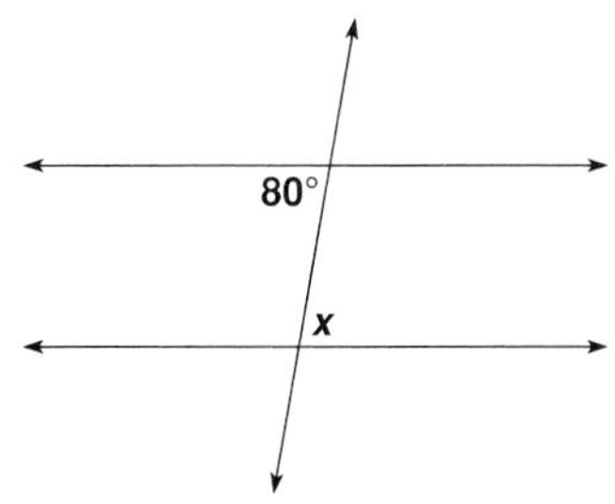

b.

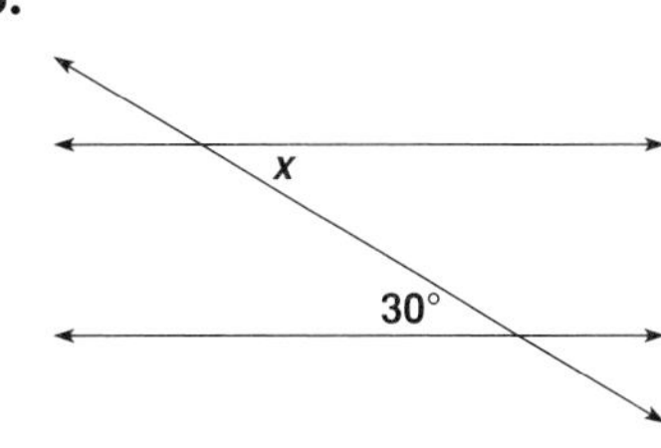

c.

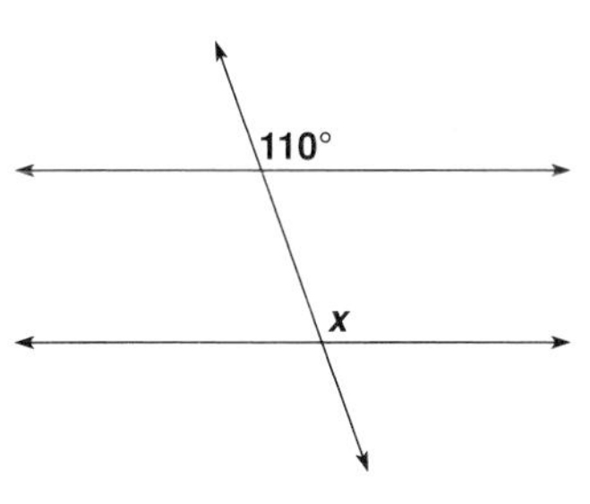

d.

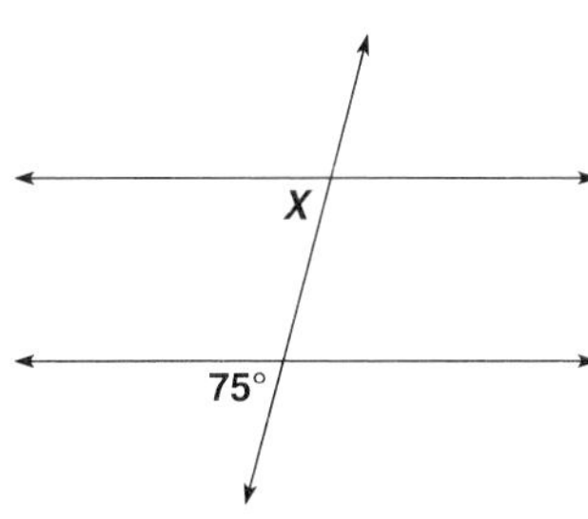

8. In each of these diagrams, parallel lines are cut by a transversal. Find the measures of the pair of angles whose representations are shown.

a.

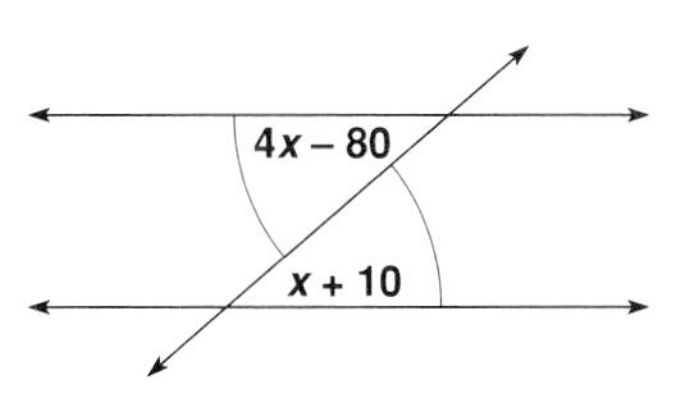

b.

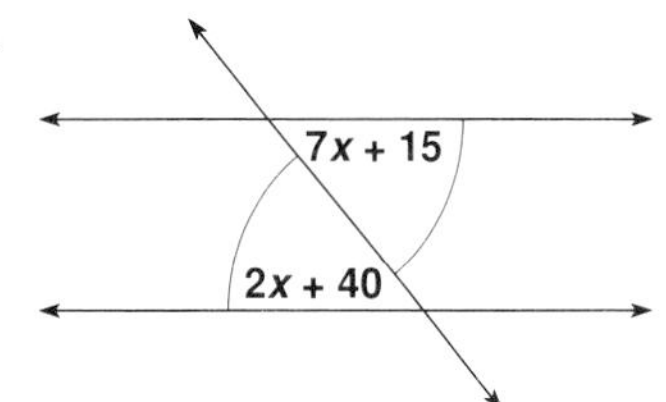

c.

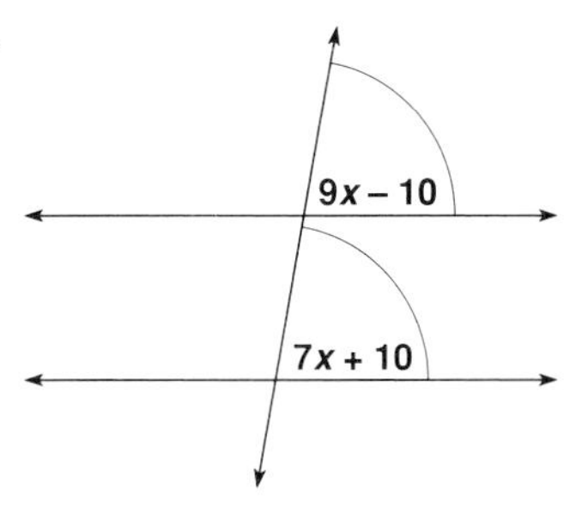

d.

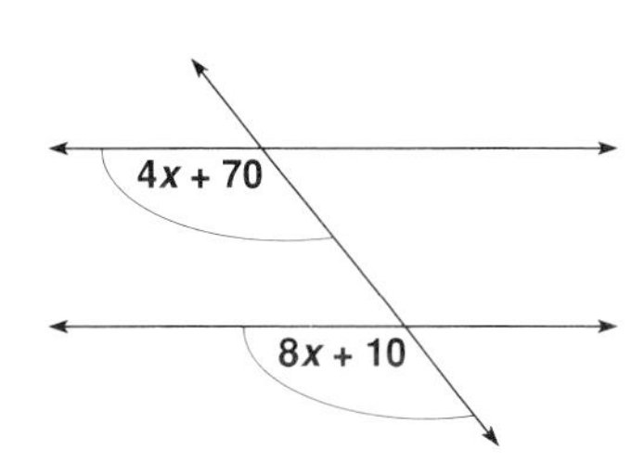

9. In the diagram, pairs of lines are cut by transversals. Name the lines that must be parallel if:

a. $\angle 2 \cong \angle 3$

b. $\angle 1 \cong \angle 4$

c. $\angle 5 \cong \angle 6$

d. $\angle 7 \cong \angle 8$

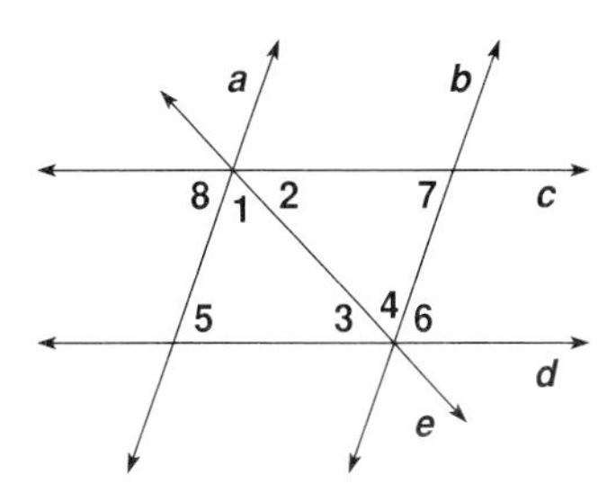

10. Lines p and q are cut by transversal t.

a. If $\angle 1$ and $\angle 2$ together form a straight angle, what is the sum of their measures?

b. If lines p and q are parallel, and $m\angle 1 = 50°$, find the measures of all the other angles.

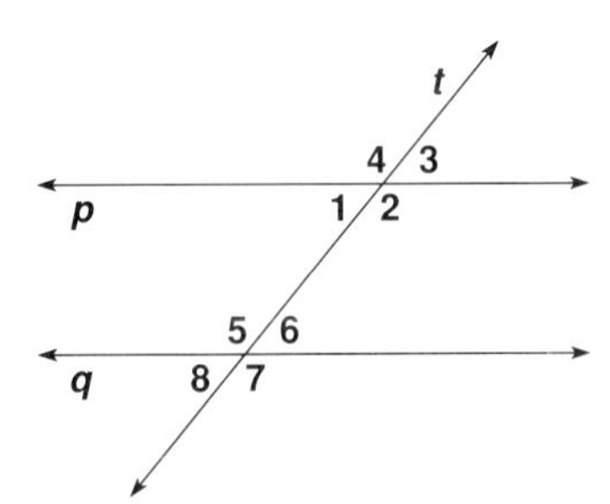

11. Five line segments intersect as shown.

a. If $m\angle BAC = 60°$ and $m\angle ACD = 4x - 40$, find the value of x that makes $\overline{AB} \parallel \overline{DC}$.

b. If $m\angle DAC = 4y + 1$ and $m\angle ACB = y + 34$, find the value of y that makes $\overline{DA} \parallel \overline{CB}$.

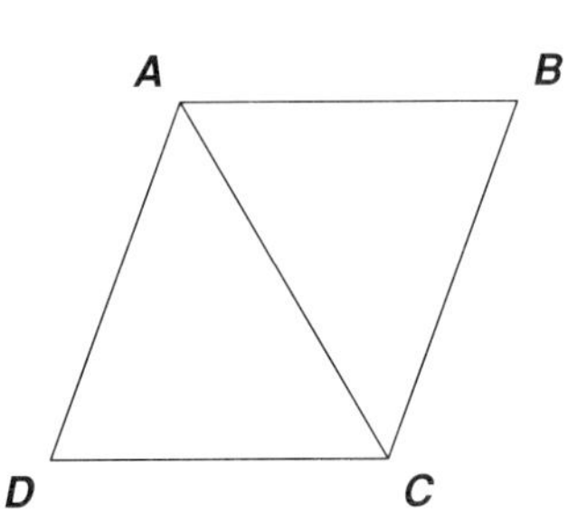

12. $\overleftrightarrow{AB}$ and $\overleftrightarrow{CD}$ are each perpendicular to $\overleftrightarrow{EF}$.

a. Draw a diagram.

b. Which is true?

(1) $\overleftrightarrow{AB} \perp \overleftrightarrow{CD}$ (2) $\overleftrightarrow{AB} \parallel \overleftrightarrow{CD}$

(3) both (4) neither

EXPLORATIONS

1. Here's how to raise a perpendicular at a point on a line.

This is the given line ℓ and a point P on it.

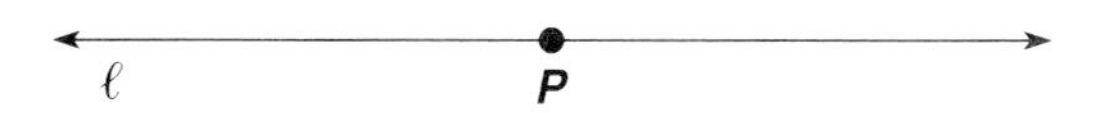

(1) With any setting on the compass, use P as the center and draw an arc that intersects ℓ. Label the points of intersection A and B.

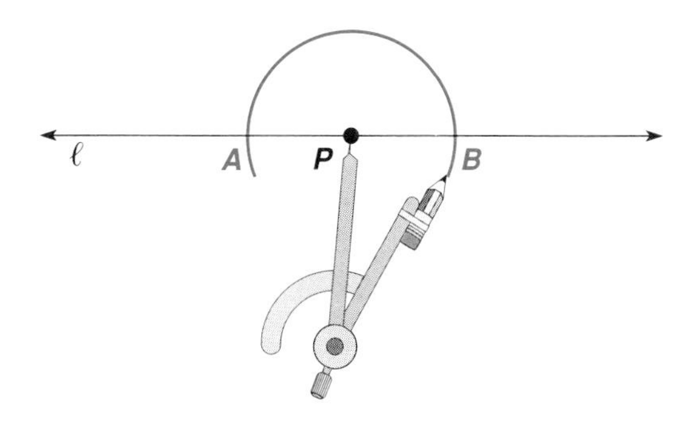

(2) Open the compass to a setting that is more than half the length of $\overline{AB}$.

Using this setting and A as the center, draw an arc above ℓ.

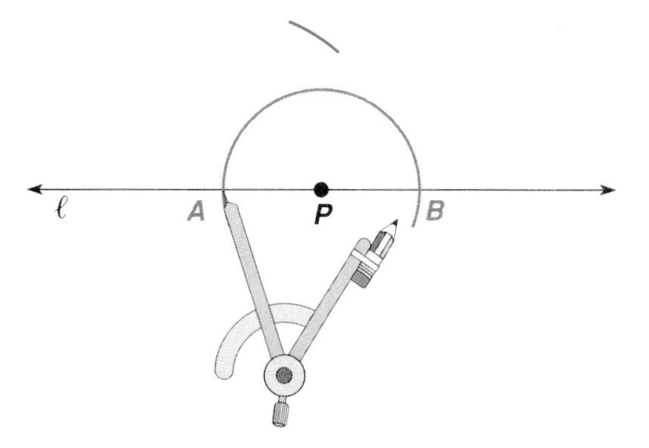

(3) With the same setting and B as the center, draw an arc above ℓ.

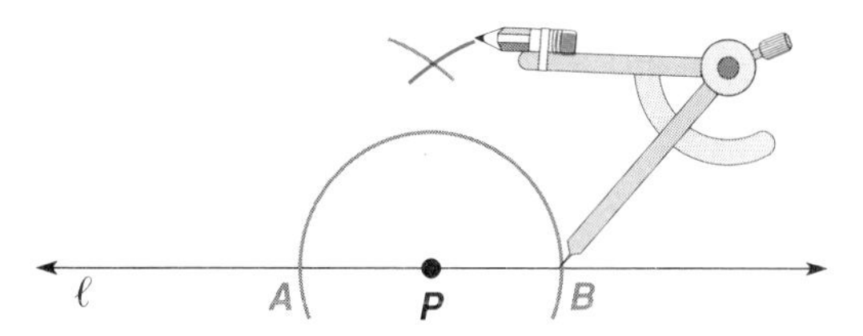

(4) With a straightedge, draw a ray from P through the point where the arcs cross.

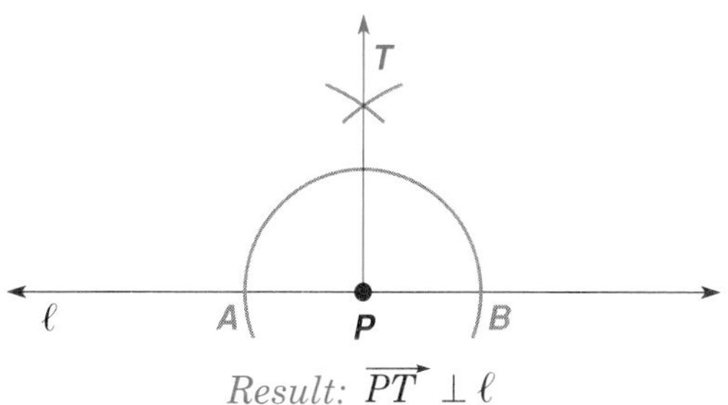

Result: $\overrightarrow{PT} \perp \ell$

Practice this construction by copying these diagrams onto separate paper.

a. Raise a perpendicular to line m at point Q.

b. **(1)** Use this construction to raise two perpendiculars to $\overleftrightarrow{AB}$, one at A and the other at B.

(2) On the perpendicular from A, mark off any length AC. Use this same length to mark off BD on the perpendicular from B. With a straightedge, draw $\overline{CD}$.

What sort of figure is $ABDC$? Explain.

2. Here's how to construct a line parallel to a given line through a particular point.

This is the given line, $\overleftrightarrow{AB}$, and a point P that is not on it.

① With a straightedge, draw a line through P to cross $\overleftrightarrow{AB}$. Label the point of intersection Q. $\overleftrightarrow{PQ}$ will be the transversal.

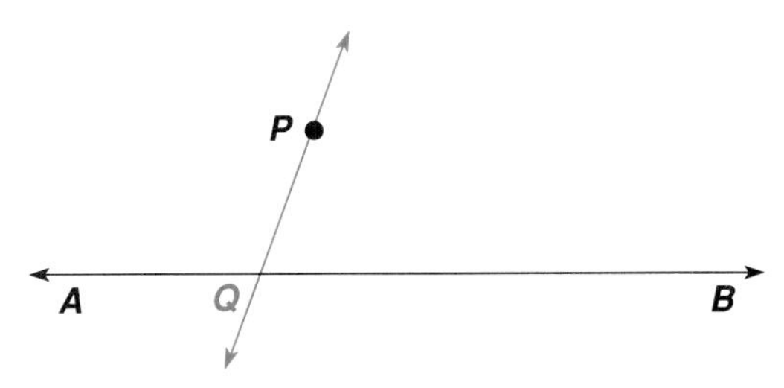

② The object is to copy an angle with vertex at Q over again at P, thereby creating a pair of congruent corresponding angles.

With any setting on the compass and Q as the center, draw an arc.

Using the same setting and P as the center, draw an arc. Use R to label the point of intersection of this arc and the transversal $\overleftrightarrow{PQ}$.

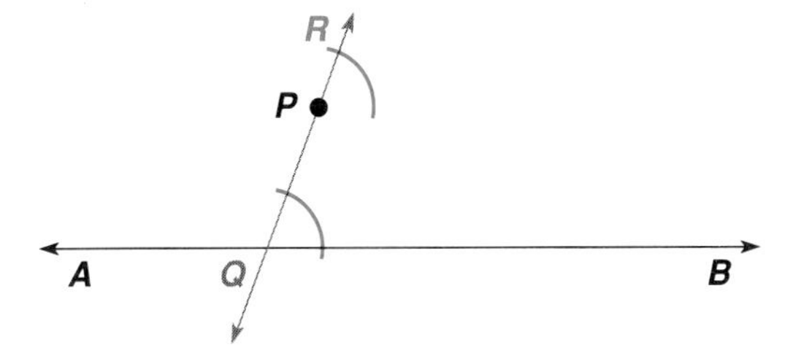

③ With the compass, measure $\angle PQB$ along its arc:

Put the point of the compass where the arc crosses $\overleftrightarrow{PQ}$ and open the compass until the pencil is where the arc crosses $\overleftrightarrow{AB}$.

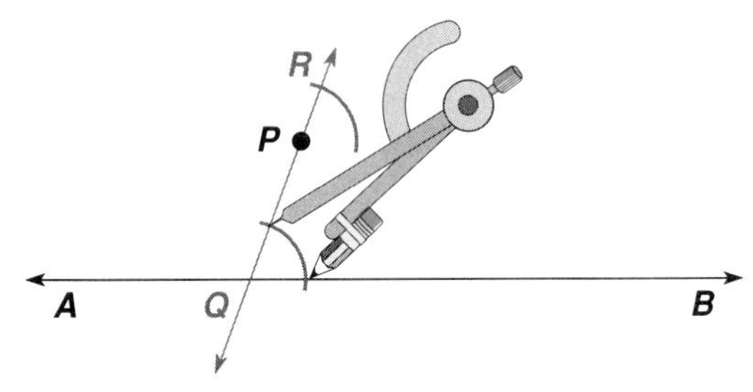

④ Copy the measure of $\angle PQB$ onto the arc with center at P:

Keep the setting on the compass. With R as the center, draw an arc that intersects the arc already at R. Use T to label this point of intersection.

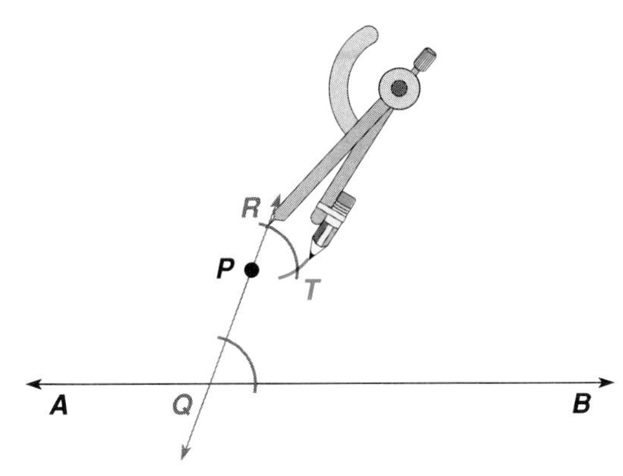

⑤ With a straightedge, draw a line through points P and T.

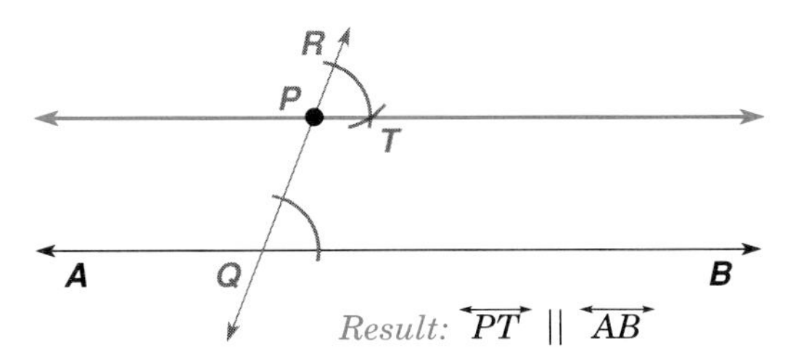

Result: $\overleftrightarrow{PT} \parallel \overleftrightarrow{AB}$

a. Practice this construction using each of these situations. Work on separate paper.

(1) Q

(2) A P B

b. **(1)** Construct a line parallel to $\overleftrightarrow{AB}$ through point P.

(2) Now construct a line parallel to the transversal you drew in part (1).

(3) What sort of figure is cut off by the two sets of parallel lines?

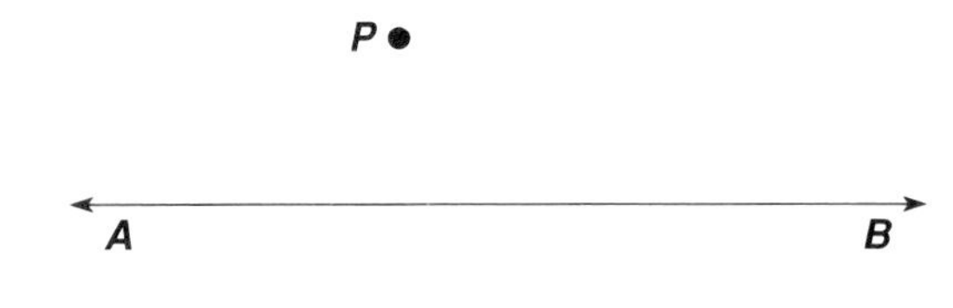

8.3 Pairs of Angles

Vertical Angles

Open and close a pair of scissors to observe the behavior of the angles formed by the blades and handles.

Pairs of opposite angles formed by intersecting lines are called ***vertical angles***.

Draw two intersecting lines. What can you say about the pairs of vertical angles? Use a protractor to verify your conclusion.

When two lines intersect, the vertical angles are congruent.
$\overleftrightarrow{AB}$ and $\overleftrightarrow{CD}$ intersect to form two pairs of vertical angles. These pairs of vertical angles are congruent.
In a diagram, using the same arc markings on a pair of angles shows that they are congruent.

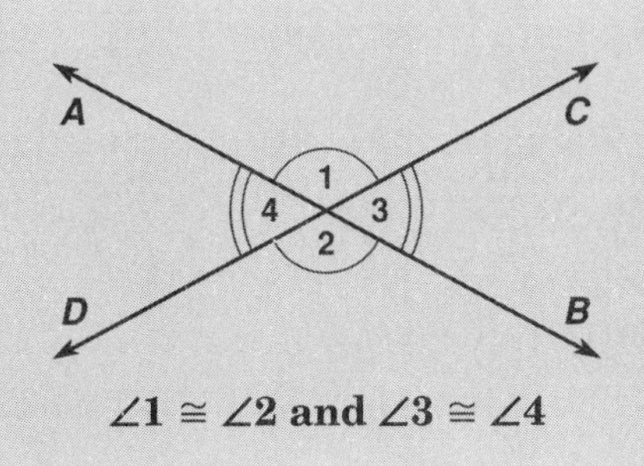

$\angle 1 \cong \angle 2$ and $\angle 3 \cong \angle 4$

Example 1 In each figure, $\overleftrightarrow{AB}$ and $\overleftrightarrow{CD}$ intersect at F. Find the value of x.

a.

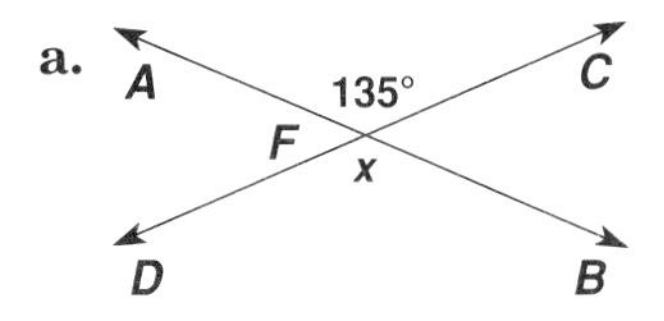

Since vertical $\angle$s are congruent:

$m\angle AFC = m\angle DFB$

$x = 135$

b.

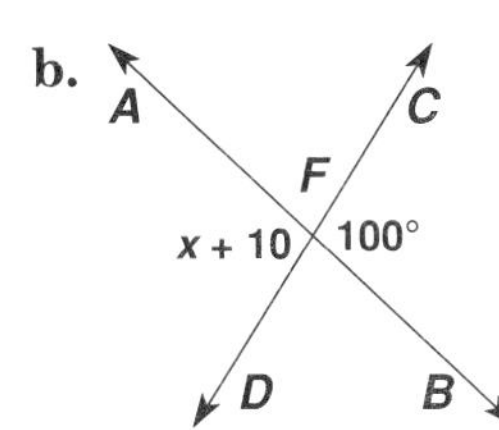

Since vertical $\angle$s are congruent:

$m\angle AFD = m\angle CFB$

$x + 10 = 100$

$x + 10 - 10 = 100 - 10$

$x = 90$

Try These *(For solutions, see page 348.)*

1. In this diagram, $\overleftrightarrow{BD}$, $\overleftrightarrow{EC}$, and $\overrightarrow{OA}$ intersect at point O.
Name a pair of vertical angles.

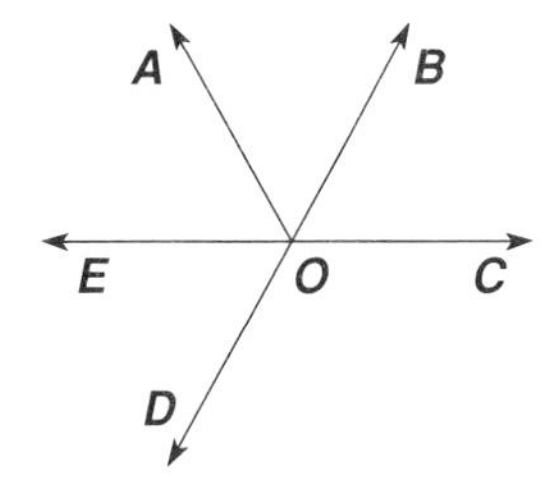

2. In this diagram, $\overleftrightarrow{SM}$ and $\overleftrightarrow{VT}$ intersect at point N.
If $m\angle MNV = 3x - 48$ and $m\angle TNS = x + 22$, find $m\angle MNV$ and $m\angle TNS$.

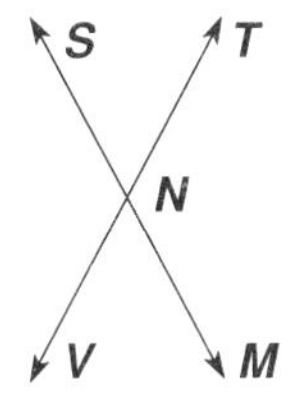

Adjacent Angles

Describe the difference, in these two diagrams, in the way $\angle 1$ and $\angle 2$ are formed.

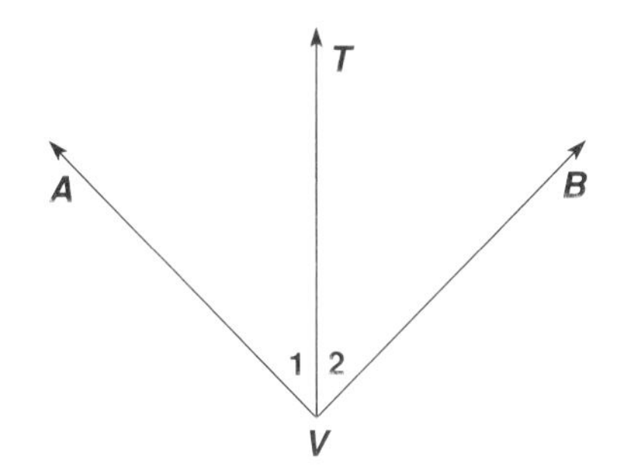

Here, $\angle 1$ and $\angle 2$ have the same vertex V and they have one side in common, $\overrightarrow{VT}$.

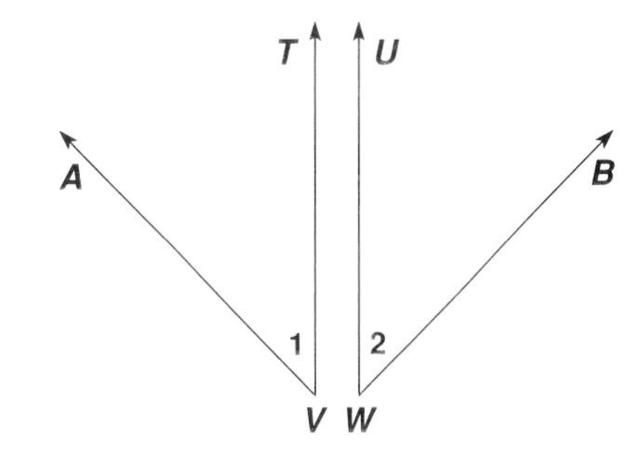

Here, $\angle 1$ and $\angle 2$ have two different vertices and different sides.

Two angles that have the same vertex, share a common side, and do not overlap are called *adjacent angles*.

Example 2 For this figure, name two pairs of adjacent angles and a pair of nonadjacent angles.

Give a reason for your answer.

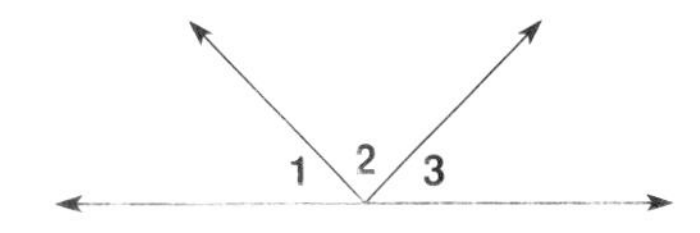

Since they have the same vertex, share a common side between them, and do not overlap:

$\angle 1$ and $\angle 2$ are adjacent angles $\quad$ $\angle 2$ and $\angle 3$ are adjacent angles

Even though they have the same vertex, $\angle 1$ and $\angle 3$ are nonadjacent angles because they do not share a common side between them.

Try These (For solutions, see page 348.)

1. For this figure, name 4 pairs of adjacent angles and 2 pairs of nonadjacent angles.

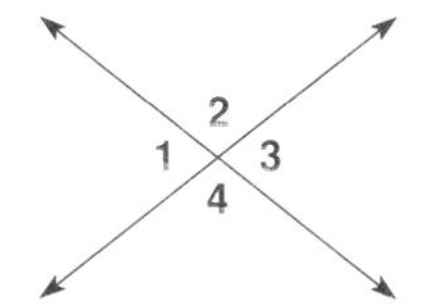

2. Are angles 1 and 2 in this diagram adjacent or nonadjacent? Explain.

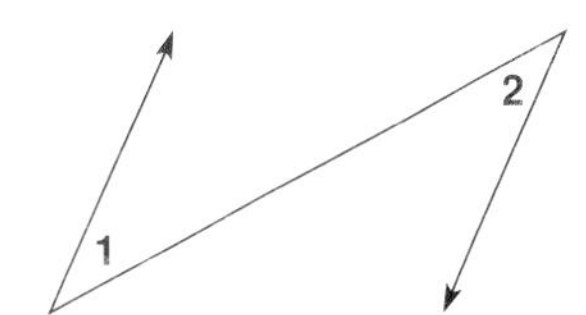

Complementary Angles

In this diagram, $\overrightarrow{BA} \perp \overrightarrow{BC}$. What do you know about the measure of $\angle ABC$?

If, together, $\angle 1$ and $\angle 2$ make up $\angle ABC$, what can you say about the sum of the measures of $\angle 1$ and $\angle 2$?

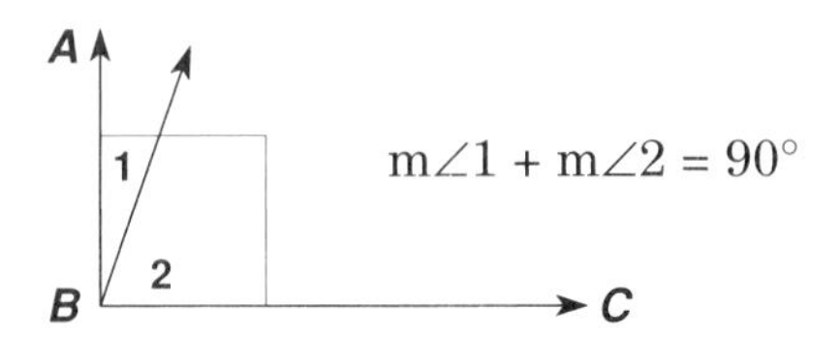

If the sum of the measures of two angles is 90°, the angles are called *complementary angles*.

Complementary angles may be adjacent angles or nonadjacent angles.

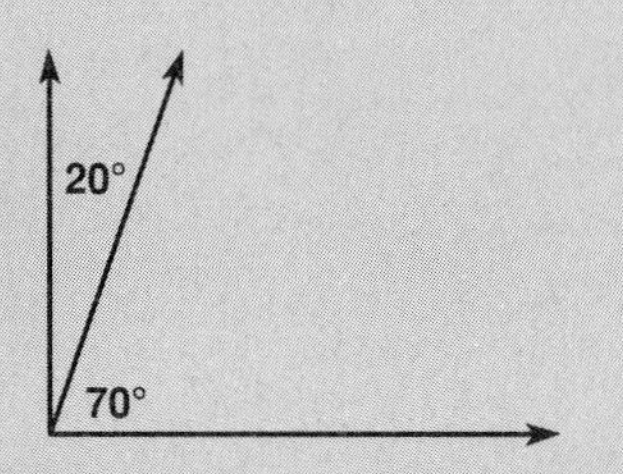

These complementary angles are adjacent.

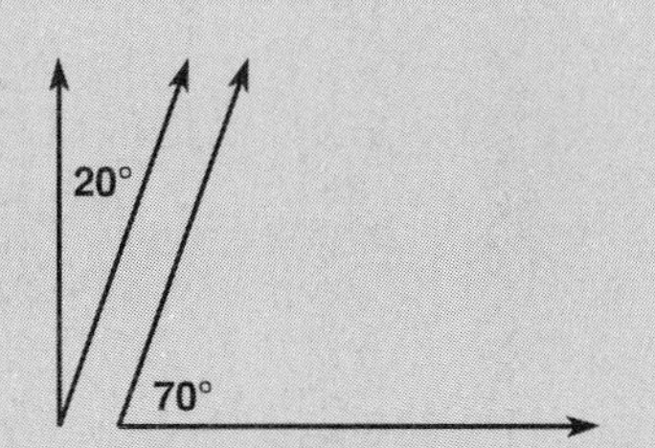

These complementary angles are nonadjacent.

Example 3 The measure of the larger of two complementary angles is twice that of the smaller. Find the number of degrees in the measure of each angle.

1 ***Read the problem for information.***
You know that two angles are complementary, meaning that the sum of their measures is 90°.
You know that the measure of the larger angle is twice that of the smaller.

2 ***Make a plan.***
Try to guess and check before writing an equation.
Let x = the measure of the smaller ∠.
Then $2x$ = the measure of the larger ∠.
$x + 2x = 90$ The angles are complementary.

3 ***Solve the equation.***
$x + 2x = 90$
$3x = 90$ Combine like terms.
$\frac{3x}{3} = \frac{90}{3}$ Divide both sides by 3.
$x = 30$
$2x = 2(30) = 60$ Substitute 30 for x.

4 ***Check***
If the ∠s measure 30° and 60°, are the ∠s complementary?
$30 + 60 \stackrel{?}{=} 90$
$90 = 90$ ✔

Answer: The ∠ measures are 30° and 60°.

Try These (For solutions, see page 348.)

1. Write the number of degrees in the complement of an angle that measures:
a. 20° **b.** 70° **c.** 45° **d.** $x°$

2. Use an equation to solve this problem. First, try guessing at an answer.
The measure of the larger of two complementary angles is 10° more than the measure of the smaller. Find the number of degrees in the measure of each of the complementary angles.

Supplementary Angles

In this diagram, $\overrightarrow{BA}$ and $\overrightarrow{BC}$ form a straight line. What do you know about the measure of $\angle ABC$?

If, together, $\angle 1$ and $\angle 2$ make up $\angle ABC$, what can you say about the sum of the measures of $\angle 1$ and $\angle 2$?

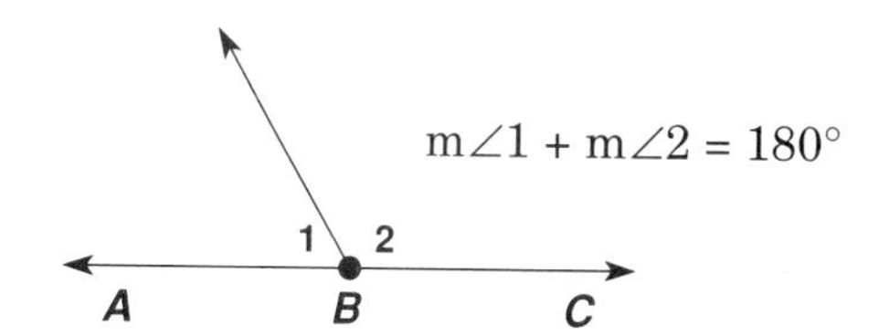

If the sum of the measures of two angles is 180°, the angles are called *supplementary angles*.

Supplementary angles may be adjacent angles or nonadjacent angles.

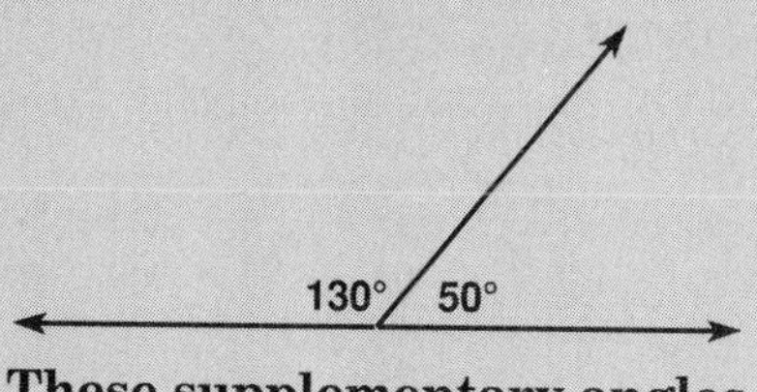

These supplementary angles are adjacent.

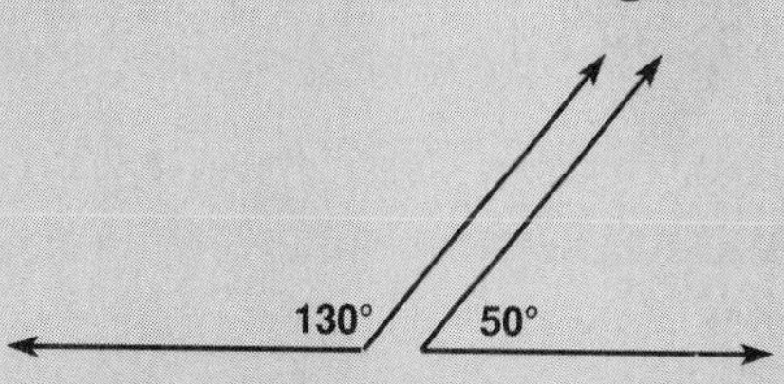

These supplementary angles are nonadjacent.

Example 4 The measure of the larger of two supplementary angles is 40° more than the measure of the smaller. Find the number of degrees in the measure of each angle.

1 ***Read the problem for information.***

You know that two angles are supplementary, meaning that the sum of their measures is 180°.

You know that the measure of the larger angle is 40° more than that of the smaller.

2 ***Make a plan.***

Try to guess and check before writing an equation.

Let x = the measure of the smaller $\angle$.
Then $x + 40$ = the measure of the larger $\angle$.
$x + x + 40 = 180$ The angles are supplementary.

3 ***Solve the equation.***

$x + x + 40 = 180$
$2x + 40 = 180$ Combine like terms.
$2x + 40 - 40 = 180 - 40$ Subtract 40 (both sides).
$2x = 140$
$\frac{2x}{2} = \frac{140}{2}$ Divide both sides by 2.
$x = 70$
$x + 40 = 70 + 40 = 110$ Substitute 70 for x.

4 ***Check.***

If the $\angle$s measure 70° and 110°, are the $\angle$s supplementary?

$70 + 110 \stackrel{?}{=} 180$
$180 = 180$ ✔

Answer: The $\angle$ measures are 70° and 110°.

Try These *(For solutions, see page 348.)*

1. Write the number of degrees in the supplement of an angle that measures:
a. 20° **b.** 160° **c.** 90° **d.** $x°$

2. Use an equation to solve this problem. First, try guessing at an answer.
The measure of the smaller of two supplementary angles is 40° less than the measure of the larger. Find the number of degrees in the measure of each of the supplementary angles.

SELF-CHECK: SECTION 8.3

1. What kind of angle pair is formed by two intersecting lines?

   Answer: vertical angles

2. What is true about the measures of vertical angles?

   Answer: Pairs of vertical angles are equal in measure, or congruent.

3. In addition to sharing the same vertex, what else must adjacent angles share?

   Answer: a side, between them

4. If the sum of the measures of two angles is 90°, the angles are called _?_.

   Answer: complementary angles

5. The sum of the measures of supplementary angles is _?_.

   Answer: 180°

6. Must complementary angles also be adjacent angles?

   Answer: No, if the sum of the measures of two angles is 90°, the angles are complementary. Complementary angles may be adjacent angles or nonadjacent angles.

EXERCISES: SECTION 8.3

1. In each diagram, tell if $\angle 1$ and $\angle 2$ are a pair of vertical angles. Give a reason for your answer.

a.

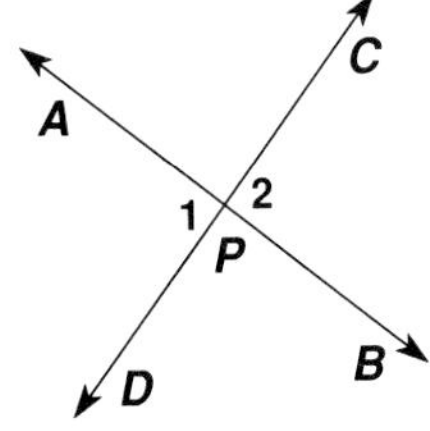

b.

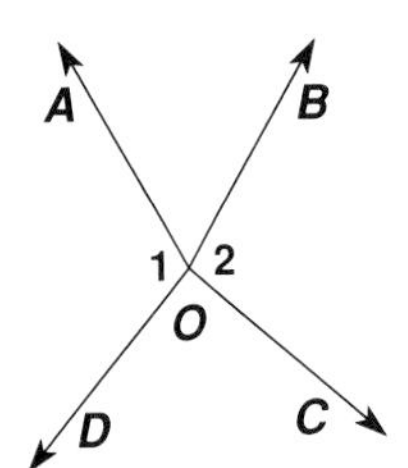

c.

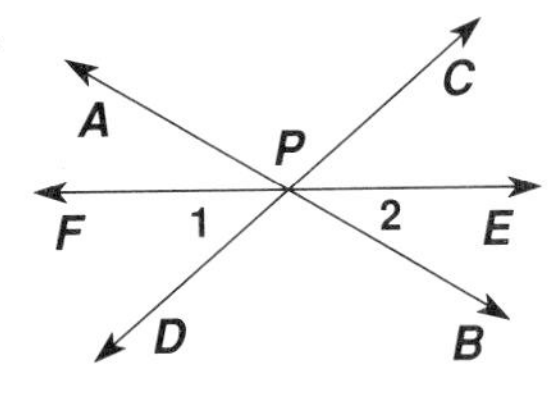

2. In each diagram, tell if $\angle 1$ and $\angle 2$ are a pair of adjacent angles. Give a reason for your answer.

a.

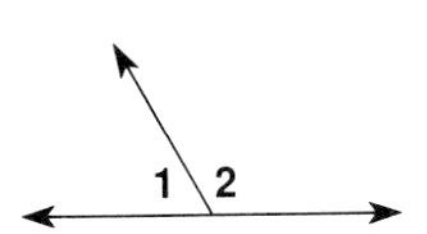

b.

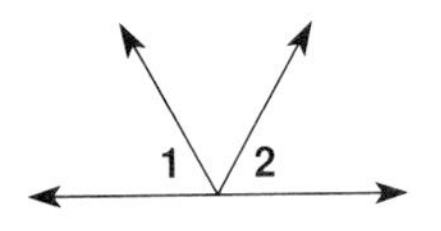

c.

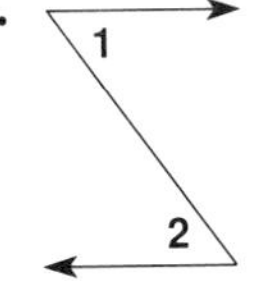

d.

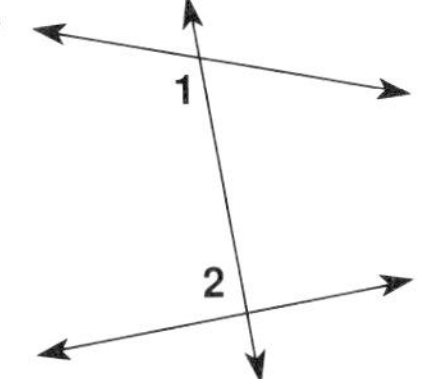

3. In this diagram, $\overleftrightarrow{EF}$ intersects perpendicular lines $\overleftrightarrow{AB}$ and $\overleftrightarrow{CD}$ at point G. Complete each statement.

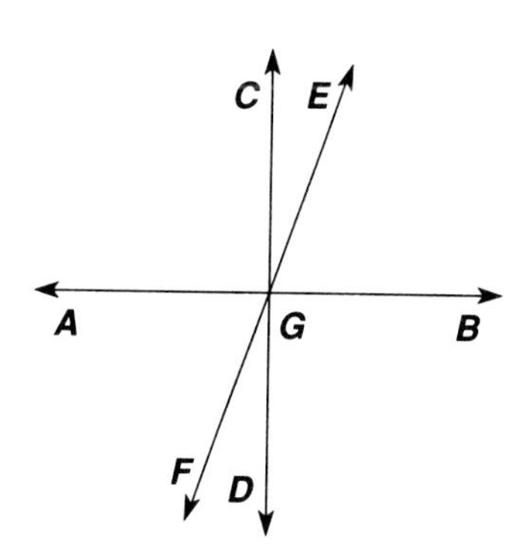

a. $\angle CGE$ and __?__ are vertical angles.
b. $\angle CGB$ and __?__ are vertical angles.
c. $\angle AGF$ and __?__ are complementary angles.
d. $\angle BGE$ and __?__ are complementary angles.
e. $\angle AGE$ and __?__ are supplementary angles.
f. $\angle DGB$ and __?__ are supplementary angles.
g. $\angle CGE$ and __?__ are supplementary angles.

4. Tell what special pair of angles is illustrated in each diagram.

a.

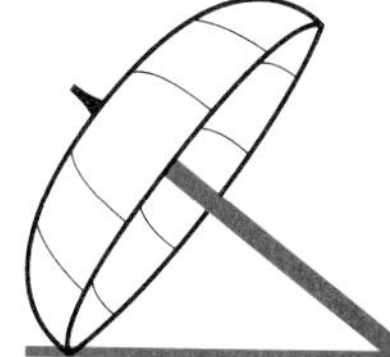

The stick of a leaning umbrella and the floor.

b.

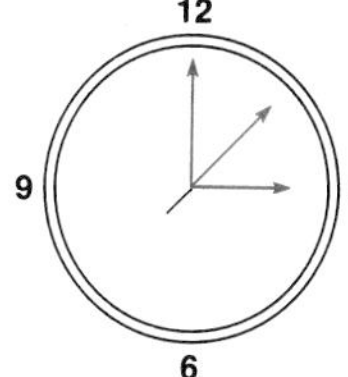

The second hand of a clock and the hour and minute hands.

5. Tell if each pair of values could represent the number of degrees in a pair of complementary angles. Give a reason for your answer.

a. 40, 50 **b.** 59, 21 **c.** x, $90 - x$ **d.** x, $90 + x$

6. Find the number of degrees in the complement of an angle whose degree measure is:

a. 15 **b.** 45 **c.** 89 **d.** n

7. Tell if each pair of values could represent the number of degrees in a pair of supplementary angles. Give a reason for your answer.

a. 40, 140 **b.** 169, 21 **c.** x, $180 - x$ **d.** x, $180 + x$

8. Find the number of degrees in the supplement of an angle whose degree measure is:

a. 45 **b.** 90 **c.** 120 **d.** n

9. Tell if each statement is true or false. If false, rewrite the statement so that it is true.

a. Two right angles are always supplementary.
b. If one of a pair of vertical angles is acute, the other must be acute.
c. If one of a pair of supplementary angles is acute, the other must be acute.
d. If two angles are complementary, the sum of their measures is equal to the measure of a straight angle.

10. Given angle measures as shown in the diagram, use letters to name:

a. two pairs of complementary angles

b. two pairs of supplementary angles

11.

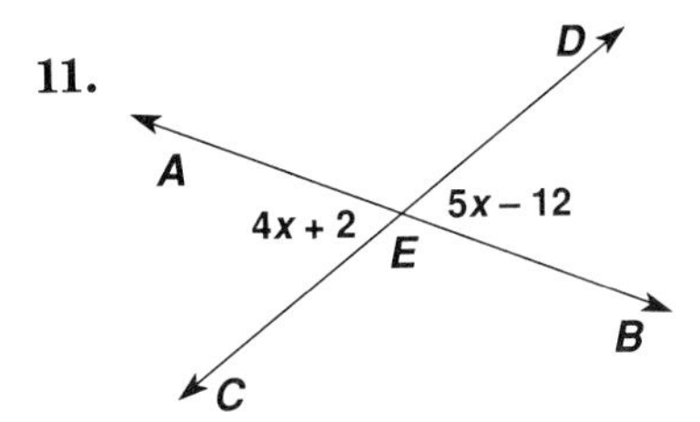

$\overleftrightarrow{AB}$ and $\overleftrightarrow{CD}$ intersect at E.
If $m\angle DEB = 5x - 12$ and $m\angle AEC = 4x + 2$, find:

a. the value of x

b. $m\angle AEC$ and $m\angle DEB$

c. $m\angle AED$ and $m\angle CEB$

12. $\overrightarrow{DC} \perp \overrightarrow{DB}$.
If $m\angle CDE = 5x + 6$ and $m\angle EDB = 2x$, find:

a. the value of x

b. $m\angle CDE$

c. $m\angle EDB$

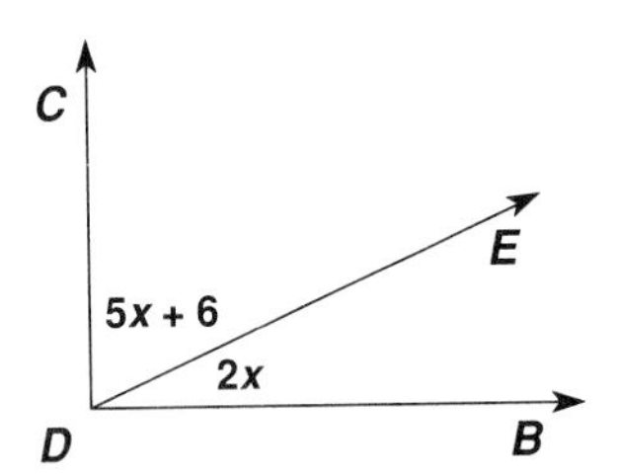

13.

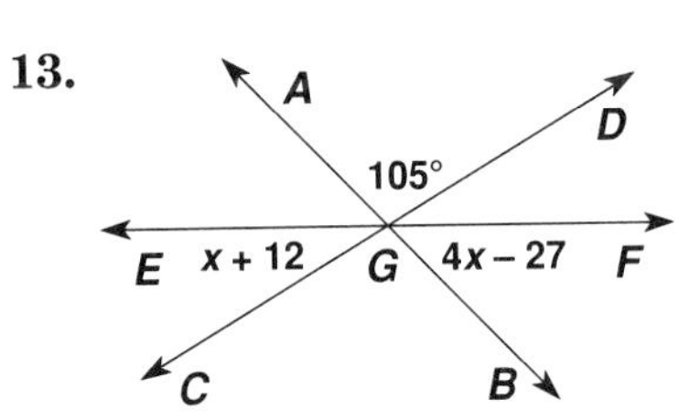

$\overleftrightarrow{AB}$, $\overleftrightarrow{CD}$ and $\overleftrightarrow{EF}$ intersect at G.
If $m\angle AGD = 105°$, $m\angle EGC = x + 12$, $m\angle FGB = 4x - 27$, find:

a. $m\angle CGB$

b. the value of x

c. $m\angle EGC$

d. $m\angle FGB$

e. $m\angle AGE$

f. $m\angle DGF$

14. Use an equation to solve each problem. First, try guesstimating an answer.

a. The measure of the larger of two supplementary angles is twice that of the smaller. Find the number of degrees in the measure of each angle.

b. The measure of the larger of two complementary angles is 20° more than the measure of the smaller. Find the number of degrees in the measure of each angle.

c. The measure of the smaller of two complementary angles is half the measure of the larger. Find the number of degrees in the measure of each angle.

d. The measure of the smaller of two supplementary angles is 30° less than the measure of the larger. Find the number of degrees in the measure of each angle.

e. The measure of the larger of two complementary angles is 12° more than twice the measure of the smaller. Find the number of degrees in the measure of each angle.

EXPLORATIONS

1. **a.** Copy and complete this table to list the measures of the complement and supplement of a given angle.

Angle Measure	Measure of the Complement	Measure of the Supplement
20°	?	?
40°	?	?
60°	?	?
$x°$	?	?

   b. Use the results of the table to describe a relationship between the supplement and the complement of an angle.

2. **a.** If x represents the degree measure of an angle, represent in terms of x the degree measure of:

   (1) the complement of the angle

   (2) the supplement of the angle

   b. Using the representations written in part *a*, write an equation to solve each of the following problems. First, guesstimate an answer.

   (1) The sum of the measures of the complement and the supplement of a certain angle is 180°. Find the measure of the angle.

   (2) The measure of the supplement of a certain angle is four times the measure of its complement. Find the measure of the angle.

(Explorations continue)

3. Here's how to construct the *bisector of an angle* (the ray that divides the angle into two halves).

① Using any setting on the compass and vertex V of the angle as the center, draw an arc that intersects both sides of the angle. Label the points of intersection A and B.

② Open the compass to a setting that is more than half the measure of the angle. With this setting, use A as the center and draw an arc. With the same setting, use B as the center and draw an arc. Use P to label this intersection.

③ With a straightedge, draw the ray from V through P. This ray bisects the angle.

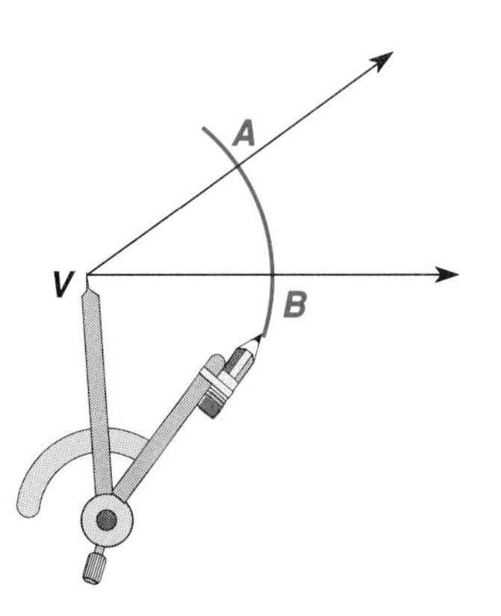

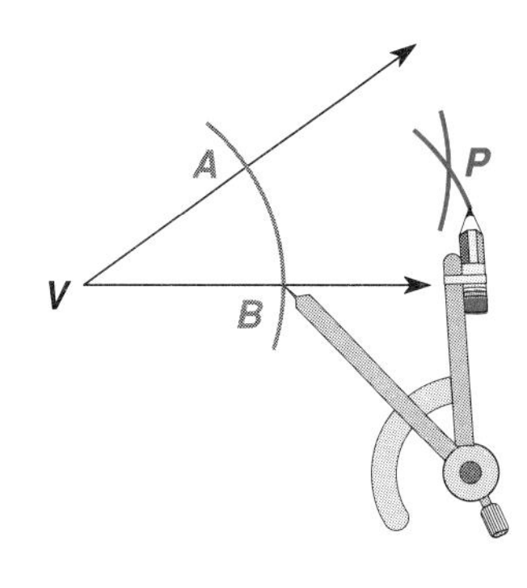

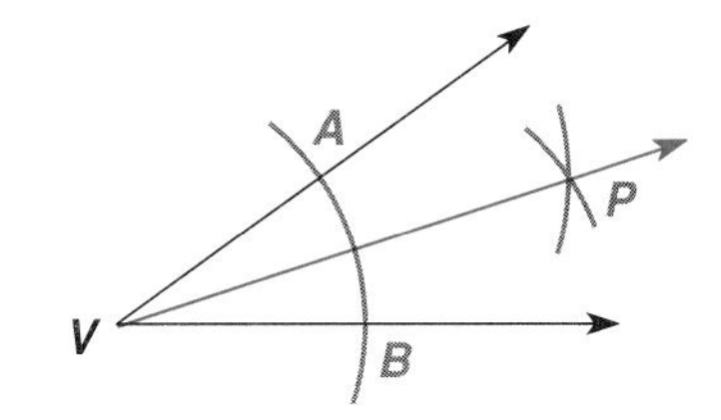

Result: $\overrightarrow{VP}$ is the bisector of $\angle AVB$.

a. On separate paper, practice this construction by finding the angle bisector for each of the following angles.

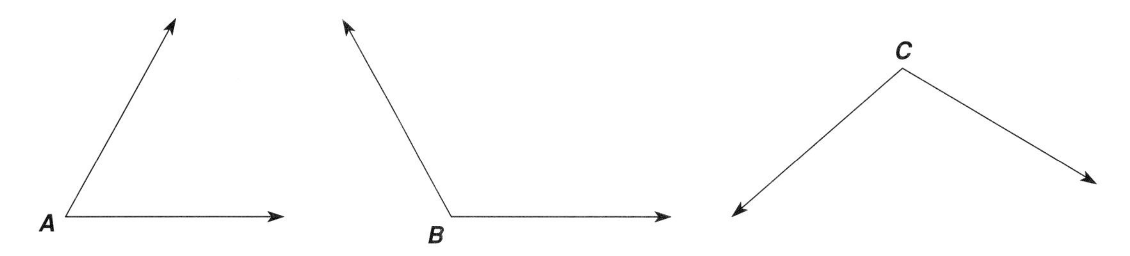

b. (1) Now that you know how to raise a perpendicular at a point on a line (see Exploration 1 on page 307) and how to bisect an angle, describe how you would construct an angle whose measure is 45°. Do the construction and then use a protractor to check the accuracy of your work.

(2) At the same time that you constructed an angle of 45°, you constructed another angle. What is its measure? Explain.

8.4 Triangles

Classifying a Triangle

The shape *triangle* is often used in construction as a support. Look for the triangular shapes in the frame of this bicycle.

Today, as you go about your usual activities, look for triangular shapes.

As suggested by its name, a ***triangle*** has 3 angles and 3 sides. The symbol for triangle is $\triangle$.

Special names are given to triangles according to the number of sides that are equal in measure.

Scalene Triangle

No sides have the same length.

Isosceles Triangle

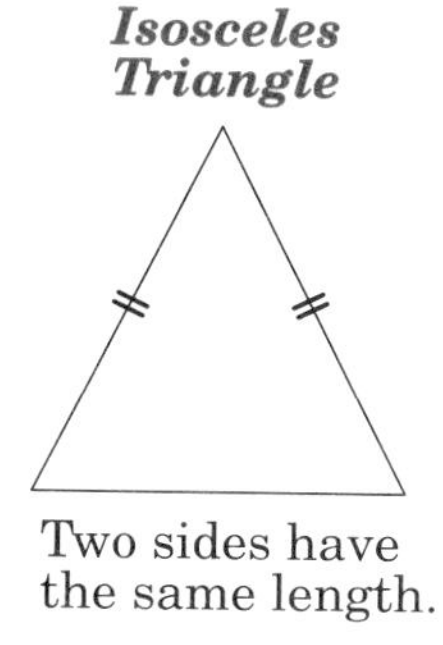

Two sides have the same length.

Equilateral Triangle

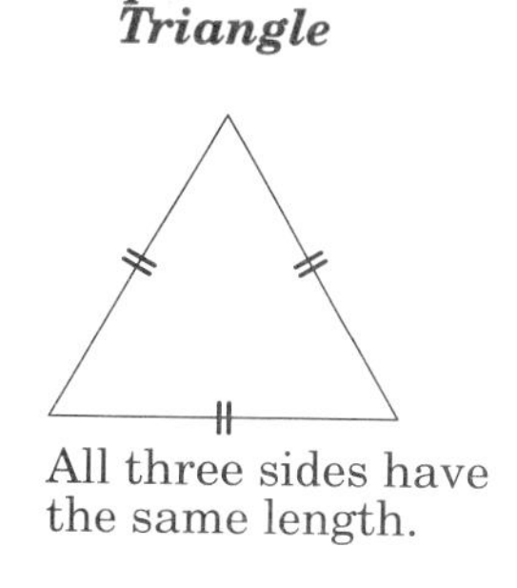

All three sides have the same length.

Study the angles of the scalene triangle. Do you think any of them are equal in measure? Use a protractor to verify your conclusion.

What do you think is true about the measure of the angles in an isosceles triangle? in an equilateral triangle? Use a protractor to verify your conclusions.

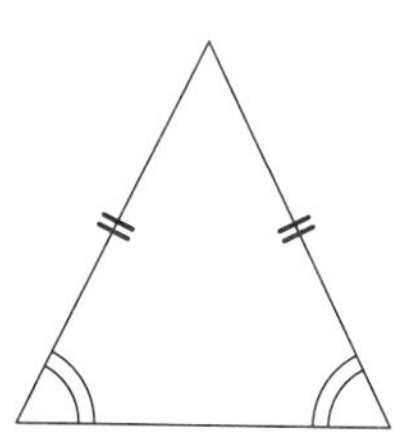

In an isosceles triangle, two angles have the same measure. The congruent angles are located opposite the congruent sides, and are called ***base angles***.

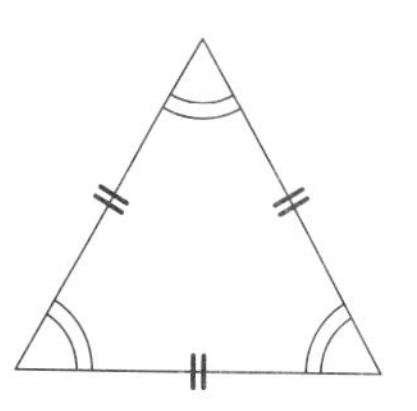

In an equilateral triangle, all three angles have the same measure. An equilateral triangle is also called ***equiangular***.

Example 1 $\triangle ABC$ is isosceles. If $AB = 2x + 4$ and $AC = 10$, find the value of x.

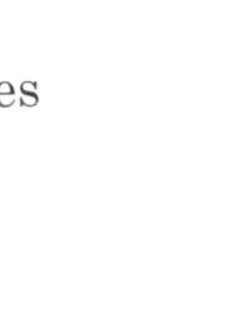

1 ***Read the problem for information.***
Since the triangle is isosceles, two of the sides are equal in measure.

2 ***Make a plan.***
Write an equation.

$$AB = AC$$
$$2x + 4 = 10$$

3 ***Solve the equation.***

$$2x + 4 = 10$$
$$2x + 4 - 4 = 10 - 4 \quad \text{Subtract 4 from both sides.}$$
$$2x = 6$$
$$\frac{2x}{2} = \frac{6}{2} \quad \text{Divide both sides by 2.}$$
$$x = 3$$

4 ***Check*** If $x = 3$, is the length of $\overline{AB}$ also equal to 10?

$$2x + 4 = 10$$
$$2(3) + 4 \stackrel{?}{=} 10$$
$$6 + 4 \stackrel{?}{=} 10$$
$$10 = 10 \checkmark$$

Answer: $x = 3$

Try These (For solutions, see page 349.)

1. Use *scalene*, *isosceles*, or *equilateral* to describe each triangle. Explain.

a.

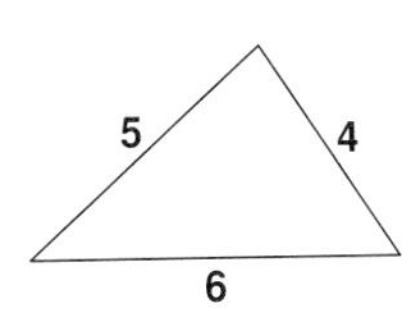

b.

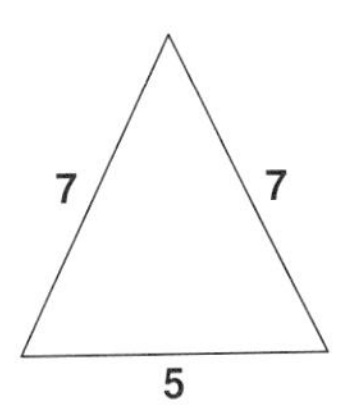

c.

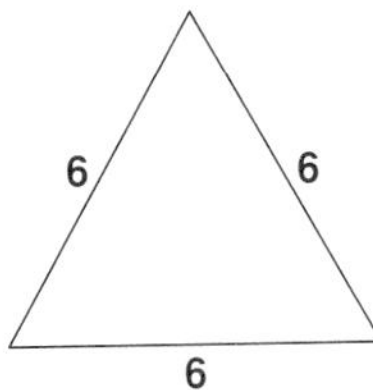

d.

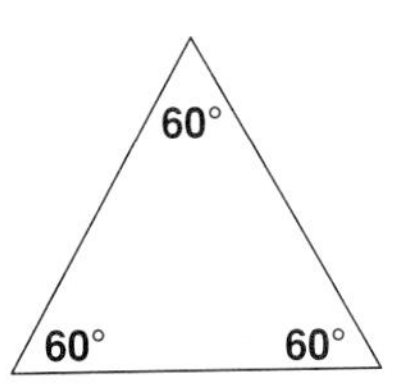

e.

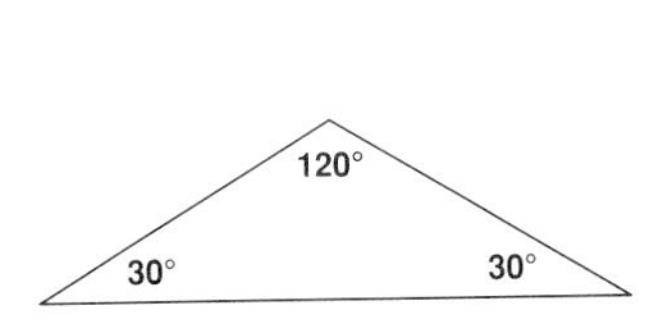

f.

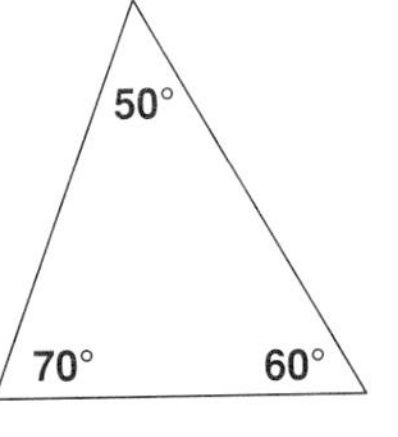

2. In equilateral $\triangle ABC$, side $AB = 3x - 2$ and sides AC and BC are each 16 cm. Find the value of x.

Angle Measure in a Triangle

Make a paper triangle. Tear the corners from your paper triangle and fit them together so that the vertices meet at a point. What can you say about the measures of the 3 angles together?

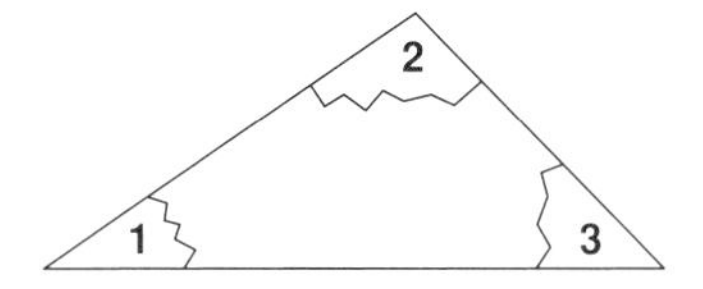

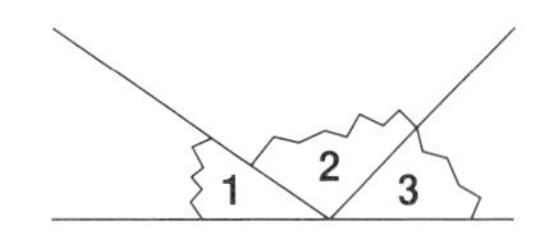

The sum of the measures of the 3 angles of a triangle is 180°.

Example 2 Find the measure of $\angle A$ in each triangle. Before studying the solution shown, see if you can calculate the answer.

a.

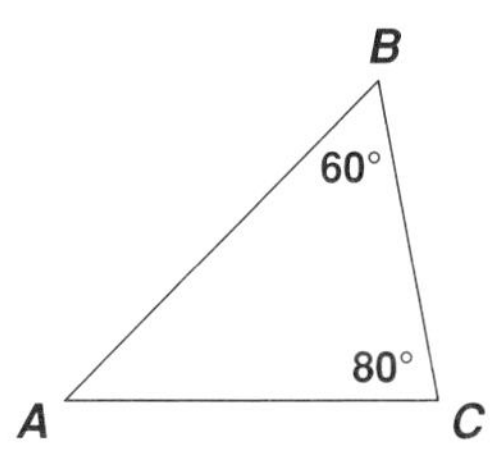

$$m\angle A + m\angle B + m\angle C = 180$$
$$m\angle A + 60 + 80 = 180$$
$$m\angle A + 140 = 180$$
$$m\angle A = 40°$$

b. $\triangle ABC$ is isosceles. In isosceles $\triangle ABC$, $m\angle C = m\angle B = 75°$.

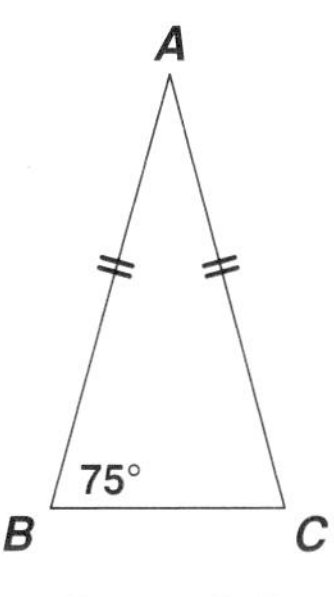

$$m\angle A + m\angle B + m\angle C = 180$$
$$m\angle A + 75 + 75 = 180$$
$$m\angle A + 150 = 180$$
$$m\angle A = 30°$$

c. $\triangle ABC$ is equilateral. In equilateral $\triangle ABC$, $m\angle A = m\angle B = m\angle C = x$

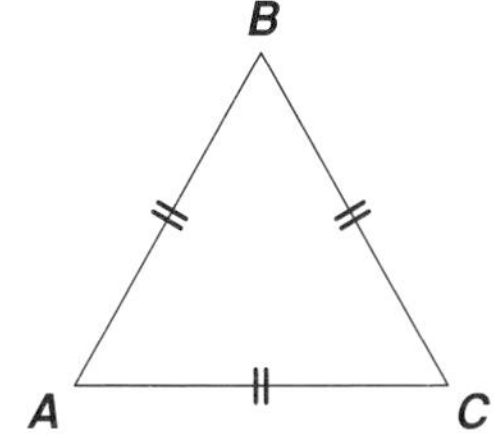

$$m\angle A + m\angle B + m\angle C = 180$$
$$x + x + x = 180$$
$$3x = 180$$
$$\frac{3x}{3} = \frac{180}{3}$$
$$x = 60$$
$$m\angle A = 60°$$

Try These (For solutions, see page 349.)

1. Find the number of degrees in the missing angle(s) of each triangle.

   a. scalene triangle

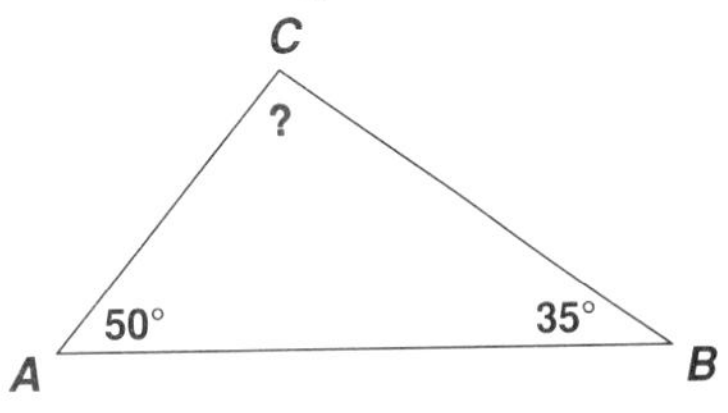

   b. isosceles triangle

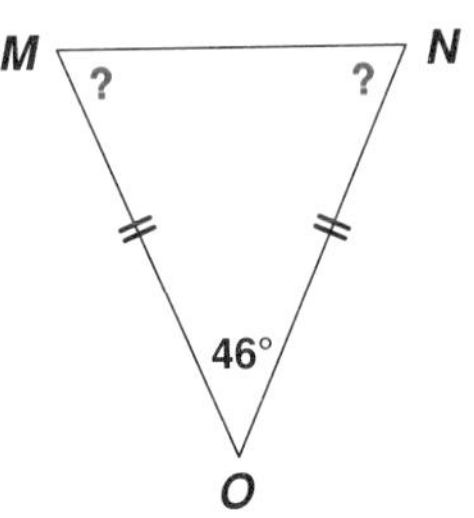

2. **a.** Draw a picture of an isosceles right triangle.

   b. How many degrees are there in the measure of each base angle?

Perimeter

Using red tape, Joan made a triangular design on her notebook cover. If the lengths of the sides of the triangle are as shown in the diagram, how many inches of tape did Joan use?

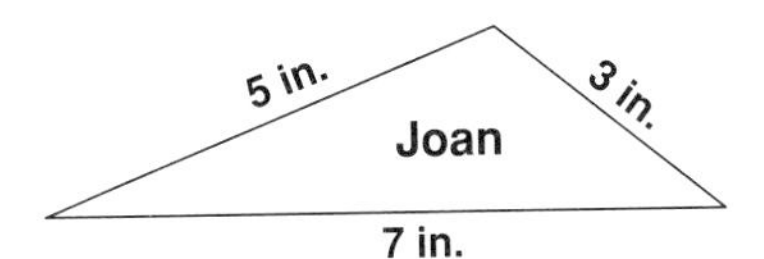

Joan used 5 + 7 + 3 or 15 inches of tape.

> **The sum of the lengths of the sides of a triangle is called the *perimeter* of the triangle.**

Example 3 Find the perimeter of each triangle.

a. $\triangle ABC$ is isosceles, $AB = AC$.

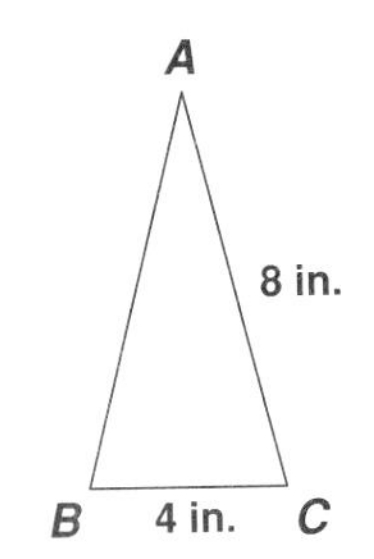

Since $\triangle ABC$ is isosceles, $AB = AC = 8$ in.

$$\text{Perimeter} = 8 + 8 + 4$$
$$= 20 \text{ in.}$$

In an isosceles triangle, the congruent sides are called ***legs*** and the third side is the ***base***. Using these words, write a rule for finding the perimeter of any isosceles triangle.

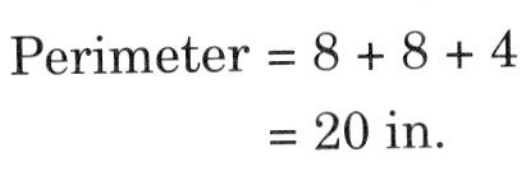

perimeter of an isosceles triangle = leg + leg + base

= 2(leg) + base

b. $\triangle PQR$ is equilateral.

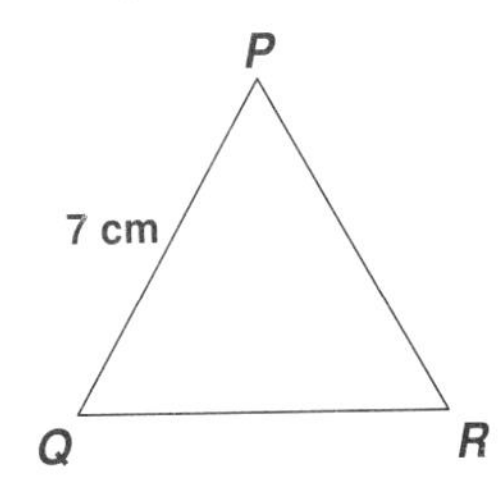

Since $\triangle PQR$ is equilateral, $PQ = PR = QR = 7$ cm

$$\text{Perimeter} = 7 + 7 + 7$$
$$= 21 \text{ cm}$$

If each of the 3 sides of an equilateral triangle has length s, write a rule in terms of s for finding the perimeter of any equilateral triangle.

perimeter of an equilateral triangle $= s + s + s$

$= 3s$

Try These (For solutions, see page 349.)

1. In an isosceles triangle, the measure of the base is 6 inches. If the measure of each of the congruent legs is twice the measure of the base, find the perimeter of the triangle.

2. The perimeter of an equilateral triangle is 27 cm. Find the measure of each of the sides.

The Triangle Inequality

Suppose you want to get from a place represented by point A to the place represented by C. Describe the shortest path.

If you first went from A to B and then to C, what would be the shape of this path? Is this way longer or shorter than going directly from A to C?

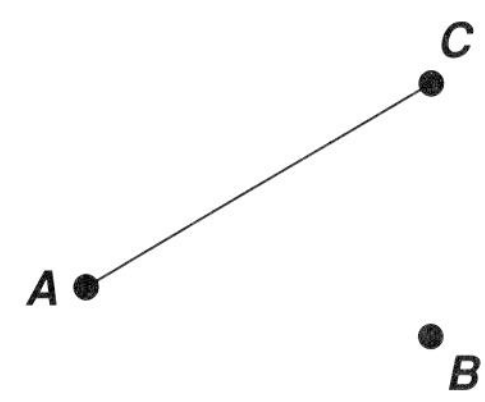

The shortest distance between two points is along a line segment.

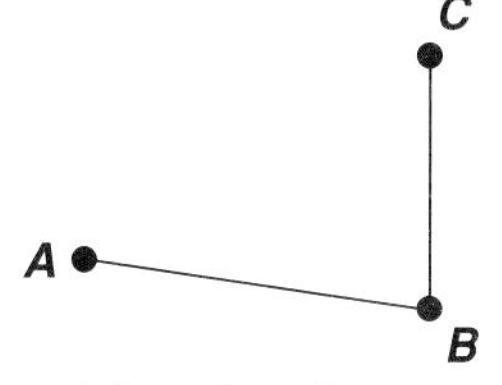

Since AC is the shortest way from A to C, any other way—such as $AB + BC$—is longer.

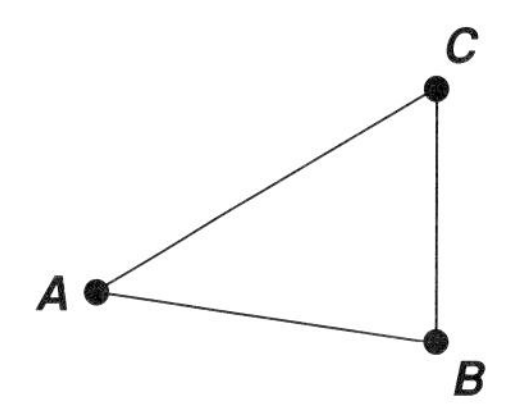

$AB + BC > AC$
$AB + AC > BC$
$AC + BC > AB$

In a triangle, the sum of the measures of any two sides is greater than the measure of the third side.

In any triangle, three different inequalities involving the measures of the sides are true.

Example 4 Tell whether three line segments of the lengths given can be the sides of a triangle. Using a ruler, try to make a triangle.

a. 5 cm, 4 cm, 3 cm

$5 + 4 \overset{?}{>} 3$ ✔

$5 + 3 \overset{?}{>} 4$ ✔

$4 + 3 \overset{?}{>} 5$ ✔

All 3 inequalities are true.

Answer: Yes, a triangle is possible.

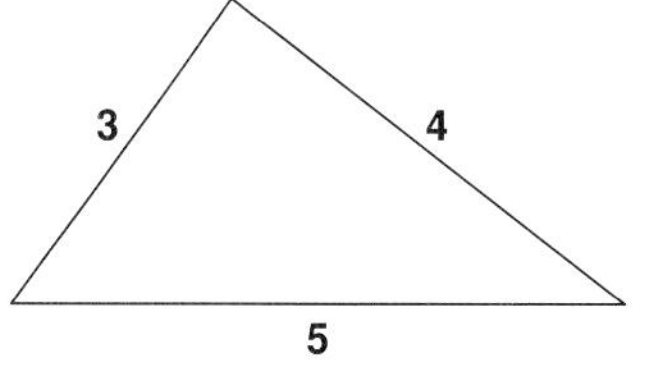

b. 4 in., 2 in., 1 in.

$4 + 2 \overset{?}{>} 1$ ✔

$4 + 1 \overset{?}{>} 2$ ✔

$2 + 1 \overset{?}{>} 4$ no

Not all of the 3 inequalities are true.

Answer: No triangle is possible.

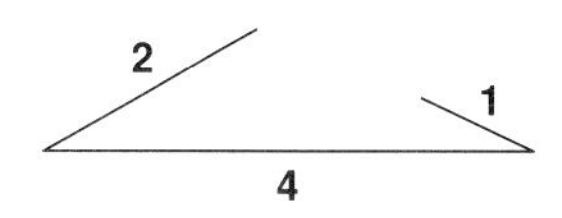

Try These *(For solutions, see page 349.)*

Tell whether these sets of 3 measures could be the lengths of the sides of a triangle. Give a reason for your answer.

1. 3, 4, 6 **2.** 2, 4, 9 **3.** 3, 2, 1

SELF-CHECK: SECTION 8.4

1. How many sides in an isosceles triangle have the same length?
Answer: 2 sides

2. What special property do the angles of an isosceles triangle have?
Answer: The base angles are equal in measure.

3. What name is given to a triangle that has all 3 sides equal in measure?
Answer: equilateral

4. What special property do the angles of an equilateral triangle have?
Answer: All 3 angles are equal in measure. An equilateral triangle is also equiangular.

5. What is true about the sum of the measures of the 3 angles of any triangle?
Answer: The sum of the measures of the angles of a triangle is 180°.

6. Describe the *perimeter* of a triangle or any polygon.
Answer: The perimeter is the sum of the lengths of the sides.

7. In order for a triangle to be formed, what must be true about the lengths of the 3 sides?
Answer: The sum of the lengths of any 2 sides must be greater than the length of the 3rd side.

EXERCISES: SECTION 8.4

1. Cut out a paper circle and fold it into an equilateral triangle. Explain your method.

2. Match each triangle name in *Column A* to the number of sides of equal measure shown in *Column B*.

Column A	*Column B*
a. scalene triangle	**(1)** all sides equal in measure
b. isosceles triangle	**(2)** no sides equal in measure
c. equilateral triangle	**(3)** two sides equal in measure

3. Find the missing angle measure in each triangle.

a.

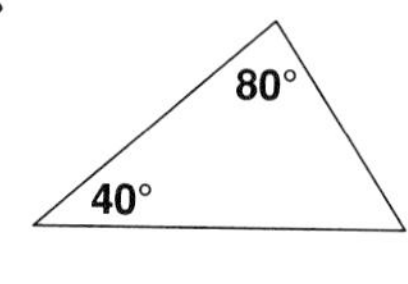

b.

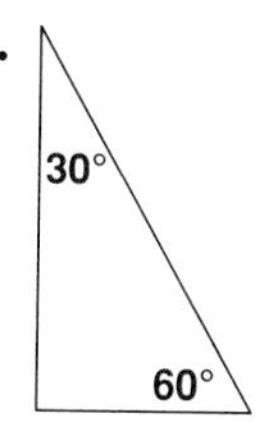

c.

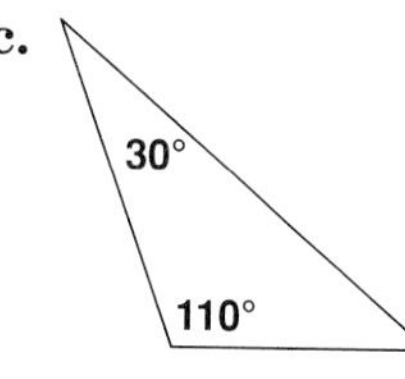

d. In $\triangle JKL$, if $\angle J \cong \angle K$ and $m\angle L = 80°$, what is $m\angle J$?

e. In equilateral $\triangle PQR$, what is the measure of $\angle P$?

f. In isosceles $\triangle RST$, if $m\angle R = 100°$, what is $m\angle S$?

4. Match each triangle with one of these names:
right, equilateral, isosceles, or scalene

a.

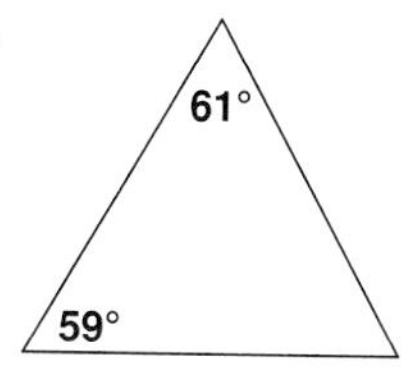

b.

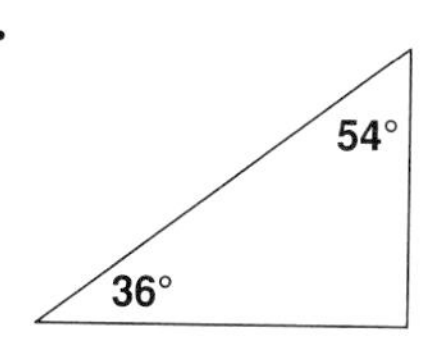

c.

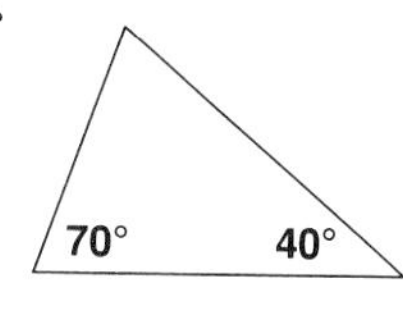

d.

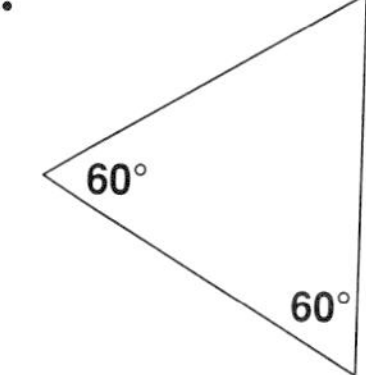

5. Do you agree or disagree with each statement? Explain.

a. Loni said there could be no more than one right angle in a triangle.

b. Lorri said there could be two obtuse angles in a triangle.

c. Larry said that if a triangle is isosceles, it could also have a right angle.

d. Lacey said that if a triangle is equilateral, it could also have a right angle.

e. Lenny said that in a triangle with a right angle, the sum of the measures of the two other angles is always 90°.

Note: A triangle that has a right angle is called a ***right triangle***.

6. For each triangle, the angle measures are represented by algebraic expressions.

- Find the value of x.
- Find the number of degrees in the measure of each angle of the triangle.
- Tell whether the triangle is right, isosceles, equilateral, or just scalene, and draw a picture of the triangle.

a. For $\triangle ABC$: $m\angle A = x$ $m\angle B = 2x$ $m\angle C = 3x$

b. For $\triangle MNO$: $m\angle M = x + 2$ $m\angle N = 3x - 5$ $m\angle O = x + 18$

c. For $\triangle JKL$: $m\angle J = x + 35$ $m\angle K = 2x + 10$ $m\angle L = 3x - 15$

d. For $\triangle RST$: $m\angle R = 4x + 7$ $m\angle S = 5x - 5$ $m\angle T = 6x - 2$

7. Find the perimeter of each triangle.

a. scalene △

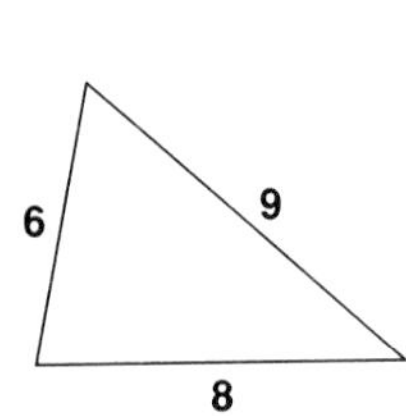

b. isosceles △

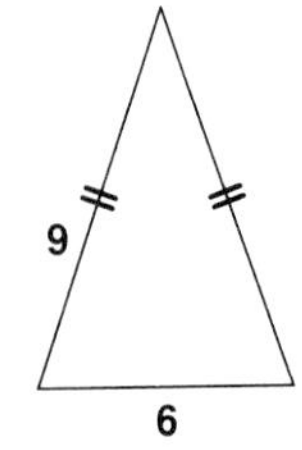

c. equilateral △

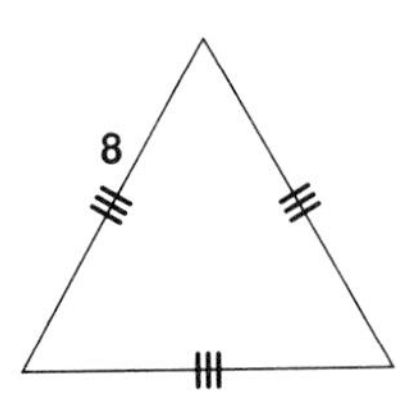

8. Use the representations shown on the sides of each triangle and the given perimeter to find the measure of the side asked for.

a. Perimeter = 43 inches.
Find AB.

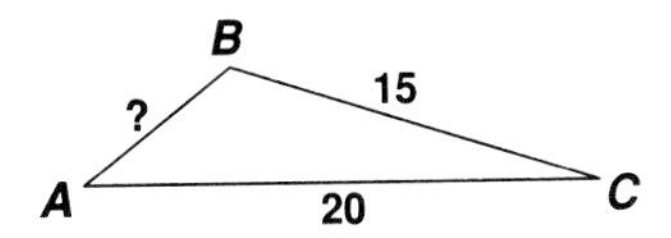

b. Perimeter = 82 centimeters.
Find the measure of the longest side.

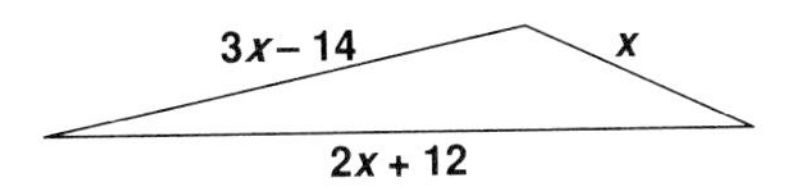

c. Perimeter = 34 feet.
Find the length of the shortest side.

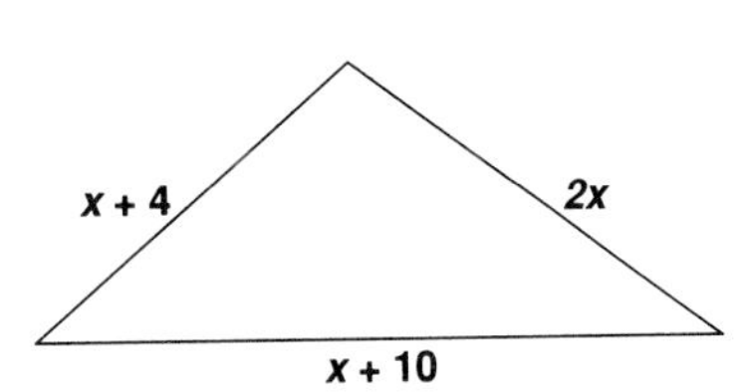

d. $\triangle DEF$ is equilateral.
Perimeter = 30 meters.
Find the value of x.

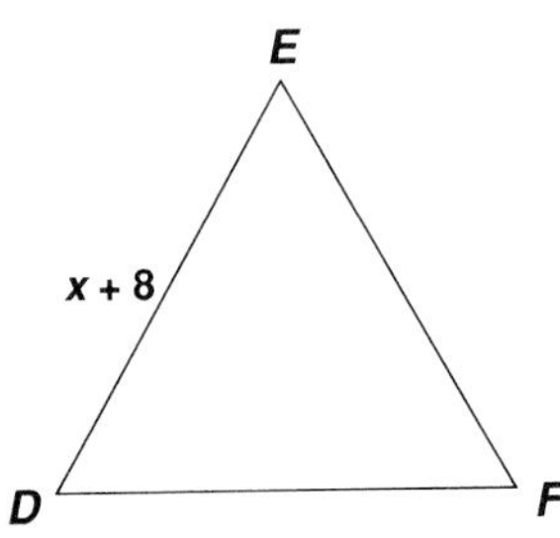

9. 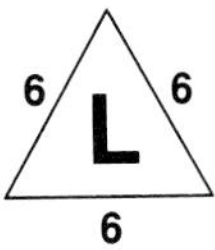

A piece of cardboard has a shape of an equilateral triangle, with each side 6 cm in length. The letter L is written inside.

a. If the triangle is rolled to the right along a table top until the letter L is again in the upright position, what distance has the triangle rolled?

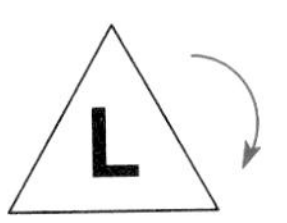

b. If the triangle is rolled to the right a number of times and stops so that the letter L is again upright, which of the following distances could the triangle have rolled? Explain.

(1) 24 cm (2) 30 cm (3) 60 cm (4) 90 cm

10. Give a reason for each answer.

a. Why could the lengths of the sides of a triangle not be 2, 3, 6?

b. The lengths of two sides of $\triangle DEF$ are 5 cm and 8 cm. The length of the third side could be

(1) 2 cm (2) 3 cm (3) 4 cm (4) 14 cm

c. In $\triangle LMN$, if $LM = 12$ in. and $MN = 15$ in., then the length of $\overline{LN}$ must be greater than _?_ and less than _?_ .

EXPLORATIONS

1. a. Copy $\triangle ABC$ onto graph paper.

(1) Draw the image $\triangle ABC'$ when $\triangle ABC$ is reflected over the vertical axis.

(2) What kind of triangle is $\triangle ACC'$? Explain.

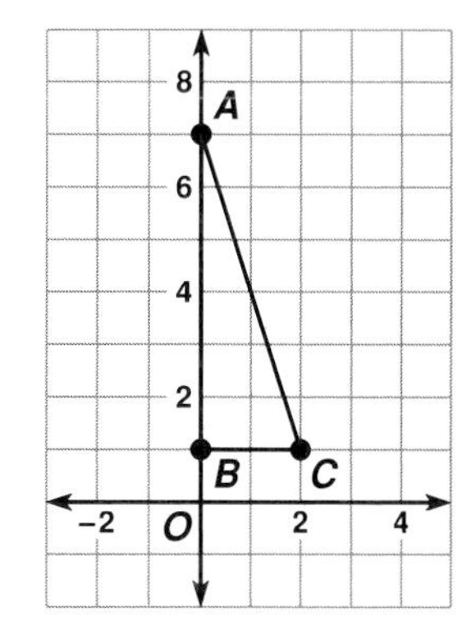

b.

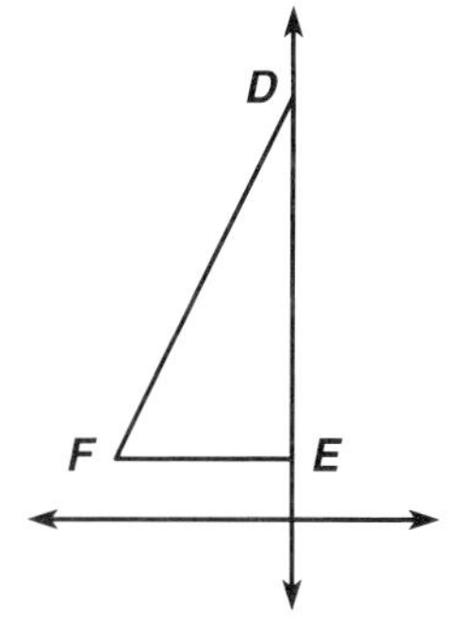

Copy $\triangle DEF$ onto plain paper, using a protractor to make sure that $m\angle F = 60°$.

(1) Draw the image $\triangle DEF'$ when $\triangle DEF$ is reflected over the vertical axis.

(2) What kind of triangle is $\triangle DFF'$? Verify your answer by measuring the angles with a protractor.

c. Copy $\triangle PQR$ onto graph paper.

(1) Draw the image $\triangle P'Q'R'$ when $\triangle PQR$ is translated 2 units right and 1 unit down.

(2) Make an observation about the relationship between the original triangle and its image.

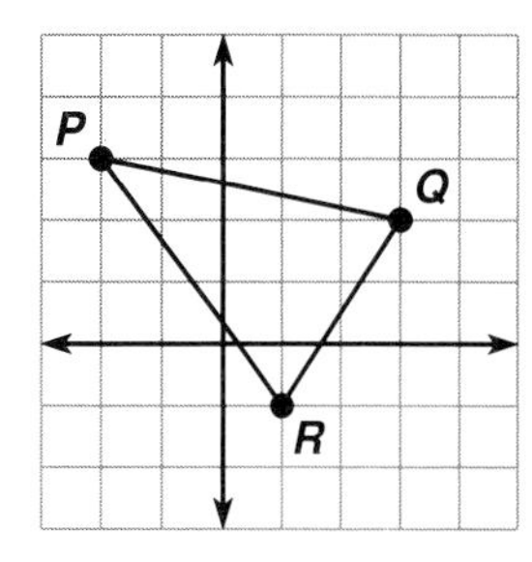

d. 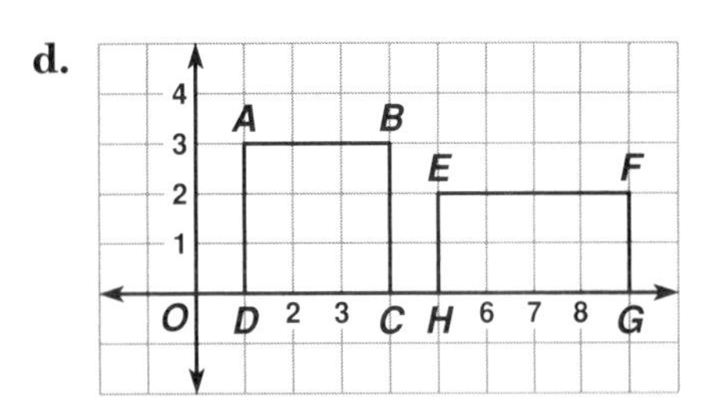

Square $ABCD$ and rectangle $EFGH$ are reflected over the horizontal axis.

(1) What are the coordinates of A' and B', the images of points A and B?

(2) What kind of figure is $ABB'A'$?

(3) What are the coordinates of E' and F', the images of points E and F?

(4) Is $EFF'E'$ the same kind of figure as $ABB'A'$? Explain.

2. In a rectangle, both pairs of opposite sides are equal in measure.

The longer side is called the *length* and the shorter side is called the *width*.

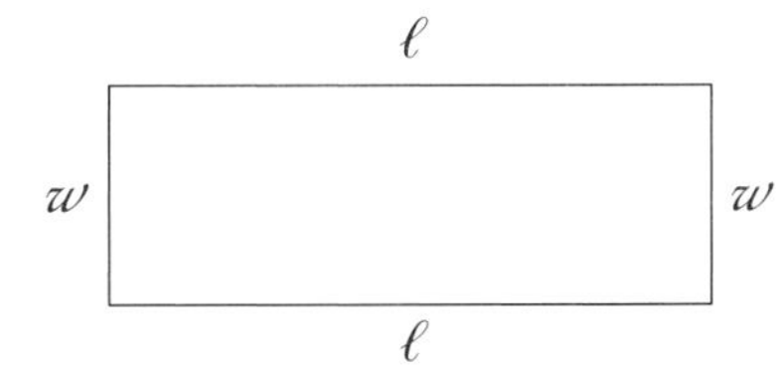

a. Using the variables shown in the figure, write a rule to find the perimeter of a rectangle.

b. If the length of the rectangle measures twice the width, and the perimeter is 72 feet, find the *dimensions* (length and width) of the rectangle.

c. If the side of a square has the same measure as the width of the rectangle in part b, find the perimeter of the square.

d. Describe a rule for finding the perimeter of a square if you know the measure of a side of the square.

3. Here's how to construct an equilateral triangle, each of whose sides measures this length: •——s——•

① Copy the given length onto a ray:

With a straightedge, draw a ray that starts at A.

Open the compass to the setting s. With A as the center, mark off an arc equal to s.

Use B to label where the arc crosses the ray.

② With the same setting s on the compass, use A as the center and draw an arc above the ray.

With the same setting s on the compass, use B as the center and draw an arc above the ray.

Use C to label the intersection of these arcs.

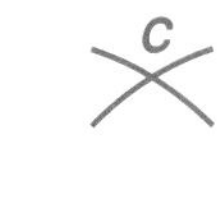

③ With a straightedge, draw a segment from A to C and from B to C.

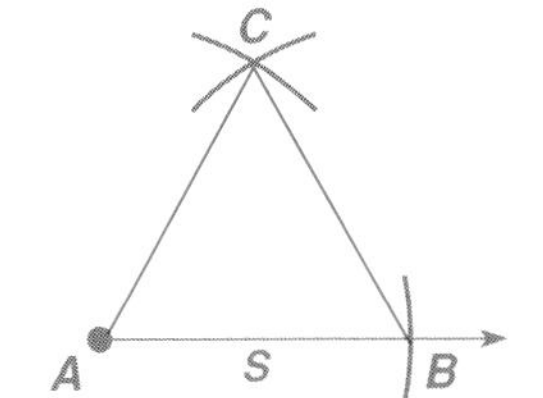

Result: $\triangle ABC$ is equilateral, with each side = s.

a. Practice this construction by using each of these lengths as a side for the equilateral triangle.

b. **(1)** Explain how you could use this construction to construct a 60° angle.

(2) Construct a 60° angle whose vertex is at the point V.

• V

(3) Explain how you could use the construction of a 60° angle to construct a 30° angle.

Construct a 30° angle and then use a protractor to check.

(4) Explain how you could use the construction of a 60° angle to construct an angle of 120°.

Construct an angle of 120° and then use a protractor to check.

(5) Name another angle measure that can be constructed beginning with an angle of 60°. Explain the construction.

8.5 More About Triangles

Special Lines in a Triangle

On their project, three construction workers are assigned to different parts of the building. Al and Chris are both on the ground level, and Bo works upstairs.

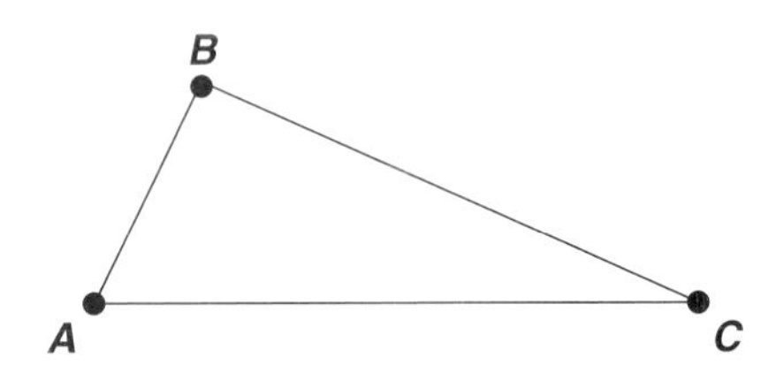

If the diagram represents the locations of these workers, tell how you would measure to find how high above the ground Bo is.

To measure Bo's height above the ground, you would have to measure along the perpendicular from B to $\overline{AC}$.

In a triangle, the line segment from a vertex, perpendicular to the opposite side, is called an *altitude*.

The altitude is also the height (or distance) from a vertex to the opposite side.

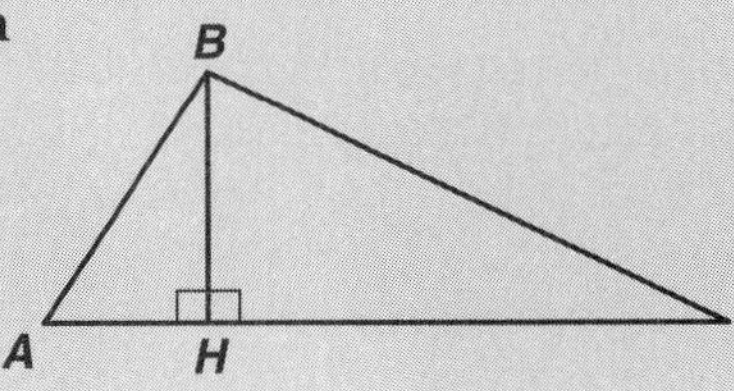

$\overline{BH}$ is the altitude from B to $\overline{AC}$.

$\overline{BH} \perp \overline{AC}$

Example 1 Using a ruler and protractor, draw the 3 altitudes of $\triangle RST$.

What do you notice when the 3 altitudes are drawn?

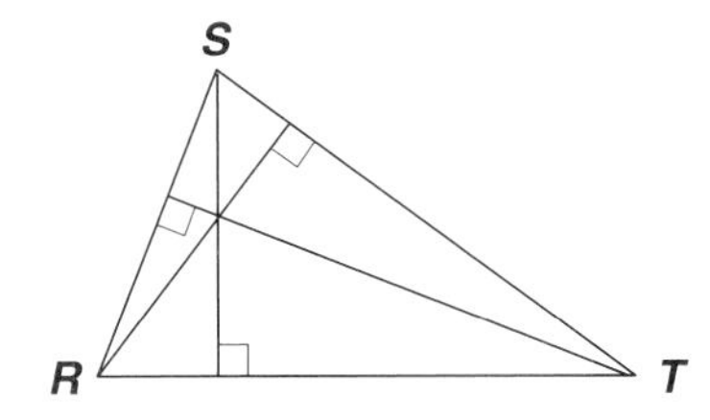

The 3 altitudes of a triangle meet at a point.

Other special lines in a triangle concern *bisectors*, which divide figures into two halves.

An *angle bisector* of a triangle is a line segment that divides an angle of a triangle into two halves.

An angle bisector of a triangle begins at the vertex of the angle and ends in the opposite side.

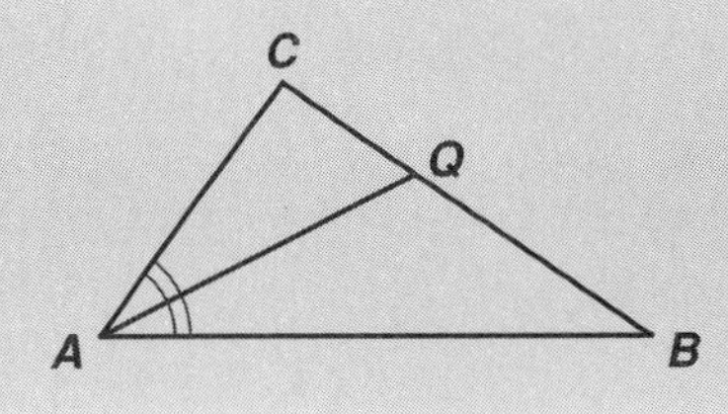

$\overline{AQ}$ is the bisector of $\angle A$.

$\angle CAQ \cong \angle QAB$

A line segment from a vertex of a triangle to the midpoint of the opposite side is called a *median*.

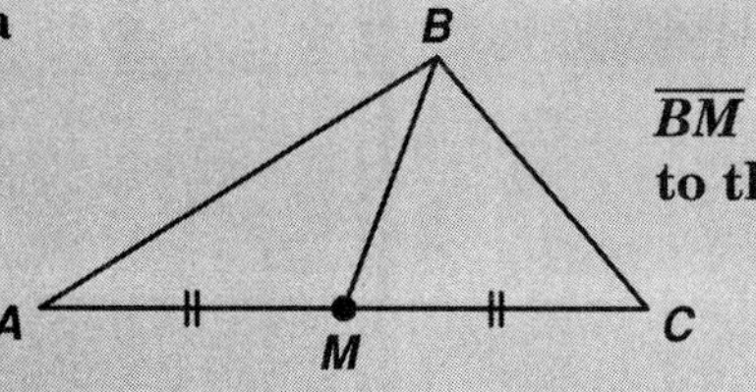

$\overline{BM}$ is the median from B to the midpoint of $\overline{AC}$.
$\overline{AM} \cong \overline{MC}$

Example 2 For $\triangle XYZ$, use a ruler and protractor to draw from Y the altitude, the angle bisector, and the median.

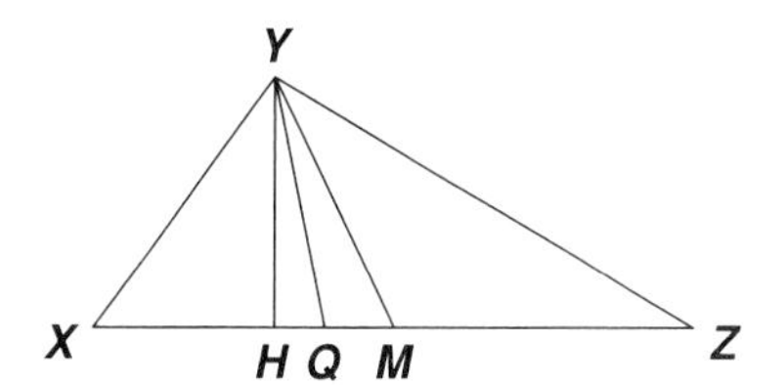

$\overline{YH}$ is the altitude from Y. $\quad \overline{YH} \perp \overline{XZ}$
$\overline{YQ}$ is the bisector of $\angle XYZ$. $\quad \angle QYX \cong \angle QYZ$
$\overline{YM}$ is the median from Y. $\quad \overline{XM} \cong \overline{MZ}$

Example 3 In $\triangle RST$, $\overline{RP}$ bisects $\angle R$. Find $m\angle S$.

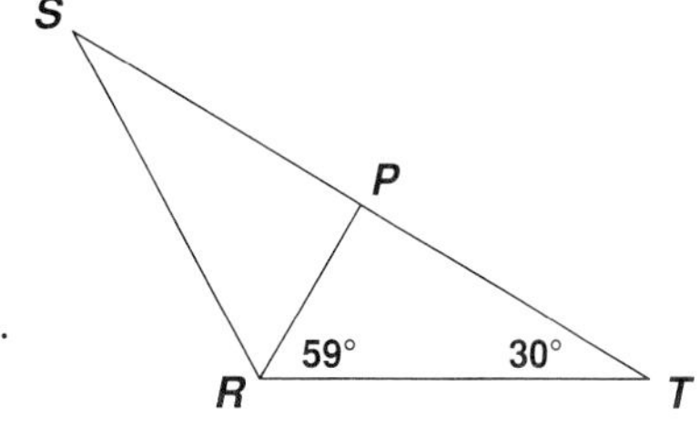

1 ***Read the problem for information.***

Since $\overline{RP}$ bisects $\angle R$, 59° is half the measure of $\angle R$. Thus, $m\angle R = 2(59) = 118°$. $\quad m\angle T = 30°$.

2 ***Make a plan.***

Write an equation using the fact that the sum of the measures of the angles of a triangle is 180°.

$$m\angle R + m\angle S + m\angle T = 180$$
$$118 + m\angle S + 30 = 180$$

3 ***Solve the equation.***

$$118 + m\angle S + 30 = 180$$
$$m\angle S + 148 = 180 \quad \text{Combine like terms.}$$
$$m\angle S = 32 \quad \text{Subtract 148 from both sides.}$$

4 ***Check***

If $m\angle S = 32°$, is the sum of the measures of the 3 angles of the triangle equal to 180°?

$$118 + 30 + 32 \stackrel{?}{=} 180$$
$$180 = 180 \checkmark$$

Answer: $m\angle S = 32°$

Try These (For solutions, see page 350.)

1. Find each of the following segments in the diagram.
   Use a protractor and ruler to verify your answers.
   a. an altitude in $\triangle ABC$
   b. a median in $\triangle ABC$
   c. an angle bisector in $\triangle ABC$

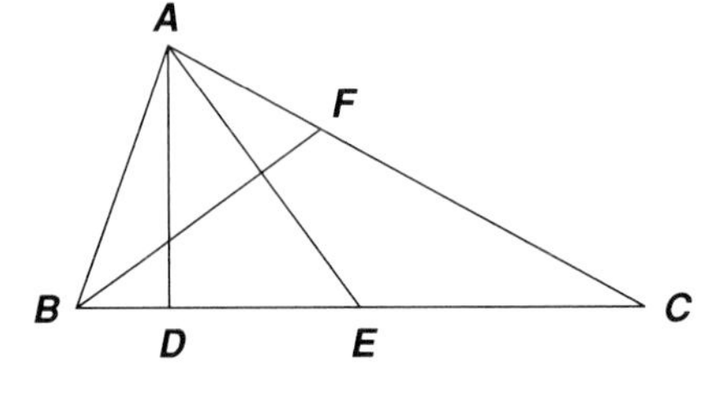

2.

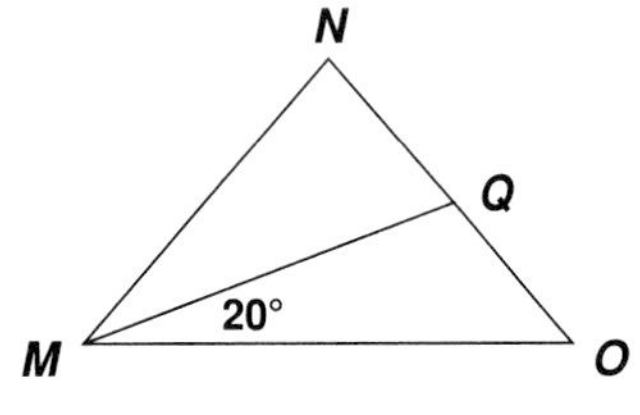

In isosceles $\triangle MNO$, $MN = ON$ and $\overline{QM}$ bisects $\angle NMO$.
If $m\angle QMO = 20°$, find:

a. $m\angle NMO$
b. $m\angle NOM$
c. $m\angle N$

Congruent Triangles

Just by looking at them, what comment would you make about $\triangle ABC$ and $\triangle A'B'C'$?

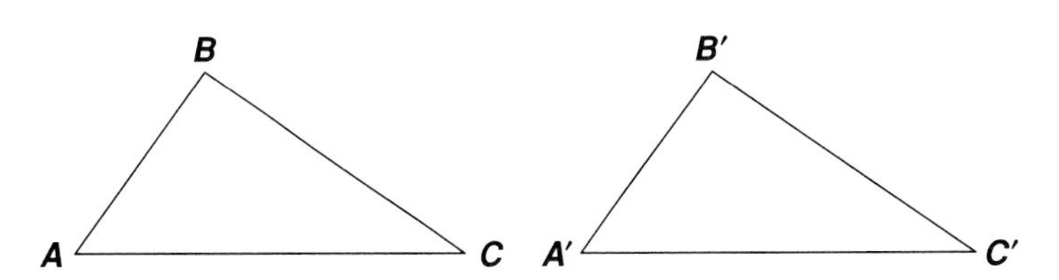

Congruent triangles have the same size and the same shape.

The measures of the angles of a triangle give the triangle its shape. In these congruent triangles, tell which angles are equal in measure.

$\angle A \cong \angle A'$ $\qquad$ $\angle B \cong \angle B'$ $\qquad$ $\angle C \cong \angle C'$

Pairs of angles of congruent triangles are equal in measure. These pairs are called ***corresponding angles*** of the triangles.

Notice that the corresponding angles of the two triangles are located opposite the sides that are equal in measure. These pairs of sides are called ***corresponding sides*** of the triangles.

$\overline{AB} \cong \overline{A'B'}$ $\qquad$ $\overline{BC} \cong \overline{B'C'}$ $\qquad$ $\overline{AC} \cong \overline{A'C'}$

Example 4 In each pair of congruent triangles, name the pairs of corresponding angles and corresponding sides.

Tell which angles lie opposite which sides.

a.

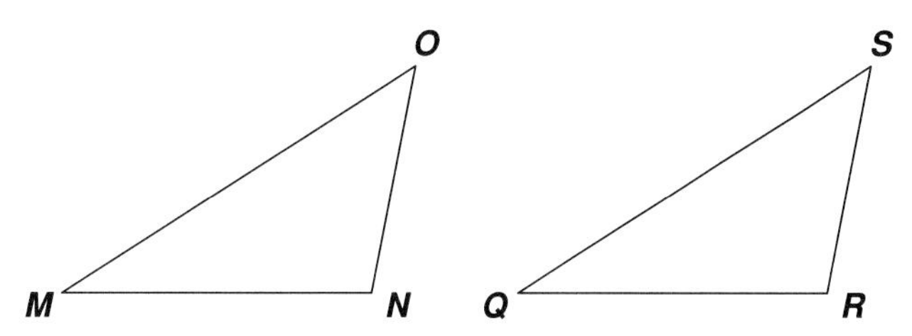

The corresponding angles are: $\angle M \cong \angle Q$ $\angle N \cong \angle R$ $\angle O \cong \angle S$

The corresponding sides are: $\overline{ON} \cong \overline{SR}$ $\overline{OM} \cong \overline{SQ}$ $\overline{MN} \cong \overline{QR}$

$\angle M$ is opposite $\overline{ON}$ and $\angle Q$ is opposite $\overline{SR}$.

$\angle N$ is opposite $\overline{OM}$ and $\angle R$ is opposite $\overline{SQ}$.

$\angle O$ is opposite $\overline{MN}$ and $\angle S$ is opposite $\overline{QR}$.

b.

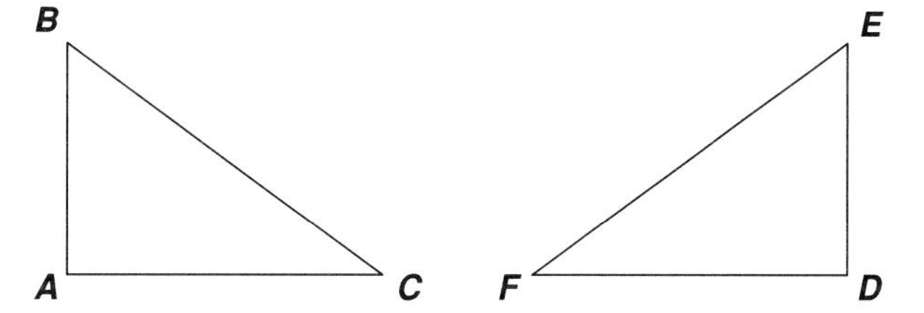

The corresponding angles are: $\angle A \cong \angle D$ $\angle B \cong \angle E$ $\angle C \cong \angle F$

The corresponding sides are: $\overline{BC} \cong \overline{EF}$ $\overline{AC} \cong \overline{DF}$ $\overline{AB} \cong \overline{DE}$

$\angle A$ is opposite $\overline{BC}$ and $\angle D$ is opposite $\overline{EF}$.

$\angle B$ is opposite $\overline{AC}$ and $\angle E$ is opposite $\overline{DF}$.

$\angle C$ is opposite $\overline{AB}$ and $\angle F$ is opposite $\overline{DE}$.

A convenient way to show corresponding sides and angles of congruent triangles is to use matching marks on the figures.

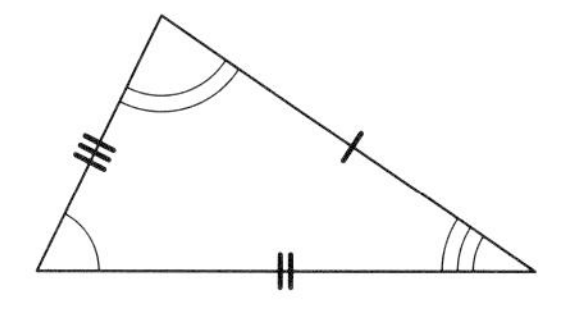

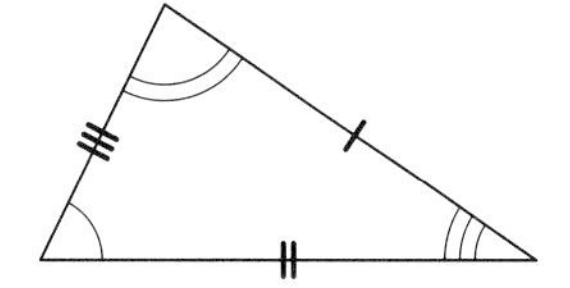

When Are Triangles Congruent?

Cut strips of paper that are the exact lengths of these three line segments.

Fit the strips together to form a triangle. Keep this triangle.

Cut three more strips of paper that are the exact lengths of these same line segments. Fit these strips together to form a triangle. What appears to be true about the two triangles? Draw a conclusion about what makes two triangles congruent.

Two triangles are congruent if the three sides of one triangle are equal in measure to the three sides of the other triangle.

In symbols, where *s* represents a side:

If $sss \cong sss$, then the triangles are congruent.

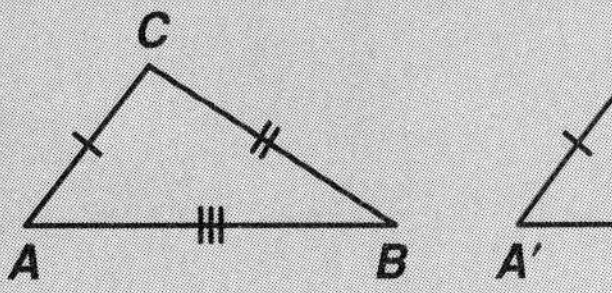

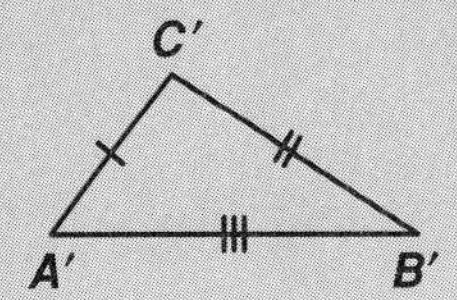

If $\overline{AC} \cong \overline{A'C'}$, $\overline{CB} \cong \overline{C'B'}$, $\overline{BA} \cong \overline{B'A'}$, then $\triangle ABC \cong \triangle A'B'C'$.

Do you think two triangles must be congruent if their three pairs of angles are equal in measure?

These two triangles that have three pairs of angles equal in measure have the same shape, but since they do not have the same size, they are not congruent.

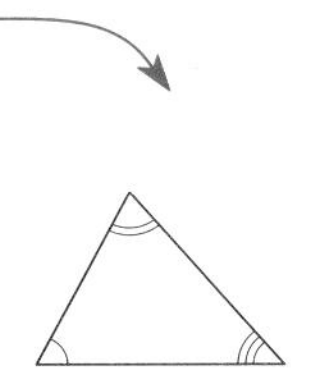

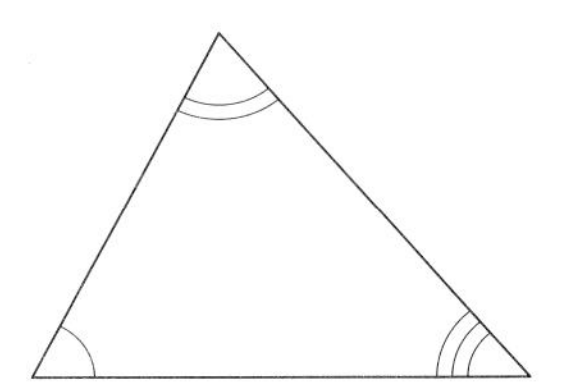

Will combinations of congruent sides and congruent angles make congruent triangles? Let's see.

Example 5 With a ruler, draw a line segment 2 inches long. Call it $\overline{AB}$. With a protractor, use A as the vertex and draw an angle of 40°. With a protractor, use B as the vertex and draw an angle of 60°. Extend the sides of the angles you drew until they intersect at C.

Start over with a new 2-inch line segment, $\overline{MN}$. With a protractor, use M as the vertex and draw an angle of 40°. With a protractor, use N as the vertex and draw an angle of 60°. Extend the sides of the angles you drew until they intersect at O.

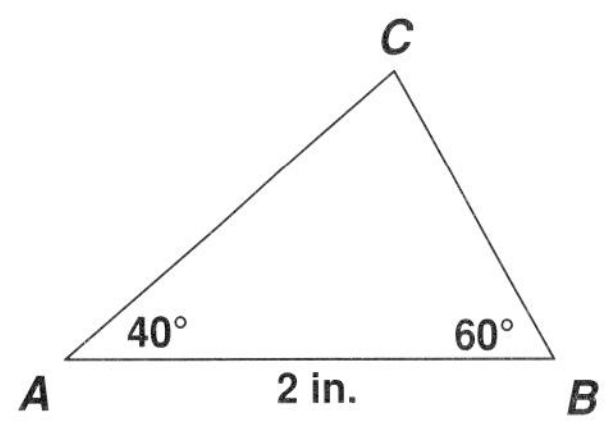

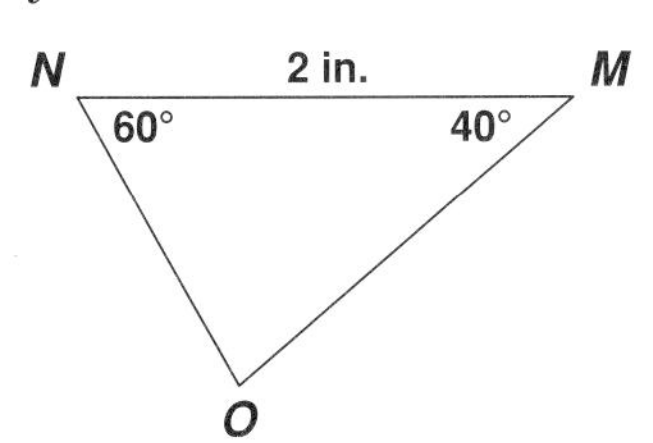

What appears to be true about $\triangle ABC$ and $\triangle MNO$? Verify your conclusion by using a protractor to measure $\angle C$ and $\angle O$, and a ruler to measure the 3 pairs of sides of the triangles.

From this example, what conditions can you say will make two triangles congruent?

Two triangles are congruent if two angles and the included side of one triangle are equal in measure to two angles and the included side of the other triangle.

In symbols, where *s* represents a side and *a* represents an angle:

If *asa* ≅ *asa*, then the triangles are congruent.

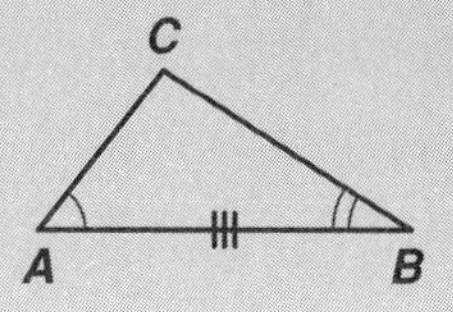

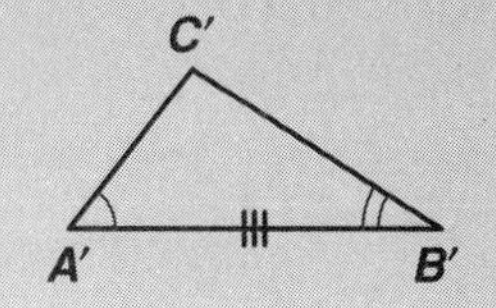

If $\angle A \cong \angle A'$, $\overline{AB} \cong \overline{A'B'}$, $\angle B \cong \angle B'$, then $\triangle ABC \cong \triangle A'B'C'$.

Another combination of congruent sides and angles that works is two angles and a side opposite one of them.

If *aas* ≅ *aas*, then the triangles are congruent.

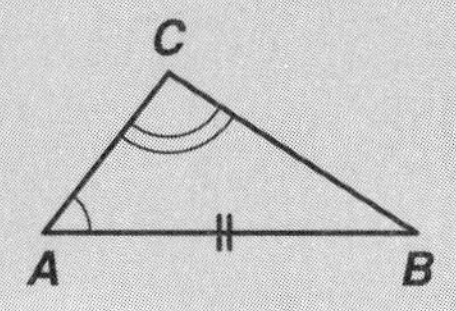

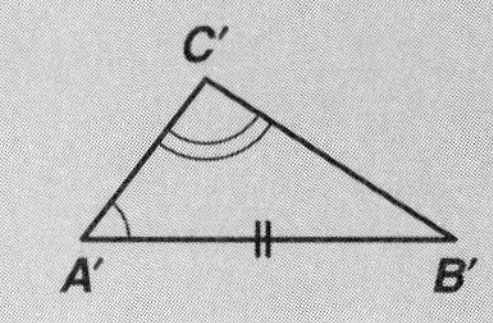

If $\angle A \cong \angle A'$, $\angle C \cong \angle C'$, $\overline{AB} \cong \overline{A'B'}$, then $\triangle ABC \cong \triangle A'B'C'$.

The combinations of sides and angles that remain deal with two pairs of sides. One of these combinations always works:

If *sas* ≅ *sas*, then the triangles are congruent.

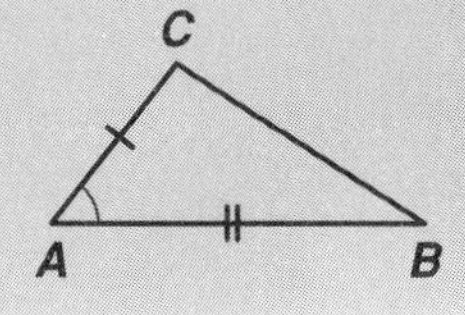

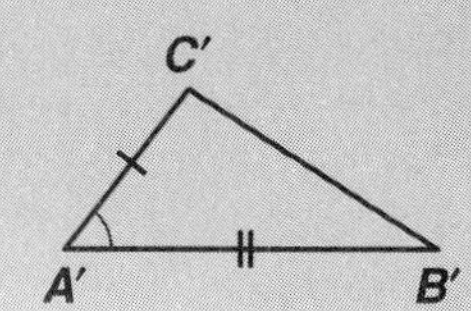

If $\overline{CA} \cong \overline{C'A'}$, $\angle A \cong \angle A'$, $\overline{BA} \cong \overline{B'A'}$, then $\triangle ABC \cong \triangle A'B'C'$.

Two triangles are congruent if these pairs of corresponding parts are equal in measure:

sss	3 sides
sas	2 sides and the included angle
asa	2 angles and the included side
aas	2 angles and the side opposite one angle

Example 6 From the markings on the pairs of triangles, tell why the triangles are congruent.

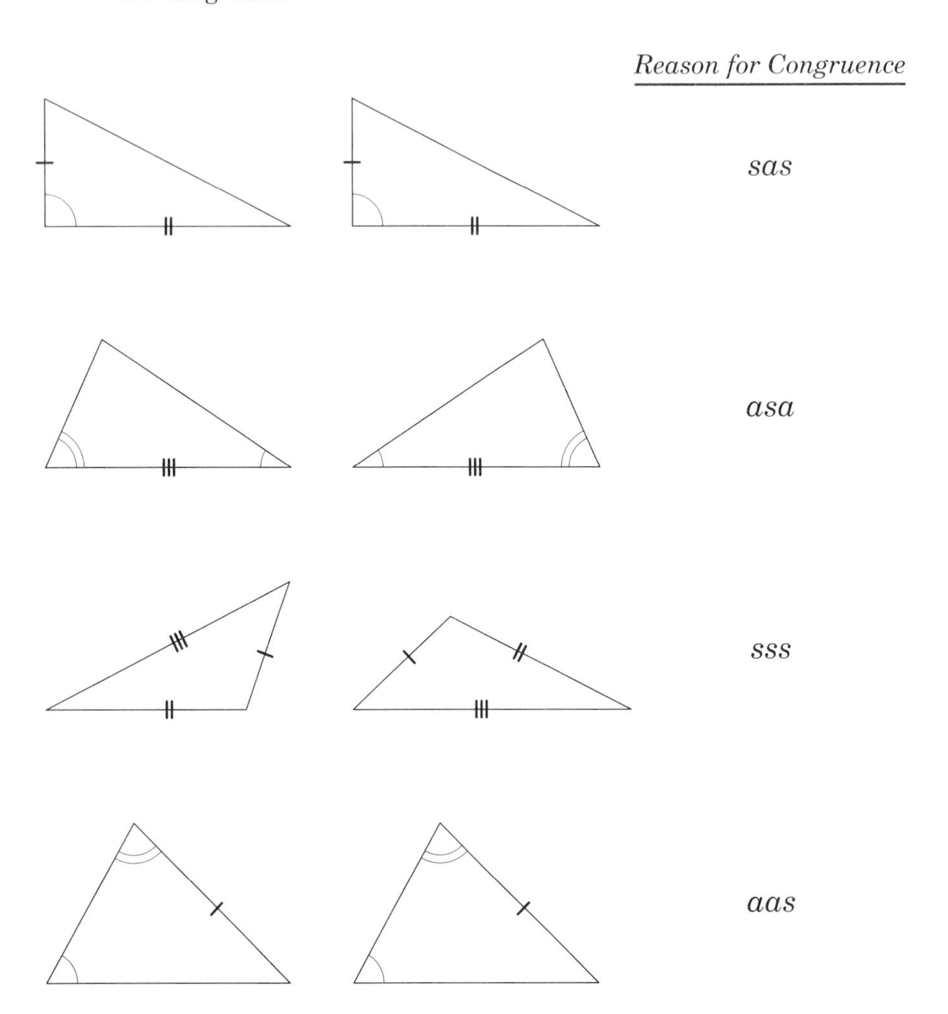

Try These (For solutions, see page 350.)

1. In the diagram, $\triangle ABC \cong \triangle RST$ with corresponding parts marked as shown.

   a. Which angle of $\triangle RST$ corresponds to $\angle C$ of $\triangle ABC$?

   b. Name another pair of corresponding angles.

   c. Which side of $\triangle ABC$ corresponds to side $\overline{RS}$ of $\triangle RST$? Name another pair of corresponding sides.

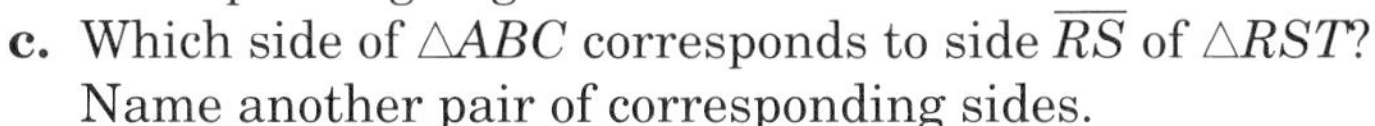

2. From the marking on the pairs of triangles, tell why triangles I and II are congruent.

a.

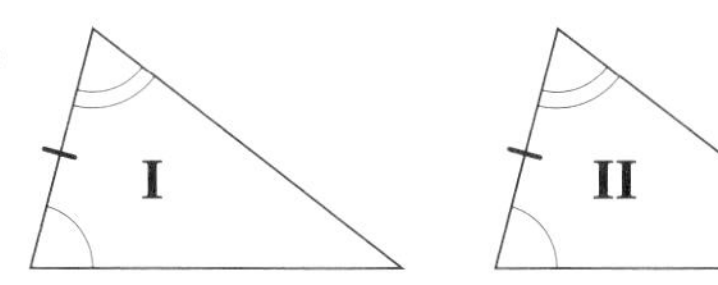

b.

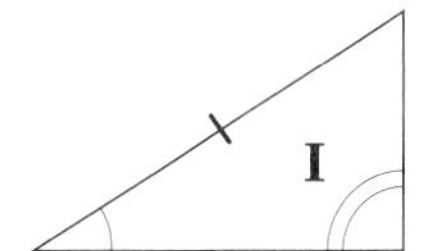

c.

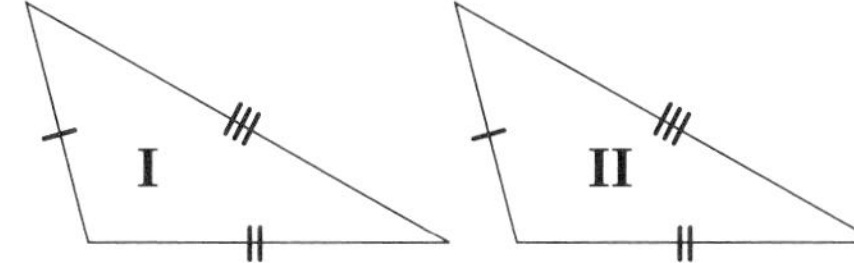

d.

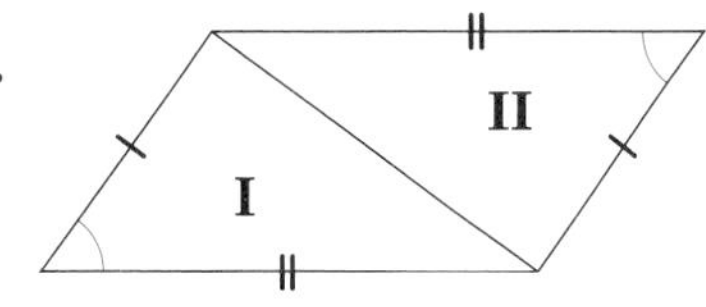

SELF-CHECK: SECTION 8.5

1. An altitude of a triangle is __?__ to the opposite side.

   Answer: perpendicular

2. A median of a triangle connects a vertex to the __?__ of the opposite side.

   Answer: midpoint

3. The line segment that divides an angle of a triangle into two congruent angles is called an __?__.

   Answer: angle bisector of the triangle

4. Triangles that have the same size and shape are called __?__.

   Answer: congruent

5. In two congruent triangles, name parts that are equal in measure.

   Answer: 3 pairs of angles and 3 pairs of sides

6. To know that two triangles are congruent, how many pairs of parts must be equal in measure? Name these parts and their order.

   Answer: 3 pairs of parts: *sss*, *sas*, *asa*, *aas*

EXERCISES: SECTION 8.5

1. Cut out a large acute triangle.
Mark the 3 vertices A, B, C, as shown.

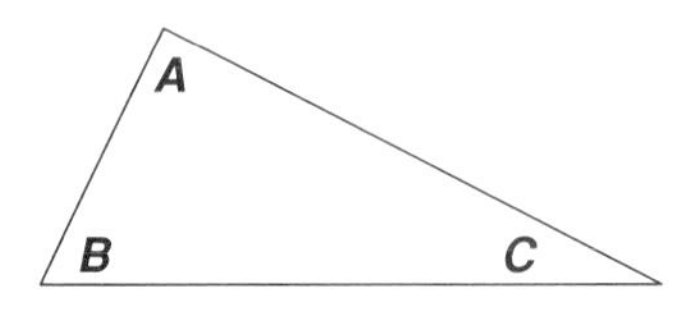

a. To draw the median from A to $\overline{BC}$, you need to find the midpoint of $\overline{BC}$. Fold B to C to find M, the midpoint of $\overline{BC}$. Explain how to draw the median.

b. If you fold side $\overline{AB}$ of the triangle onto side $\overline{AC}$, what special line of the triangle is along this fold?

c. If you fold A to some point E on $\overline{BC}$ so that $\angle AEB$ is a right angle, what special line of the triangle is along this fold?

2. Find each of the following segments in the diagram. Use a protractor and ruler to verify your answers.

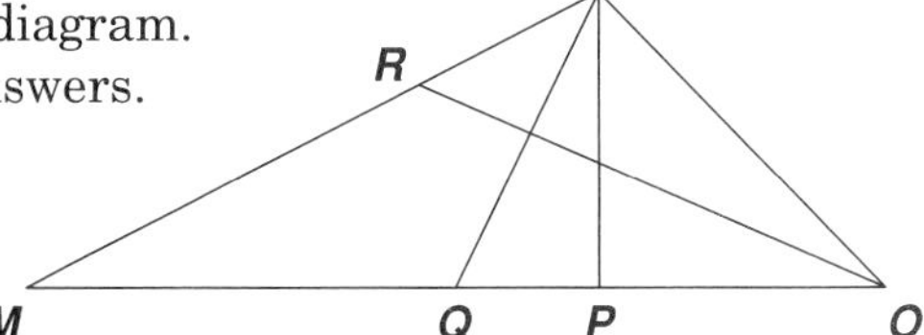

a. an altitude in $\triangle MNO$

b. a median in $\triangle MNO$

c. an angle bisector in $\triangle MNO$

3. Draw a large isosceles triangle ABC, with $AB = AC$.

a. Use a protractor and ruler to draw altitude $\overline{AH}$ from A to $\overline{BC}$ and median $\overline{AM}$ from A to $\overline{BC}$.

b. What do you observe about the altitude and median from A to $\overline{BC}$ in this isosceles triangle?

c. Do you think this observation will hold true for the altitude and median from B to $\overline{AC}$? Verify your prediction by using a protractor and ruler to draw these lines.

d. What do you predict about the altitude and median from C to $\overline{AB}$? Verify your prediction by using a protractor and ruler to draw these lines.

e. Summarize your findings about the altitudes and medians in an isosceles triangle.

f. What do you predict about the altitudes and medians in an equilateral triangle? Explain. Use a protractor and ruler to verify your conclusions.

4. a. Draw a large isosceles triangle ABC.

(1) Fold $\triangle ABC$ so that vertex B falls on vertex C, then unfold the paper. What appears to be true about the 2 triangles created?

(2) The fold in the paper is a ***line of symmetry***, since the figure on one side of the fold is the same as on the other. Do you think there are other lines of symmetry in isosceles triangle ABC? Fold the paper to verify your conclusions.

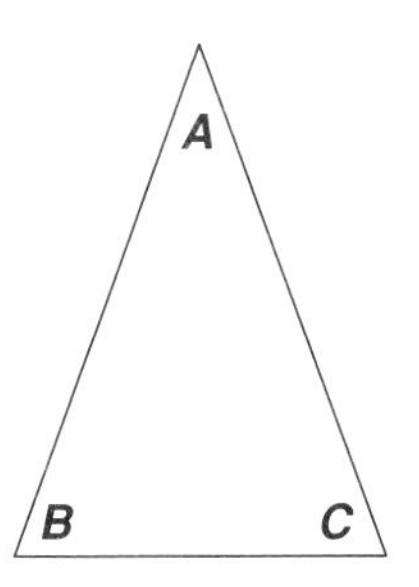

b. Draw a large equilateral triangle *DEF*.

(1) Fold $\triangle DEF$ so that vertex *E* falls on vertex *F*, then unfold the paper. What appears to be true about the 2 triangles created?

(2) Do you think there are other lines of symmetry in equilateral triangle *DEF*? Fold the paper to verify your conclusions.

c. Summarize your findings regarding the number of lines of symmetry in an isosceles triangle and in an equilateral triangle.

d. How many lines of symmetry would you expect to find in a rectangle? in a square? Fold paper figures to verify your conclusions.

5. **a.** Draw a large right $\triangle ABC$, with the right angle at *C*.

b. Use a ruler and protractor to draw the altitude from *C* to $\overline{AB}$.

c. Try to draw the altitude from *A* to $\overline{BC}$. What do you notice?

d. Try to draw the altitude from *B* to $\overline{AC}$. What do you notice?

e. Are there 3 altitudes in a right triangle? Explain.

f. Do the 3 altitudes of a right triangle meet at a point? Where is that point?

6. **a.** Draw a large obtuse $\triangle ABC$, with the obtuse angle at *C*.

b. Use a ruler and protractor to draw altitude $\overline{CH}$ from *C* to $\overline{AB}$.

c. Draw the altitude $\overline{AJ}$ from *A* to $\overline{CB}$. What must you do to $\overline{CB}$ so that the altitude intersects it?

d. Draw the altitude $\overline{BK}$ from *B* to $\overline{AC}$. What must you do to $\overline{AC}$ so that the altitude intersects it?

e. Here is what your diagram should look like so far.

f. If you extend altitude $\overline{HC}$ through *C*, altitude $\overline{BK}$ through *K*, and altitude $\overline{AJ}$ through *J*, what do you think will happen to the 3 altitudes? Describe the location of the point at which the 3 altitudes of this obtuse triangle meet.

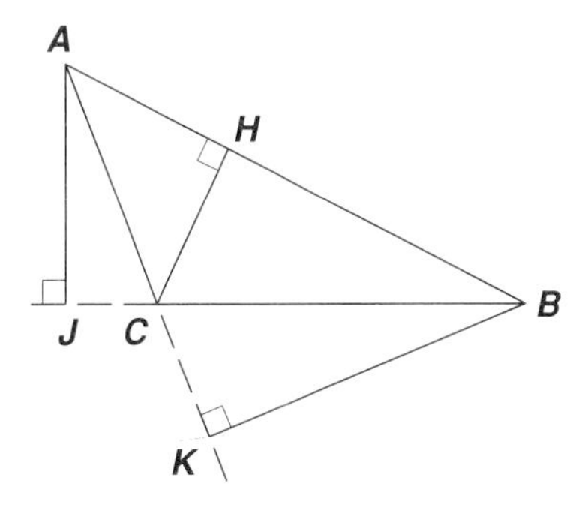

7. Draw a diagram to help you see what each situation is about. Use an equation to solve.

a. $\overline{AD}$ is a median in scalene $\triangle ABC$. If $BD = y + 7$ and $CD = 2y - 1$, find the length of $\overline{BC}$.

b. $\overline{PQ}$ is an angle bisector in scalene $\triangle MNP$. If $m\angle MPQ = x + 4$ and $m\angle NPQ = 2x - 24$, find $m\angle MPN$.

c. $\overline{JH}$ is an altitude in scalene $\triangle JKL$. If $m\angle JHL = 6z + 24$, find the value of *z*.

8. In isosceles $\triangle ABC$, $\overline{BQ}$ bisects $\angle ABC$ and $\overline{CQ}$ bisects $\angle ACB$, to form isosceles $\triangle BQC$. If $m\angle ABQ = 25°$, find:

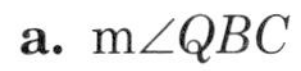

a. $m\angle QBC$ **b.** $m\angle QCB$

c. $m\angle Q$ **d.** $m\angle A$

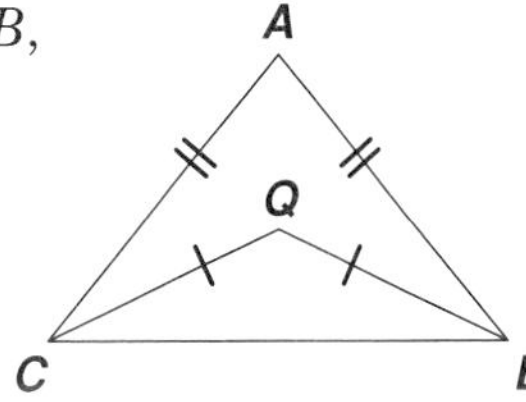

9. 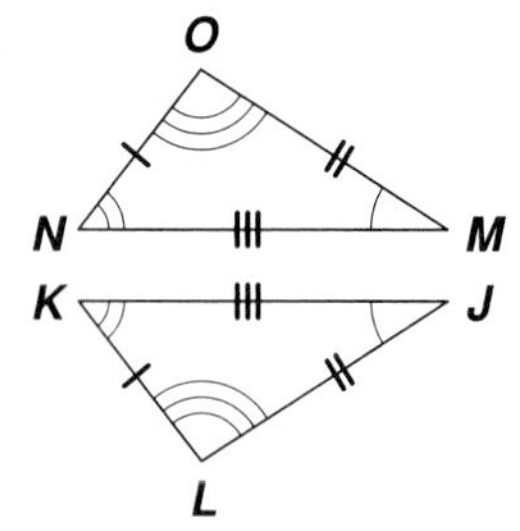

In the diagram, $\triangle MNO \cong \triangle JKL$ with corresponding parts marked as shown.

a. Which angle of $\triangle MNO$ corresponds to $\angle K$ of $\triangle JKL$?

b. Which side of $\triangle JKL$ corresponds to $\overline{OM}$ of $\triangle MNO$?

c. If $m\angle O = 80°$ and $m\angle J = 40°$, find $m\angle N$.

10. From the markings on the diagrams, tell why triangles I and II are congruent.

a. 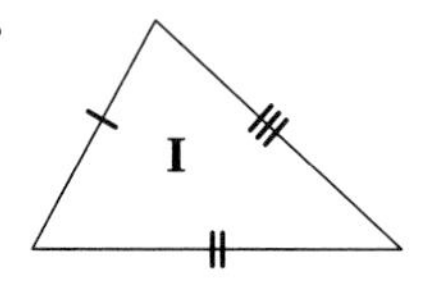

II

b. 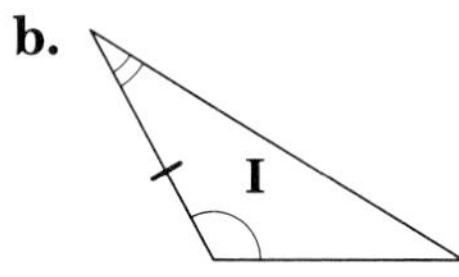

II

c. 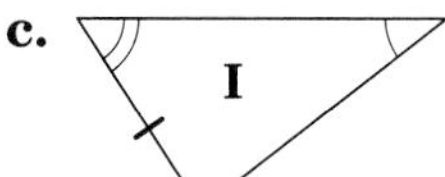

II

d. 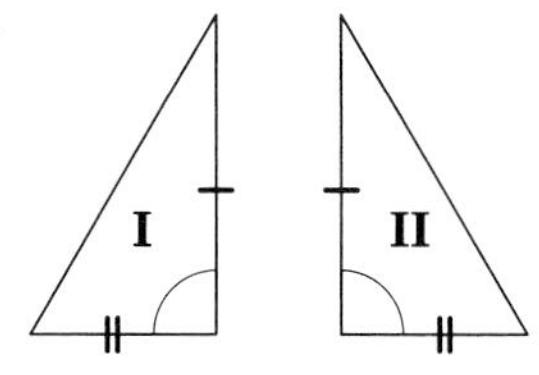

11. In each case, according to the markings shown, which pair of triangles is not necessarily congruent? Explain.

a. 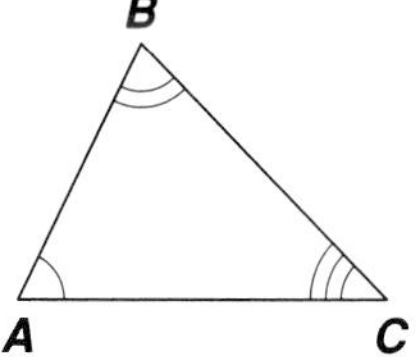

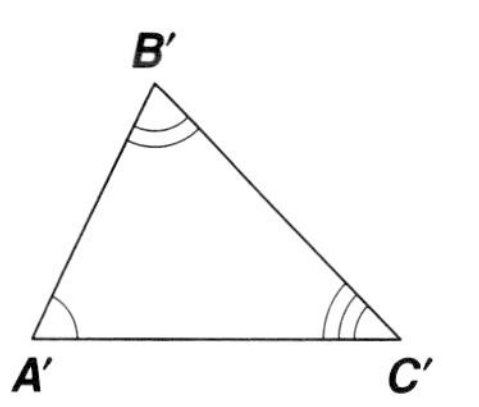

or

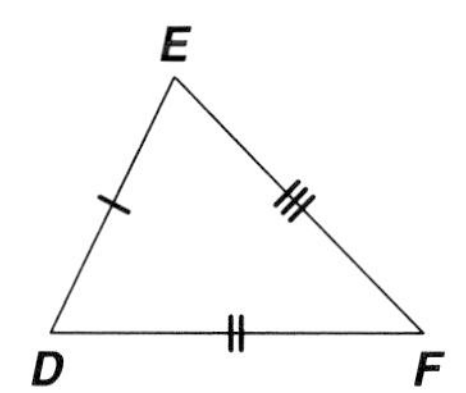

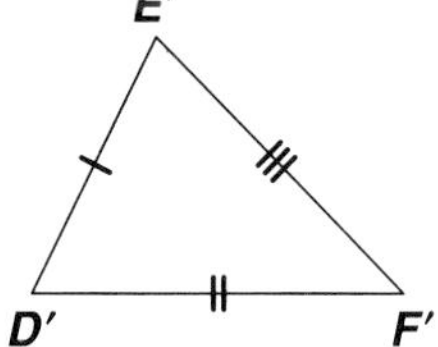

b. 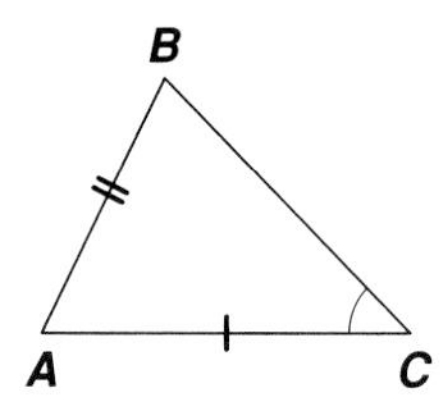

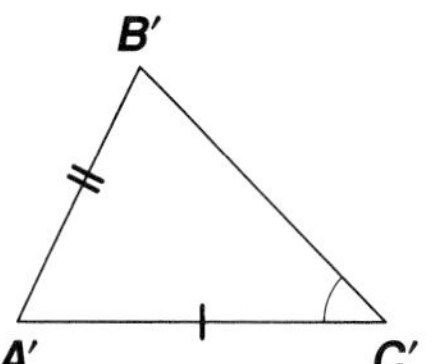

or

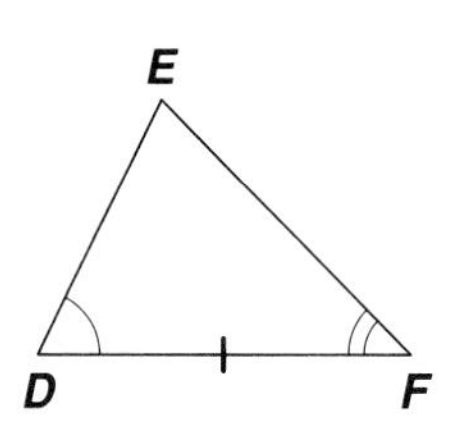

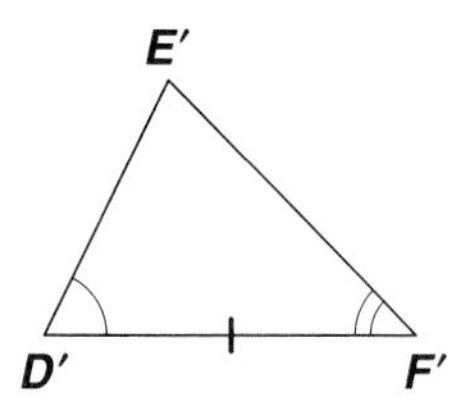

c. 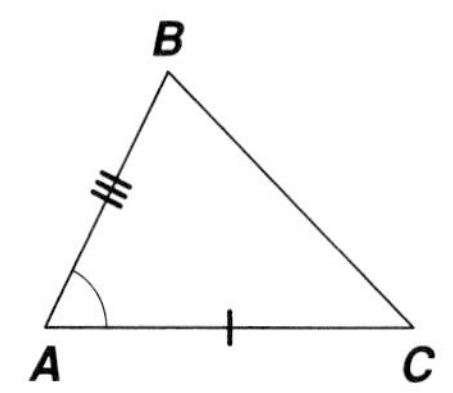

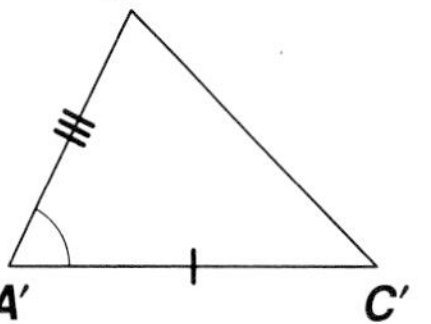

or

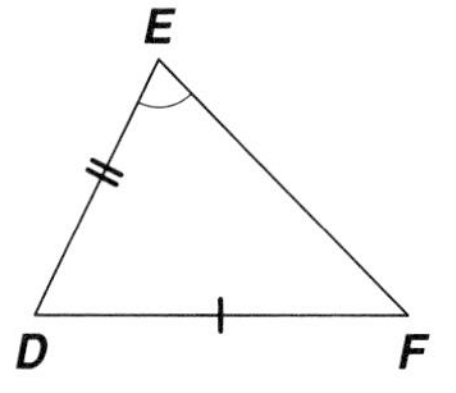

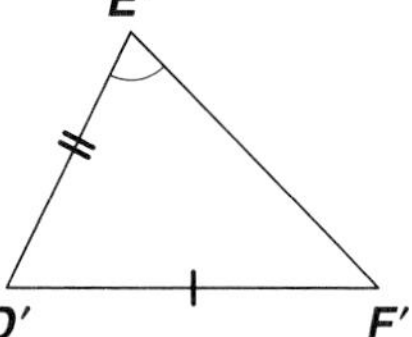

12. Determine from the information in the diagrams if each pair of triangles must be congruent. If the triangles must be congruent, name the method of congruence.

a.

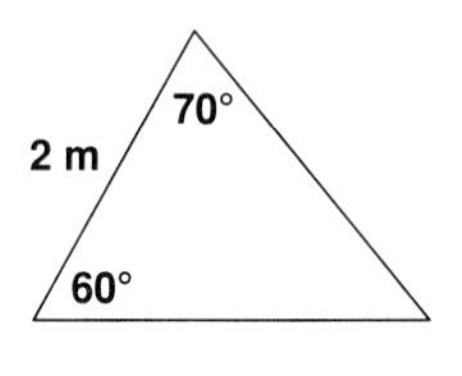

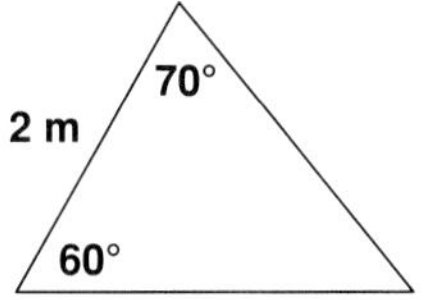

b.

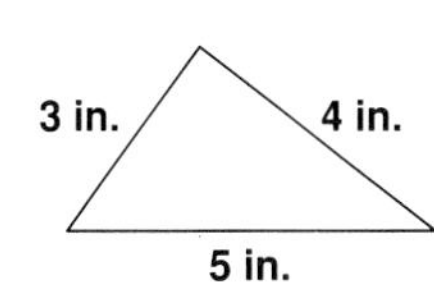

4 in. 3 in. 5 in.

c.

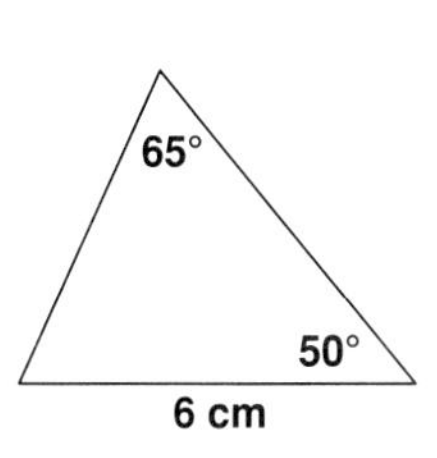

50° 6 cm 65°

d.

2 ft. 3 ft. 60° 30°

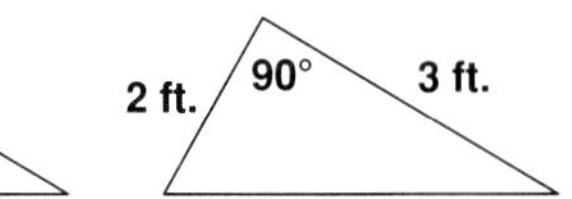

e.

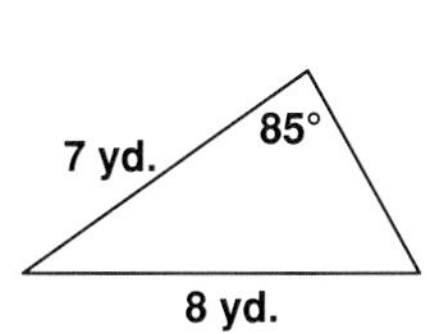

8 yd. 85° 7 yd.

f.

63° 55° 5 m

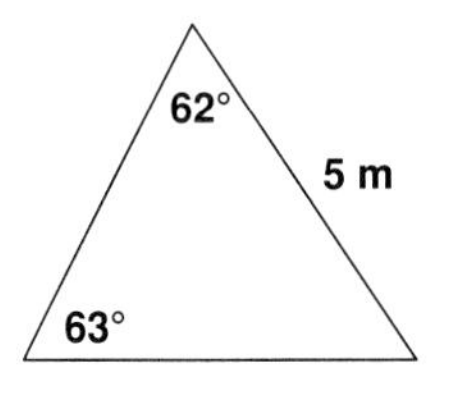

13. Each diagram shows a pair of triangles that can be shown to be congruent. In each case, only two pairs of parts are given as congruent. For each case:

- Name a third pair of parts that you know must be congruent. Explain.
- Name the method that makes the triangles congruent.

a.

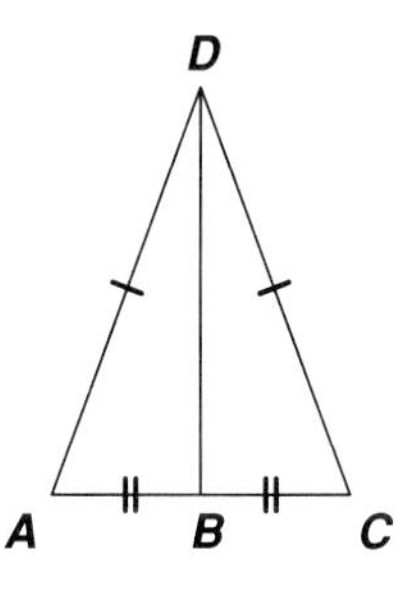

b.

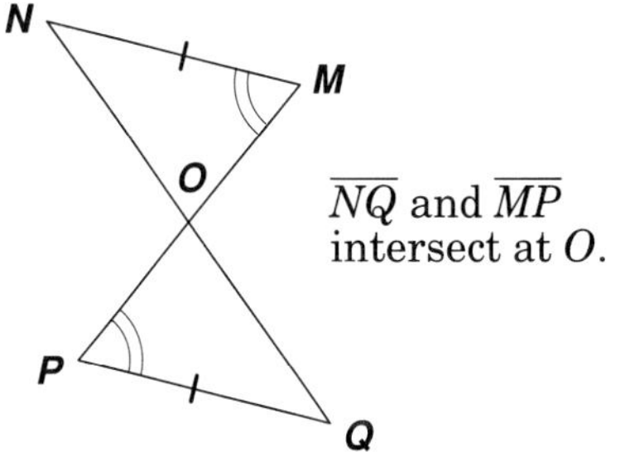

$\overline{NQ}$ and $\overline{MP}$ intersect at O.

c.

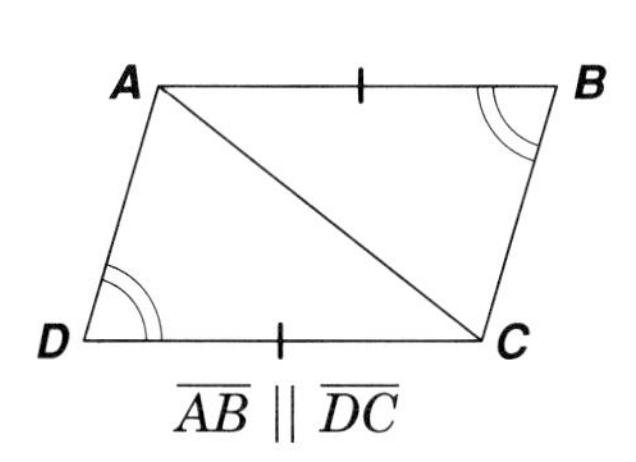

$\overline{AB} \parallel \overline{DC}$

d.

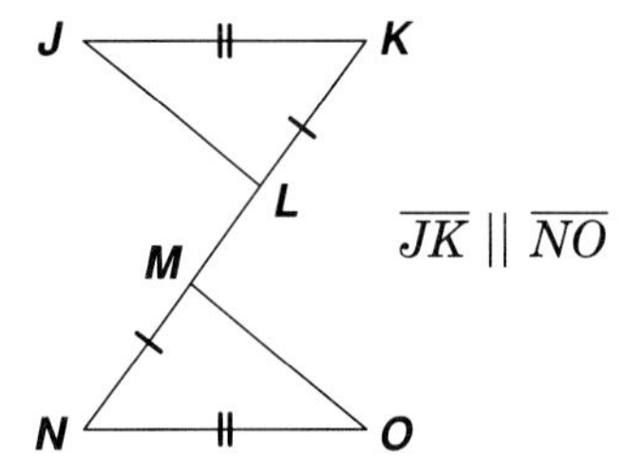

$\overline{JK} \parallel \overline{NO}$

EXPLORATIONS

1. With a compass and straightedge, you have seen how to copy the measure of a line segment (see Exploration 2, page 292) and the measure of an angle (see Exploration 3, page 293). Using these basic constructions:

a. Construct a triangle congruent to $\triangle ABC$ by copying, in order, $\angle A$, $\overline{AB}$, and $\angle B$.

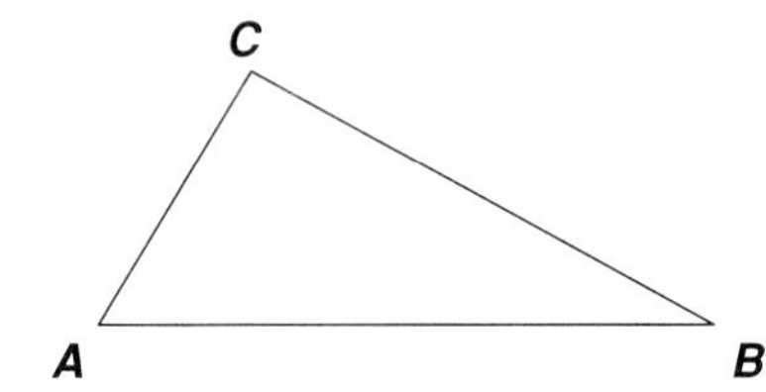

b. Construct a triangle congruent to $\triangle MNO$ by copying, in order, $\overline{ON}$, $\angle N$, and $\overline{NM}$.

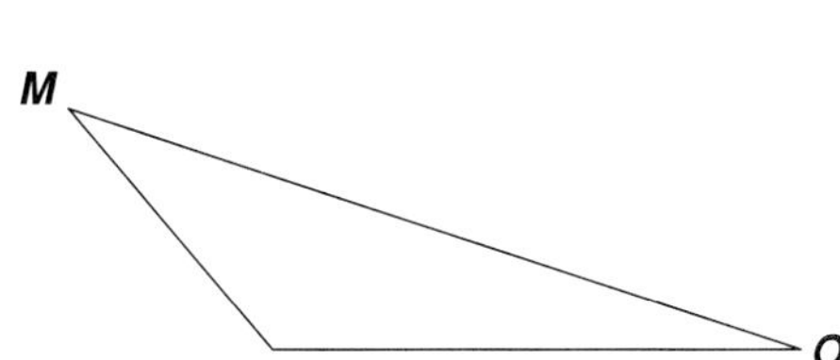

c. Construct a triangle congruent to $\triangle JKL$ by copying the 3 sides. To see how to construct a triangle using side lengths, refer to the construction of an equilateral triangle (see Exploration 3 on page 328).

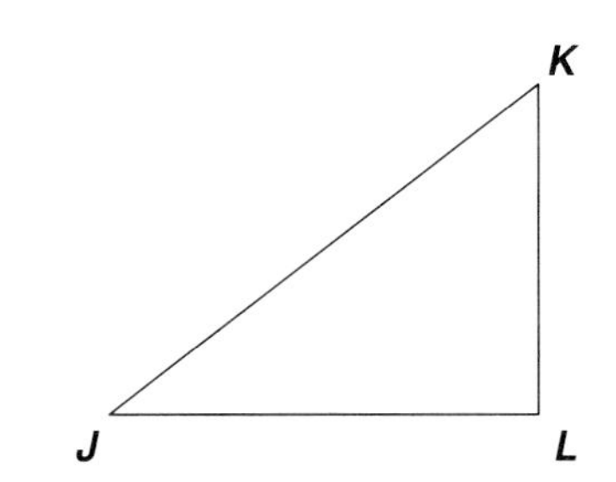

2. To construct a median of a triangle, what point must you first find?

(Refer to Exploration 4 on page 294).

Copy $\triangle ABC$, and construct the median from A to $\overline{BC}$.

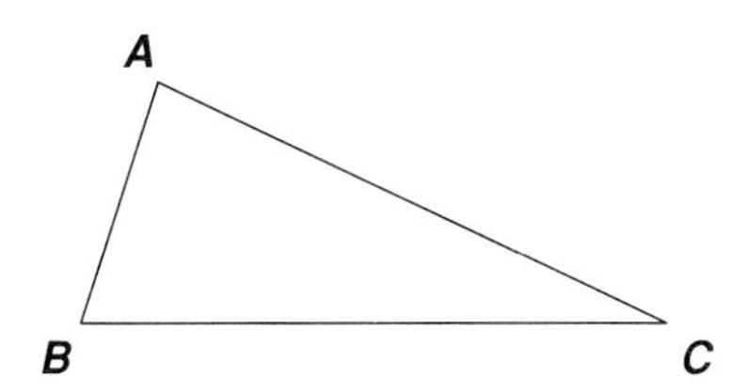

3. a. Follow these steps to construct an altitude from A to $\overline{BC}$ of $\triangle ABC$.

① Open the compass to a setting that is slightly greater than the distance from A to $\overline{BC}$.

Using this setting and A as the center, draw an arc below $\overline{BC}$ that intersects $\overline{BC}$ in 2 places, P and Q.

② Open the compass to a setting that is more than half the length of $\overline{PQ}$.

Using this setting and P as the center, draw an arc below $\overline{BC}$.

With the same setting and Q as the center, draw an arc below $\overline{BC}$.

These arcs intersect.

③ With a straightedge, draw a ray from A through the point where the arcs intersect.

Use T to label the point where this ray crosses $\overline{BC}$.

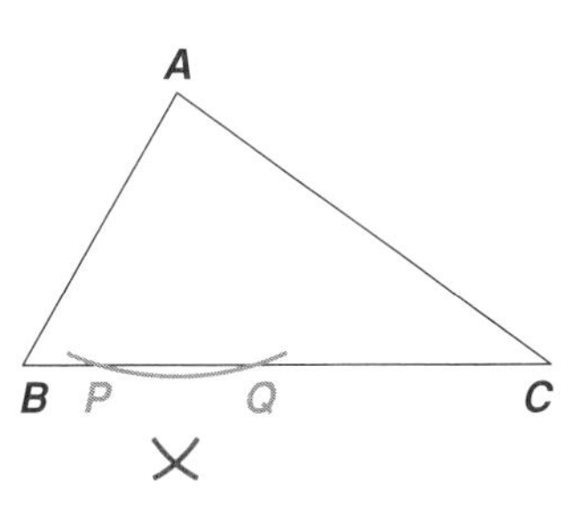

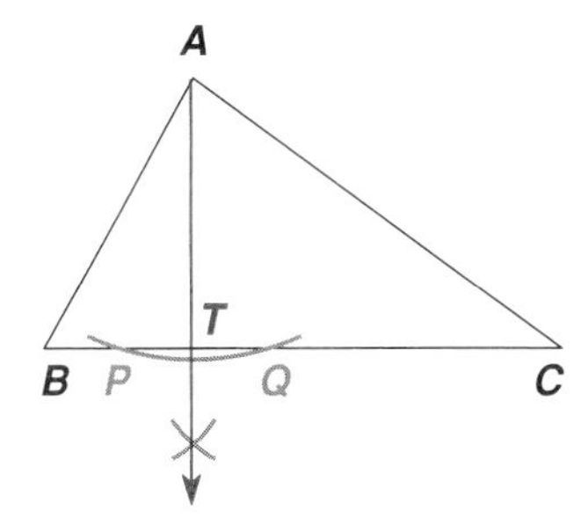

Result: $\overline{AT} \perp \overline{BC}$
$\overline{AT}$ is an altitude.

b. Practice this construction by copying each triangle and constructing the altitude from P to $\overline{SR}$.

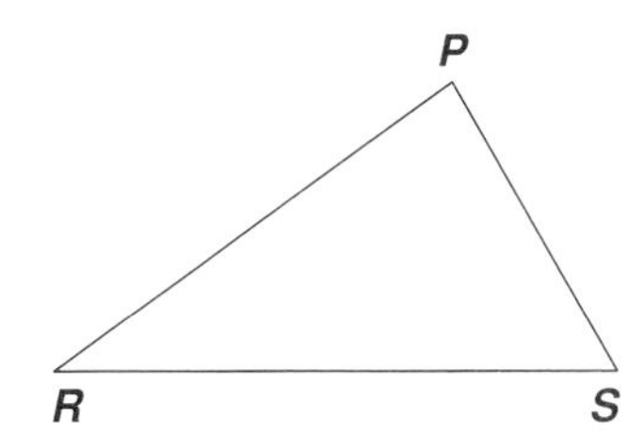

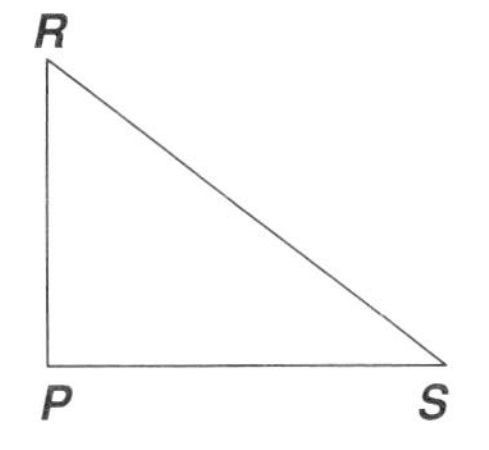

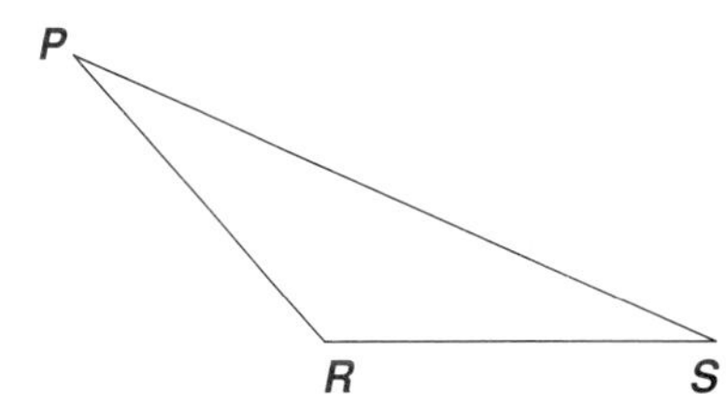

First extend $\overline{SR}$ through R.

Things You Should Know After Studying This Chapter

KEY SKILLS

8.1 Use symbols to name lines, line segments, rays, and angles.
Find the length of a line segment by using an equation.
Find the measure of an angle by using a protractor and by using an equation.
Identify an angle as acute, right, obtuse, or straight.

8.2 Identify perpendicular and parallel lines.
Identify alternate interior angles and corresponding angles formed by two lines and a transversal.
Find the measures of angles formed by two parallel lines and a transversal.

8.3 Identify angle pairs: vertical, adjacent, complementary, and supplementary.
Use an equation to find the measures of angle pairs.

8.4 Apply the properties of isosceles and equilateral triangles.
Find the measures of angles in a triangle.
Find the perimeter of a triangle or of any polygon.
Write rules for the perimeters of special polygons.
Apply the triangle inequality.

8.5 Identify the altitudes, medians, and angle bisectors of a triangle.
Identify the corresponding sides and angles of congruent triangles.
Apply *sss*, *sas*, *asa*, or *aas* to determine if two triangles are congruent.

KEY TERMS AND SYMBOLS

8.1 point (P) • line ($\overleftrightarrow{AB}$) • line segment ($\overline{AB}$) • ray ($\overrightarrow{AB}$) • end point • midpoint • angle (vertex, side, size) • degree • protractor • classes of angles (acute, right, obtuse, straight) • congruent figures

8.2 perpendicular ($\perp$) • perpendicular bisector • parallel ($\parallel$) • transversal • angles formed by two lines and a transversal (alternate interior, corresponding)

8.3 angle pairs (vertical, adjacent, complementary, supplementary)

8.4 triangle ($\triangle$) • scalene • isosceles (legs, base) • equilateral/equiangular • perimeter

8.5 altitude • angle bisector • median • congruent triangles (*sss*, *sas*, *asa*, *aas*) • line of symmetry

CHAPTER REVIEW EXERCISES

1. Find each of the following in the diagram, and name each in mathematical notation.

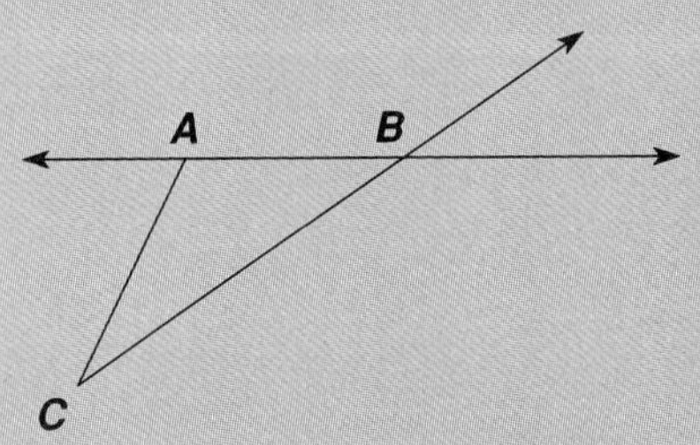

a. a line **b.** a line segment
c. a ray **d.** an acute angle
e. a triangle **f.** an obtuse angle

2. On a number line, point A has the coordinate -5 and point B has the coordinate -1.

a. Draw a diagram.
b. Count units to find the distance AB.
c. Write an absolute-value expression to find the distance AB.

3. C 5x – 7 M 3x + 11 D

M is the midpoint of $\overline{CD}$.
If $CM = 5x - 7$ and $DM = 3x + 11$, find:

a. the value of x
b. the lengths of $\overline{CM}$ and $\overline{DM}$
c. the length of $\overline{CD}$

4. By observation, name the angle that has $\overrightarrow{AB}$ as one side and whose measure is about:

a. 80° **b.** 120° **c.** 35° **d.** 160°

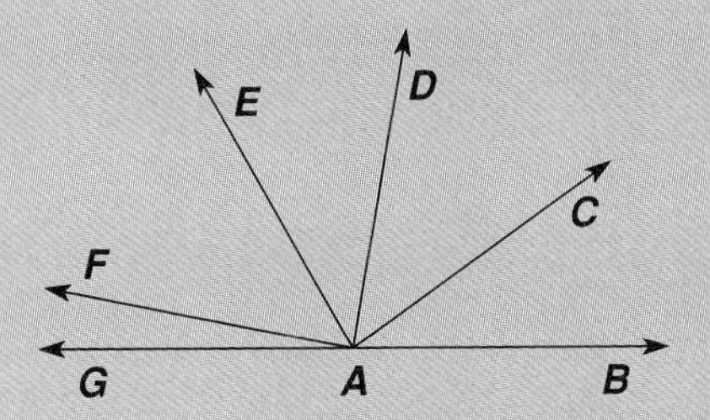

5. 6x + 10 A 4x + 50 B

If $\angle A \cong \angle B$ and their degree measures are as represented in the diagram, use an equation to find their degree measures.

6. According to the markings in the diagram, which of the following statements is true?

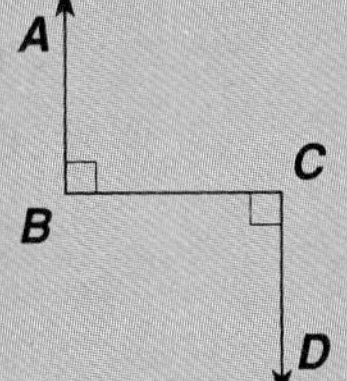

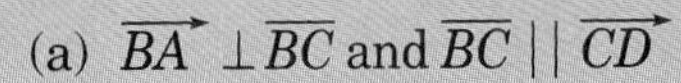

(a) $\overrightarrow{BA} \perp \overline{BC}$ and $\overline{BC} \parallel \overrightarrow{CD}$
(b) $\overrightarrow{BA} \parallel \overline{BC}$ and $\overline{BC} \perp \overrightarrow{CD}$
(c) $\overrightarrow{BA} \perp \overline{BC}$ and $\overline{BC} \perp \overrightarrow{CD}$
(d) $\overrightarrow{BA} \parallel \overline{BC}$ and $\overline{BC} \parallel \overrightarrow{CD}$

7.

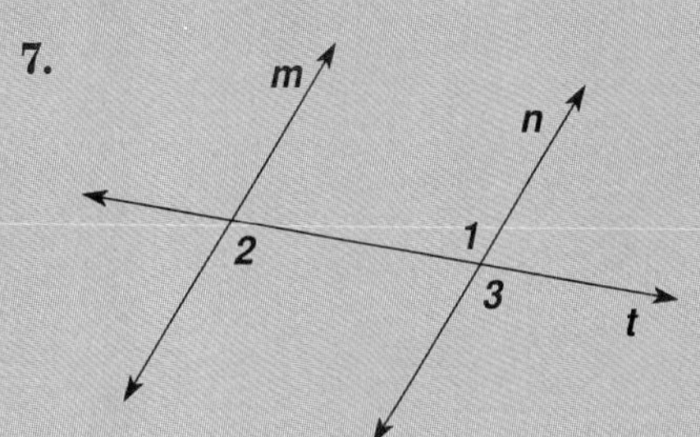

Parallel lines m and n are cut by transversal t. If $m\angle 1 = 112°$, find the measures of angles 2 and 3. Give reasons for your answers.

8. Transversal $\overleftrightarrow{EF}$ intersects parallel lines $\overleftrightarrow{AB}$ and $\overleftrightarrow{CD}$ in points G and H respectively.

a. Draw a diagram.
b. If $m\angle AGH = 4x + 3$ and $m\angle GHD = 6x - 33$, find the number of degrees in $\angle AGH$.

9. If $m\angle R = 40°$, what is the measure of the complement of $\angle R$?

CHAPTER REVIEW EXERCISES

10. If m$\angle A = 6x - 18$ and m$\angle B = 3x + 27$, find the number of degrees of each angle if $\angle A$ and $\angle B$ are:

a. vertical angles

b. complementary angles

c. supplementary angles

11. In isosceles $\triangle PQR$, $\overline{PQ} \cong \overline{PR}$. If m$\angle P = 100°$, find m$\angle Q$.

12. In $\triangle ABC$, m$\angle A = y + 30$, m$\angle B = 2y + 5$, and m$\angle C = 3y - 65$. Find the degree measure of each angle.

13. In an isosceles triangle, the measure of each of the congruent legs is 4.5 cm and the measure of the base is 1.7 cm. Find the perimeter of the triangle.

14. In equilateral $\triangle DEF$, $DE = 6x - 6$ and $EF = 2x + 10$. Find the perimeter of the triangle.

15. In a triangle, the sum of the lengths of any two sides is __?__ the length of the third side.

(a) equal to (b) less than
(c) greater than (d) unrelated to

16. Which of these sets of values could be the lengths of the sides of a triangle?

(a) 11, 5, 5 (b) 11, 11, 5
(c) 16, 11, 5 (d) 11, 5, 4

17. If $\overline{MP}$ is an altitude in $\triangle MNO$, which statement must be true?

(a) $\overline{MP} \perp \overline{MO}$ (b) $\overline{MP}$ bisects $\overline{NO}$
(c) $\overline{MP} \perp \overline{NO}$ (d) $\overline{MP}$ bisects $\angle M$

18. $\overline{AD}$ is a median in $\triangle ABC$. If $BD = x + 5$ and $DC = 2x - 2$, find:

a. the value of x

b. the length of $\overline{BD}$

c. the length of $\overline{BC}$

19. From the information in each diagram, tell why triangles I and II are congruent.

a.

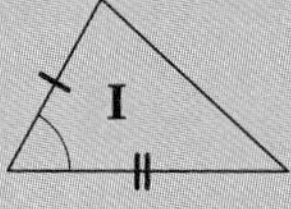

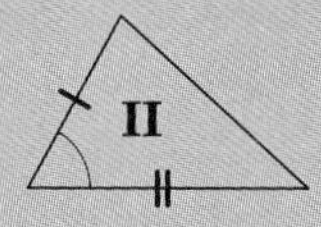

b.

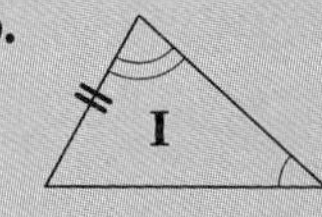

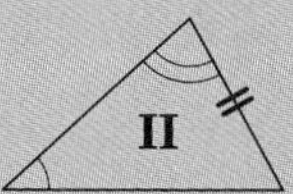

c.

70°
I
80°
5 cm

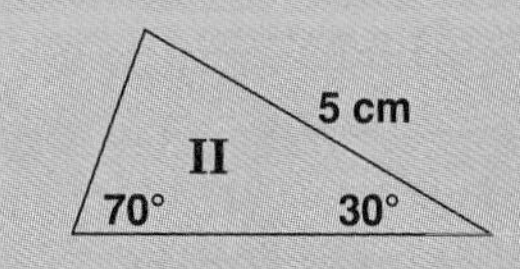

d.

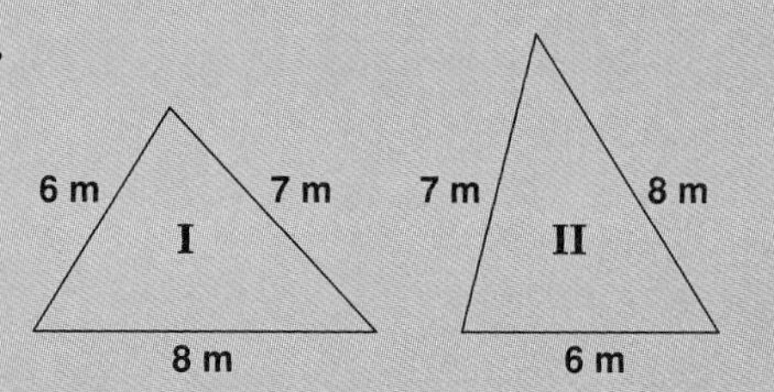

SOLUTIONS FOR TRY THESE

8.1 Basic Geometric Figures

Page 280

1. line segment AB

2. line CD

Page 282

1. a.

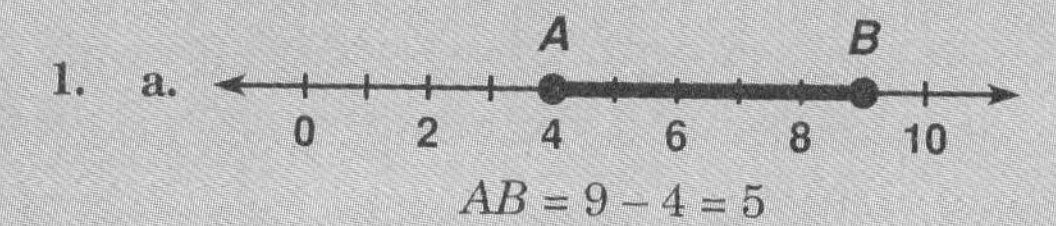

$AB = 9 - 4 = 5$

b.

R S
–4 –2 0 2 4 6

$RS = 4 - (-3) = 4 + 3 = 7$

2. a.

$$AM = MB$$
$$4x - 2 = 3x + 1$$

b.

$$4x - 2 = 3x + 1$$
$$4x - 3x - 2 = 3x - 3x + 1 \quad \text{Subtract 3x from both sides.}$$
$$x - 2 = 1$$
$$x - 2 + 2 = 1 + 2 \quad \text{Add 2 to both sides.}$$
$$x = 3$$

c.

$$AM = 4x - 2 = 4(3) - 2 = 12 - 2 = 10$$
$$MB = 3x + 1 = 3(3) + 1 = 9 + 1 = 10$$
$$AB = AM + MB = 10 + 10 = 20$$

Page 282

1. $\angle M, \angle T, \angle C, \angle R, \angle MYC, \angle TYR$

2. There is more than one angle with vertex at point Y.

3. $\angle MYC, \angle CYR, \angle RYT, \angle TYM, \angle MYR, \angle TYC,$
$\angle MYT, \angle CYR, \angle TYC, \angle MYR$

Page 284

1. $m\angle RST = 90°$

2. $m\angle UVW = 25°$

Page 285

1. $\angle ABC$ is obtuse.

2. $\angle DEF$ is acute.

3. $\angle MNO$ is a right angle.

Page 286

$$m\angle ABC = m\angle DEF$$
$$2x + 5 = x + 25$$
$$2x - x + 5 = x - x + 25 \quad \text{Subtract } x \text{ from both sides.}$$
$$x + 5 = 25$$
$$x + 5 - 5 = 25 - 5 \quad \text{Subtract 5 from both sides.}$$
$$x = 20$$
$$m\angle ABC = 2x + 5 = 2(20) + 5 \quad \text{Substitute } x = 20.$$
$$= 40 + 5 = 45$$
$$m\angle DEF = x + 25 = 20 + 25 = 45$$

Check: $m\angle ABC = m\angle DEF = 45°$ ✔

8.2 Pairs of Lines

Page 297

1. $\overline{AB}$ does not go through the midpoint of $\overline{RS}$.
$\overline{KL}$ does go through the midpoint of $\overline{CD}$.
$\overline{KL}$ is the perpendicular bisector of $\overline{CD}$.

2.

$$m\angle PQR = 90 \quad \text{Perpendiculars meet at right angles.}$$
$$5x - 10 = 90$$
$$5x - 10 + 10 = 90 + 10 \quad \text{Add 10 to both sides.}$$
$$5x = 100$$
$$\frac{5x}{5} = \frac{100}{5} \quad \text{Divide both sides by 5.}$$
$$x = 20$$

Check: If $x = 20$, does $5x - 10$ equal 90?

$$5x - 10 \stackrel{?}{=} 90$$
$$5(20) - 10 \stackrel{?}{=} 90$$
$$100 - 10 \stackrel{?}{=} 90$$
$$90 = 90 \checkmark$$

Answer: $x = 20$

SOLUTIONS FOR TRY THESE

Page 302

1. $\angle x$ and $\angle y$ are alternate interior angles of the parallel lines.

$$m\angle x = m\angle y = 110°$$

$\angle z$ and $\angle y$ are corresponding angles of the parallel lines.

$$m\angle z = m\angle y = 110°$$

2.
$$
\begin{aligned}
m\angle AGH &= m\angle DHG && \text{Alternate interior angles of parallel lines are equal in measure.}\\
8x - 20 &= 4x + 44 \\
8x - 4x - 20 &= 4x - 4x + 44 && \text{Subtract } 4x \text{ from both sides.}\\
4x - 20 &= 44 \\
4x - 20 + 20 &= 44 + 20 && \text{Add 20 to both sides.}\\
4x &= 64 \\
\frac{4x}{4} &= \frac{64}{4} && \text{Divide both sides by 4.}\\
x &= 16
\end{aligned}
$$

$$m\angle AGH = 8x - 20 = 8(16) - 20 = 128 - 20 = 108$$
$$m\angle DHG = 4x + 44 = 4(16) + 44 = 64 + 44 = 108$$

Check: If $x = 16$, are the two angles equal in measure?

$$108 = 108 \checkmark$$

Answer: $m\angle AGH = m\angle DHG = 108°$

8.3 Pairs of Angles

Page 310

1. Vertical angles are formed by intersecting lines. When $\overleftrightarrow{BD}$ and $\overleftrightarrow{EC}$ intersect, vertical angles $\angle BOC$ and $\angle EOD$ are formed.

2.
$$
\begin{aligned}
m\angle MNV &= m\angle TNS && \text{Vertical angles are congruent.}\\
3x - 48 &= x + 22 \\
3x - x - 48 &= x - x + 22 && \text{Subtract } x \text{ from both sides.}\\
2x - 48 &= 22 \\
2x - 48 + 48 &= 22 + 48 && \text{Add 48 to both sides.}\\
2x &= 70 \\
\frac{2x}{2} &= \frac{70}{2} && \text{Divide both sides by 2.}\\
x &= 35
\end{aligned}
$$

$$
\begin{aligned}
m\angle MNV &= 3x - 48 = 3(35) - 48 && \text{Substitute } x = 35.\\
&= 105 - 48 = 57 \\
m\angle TNS &= x + 22 = 35 + 22 \\
&= 57
\end{aligned}
$$

Check: If $x = 35$, are the measures of the $\angle$s equal?

$$57 = 57 \checkmark$$

Answer: $m\angle MNV = m\angle TNS = 57°$

Page 311

1. adjacent angles: $\angle 1$ and $\angle 2$, $\angle 1$ and $\angle 4$, $\angle 2$ and $\angle 3$, $\angle 4$ and $\angle 3$

   nonadjacent angles: $\angle 1$ and $\angle 3$, $\angle 2$ and $\angle 4$

2. Angles 1 and 2 are nonadjacent because they do not share the same vertex.

Page 312

1. **a.** $90° - 20°$, or $70°$ **b.** $90° - 70°$, or $20°$
   c. $90° - 45°$, or $45°$ **d.** $90° - x°$

2. Let x = the measure of the smaller angle.
   Then $x + 10$ = the measure of the larger angle.

$$
\begin{aligned}
x + x + 10 &= 90 && \text{The angles are complementary.}\\
2x + 10 &= 90 \\
2x + 10 - 10 &= 90 - 10 && \text{Subtract 10 from both sides.}\\
2x &= 80 \\
\frac{2x}{2} &= \frac{80}{2} && \text{Divide both sides by 2.}\\
x &= 40 \\
x + 10 &= 40 + 10 = 50 && \text{Substitute } x = 40.
\end{aligned}
$$

Check: If the measures of the angles are 40° and 50°, are the angles complementary?

$$40 + 50 = 90 \checkmark$$

Answer: The measures of the angles are 40° and 50°.

Page 313

1. **a.** $180° - 20°$, or $160°$ **b.** $180° - 160°$, or $20°$
   c. $180° - 90°$, or $90°$ **d.** $180° - x°$

2. Let x = the measure of the larger angle.
   Then $x - 40$ = the measure of the smaller angle.

$$
\begin{aligned}
x + x - 40 &= 180 && \text{The angles are supplementary.}\\
2x - 40 &= 180 \\
2x - 40 + 40 &= 180 + 40 && \text{Add 40 to both sides.}\\
2x &= 220 \\
\frac{2x}{2} &= \frac{220}{2} && \text{Divide both sides by 2.}\\
x &= 110 \\
x - 40 &= 110 - 40 = 70 && \text{Substitute } x = 110.
\end{aligned}
$$

Check: If the measures of the angles are 110° and 70°, are the angles supplementary?

$$110 + 70 = 180 \checkmark$$

Answer: The measures of the angles are 110° and 70°.

SOLUTIONS FOR TRY THESE

8.4 *Triangles*

Page 320

1.

	Name	Explanation
a.	scalene	No sides equal are in measure.
b.	isosceles	Two sides are equal in measure.
c.	equilateral	All three sides are equal in measure.
d.	equilateral	All three angles equal in measure means that the three sides are equal in measure.
e.	isosceles	Two angles equal in measure means that two sides are equal in measure.
f.	scalene	No angles equal in measure means that no sides are equal in measure.

2.

$AB = AC = BC$ — Sides of equilateral △.

$3x - 2 = 16$ — Substitute given values.

$3x - 2 + 2 = 16 + 2$ — Add 2 to both sides.

$3x = 18$

$\frac{3x}{3} = \frac{18}{3}$ — Divide both sides by 3.

$x = 6$

Check: If $x = 6$, does $AB = 16$?

$3(6) - 2 \stackrel{?}{=} 16$

$18 - 2 \stackrel{?}{=} 16$

$16 = 16$ ✔

Answer: $x = 6$

Page 322

1. The sum of the measures of the 3 angles of any triangle is 180°.

a. $m\angle A + m\angle B + m\angle C = 180$

$50 + 35 + m\angle C = 180$

$85 + m\angle C = 180$

$85 - 85 + m\angle C = 180 - 85$

$m\angle C = 95°$

b. Let $x = m\angle M$.
Then $x = m\angle N$.

$m\angle M + m\angle N + m\angle O = 180$

$x + x + 46 = 180$

$2x + 46 = 180$

$2x + 46 - 46 = 180 - 46$

$2x = 134$

$\frac{2x}{2} = \frac{134}{2}$

$x = 67°$

Answer: $m\angle M = m\angle N = 67°$

2. a.

b. If the measure of each of the equal base angles is x, then:

$x + x + 90 = 180$

$2x = 90$

$x = 45°$

Page 323

1. In this isosceles triangle, the measure of each of the congruent legs is 2(base) or 2(6) or 12 inches.
The perimeter of this isosceles triangle is 12 + 12 + 6, or 30 inches.

2. In an equilateral triangle, the 3 sides are equal in measure.
Since the perimeter is 27 cm, the length of each side is 27 ÷ 3, or 9 cm.

Page 324

To be a △, the sum of the measures of any 2 sides must be greater than the measure of the 3rd side.

1. $3 + 4 \stackrel{?}{>} 6$ ✔
$3 + 6 \stackrel{?}{>} 4$ ✔
$4 + 6 \stackrel{?}{>} 3$ ✔
Answer: Yes, a △.

2. $2 + 4 \stackrel{?}{>} 9$ no
Answer: no △

3. $1 + 2 \stackrel{?}{>} 3$ no
Answer: no △

SOLUTIONS FOR TRY THESE

8.5 More About Triangles

Page 332

1. In $\triangle ABC$: **a.** $\overline{AD}$ is an altitude **b.** $\overline{AE}$ is a median
   c. $\overline{BF}$ is an angle bisector
2. **a.** $m\angle NMO = 2(m\angle QMO)$ — $\overline{QM}$ is an angle bisector.
   $= 2(20)$
   $= 40°$

   b. $m\angle NOM = m\angle NMO$ — Base angles of an isosceles triangle are equal in measure.
   $= 40°$

   c. $m\angle NMO + m\angle NOM + m\angle N = 180$ — Sum of the measures of the 3 angles of a triangle = 180°.
   $40 + 40 + m\angle N = 180$
   $80 + m\angle N = 180$
   $m\angle N = 100°$

Page 337

1. **a.** $\angle C$ and $\angle T$ are corresponding angles.
   b. $\angle A$ corresponds to $\angle R$, and $\angle B$ corresponds to $\angle S$.
   c. $\overline{RS}$ and $\overline{AB}$ are corresponding sides.
   d. $\overline{RT}$ corresponds to $\overline{AC}$, and $\overline{ST}$ corresponds to $\overline{BC}$.
2. **a.** *asa* makes $\triangle\text{I} \cong \triangle\text{II}$.
   b. *aas* makes $\triangle\text{I} \cong \triangle\text{II}$.
   c. *sss* makes $\triangle\text{I} \cong \triangle\text{II}$.
   d. *sas* makes $\triangle\text{I} \cong \triangle\text{II}$.

Geometric Measures 9

ACTIVITY

Using unlined paper, ruler, and scissors, make 4 figures of Shape *A* and 2 of Shape *B*, with the dimensions shown.

Form a quadrilateral using 2 of shape *A*.

Form a quadrilateral using 2 of shape *B*.

Form a quadrilateral using 4 of shape *A*.

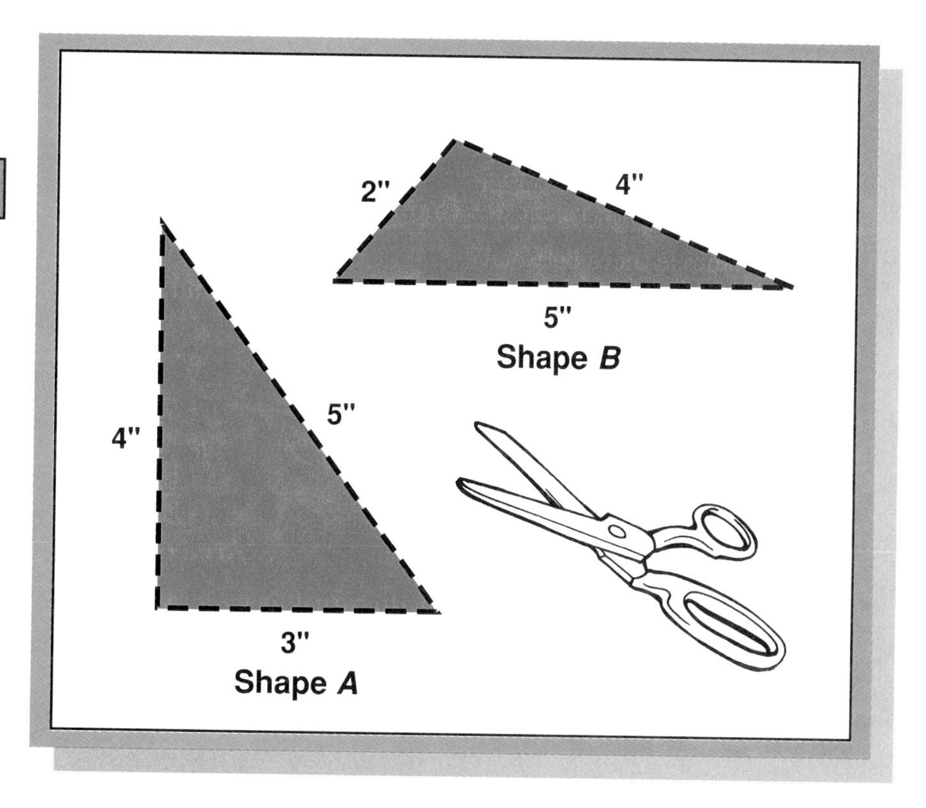

9.1 Area

Rectangle

How many 1" by 1" squares will fit inside a 3" by 5" rectangle?

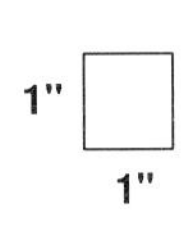

A 1" by 1" square, called 1 square inch.

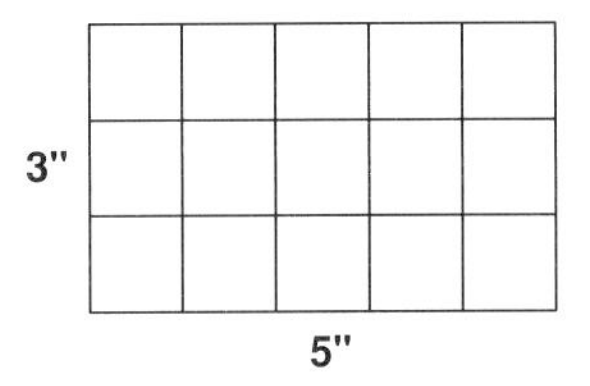

A 3" by 5" rectangle contains 15 square inches.

The number of square inches that can fit inside a figure is the *area* of the figure.

What relationship is there between
the dimensions of the rectangle, 3 inches by 5 inches, and
the area of the rectangle, 15 square inches?

3 inches × 5 inches = 15 square inches

Write a rule to find the area of a rectangle whose base measures b inches and whose height measures h inches.

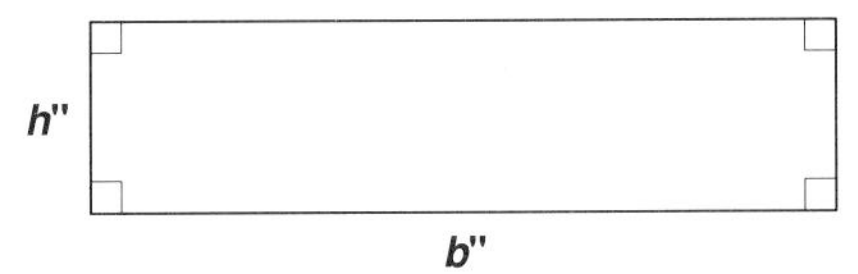

Area of Rectangle = $b \times h$ square inches

Example 1 Mrs. Prah is planning to carpet Jason's room, which is a rectangular shape of length 14 feet and width 12 feet.

a. How many square feet are in the area of Jason's room?

Apply the rule for finding the area of a rectangle.

14 feet × 12 feet = 168 square feet

There are 168 square feet in the area of Jason's room.

b. Carpeting is sold by the square yard. How many square yards are there in the area of Jason's room?

3 feet = 1 yard
9 square feet = 1 square yard

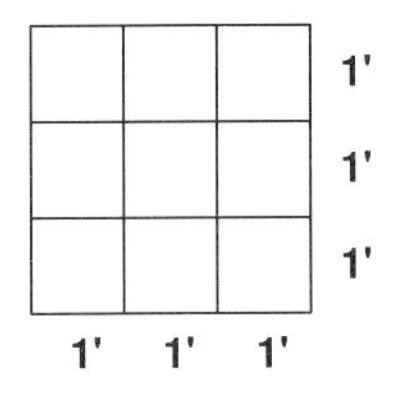

A square yard contains 9 square feet.

To change from square feet to square yards, multiply by the conversion fraction $\frac{1 \text{ square yard}}{9 \text{ square feet}}$, whose value is 1.

$$\overset{56}{\cancel{168}} \cancel{\text{square feet}} \times \frac{1 \text{ square yard}}{\underset{3}{\cancel{9}} \cancel{\text{square feet}}} = \frac{56}{3} \text{ or } 18\tfrac{2}{3} \text{ square yards}$$

There are $18\frac{2}{3}$ square yards in the area of Jason's room.

c. The number of square yards of carpeting that must be purchased depends on the width of the roll on which the carpet comes.

If the carpet that Jason chose comes on rolls that are 12 feet wide, and lengths must be purchased in full yards, how many square yards of carpet must Mrs. Prah buy?

How much extra carpet will there be?

Draw a diagram to model the problem.

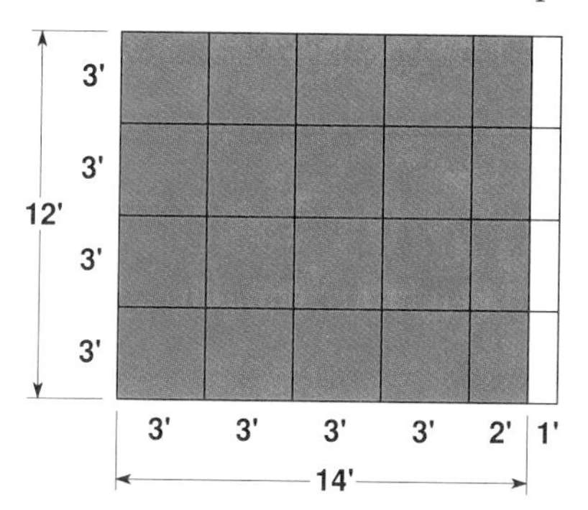

Because the carpet is sold in full yards, Mrs. Prah will have to buy 20 square yards.

After cutting off the carpet needed for the room, the extra piece of carpet will measure 12 feet by 1 foot.

Example 2 The vertices of a rectangle are $A(2, 0)$, $B(7, 0)$, $C(7, 4)$, $D(2, 4)$. Find the area of the rectangle.

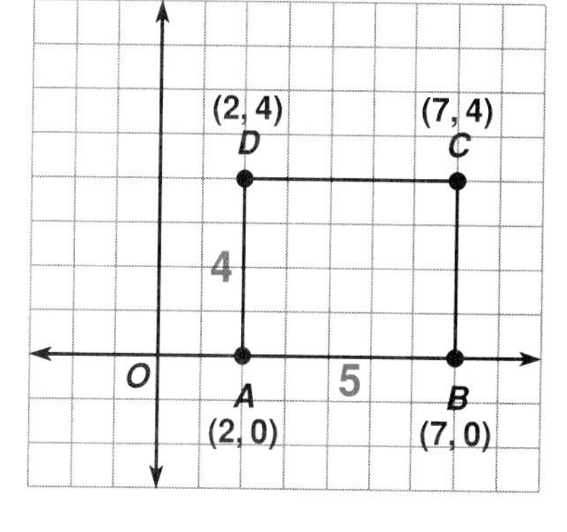

To find the dimensions of the rectangle, count boxes or subtract coordinates.

base $AB = 7 - 2 = 5$ units

height $DA = 4 - 0 = 4$ units

Area = base × height

= 5 units × 4 units

= 20 square units

To verify this result, count the number of square units on the graph that are inside the rectangle.

Try These *(For solutions, see page 417.)*

1. The dimensions of a rectangle are 9 feet by 12 feet.
   Find: **a.** the number of square feet in the area of the rectangle
   b. the number of square yards in the area of the rectangle
   c. the number of feet in the perimeter of the rectangle
2. The vertices of a rectangle are $A(1, 3)$, $B(5, 3)$, $C(5, 0)$, $D(1, 0)$. Draw this rectangle on graph paper and find its area.

Square

When the base and height of a rectangle are equal in measure, what special kind of rectangle is this figure?

Example 3 Sara Lee has 20 feet of fencing with which she wants to lay out a square patch for growing herbs. What will be the area of the patch?

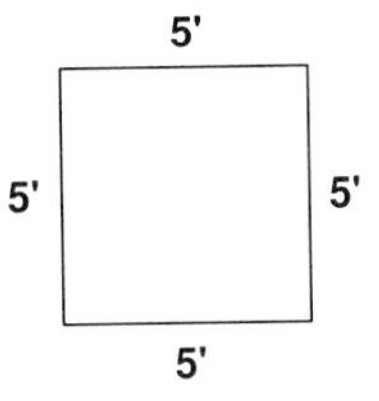

The fencing will be the perimeter of the square. To find the length of each of the 4 equal sides, divide 20 by 4.

20 feet ÷ 4 = 5 feet ← length of side

Since a square is a rectangle in which the base and height are equal in measure, the area of the square is

5 feet × 5 feet or 25 square feet

How would you square the number 5 on a calculator?

You could enter: 5 [×] 5 [=]

You could also make use of the constant feature of the calculator.

Enter: 5 [×] [=] *Display:*

Using the fact that the base and height of a square, both called the *sides,* are equal in measure, write a rule to find the area of a square whose side measures s inches.

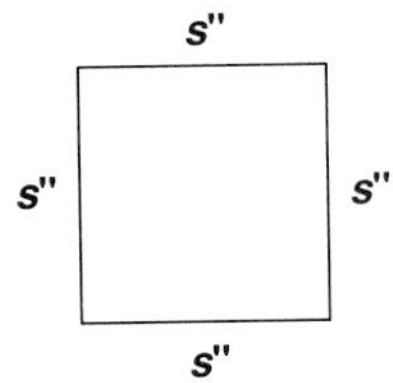

Area of Square = s^2 square inches

Example 4 Raoul has made a square wooden cutting board for his mom. If the square board has 144 square inches of area, what is the measure of each side?

Apply the rule for the area of a square.

Area of Square = s^2

$144 = s^2$ The area is 144.

$\sqrt{144} = s$ To find a side, undo the operation of squaring. This inverse operation is called ***square root***, symbol $\sqrt{\ }$.

$12 = s$ Look for 2 identical factors of 144.

The side of the square measures 12 inches.

To find the square root of a number on a calculator,

Enter: 144 [$\sqrt{\ }$] *Display:* 12.

Try These *(For solutions, see page 417.)*

1. Each side of a square measures 15 feet.
   Find: **a.** the number of square feet in the area of the square
   b. the number of square yards in the area of the square
2. The vertices of a square are $J(-2, 4)$, $K(2, 4)$, $L(2, 0)$, $M(-2, 0)$. Draw the square on graph paper and find its area.
3. The area of a square is 36 square inches.
   Find: **a.** the measure of a side of the square
   b. the perimeter of the square
4. Write a key sequence for evaluating each expression on a calculator. Carry out your key sequence and write the display.
   a. 17^2 **b.** $\sqrt{1{,}156}$

Right Triangle

Cut out a 5" by 12" rectangle.

Draw a diagonal of the rectangle.

Cut along the diagonal to divide the rectangle into 2 figures. Describe these figures.

How do you think the area of one of these figures compares to the area of the original rectangle?

What is the area of the original rectangle?

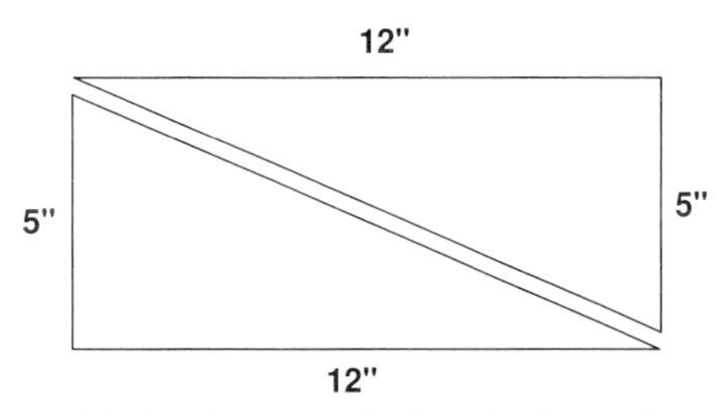

Cutting the rectangle along its diagonal results in two congruent right triangles.

The area of each triangle is half the area of the rectangle.

What is the area of one right triangle cut from the original rectangle?

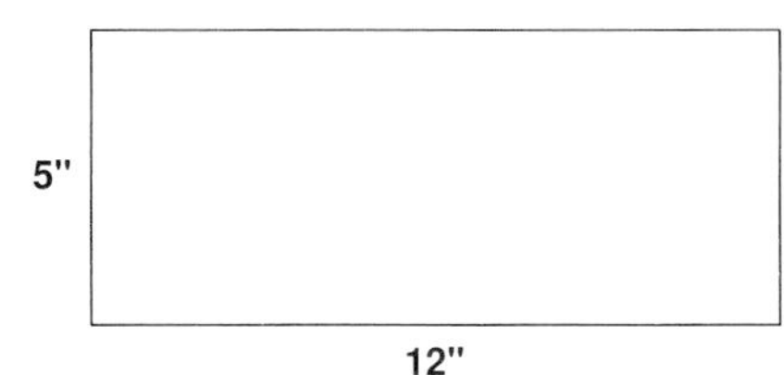

Area of Rectangle = 5" × 12"
= 60 square inches

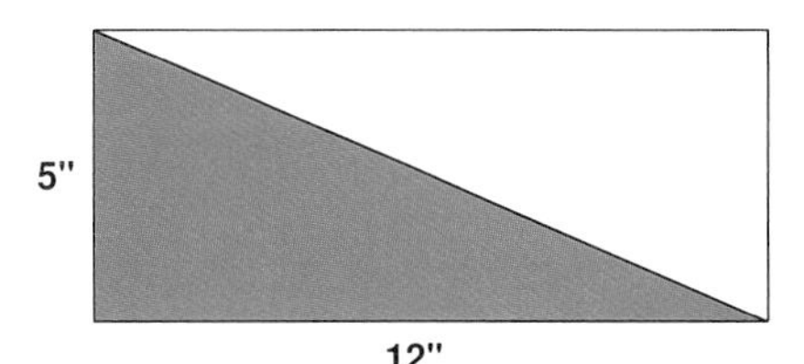

Area of Right Triangle $= \frac{1}{2}(5" \times 12")$
$= \frac{1}{2}(60)$
= 30 square inches

The area of a right triangle is one-half the area of a rectangle of the same dimensions.

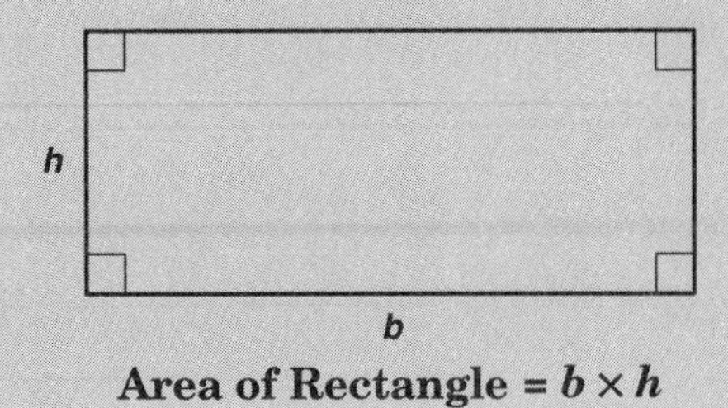

Area of Rectangle = $b \times h$

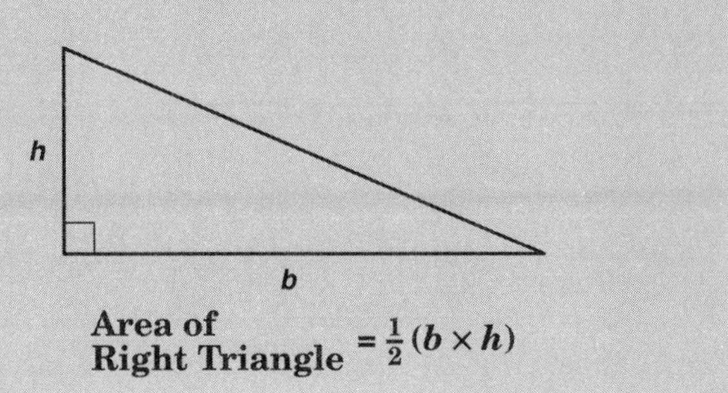

Area of Right Triangle $= \frac{1}{2}(b \times h)$

The perpendicular sides of a right triangle are called the *legs* of the triangle.

The third side of a right triangle, the side opposite the right angle, is called the *hypotenuse*.

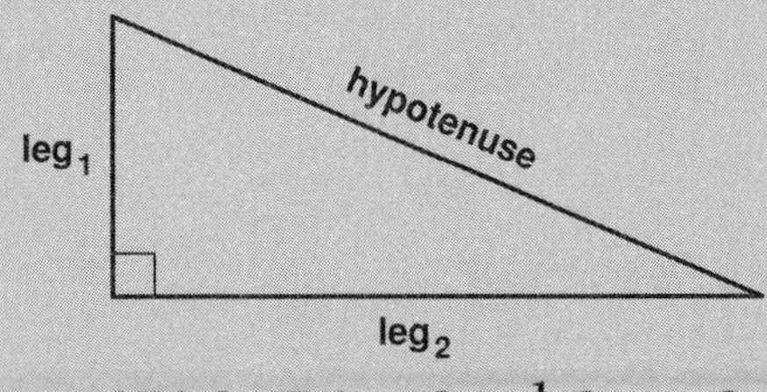

Area of Right Triangle $= \frac{1}{2}(\text{leg}_1 \times \text{leg}_2)$

Example 5 The vertices of a right triangle are $A(1, 2)$, $B(7, 2)$, $C(1, 5)$.

a. Which sides are the legs of the triangle?

From the graph, you can see that the perpendicular sides are $\overline{AB}$ and $\overline{CA}$.

b. How many square units are in the area of the triangle?

To find the measures of the legs, count boxes or subtract coordinates.

$$\text{Area} = \tfrac{1}{2}(\text{leg}_1 \times \text{leg}_2)$$

$$= \tfrac{1}{2}(6 \text{ units} \times 3 \text{ units})$$

$$= \tfrac{1}{2}(18 \text{ square units}) = 9 \text{ square units}$$

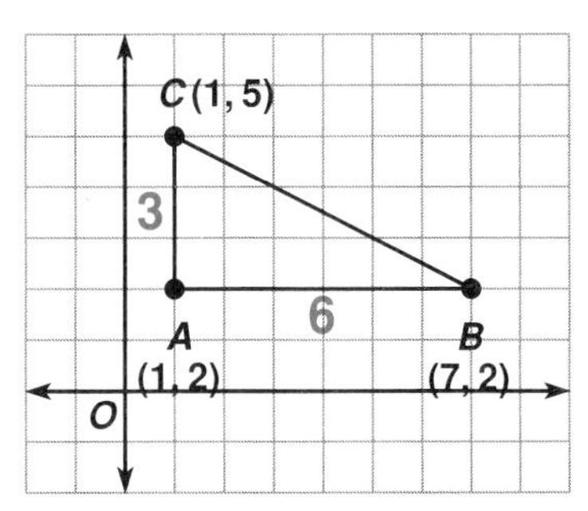

leg $AB = 7 - 1 = 6$ units
leg $CA = 5 - 2 = 3$ units

Example 6 The area of a right triangle is 40 square centimeters.

If one leg of the triangle measures 4 centimeters, what is the measure of the other leg?

Apply the rule for the area of a right triangle, substituting the values you know.

$$\text{Area of Right Triangle} = \tfrac{1}{2}(\text{leg}_1 \times \text{leg}_2)$$

$40 = \tfrac{1}{2}(4 \times \ell)$ — Area = 40, leg = 4, other leg = ℓ

$40 = \tfrac{\cancel{1}}{\cancel{2}_1} \cdot \cancel{4}^{2}\ell$ — Cancel.

$40 = 2\ell$

$\frac{40}{2} = \frac{2\ell}{2}$ — Divide both sides by 2.

$20 = \ell$

Check: If the legs measure 20 and 4, does the area = 40?

$\text{area} = \tfrac{1}{2}(20 \times 4)$

$= \tfrac{1}{2}(80)$

$= 40$ ✓

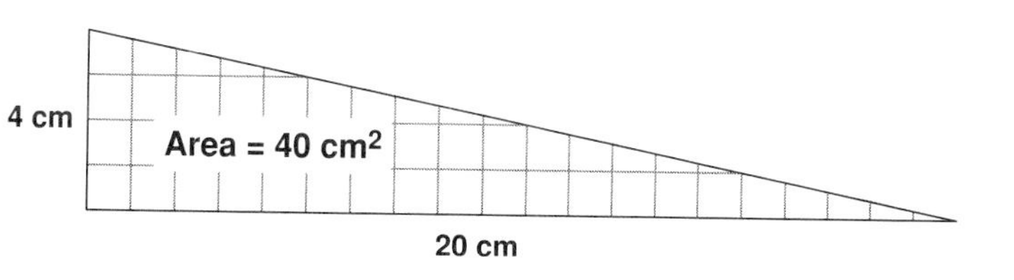

Answer: The other leg of the triangle measures 20 centimeters.

What would be the area of the rectangle that has the same dimensions as this right triangle?

The rectangle with dimensions 20 cm and 4 cm is twice the area of the right triangle.

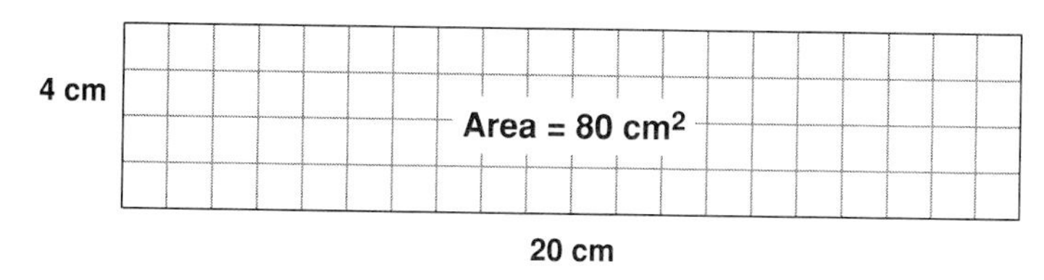

Try These (For solutions, see page 418.)

(For solutions, see page 418.)

1. Find the area of this right triangle.

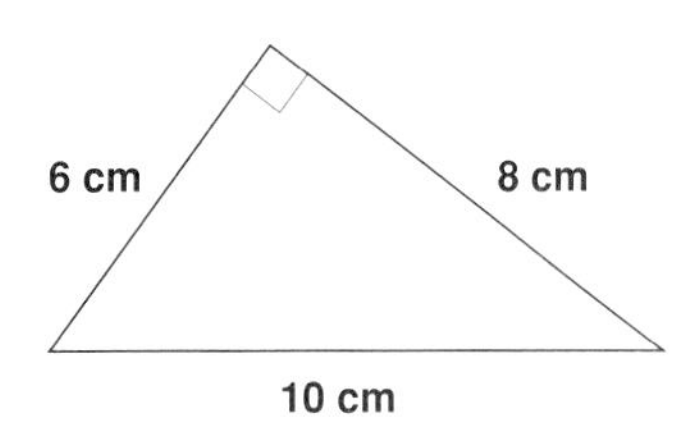

2. The area of a right triangle is 66 square inches.

   If one leg of the triangle measures 12 inches, what is the measure of the other leg?

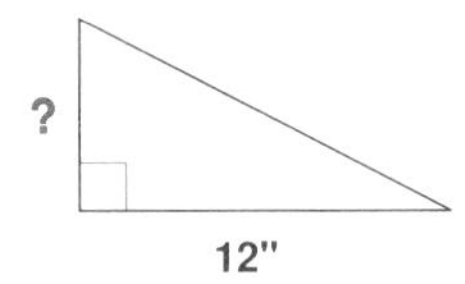

SELF-CHECK: SECTION 9.1

1. What do we call a square each of whose sides measures 1 inch?

   Answer: a square inch

2. What is the meaning of *area*?

   Answer: The area of a figure is the number of square units — square inches, square feet, square meters, etc. — that can fit inside the figure.

3. If the dimensions of a rectangle are called the base b and height h, what is a rule for finding the area of a rectangle?

   Answer: Area of Rectangle = $b \times h$

4. What is a rule for finding the area of a square from the measure of its side s?

   Answer: Area of Square = s^2

5. What is the relationship between the area of a right triangle and the area of a rectangle of the same dimensions?

   Answer: The area of the right triangle is one-half the area of the rectangle.
   Area of Right Triangle = $\frac{1}{2}(b \times h)$

6. The perpendicular sides of a right triangle are called the *legs* of the triangle. Write a rule for finding the area of a right triangle using the measures of the legs.

   Answer: Area of Right Triangle = $\frac{1}{2}(\text{leg}_1 \times \text{leg}_2)$

EXERCISES: SECTION 9.1

1. Find the number of square units in the area of each figure.

a.

b. 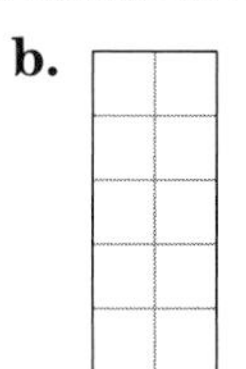

c. 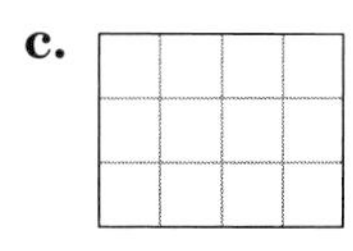

d. 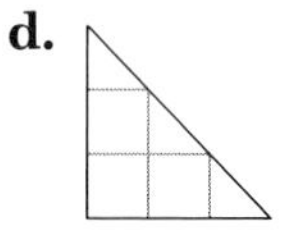

e. 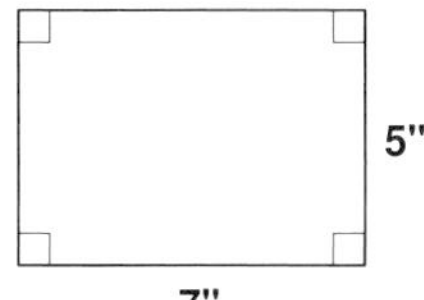

f. 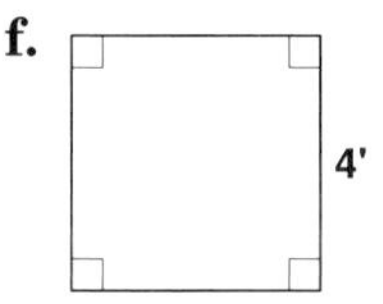

g. 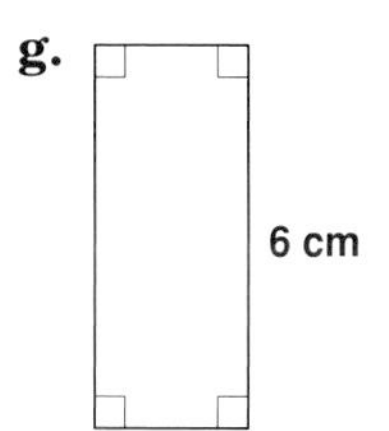

h. 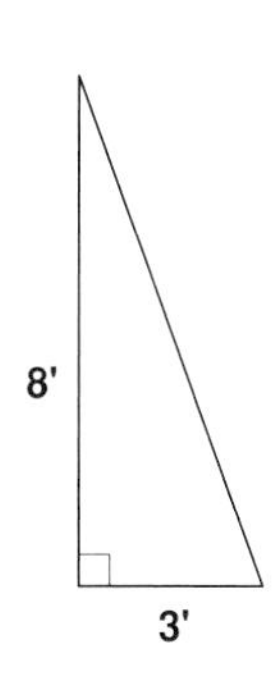

2. **a.** Find the area of the square whose side measures:

(1) 4 in. **(2)** 20 m **(3)** 6.5 cm **(4)** $8\frac{1}{2}$ ft.

b. Find the measure of the side of the square whose area is:

(1) 100 ft.^{2} **(2)** 81 cm^2 **(3)** 121 yd.^{2} **(4)** 225 m^2

c. A large square is filled with small squares, each with an area of 1 square inch. Is it possible for the large square to contain 40 of the small squares? Explain.

3. The letter in each diagram is made up of congruent squares. Each side of the square measures 3 inches.

a.

b. 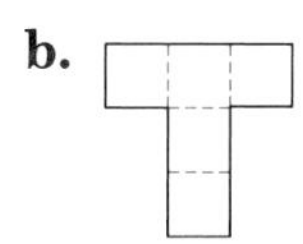

c. 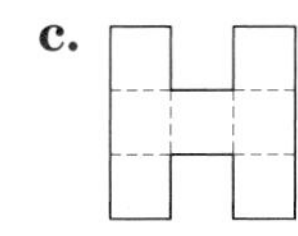

For each letter, find: **(1)** the perimeter **(2)** the area

4. Write a key sequence for evaluating each expression on a calculator. Carry out your key sequence and write the display.

a. 32^2 **b.** 41^2 **c.** 65^2

d. $\sqrt{2{,}304}$ **e.** $\sqrt{2{,}809}$ **f.** $\sqrt{11{,}025}$

5. Find the area of the rectangle with the given dimensions.

	base	*height*		*base*	*height*
a.	3 in.	7 in.	**b.**	6 ft.	4 ft.
c.	2.5 m	4 m	**d.**	5.2 cm	3.4 cm
e.	8 yd.	$2\frac{1}{4}$ yd.	**f.**	$3\frac{1}{2}$ in.	$1\frac{1}{8}$ in.
g.	2 ft.	6 in.	**h.**	5 m	50 cm

6. Find the area of the right triangle whose legs measure:

a. 4 in. and 6 in. **b.** 5 m and 3 m **c.** 2.4 cm and 5 cm

d. 2 yd. and $\frac{3}{4}$ yd. **e.** 10 m and 50 cm **f.** 2 ft. and 6 in.

7. Two congruent right triangles are fitted together to form a third triangle

a. Make an observation about the third triangle formed.

b. Find the area of the third triangle.

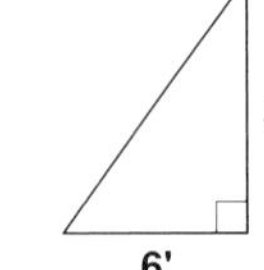

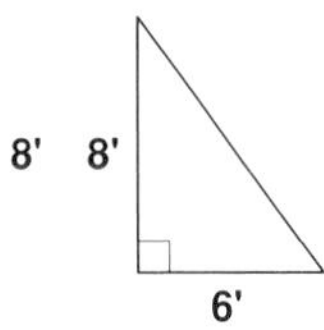

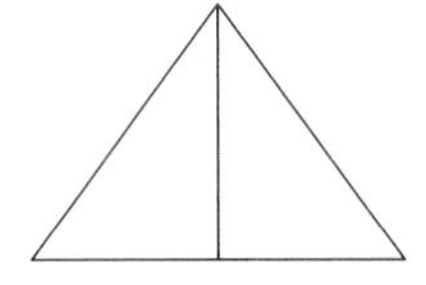

8. On graph paper, draw each figure with the given vertices, and find the area of the figure

a. $ABCD$ with $A(0, 0)$ $B(4, 0)$ $C(4, 8)$ $D(0, 8)$

b. $JKLM$ with $J(1, 1)$ $K(8, 1)$ $L(8, 6)$ $M(1, 6)$

c. $QRST$ with $Q(-4, 0)$ $R(3, 0)$ $S(3, 4)$ $T(-4, 4)$

d. $WXYZ$ with $W(0, 0)$ $X(7, 0)$ $Y(7, 7)$ $Z(0, 7)$

e. $DEFG$ with $D(3, -3)$ $E(3, 3)$ $F(-3, 3)$ $G(-3, -3)$

f. $PQRS$ with $P(4, -5)$ $Q(4, 5)$ $R(-4, 5)$ $S(-4, -5)$

g. $TUVW$ with $T(4, -3)$ $U(4, 0)$ $V(-4, 0)$ $W(-4, -3)$

h. ABC with $A(0, 0)$ $B(6, 0)$ $C(0, 8)$

i. DEF with $D(-4, 0)$ $E(0, 0)$ $F(0, 7)$

j. XYZ with $X(0, -4)$ $Y(7, -4)$ $Z(0, 5)$

9. Find a fourth pair of coordinates that, together with the 3 pairs of coordinates given, will form a rectangle or square. Find the area of each figure.

a. $A(0, 0)$ $B(0, 4)$ $C(5, 4)$ $D(5, ?)$

b. $W(3, 5)$ $X(0, 5)$ $Y(0, -5)$ $Z(?, -5)$

c. $P(-3, 0)$ $Q(-3, 3)$ $R(5, 3)$ $S(?, ?)$

d. $G(-4, 4)$ $H(4, 4)$ $I(4, -4)$ $J(?, ?)$

e. $S(5, -7)$ $T(5, 2)$ $U(-4, 2)$ $V(?, ?)$

10. Use graph paper to draw the figures.

a. **(1)** Draw 4 rectangles, all with different shapes, and all with an area of 36 square units.

(2) Find the perimeter of each rectangle.

(3) Describe the rectangle with area 36 square units that has the least possible perimeter.

(4) Describe a situation in which it would be good to know about a rectangle of a certain area that has the least possible perimeter.

b. **(1)** Draw 4 rectangles, all with different shapes, and all with a perimeter of 32 units.

(2) Find the area of each rectangle.

(3) Describe the rectangle with perimeter 32 units that has the greatest possible area.

(4) Describe a situation in which it would be good to know about a rectangle of a certain perimeter that has the greatest possible area.

11. A rectangular board covers a box that measures 14 inches long by 3 inches wide by 4 inches deep. If the board extends 1 inch uniformly over each edge of the box, as shown, find the area of the board.

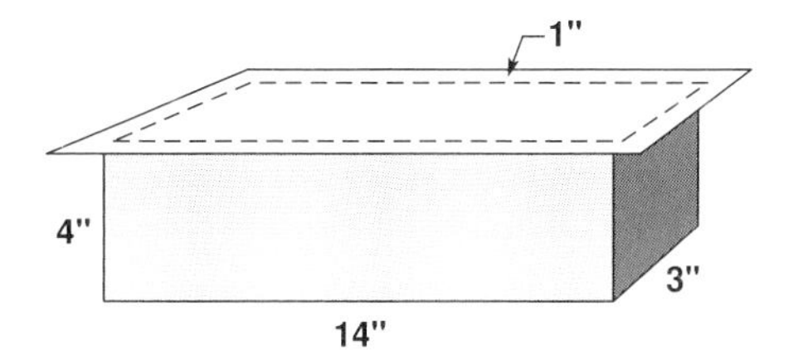

12. Find the area of each figure described.

a. a square with perimeter 24 yards

b. a rectangle with perimeter 34 inches and base 15 inches

c. a square whose perimeter and area have the same numerical value

d. an isosceles right triangle with perimeter 34 inches and hypotenuse 14 inches.

13. Use an equation to find the value of the variable shown.

a. The area of this rectangle is 120 square units.

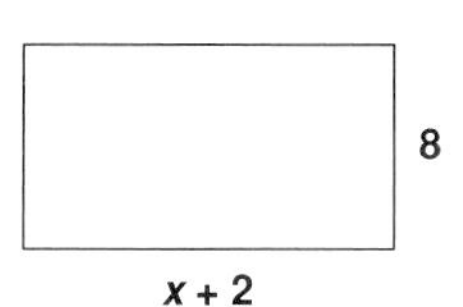

Find the value of x.

b. The area of this rectangle is 360 square units.

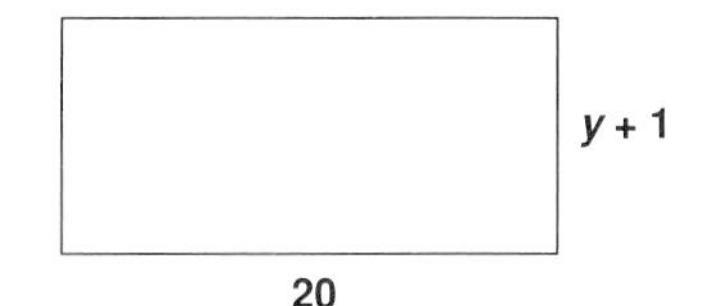

Find the value of y.

c. The area of this right triangle is 100 square units.

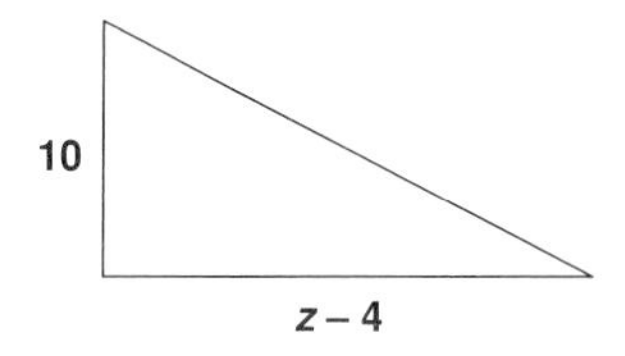

Find the value of z.

14. Use the area rule for each figure to find the missing dimension.

a. The area of a rectangle is 42 square feet, and the base measures 7 feet. Find the measure of the height.

b. The area of a rectangle is 108 square centimeters, and the height measures 9 centimeters. Find the measure of the base.

c. The area of a square is 144 square meters. Find the measure of the side.

d. The area of a right triangle is 84 square inches. If one leg measures 14 inches, find the measure of the other leg.

15. Guesstimate an answer before solving each problem.

a. **(1)** Nancy wants to have a rectangular rug 8 feet by 11 feet in her dining room.
How many square feet are in the area covered by the rug?

(2) Nancy is having the rug cut from carpet that comes on a roll that is 12 feet wide, and the carpet is sold in lengths of full yards.
How many square yards of the carpet must she buy to have the rug made?
After the rug is made, how many square feet of carpet will be left over?

b.

The triangular sail on Janet Finn's boat measures 6 feet across and 10 feet high.

(1) How many square feet are in the area of the sail?

(2) How many square yards of cloth are needed for Janet's sail?

c. A baseball diamond is a square measuring 90 feet on a side.

(1) What is the area of the infield (inside the diamond)?

(2) Hank just hit a home run. How far must he run to touch each base and get back to home plate?

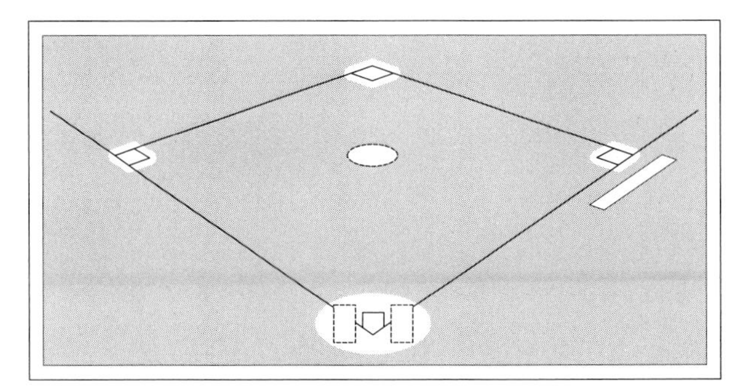

d. Simon has 100 feet of fencing with which to lay out a square pen in his yard for Tim's guide dog.

(1) What will be the measure of each side of the square pen?

(2) What will be the area of the square pen?

e. Sheffield Township is building a new road 1,380 meters long. Each of the 4 lanes of the road is 3 meters wide. How many square meters of roadway must be paved?

f. The Burtons are laying new floor covering in their kitchen, which measures 9 feet by 15 feet.

(1) If they use 1-foot-square tiles that cost $1 each, what will be the total cost of the tiles?

(2) If they choose linoleum costing $7 a square yard, what will they have to pay?

(3) Which covering is less expensive? How much money can the Burtons save by using the less expensive covering?

g. Jim is buying wood molding to frame a 5" by 7" photo for his parents' anniversary. Before framing the photo, Jim intends to paste the photo on an 11" by 13" colored mat.

(1) What is the area of the photo?

(2) After the photo is centered and pasted on the mat, what is the area of the mat that will show?

(3) What is the length of molding Jim needs for the frame?

h. Mrs. Cuervo used some empty cartons from the supermarket to mail clothes to her daughter at college. She measured the cartons to find one that met the post-office requirement that the length and girth (perimeter of an end) combined should be no more than 108 inches.

(1) Which dimensions satisfy the requirements?

(a) 16 in. by 22 in. by 40 in.

(b) 25 in. by 22 in. by 30 in.

(c) 15 in. by 20 in. by 36 in.

(d) 18 in. by 18 in. by 38 in.

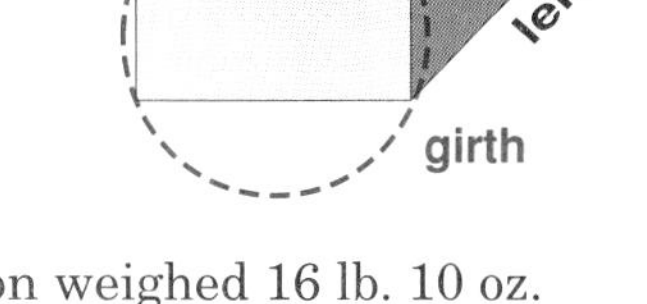

(2) When it was ready to be mailed, the carton weighed 16 lb. 10 oz. If it was sent fourth-class to Zone 7, use the accompanying table to find the postal rate.

Weight. up to but not exceeding— (pounds)	Zones							
	Local	1 & 2	3	4	5	6	7	8
2	$1.63	$1.69	$1.81	$1.97	$2.24	$2.35	$2.35	$2.35
3	1.68	1.78	1.95	2.20	2.59	2.98	3.42	4.25
4	1.74	1.86	2.10	2.42	2.94	3.46	4.05	5.25
5	1.79	1.95	2.24	2.65	3.29	3.94	4.67	6.25
6	1.85	2.04	2.39	2.87	3.64	4.43	5.30	7.34
7	1.91	2.12	2.53	3.10	4.00	4.91	5.92	8.30
8	1.96	2.21	2.68	3.32	4.35	5.39	6.55	9.26
9	2.02	2.30	2.82	3.55	4.70	5.87	7.17	10.22
10	2.07	2.38	2.97	3.78	5.05	6.35	7.79	11.18
11	2.13	2.47	3.11	4.00	5.40	6.83	8.42	12.14
12	2.19	2.56	3.25	4.22	5.75	7.30	9.03	13.09
13	2.24	2.64	3.40	4.44	6.10	7.78	9.65	14.03
14	2.28	2.69	3.48	4.56	6.27	8.02	9.96	14.50
15	2.32	2.75	3.55	4.67	6.44	8.24	10.24	14.94
16	2.35	2.79	3.63	4.78	6.60	8.45	10.52	15.35
17	2.39	2.84	3.70	4.88	6.75	8.66	10.77	15.74
18	2.42	2.89	3.76	4.98	6.90	8.85	11.02	16.11
19	2.46	2.93	3.83	5.07	7.03	9.03	11.25	16.45
20	2.49	2.98	3.89	5.16	7.16	9.20	11.47	16.79

EXPLORATIONS

Using graph paper, cut out 3 squares, one each whose side measures 3 units, 4 units, and 5 units.

Using the same unit that is on your graph paper, cut out from unlined paper a right triangle with legs that measure 3 units and 4 units, and a hypotenuse of 5 units.

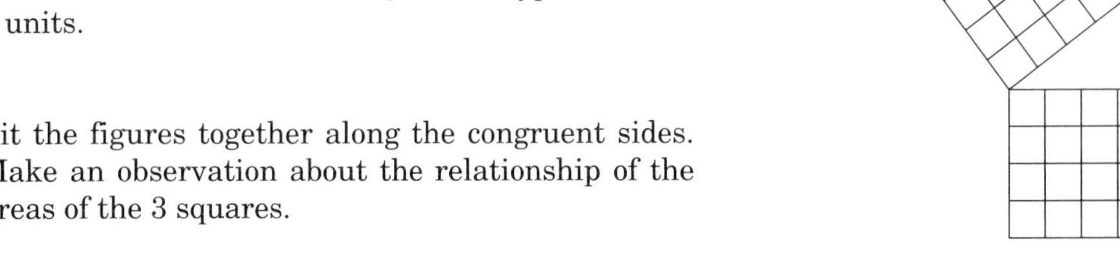

Fit the figures together along the congruent sides. Make an observation about the relationship of the areas of the 3 squares.

The relationship among the areas of the squares that can be formed on the sides of a right triangle is called the *Pythagorean Relation*, after the Greek mathematician Pythagoras who recorded it more than 2,000 years ago.

The area of the square on the hypoteneuse is equal to the sum of the areas of the squares on the legs.

a. (1) According to the Pythagorean Relation, what do you think the measure of the hypotenuse must be if the legs of a right triangle measure 6 units and 8 units? Verify your prediction by making a model from graph paper. Which of the following sentences summarizes the model?

(a) $6 + 8 = 10$ (b) $6^2 + 8^2 = 14^2$ (c) $6^2 + 8^2 = 10^2$

(2) According to the Pythagorean Relation, what do you think the measure of the hypotenuse must be if the legs of a right triangle measure 5 units and 12 units? Verify your prediction by making a model from graph paper. Write a sentence that relates the numbers 5, 12, 13.

(3) According to the Pythagorean Relation, what do you think the measure of the hypotenuse must be if the legs of a right triangle measure 8 units and 15 units? Verify your prediction by making a model from graph paper. Write a sentence that relates the numbers 8, 15, 17.

(4) From the sentences you have written relating the measures of the 3 sides of a right triangle, write a general sentence using a and b as the measures of the legs of the right triangle and c as the measure of the hypotenuse.

b. The Pythagorean Relation among the measures of the sides of a right triangle is:

$(\text{leg})^2 + (\text{leg})^2 = (\text{hypotenuse})^2$

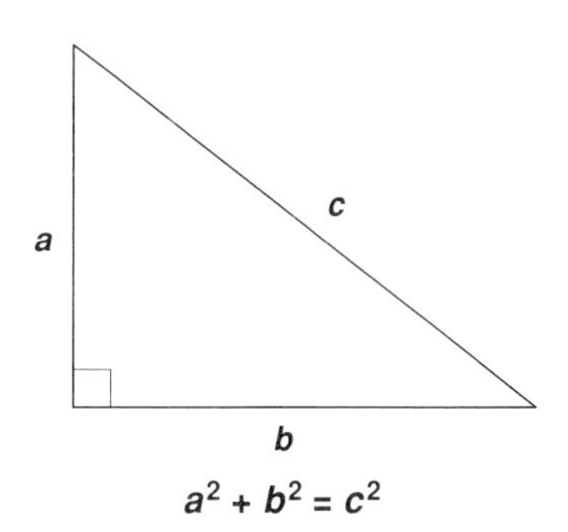

If you know any 2 of the 3 measures, you can use the Pythagorean Relation to find the missing measure.

Example: The legs of a right triangle measure 9 units and 12 units. What is the measure of the hypotenuse?

Using a diagram helps keep track of which numbers belong to the legs and which to the hypotenuse.

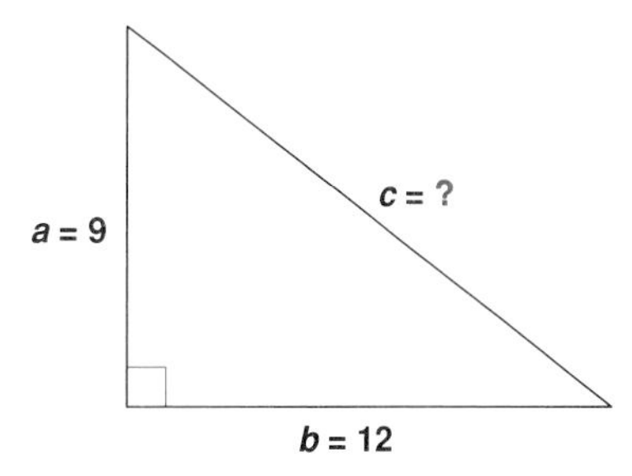

$a^2 + b^2 = c^2$	The Pythagorean Relation.
$9^2 + 12^2 = c^2$	You are looking for the value of *c*.
$81 + 144 = c^2$	Do the squares.
$225 = c^2$	Do the addition.
$\sqrt{225} = c$	To find *c*, you need square roots.
$15 = c$	$225 = 15 \times 15$

Check: Does $9^2 + 12^2 = 15^2$?

$81 + 144 \stackrel{?}{=} 225$

$225 = 225$ ✓

Answer: In a right triangle whose legs measure 9 units and 12 units, the hypotenuse measures 15 units.

Use the Pythagorean Relation to find the missing measures of the sides of a right triangle. Check your results.

a	b	c
3	4	?
5	12	?
?	8	10
8	15	?
10	?	26

c. Three numbers that are connected by the Pythagorean Relation are called *Pythagorean Triples*. These special number triples show interesting number patterns.

Let's examine 5 rows of Pythagorean Triples

	a	*b*	*c*
row 1	3	4	5
row 2	5	12	13
row 3	7	24	25
row 4	9	40	41
row 5	11	60	61

(1) Make an observation about the numbers in the a-column.

(2) How are the numbers in the c-column related to the numbers in the b-column?

(3) Carry out this calculation for each row of numbers with these column headings: row $\times\, a$ + row

Example: $1 \times 3 + 1$

Compare your results to the numbers in the columns, and draw a conclusion.

(4) How do the squares of the numbers in the a-column relate to the numbers in the b-column and c-column?

(5) Using the relationships you have seen, write the numbers in row 6.

Use a calculator to verify your result.

d. Examine these 4 rows of Pythagorean Triples.

	a	*b*	*c*
row 1	4	3	5
row 2	8	15	17
row 3	12	35	37
row 4	16	63	65

(1) Make an observation about the numbers in the a-column.

(2) How are the numbers in the c-column related to the numbers in the b-column?

(3) Carry out this calculation for each row of numbers with these column headings: row $\times\, a - 1$

Example: $1 \times 4 - 1$

Describe the numbers you get.

(4) For each row of numbers, carry out this calculation: $(a \div 2)^2$
Describe the numbers you get.

(5) Using the relationships you have seen, write the numbers in row 5.

Use a calculator to verify your result.

9.2 More About Area

Parallelogram

Cut out a parallelogram of
height 5 inches and base 12 inches.

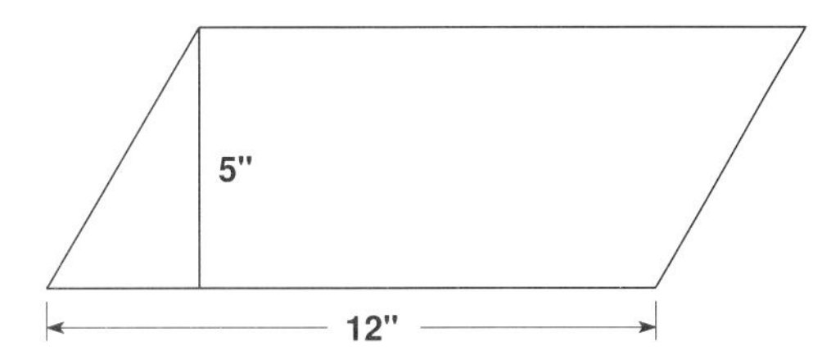

From the parallelogram, cut off a triangle at the left, along the altitude.

Reattach the triangle at the right. What kind of figure results?

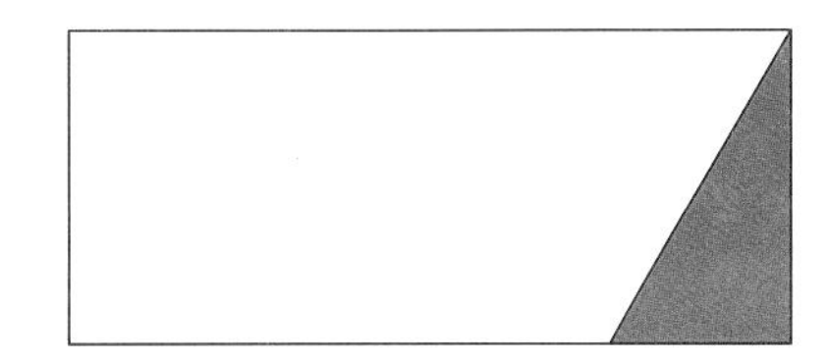

After reattaching the triangle, the result is a rectangle.

How do you think the area of the newly formed rectangle compares to the area of the original parallelogram?

The areas must be the same since you used the same pieces to form this new figure from the original.

The area of a parallelogram is the same as the area of a rectangle of the same dimensions.

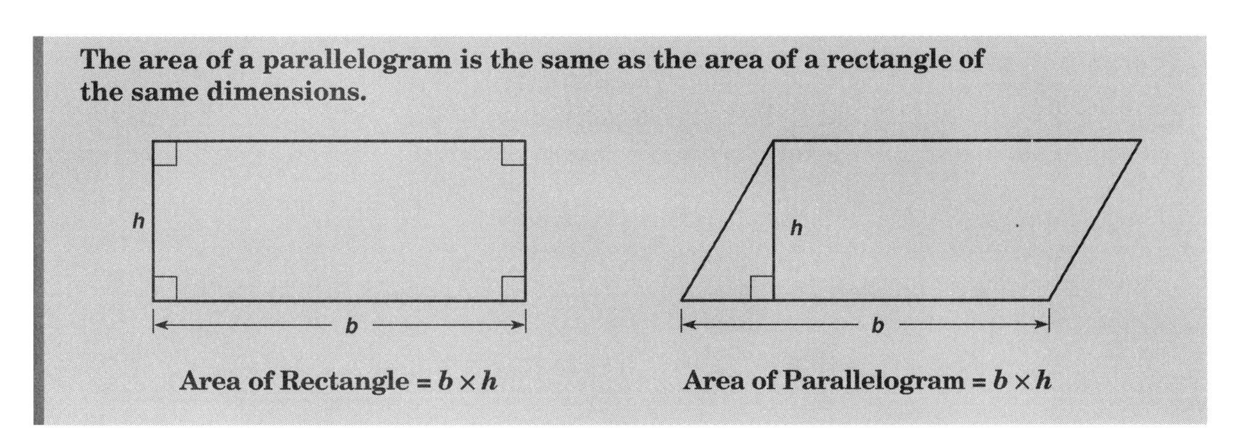

Example 1 Compare the areas of these 3 figures.

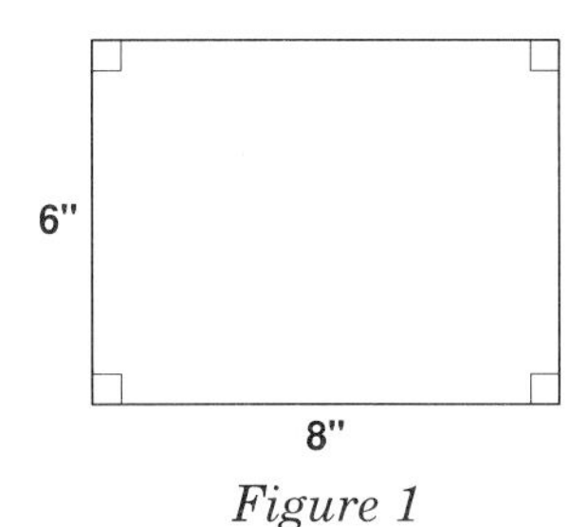

Figure 1

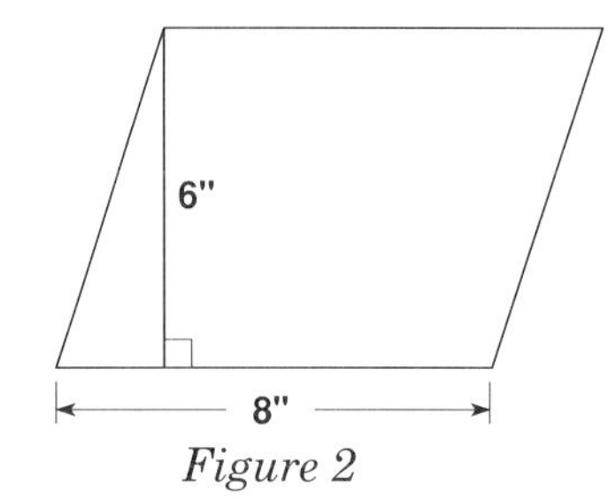

Figure 2

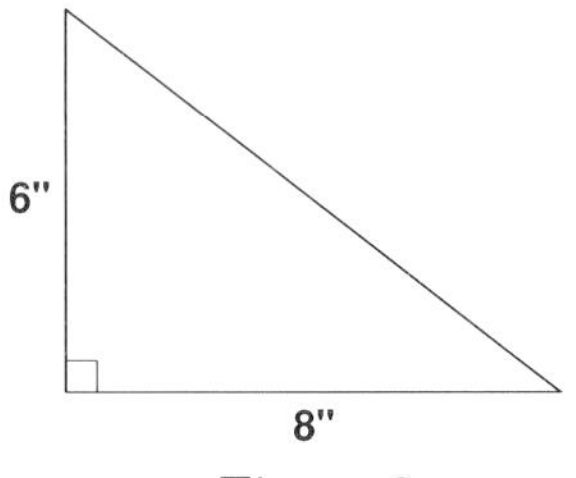

Figure 3

Figure 1 is a rectangle.	Its area is 6" × 8" or 48 square inches.
Figure 2 is a parallelogram.	Its area is 6" × 8" or 48 square inches.
Figure 3 is a right triangle.	Its area is $\frac{1}{2}$(6" × 8") or 24 square inches.

Figures 1 and 2 are equal in area.
Figure 3 has half the area of Figures 1 or 2.

Example 2 The vertices of a parallelogram are $M(7, 4)$, $N(2, 4)$, $O(0, 0)$, $P(5, 0)$.
Find the area of the parallelogram.

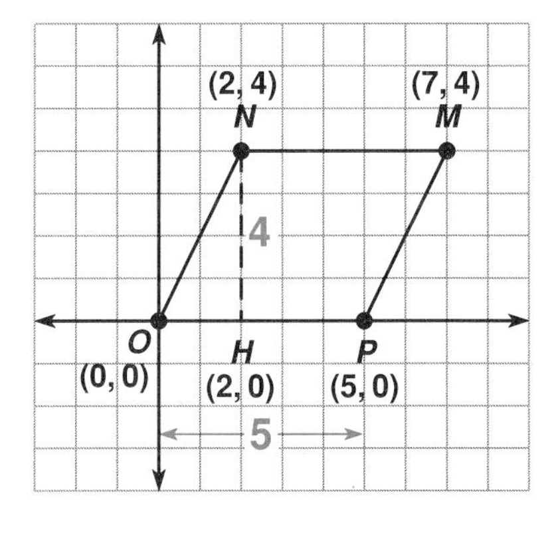

To find the length of base $\overline{OP}$, count boxes or subtract coordinates.

base $OP = 5 - 0 = 5$ units

To find the height of the parallelogram, draw $\overline{NH}$, the altitude from N to base $\overline{OP}$.

From the graph, you can see that point H has the coordinates (2, 0).

altitude $NH = 4 - 0 = 4$

Area of Parallelogram = base × height

= 5 units × 4 units = 20 square units

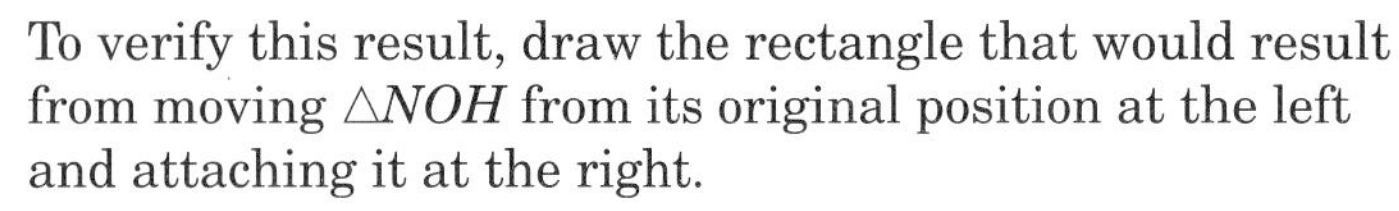

To verify this result, draw the rectangle that would result from moving $\triangle NOH$ from its original position at the left and attaching it at the right.

Count the boxes in the shaded rectangle.

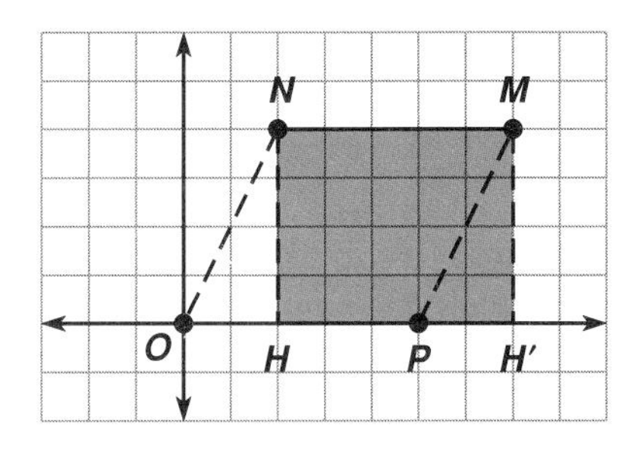

Try These (For solutions, see page 418.)

1. Find the area of each figure.

a.

5'

12'

b.

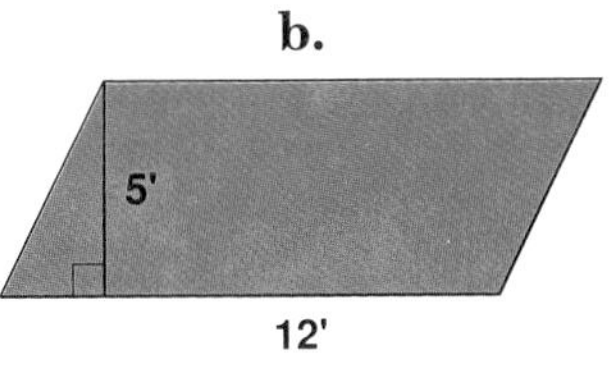

c.

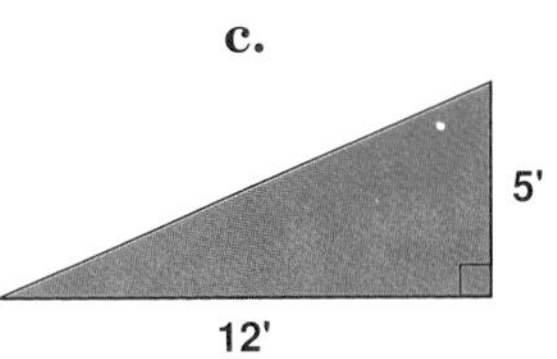

2. **a.** On graph paper, draw parallelogram $ABCD$ with vertices $A(1, 0), B(9, 0), C(12, 7), D(4, 7)$.
   b. Find the length of base $\overline{AB}$.
   c. From D, draw altitude $\overline{DF}$ to base $\overline{AB}$.
   What are the coordinates of F? Find the length of altitude $\overline{DF}$.
   d. Find the area of parallelogram $ABCD$.

Triangle

Cut out a parallelogram of height 5" and base 12". Draw a diagonal of the parallelogram. Cut along the diagonal to divide the parallelogram into 2 figures. Describe these figures.

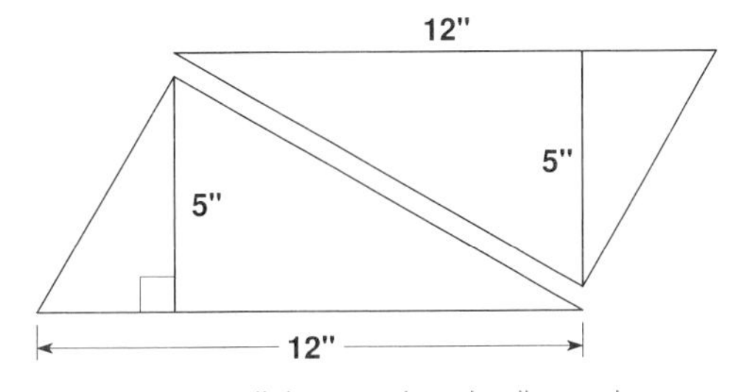

Cutting the parallelogram along its diagonal results in 2 congruent scalene triangles.

How do you think the area of one of these figures compares to the area of the original parallelogram?

What is the area of the original parallelogram?

What is the area of one scalene triangle cut from the parallelogram?

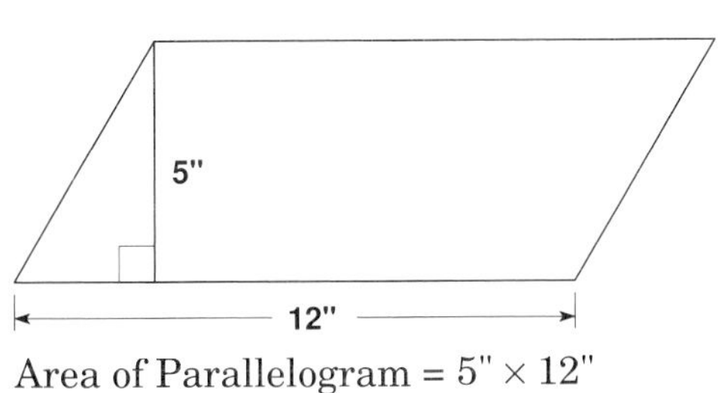

Area of Parallelogram = 5" × 12"
= 60 square inches

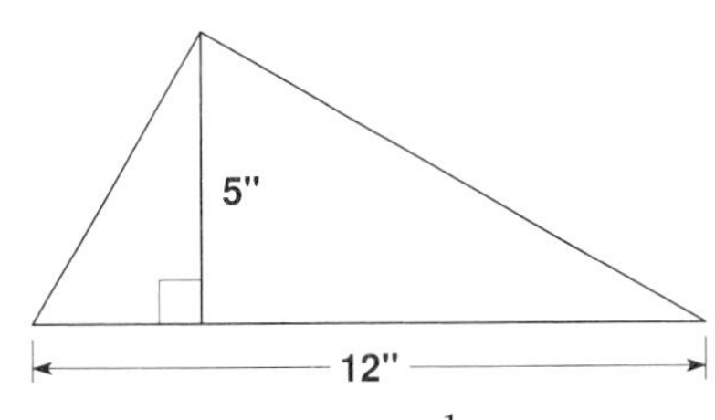

Area of Triangle = $\frac{1}{2}$(5" × 12")
= $\frac{1}{2}$(60)
= 30 square inches

The area of a triangle is one-half the area of a parallelogram of the same dimensions.

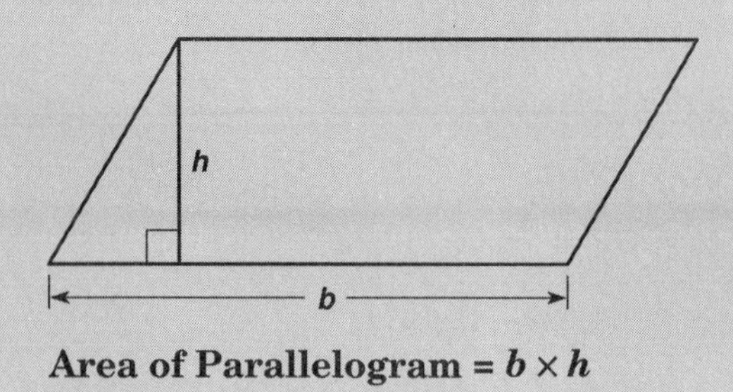

Area of Parallelogram = $b \times h$

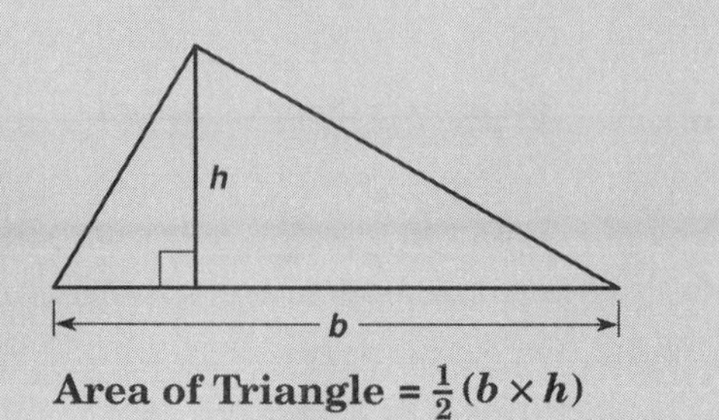

Area of Triangle = $\frac{1}{2}(b \times h)$

Example 3 Find the area of each triangle.

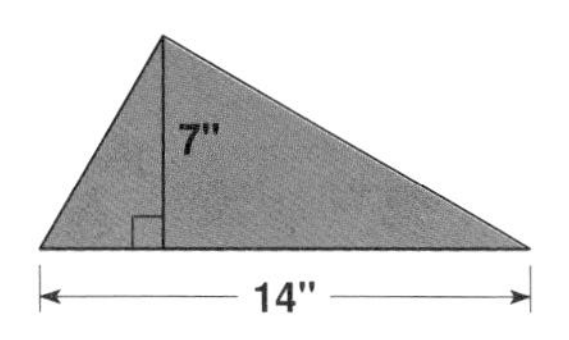

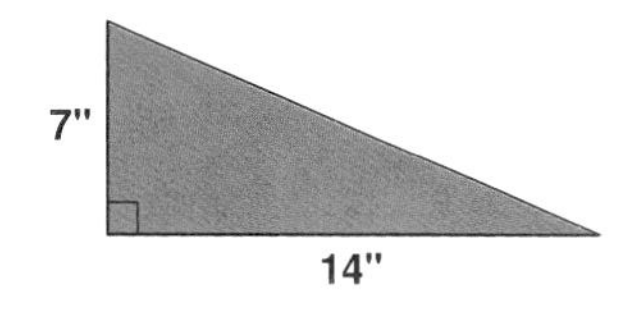

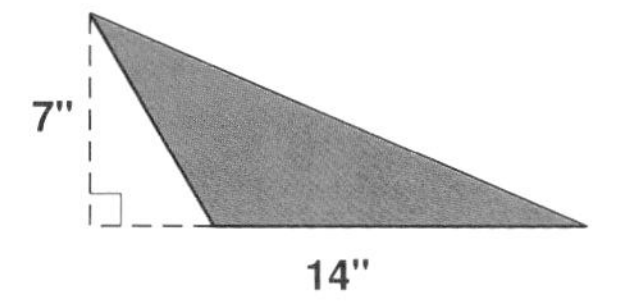

The rule for finding the area of any kind of triangle is always the same: one-half the product of the measures of the base and altitude.

Note the different positions that the altitude takes depending upon the type of triangle, and which side is the base.

In an acute triangle, the altitude falls inside the triangle.
In a right triangle, the altitude is a leg of the triangle.
In an obtuse triangle, the altitude may fall outside the triangle.

Since these three triangles all have the same dimensions,they will all have the same area.

Area of Triangle $= \frac{1}{2}(b \times h)$
$= \frac{1}{2}(14" \times 7")$
$= \frac{1}{2}$(98 square inches) = 49 square inches

Example 4 The vertices of a triangle are $A(1,0)$, $B(8, 0)$, $C(3, 4)$. Find the area of the triangle.

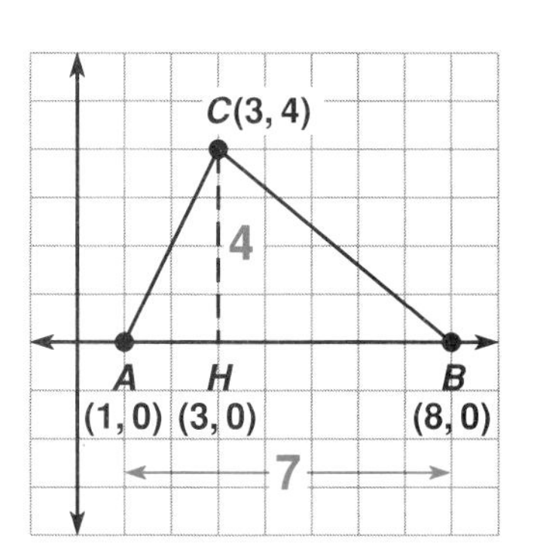

To find the length of base $\overline{AB}$, count boxes or subtract coordinates.

base $AB = 8 - 1 = 7$ units

To find the height of the triangle, draw $\overline{CH}$, the altitude from C to base $\overline{AB}$.

From the graph, you can see that point H has coordinates $(3, 0)$.

altitude $CH = 4 - 0 = 4$ units

Area of Triangle $= \frac{1}{2}(b \times h)$
$= \frac{1}{2}$(7 units × 4 units)
$= \frac{1}{2}$(28 square units) = 14 square units

Try These *(For solutions, see page 418.)*

1. Find the area of each shaded triangle, using the dimensions shown on the diagram.

a.

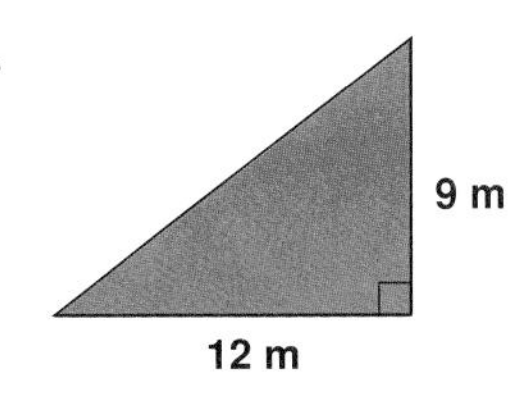

b.

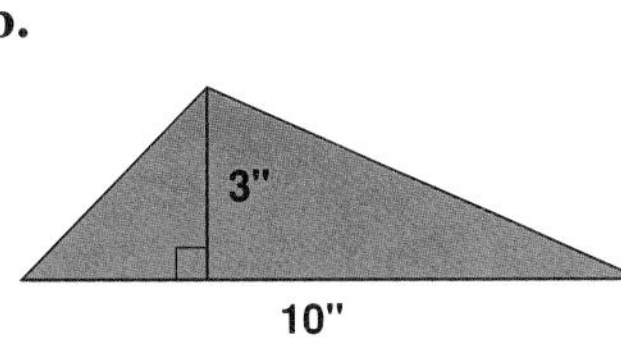

c.

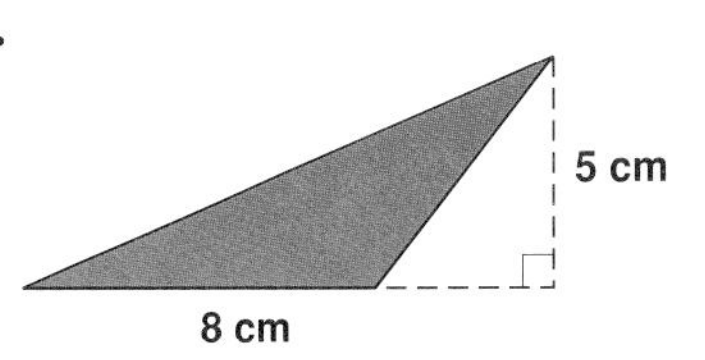

2. **a.** On graph paper, draw triangle PQR with vertices $P(2, 0)$, $Q(14, 0)$, $R(5, 5)$.

   b. Find the length of base $\overline{PQ}$.

   c. From R, draw altitude $\overline{RH}$ to base $\overline{PQ}$. What are the coordinates of H? Find the length of altitude $\overline{RH}$.

   d. Find the area of triangle PQR.

Trapezoid

Cut out 2 scalene triangles that are not congruent, but have the same height, and the measure of one side the same.

Triangle 1

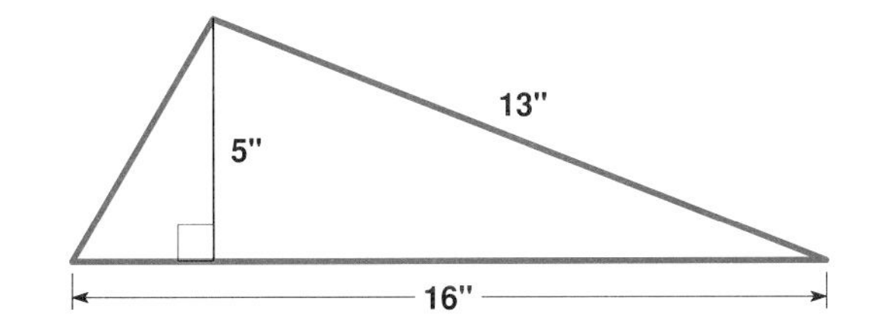

Triangle 2

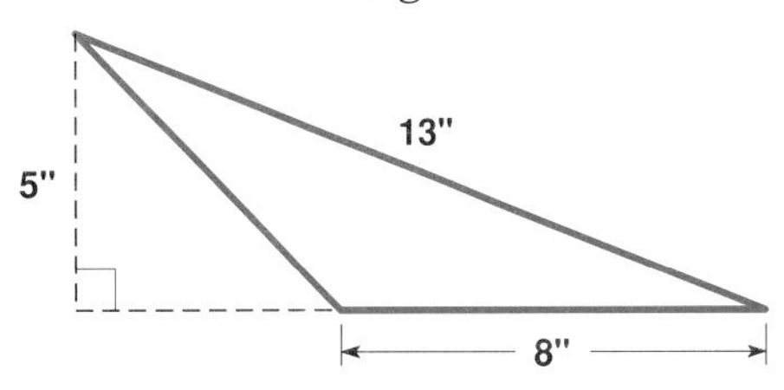

Put the triangles together so that they come together along the sides that are equal in measure.

Describe the figure formed.

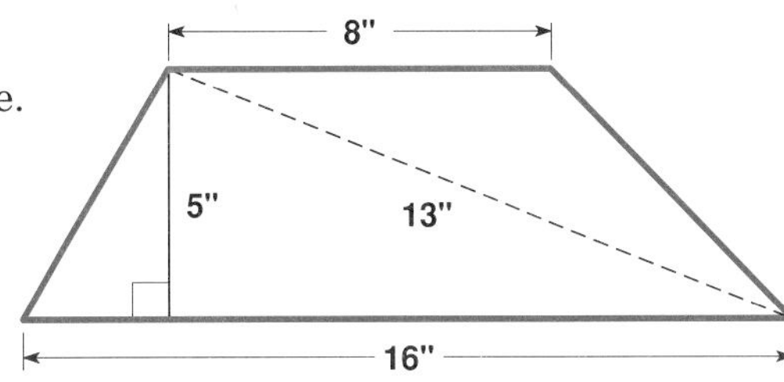

The 2 triangles form a trapezoid.

How does the area of the newly formed trapezoid compare to the areas of the original triangles?

$$
\begin{aligned}
\text{Area of Trapezoid} &= \text{Area of Triangle 1} &&+ \text{Area of Triangle 2} \\
&= \tfrac{1}{2}(b_1 \times h) &&+ \tfrac{1}{2}(b_2 \times h) \\
&= \tfrac{1}{2}(16 \times 5) &&+ \tfrac{1}{2}(8 \times 5) \\
&= \tfrac{1}{2}(80) &&+ \tfrac{1}{2}(40) \\
&= 40 &&+ 20 \\
&= &&60 \text{ square inches}
\end{aligned}
$$

The area of a trapezoid is the sum of the areas of 2 triangles. The triangles have the same height.

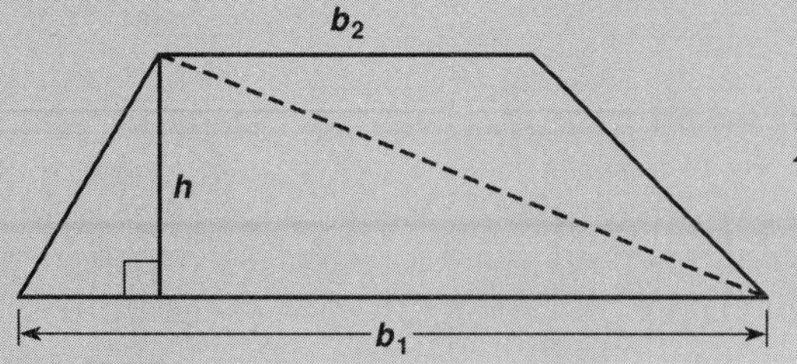

Area of Trapezoid $= \frac{1}{2}(b_1 \times h) + \frac{1}{2}(b_2 \times h)$

Example 5 Find the area of trapezoid *ABCD*.

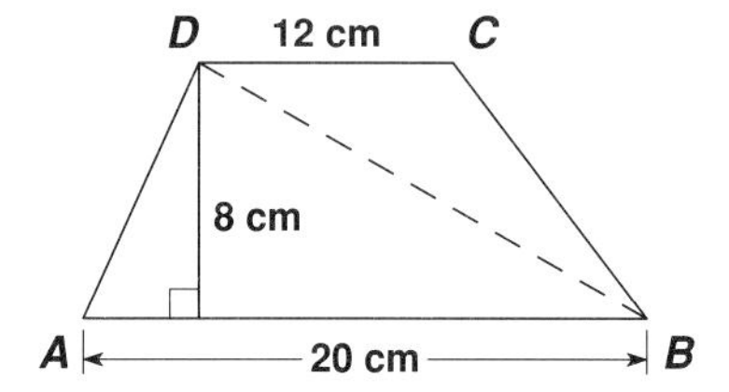

The distance between the parallel sides of this trapezoid is always 8 cm.

This height of the trapezoid is also the height of the 2 triangles that form the trapezoid.

The area of the trapezoid is the sum of the areas of the 2 triangles.

$$\begin{aligned}
\text{Trapezoid } ABCD &= \text{Triangle } ABD + \text{Triangle } BCD \\
&= \tfrac{1}{2}(20 \text{ cm} \times 8 \text{ cm}) + \tfrac{1}{2}(12 \text{ cm} \times 8 \text{ cm}) \\
&= \tfrac{1}{2}(160 \text{ cm}^2) + \tfrac{1}{2}(96 \text{ cm}^2) \\
&= 80 \text{ cm}^2 + 48 \text{ cm}^2 \\
&= 128 \text{ cm}^2
\end{aligned}$$

Study the calculation of the area.

Note that each product has the factor $\frac{1}{2}$, and also each product has the factor 8.

Using the distributive property, common factors can be used just once, to simplify the calculation.

$$\begin{aligned}
&\tfrac{1}{2} \times 8 \times 20 + \tfrac{1}{2} \times 8 \times 12 \\
=\ &\tfrac{1}{2} \times 8(20 + 12) \\
=\ &4(32) \\
=\ &128
\end{aligned}$$

The rule for calculating the area of a trapezoid can be written in a more compact form.

Area of Trapezoid $= \frac{1}{2}(b_1 \times h) + \frac{1}{2}(b_2 \times h)$ **becomes** $\frac{1}{2}h(b_1 + b_2)$

Example 6 Find the area of the trapezoid whose vertices are $M(8, 4)$, $N(3,4)$, $O(0, 0)$, $P(10,0)$.

To apply the area rule for a trapezoid, you need to know 3 measures: the measure of the altitude, and the measure of each of the 2 bases of the trapezoid.

The height, NH, is the distance between the parallel bases.

height NH = 4 units
base OP = 10 − 0 = 10 units
base NM = 8 − 3 = 5 units

Trapezoid $MNOP = \frac{1}{2}h(b_1 + b_2)$

$= \frac{1}{2} \times 4(10 + 5)$ $h = 4$, $b_1 = 10$, $b_2 = 5$

$= \frac{1}{2} \times 4(15)$ First, work in the parentheses.

$= 2(15)$ Do the multiplication at the left.

$= 30$ Do the next multiplication.

Answer: The area of the trapezoid is 30 square units.

Try These *(For solutions, see page 419.)*

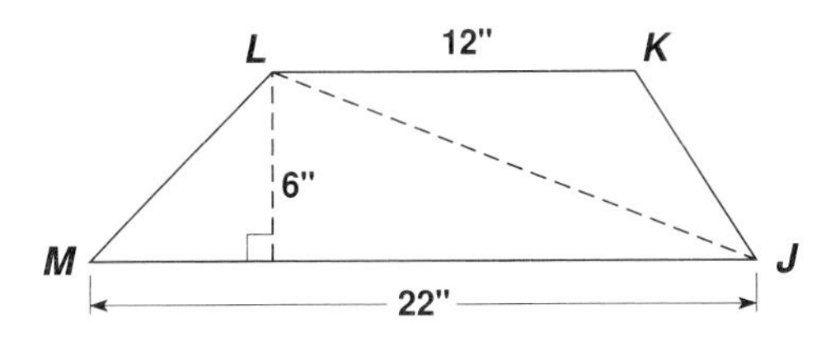

1. **a.** When diagonal $\overline{LJ}$ is drawn in trapezoid $JKLM$, name the 2 triangles that make up the trapezoid.
   b. What is the measure of the altitude to base $\overline{MJ}$ of $\triangle MLJ$?
   What is the measure of the altitude to base $\overline{LK}$ of $\triangle JKL$?
   c. What are the measures of $\overline{MJ}$ and $\overline{LK}$?
   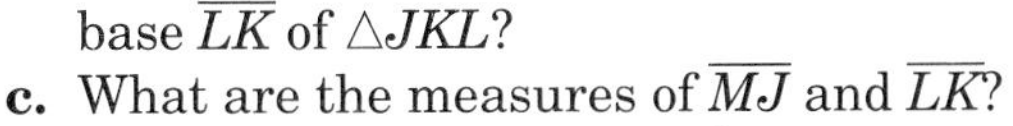
   d. What is the area of $\triangle MLJ$?
   What is the area of $\triangle JKL$?
   e. What is the area of trapezoid $JKLM$?
2. **a.** On graph paper, draw trapezoid $ABCD$ with vertices $A(3, 0)$, $B(12, 0)$, $C(9, 6)$, $D(5, 6)$.
   b. What is the distance between the parallel sides?
   c. What are the measures of the two bases of the trapezoid?
   d. Apply the rule $\frac{1}{2}h(b_1 + b_2)$ to find the area of trapezoid $ABCD$.

SELF-CHECK: SECTION 9.2

1. How is the area of a parallelogram related to the area of a rectangle of the same dimensions?

Answer: A parallelogram and rectangle of the same dimensions are equal in area.

2. If the dimensions of a parallelogram are called the base b and the height h, what is a rule for finding the area?

Answer: Area of Parallelogram = $b \times h$

3. **a.** When a parallelogram is cut along a diagonal, describe the figures formed.

Answer: 2 congruent triangles

b. How is the area of one of these triangles related to the area of the parallelogram?

Answer: The area of the triangle is one-half the area of the parallelogram.

c. If the dimensions of the parallelogram are called the base b and the height h, what is a rule for finding the area of the triangle cut from the parallelogram?

Answer: Area of Triangle = $\frac{1}{2}(b \times h)$

4. **a.** If a diagonal of a trapezoid is drawn, describe the figures formed.

Answer: 2 triangles that are not congruent, but they have the same height

b. How is the area of the trapezoid related to the areas of these 2 triangles?

Answer: The area of the trapezoid is equal to the sum of the areas of the 2 triangles.

c. If the 2 triangles have the same height h drawn to the different bases b_1 and b_2, write a rule for finding the area of each triangle, and the area of the trapezoid.

Answer: Area of Triangle 1 = $\frac{1}{2}(b_1 \times h)$

Area of Triangle 2 = $\frac{1}{2}(b_2 \times h)$

Area of Trapezoid = $\frac{1}{2}(b_1 \times h) + \frac{1}{2}(b_2 \times h)$

d. Use the common factors in this sum to write the rule for the area of a trapezoid in a more compact form.

Answer: Area of Trapezoid = $\frac{1}{2}h(b_1 + b_2)$

EXERCISES: SECTION 9.2

1. **a.** Using the 2 shapes shown, draw:

(1) a parallelogram

(2) a rectangle

b. What must be true about the areas of the parallelogram and rectangle you have drawn? Explain.

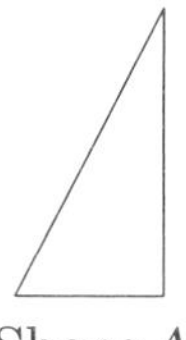
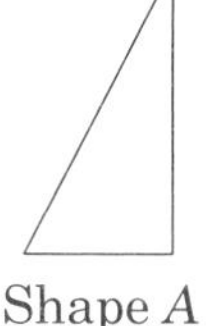

Shape A

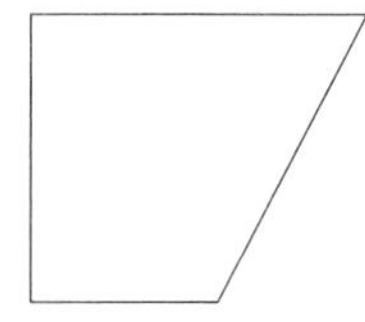

Shape B

2. Find the number of square units in the area of each figure

a. 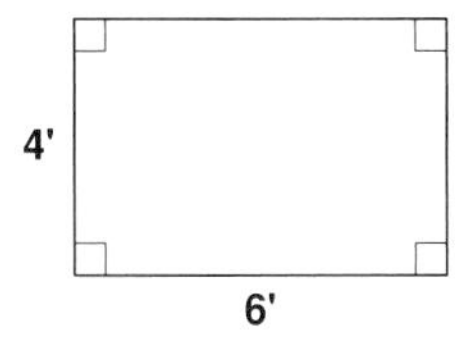

b. 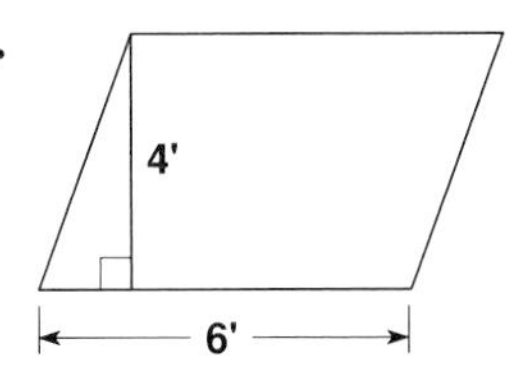

c. 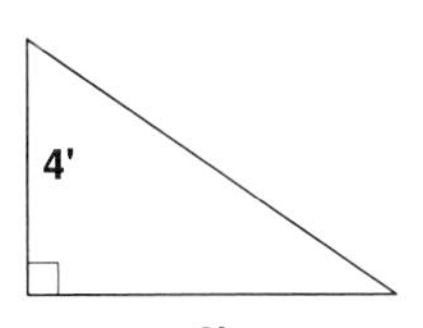

d. 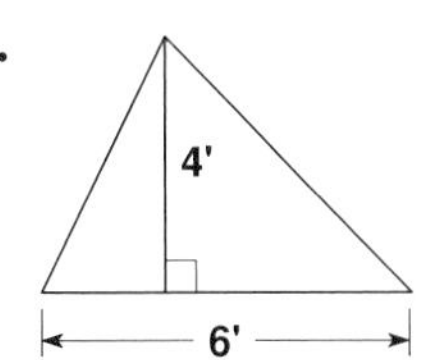

e. a rectangle with base $8\frac{1}{2}$ cm and height 6 cm
f. a square with side 9'
g. a right triangle with legs 5 yards and 12 yards
h. a parallelogram with base 7.6" and height 12"
i. a triangle with base 10 m and height 6.4 m

3. Find the area and perimeter of each figure

a. 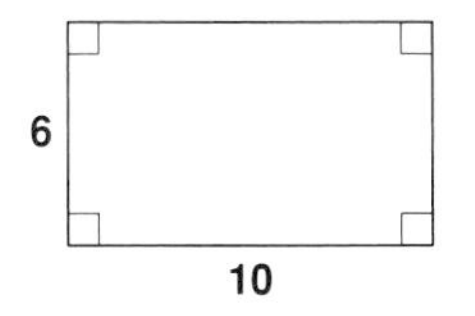

b. 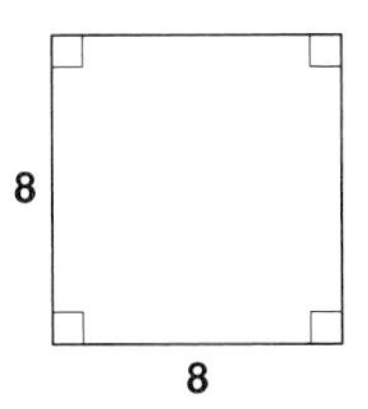

c. 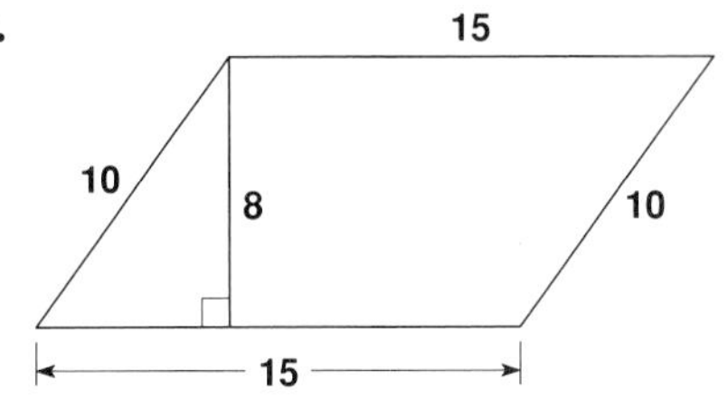

d. 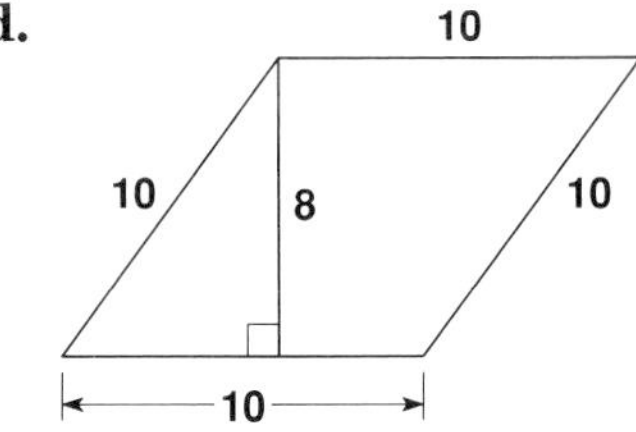

e. 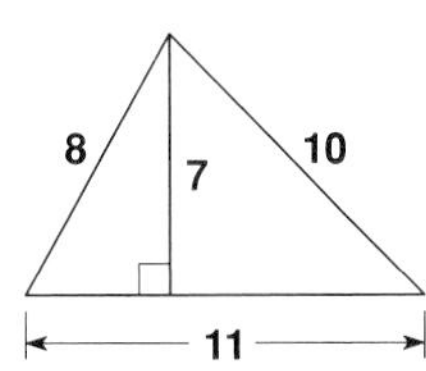

f. 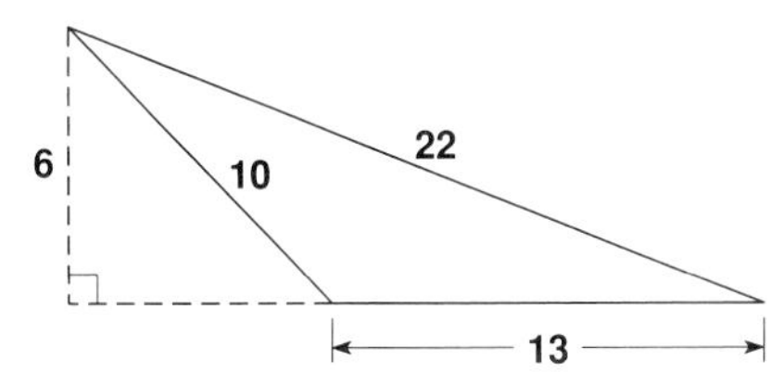

4. 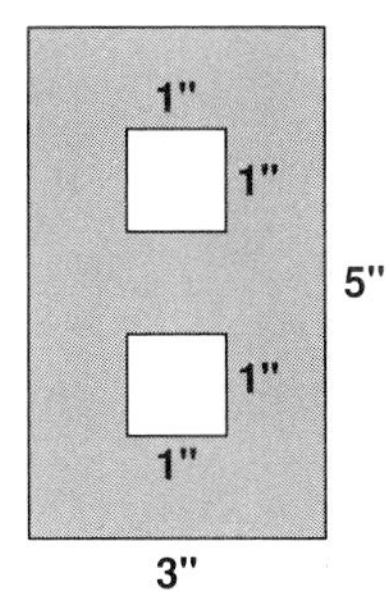

The design for a special metal plate to cover electrical outlets is made up of 2 congruent squares cut out from a rectangle. Using the dimensions shown, find the area covered by the metal.

5. The design for a wooden hot pad is made up of 6 congruent parallelograms placed side by side around a center point. Each parallelogram has a base of 2.5" and a height of 2.2".
   a. Find the area of the pad.
   b. Find the perimeter of the pad.
   c. How many degrees are there in the measure of each angle with a vertex at the center of the pad?

6. Each of these oddly-shaped figures is made up of familiar shapes for which you know how to find area.

Find the area of each figure.

a.

16 ft.
12 ft.
8 ft.
6 ft.

b.

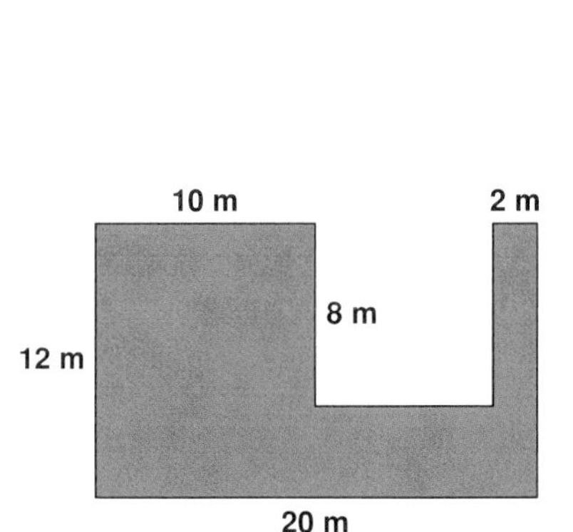

c.

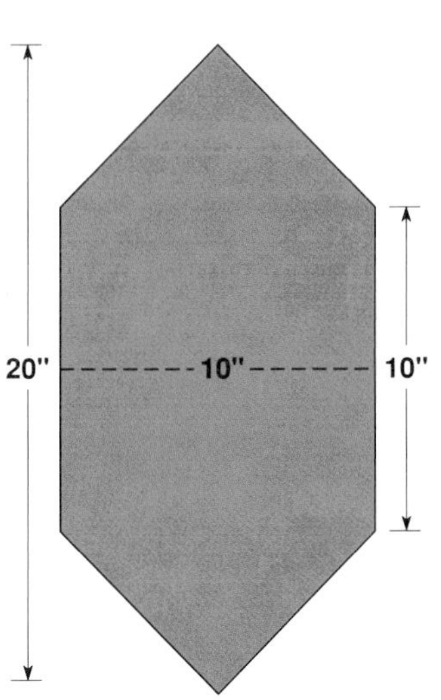

7. The coordinates of the vertices of a parallelogram *MNOP* are *M*(10, 5), *N*(3, 5), *O*(0, 0), *P*(7, 0).

a. Draw the figure on graph paper

b. How long is base $\overline{OP}$?

c. Draw altitude $\overline{NH}$ from *N* to $\overline{OP}$. What are the coordinates of *H*? How long is altitude $\overline{NH}$?

d. Find the area of parallelogram *MNOP*.

8. On graph paper, draw each parallelogram with the given vertices, and find the area of the parallelogram.

a. *MNOP* with	*M*(10, 5)	*N*(2, 5)	*O*(0, 0)	*P*(8, 0)
b. *JKLM* with	*J*(3, 0)	*K*(8, 0)	*L*(10, 5)	*M*(5, 5)
c. *PQRS* with	*P*(–4, 0)	*Q*(9, 0)	*R*(15, 4)	*S*(2, 4)
d. *ABCD* with	*A*(–4, –2)	*B*(8, –2)	*C*(10, 3)	*D*(–2, 3)
e. *TUVW* with	*T*(3, 6)	*U*(0, 8)	*V*(0, 2)	*W*(3, 0)

9. Find a fourth pair of coordinates that, together with the 3 pairs of coordinates given, will form a parallelogram. Find the area of each parallelogram.

a. *A*(1, 1)	*B*(2, 3)	*C*(6, 3)	*D*(5, ?)
b. *J*(–1, 4)	*K*(2, 4)	*L*(4, –1)	*M*(?, –1)
c. *E*(–1, 2)	*F*(1, 3)	*G*(1, –2)	*H*(–1, ?)
d. *R*(8, 0)	*S*(–3, 0)	*T*(0, 4)	*U*(?, ?)

10. On graph paper, draw each triangle with the given vertices, and find the area of the triangle.

a. *MNO* with $M(0, -8)$ $N(10, 0)$ $O(0, 0)$

b. *OPQ* with $O(0, 0)$ $P(14, 0)$ $Q(5, 5)$

c. *ABC* with $A(-4, 0)$ $B(0, 6)$ $C(8, 0)$

d. *RST* with $R(-4, 0)$ $S(3, -5)$ $T(8, 0)$

e. *NOP* with $N(0, 8)$ $O(0, 0)$ $P(3, 6)$

11. a. Choose the numbers you need from the numbers shown to find the area of each shaded triangle.

Triangle I

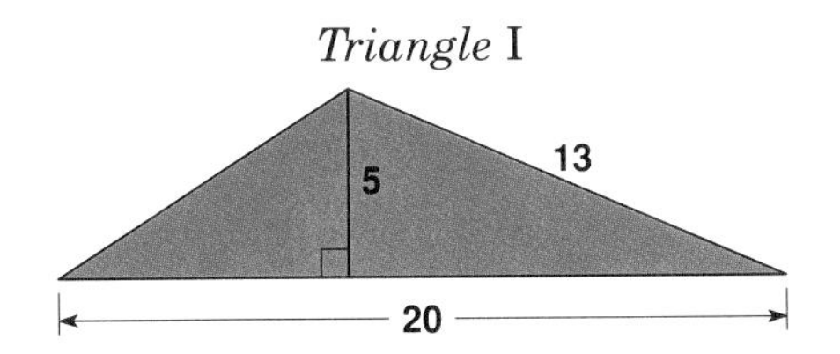

Triangle II

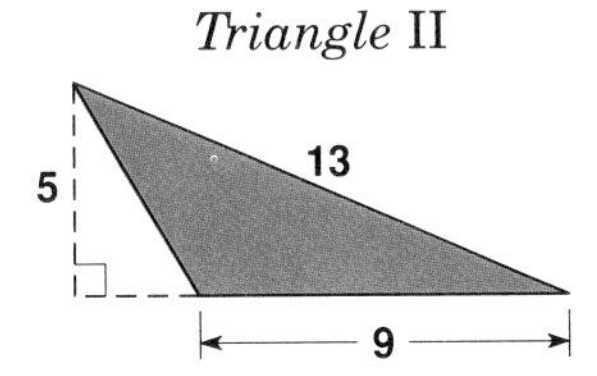

b. Explain how you could fit the 2 shaded triangles together to form a trapezoid. Draw the trapezoid, showing the numbers you know.

c. Use the areas you found for the triangles to determine the area of the trapezoid.

d. Use the rule $\frac{1}{2}h(b_1 + b_2)$ to determine the area of the trapezoid. Does this result agree with the result you found by using the triangles to find the area of the trapezoid?

12. Find the number of square units in the area of each trapezoid.

a.

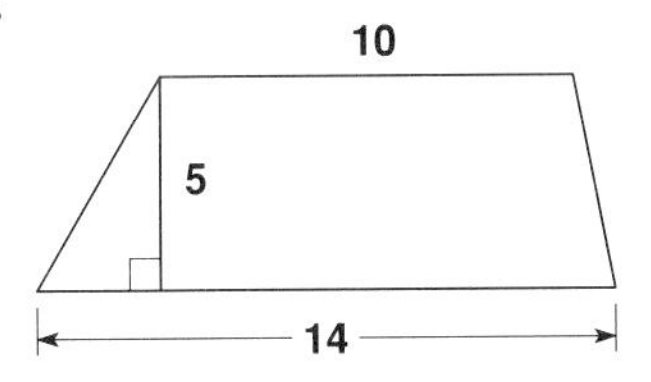

b

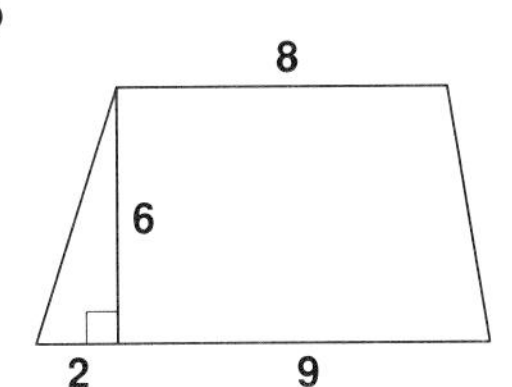

c.

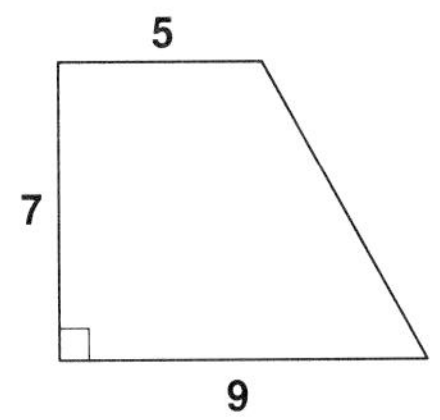

d. a trapezoid with height 12.8 cm and bases 10 m and 30 cm

e. a trapezoid with height 9 inches and bases 8 feet and 12 feet

13. The coordinates of the vertices of trapezoid *MNOP* are $M(8, 5)$, $N(2, 5)$, $O(0, 0)$, $P(12, 0)$.

a. Draw the figure on graph paper.

b. How long is base $\overline{OP}$? base $\overline{NM}$?

c. Draw altitude $\overline{NH}$ from N to $\overline{OP}$. What are the coordinates of H? How long is altitude $\overline{NH}$?

d. Find the area of trapezoid *MNOP*.

14. On graph paper, draw each trapezoid with the given vertices, and find the area of the trapezoid.

a. *ABCD*	with	*A*(2, 0)	*B*(14, 0)	*C*(10, 7)	*D*5, 7)
b. *RSTU*	with	*R*(–4, 0)	*S*(12, 0)	*T*(8, 8)	*U*(–2, 8)
c. *GIJK*	with	*G*(–5, –5)	*I*(10, –5)	*J*(8, 7)	*K*(–2, 7)
d. *EFGH*	with	*E*(3, 5)	*F*(3, 0)	*G*(11, 0)	*H*(8, 5)
e. *KLMN*	with	*K*(0, 8)	*L*(4, 8)	*M*(4, –2)	*N*(0, –5)

15. Use an equation to find the value of the variable shown.

a. The area of this parallelogram is 120 square units.

$x - 3$

12

Find the value of x.

b. The area of this triangle is 77 square units.

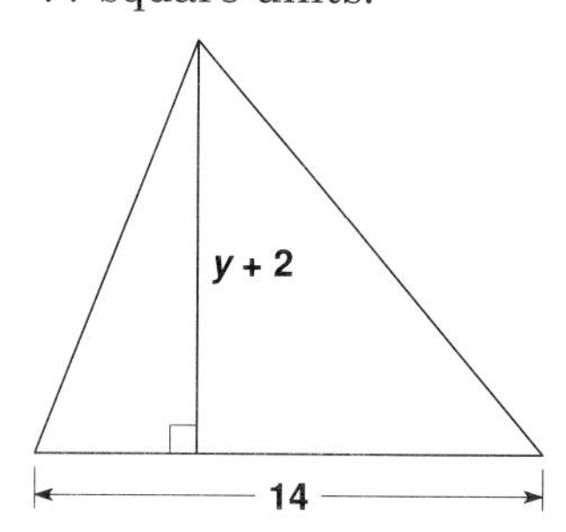

Find the value of y.

c. The area of this trapezoid is 100 square units.

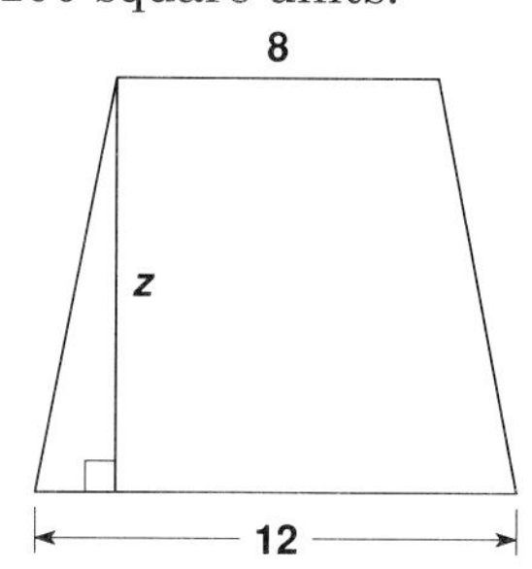

Find the value of z.

16. Use the area rule for each figure to find the missing dimension.

a. The area of a parallelogram is 41 square meters, and the base measures 8.2 meters. Find the measure of the height.

b. The area of a triangle is 108 square feet and a base measures 12 feet. Find the measure of the altitude drawn to that base.

c. The area of a trapezoid is 24 square yards, and the measures of the bases are 3 yards and 9 yards. Find the measure of the altitude.

d. The area of a trapezoid is 36 square inches, the height is 6 inches, and the shorter base measures 5 inches. Find the measure of the longer base.

EXPLORATIONS

On graph paper, draw $\triangle ABC$ with vertices $A(2, 2)$, $B(10, 4)$, $C(4, 8)$.
Explain why you cannot easily find the area of $\triangle ABC$ using the area rule.

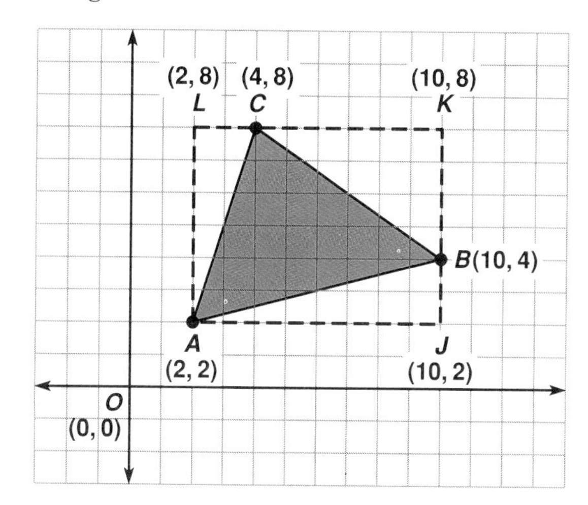

Since you need to work with horizontal and vertical lines to find area, let's draw a rectangle through the vertices of $\triangle ABC$.

Explain how you would find the area of rectangle $AJKL$.

Rectangle $AJKL$ is made up of 4 triangles. Name them.

Explain why it is easy to find the areas of 3 of these triangles: $\triangle ALC$, $\triangle CKB$, $\triangle ABJ$.

What calculations would you do with the area of the rectangle and the areas of the 3 right triangles to be left with the area of $\triangle ABC$?

Here's how to organize your work.

1. From their positions on the graph, complete the coordinates of the vertices of the rectangle. $A(2, 2)$ $J(10, 2)$ $K(10, 8)$ $L(2, 8)$
2. Find the area of the rectangle.

   base $AJ = 10 - 2 = 8$

   height $AL = 8 - 2 = 6$

   Area of Rectangle $AJKL$ = base × height

   $= 8 \times 6 = 48$ square units
3. Find the area of the 3 right triangles.

   Legs of Right Triangle ALC are $AL = 6$ and $LC = 2$.

   Area of Right Triangle $ALC = \frac{1}{2}(6 \times 2) = \frac{1}{2}(12) = 6$ square units

   Legs of Right Triangle CBK are $CK = 6$ and $KB = 4$.

   Area of Right Triangle $CBK = \frac{1}{2}(6 \times 4) = \frac{1}{2}(24) = 12$ square units

   Legs of Right Triangle ABJ are $AJ = 8$ and $BJ = 2$

   Area of Right Triangle $ABJ = \frac{1}{2}(8 \times 2) = \frac{1}{2}(16) = 8$ square units
4. Subtract the sum of the areas of the 3 right triangles from the area of the rectangle to give the area of the original triangle.

   $\triangle ABC$ = Rectangle $AJKL - (\triangle ALC + \triangle CKB + \triangle ABJ)$

   $= 48 - (6 + 12 + 8)$

   $= 48 - (26) = 22$ square units

Match this answer against what you see on the grid. Does the answer seem reasonable?

Use this method of building a rectangle through the vertices of a triangle to find the areas of these triangles:

a. *MNO* with $M(7, 2)$ $N(2, 5)$ $O(0, 0)$

b. *NOP* with $N(2, 4)$ $O(0, 0)$ $P(6, 1)$

c. *GHI* with $G(-2, 2)$ $H(2, 6)$ $I(8, 5)$

d. *UVW* with $U(2, 5)$ $V(6, -6)$ $W(-2, -3)$

9.3 The Circle

Circumference

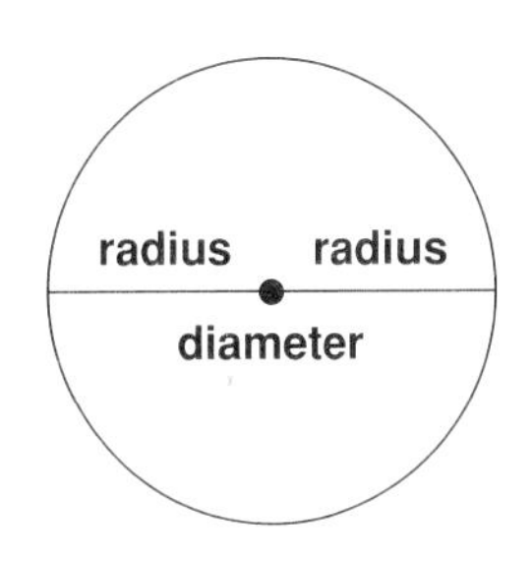

Use a compass to draw a circle of radius 2 inches.

How long is the diameter of this circle?

The diameter of this circle is 4 inches, twice the measure of the radius.

The measure of a diameter of a circle is twice the measure of a radius. $d = 2r$

Carefully lay out a piece of string along the perimeter of the circle, which is called the ***circumference***. Cut the string to the length of the circumference. Use a ruler to measure the length of the string.

How does the length of the circumference compare to the length of the diameter?

The circumference of this circle measures about 12.5 inches, just a little more than 3 times the length of the diameter.

The ratio of the measures is just more than 3. $\frac{\text{circumference}}{\text{diameter}} = 3+$

Repeat this experiment with a circle of a different size radius. What general result appears to be true?

The ratio of the measure of the circumference and diameter of a circle is always the same, just more than 3.

$$\frac{C}{d} = 3+$$

The value of this constant ratio is represented by the Greek letter pi, π.

$$\frac{C}{d} = \pi$$

Close approximations of π are 3.14 or $\frac{22}{7}$.

Example 1 Find the length of the circumference of a circle whose diameter measures 5 inches.

Apply the relationship between the measures of the circumference and diameter of a circle. $\frac{C}{d} = \pi$

Substitute the value of the diameter. $\frac{C}{5''} = \pi$

Cross multiply. $C = 5\pi$ inches

The measure of the circumference may be left in terms of π or it may be evaluated by substituting a value for π.

$C = 5(3.14)$

$= 15.7$ inches

To find the measure of the circumference of a circle when the diameter is known, use the rule $C = \pi d$.

Example 2 Find the number of feet in the circumference of the circle whose radius measures 14 feet.

To apply the rule, you need to know the length of the diameter.

$$d = 2r$$
$$= 2(14') = 28'$$

$$C = \pi d \quad \text{The rule for circumference.}$$
$$= \frac{22}{7} \cdot 28 \quad d = 28' \text{ Use: } \pi = \frac{22}{7}$$
$$= \frac{22}{7} \cdot \frac{28}{1} \quad \text{Rewrite the whole number.}$$
$$= \frac{22}{\cancel{7}_1} \cdot \frac{\cancel{28}^4}{1} \quad \text{Cancel.}$$
$$= 88 \quad \text{Multiply.}$$

The length of the circumference is 88 feet.

It is also useful to convert the rule for circumference from using the measure of the diameter to using the measure of the radius.

$$C = \pi d$$
$$= \pi(2r)$$
$$C = 2\pi r$$

To find the measure of the circumference of a circle, use:
$C = 2\pi r$ when you know the measure of the radius
$C = \pi d$ when you know the measure of the diameter

You can also use these rules to find the measure of the radius or diameter when the measure of the circumference is known.

Example 3 Find the measure of the radius of a circle whose circumference measures 16π feet

$$C = 2\pi r \quad \text{Apply the rule for circumference.}$$
$$16\pi = 2\pi r \quad \text{Circumference} = 16\pi$$
$$16 = 2r \quad \text{Divide both sides by } \pi.$$
$$8 = r \quad \text{Divide both sides by 2.}$$

The measure of the radius is 8 feet.

Try These *(For solutions, see page 420.)*

1. Find the circumference of the circle:
   a. whose diameter measures 10 inches Use: $\pi = 3.14$
   b. whose radius measures 7 meters Use: $\pi = \frac{22}{7}$
2. The circumference of a circle measures 16π centimeters. Find the measure of the radius.

Area

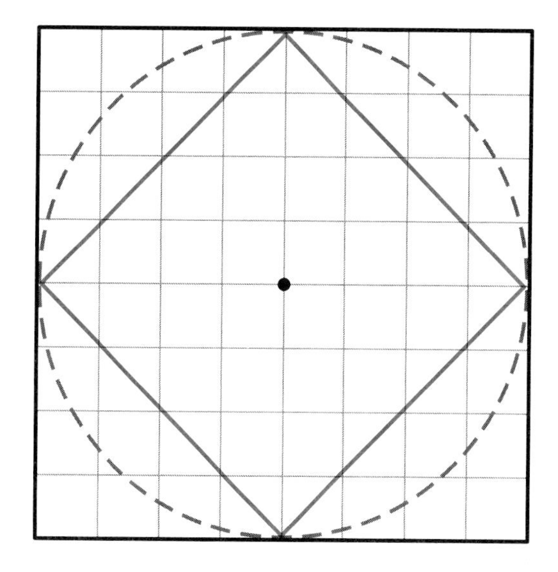

This figure shows a circle and 2 squares that "hug" the circle, all resting on a grid.

How many square units are in the area of the outer square? 64

How many square units are in the area of the inner square?

Counting whole square units and half square units, there are 32 square units in all for the inner square.

Make an observation about the number of square units in the area of the circle.

The number of square units in the area of the circle is about halfway between the areas of the 2 squares.

$$\text{Area of Circle} \approx \frac{64 + 32}{2} = 48$$

From the grid, tell the length of the radius of the circle. 4 units

How does the approximate area of the circle, 48 square units, compare to the length of the radius, 4 units? $48 = 3 \times 4^2$

Repeat these calculations with circles of different radii to verify that:

The area of a circle is about 3 times the square of the measure of the radius. $A = \pi r^2$

Example 4 Find the number of square centimeters in the area of a circle whose radius measures 7 centimeters.

Area of Circle $= \pi r^2$	Apply the rule for area.
$= \frac{22}{7}(7\text{ cm})^2$	Use $\pi = \frac{22}{7}$ and r = 7 cm.
$= \frac{22}{7}(49\text{ cm}^2)$	Square first.
$= \frac{22}{\cancel{7}_1}(\cancel{49}^{7}\text{ cm}^2)$	Cancel.
$= 154\text{ cm}^2$	Multiply.

The area of the circle is 154 square centimeters.

Example 5 If the area of a circle contains 81π square inches, how long is the radius?

Area of Circle $= \pi r^2$	Apply the rule for area.
$81\pi = \pi r^2$	Area $= 81\pi$
$81 = r^2$	Divide both sides by π.
$\sqrt{81} = r$	Take the square root.
$9 = r$	$9 \times 9 = 81$.

The measure of the radius is 9 inches.

Try These *(For solutions, see page 420.)*

1. Find the area of the circle whose radius measures 10 inches. Use: $\pi = 3.14$
2. The area of a circle contains 64π square feet. Find the measure of the radius.

SELF-CHECK: SECTION 9.3

1. What is the relationship between the measures of a diameter and radius of a circle?

   Answer: diameter $= 2 \times$ radius
2. What special name is given to the perimeter of a circle?

   Answer: circumference
3. What is the relationship between the measures of the circumference and diameter of a circle?

   Answer: The measure of the circumference is always just more than 3 times the measure of the diameter.
4. **a.** What ratio does the Greek letter π represent?

   Answer: π represents the constant ratio between the measures of the circumference and diameter of a circle.

   b. What are approximate values of π?

   Answer: $\pi \approx 3.14$ or $\pi \approx \frac{22}{7}$
5. Using symbols, write the rule that tells the relationship between the measures of:

   a. the circumference and diameter of a circle

   Answer: $C = \pi d$

   b. the circumference and radius of a circle

   Answer: $C = 2\pi r$
6. What is the rule for finding the area of a circle?

   Answer: Area of Circle $= \pi r^2$

EXERCISES: SECTION 9.3

1. For each circle, write the length of a radius and a diameter.

a.

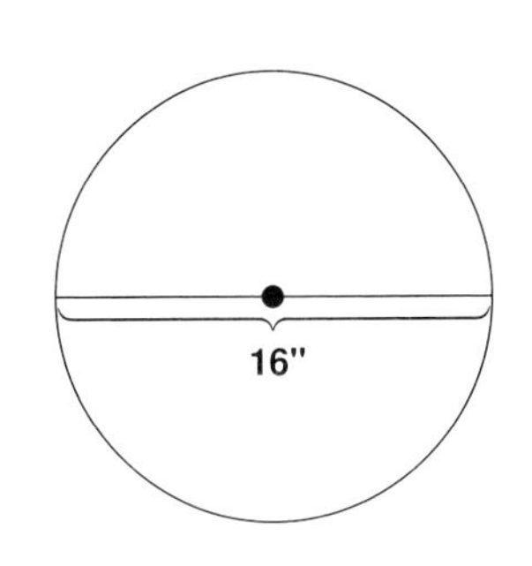

b.

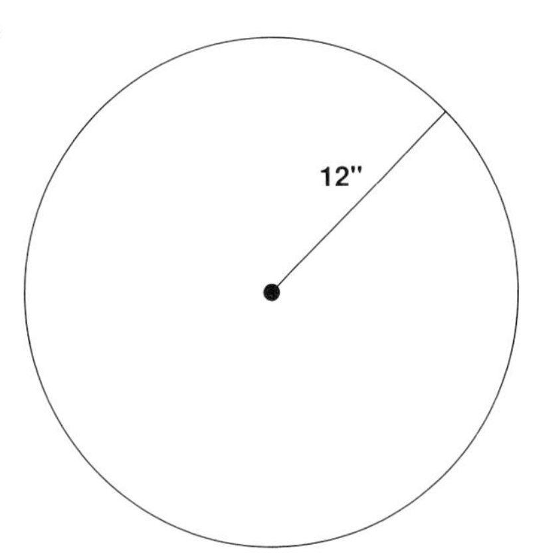

c.

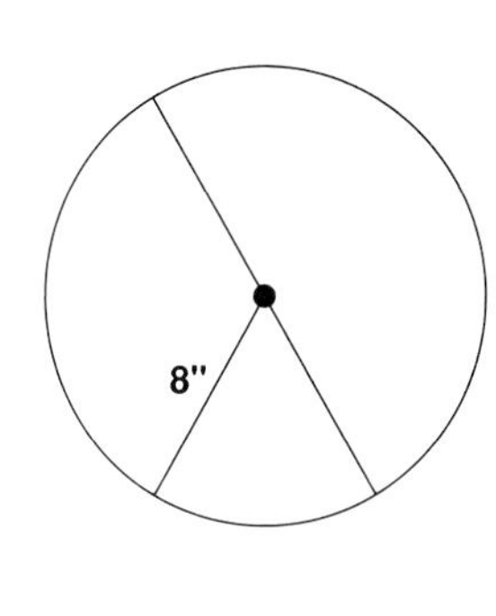

d.

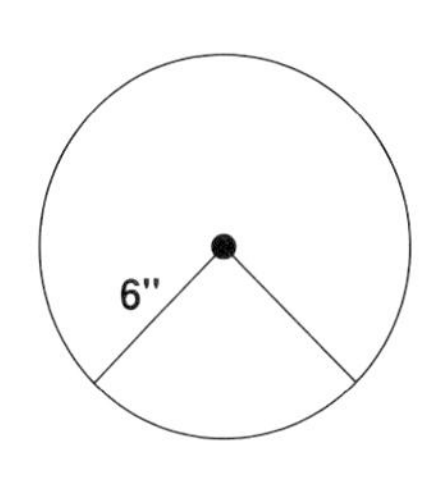

e.

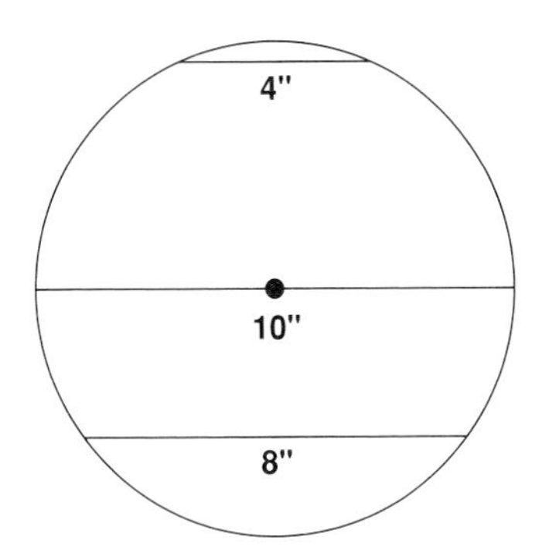

f.

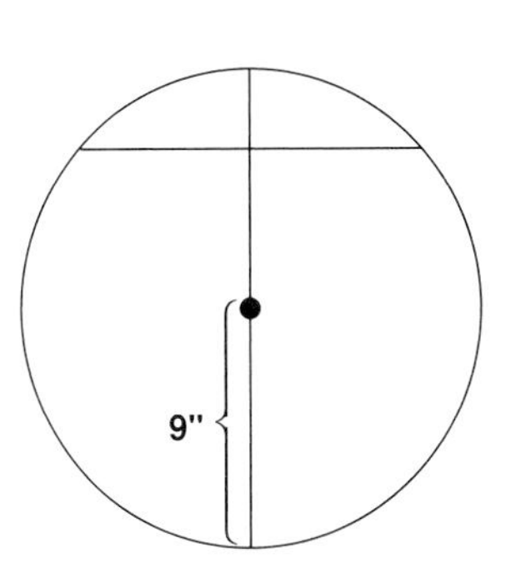

2. Find the circumference, in terms of π, of a circle:

a. with diameter 4 yards

b. with diameter 5 inches

c. with radius 2 meters

d. with radius 3.5 centimeters

e. with diameter $6\frac{1}{2}$ feet

f. with radius 120.3 millimeters

3. Use the given approximation for π to find the circumference of each circle.

a. $\pi = 3.14$ and **(1)** diameter = 20 cm **(2)** radius = 15 mm
(3) diameter = 1.1 m **(4)** radius = 0.2 yd.

b. $\pi = \frac{22}{7}$ and **(1)** diameter = 14 ft. **(2)** radius = 28 cm
(3) diameter = 49 mm **(4)** radius = 0.7 yd.

4. Find the diameter of the circle with the given circumference:

a. when the circumference is in terms of π

(1) $C = 20\pi$ cm **(2)** $C = 14\pi$ ft. **(3)** $C = 3.8\pi$ yd.

b. when an approximation has been used for π

(1) $C = 22$ m when $\pi = \frac{22}{7}$ **(2)** $C = 31.4$ cm when $\pi = 3.14$

(3) $C = 44''$ when $\pi = \frac{22}{7}$ **(4)** $C = 628$ mm when $\pi = 3.14$

5. In each figure, the point represents the center of a circle. Using the dimensions shown, find the perimeter of the figure. Use: $\pi = 3.14$

a.

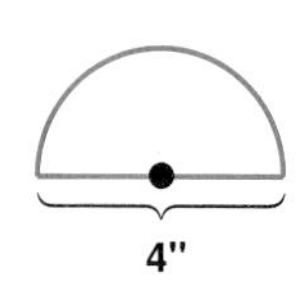

b.

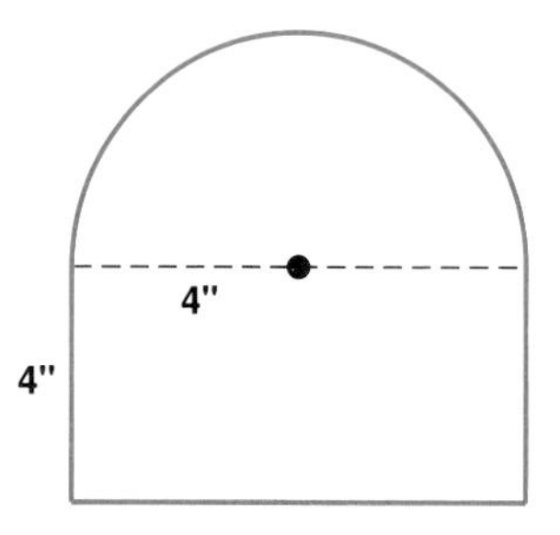

c.

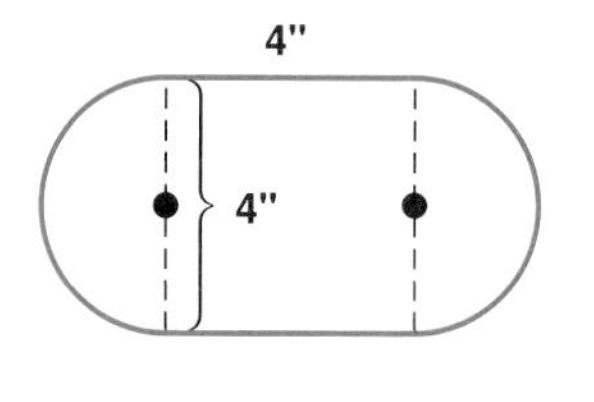

6. A rectangular doorway is 6 feet wide and 8 feet high. Could a circular table top of diameter 9 feet pass through this doorway? Explain.

7. Find the area, in terms of π, of a circle:

a. with radius 6 in. **b.** with radius 4 m

c. with radius $\frac{1}{4}$ ft. **d.** with radius 1.2 cm

e. with diameter 8 yd. **f.** with diameter 2.2 m

8. Use the given approximation for π to find the area of each circle.

a. $\pi = 3.14$ and **(1)** radius = 10 in. **(2)** radius = 20 mm

(3) diameter = 100 m **(4)** diameter = 2.4 ft.

b. $\pi = \frac{22}{7}$ and **(1)** radius = 7 ft. **(2)** radius = 14 m

(3) diameter = 28 cm **(4)** diameter = 42 in.

9. Find the radius of the circle with the given area:

a. when the area is in terms of π

(1) $A = 25\pi$ sq. in. **(2)** $A = 100\pi$ sq. yd. **(3)** $A = 225\pi$ m^2

b. when an approximation has been used for π

(1) $A = 314$ sq. in. when $\pi = 3.14$

(2) $A = 154$ sq. ft. when $\pi = \frac{22}{7}$

(3) $A = 1{,}256$ cm^2 when $\pi = 3.14$

(4) $A = 616$ m^2 when $\pi = \frac{22}{7}$

10. The shocks from an earthquake move in circular motion. From a quake on June 12, with epicenter outside of San Leandro, the greatest circumference recorded measured 110 π miles.

If the town of San Luis is 50 miles from the epicenter, was this town within range of the quake? Explain.

11. In each figure, the point represents the center of a circle. Using the dimensions shown, find the area of the shaded figure. Use: $\pi = 3.14$

a.

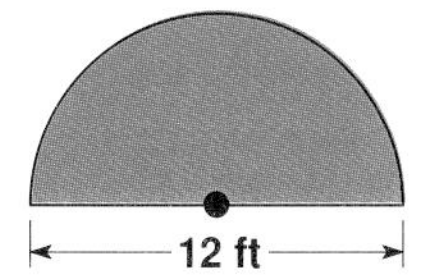

b

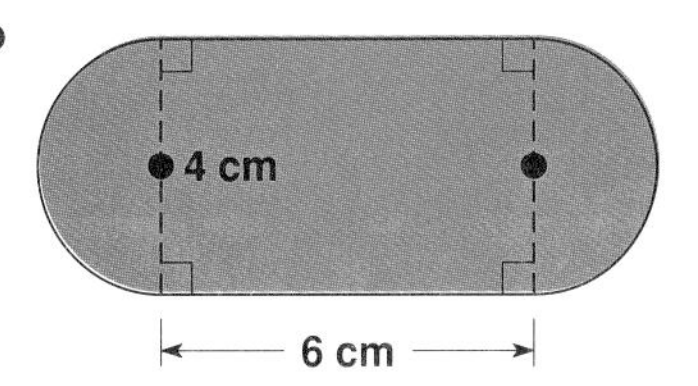

c

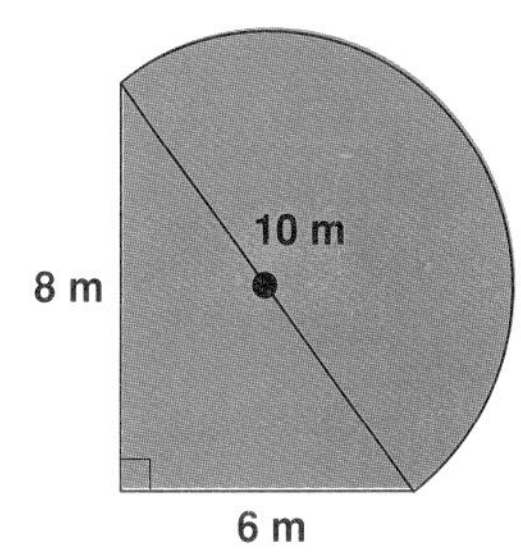

d.

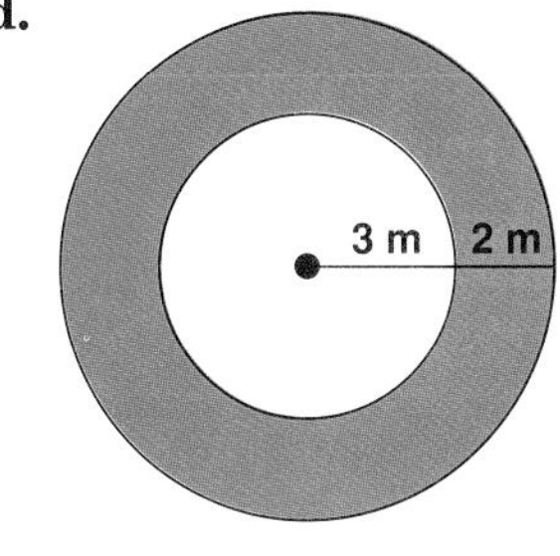

e.

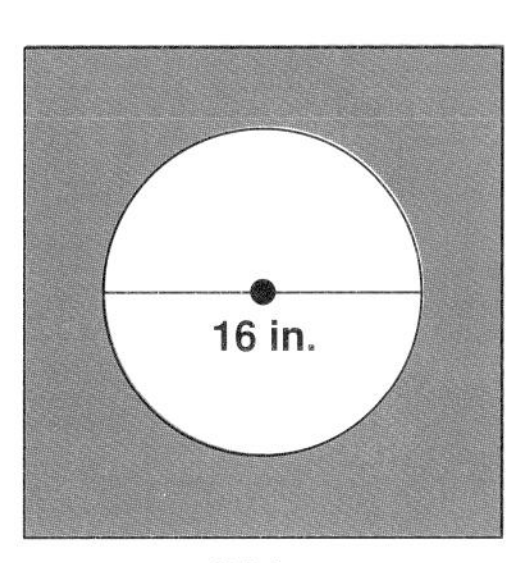

f.

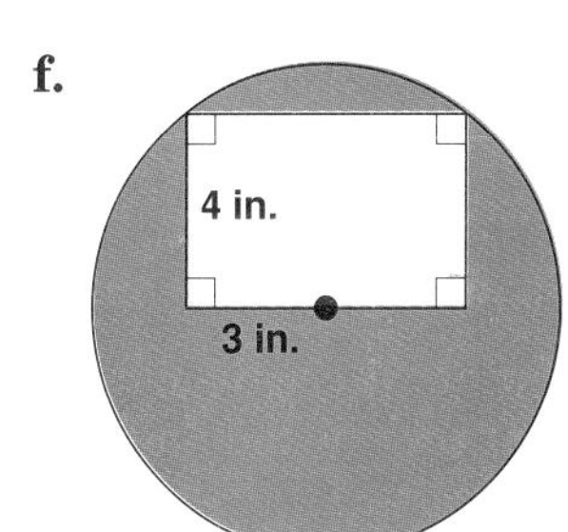

12. Guesstimate an answer before solving each problem. Answers may be left in terms of π.

a. An electric fan has blades that rotate. If the measure of a blade from the center of the fan to the tip of the blade is 5.5", what is the area of the region that the blades sweep through?

b. Each of the 2 windshield wipers on Theo's car sweeps through a semicircular region. If a wiper blade is 16 inches long, what is the total area of the windshield that is cleared?

c. The 9 rectangular panes of glass in the Edison's new windows combine to form a large rectangle 30 inches across and 48 inches high. The remaining 5 panes of glass at the top of the window form a semicircle. What is the total area of all the glass?

d. The radius of Tallulah's country-western CD measures 6 cm. The disc has 3 bands. The clear inner band has a radius of 2.5 cm, and the clear band at the outer rim is 0.5 cm wide. Find the area of the middle band, which contains the music.

EXPLORATIONS

1. Here's another way to come up with the area rule for a circle.

Cut out a paper circle about 5" in diameter.

Fold the circle in half 3 consecutive times to create 8 congruent parts.

Shade every other part.

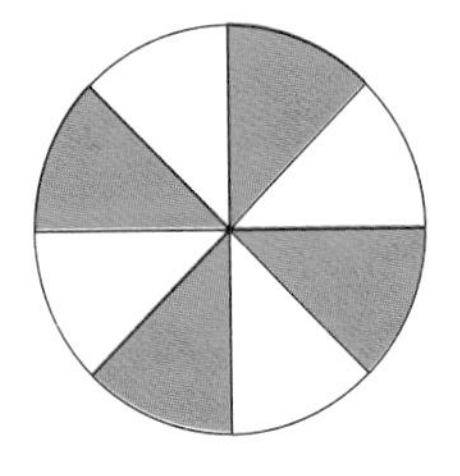

Cut out these 8 parts, and lay them out so that the shaded and unshaded parts alternate, as shown.

When the 8 parts of the circle are laid out this way, what does the resulting figure remind you of?

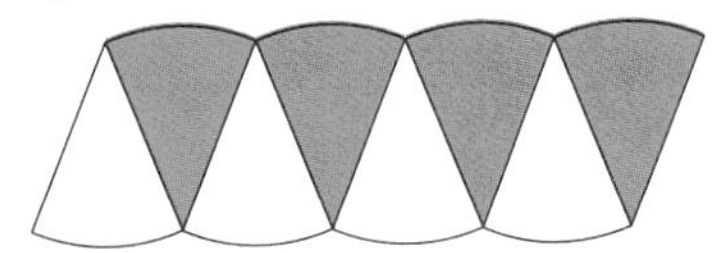

This figure is not quite a parallelogram.

How does the area of this "parallelogram" compare to the area of the circle?

The area of the circle must be the same as the area of this "parallelogram" since the same parts were used.

To calculate the area of this "parallelogram," what measures do you need to know?

You need to know the measures of the base and height.

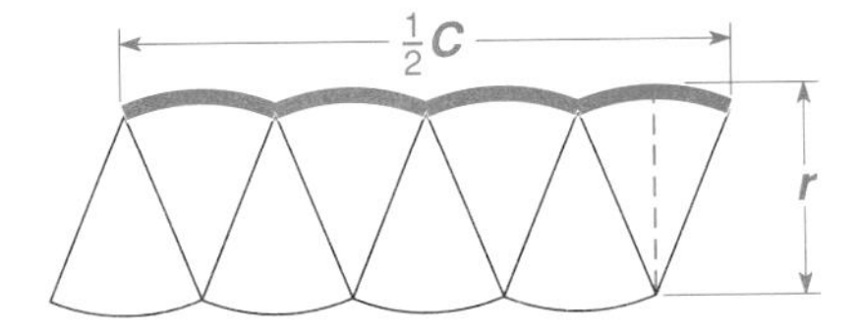

The base is as long as 4 arcs of the circle, or one-half the measure of the circumference.

The height is equal to the measure of the radius.

Area of Circle ≈ Area of Parallelogram

$= \text{base} \times \text{height}$

$= \frac{1}{2}C \times r$

$= \frac{1}{2}(2\pi r) \times r$

Area of Circle $= \pi r^2$

2. Tracing Puzzle

Each of these figures is made up of points, called *vertices*, and strokes that connect the vertices. The strokes may be straight or curved.

Lay a thin sheet of paper over a figure, and try to trace the figure WITHOUT LIFTING YOUR PENCIL OR RETRACING ANY STROKE.

Begin at a vertex. If you cannot complete the trace, try a different vertex as a starting point.

For example, in this figure, if you begin at vertex A, you cannot complete the trace.

But, begin at vertex B or at vertex C, and it works!

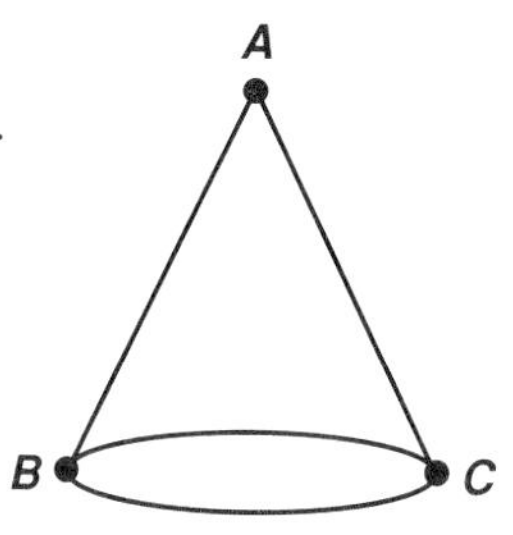

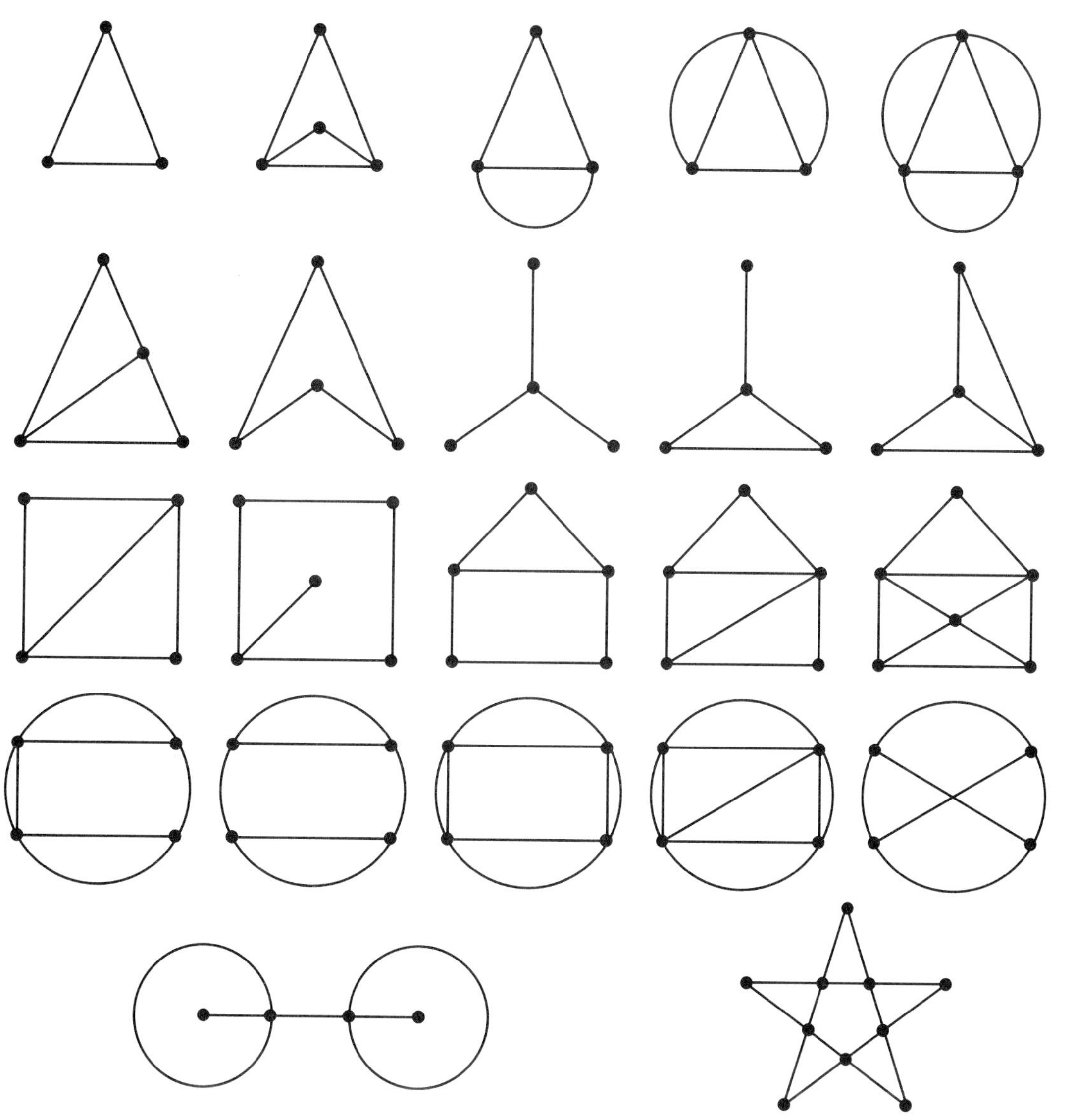

9.4 Three Dimensions

3-Dimensional Figures

Draw a 1-inch square at each corner of a piece of notebook paper.

From each outside corner of the paper, draw the diagonal of the square.

Cut each corner of the paper along the diagonal of the square.

Fold the paper up (along the dotted lines shown in the diagram) to create an open box.

You have just gone from 2 dimensions to 3 dimensions.

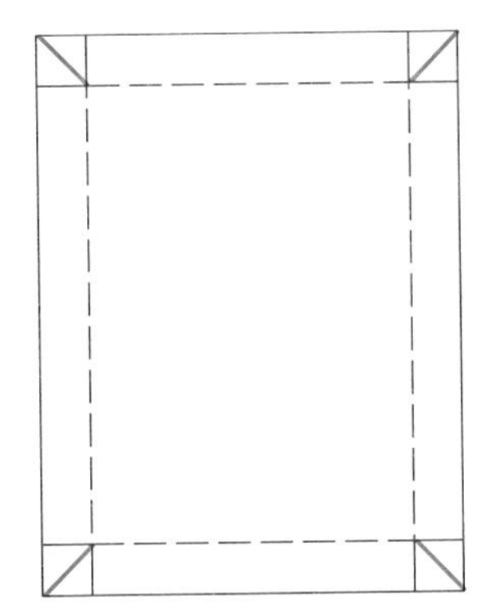

Here are some familiar 3-dimensional figures, or ***solids***.

A ***prism*** is a special 3-dimensional figure with bases that are congruent.

A ***right prism*** has sides perpendicular to the bases.

A prism may be named according to the shape of its bases.

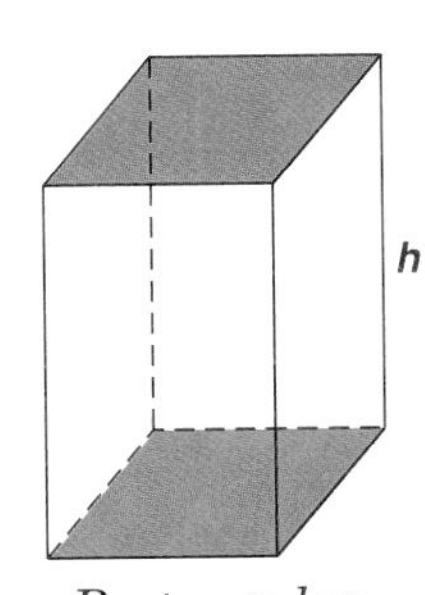

Rectangular Right Prism

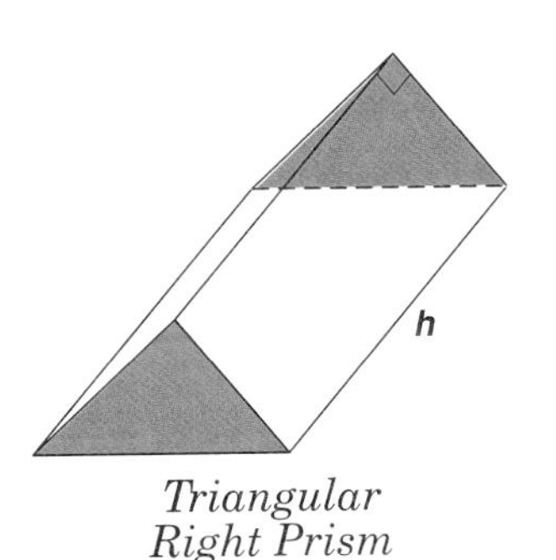

Triangular Right Prism

A ***pyramid*** has triangular faces on a base such as a square.

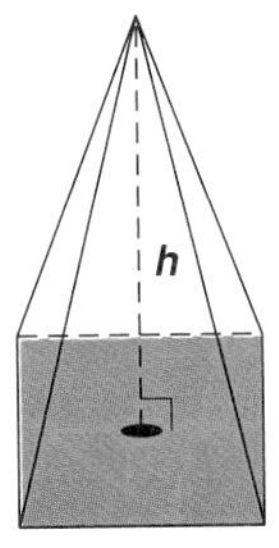

Pyramid

Other solids are related to a circle.

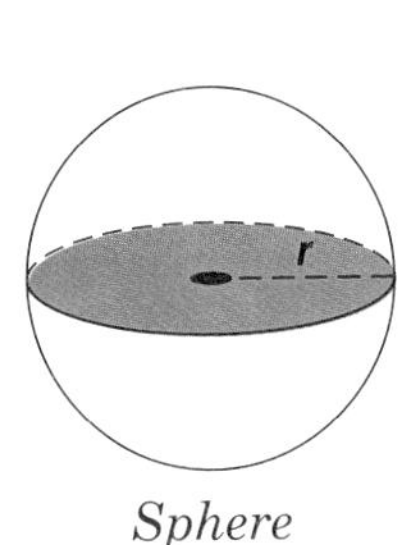

Sphere

Cylinder

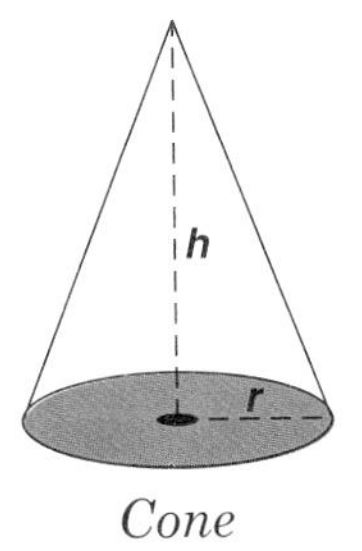

Cone

Example 1 Cut out and mark a piece of paper in this pattern.

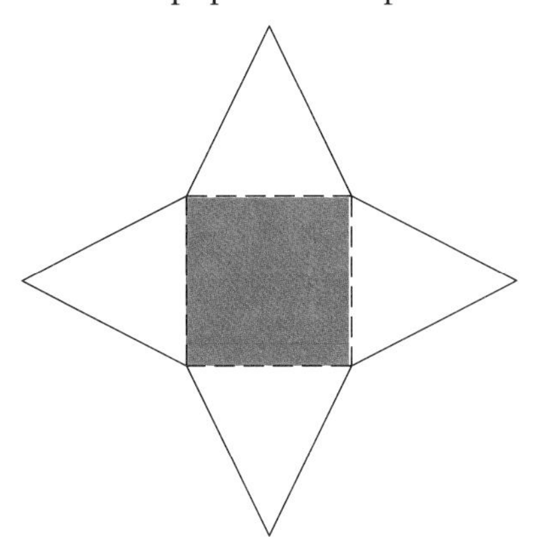

Fold the paper model along the dotted lines. What kind of 3-dimensional figure is formed?

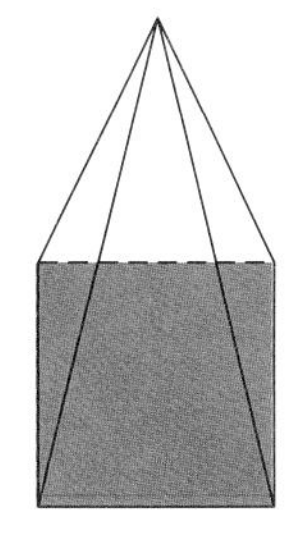

The figure is a square pyramid.

Try This *(For solution, see page 420.)*

Cut out and mark a piece of paper in this pattern.

Fold the paper model along the dotted lines.

What kind of 3-dimensional figure is formed?

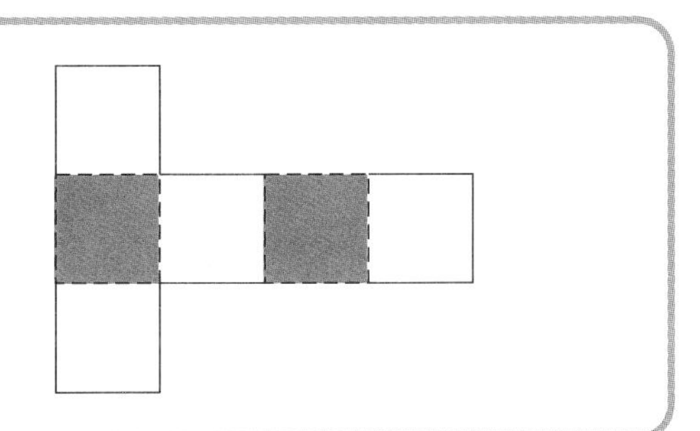

Surface Area

From a sheet of plain paper, cut out a 5" by 8" piece, and mark off this piece in 1" squares.

From each outside corner of the paper, draw the diagonal of the corner square.

Cut each corner of the paper along the diagonals of the squares.

Fold the paper up to create an open box.

How many sides (called ***faces***) does the open box have?

How many square inches are in each face?

Make a face to exactly cover the top of this box. How many square inches are in this face?

What is the geometric name of this box?

What are the measures of the 3 dimensions of this box?

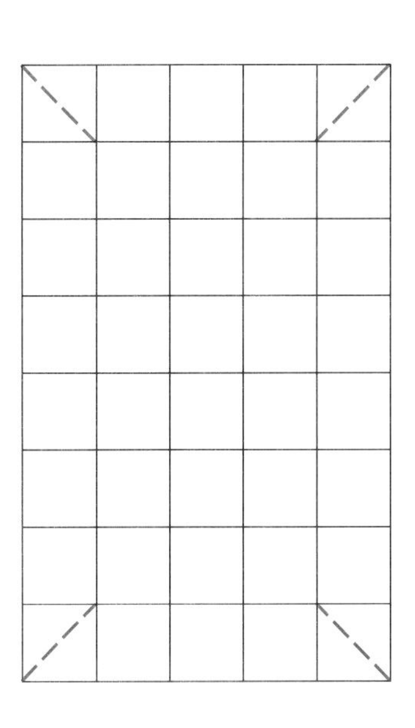

(Continues on next page)

Your covered box is a rectangular right prism with 6 faces. The prism is 3" wide, 6" long, and 1" high. Here are the areas of the 6 rectangular faces of the prism. In all, the 6 faces contain 54 square inches.

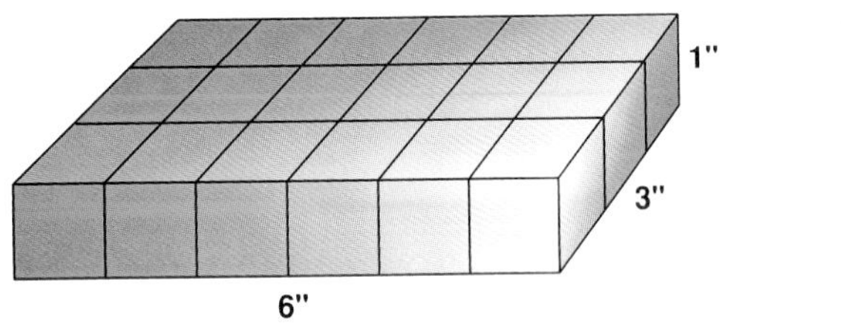

Face	*Area*	*in.*^{2}
front	6×1	6
back	6×1	6
top	6×3	18
bottom	6×3	18
left	3×1	3
right	3×1	3
		54

The sum of the areas of the 6 faces of a rectangular prism is called the *surface area*.

Example 2 In shop class, Alice made a wooden paperweight, which she decorated by covering it with 1-inch-square tiles. How many tiles did Alice use?

The paperweight is a rectangular prism. Think of unfolding the prism to get a better look at the 6 faces.

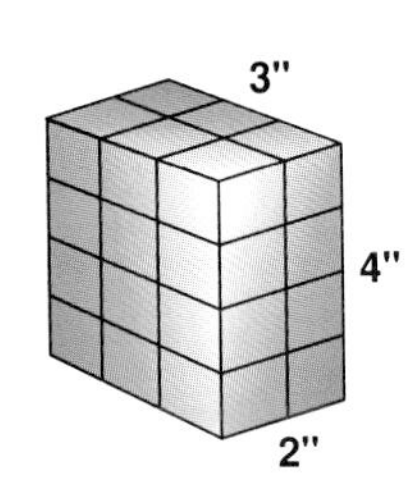

Face	*Area*	*in.*^{2}
front	3×4	12
back	3×4	12
top	3×2	6
bottom	3×2	6
left	2×4	8
right	2×4	8
		52

Answer: Alice used 52 one-inch tiles to cover all the faces of the paperweight.

Example 3 Find the surface area of this triangular prism.

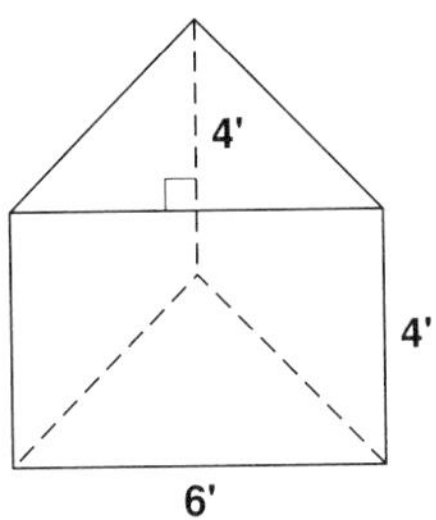

Unfold the prism to examine the dimensions of the 5 faces.

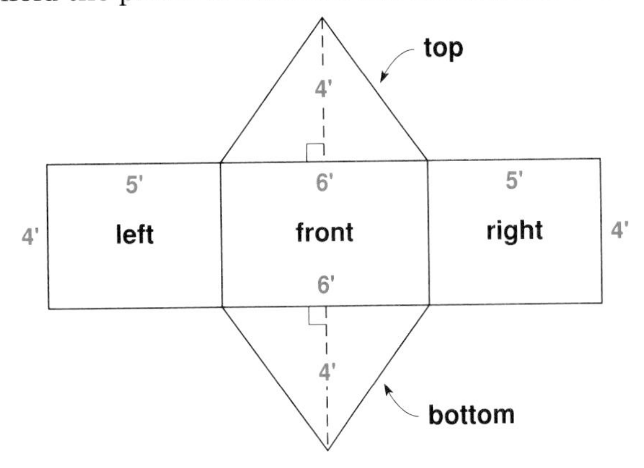

Face	*Area*	*ft.*^{2}
front	6×4	24
top	$\frac{1}{2}(6 \times 4)$	12
bottom	$\frac{1}{2}(6 \times 4)$	12
left	5×4	20
right	5×4	20
		88

Answer: The surface area of the prism is 88 square feet.

Try This *(For solution, see page 420.)*

Find the surface area of this rectangular prism.

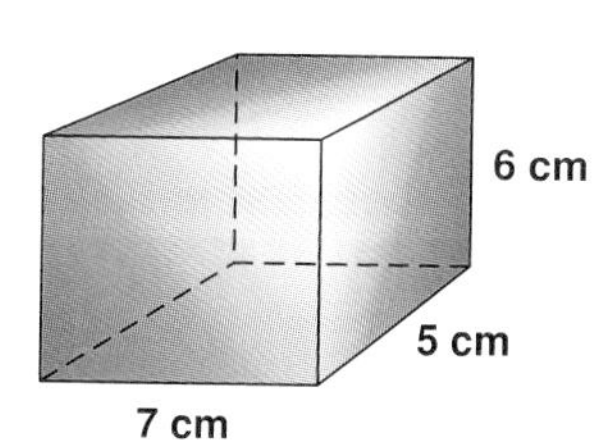

SELF-CHECK: SECTION 9.4

1. Describe a difference between a 2-dimensional figure and a 3-dimensional figure.

   Answer: A 2-dimensional figure, such as a rectangle, is *flat*, while a 3-dimensional figure, such as a prism, is *solid*.

2. What geometric measure are you using when you talk about:

   a. the number of square units that fills a rectangle?

   Answer: area

   b. the number of units it takes to go completely around the edges of a rectangle?

   Answer: perimeter

   c. the number of square units that covers all the faces of a prism?

   Answer: surface area

3. How do you calculate the surface area of a solid figure?

   Answer: Find the area of each of the faces, and add all these areas.

EXERCISES: SECTION 9.4

1. Mark a piece of paper in each pattern, and cut them out. Make your patterns bigger for easier folding. Fold the paper along the dotted lines, and tell the kind of 3-dimensional figure that is formed.

   a.

   b.

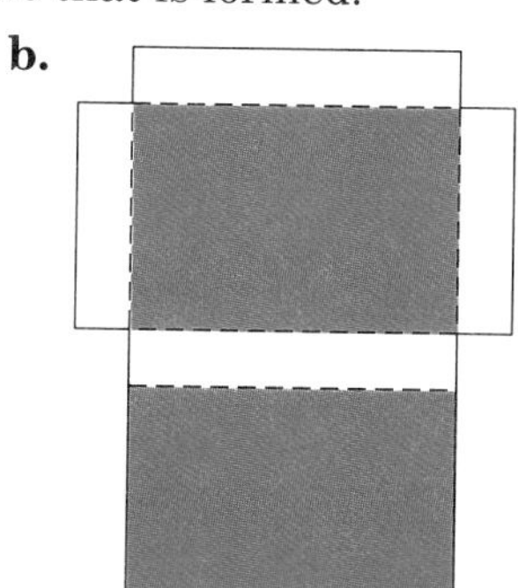

   c.

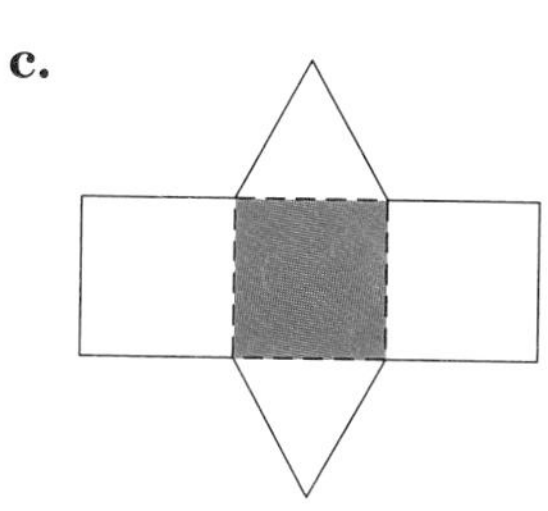

d.

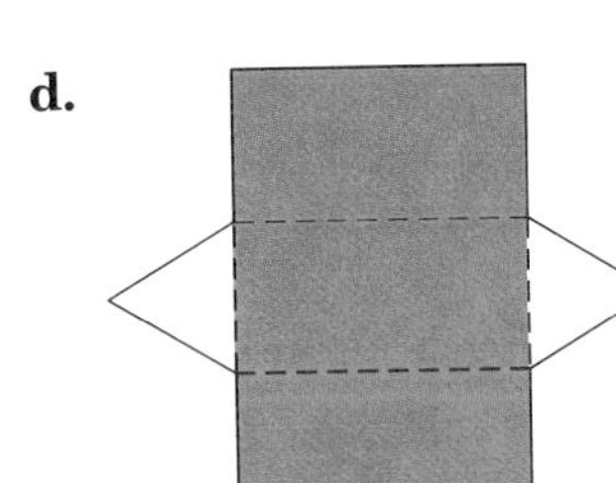

e.

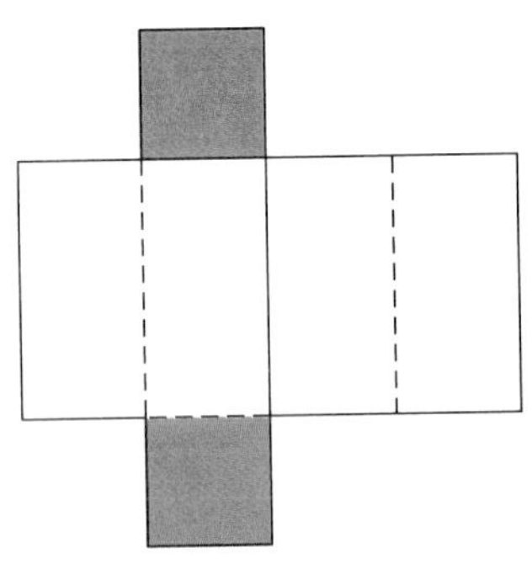

f.

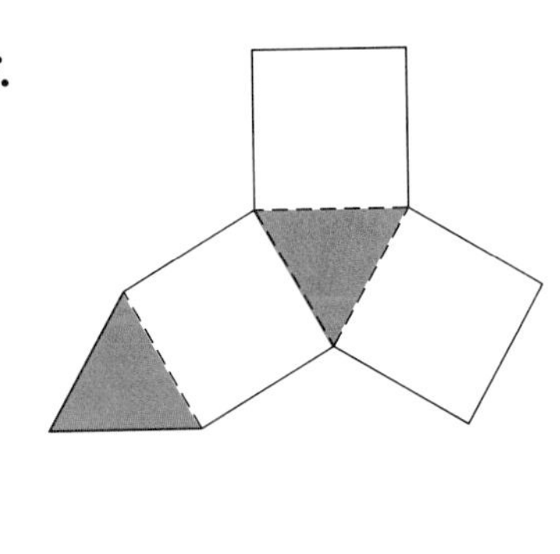

2. Think of folding each of these patterns to make a cube. Which face do you think will be opposite face A? opposite face B? To verify your predictions, make these patterns, cut them out, and fold them.

a.

A	B	C
	D	
	E	
	F	

b.

	A	B
C	D	
	E	
	F	

c.

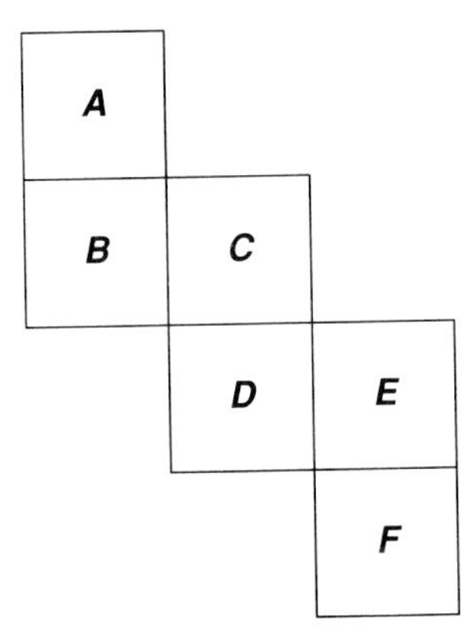

3. Draw a flat pattern for each solid figure.

a.

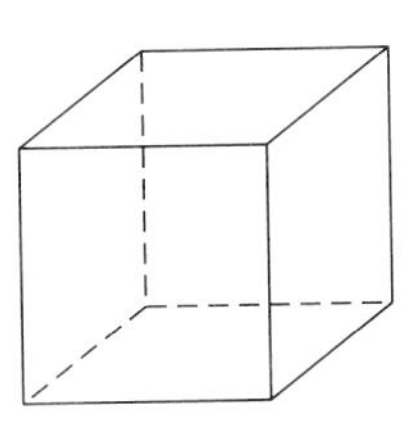

b.

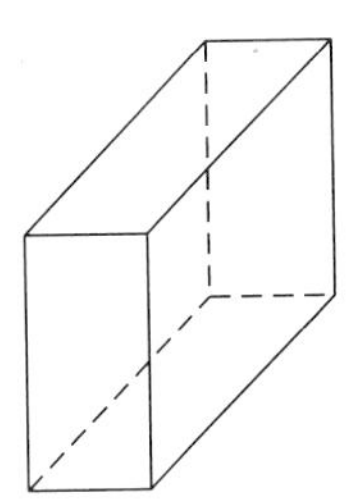

c.

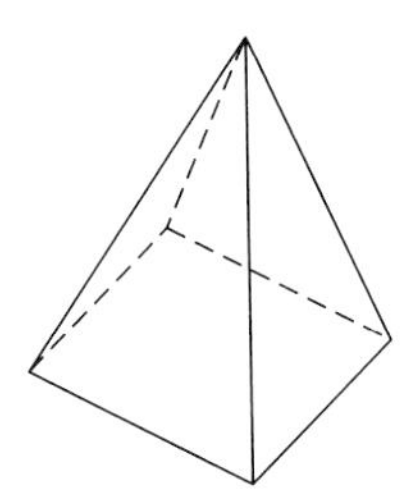

d.

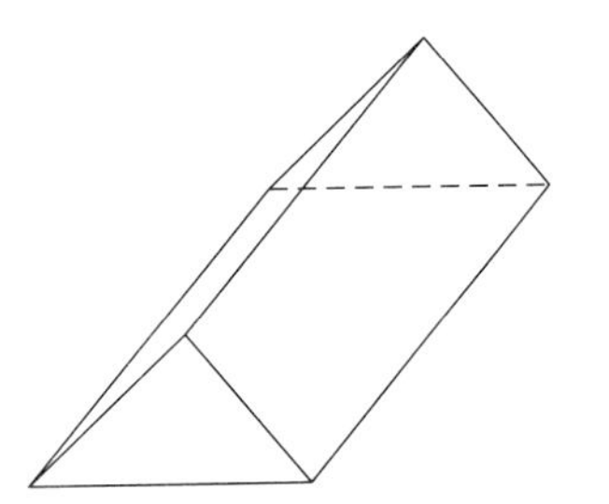

e.

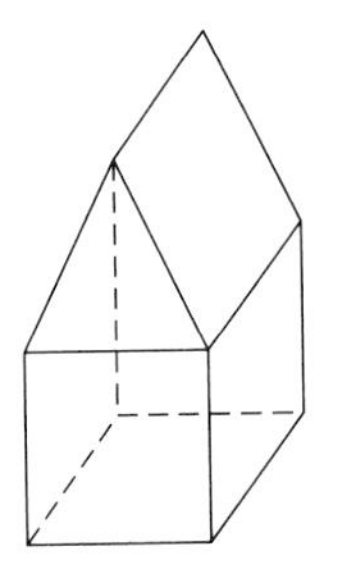

f.

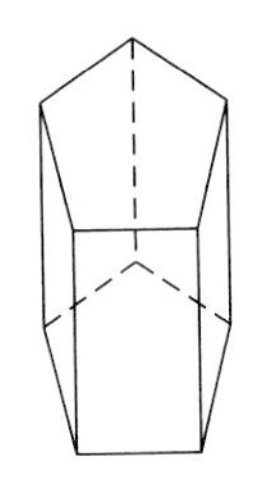

4. When all the faces of a solid figure are flat, the figure is called a ***polyhedron***. You can practice drawing polyhedrons by working on graph paper.

a. (1) Graph the points $A(3, 1)$, $B(6, 2)$, $C(5, 3)$, $D(2, 2)$.

(2) Each point A, B, C, D is now translated 4 units up on the graph to image points A', B', C', D'.

Graph these image points.

(3) Connect the points A, B, C, D to form the figure $ABCD$.

Connect the points A', B', C', D' to form $A'B'C'D'$.

Draw $\overline{AA'}$, $\overline{BB'}$, $\overline{CC'}$, $\overline{DD'}$.

Describe the resulting polyhedron.

b. (1) Graph the points $P(3, 7)$, $Q(7, 7)$, $R(5, 5)$, $S(1, 5)$.

(2) Each point P, Q, R, S is now translated 4 units down on the graph to image points P', Q', R', S'.

Graph these image points.

(3) Connect the points P, Q, R, S to form the figure $PQRS$.

Connect the points P', Q', R', S' to form $P'Q'R'S'$.

Draw $\overline{PP'}$, $\overline{QQ'}$, $\overline{RR'}$, $\overline{SS'}$.

Describe the resulting polyhedron.

c. (1) Graph the points $X(6, 6)$, $Y(9, 4)$, $Z(7, 2)$.

(2) Each point X, Y, Z, *is* now translated 5 units to the left on the graph to image points X', Y', Z'.

Graph these image points.

(3) Connect the points X, Y, Z to form the figure XYZ.

Connect the points X', Y', Z' to form $X'Y'Z'$.

Draw $\overline{XX'}$, $\overline{YY'}$, $\overline{ZZ'}$.

Describe the resulting polyhedron.

d. (1) Graph the points $E(-2, 1)$, $F(-1, -1)$, $G(-1, 3)$.

(2) Each point E, F, G is now translated 3 units to the right on the graph to image points E', F', G'.

Graph these image points.

(3) Connect the points E, F, G to form the figure EFG.

Connect the points E', F', G' to form $E'F'G'$.

Draw $\overline{EE'}$, $\overline{FF'}$, $\overline{GG'}$.

Describe the resulting polyhedron.

5. From a sheet of plain paper, cut out an 8" by 10" piece and mark off this piece in 1" squares.

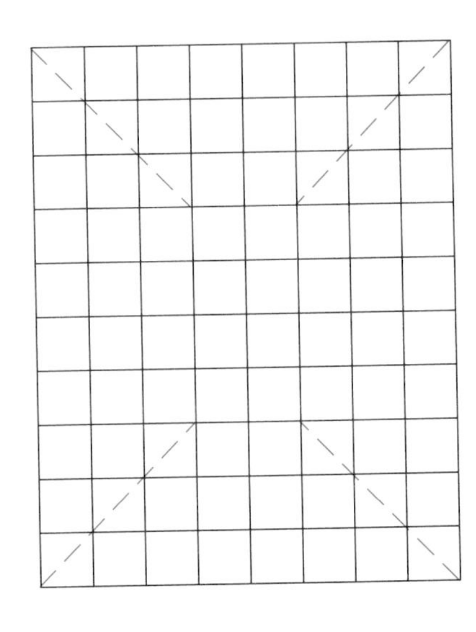

   a. From each outside corner of the paper, mark off a square block of 9 squares, and draw the diagonal.

      Cut each corner of the paper along these diagonals.

      Fold the paper up to create an open box.

      Make a face to exactly cover the top of this box.

      How many faces does this covered box have?

      How many square inches are in each face?

      What is the surface area of this covered box?

   b. Unfold the paper.
      Now create an open box by folding up only 2 squares.
      Make a face to exactly cover the top of this box.
      Find the surface area of this covered box.

   c. Unfold the paper.
      Now create an open box by folding the paper up just one square.
      Make a face to exactly cover the top of this box.
      Find the surface area of this covered box.

6. Find the surface area of each solid figure.

   a.

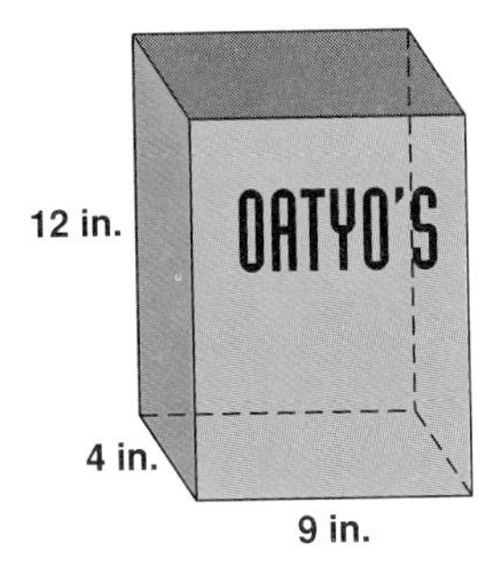

   b.

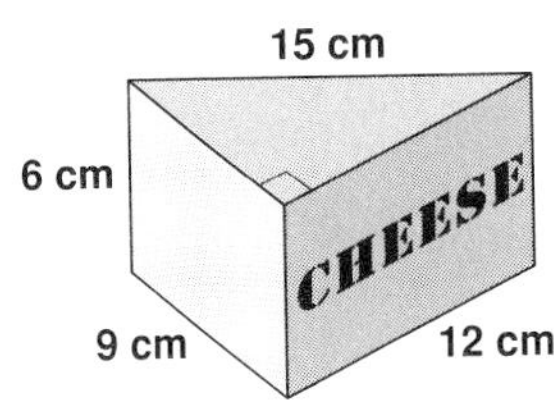

   c.

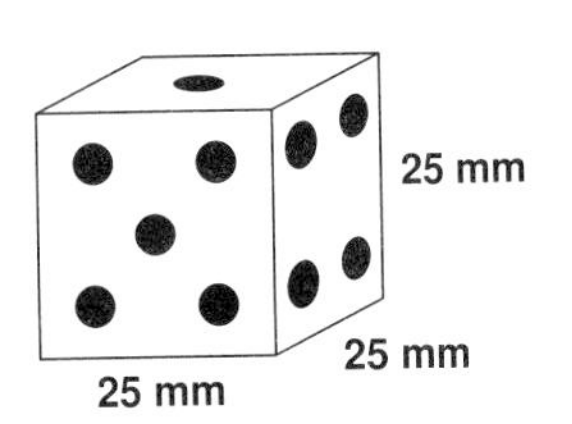

   d.

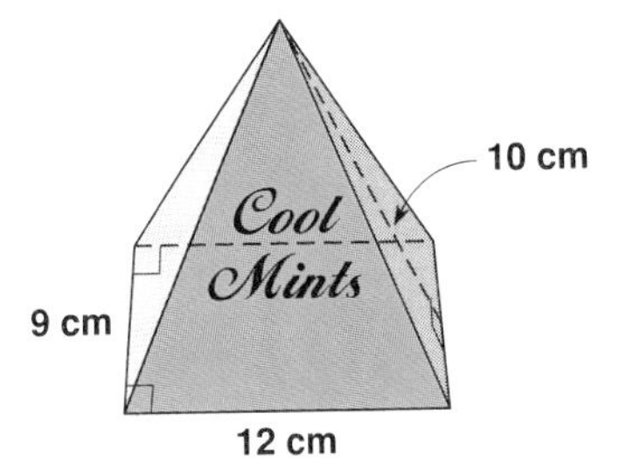

7. **a.** How many faces does a cube have? What is true about the size of all the faces of the cube?

b. Use the variable e (for edge) to write an expression for the area of face I of this cube.

c. Using the fact that all 6 faces of a cube are congruent, write a rule (in terms of the edge e) for finding the surface area of a cube.

d. Use your rule to find the surface area of each of these cubes:

(1) $e = 3$ in. **(2)** $e = 2$ cm **(3)** $e = 3.5$ m **(4)** $e = \frac{2}{3}$ ft.

8. **a.** Using the variables shown on the diagram of a rectangular prism, write an expression to represent the area of:

(1) face I **(2)** face II **(3)** face III

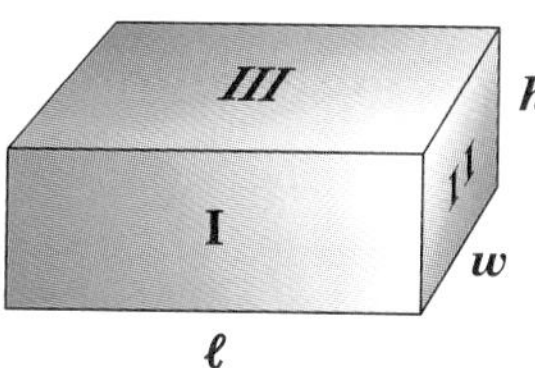

b. Include the hidden faces that are congruent to faces I, II, III, and write a rule for the surface area of the rectangular prism.

c. Using this rule for surface area $S = 2\ell w + 2wh + 2\ell h$, find the surface area of each of these rectangular prisms:

(1) $\ell = 5$ ft. $w = 2$ ft. $h = 4$ ft.

(2) $\ell = 2$ in. $w = 1$ in. $h = 6$ in.

(3) $\ell = 3$ yd. $w = 3$ yd. $h = \frac{2}{3}$ yd.

(4) $\ell = 10$ cm $w = 8.5$ cm $h = 5$ cm

9. Have you ever cut the label off a soup can, perhaps to clip a coupon or recipe, or just to recycle the can? If so, you have seen that the shape of the label is a rectangle. What is the geometric name for the shape of the can?

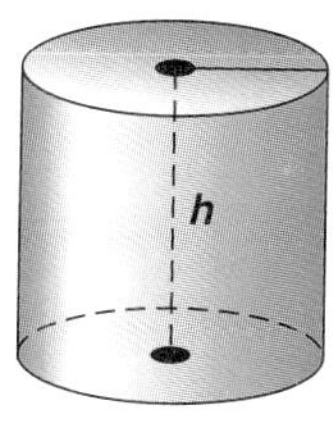

a. Describe the pattern you would get if you cut open the cylinder and laid it flat.

b. The height of the rectangular label is about the height of the cylindrical can.

(1) How is the base of the rectangle related to the circular top of the can?

(2) How would you get the area of the rectangle?

(3) How would you get the area of each circular base of the can?

(4) Using r for radius and h for height, write a rule for the surface area of a cylinder.

c. Using this rule for surface area $S = 2\pi r^2 + 2\pi rh$, find the surface area of each of these cylinders, leaving answers in terms of π:

(1) $r = 3$ in. $h = 7$ in. **(2)** $r = 6$ cm $h = 10$ cm

(3) $r = 8$ ft. $h = \frac{1}{2}$ ft. **(4)** $r = 1.1$ ft. $h = 1$ ft.

10. Predict an answer to each situation.

a. What would happen to the surface area of a cube if the length of each edge were doubled?

(1) To verify your prediction, do these calculations using a flat model to find the surface areas:

Find the surface area of the cube that measures 6 in. by 6 in. by 6 in.
Then find the surface area of the cube that measures 12 in. by 12 in. by 12 in.
Compare the two surface areas.

(2) Do the calculations again, using the rule: $S = 6e^2$

b. What would happen to the surface area of a rectangular prism if each dimension were doubled?

(1) To verify your prediction, do these calculations using a flat model to find the surface areas:

Find the surface area of the rectangular prism with dimensions 6 in. by 8 in. by 10 in.
Then find the surface area of the rectangular prism with dimensions 12 in. by 16 in. by 20 in.
Compare the two surface areas.

(2) Do the calculations again, this time using the rule: $S = 2\ell w + 2wh + 2\ell h$

c. Which would cause the greater increase in the surface area of a cylinder, doubling the height or doubling the radius of the base?

(1) To verify your prediction, do these calculations using a flat model to find the surface areas:

Find the surface area of a cylinder of radius 6 in. and height 10 in.
Find the surface area of a cylinder of radius 6 in. and height 20 in.
Find the surface area of a cylinder of radius 12 in. and height 10 in.
Compare the second and third surface areas to the first.

(2) Do the calculations again, this time using the rule: $S = 2\pi r^2 + 2\pi rh$

11. Guesstimate an answer before solving each problem.

a. Maseo has collected 250 postage stamps that are 1-inch square. He wants to cover all the faces of a 9-inch cube with the stamps, with no stamps overlapping. Does he have enough stamps? Explain.

b. To keep down the cost of packaging material, the surface area of a dog food box is to be no more than 166 square inches. If the length of the box is 6 inches and the width is 8 inches, what is the maximum height of the box?

c. Dana is planning to use blue paint on the walls and ceiling of her room, which measures 12 feet by 14 feet and is 8 feet high.

(1) Including the door, but omitting the two 3 ft.-by-4 ft. windows, what is the total area to be painted?

(2) If a gallon of paint covers about 400 square feet, how many gallons should Dana buy?

d. Marsha wants to wrap a gift that is in a box measuring 12" by 16" by 4". She has a sheet of wrapping paper that measures 18" by 36".

(1) Will Marsha be able to wrap the package? Explain.

(2) Draw a diagram to show how to place the package on the paper in order to wrap it. Is everything OK? Explain.

EXPLORATIONS

Here are prisms with bases that have 3, 4, 5, and 6 sides.

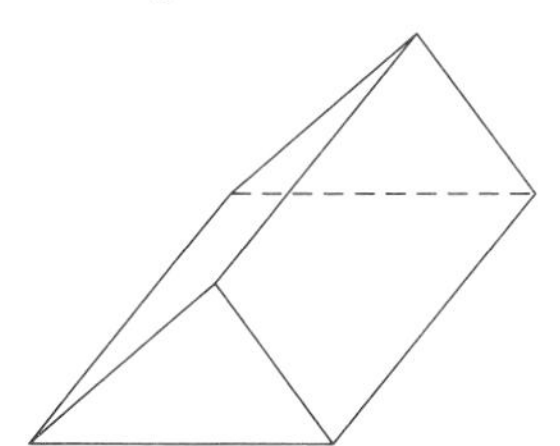

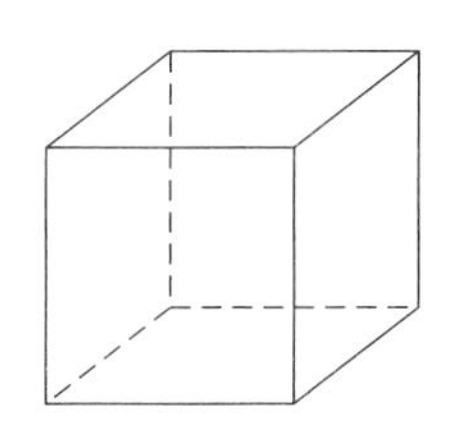

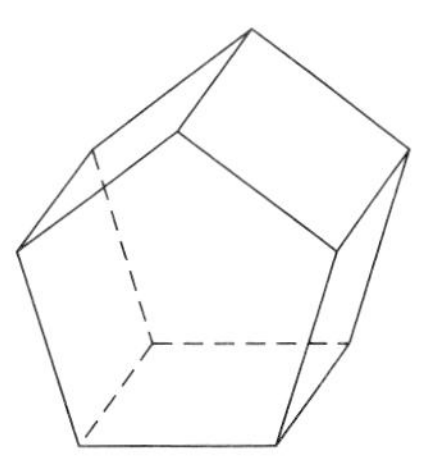

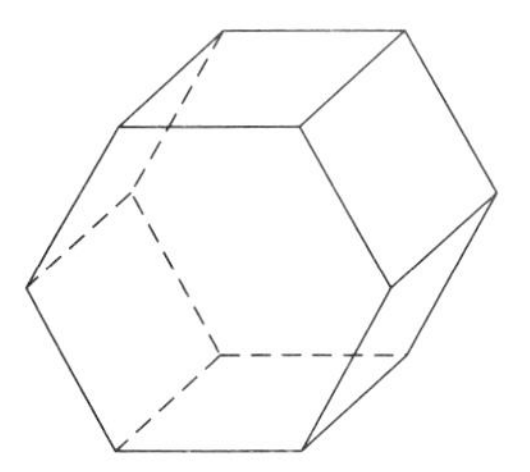

The line segments where the faces meet are *edges*, and the points where the edges meet are *vertices*.

	Number of sides in a base				
	3	4	5	6	n
Vertices	?	?	?	?	?
Faces	?	?	?	?	?
Edges	?	?	?	?	?

a. Copy the table and fill in the number of vertices for the figures with bases of 3, 4, 5, and 6 sides.
Look for a pattern.
How many vertices do you think there would be if the base of the prism had 7 sides? n sides?

b. Fill in the row labeled *Faces*.
A pattern should show you how to fill in the column headed n.

c. In the same way, fill in the row labeled *Edges*.

d. Now look at the column headed 3 in your table.
Look for a pattern that relates the numbers of vertices, faces, and edges in this column.
Does your rule work in the column headed 4? 5? 6?
Write your rule as a general statement, using v for the number of vertices, f for the number of faces, and e for the number of edges.

9.5 Volume

Prism

Let's make another box.

From a sheet of plain paper, cut out a 5" by 8" piece, and mark off this piece in 1" squares.

From each outside corner of the paper, draw the diagonal of the corner square.

Cut each corner of the paper along the diagonals of the squares.

Fold the paper up to create an open box.

What are the dimensions of this box?

Think of filling the box with one-inch cubes. How many cubes would you need?

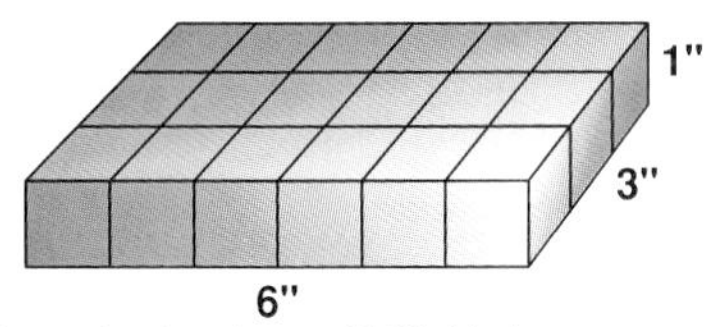

18 one-inch cubes will fill this box.

Suppose the box were 2 inches high. How many 1-inch cubes would you need to fill the box?

What relationship is there between the dimensions of the box and the number of one-inch cubes needed to fill it?

6 inches × 3 inches × 2 inches = 36 cubic inches

area of the base × height

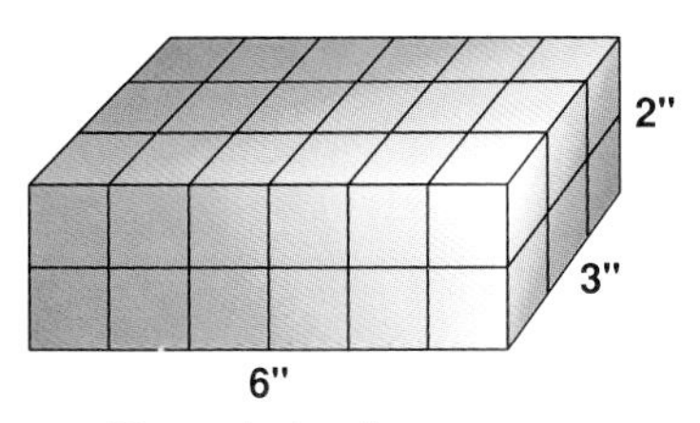

36 one-inch cubes will fill this box.

> **The number of cubic inches that fills a solid figure is the *volume* of the figure.**
>
> **Volume of Prism = Area of the Base × Height of the Prism**
>
> $V = B \times h$

Example 1 Find the volume of each prism.

a. rectangular prism

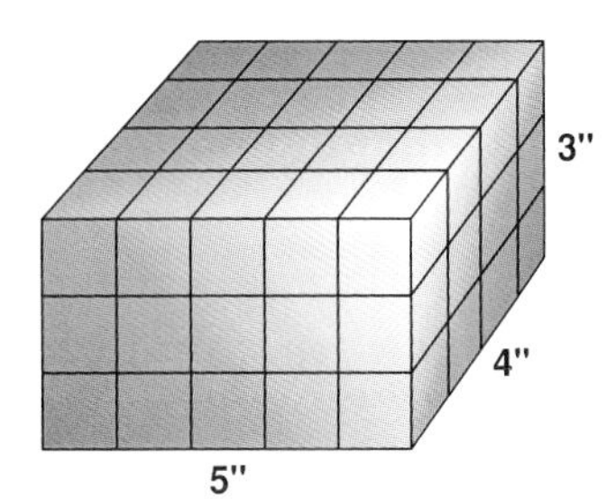

$V = B \times h$ Apply the volume rule.

$V = \ell \times w \times h$ Base is a rectangle.

$= 5" \times 4" \times 3"$

$=$ 60 cubic inches

b. cube

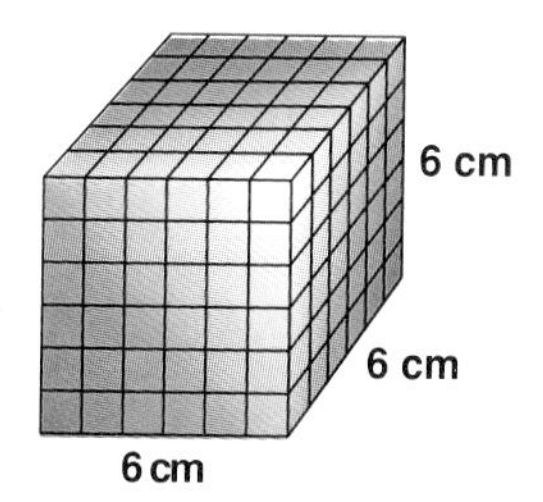

$V =$	B	$\times\ h$	Apply the volume rule.
$V =$	s^2	$\times\ h$	Base is a square.
$=$	$(6 \text{ cm})^2$	$\times$ 6 cm	
$=$	36 cm^2	$\times$ 6 cm	
$=$	216 cm^3		

Try These *(For solutions, see page 421.)*

Find the volume of each prism.

1. Rectangular Prism

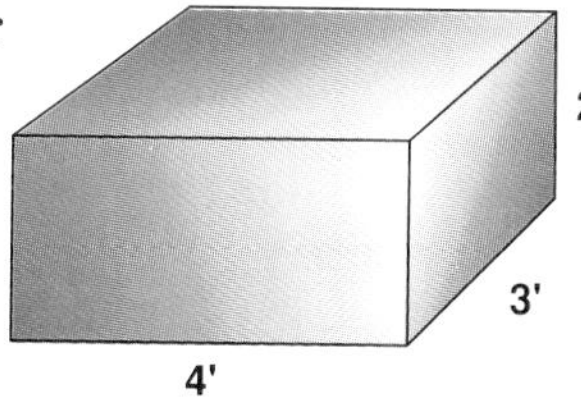

2. Cube

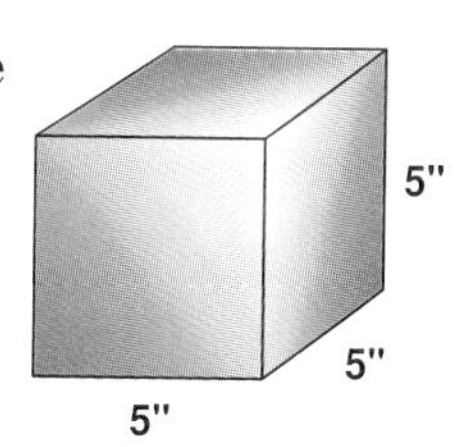

Cylinder

How would you fill this cylinder with one-inch cubes?

As with a prism, first cover the base with cubes (or parts of cubes), and then build layers until you have reached the height of the cylinder.

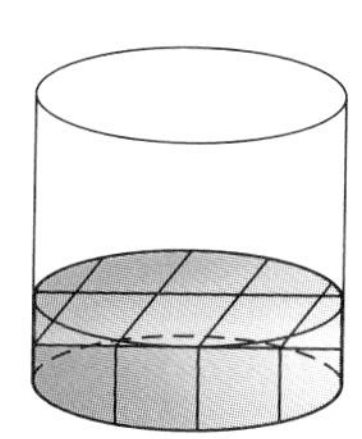

Volume of Cylinder = Area of the Base × Height

$= \pi r^2 \times h$ Base is a circle.

Volume of Cylinder = $\pi r^2 h$

Example 2 Find the volume of the cylinder whose base has a radius of 4 inches, and whose height measures 7 inches.

Volume of Cylinder $= \pi r^2 h$	Apply the volume rule.
$= \pi(4 \text{ in.})^2 \times 7 \text{ in.}$	Substitute for r and h.
$= \pi(16 \text{ in.}^2) \times 7 \text{ in.}$	Square first.
$= \pi(112 \text{ in.}^3)$	Do the multiplication.
$= 112\pi \text{ in.}^3$	Rearrange the position of π.

Answer: The volume of this cylinder is 112π cubic inches.
Replacing π by 3.14, the volume is 112(3.14), or 351.68 cubic inches.

Try This *(For solution, see page 421.)*

Find the volume of the cylinder whose base has a radius of 5 inches, and whose height measures 3 inches. Use: $\pi = 3.14$.

Pyramid, Cone

Two different shapes that have congruent bases and equal altitudes may have a relationship in their volume.

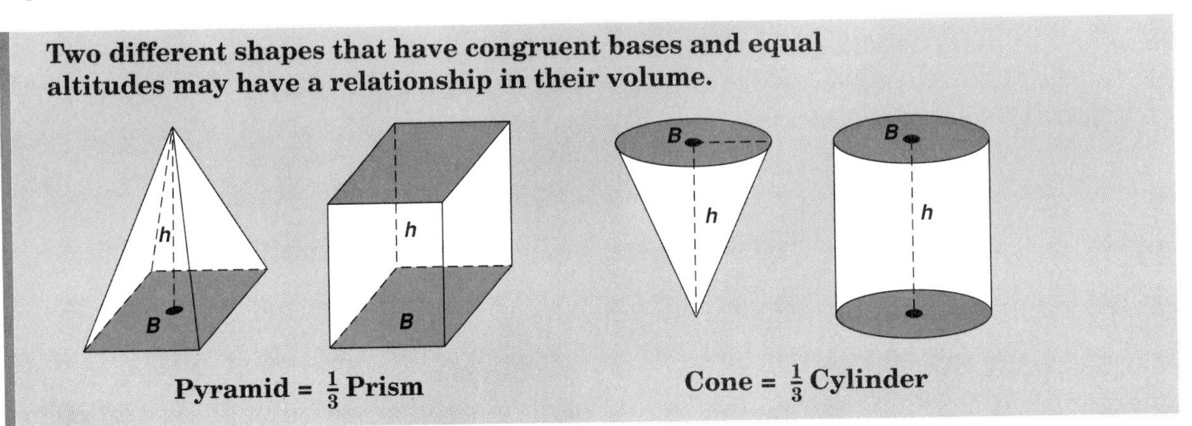

Example 3 A pyramid has a 3-inch square base and a height of 5 inches. Find the volume of the pyramid.

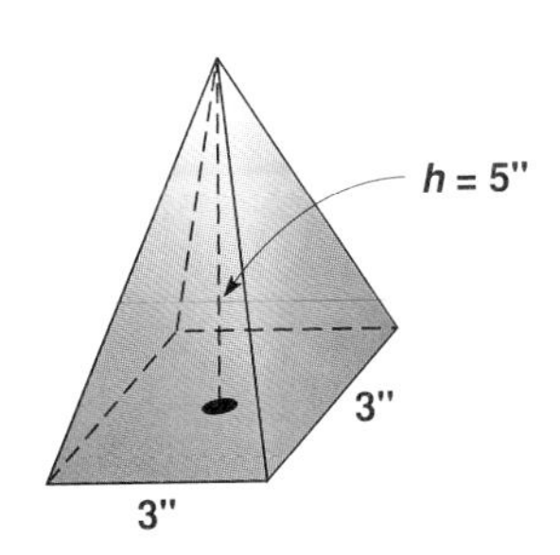

Volume of Pyramid $= \frac{1}{3}$(Volume of Prism)	Pyramid is related to prism.
$= \frac{1}{3}$(Area of Base $\times$ Height)	Insert volume rule for prism.
$= \frac{1}{3}(3^2 \times 5)$	Substitute.
$= \frac{1}{3}(9 \times 5)$	Square first.
$= \frac{1}{3}(45)$	Work in parentheses.
$= 15$	Do the multiplication.

Answer: The volume of the pyramid is 15 cubic inches.

Example 4 A cone has a circular base of radius 6 centimeters. The height of the cone is 8 centimeters.

Find the volume of the cone, in terms of π.

Using $\pi = 3.14$, find the volume of the cone, to the nearest cubic centimeter.

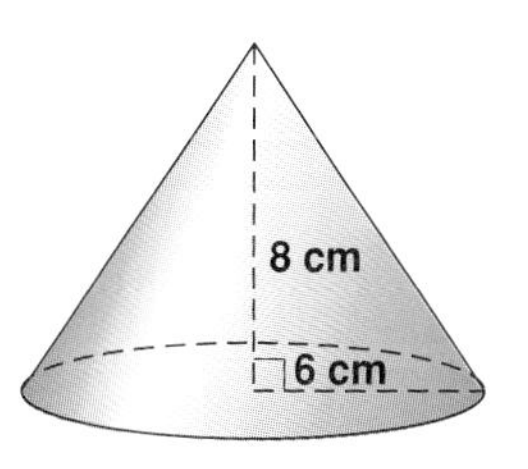

Volume of Cone $= \frac{1}{3}$(Volume of Cylinder)	Cone is related to cylinder.
$= \frac{1}{3}(\pi r^2 h)$	Insert volume rule for cylinder.
$= \frac{1}{3}\pi(6^2 \times 8)$	Substitute for r and h.
$= \frac{1}{3}\pi(36 \times 8)$	Square first.
$= \frac{1}{3}\pi(288)$	Work in parentheses.
$= 96\pi$	Do the multiplication.

The volume of the cone is 96π cubic centimeters.

Substituting for π: $96(3.14) = 301.44$

Answer: Rounded to the nearest integer, the volume of the cone is 301 cubic centimeters.

Try These *(For solutions, see page 421.)*

1. Find the volume of this prism and of a pyramid of the same dimensions.

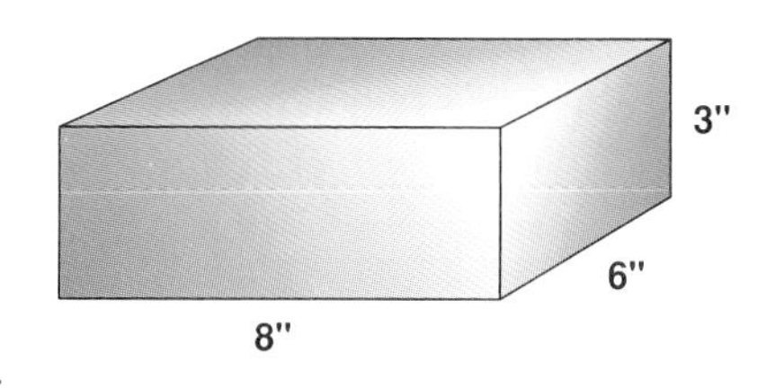

2. A cone has a circular base of radius 7 inches.
   The height of the cone is 6 inches.
   Find the volume of the cone, in terms of π.
   Find the volume of the cone, to the nearest cubic inch.
   Use: $\pi = \frac{22}{7}$

SELF-CHECK: SECTION 9.5

1. What is the meaning of *volume*?

   Answer: The volume of a solid figure is the number of cubic units —cubic inches, cubic feet, etc.—that can fit inside the figure.

2. If you know how many square units there are in the area of the base of a prism, how can you find the volume of the prism?

   Answer: Build layers of this area until you reach the height of the prism.
   This is the same as multiplying the area of the base by the height of the prism.

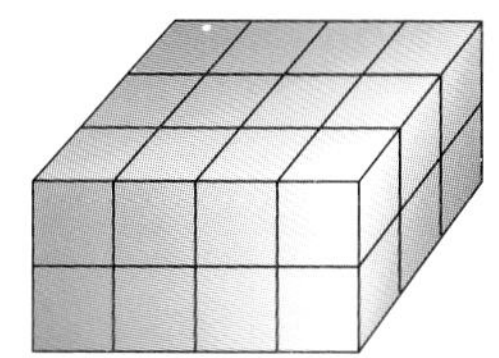

3. **a.** How can you apply the basic rule Volume = Area of Base × Height to find the volume of a rectangular prism?

Answer: Find the area of the rectangular base (by multiplying its 2 dimensions), and then multiply by the height of the prism.

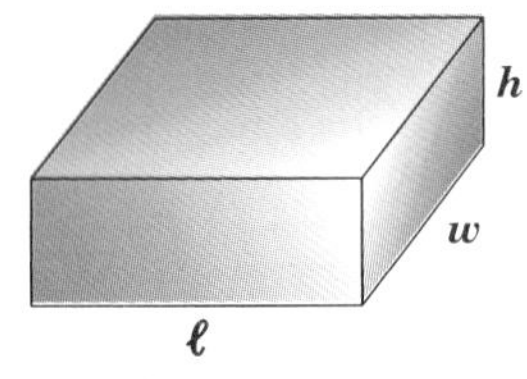

b. How does the volume calculation change for a cube?

Answer: You still find the area of the base first, but this time the base is a square. Then you still multiply by the height of the figure, but this time the height is the same measure as a side of the square.

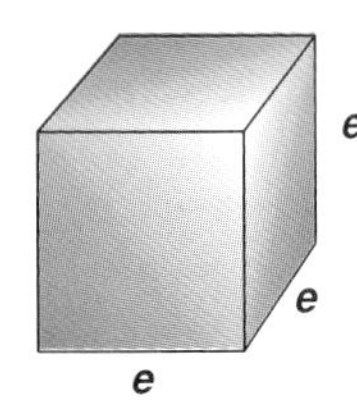

$V = e \times e \times e$
or $V = e^3$

c. How does the volume calculation change for a cylinder?

Answer: You still find the area of the base first, but this time the base is a circle. Then you still multiply by the height of the figure.

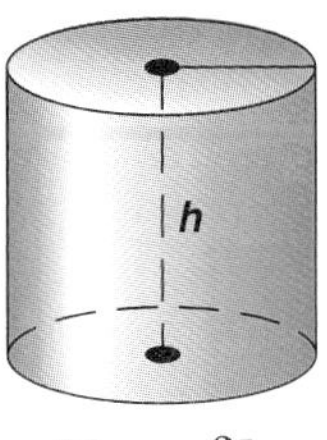

$V = \pi r^2 h$

4. You are pouring water from a cup in the shape of a cone into a cylinder of the same dimensions of the cone. How many conical cups will fill the cylinder?

Answer: Exactly 3 cups; the volume of a cone equals one-third the volume of a cylinder of the same dimensions.

EXERCISES: SECTION 9.5

1. Use the rule $V = \ell \cdot w \cdot h$ to find the volume of each rectangular prism.

a. $\ell = 6$ in. $w = 5$ in. $h = 13$ in.

b. $\ell = 12$ cm $w = 7$ cm $h = 4$ cm

c. $\ell = 1.8$ m $w = 3.5$ m $h = 4$ m

d. $\ell = \frac{3}{4}$ ft. $w = \frac{1}{2}$ ft. $h = 1$ ft.

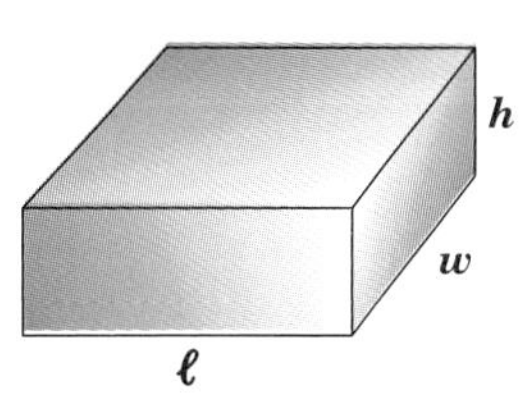

2. Substitute the given measures into the rule $V = \ell \cdot w \cdot h$ to find the missing measure.

a. $V = 240$ ft.^{3} $\ell = 16$ ft. $w = 5$ ft. Find h.

b. $V = 672$ in.^{3} $w = 8$ in. $h = 7$ in. Find ℓ.

c. $V = 882$ cm^3 $\ell = 14$ cm $h = 9$ cm Find w.

3. Use the rule $V = e^3$ to find the volume of each cube.

a. $e = 6$ ft. **b.** $e = 10$ in.

c. $e = 1.5$ cm **d.** $e = \frac{1}{2}$ yd.

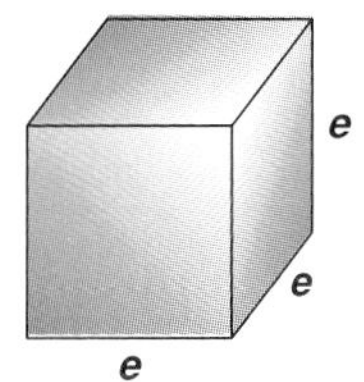

4. Find the measure of the edge of each cube if:

a. $V = 125$ cu. in. **b.** $V = 1{,}000$ cu. ft. **c.** $V = 512$ cm^3

5. Use the rule $V = \pi r^2 h$, with $\pi = 3.14$, to find the volume of each cylinder.

a. $r = 10$ cm $h = 10$ cm

b. $r = 8$ in. $h = 12$ in.

c. $r = 3$ ft. $h = \frac{1}{3}$ ft.

d. $r = 1.2$ yd. $h = 4$ yd.

6. Substitute the given measures into the rule $V = \pi r^2 h$ to find the missing measure.

a. $V = 1{,}130.4$ in.^{3} $\pi = 3.14$ $r = 6$ in. Find h.

b. $V = 3{,}080$ cm^3 $\pi = \frac{22}{7}$ $r = 14$ cm Find h.

c. $V = 792$ yd.^{3} $\pi = \frac{22}{7}$ $h = 7$ yd. Find r.

d. $V = 3{,}768$ ft.^{3} $\pi = 3.14$ $h = 12$ ft. Find r.

7. Use the rule $V = \frac{1}{3}\pi r^2 h$, with $\pi = 3.14$, to find the volume of each cone.

a. $r = 8$ cm $\quad h = 12$ cm

b. $r = 10$ in. $\quad h = 6$ in.

c. $r = 9$ ft. $\quad h = 3.3$ ft.

d. $r = 0.4$ yd. $\quad h = 0.9$ yd.

8. Substitute the given measures into the rule $V = \frac{1}{3}\pi r^2 h$ to find the missing measure.

a. $V = 924$ in.^{3} $\quad \pi = \frac{22}{7}$ $\quad r = 7$ in. $\quad$ Find h.

b. $V = 1{,}017.36$ cm^3 $\quad \pi = 3.14$ $\quad r = 9$ cm $\quad$ Find h.

c. $V = 550$ ft.^{3} $\quad \pi = \frac{22}{7}$ $\quad h = 21$ ft. $\quad$ Find r.

d. $V = 763.02$ m^3 $\quad \pi = 3.14$ $\quad h = 9$ m $\quad$ Find r.

9. Use the rule $V = \frac{4}{3}\pi r^3$ to find the volume of each sphere.

a. $r = 5$ in. $\quad$ Leave π.

b. $r = 7$ ft. $\quad$ Use: $\pi = \frac{22}{7}$

c. $r = 30$ mm $\quad$ Use: $\pi = 3.14$

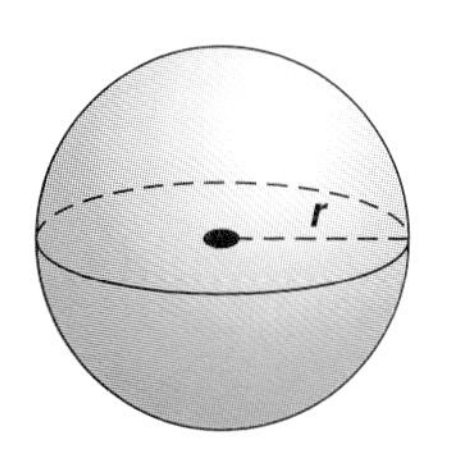

10. For each prism described, find the volume by using the rule $V = B \times h$, where B is the area of a base, and h is the height of the prism.

	Base of Prism	*Height of Prism*
a.	a 3-inch square	6 inches
b.	a right triangle, with legs 2 ft. and 4 ft.	3 ft.
c.	a parallelogram, with base 3.2 cm and height 2.5 cm	10 cm
d.	a trapezoid, with bases 10 in. and 12 in., and height 6 in.	8 in.

11. Find the volume of each solid figure.

a.

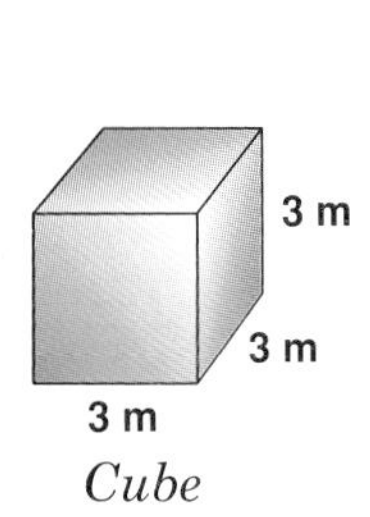

Cube

b.

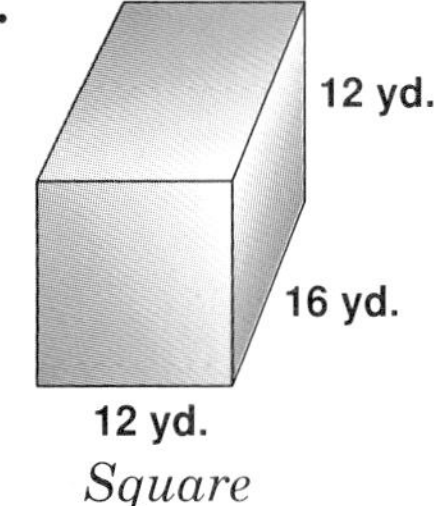

Square Base

c.

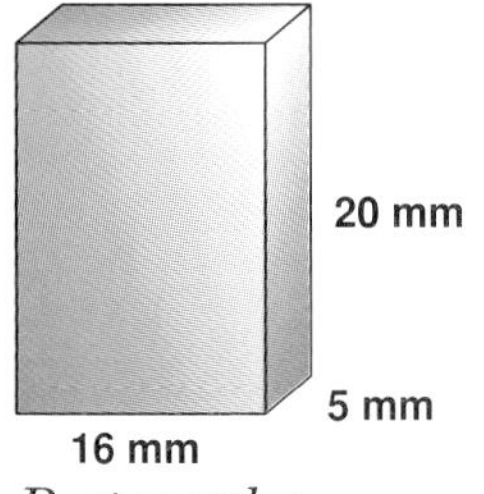

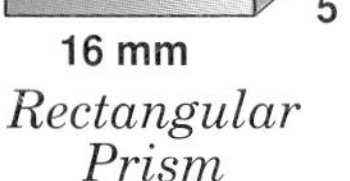

Rectangular Prism

d.

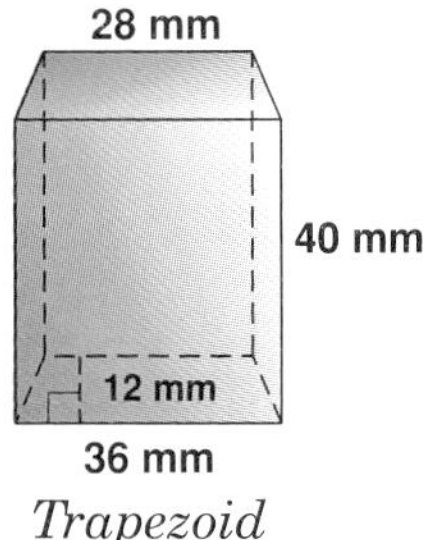

Trapezoid Base

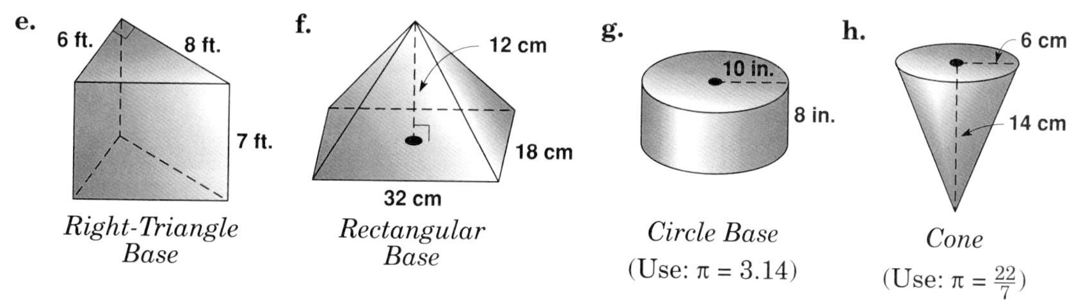

12. Predict an answer to each situation.

a. What would happen to the volume of a cube if the length of each edge were doubled?

To verify your prediction, do these calculations using the rule $V = e^3$:

Find the volume of the cube whose edge measures 3 cm.

Find the volume of the cube whose edge measures 6 cm.

Compare the two volumes.

b. What would happen to the volume of a rectangular prism if each dimension were doubled?

To verify your prediction, do these calculations using the rule $V = \ell \cdot w \cdot h$:

Find the volume of the rectangular prism with dimensions 3 in. by 5 in. by 8 in.

Find the volume of the rectangular prism with dimensions 6 in. by 10 in. by 16 in.

Compare the two volumes.

c. Which would cause the greater increase in the volume of a cylinder, doubling the radius or doubling the height?

To verify your prediction, do these calculations using the rule $V = \pi r^2 h$:

Find the volume of a cylinder of radius 5 in. and height 7 in.

Find the volume of a cylinder of radius 10 in. and height 7 in.

Find the volume of a cylinder of radius 5 in. and height 14 in.

Compare the second and third volumes to the first.

13. Guesstimate an answer before solving each problem. When working with π, use: $\pi = 3.14$

a. Students in the Adult Ed sculpture class will be working with modeling clay.

If each of 20 students is to be given a cube of clay measuring 15 cm on an edge, how much clay is needed altogether?

b. A construction company has to fill in a hole in the ground, which measures 72 feet long, 30 feet wide, and 8 feet deep. How many cubic yards of earth are needed to fill the hole? (1 cubic yard = 27 cubic feet)

c. A steel block coming off an assembly line measures 35 inches by 18 inches by 0.5 inch. If steel weighs 0.28 pound per cubic inch, what is the weight of the block?

d. Stuart works in shipping at Blake's Auto Supply. How many cartons each measuring 8 in. by 12 in. by 16 in. can he pack in a crate that measures 2 ft. by 3 ft. by 4 ft.?

e. When it was built, nearly 5,000 years ago, the height of the Great Pyramid was 481.4 feet. Today the pyramid stands 451.4 feet tall.

The measure of each side of the square base remains at 755 feet.

By how much has the volume of the pyramid decreased?

f. Cut 2 pieces of notebook paper, each to size 8" by 10".

Roll one paper "the long way" into the shape of a cylinder.

Roll the other paper "the short way" into the shape of a cylinder.

(1) Which cylinder has the greater volume? Explain.

(2) Make an observation about the surface areas of the two cylinders.

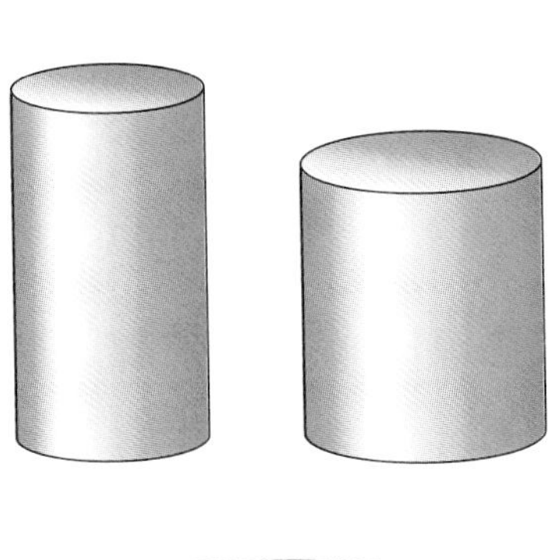

g. A styrofoam insulator is in the shape of a hollow cylinder with an outer diameter of 10 feet and an inner diameter of 8 feet.

The insulator is 5 feet deep.

If one pound of styrofoam can fill 4 cubic feet, how many pounds of styrofoam are needed?

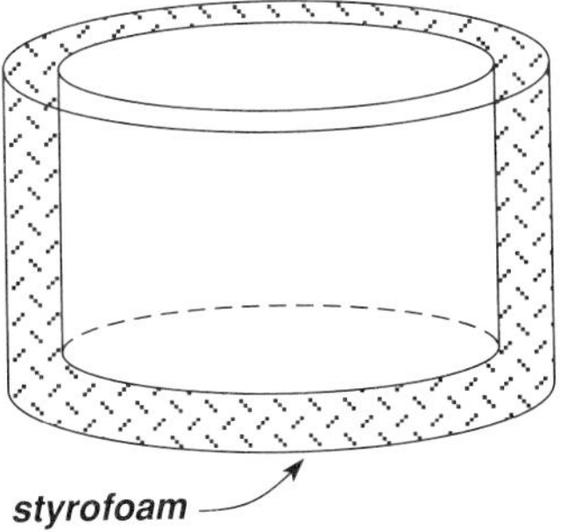

EXPLORATIONS

1. **a.** A flat figure with 2 dimensions is drawn on a flat surface called a *plane*.
A solid figure with 3 dimensions is drawn in *space*.
Match the geometric figures in *Column A* with a description in *Column B*.

Column A	*Column B*
(1) perimeter	(a) the number of square units that can fit inside a plane figure
(2) area	(b) the number of cubic units that can fit inside a solid figure
(3) surface area	(c) the number of units in the distance around a plane figure
(4) volume	(d) the sum of the areas of all the faces of a solid figure

b. Perimeter and area are different geometric measures of a plane figure but, for special figures, the perimeter and area can have the same numerical value.

(1) Find the length of the side of the square for which the perimeter has the same numerical value as the area.

(2) Find the length of the radius of the circle for which the circumference has the same numerical value as the area.

c. Find the length of the edge of the cube for which the surface area has the same numerical value as the volume.

2. **a.** How many different rectangular prisms can you create, each of which has a volume of 24 cubic units? What are the dimensions?

To verify your result, copy and complete this chart showing dimensions of rectangular prisms with volume 24 cubic units.

If the dimensions are, say 1, 2, 12, how many different prisms can be formed?

ℓ	w	h
1	1	?
1	2	?
1	3	?
1	4	?
2	2	?
2	3	?

b. Use the dimensions in the completed chart to find the surface area of each prism shown.

c. **(1)** Which of the prisms has the greatest surface area?

(2) Which has the least surface area?

(3) Do any of these prisms have the same surface area?

Things You Should Know After Studying This Chapter

KEY SKILLS

9.1 Find the area of a square, rectangle, and right triangle.

9.2 Find the area of a parallelogram, trapezoid, and any triangle.

9.3 Find the circumference and area of a circle.

9.4 Recognize special 3-dimensional figures, such as a prism, pyramid, cylinder, and cone.
Find surface area.

9.5 Find volume.

KEY TERMS

9.1 area • square unit (square inch, square foot, etc.) • rectangle (base, height) • square (side) • right triangle (legs, hypotenuse) • square root

9.2 parallelogram • triangle • trapezoid

9.3 circle • radius • diameter • circumference

9.4 polyhedron • prism • pyramid • sphere • cylinder • cone • face • edge • surface area

9.5 volume • cubic unit

Things You Should Know After Studying This Chapter

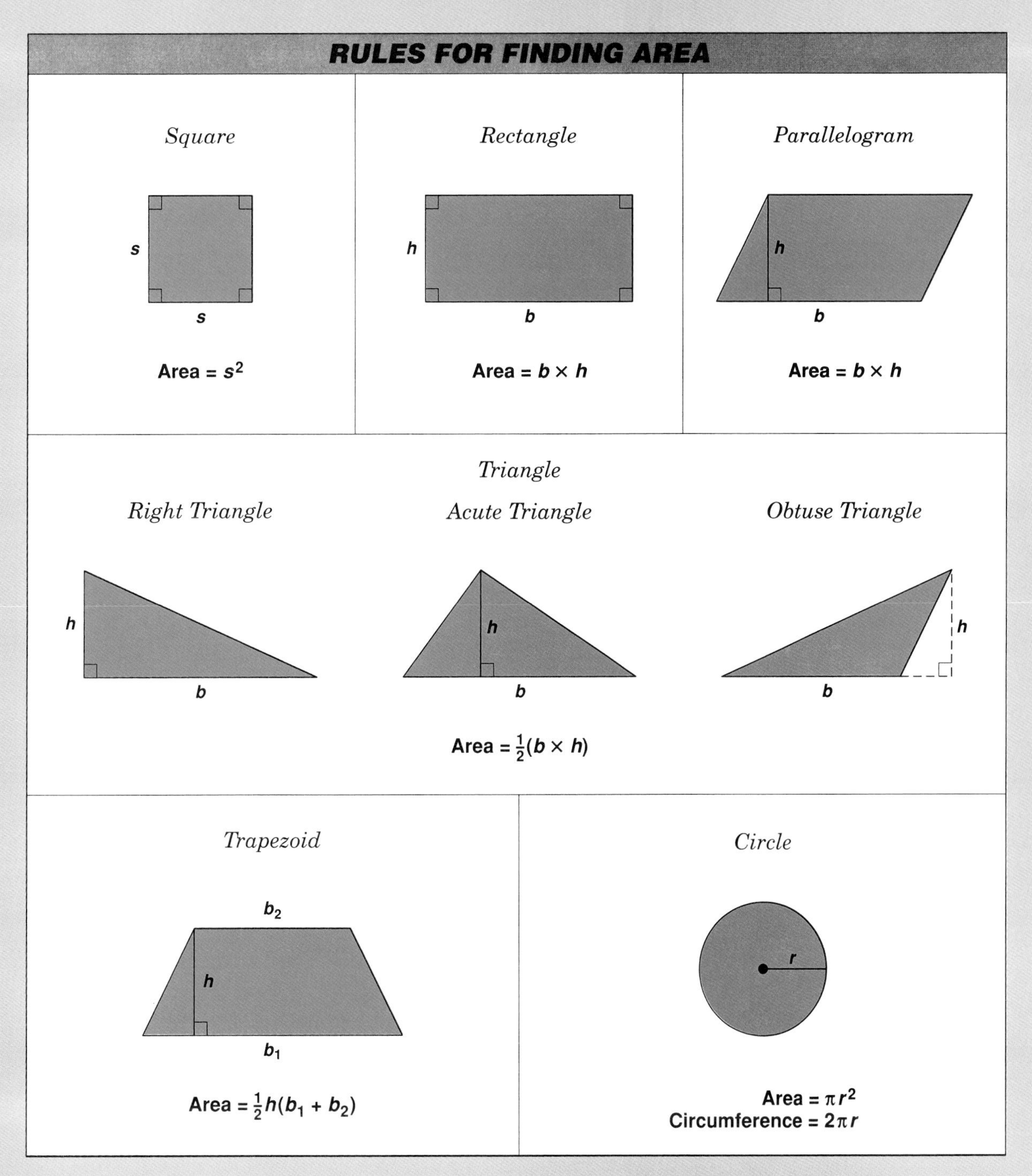

Things You Should Know After Studying This Chapter

RULES FOR FINDING VOLUME		
Volume of Prism = Area of Base × Height of Prism		
Cube	*Rectangular Prism*	*Pyramid*
$V = e^3$	$V = \ell \cdot w \cdot h$	$V = \frac{1}{3}(B \cdot h)$
Cylinder	*Cone*	*Sphere*
$V = \pi r^2 h$	$V = \frac{1}{3}\pi r^2 h$	$V = \frac{4}{3}\pi r^3$

CHAPTER REVIEW EXERCISES

1. From the moment an infant is weighed at birth, measurements are important in our lives. Match each item in *Column A* with an appropriate measure from *Column B*.

Column A *Item*	*Column B* *Approximate Measure*
a. a looseleaf page	(1) 5 square centimeters
b. the diameter of Earth	(2) 93.5 square inches
	(3) 100 square centimeters
c. a coffee can	(4) 3,000 miles
d. an ice cube	(5) 4,900 kilometers
e. an ocean depth	(6) 26 miles
f. a dollar bill	(7) 200 cubic inches
g. California to New York	(8) 6 cubic meters
	(9) 2.5 cubic inches
h. airspace per student	(10) 5 fathoms
i. a postage stamp	
j. marathon distance	

2. Find each shaded area.

a.

b.

8 m

8 m

c.

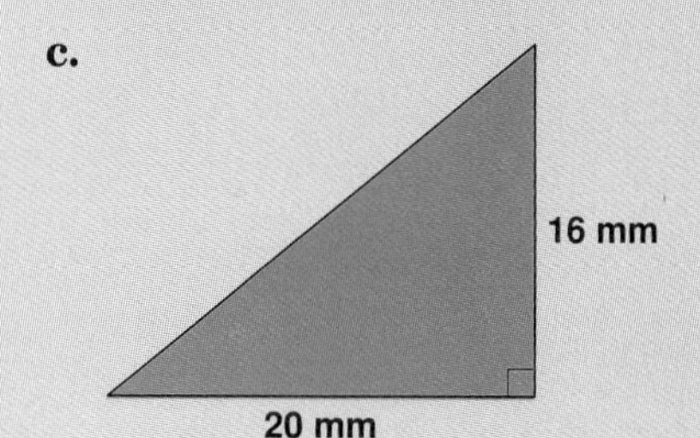

d.

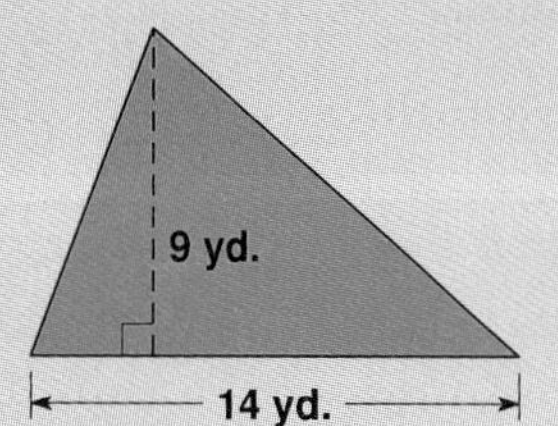

e.

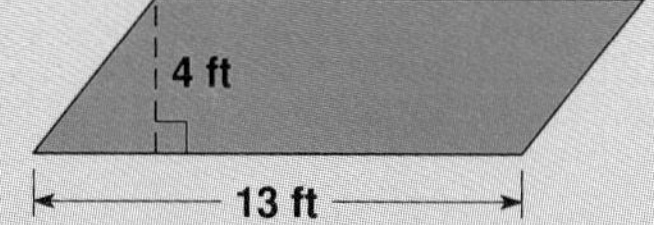

f.

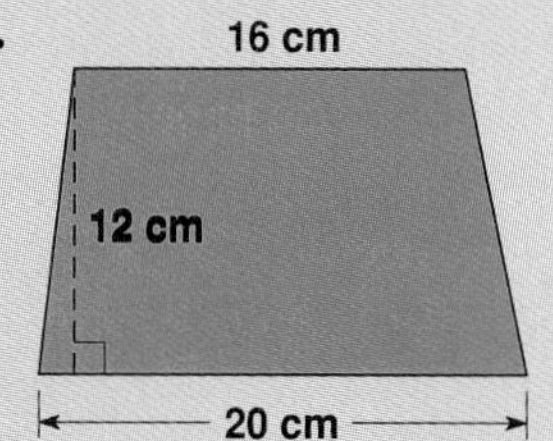

g.

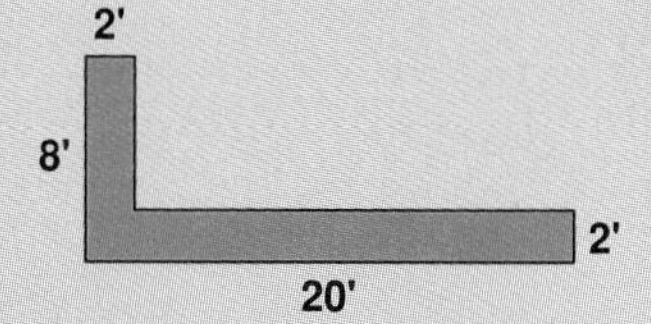

h.

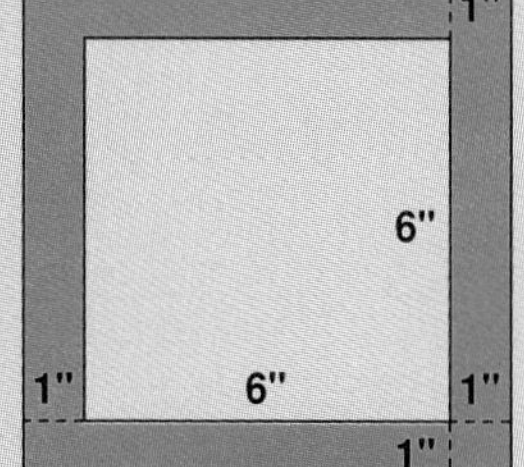

CHAPTER REVIEW EXERCISES

3. Draw a diagram for each situation and find the missing measures.
   a. In rectangle $ABCD$, $AB = 5$ in. and the area is 45 sq. in. Find BC.
   b. In rectangle $EFGH$, $EH = 14$ cm and the perimeter is 40 cm.
      Find: **(1)** EF **(2)** the area
   c. The perimeter of square $PQRS$ is 24 mm. Find the area.
   d. Right triangle JKL, with a right angle at K, has an area of 60 sq. ft. If $JK = 15$ ft., find KL.
   e. The shorter base $\overline{WX}$ of trapezoid $WXYZ$ measures 9 in, and the area of the trapezoid is 66 sq. in. If the height is 6 in., find the length of base $\overline{YZ}$.
4. Draw each figure on graph paper, and find the area.
   a. $MNOP$ with vertices
      $M(5, 6)$ $N(0, 6)$ $O(0, 0)$ $P(5, 0)$
   b. $DEFG$ with vertices
      $D(0, 1)$ $E(0, 6)$ $F(3, 8)$ $G(3, 3)$
   c. ABC with vertices
      $A(-2, 5)$ $B(4, -2)$ $C(-2, -2)$
   d. RST with vertices
      $R(-2, 0)$ $S(0, 5)$ $T(6, 0)$
   e. $JKLM$ with vertices
      $J(-1, -1)$ $K(1, 3)$ $L(5, 3)$ $M(8, -1)$
5. Find the circumference and area of each circle.
   a. radius = 4 cm Leave π.
   b. radius = 10 in. Use: $\pi = 3.14$
   c. diameter = 14 ft. Use: $\pi = \frac{22}{7}$
6. Find the surface area of each solid figure.
   a.

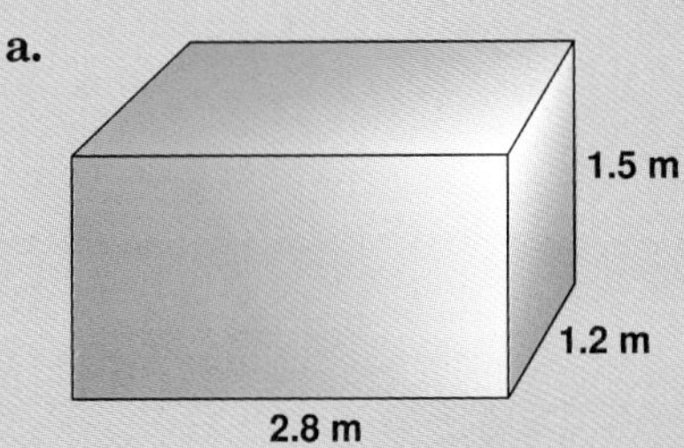

   b.

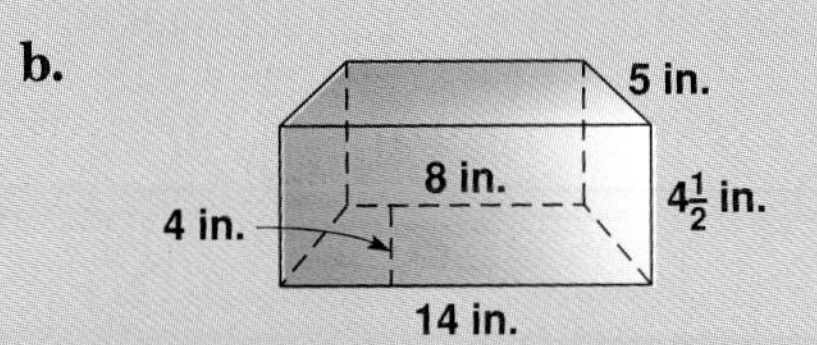

   c.

Use: $\pi = 3.14$

   d.

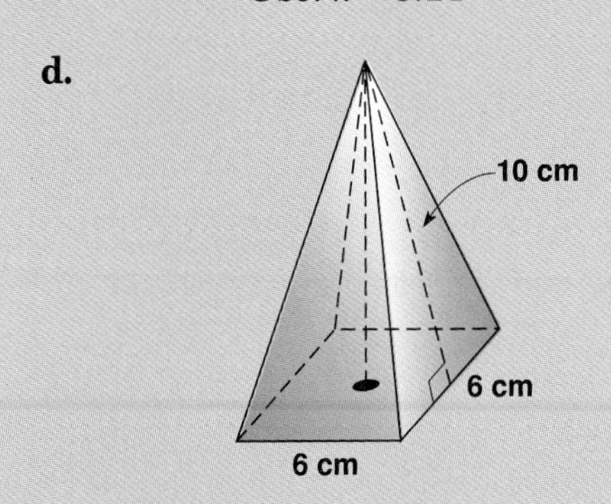

7. Find the volume of each solid figure.
   a.

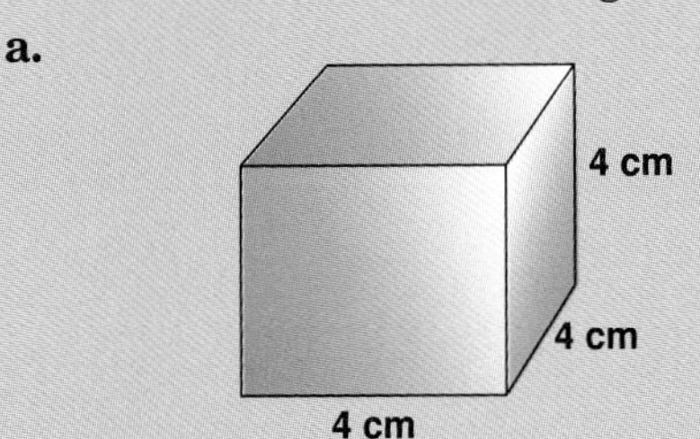

   b.

c.

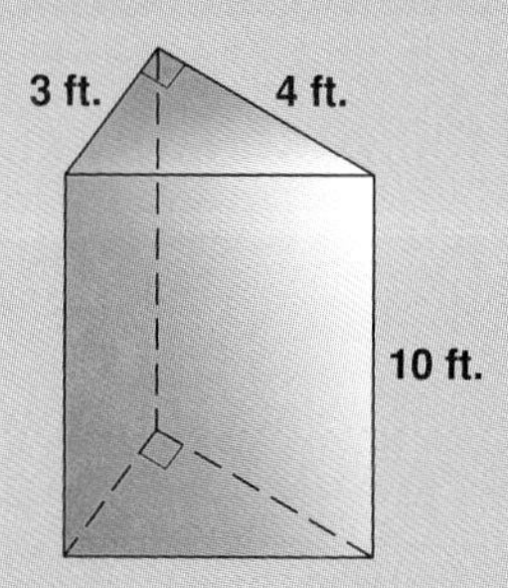

d.

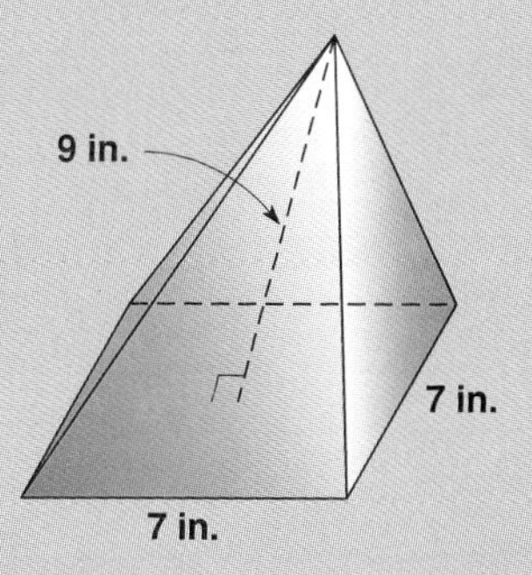

e.

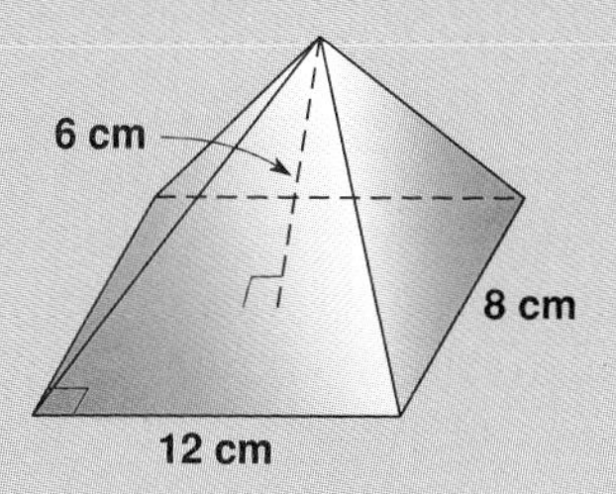

f.

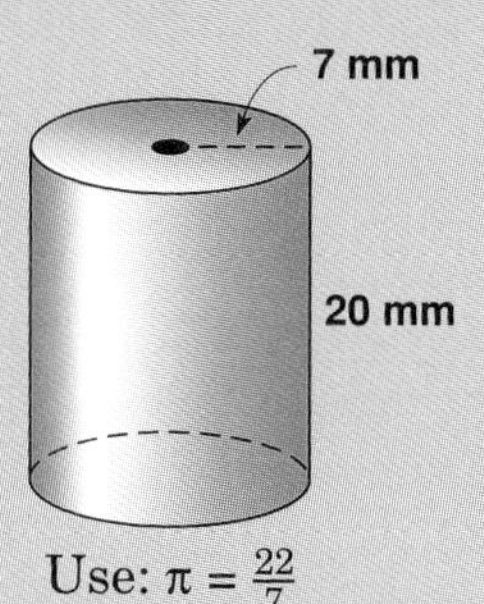

Use: $\pi = \frac{22}{7}$

g.

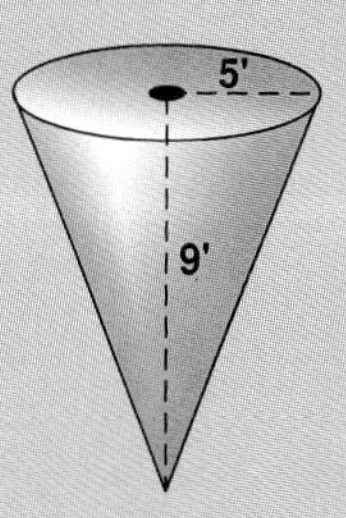

Use: $\pi = 3.14$

h.

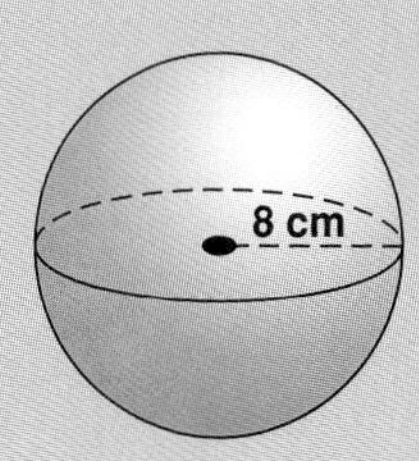

Use: $\pi = 3.14$

8. Guesstimate an answer before solving each problem.

a. A Boy Scout scarf is in the shape of a triangle, with a base of 28 inches and a height of 16.5 inches. How many square inches of material are in the scarf?

b. A 240-meter length of the East River borders one side of the River Housing Community. The other 3 sides of the rectangular region are surrounded by roads totaling 560 meters in length.

(1) What are the dimensions of the rectangular region?

(2) What is the area?

c. A bicycle tire has a diameter of 26 inches. How many feet will the bicycle travel in one complete rotation of the tire? (Use: $\pi = 3.14$)

d. If King Arthur's Round Table could seat 22 knights, with each knight occupying $7\frac{7}{11}\pi$ inches of the circumference, what was the radius of the table?

e. Sandra wants to decorate a wooden cube whose edge measures 8 inches with 1-inch squares made of colored foil. How many squares will she need?

f. In his Physics class, Howard demonstrated how a prism is used to see that white light is made up of all the colors of the spectrum. The height of the prism is 8 cm, and its triangular base is 4 cm across, with an altitude of 3 cm. What is the volume of the prism?

SOLUTIONS FOR TRY THESE

9.1 Area

Page 353

1. **a.** area = 9 feet × 12 feet

   = 108 square feet

   b. Let x = the number of square yards in the area.

$$\frac{1 \text{ square yard}}{9 \text{ square feet}} = \frac{x \text{ square yards}}{108 \text{ square feet}}$$

$$9x = 108$$

$$\frac{9x}{9} = \frac{108}{9}$$

$$x = 12$$

Answer: There are 12 square yards in the area of the rectangle.

   c. The perimeter of the rectangle is the sum of the measures of the sides.

   Of the 4 sides, 2 bases are equal in measure and 2 heights are equal in measure.

$P = 2b + 2h$

$= 2(9 \text{ feet}) + 2(12 \text{ feet})$

$= 18 \text{ feet} + 24 \text{ feet}$

$= 42 \text{ feet}$

2. Find the dimensions of the rectangle by counting boxes or by subtracting coordinates.

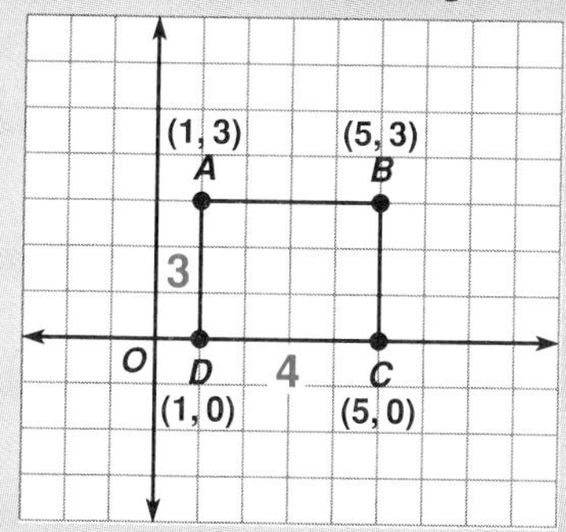

base $DC = 5 - 1 = 4$

height $AD = 3 - 0 = 3$

Area = base × height

= 4 × 3

= 12 square units

Page 355

1. **a.** area = $15^2 = 15 \times 15 = 225$ square feet

   b. Let x = the number of square yards in the area.

$$\frac{1 \text{ square yard}}{9 \text{ square feet}} = \frac{x \text{ square yards}}{225 \text{ square feet}}$$

$$9x = 225$$

$$\frac{9x}{9} = \frac{225}{9}$$

$$x = 25$$

Answer: There are 25 square yards in the area of the square.

2. Find the measure of the side of the square by counting boxes or by subtracting coordinates.

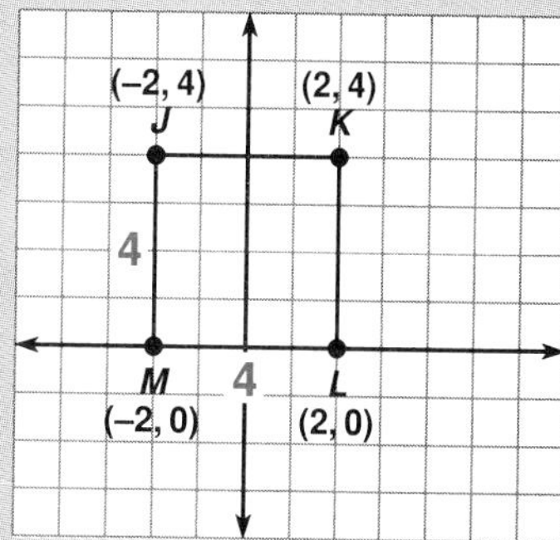

side $ML = 2 - (-2) = 2 + 2 = 4$

side $JM = 4 - 0 = 4$

Area = side^2

$= 4^2 = 4 \times 4$

= 16 square units

3. **a.** Area of Square = s^2 — The rule for area of a square.

$36 = s^2$ — The area is 36.

$\sqrt{36} = s$ — The measure of the side is the square root of the area.

$6 = s$

Answer: Each side of the square measures 6 inches.

   b. perimeter = $4s$

   = 4(6 inches) = 24 inches

4. **a.** 17 [×] [=] *Display:* 289.

   b. 1156 [√] *Display:* 34.

SOLUTIONS FOR TRY THESE

Page 358

1. Area of Right Triangle $= \frac{1}{2}(\text{leg}_1 \times \text{leg}_2)$
$= \frac{1}{2}(6 \text{ cm} \times 8 \text{ cm})$
$= \frac{1}{2}(48 \text{ cm}^2)$
$= 24 \text{ cm}^2$

2. Apply the rule for the area of a right triangle, substituting the numbers you know.

Area $= \frac{1}{2}(\text{leg}_1 \times \text{leg}_2)$

$66 = \frac{1}{2}(12 \times \ell)$ Area = 66, one leg = 12, other leg = ℓ

$66 = \frac{1}{2}(\overset{6}{\cancel{12}} \times \ell)$ Cancel.

$66 = 6\ell$

$\frac{66}{6} = \frac{6\ell}{6}$ Divide both sides by 6.

$11 = \ell$

Answer: The measure of the other leg is 11 inches.

2. a.

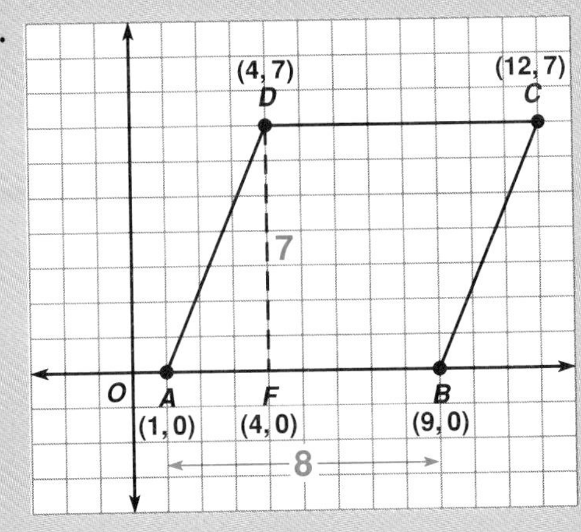

b. base $AB = 9 - 1 = 8$ units

c. $F(4, 0)$; height $DF = 7 - 0 = 7$ units

d. Area of Parallelogram = base × height
= 8 units × 7 units
= 56 square units

9.2 More About Area

Page 369

1. a. Area of Rectangle $= b \times h$
$= 12' \times 5'$
$= 60$ square feet

b. Area of Parallelogram $= b \times h$
$= 12' \times 5'$
$= 60$ square feet

c. Area of Right Triangle $= \frac{1}{2}(b \times h)$
$= \frac{1}{2}(12' \times 5')$
$= \frac{1}{2}(60 \text{ square feet})$
$= 30$ square feet

Page 371

1. a. This is a right triangle. The legs of the triangle are a base and the altitude to that base.
Area $= \frac{1}{2}(b \times h)$
$= \frac{1}{2}(12 \text{ m} \times 9 \text{ m})$
$= \frac{1}{2}(108 \text{ m}^2) = 54 \text{ m}^2$

b. This is an acute triangle.
Area $= \frac{1}{2}(b \times h)$
$= \frac{1}{2}(10'' \times 3'')$
$= \frac{1}{2}(30 \text{ square inches})$
$= 15$ square inches

c. This is an obtuse triangle, for which the altitude shown falls outside the triangle.
Area $= \frac{1}{2}(b \times h)$
$= \frac{1}{2}(8 \text{ cm} \times 5 \text{ cm})$
$= \frac{1}{2}(40 \text{ cm}^2) = 20 \text{ cm}^2$

SOLUTIONS FOR TRY THESE

2. a.

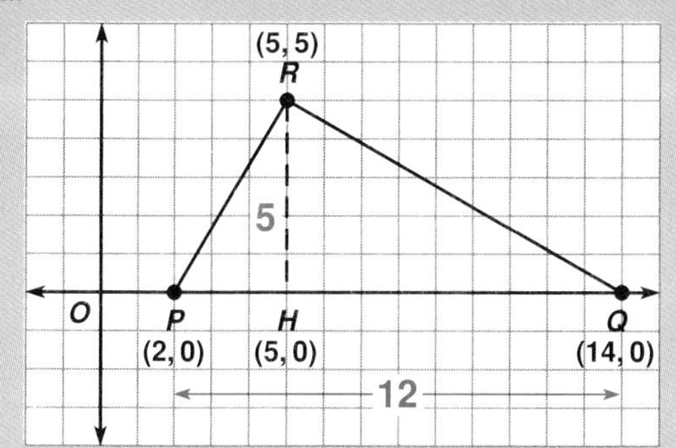

b. base $PQ = 14 - 2 = 12$ units

c. $H(5, 0)$; height $RH = 5 - 0 = 5$ units

d. Area $= \frac{1}{2}(b \times h)$

$= \frac{1}{2}(12 \text{ units} \times 5 \text{ units})$

$= \frac{1}{2}(60 \text{ square units})$

$= 30$ square units

Page 373

1. a. Trapezoid $JKLM$ is made up of $\triangle MLJ$ and $\triangle JKL$.

b. The height of each triangle is 6".

c. $MJ = 22"$ $LK = 12"$

d. Area of $\triangle MLJ = \frac{1}{2}(22" \times 6")$

$= \frac{1}{2}(132 \text{ sq. in.})$

$= 66$ sq. in.

Area of $\triangle JKL = \frac{1}{2}(12" \times 6")$

$= \frac{1}{2}(72 \text{ sq. in.})$

$= 36$ sq. in.

e. Area of Trapezoid $JKLM$ = Area of $\triangle MLJ$ + Area of $\triangle JLK$

66 sq. in. + 36 sq. in.

102 sq. in.

2. a.

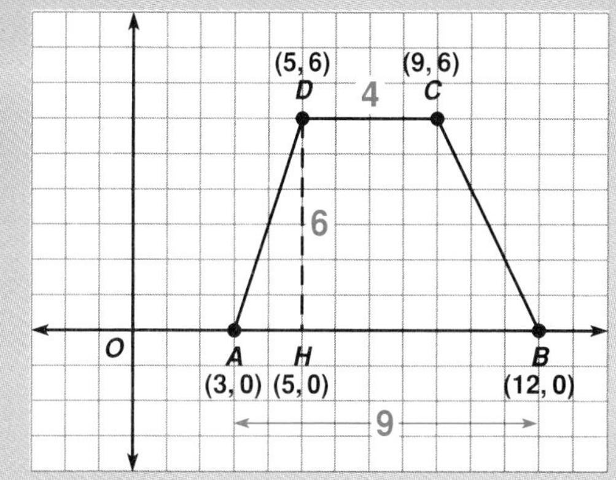

b. height $DH = 6$ units

c. base $AB = 12 - 3 = 9$ units

base $DC = 9 - 5 = 4$ units

d. Area of Trapezoid $= \frac{1}{2}h(b_1 + b_2)$

$= \frac{1}{2} \times 6(9 + 4)$ $h = 6,\ b_1, = 9,\ b_2 = 4$

$= \frac{1}{2} \times 6(13)$ First, work in the parentheses.

$= 3(13)$ Do the multiplication at the left.

$= 39$ Do the next multiplication.

The area of the trapezoid is 39 square units.

(Solutions continue)

SOLUTIONS FOR TRY THESE

9.3 The Circle

Page 382

1. **a.** $C = \pi d$ — Use the circumference rule with diameter.

$= 3.14 \times 10''$ — Substitute.

$= 31.4''$ — Multiply.

Answer: The circumference measures 31.4 inches.

b. $C = 2\pi r$ — Use the circumference rule with radius.

$= 2 \times \frac{22}{7} \times 7\text{m}$ — Substitute.

$= 2 \times \frac{22}{\cancel{7}} \times \overset{1}{\cancel{7}}\text{m}$ — Cancel.

$= 44\text{ m}$ — Multiply the remaining factors.

Answer: The circumference measures 44 meters.

2. $C = 2\pi r$ — Use the circumference rule with radius.

$16\pi = 2\pi r$ — Substitute.

$16 = 2r$ — Divide both sides by π.

$8 = r$ — Divide both sides by 2.

Answer: The radius measures 8 centimeters.

Page 384

1. $A = \pi r^2$ — Apply the rule for area.

$= 3.14(10'')^2$ — Substitute.

$= 3.14(100\text{ sq. in.})$ — Square first.

$= 314\text{ sq. in.}$ — Multiply.

Answer: The area is 314 square inches.

2. $A = \pi r^2$ — Apply the rule for area.

$64\pi = \pi r^2$ — Substitute.

$64 = r^2$ — Divide both sides by π.

$\sqrt{64} = r$ — Take the square root.

$8 = r$ — $8 \times 8 = 64$.

Answer: The radius measures 8 feet.

9.4 Three Dimensions

Page 391

The folded pattern forms a rectangular prism with equal dimensions.

This figure is called a *cube*.

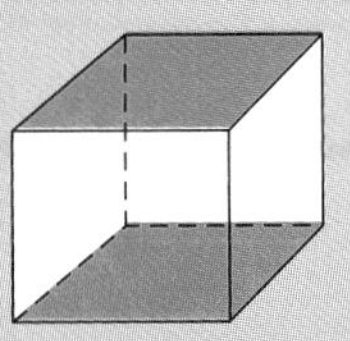

Page 393

Unfold the prism to examine the dimensions of the 6 faces.

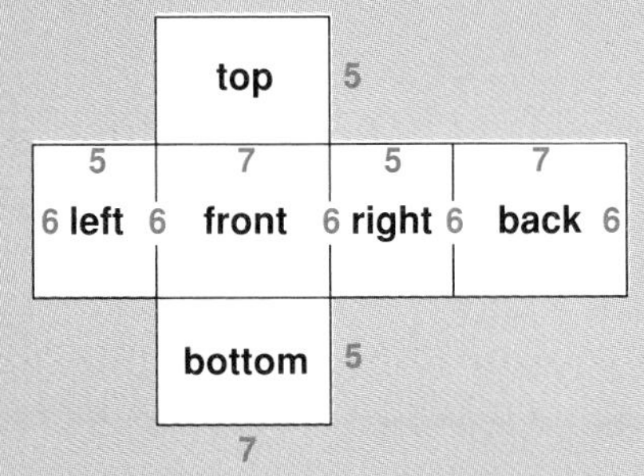

Face	Area	cm.^{2}
front	7×6	42
back	7×6	42
top	7×5	35
bottom	7×5	35
left	5×6	30
right	5×6	30
	Total:	214

The surface area of the prism is 214 square feet.

(Solutions continue)

SOLUTIONS FOR TRY THESE

9.5 Volume

Page 401

1. $V = B \times h$ Apply the volume rule.

$V = \ell \times w \times h$ Base is a rectangle.

$= 4' \times 3' \times 2'$

$= 24$ cubic feet

2. $V = B \times h$ Apply the volume rule.

$V = s^2 \times h$ Base is a square.

$= (5 \text{ in.})^2 \times 5 \text{ in.}$

$= 25 \text{ in.}^2 \times 5 \text{ in.}$

$= 125 \text{ in.}^3$

Page 402

Volume of Cylinder $= \pi r^2 h$ Apply the volume rule.

$= \pi(5 \text{ in.})^2 \times 3 \text{ in.}$ Substitute for r and h.

$= \pi(25 \text{ in.}^2) \times 3 \text{ in.}$ Square first.

$= \pi(75 \text{ in.}^3)$ Do the multiplication.

$= 3.14(75 \text{ in.}^3)$ Substitute for π.

$= 235.5 \text{ in.}^3$

The volume of this cylinder is 235.5 cubic inches

Page 403

1. Volume of Prism $= l \cdot w \cdot h$

$= 8 \cdot 6 \cdot 3$

$= 144$ cubic inches

Volume of Pyramid $= \frac{1}{3}$(Volume of Prism)

$= \frac{1}{3}$(144 cubic inches)

$= 48$ cubic inches

2. Volume of Cone $= \frac{1}{3}\pi r^2 h$ Apply the volume rule.

$= \frac{1}{3}\pi(7^2 \times 6)$ Substitute for r and h.

$= \frac{1}{3}\pi(49 \times 6)$ Square first.

$= \frac{1}{3}\pi(294)$ Work in parentheses.

$= 98\pi$ Do the multiplication.

The volume of the cone is 98π cubic inches.
Substituting for π: $\overset{14}{\cancel{98}}\left(\frac{22}{\underset{1}{\cancel{7}}}\right) = 14 \times 22$

$= 308$ cubic inches

Applying Ratio and Proportion 10

ACTIVITY

Using a 4 by 4 grid, create a design by shading some of the boxes and leaving some of the boxes unshaded. Make enough copies of your design so that you can view different groups of the same design—groups of 1, 2, 3, 4, 5, and so on.

In each group, compare the total number of shaded boxes to the total number of unshaded boxes.

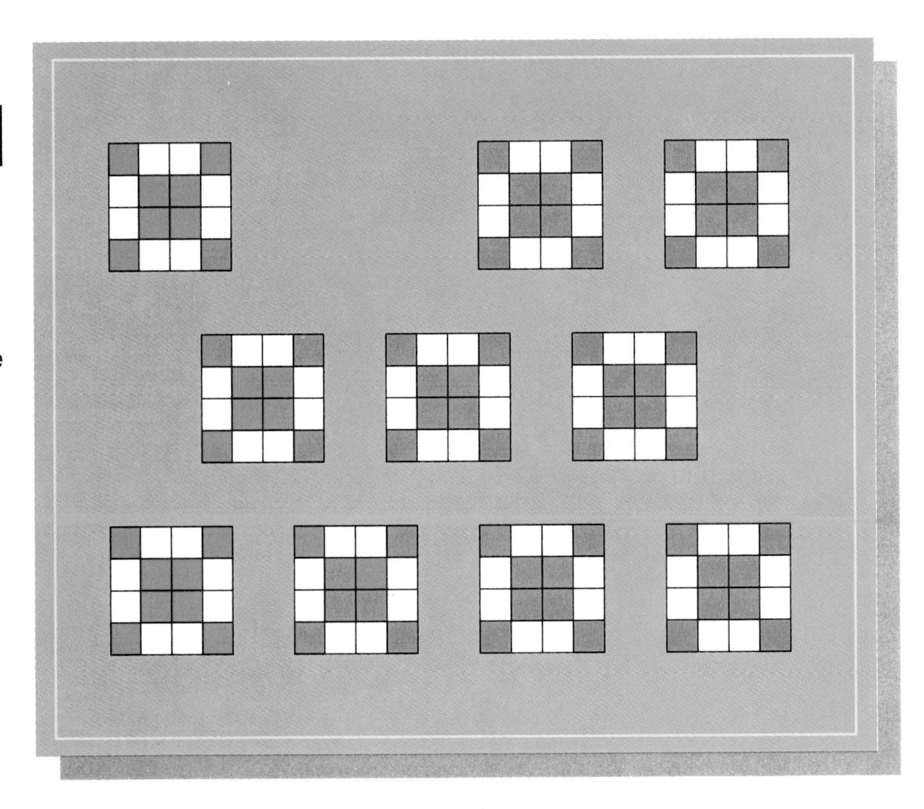

10.1 Ratio

To fit through the door of Wonderland, 4-foot-tall Alice drank a magic potion that shrank her to 1 foot.

How can you compare Alice's heights?

You can use subtraction to compare the two heights, saying that her original height has been decreased by 3 feet.

You can also use division to compare the two heights, saying

$$\frac{\text{new}}{\text{old}} = \frac{1}{4} \quad \text{or} \quad \frac{\text{old}}{\text{new}} = \frac{4}{1}$$

A *ratio* compares two numbers by division.

1 to 4 or 1 : 4 or $\frac{1}{4}$ ← ways to write the ratio that compares the numbers 1 and 4

Using a fraction bar to write a ratio shows that the familiar number one-fourth is also the ratio of the integer 1 to the integer 4.

Numbers that can be written as the *ratio* of two integers are called *rational numbers*.

Example 1

a. Name some rational numbers that are in the ratio $\frac{1}{4}$.

$\frac{1 \times 2}{4 \times 2} = \frac{2}{8}$ $\quad$ $\frac{2}{8}$ is in the ratio $\frac{1}{4}$

$\frac{1 \times 3}{4 \times 3} = \frac{3}{12}$ $\quad$ $\frac{3}{12}$ is in the ratio $\frac{1}{4}$

$\frac{1 \times 4}{4 \times 4} = \frac{4}{16}$ $\quad$ $\frac{4}{16}$ is in the ratio $\frac{1}{4}$

To get other rational numbers in the ratio $\frac{1}{4}$, multiply numerator and denominator by the same factor.

b. Is $\frac{6}{36}$ in the ratio $\frac{1}{4}$?

$\frac{6 \div 6}{36 \div 6} = \frac{1}{6}$ $\quad$ No, $\frac{6}{36}$ is in the ratio $\frac{1}{6}$, not $\frac{1}{4}$.

To see if a rational number reduces to the ratio $\frac{1}{4}$, divide numerator and denominator by the greatest common factor.

Ratio

- A ratio is a comparison of two numbers by division.
- The ratio 1 to 4 is written 1 : 4 or $\frac{1}{4}$.
- The ratio 4 to 1 is not the same as the ratio 1 to 4.
- Other ratios equal to $\frac{1}{4}$ are found by multiplying numerator and denominator by the same factor.

  $\frac{1 \times 3}{4 \times 3} = \frac{3}{12}$ $\frac{3}{12}$ is in the ratio $\frac{1}{4}$

- To reduce a ratio:

  Divide numerator and denominator by the greatest common factor. or Factor and cancel.

  $\frac{6 \div 6}{24 \div 6} = \frac{1}{4}$ $\frac{6}{24} = \frac{1 \times \cancel{6}}{4 \times \cancel{6}} = \frac{1}{4}$ $\frac{6}{24}$ is in the ratio $\frac{1}{4}$

Example 2 Rosa and Rita shared a baby-sitting job that paid \$91 over a week's time. Since Rita spent more time on the job than Rosa, the girls agreed to split the money in the ratio 5 to 8, with Rita getting the larger amount. How much money did each girl get?

1 ***Read the problem for information.***

The amounts of money are in the ratio 5 to 8.

The total amount of money is \$91.

2 ***Make a plan.***

Before writing an equation to model the problem, guesstimate.

The girls' shares are in the ratio $\frac{5}{8}$ but you don't know the factor that multiplies the numerator and denominator of the ratio.

Say this ratio factor is n, as in $\frac{5n}{8n}$.

Then Rosa's share is $5n$ and Rita's share is $8n$.

$8n + 5n = 91$ The total money is \$91.

3 ***Solve the equation.***

$8n + 5n = 91$

$13n = 91$ Combine like terms.

$\frac{13n}{13} = \frac{91}{13}$ Undo × by ÷.

$n = 7$

Rosa's share = $5n = 5(7) = 35$

Rita's share = $8n = 8(7) = 56$

4 ***Check your result.***

Are the amounts in the ratio $\frac{5}{8}$?

$\frac{35}{56} = \frac{35 \div 7}{56 \div 7} = \frac{5}{8}$ ✓

Does the sum of the amounts equal \$91?

$35 + 56 \stackrel{?}{=} 91$

$91 = 91$ ✓

Answer: Rosa got \$35 and Rita got \$56.

Try These *(For solutions, see page 464.)*

1. Which of these rational numbers is in the ratio $\frac{4}{5}$? Explain.

   (a) $\frac{5}{4}$ (b) $\frac{8}{15}$ (c) $\frac{12}{10}$ (d) $\frac{8}{10}$

2. Reduce the ratio $\frac{15}{21}$ to simplest form.

3. After guesstimating, use an equation to solve this problem.

   Find two numbers in the ratio 3 : 5 that have a sum of 32.

Using Ratio to Compare Measurements

Ryan spent 45 minutes typing his history report and Jon spent 1 hour typing his.

Compare the times that the boys spent typing.

Since the times are in different units, you must first change one unit to be the same as the other.

$$\frac{\text{Ryan}}{\text{Jon}} = \frac{45 \text{ minutes}}{60 \text{ minutes}}$$ 1 hour = 60 minutes

$$= \frac{3 \times \cancel{15 \text{ minutes}}}{4 \times \cancel{15 \text{ minutes}}}$$ Cancel the common factor. The common unit also cancels.

$$= \frac{3}{4}$$ The ratio has no unit.

The ratio of Ryan's time to Jon's time was 3 to 4. In other words, Ryan spent 3 minutes typing material that took Jon 4 minutes to type.

Comparing Measurements

- **The two measures must be in the same unit.**
- **The common unit cancels, as does a common ratio factor.**

Example 3 Write the ratio of 75 cents to $2.25 in simplest form.

$$\frac{75 \text{ cents}}{\$2.25} = \frac{75 \text{ cents}}{225 \text{ cents}} = \frac{1 \times \cancel{75 \text{ cents}}}{3 \times \cancel{75 \text{ cents}}} = \frac{1}{3}$$

Try This *(For solution, see page 464.)*

Write a ratio to compare the measures 2 pounds to 8 ounces.
Express the ratio in simplest form.

Using Ratio as a Rate

Describe the posted speed limit in words.
Explain the meaning.

55 miles per hour
means
driving a distance of 55 miles in 1 hour

This speed, or ***rate***, at which a car travels may be written as a ratio. Notice that two different units of measure (miles and hours) are used in this rate.
A rate is usually written so that the number in the denominator is 1.

$$\frac{55 \text{ miles}}{1 \text{ hour}}$$

Example 4 If Jamie earns \$19.50 for 3 hours of work, what is his rate of pay?

$\frac{\$19.50}{3 \text{ hours}}$ Write a ratio.

$\frac{\$6.50}{1 \text{ hour}}$ Divide so that the value in the denominator is 1.

Answer: Jamie's rate of pay is \$6.50 per hour.

Ratio as a Rate

- **There are two different units of measure, which must be mentioned in the rate.**
- **Reduce the numbers of the ratio until the denominator is 1.**

Try This *(For solution, see page 464.)*

If Isabel can type 250 words in 5 minutes, what is her typing rate per minute?

SELF-CHECK: SECTION 10.1

1. What operation does a ratio use to compare two numbers?

   Answer: division
2. Write the ratio 2 to 3 in two other ways.

   Answer: 2 : 3 or $\frac{2}{3}$
3. Since it can be expressed as the ratio of two integers, a number like $\frac{2}{3}$ is called a __?__.

   Answer: rational number

4. **a.** Starting with the ratio $\frac{2}{3}$, how do you find other rational numbers that have the same ratio?

Answer: Multiply numerator and denominator of the ratio $\frac{2}{3}$ by the same ratio factor. For example:

$$\frac{2 \times 5}{3 \times 5} = \frac{10}{15} \qquad \frac{10}{15} \text{ is in the ratio } \frac{2}{3}.$$

b. Name three other rational numbers that are in the ratio $\frac{2}{3}$.

Answer: $\frac{4}{6}, \frac{6}{9}, \frac{8}{12}$ are each in the ratio $\frac{2}{3}$.

c. If n is the ratio factor, write a general rational number that is in the ratio $\frac{2}{3}$.

Answer: $\frac{2n}{3n}$

5. What would you do to reduce the ratio $\frac{6}{12}$ to simplest form?

Answer: Divide numerator and denominator by the greatest common factor.

$$\frac{6 \div 6}{12 \div 6} = \frac{1}{2}$$

or

Factor, and cancel common factors.

$$\frac{6}{12} = \frac{1 \times \cancel{6}}{2 \times \cancel{6}} = \frac{1}{2}$$

6. What must you do first in order to write a ratio comparing the two measures 3 feet and 9 inches?

Answer: Get the same unit for the two measures. Change feet to inches.

7. What does the rate 40 miles per hour mean?

Answer: You can go a distance of 40 miles in 1 hour.

8. To write a rate, you should reduce the numbers of the ratio until the number in the denominator is __?__.

Answer: 1

EXERCISES: SECTION 10.1

1. In Mrs. Klein's math class, there are 12 boys and 14 girls. Write the ratio of:

a. boys to girls **b.** girls to boys

c. boys to the total number of students in the class

2. A survey of 30 car owners found that 10 had red cars, 12 had white cars, 5 had blue cars, and 3 had black cars. Write the ratio of:

a. the number of white cars to the total number of cars

b. the number of black cars to the number of white cars

c. the number of white cars to the number of red cars

d. the number of blue cars to the total number of red cars and white cars

e. the number of green cars to the total number of cars

3. Write three ratios equivalent to each of these ratios:

a. $\frac{1}{5}$ **b.** $\frac{3}{4}$ **c.** $\frac{7}{2}$

d. $3 : 2$ **e.** $\frac{2.5}{1}$ **f.** $1\frac{1}{2} : 1$

4. Write each ratio in simplest form.

a. $\frac{4}{10}$ **b.** $\frac{18}{27}$ **c.** $\frac{70}{20}$

d. $7 : 42$ **e.** $24 : 32$ **f.** $10.2 : 0.2$

5. Do you agree or disagree? Explain.

a. Tom said that *rational number* is a good way to describe a number that can be expressed as the ratio of two integers.

b. Toni said that the ratio 5 : 1 is the same as the ratio 1 : 5 because the two ratios have the same numbers.

c. Tim said that the ratio $\frac{8}{9}$ is equivalent to the ratio $\frac{2}{3}$ because you can divide 8 by 4 and 9 by 3.

d. Tanya said that the rational number with denominator 21 that is equivalent to the rational number $\frac{2}{3}$ is $\frac{14}{21}$.

6. Write a ratio to compare each pair of measures. Express the ratio in simplest form.

a. 4 inches to 1 foot **b.** 3 months to 1 year

c. 10 hours to 1 day **d.** 1 dollar to 1 dime

e. 3 quarts to 1 gallon **f.** 6 centimeters to 1 meter

g. 2 pounds to 4 ounces **h.** 3 weeks to 3 days

i. Using ratio to compare 3 nickels to 2 dimes, Linda wrote 3 : 2 and Liza wrote 3 : 4. With whom do you agree? Explain.

7. Use an equation to solve each problem. First guesstimate an answer.

a. Find two numbers in the ratio 4 : 7 that have a sum of 55.

b. Jon and Jessie divide $34 in the ratio of 9 : 8. If Jon gets the larger share, how much does each get?

c. The ratio of boys to girls in the sophomore class at Burns High is 5 : 7. If there are 252 students in the class, how many boys are there?

d. Hopedale Village outlines its parking lot so that the ratio of spaces reserved for handicapped to other parking is 2 : 18. If there are 340 parking spaces in the lot, how many are marked for the handicapped?

e. The measures of two angles are in the ratio 2 : 1. If the angles are supplementary, find the measure of each angle.

f. The length and width of a rectangular picture frame are in a ratio of 3 : 2. If the perimeter of the frame is 70 inches, find the dimensions.

8. Write each rate in simplest form.

a. a motorbike traveling 175 miles in 5 hours

b. a worker earning $65 in 4 hours

c. a motor making 240 revolutions every 6 minutes

d. 96 tourists equally distributed among 3 buses

e. a school program that has 200 students with 8 teachers

f. a snack that has $1\frac{1}{2}$ cups of nuts to $\frac{1}{2}$ cup of raisins

9. Be a comparison shopper and tell which is the better buy in each situation. Explain your reasoning.

a. Klinger's is selling 3 cans of baked beans for 79¢, and Friendly Foods is selling the same item at 4 for 99¢.

b. A and C Stationers has the same kind of looseleaf paper in two different size packages: 150 sheets for 39¢ or 200 sheets for 59¢.

c. Hector wants 16 bagels. The Bagel Bin sells 4 bagels for $1.59. At Deli Delite, a dozen bagels cost $5.16 with a 13th bagel free.

d. Maggie needs 500 advertising fliers printed. Gotham Printers charges $7 for the first 100 copies, and $3 for each additional 100 copies. Adams Printing charges $10 for the first 250 copies, and $2 for each additional 50 copies.

10. Choose the best answer for each question.

a. At Circus Spectacular, the ratio of adults' to children's ticket prices is 5 to 2. For a benefit performance, when all prices were doubled, the ratio was

(1) 10 : 2 (2) 5 : 4 (3) 5 : 1 (4) 5 : 2

b. The ratio $4\frac{1}{2} : 27$ is equal to

(1) $\frac{1}{2}$ (2) $\frac{1}{3}$ (3) $\frac{1}{5}$ (4) $\frac{1}{6}$

c. If a and b are positive integers and $a : b$ is $1 : 2$, then $a : 3b$ is

(1) 1 : 5 (2) 1 : 1 (3) 1 : 6 (4) 6 : 1

11. A ratio can be extended to compare more than two numbers. For example, 1 : 2 : 4. Use an equation to solve each problem. First guesstimate an answer.

a. Find three numbers in the ratio 2 : 4 : 5 that have a sum of 44.

b. The weights of the three Cleaver children are in the ratio 1 : 2 : 5. If their combined weight is 200 pounds, what are the three weights?

c. The lengths of the sides of a triangle are in the ratio 3 : 4 : 5. If the perimeter of the triangle is 96 cm, find the measure of each side.

d. The measures of the angles of a triangle are in the ratio 2 : 3 : 7. Find the measure of each angle.

e. The measures of the angles of a triangle are in the ratio 1 : 2 : 3. What kind of triangle is this?

12. For each of these pairs of similar figures:

a. *squares* **b.** *rectangles* **c.** *right triangles*

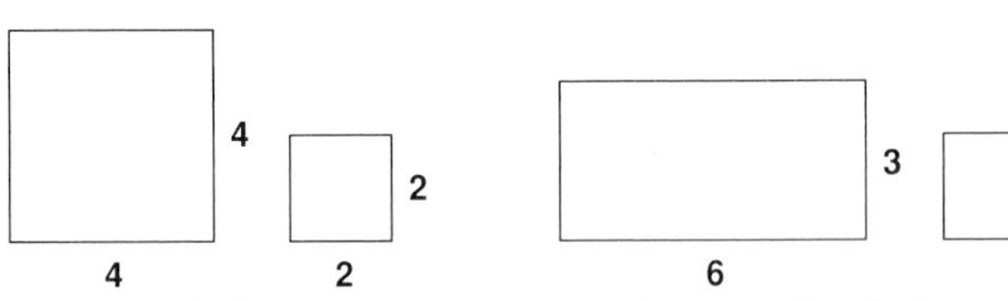

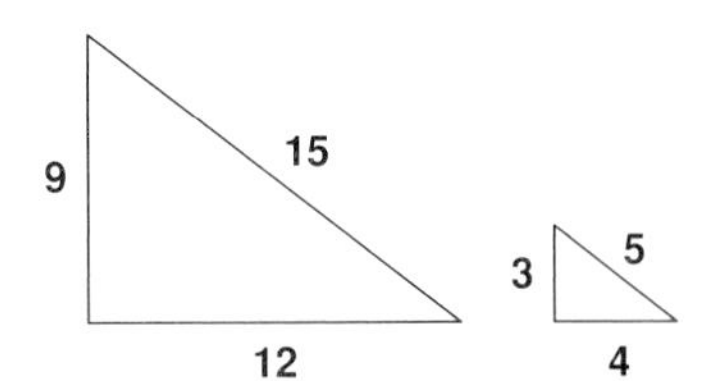

Answer the following questions, writing ratios in lowest terms.

(1) What is the ratio of the lengths of a pair of corresponding sides?

(2) Find the perimeter of each figure.

(3) Find the ratio of the perimeters.

(4) How does the ratio of the perimeters compare to the ratio of the lengths of corresponding sides?

(5) Find the area of each figure.

(6) Find the ratio of the areas.

(7) How does the ratio of the areas compare to the ratio of the lengths of corresponding sides?

EXPLORATIONS

1. Some body builders think that your neck, your biceps muscles, and your calf should be about the same size.

Work with a partner. Use a piece of string to measure these three parts of your body. Then use a ruler to determine the string measures in metric or customary units. Record these values.

Write ratios of your measurements: neck to biceps, neck to calf, and biceps to calf.

Use a calculator to rewrite each ratio as a decimal.

How close are your ratios to the ideal value 1.00?

2. Here is the record of how the school teams at Burns High did this season. The Student Council wants to rank the teams on their performances.

Team	Won	Lost	Played
Alps	6	3	9
Bees	4	2	6
Colts	6	4	10
Darts	9	3	12

Here are statements made by council members, who offered reasons for ranking the teams.

Statement	*Reason*
(1) Darts are in first place.	They won the most games.
(2) Bees are in last place.	?
(3) Colts are in last place.	?
(4) Alps and Colts are tied.	?
(5) Alps and Darts are tied.	?
(6) Alps and Bees are tied.	?

a. Write a reason for statements 2–6.

b. What do you think would be a fair way to rank the teams? After describing your method, tell what your ranking of the teams would be.

10.2 Proportion

Gena has read 50 pages of a 100-page book, and Joan has read 100 pages of a 200-page book.

Compare the amounts read.

One way to compare these amounts is to write a ratio for each.

$\frac{50}{100}$ or $\frac{1}{2}$ $\qquad$ $\frac{100}{200}$ or $\frac{1}{2}$

Since each of the two ratios is equal to the same value, $\frac{1}{2}$, these two ratios are equivalent. $\quad \frac{50}{100} = \frac{100}{200}$

An equation that says two ratios are equivalent is called a *proportion*.

Example 1 Can these pairs of ratios form a proportion? Explain.

a. $\frac{3}{9}$ and $\frac{6}{18}$ $\quad$ When reduced, both ratios are equal to the same value, $\frac{1}{3}$. Thus, these ratios can form a proportion.

b. $\frac{4}{8}$ and $\frac{5}{15}$ $\quad$ When reduced, these ratios are equal to different values $\left(\frac{4}{8} = \frac{1}{2}\text{ while }\frac{5}{15} = \frac{1}{3}\right)$. Thus, these ratios cannot form a proportion.

c. $\frac{1}{2}$ and $\frac{x}{4}$ $\quad$ If x is replaced by 2 so that the ratio reduces to $\frac{1}{2}$, these ratios can form a proportion. If x is replaced by a value other than 2, these ratios cannot form a proportion.

Since a ratio contains two terms, a proportion has four terms.

Read this proportion this way:

1 is to 2 as 3 is to 6

The "outer" terms of the proportion are called ***extremes*** and the "inner" terms are called ***means***.

extreme (1) mean (3)

$$\frac{1}{2} = \frac{3}{6}$$

mean (2) extreme (6)

Consider the product of the means in this proportion, 2×3, and the product of the extremes, 1×6. What is true about these ***cross products***?
Try other proportions. Are the cross products always equal?
If the statement is not a proportion, are the cross products still equal?

When two ratios form a proportion, the product of the means is equal to the product of the extremes.

Example 2 Can these pairs of ratios form a proportion? Explain.

a. $\frac{5}{7}$ and $\frac{15}{21}$

$$\frac{5}{7} \stackrel{?}{=} \frac{15}{21}$$

To form a proportion, the cross products must be equal.

$$7 \times 15 \stackrel{?}{=} 5 \times 21$$

$$105 \stackrel{?}{=} 105 \quad \text{true}$$

Answer: Since the cross products are equal, these ratios can form a proportion.

b. $\frac{3}{11}$ and $\frac{27}{88}$

$$\frac{3}{11} \stackrel{?}{=} \frac{27}{88}$$

To form a proportion, the cross products must be equal.

$$11 \times 27 \stackrel{?}{=} 3 \times 88$$

$$297 \stackrel{?}{=} 264 \quad \text{false}$$

Answer: Since the cross products are not equal, these ratios cannot form a proportion.

Property of Proportion

- **If two ratios form a proportion, then the cross products are equal.**
- **If the cross products of two ratios are equal, then the ratios form a proportion.**

Example 3 Find the value of the variable that makes the open sentence a proportion.

$$\frac{n}{8} = \frac{15}{24}$$

To form a proportion, the cross products must be equal.

$$24n = 8 \cdot 15$$

Either cross product may be written first.

$$24n = 120$$

$$\frac{24n}{24} = \frac{120}{24}$$

Undo × by ÷.

$$n = 5$$

Check:

$$\frac{n}{8} = \frac{15}{24}$$

Replace n by 5 in the original sentence.
Then reduce each ratio to see if they are equal.

$$\frac{5}{8} \stackrel{?}{=} \frac{15}{24}$$

Do not repeat the cross-product method.

$$\frac{5}{8} \stackrel{?}{=} \frac{5 \cdot \cancel{3}}{8 \cdot \cancel{3}}$$

$$\frac{5}{8} = \frac{5}{8} \checkmark$$

Answer: $n = 5$

Try These (For solutions, see page 464.)

1. Which of these sentences is a proportion? Give a reason for your answer.
   a. $\frac{7}{8} \stackrel{?}{=} \frac{14}{16}$ b. $\frac{5}{3} \stackrel{?}{=} \frac{25}{18}$
2. Find the value of the variable that makes the open sentence a proportion.
   a. $\frac{x}{6} = \frac{20}{30}$ b. $\frac{7}{y} = \frac{21}{27}$

Using a Proportion to Solve a Problem

Example 4 Two out of three students in a math class said that math is their favorite subject. If there are 21 students in the class, how many said math is their favorite?

1 ***Read the problem for information.***

A ratio is given: $\frac{2}{3}$ ← number choosing math / ← number asked

Of the 21 students asked, you need to find the number that said math is their favorite.

2 ***Make a plan.***

Before writing an equation to model the problem, try to guess and check.

To write a proportion, you must write the ratios in the same order. $\frac{\text{number choosing math}}{\text{number asked}}$

Let n = the number of students who chose math.

Then: $\frac{n}{21} = \frac{2}{3}$

3 ***Solve the proportion.***

$\frac{n}{21} = \frac{2}{3}$ $\frac{\text{number math}}{\text{number asked}} = \frac{\text{number math}}{\text{number asked}}$

$3n = 2 \cdot 21$ The cross products must be equal.

$3n = 42$

$\frac{3n}{3} = \frac{42}{3}$ Undo × by ÷.

$n = 14$

4 ***Check your result.***

If 14 students chose math, is the ratio $\frac{\text{number choosing math}}{\text{number asked}}$ equal to $\frac{2}{3}$?

$\frac{14}{21} \stackrel{?}{=} \frac{2}{3}$

$\frac{\cancel{7} \cdot 2}{\cancel{7} \cdot 3} \stackrel{?}{=} \frac{2}{3}$

$\frac{2}{3} = \frac{2}{3}$ ✔

Answer: Of the 21 students asked, 14 students chose math as their favorite subject.

Example 5 Sonee Electronics can produce 75 TV sets in 4 hours.
At this rate, how long would it take to produce 375 TV sets?

1 ***Read the problem for information.***

A rate is given: $\frac{75 \text{ TV sets}}{4 \text{ hours}}$

2 ***Make a plan.***

Write a second ratio in the same order, where x represents the number of hours it takes to produce 375 TV sets.

$$\frac{375 \text{ TV sets}}{x \text{ hours}}$$

Write a proportion from the two rates.

$$\frac{75}{4} = \frac{375}{x}$$

3 ***Solve the proportion.***

Solve the proportion, by making the cross products equal.

$$75x = 4 \cdot 375$$
$$75x = 1{,}500$$
$$\frac{75x}{75} = \frac{1{,}500}{75}$$
$$x = 20$$

4 ***Check your result.***

If it takes 20 hours to produce 375 TV sets, is the rate of production equal to 75 TV sets in 4 hours?

$$\frac{375}{20} \stackrel{?}{=} \frac{75}{4}$$
$$\frac{\cancel{5} \cdot 75}{\cancel{5} \cdot 4} \stackrel{?}{=} \frac{75}{4}$$
$$\frac{75}{4} = \frac{75}{4} \checkmark$$

Answer: At the given rate of production, it will take 20 hours to produce 375 TV sets.

Try This *(For solution, see page 465.)*

Use a proportion to solve this problem. First try a guesstimate.

Mike's car used 7 gallons of gas for a trip of 168 miles.
At this rate, how much gas would Mike use for a trip of 840 miles?

Direct Variation

x	2	3	4	5	6
y	4	6	8	10	?

Examine the numbers in this table.

What number would you replace ? by to stay in the pattern? Explain.

The missing number should be 12.

You may have noticed that as the x-values are going up by 1, the y-values are going up by 2.

Observe that each y-value is twice its x-value. $y = 2x$

Another way to express this relationship between y and x is to say that the ratio of y to x is always 2 to 1.

$$\frac{y}{x} = \frac{2}{1}$$

Using this same ratio, what would be the value of y when $x = 100$?

To keep the ratio $\frac{y}{x}$ as $\frac{2}{1}$, y would be 200 when x is 100.

Example 6 For this table of values:

a	2	5	6	8	10
b	6	15	18	24	?

a. Is the ratio $\frac{b}{a}$ constant? Explain.

Use the given values to find the ratio of each pair.

$$\frac{6}{2} = \frac{3}{1} \quad \frac{15}{5} = \frac{3}{1} \quad \frac{18}{6} = \frac{3}{1} \quad \frac{24}{8} = \frac{3}{1}$$

Since the ratio for all the pairs is the same value, $\frac{3}{1}$, the ratio $\frac{b}{a}$ is constant.

b. Using the constant ratio, write a rule that relates b to a.

$$\frac{b}{a} = \frac{3}{1} \text{ or } b = 3a$$

c. Using this constant ratio, what would be the value of b when $a = 50$?

$b = 3a$
$b = 3(50) = 150$

Two values that vary so that their ratio is constant are in *direct variation*.

Example 7 The perimeter of a square varies directly as the length of a side. If the perimeter is 28 cm when the side measures 7 cm, find the perimeter when the side measures 12 cm.

First find the constant ratio from the pair of values.

Let p represent perimeter, and s the length of a side.	$\frac{p}{s} = \frac{28}{7} = \frac{4}{1}$ or 4
Use the constant ratio to write a rule that relates p to s.	$\frac{p}{s} = \frac{4}{1}$ or $p = 4s$
Use the rule to find the value of p when $s = 12$ cm.	$p = 4s$ $p = 4(12) = 48$

Answer: $p = 48$ cm when $s = 12$ cm

Direct Variation

To solve a problem involving direct variation:

1. **Use a given pair of numbers to find the constant ratio.**
2. **Use the constant ratio to write a rule relating the variables.**
3. **Use the rule that relates the variables to find the value that corresponds to a known value.**

Example 8 s varies directly as t. If $s = 42$ when $t = 7$, find s when $t = 4$.

First find the constant ratio from the given pair of values. $\frac{s}{t} = \frac{42}{7} = \frac{6}{1}$ or 6

Use the constant ratio to write a rule relating the variables. $\frac{s}{t} = \frac{6}{1}$ or $s = 6t$

Use the rule that relates the variables to find the value of s that corresponds to $t = 4$.

$s = 6t$
$s = 6(4) = 24$

Check: If $s = 24$ when $t = 4$, are these numbers in the constant ratio $\frac{6}{1}$?

$$\frac{24}{4} \stackrel{?}{=} \frac{6}{1}$$

$$\frac{\cancel{4} \cdot 6}{\cancel{4} \cdot 1} \stackrel{?}{=} \frac{6}{1}$$

$$\frac{6}{1} = \frac{6}{1} \checkmark$$

Answer: $s = 24$ when $t = 4$

Try This *(For solution, see page 465.)*

m varies directly as n. If $m = 36$ when $n = 9$, find m when $n = 5$.

SELF-CHECK: SECTION 10.2

1. What statement does a proportion make?

Answer: that two ratios are equivalent

2. What names are given to the terms of a proportion?

Answer: the means and the extremes

3. When two ratios form a proportion, what is true about the cross products?

Answer: In a proportion, the product of the means is equal to the product of the extremes.

4. Consider this open sentence: $\frac{2}{3} = \frac{n}{24}$

   a. To find the value of n that makes this sentence a proportion, what is the first step?

   Answer: Set the cross products equal. $3n = 2 \cdot 24$

   b. How will you check that $n = 16$ is the correct answer?

   Answer: Substitute into the original equation.
   Reduce the ratios on each side to simplest form.

5. When writing a proportion to solve a problem, how must the numbers be arranged in the ratios?

   Answer: The ratios must be in the same order. For example:

   $$\frac{\text{number of miles}}{\text{number of hours}} = \frac{\text{number of miles}}{\text{number of hours}}$$

6. If the ratio $\frac{y}{x}$ is the constant value 2, write a rule to describe the relationship between y and x.

   Answer: $\frac{y}{x} = \frac{2}{1}$ or $y = 2x$

EXERCISES: SECTION 10.2

1. Reduce each pair of ratios to determine if the two ratios can form a proportion.

   a. $\frac{1}{2}$ and $\frac{5}{10}$ b. $\frac{2}{3}$ and $\frac{4}{12}$ c. $\frac{5}{25}$ and $\frac{20}{100}$

2. Arrange the 4 numbers in each group to form a proportion.

   a. 1, 2, 3, 6 b. 3, 2, 6, 9 c. 100, 1, 2, 50

3. In each proportion, tell which terms are the means and which are the extremes.

   a. $\frac{4}{7} = \frac{12}{21}$ b. $9 : 5 = 54 : 30$ c. $\frac{3}{x} = \frac{9}{15}$

4. Use cross products to determine if the two ratios can form a proportion.

   a. $\frac{4}{8}$ and $\frac{10}{20}$ b. $\frac{18}{6}$ and $\frac{3}{2}$ c. $\frac{1.5}{3}$ and $\frac{9}{18}$

   d. $3 : 2$ and $18 : 12$ e. $4 : 17$ and $12 : 51$ f. $12 : 48$ and $240 : 60$

5. Find the value of the variable that makes the open sentence a proportion. Check your result.

   a. $\frac{x}{6} = \frac{10}{12}$ b. $\frac{9}{n} = \frac{3}{4}$ c. $\frac{20}{36} = \frac{z}{9}$

   d. $\frac{8}{100} = \frac{4}{t}$ e. $\frac{x}{5} = \frac{11}{55}$ f. $\frac{20}{y} = \frac{2}{9}$

   g. $6 : 30 = 4 : x$ h. $9 : y = 3 : 4$ i. $8 : 100 = w : 50$

   j. $\frac{2x}{5} = \frac{32}{2}$ k. $\frac{16}{2n} = \frac{5}{10}$ l. $\frac{4}{x+2} = \frac{1}{3}$

6. Use a proportion to find each amount.

a. If 1 pound = 16 ounces, how many ounces are in:

(1) 3 pounds **(2)** $1\frac{3}{4}$ pounds **(3)** 4.5 pounds

b. If 1 yard = 3 feet, how many feet are in:

(1) 4 yards **(2)** $2\frac{2}{3}$ yards **(3)** 3.2 yards

c. If 1 mile = 1.6 kilometers, how many kilometers are in:

(1) 6 miles **(2)** $2\frac{1}{2}$ miles **(3)** 6.5 miles

7. Use a proportion to solve each problem. First try a guesstimate.

a. Amy got 3 hits for 8 times at bat. At this rate, how many hits can she expect in 32 times at bat?

b. At Sunrise Beach, the charge for renting a rowboat is \$7 for 2 hours. At this rate, what would be the charge for renting a boat for 5 hours?

c. Reliable Day Care provides 3 teachers for every 12 children. If 132 children are enrolled, how many teachers are needed?

d. At Saturday's football game, for every 48 tickets sold, 3 were given away. If 600 tickets were given away, how many tickets were sold?

e. An airplane travels 1,232 miles in 4 hours. At this rate, how many hours will it take for the plane to travel 11,088 miles?

f. The cost of going 50 miles on a turnpike is \$2.00. What can a driver expect to pay for going 300 miles on this turnpike?

g. Using 7 gallons of gas, Will drove his car 161.7 miles. At this same rate of gas usage, how far can he expect to drive on 10 gallons of gas?

h. Jane can read 10 pages of her book in 25 minutes. At this rate, how many pages can she read in 1 hour?

i. If a manufacturer uses $2\frac{1}{2}$ pounds of cotton in the production of 2 dozen sweaters, how many pounds of cotton would be needed to produce 5 dozen of these sweaters?

j. A recipe calls for $1\frac{3}{4}$ cups of oil for a cake to serve 9 people. How much oil is needed if the recipe is increased to serve 18 people?

k. If a traffic signal blinks 52 times a minute, about how many times will it blink in an hour?

(1) 50 (2) 60 (3) 310 (4) 3,100

8. What number can replace ? in each table to stay in the pattern?

a.

x	2	5	8	11
y	6	15	24	?

b.

r	50	40	30	10
s	5	4	3	?

c.

a	4	2	0	−2
b	2	1	0	?

9. To solve each problem of direct variation:

- Use the given pair of numbers to find the constant ratio.
- Use the constant ratio to write a rule relating the variables.
- Use the rule to find the required value.

a. y varies directly as x.
If $y = 24$ when $x = 4$, find y when $x = 5$.

b. p varies directly as t.
If $p = 6$ when $t = 5$, find p when $t = 10$.

c. r varies directly as s.
If $r = 18$ when $s = 36$, find s when $r = 20$.

d. m varies directly as n.
If $m = 0.3$ when $n = 9$, find n when $m = 8$.

10. Solve each problem of direct variation in two ways.

- Determine the constant ratio and write a rule relating the variables.
- Write a proportion

Compare the results of the two solutions.

a. The perimeter of an equilateral triangle varies directly with the length of a side. If the perimeter is 36 inches when the side measures 12 inches, find the perimeter when the side measures 5 inches.

b. The distance a spring is stretched varies directly with the weight applied to the spring. If a weight of 48 grams stretches a spring 12 cm, how far will a weight of 20 grams stretch the spring?

c. The distance a car travels varies directly with the number of hours spent driving. If a car covers a distance of 150 miles in 3 hours, how far will it go in 7 hours?

d. Over a year's time, the amount of simple interest earned on a certificate of deposit varies directly with the rate of interest. If a deposit of $1,000 earns $55 interest:

(1) How much interest would a $3,000 deposit earn?

(2) How much would you have to deposit to earn $220 in interest?

EXPLORATIONS

1. **a.** Copy this picture on your graph paper.
   On another piece of graph paper, try to draw the picture twice as large.

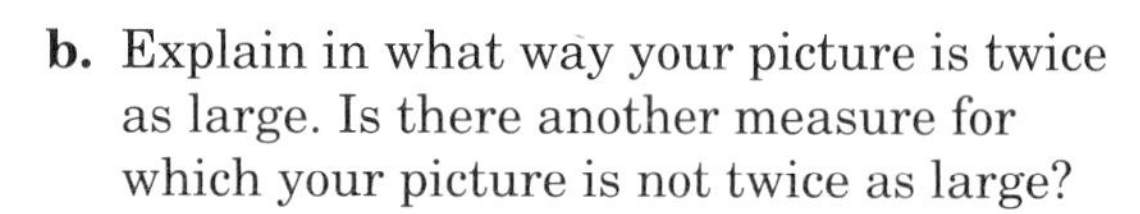

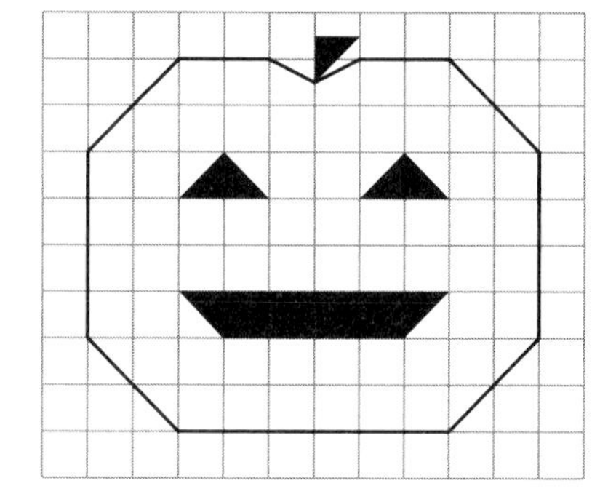

   b. Explain in what way your picture is twice as large. Is there another measure for which your picture is not twice as large?

2. To solve a proportion, rewriting the ratios so that they have a common denominator helps you to see the solution directly.

   Example: Solve the proportion:

   $\frac{x}{9} = \frac{2}{3}$

   $\frac{x}{9} = \frac{2}{3} \cdot \frac{3}{3}$ Rewrite the ratio on the right to have the same denominator as the ratio on the left.

   $\frac{x}{9} = \frac{6}{9}$ Since the denominators of the two ratios are equal, the numerators must be equal.

   $x = 6$

   a. Solve each proportion by getting the ratios to have a common denominator. Check your result in the original open sentence.

   (1) $\frac{x}{12} = \frac{2}{3}$ **(2)** $\frac{x}{30} = \frac{2}{15}$ **(3)** $\frac{21}{36} = \frac{x}{12}$ **(4)** $\frac{8}{34} = \frac{x}{17}$

   b. Explain how you could adapt this method to solve this proportion: $\frac{54}{x} = \frac{18}{24}$

   Carry out your method to get an answer. Check your result in the original open sentence.

   Solve the proportion by the method of cross products. Did you get the same result? Comment on the amount of arithmetic required by each method.

   Note that the cross-product method always works, and a calculator is available to help with the arithmetic. However, looking for shortcuts in arithmetic is a challenge, and gives you another view of the problem.

3. **a.** On graph paper, draw $\triangle ABC$ with vertices $A(4, 3)$, $B(1, 2)$, and $C(3, -1)$.

   b. Double each coordinate (for example, (4, 3) to (8, 6)) to get the image points A', B', and C'. Draw $\triangle A'B'C'$ on the same set of axes.

   c. Compare the lengths of the sides of $\triangle A'B'C'$ with those of $\triangle ABC$. Draw a conclusion relating the changes in lengths to the changes in the coordinates.

10.3 Percent

A recent poll about the election for mayor of New City showed that 43 of 100 voters intended to vote for Jane Brown.

Write the result of this poll as a ratio.

Since this fraction has a denominator of 100, in what other familiar way may the ratio be written?

$\frac{43}{100}$ or 43%

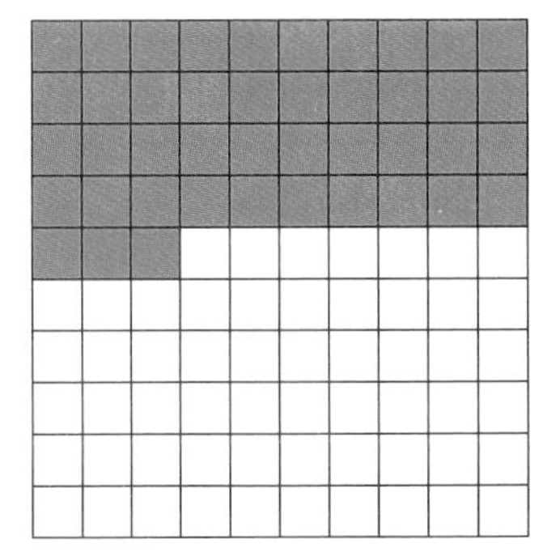

Shading 43 of the 100 boxes on a 10 by 10 grid is a way of modeling 43%.

A *percent* is a ratio where the comparison is to the number 100.

Using Proportions to Solve Percent Problems

Since a percent is a ratio, problems about percent can be solved by using a proportion.

Consider three types of questions about percent and their corresponding proportions.

All three of the following questions are related to this one proportion: $\frac{25}{100} = \frac{12}{48}$

Question	*Proportion*	*Answer*
What is 25% of 48?	$\frac{25}{100} = \frac{x}{48}$	12
What percent of 48 is 12?	$\frac{x}{100} = \frac{12}{48}$	25%
12 is 25% of what number?	$\frac{25}{100} = \frac{12}{x}$	48

Since the questions are worded in a similar way, setting up the correct proportion depends on careful reading. Study the proportions already written to pick out information that will help you write proportions for similar questions.

- What do you notice about the placement of the number 100?
- If the symbol % is used in the question, where does the number that goes with the % get placed?
- Where do numbers that go with the word *is* get placed?
- Where do numbers that go with the word *of* get placed?

Writing a Proportion to Answer a Question About Percent

- 100 is always one of the denominators of the proportion.
- Whenever the symbol % is used in the problem, the number with that symbol is the numerator that goes with 100.
- Numbers that go with the word *is* become numerators.
- Numbers that go with the word *of* become denominators.

$$\frac{\text{number that goes with symbol \%}}{100} = \frac{\text{is}}{\text{of}}$$

Example 1 Write a proportion to answer each question.

Question	*Proportion*	*Answer*
16 is 20% of what number?	$\frac{20}{100} = \frac{16}{x}$	80
What is 20% of 80?	$\frac{20}{100} = \frac{x}{80}$	16
What percent of 80 is 16?	$\frac{x}{100} = \frac{16}{80}$	20%

The basic questions about percent are the core of many problems.

Example 2 The sale price of a TV is $200, which is 80% of the regular price. What is the regular price?

The basic question: 200 is 80% of what number?

Write a proportion, letting x represent the regular price.

$$\frac{80}{100} = \frac{200}{x}$$

$$80x = 20{,}000$$

$$\frac{80x}{80} = \frac{20{,}000}{80}$$

$$x = 250$$

Check: Is 80% of $250 equal to $200?

$250(.80) \stackrel{?}{=} 200$

$200 = 200$ ✔

Answer: The regular price of the TV was $250.

Try These (For solutions, see page 465.)

1. Express 40% as:
   a. a fraction in simplest form
   b. a decimal

2. Write a proportion to answer each question.
   a. What is 40% of 70?
   b. What percent of 70 is 28?
   c. 28 is 40% of what number?

3. For each problem:
   - Write the basic question of percent that is asked.
   - Write a proportion to answer the question.
   - Solve the proportion and check your result.

   a. A 20 % discount is being offered on a jacket that regularly sells for $90. How much will Jorge save if he buys the jacket?
   b. Forty-five of 75 people interviewed said they liked orange juice with breakfast. What percent of those interviewed liked orange juice with breakfast?
   c. The sale price of a bicycle is $275, which is 75% of the regular price. What is the regular price?

Percent of Increase or Decrease

A bookstore owner buys a book for $13 and sells it for $18.20. What is the *markup* on the price of the book?

The markup, or amount of increase, is the difference between the cost to the dealer and the selling price.

$$\$18.20 - \$13 = \underset{\uparrow \text{ markup}}{\$5.20}$$

If we now compare the amount of the increase to the original cost, we will know the percent by which the cost has increased.

$$\frac{\$5.20}{\$13} = 0.4 \text{ or } 40\% \leftarrow \text{percent of increase}$$

The same thinking is applied to situations in which an amount is decreased.

Example 3 A sweater is marked down from $20 to $14. What is the percent of decrease?

First find the amount of the markdown, or the amount of decrease.

$$\$20 - \$14 = \underset{\uparrow \text{ markdown}}{\$6}$$

Then compare that amount to the original amount.

$$\frac{\$6}{\$20} = 0.3 \text{ or } 30\%$$

Answer: The sweater has been marked down by 30%.

Percent of Increase or Decrease

$$\text{Percent of Increase} = \frac{\text{Amount of Increase}}{\text{Original Amount}}$$

$$\text{Percent of Decrease} = \frac{\text{Amount of Decrease}}{\text{Original Amount}}$$

Try This *(For solution, see page 466.)*

Earl reduced his weight from 250 pounds to 200 pounds.
Find the percent of decrease in his weight.

SELF-CHECK: SECTION 10.3

1. When you express 25% as a ratio, what numbers are you comparing?

Answer: 25 to 100

2. Since 25% can be written as $\frac{25}{100}$, what is the simplest equivalent fraction?

Answer: $25\% = \frac{25}{100} = \frac{1}{4}$

3. Since 25% can be written as $\frac{25}{100}$, what is an equivalent decimal?

Answer: $25\% = \frac{25}{100} = .25$

4. Write the statement "25% of 48 is 12" as a proportion.

Answer: $\frac{25}{100} = \frac{12}{48}$

5. Write the statement "48 is 400% of 12" as a proportion.

Answer: $\frac{48}{12} = \frac{400}{100}$

6. Suppose the amount \$8 is increased to the amount \$10.

a. What is the amount of increase?

Answer: \$2 (calculation: \$10 – \$8 = \$2)

b. How would you find the percent of increase?

Answer: Compare the amount of increase, \$2, to the original amount, \$8.

c. What is the percent of increase?

Answer: 25% (calculation: $\frac{2}{8}$ or $\frac{1}{4}$ or 25%)

EXERCISES: SECTION 10.3

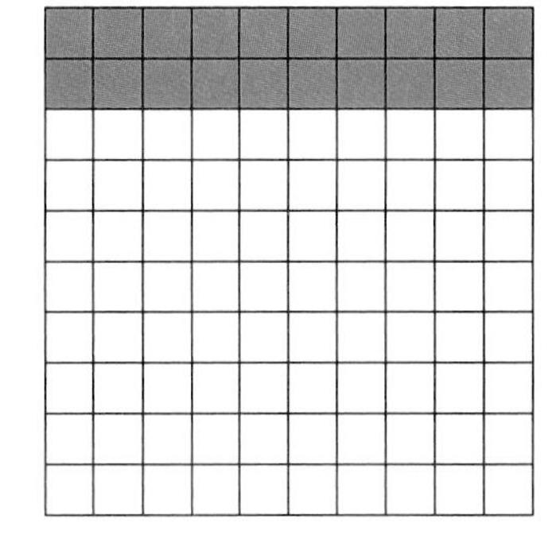

1. **a.** Write the ratio of the number of shaded boxes to the total number of boxes shown on this grid.

   b. Write this ratio as a percent.

   c. The 20 shaded boxes of the 100 boxes also show you the result of 20% of 100.

$$20\% \text{ of } 100 = \frac{20}{100} \cdot 100$$
$$= 20$$

   Explain how you could use this model to get 20% of 50.

2. Copy and complete this table to show equivalent forms of each ratio.

Ratio	Percent	Decimal	Ratio	Percent	Decimal
a. $\frac{1}{2}$	50%	?	**e.** $\frac{3}{4}$	?	?
b. ?	$33\frac{1}{3}\%$	$.\overline{33}$	**f.** ?	60%	?
c. $\frac{1}{4}$	?	?	**g.** $\frac{3}{8}$	?	?
d. ?	?	.20	**h.** ?	?	$.\overline{66}$

3. This circle graph shows the distribution of grades on a test.

   80 students took the test.

   a. How many students got:

   (1) A **(2)** B **(3)** C

   b. If C was the lowest passing grade, how many students failed the test?

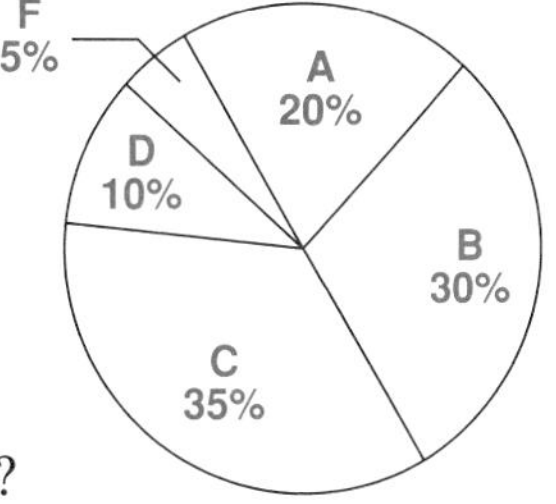

4. Write a proportion to answer each of these basic questions about percent. Solve the proportion and check your result.

   a. **(1)** What is 40% of 90? **(2)** What percent of 90 is 36? **(3)** 36 is 40% of what number?

   b. **(1)** What is 30% of 60? **(2)** What percent of 60 is 18? **(3)** 18 is 30% of what number?

   c. **(1)** 60 is 75% of what number? **(2)** What is 75% of 80? **(3)** What percent of 80 is 60?

   d. What is 20% of 64?

   e. What percent of 10 is 2?

   f. 8 is 20% of what number?

   g. What is 120% of 90?

   h. What percent of 25 is 40?

   i. 21 is 75% of what number?

5. Choose the best estimate.

a. 12 is about 50% of

(1) 15 (2) 21 (3) 25 (4) 32

b. About 25% of 59 is

(1) 12 (2) 15 (3) 18 (4) 21

c. 35 out of 48 is about

(1) 75% (2) 68% (3) 60% (4) 50%

d. 8 out of 99 is about

(1) 6% (2) 7% (3) 8% (4) 9%

6. Give a reason for your answer to each question.
In your reason, use the results of calculations.

a. Is 60% of 50 more or less than 50% of 60?

b. Which is more: 20% off $100 or another 10% off 10% off $100?

c. "Buy 2 get 1 free" is the same as a saving of ___?___ percent.

d. The Nets won 4 out of 25 games, and the Jets won 3 out of 20.
Which team has the better record?

7. Explain how each person should recognize that the answer to the question is not reasonable. Write the correct calculation, and answer.

a. To answer the question "What is 30% of 60?", Ruth calculated: $30 \times 60 = 1{,}800$

b. To answer the question "60 is what percent of 240?", Roz calculated: $\frac{240}{60} = 4\%$

c. To answer the question "75% of what number is 150?", Rob calculated: $0.75 \times 150 = 112.5$

8. For each problem:

- Write the basic question of percent.
- Write a proportion that will answer the question.
- Estimate an answer.
- Solve the proportion and check your result.

a. A house worth $92,500 is insured for 80% of its value.
If the house were destroyed by fire, how much would the owner get?

b. Four of the radios that came off an assembly line were defective.
If 200 radios were produced, what percent were defective?

c. The 3 students absent are 15% of the class. How many are in the class?

d. Trish received $2.50 in interest on her savings account.
If her bank's interest rate is 2%, how much is in her account?

e. After one year, Ossie got $30 interest on a $500 U. S. Savings Bond.
What was the rate of interest?

f. Andy earns a 6% commission on his sales.
If his sales totaled $7,000, how much did he get in commission?

g. The 30% discount on a jacket saved Luz $24.
What was the original price of the jacket?

h. Jackson Realty gets 5% commission on the sale of a house.
How much does this agency get on the sale of a $125,000 house?

i. The discounted price of a $400 computer printer is $320.
What percent of the original price is the discounted price?

9. Use a proportion in the solution of each problem.

a. A VCR that regularly sells for \$350 is on sale for 20% off. How much would a customer pay?

b. Of the 25 students in a class, 5 were absent. What percent of the class was present?

c. Ms. Carter is paid a monthly salary of \$450 plus an 8% commission on her sales. If her December sales totaled \$8,500, what were her earnings for the month?

d. On a geography test, 20% of the questions were about Asia.

If 6 questions were about Asia, how many were about other places?

e. Of the 60 children on a school trip, 70% brought their lunch and the rest brought money to buy lunch. How many children were buying lunch?

f. Rock paid 25% down on the purchase of an \$800 sofa. If the balance is to be paid in 12 equal monthly installments, how much must he pay each month?

g. Bert is buying a pair of running shoes regularly priced at \$80. Conroy's is selling these shoes at 15% off and Wade's is selling them for \$64. Where should Bert buy his shoes?

h. The 7% sales tax on Darlene's purchase of gloves and a scarf came to \$1.96. If, before tax, the gloves were priced at \$16, what was the price of the scarf?

10. Use a ratio to solve each problem.

a. A dealer buys a telescope at the wholesale cost of \$85 and marks it up to sell for \$153. What is the percent increase?

b. The price of a boom box was marked down from \$340 to \$272. What was the percent of decrease?

c. The population of Dexter County rose from 63,400 to 78,616. What was the percent of increase?

d. In one month, the number of passengers on Chuck's Charter Airline fell from 2,580 to 2,451. What was the percent of decrease?

e. Ms. Cash bought a gold chain for \$120 and sold it at a markup of 75%. How much did she sell it for?

11. In each of the following, x and y are whole numbers.

a. If x is one-tenth of y, then y is what percent of x?

(1) 10% (2) 90% (3) 100% (4) 1,000%

b. If x is one-half of y, then y is what percent of x?

(1) 50% (2) 100% (3) 150% (4) 200%

EXPLORATIONS

a. The 60 students in a physical education class were asked to choose an elective. The results of the selection are shown in the table.

Activity	Number of Students
Badminton	15
Basketball	12
Gynmastics	6
Volleyball	15
Wrestling	12

To draw a circle graph that displays this information, consider these questions:

(1) For a circle graph that displays percents:

How will you find the percent of students choosing each activity?

(2) Each section of the circle graph has an angle at the center of the circle. The measure of each angle determines the size of the section.

How will you find the angle measure for each section?

Draw the circle graph, using a protractor to measure off the angle needed for each section. Label each section with its activity and percent.

b. Conduct a survey of your own with your classmates. Display the results in a circle graph.

Sample Survey Question: When do you do the homework that is assigned on Friday?

Which breakdown of the results would make a more interesting and informative circle graph? Explain.

(1)	(2)
Friday	Friday
Saturday	Saturday, day
Sunday	Saturday, night
	Sunday, day
	Sunday, night

10.4 Similarity

Scale Drawings

The Lawsons have hired the architect Mona Aaron to draw up plans for a new den they plan to build onto their house. How might Ms. Aaron present her ideas to the Lawsons?

Ms. Aaron's plans could be presented in a diagram that is conveniently smaller than the actual size of the room, and is in proportion to the actual size.

Such an architectural diagram, usually called a ***blueprint***, will tell the ratio between the actual dimensions and the measures used in the diagram. For example, the ratio used, called the ***scale***, might be that 1 foot is represented by $\frac{1}{4}$ inch.

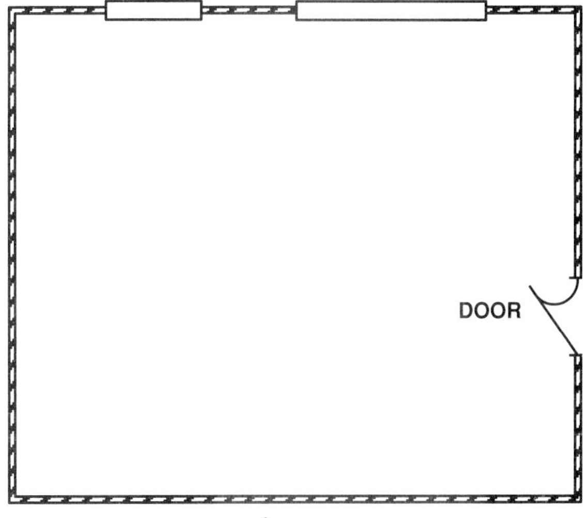

Scale: $\frac{1}{4}$ inch = 1 foot

Example 1 The scale of a blueprint of a room is $\frac{1}{4}$ inch to represent 1 foot. If the line segment that represents the width of the room measures 3 inches, what is the actual width of the room?

1 ***Read the problem for information.***
A ratio is given, $\frac{1}{4}$ inch : 1 foot.

You have to find the number of actual feet that corresponds to 3 inches in the blueprint.

2 ***Make a plan.***
First try to guesstimate an answer, then set up a proportion.

Let x = the number of actual feet represented by 3 inches in the diagram.

Be sure to keep the ratios in order, say:

$$\frac{\text{diagram inches}}{\text{actual feet}} = \frac{\text{diagram inches}}{\text{actual feet}}$$

$$\frac{\frac{1}{4}\text{ inch}}{1\text{ foot}} = \frac{3\text{ inches}}{x\text{ feet}}$$

3 ***Solve the proportion.***

$$\frac{\frac{1}{4}}{1} = \frac{3}{x}$$

$$\frac{1}{4}x = 3$$ In a proportion, the cross products are equal.

$$x = 12$$ Multiply both sides by 4.

4 ***Check your result.***
Are the two ratios equivalent?

$$\frac{\frac{1}{4}}{1} \stackrel{?}{=} \frac{3}{12}$$

$$\frac{\cancel{4}\cdot\frac{1}{4}}{\cancel{4}\cdot 1} \stackrel{?}{=} \frac{1\cdot\cancel{3}}{4\cdot\cancel{3}}$$

$$\frac{1}{4} = \frac{1}{4} \checkmark$$

Answer: The actual width of the room is 12 feet.

A blueprint is useful for making calculations and approximating costs.

Example 2 In this blueprint of the first floor of the Bell's new home, 1 inch represents 16 feet.

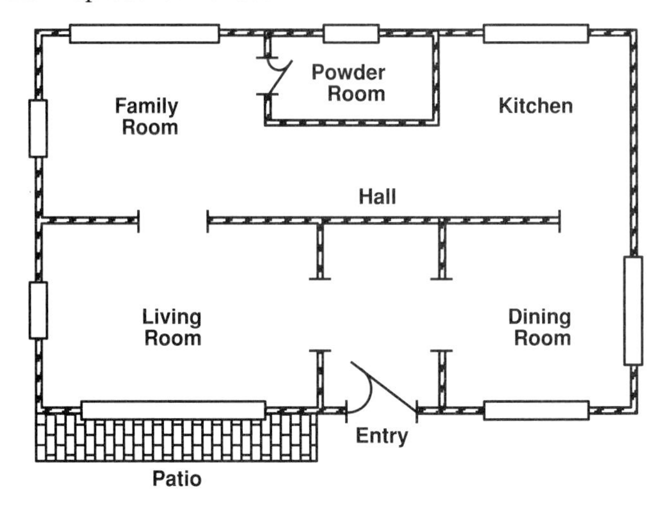

If, in the blueprint, the dimensions of the living room are 1.5 inches by 1 inch:

a. Find the actual dimensions.

Since the width of the room is represented by 1 inch and the scale is 1 inch = 16 feet, the actual width is 16 feet.

Set up a proportion to find the actual length.

$$\frac{\text{diagram inches}}{\text{actual feet}} = \frac{\text{diagram inches}}{\text{actual feet}}$$

Let x represent the actual length.

$$\frac{1 \text{ inch}}{16 \text{ feet}} = \frac{1.5 \text{ inches}}{x \text{ feet}}$$

Thus, the actual dimensions of the living room are 24 feet by 16 feet.

$$1x = 16 \times 1.5$$

$$x = 24 \quad \text{length of room}$$

b. Find the actual area of the living room.

Area of Rectangle = length × width

= 24 ft. × 16 ft. = 384 square feet

c. If the Bell's plan to carpet the living room at the cost of \$20 per square yard, what is the approximate cost?

Since carpeting is sold by the square yard and the dimensions of the room are in square feet, set up a proportion to convert to the same unit.

1 square yard = 9 square feet

$$\frac{1 \text{ sq. yd.}}{9 \text{ sq. ft.}} = \frac{n \text{ sq. yd.}}{384 \text{ sq. ft.}}$$

$$9n = 1 \times 384$$

$$\frac{9n}{9} = \frac{384}{9}$$

$$n \approx 43 \text{ sq. yd.}$$

To find the cost: 43 sq. yd. × \$20 = \$860

Try This *(For solution, see page 466.)*

In this drawing that represents a computer chip, 1 inch corresponds to 0.5 millimeter.

Find the actual width of the computer chip if the width of the chip in the drawing is 2.0 inches.

Scale: 1 in. = 0.5 mm

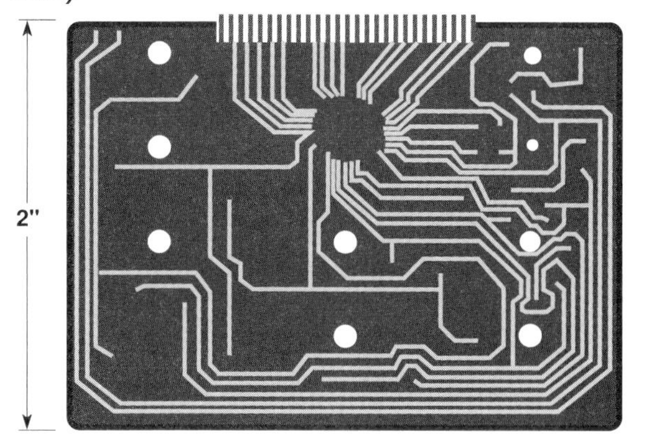

Similar Triangles

What observation can you make about these two triangles?

These triangles have the same shape, but are not the same size.

Use a protractor to verify that the 3 pairs of corresponding angles are equal in measure.

If $AC = 1.2$ and $A'C' = 0.9$, determine the ratio $AC : A'C'$. (Use a calculator.)

If $AB = 1.0$ and $A'B' = 0.75$, determine the ratio $AB : A'B'$.

If $BC = 2.0$ and $B'C' = 1.5$, determine the ratio $BC : B'C'$

What observation can you make about the ratios of the lengths of these 3 pairs of sides, if these triangles have the same shape?

Since the ratios are equal, the pairs of corresponding sides are in proportion.

Triangles that have the same shape are called *similar triangles.*
In similar triangles:
The measures of corresponding angles are equal.
The measures of corresponding sides are in proportion. $\sim$ **means *is similar to***

Try This *(For solution, see page 466.)*

$\triangle PQR \sim \triangle MNO$

From the markings in the diagram, name the

a. corresponding angles

b. corresponding sides

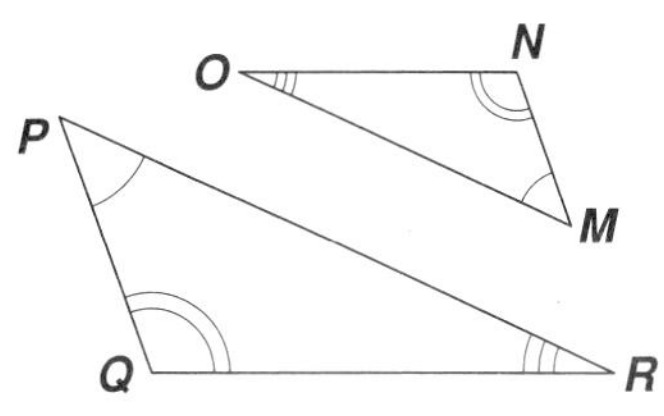

Example 3 $\triangle DEF \sim \triangle PQR$.

The measures of the sides of $\triangle DEF$ are 15 cm, 20 cm, and 25 cm. In $\triangle PQR$, $PQ = 3$ cm, and $PR = 4$ cm.

Find QR.

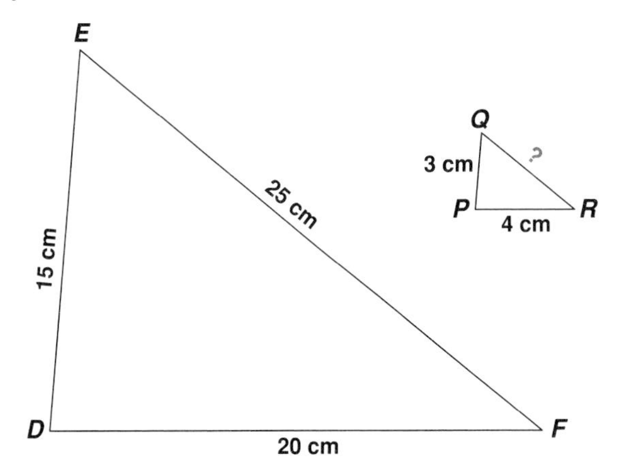

1 *Read the problem for information.*

You are told that the triangles are similar.

You are told the measures of 3 sides of one triangle and 2 sides of the other triangle.

You have to find the missing length.

Since the triangles are similar, you know that the measures of the corresponding sides are in proportion.

2 *Make a plan.*

Write a proportion.

$$\frac{DE}{PQ} = \frac{EF}{QR}$$

$$\frac{15}{3} = \frac{25}{QR}$$

3 *Solve the proportion.*

$$\frac{15}{3} = \frac{25}{QR}$$

$$15 \cdot QR = 3 \cdot 25$$

$$15QR = 75$$

$$\frac{15QR}{15} = \frac{75}{15}$$

$$QR = 5$$

4 *Check your result.*

If $QR = 5$, are the lengths of the sides in proportion?

$$\frac{DF}{PR} \stackrel{?}{=} \frac{DE}{PQ} \stackrel{?}{=} \frac{EF}{QR}$$

$$\frac{20}{4} \stackrel{?}{=} \frac{15}{3} \stackrel{?}{=} \frac{25}{5}$$

$$\frac{5}{1} = \frac{5}{1} = \frac{5}{1} \checkmark$$

Answer: $QR = 5$ cm

Try These *(For solutions, see page 467.)*

1. Since the angles of a triangle give the triangle its shape, two triangles are similar if they are alike in angle measure.
   If $\triangle ABC$ and $\triangle DEF$ are alike in measure for two pairs of angles, what must be true about the third pair of angles? Explain.

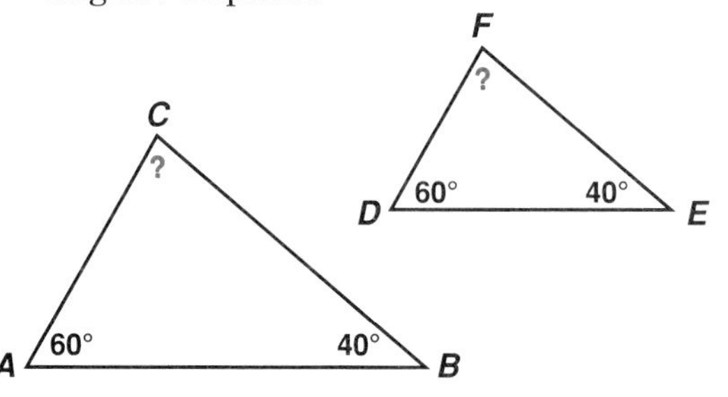

2. $\triangle ABC \sim \triangle DEF$ with the measures of the sides as shown.
   Find DF.

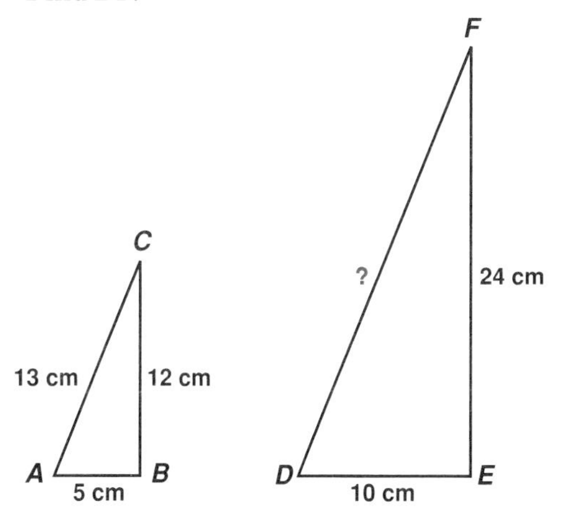

Using Similar Triangles to Solve Shadow Problems

During a hike in the woods, a forest ranger showed a group of scouts how to find the height of a tree by measuring the height of a stick pounded into the ground near the tree.

The ranger and the scouts also found the measures of the shadows of the stick and the tree. These were easy to measure, since they were on the ground.

The ranger then told the scouts about the shadow rule:

During the day, two objects that are near each other have shadows whose measures are proportional to the heights of the objects.

(Continues on next page)

Here is a drawing of the situation.
Explain why the shadow rule works.

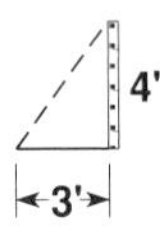

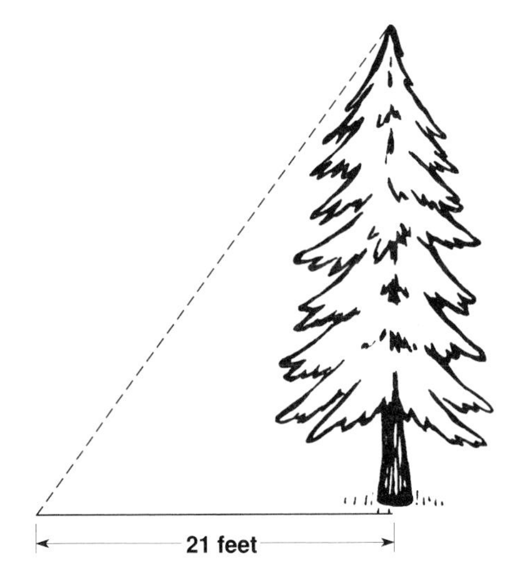

The shadow rule sets up similar triangles. When the triangles are similar, the measures of the corresponding sides are in proportion.

What special angle is included in the similar triangles that represent shadow problems?

Because the objects are standing perpendicular to the ground, these triangles include right angles.

Example 4 Calculate the height of the tree in the problem with the forest ranger.

Apply the shadow rule to write a proportion, with x as the height of the tree.

$$\frac{\text{stick's height}}{\text{stick's shadow}} = \frac{\text{tree's height}}{\text{tree's shadow}}$$

$$\frac{4 \text{ feet}}{3 \text{ feet}} = \frac{x \text{ feet}}{21 \text{ feet}}$$

$$3x = 4 \cdot 21$$

$$3x = 84$$

$$x = 28$$

Check: If the tree is 28 feet tall, are the measures in proportion?

$$\frac{\text{stick's height}}{\text{stick's shadow}} \stackrel{?}{=} \frac{\text{tree's height}}{\text{tree's shadow}}$$

$$\frac{4}{3} \stackrel{?}{=} \frac{28}{21}$$

$$\frac{4}{3} \stackrel{?}{=} \frac{4 \cdot \cancel{7}}{3 \cdot \cancel{7}}$$

$$\frac{4}{3} = \frac{4}{3} \checkmark$$

Answer: The tree is 28 feet tall.

Try This *(For solution, see page 467.)*

A man is standing near a 4-foot-tall boy. If the man's shadow measures 3 feet and the boy's shadow measures 2 feet, how tall is the man?

SELF-CHECK: SECTION 10.4

1. What is the purpose of a scale drawing?

Answer: A scale drawing is a more convenient size than the subject of the drawing. The scale drawing can be referred to with ease.

2. What is true about the dimensions represented in a scale drawing and the actual dimensions?

Answer: The dimensions in the scale drawing are in proportion to the actual dimensions.

3. What do we call triangles that have the same shape but not necessarily the same size?

Answer: similar triangles

4. What is true about the measures of the angles of similar triangles? of the sides?

Answer: In similar triangles, the measures of corresponding angles are equal, and the measures of corresponding sides are in proportion.

5. What diagram models a shadow problem?

Answer: 2 similar right triangles

EXERCISES: SECTION 10.4

1. For each of these scale drawings:

- Use a ruler to determine the length and width of the drawing.
- Use the given scale to determine the actual dimensions of the object for which the drawing is a model.

a. This is a blueprint of a room.

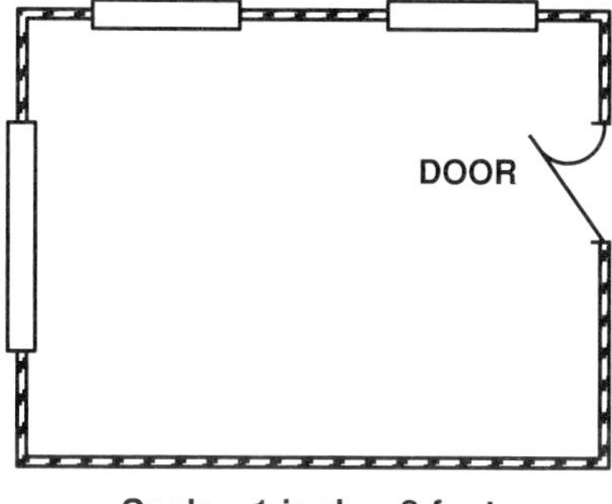

Scale: 1 inch = 8 feet

b. This is a drawing of a chip for a wristwatch.

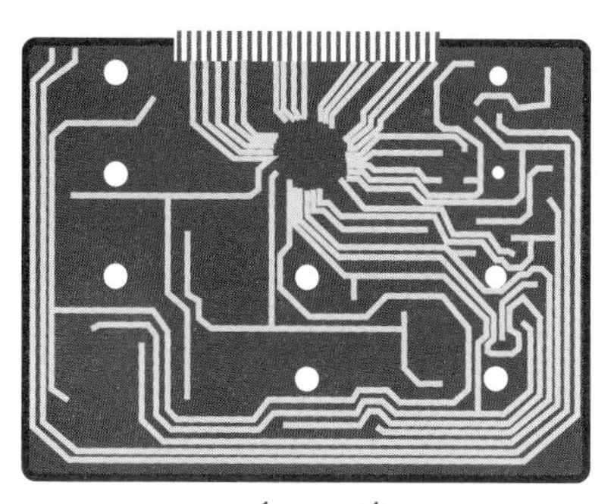

Scale: $\frac{1}{2}$ in. = $\frac{1}{10}$ mm

2. Use the distance on each map and the given scale to find the actual distance between the cities mentioned.

a. The town of Nance is west of Alba, and the distance between them on the map is 1.4 inches.

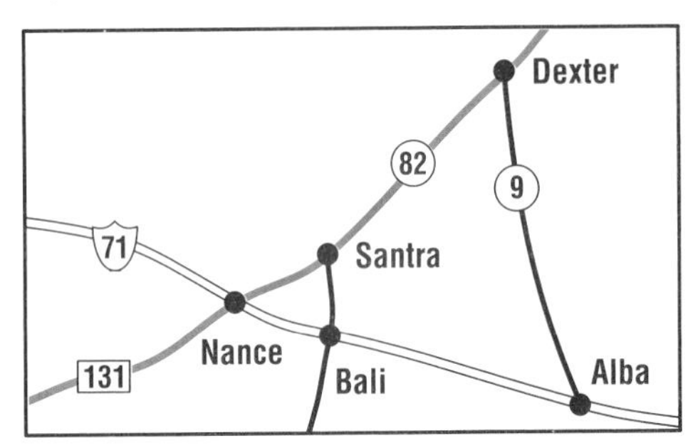

Scale: 1 inch = 150 miles

b. The city of Pueblo is south of Denver, and the distance between them on the map is three-quarters of an inch.

Scale: $\frac{1}{2}$ inch = 70 miles

3. The scale on a map is 1 inch = 80 miles.

a. Find the actual distance between two cities if the map distance is:

(1) 3 in. **(2)** 1.5 in. **(3)** $\frac{3}{4}$ in. **(4)** $2\frac{1}{4}$ in.

b. Find the map distance between two cities if the actual distance is:

(1) 160 mi. **(2)** 40 mi. **(3)** 20 mi. **(4)** 200 mi.

4. Jan made a 12-inch model of a 36-foot boat.

a. What are the actual dimensions of a deck area whose dimensions in the model are:

(1) 3 in. by 4 in. **(2)** $\frac{1}{2}$ in. by 1 in.

b. What are the dimensions in the model of a cabin whose actual dimensions are:

(1) 8 ft. by 12 ft. **(2)** 9 ft. by 16 ft.

5. A blueprint for an addition to Slocum High shows a corridor with 8 rooms and 2 flights of stairs. The 6 classrooms along the sides of the corridor are all the same size. The 2 square science labs at the end of the corridor are the same length as the classrooms.

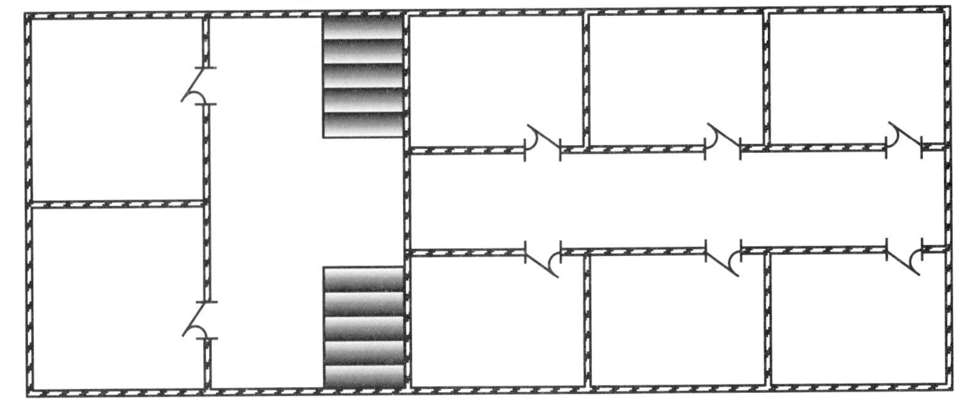

a. If the 5-inch-by-2-inch drawing represents the actual dimensions of 120 feet by 48 feet, how many actual feet are represented by 1 inch on the drawing?

b. Find the actual dimensions of each classroom and each science lab.

c. The floors of the classrooms and labs are to be tiled.

(1) What is the total area to be tiled?

(2) If the tile costs $1.50 per square foot, what is the total cost of tiling?

6. For each scale model, find the required dimension.

a. A road distance of 45 km is represented on a map by a distance of 3 cm. What road distance is represented by 5 cm on the map?

b. In a photo, the Baker County Museum is 4 inches high and 6 inches wide. If the actual width of the museum is 75 feet, what is the actual height?

c. A billboard displays a 12-foot-high blowup of a bottle of soda pop. If the bottle is really 9 inches high, with a 4-inch-high label, what is the height of the label in the blowup?

d. The drawing of an airship is on a scale of 4 cm = 5m. If the gondola cabin attached to the underside of the airship is 3 cm long in the drawing, what is its actual length?

7. For each pair of similar triangles, find the missing angle measures.

a. $\triangle ABC \sim \triangle DEF$

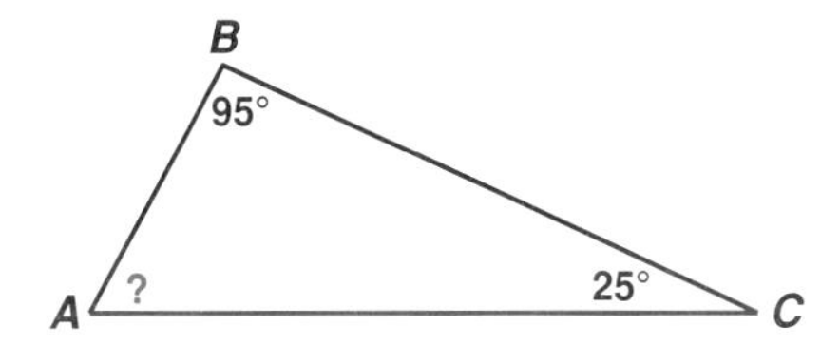

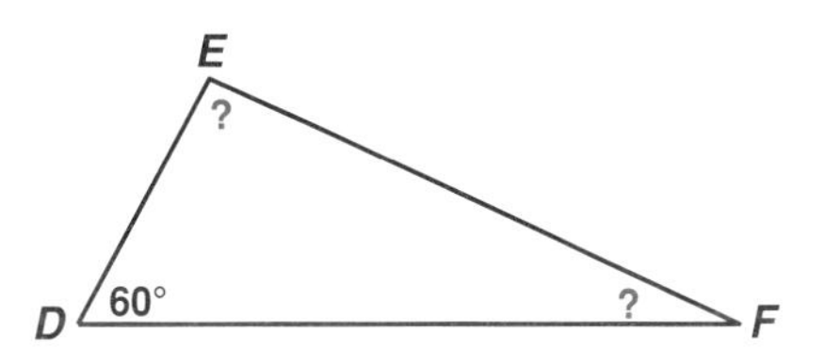

b. $\triangle GHI \sim \triangle JLK$

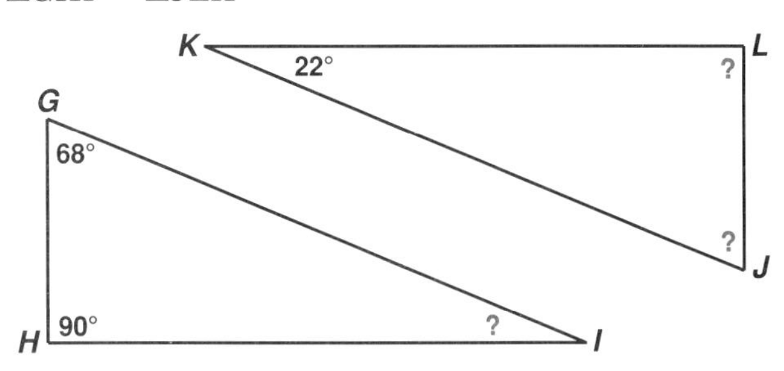

8. Use the given angle measures to tell if the triangles in each pair are similar. Answer *yes, no,* or *cannot tell*, and give a reason.

a. $\triangle ABC$ $m\angle A = 35°$ $m\angle B = 60°$ $m\angle C = 85°$

$\triangle DEF$ $m\angle D = 60°$ $m\angle E = 85°$ $m\angle F = 35°$

b. $\triangle GHJ$ $m\angle G = 40°$ $m\angle H = 50°$

$\triangle KLM$ $m\angle K = 50°$ $m\angle L = 90°$

c. $\triangle TOM$ $m\angle T = 40°$ $m\angle O = 80°$

$\triangle RAY$ $m\angle R = 60°$ $m\angle A = 70°$

d. $\triangle PQR$ $m\angle P = 60°$ $m\angle Q = 60°$

$\triangle STU$ $m\angle S = 60°$

e. $\triangle ACE$ and $\triangle BDF$ are isosceles.

$m\angle A = 100°$, and each of the two congruent angles in $\triangle BDF$ measures 40°.

f. $\triangle JKL$ and $\triangle MNP$ are right triangles.

$m\angle J = 35°$ and $m\angle N = 55°$.

9. For each pair of similar triangles, find the missing measures.

a. $\triangle CFE \sim \triangle ADB$

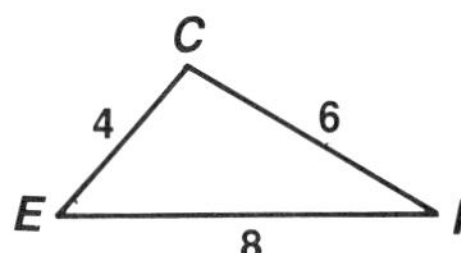

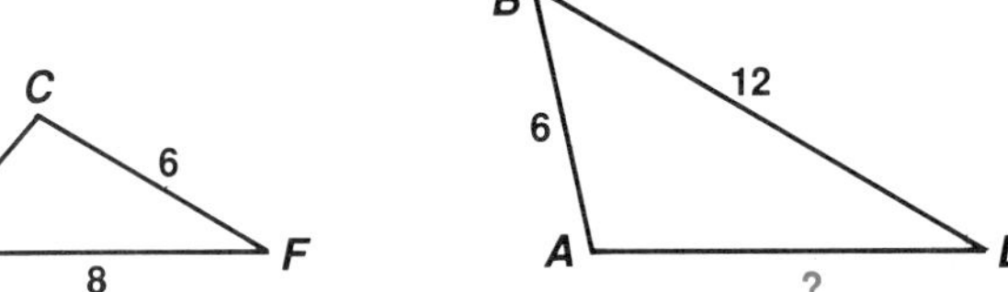

b. $\triangle FHG \sim \triangle LKJ$

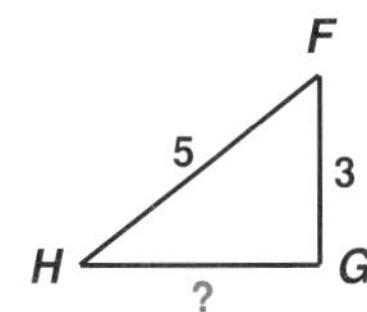

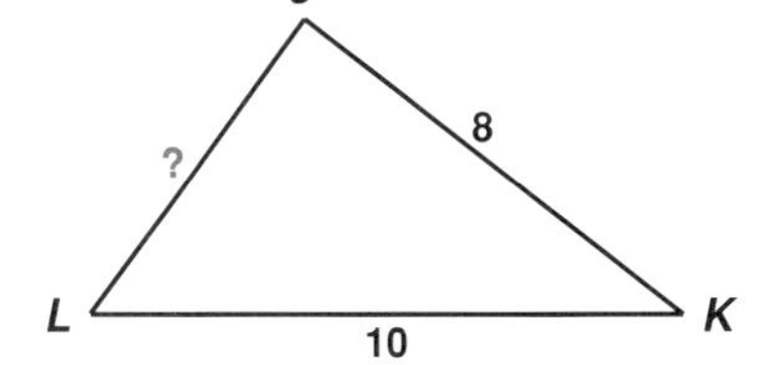

10. Explain your reasoning in each situation.

a. In the diagram, $\overline{AP}$ intersects $\overline{BQ}$ and $\overline{AB} \parallel \overline{QP}$. Draw a conclusion about $\triangle ABT$ and $\triangle PQT$.

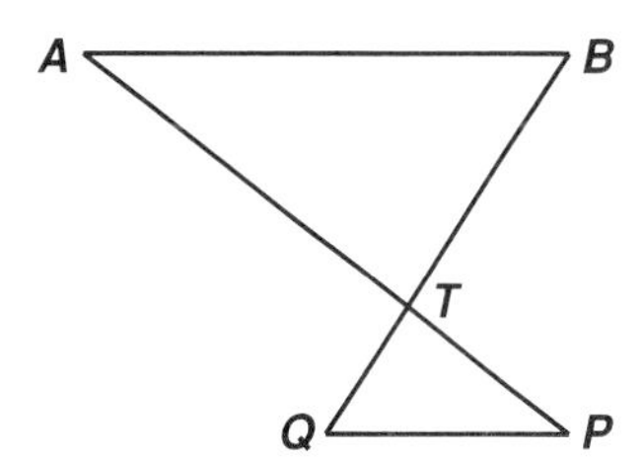

b. (Triangle ABC with AB = 14, BC = 21, AC = 28)

The set of numbers that could be the length of the sides of a triangle similar to $\triangle ABC$ is

(1) 2, 3, 6 (2) 8, 9, 12

(3) 16, 24, 32 (4) 20, 32, 40

11. Use the shadow rule to find each required length.

a. A 5-foot dogwood tree casts a 2-foot shadow, while a nearby maple tree casts a 14-foot shadow. How tall is the maple tree?

b. A 6-foot-tall man standing near a building casts a shadow 5 feet long. If the shadow of the building is 35 feet long, how tall is the building?

c. A 24-foot-high tree standing at the edge of Silver Pond has a shadow that just reaches the opposite shore of the pond. If a 9-foot-tall giraffe that is grazing casts a 15-foot shadow, how wide is Silver Pond?

EXPLORATIONS

1. At the top of a full sheet of graph paper, draw a rectangle that is 6 by 8 units. Using different places on the same paper, draw these four new rectangles that are related to the original rectangle, numbering them (1) – (4).

(1) a 9 by 11 rectangle (add 3 to each original dimension)

(2) a 2 by 4 rectangle (subtract 4 from each original dimension)

(3) a 9 by 12 rectangle (multiply each original dimension by 1.5)

(4) a 3 by 4 rectangle (divide each original dimension by 2)

a. Which of the new rectangles have the same shape as the original rectangle? Explain why you think this is so.

b. Repeat this experiment on a new sheet of graph paper.

For your original rectangle, choose dimensions less than 13.

Create four new rectangles by:

(1) adding the same number to both original dimensions

(2) subtracting the same number from both original dimensions

(3) multiplying both original dimensions by the same number

(4) dividing both original dimensions by the same number

Which of the new rectangles have the same shape as the original rectangle? Are these results the same as the results of the first trial?

c. On a new sheet of graph paper, start with a square, and repeat this experiment.

Which of the new figures are the same shape as the original square?

Draw a conclusion about when squares are similar.

d. Why do you think the results of the experiment starting with a square are different from the results of the experiment starting with a rectangle?

e. Predict the results of the experiment if the original figure is:

(1) an equilateral triangle **(2)** an isosceles triangle

f. Compare the results of the experiment when the original figure has 3 sides to the results when the original figure has 4 sides.

Predict the results if the original figure has 5 sides.

2. Here is a twist on the idea of similarity.

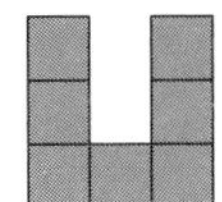

← Using this pattern, replace each box of the pattern by the pattern itself.

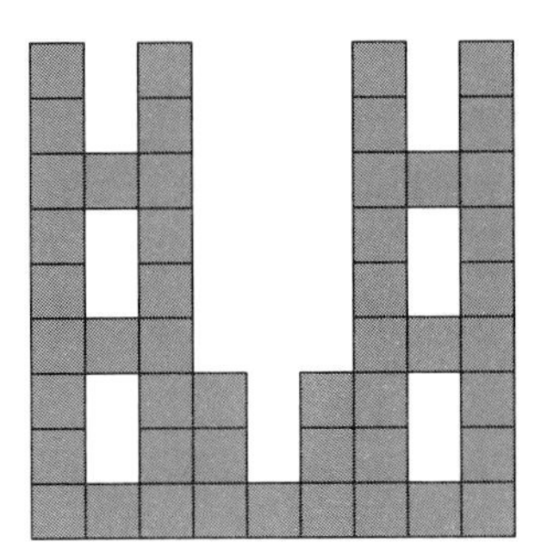

There are 7 boxes on the original pattern.

Do you see 7 of the same patterns in the result?

The new figure is *self-similar* to the original figure.

For each of the following figures, create a self-similar figure by replacing each box of the pattern by the pattern itself.

a.

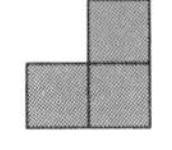

b.

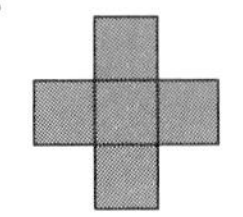

Things You Should Know After Studying This Chapter

KEY SKILLS

10.1 Write a ratio equivalent to a given ratio.
Reduce a ratio to simplest form.
Solve a problem that contains a ratio relationship.
Use ratio to compare measurements.
Use ratio to write a rate.

10.2 Determine if two ratios form a proportion.
Solve a proportion.
Use a proportion to solve a problem.
Find the missing value of a pair when the numbers are in direct variation.

10.3 Write a ratio as a percent.
Write a proportion to answer a question about percent.
Use a proportion to solve a problem about percent.

10.4 Use a scale drawing to find actual dimensions.
Use a proportion to find the length of a side in similar triangles.
Use similar triangles to solve shadow problems.

KEY TERMS AND SYMBOLS

10.1 ratio $\left(2 \text{ to } 3,\ 2:3,\ \frac{2}{3}\right)$ • rational number • rate

10.2 proportion • extremes • means • cross product • constant ratio • direct variation

10.3 percent (%) • percent of increase or decrease

10.4 scale drawing • similar triangles ($\sim$) • shadow rule

CHAPTER REVIEW EXERCISES

1. In a box of 50 marbles, there are 15 red marbles, 20 white marbles, and the remaining marbles are blue.

a. Write the ratio of each color marble to the total number of marbles in the box.

b. Which of these ratios is equivalent to $\frac{2}{5}$?

2. Reduce each ratio to simplest form.

a. $\frac{15}{60}$

b. $18 : 8$

c. $\frac{4.2}{0.6}$

3. Write a ratio to compare each pair of measures. Express the ratio in simplest form.

a. 8 inches to 1 foot

b. 6 months to 2 years

4. Write each comparison as an hourly rate.

a. 180 miles in 3 hours

b. $90 in 3 hours

5. Use an equation to solve each problem. First guesstimate an answer.

a. Find two numbers in the ratio $2 : 3$ that have a sum of 45.

b. If the measures of the angles of a triangle are in the ratio $1 : 2 : 3$, find the measure of the smallest angle.

6. Determine if each pair of ratios can form a proportion.

a. $42 : 6$ and $14 : 2$

b. $\frac{3}{4}$ and $\frac{6}{9}$

c. $\frac{5}{3}$ and $\frac{2.5}{1.5}$

7. Arrange the numbers 2, 64, 16, 8 to form a proportion.

8. Solve each proportion.

a. $\frac{14}{x} = \frac{7}{8}$

b. $\frac{n}{20} = \frac{0.5}{5}$

c. $w : 9 = 4 : 3$

9. Use a proportion to solve each problem. First guesstimate an answer.

a. If Pam can type 150 words in 3 minutes, how many words can you expect her to type in 20 minutes?

b. A survey of voters in May City showed that 7 out of 10 belong to a political party. If there are 12,500 voters, how many belong to a political party?

c. The distance a car travels varies directly with the number of hours spent driving. If a car covers a distance of 200 miles in 4 hours, how far will it go in 9 hours?

CHAPTER REVIEW EXERCISES

10. Use a proportion to solve each basic question about percent.

a. Find 20% of 42.

b. 18 is 60% of what number?

c. What percent of 120 is 90?

11. Use a proportion to solve each problem. First guesstimate an answer.

a. How much would Paul save on the purchase of a VCR that is discounted by 30% of its regular $245 price?

b. Of Paula's total income of $50,000, she had received $20,000 in commissions. What percent of her total earnings were from commissions?

c. On her driving test, Pauline answered 12 questions correctly, and received a grade of 60%. How many questions were there?

12. Use a ratio to solve each problem.

a. After a cut in tuition, the enrollment at Claghorn College jumped from 16,400 to 17,056. What was the percent of increase?

b. A $250 desk is on sale for $150. What is the percent of decrease?

c. Dana got a raise in her hourly pay rate, from $8.50 to $9.18. What is the percent of increase?

13. The scale on a map is 1 inch = 150 miles.

a. Find the actual distance between two cities if the map distance is 2.5 inches.

b. Find the map distance between two cities if the actual distance is 525 miles.

14. Each pair of triangles is similar.

a. Find the missing angle measures.

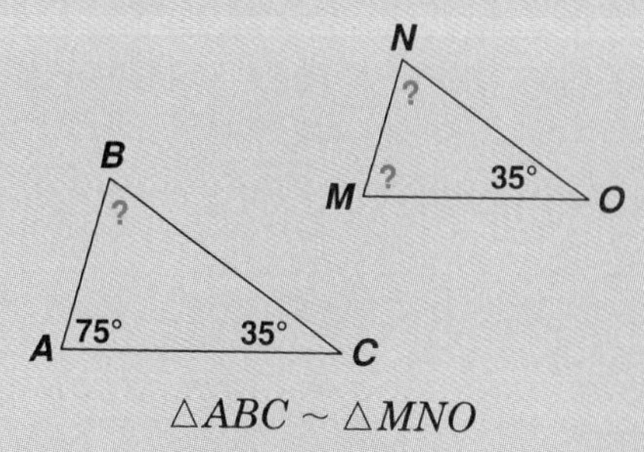

$\triangle ABC \sim \triangle MNO$

b. Find the missing side lengths.

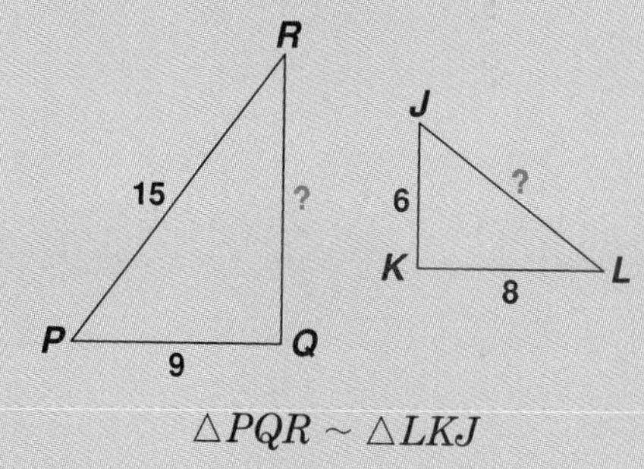

$\triangle PQR \sim \triangle LKJ$

15. Outside her home, 5-foot-tall Tanya cast a shadow 2.5 feet long as she stood near the carton in which the new washer/dryer set was delivered. If the carton cast a shadow of 3.25 feet, how tall was the carton?

SOLUTIONS FOR TRY THESE

10.1 Ratio

Page 425

1. (d) $\frac{8}{10}$ If numerator and denominator of $\frac{4}{5}$ are each multiplied by the ratio factor 2, the result is $\frac{8}{10}$.

2. $\frac{15}{21} = \frac{5 \times 3}{7 \times 3} = \frac{5}{7}$

3. If n is the ratio factor, then the two numbers are $3n$ and $5n$.

$3n + 5n = 32$ The sum of the numbers is 32.

$8n = 32$ Combine like terms.

$\frac{8n}{8} = \frac{32}{8}$ Undo × by ÷.

$n = 4$

$3n = 3(4) = 12$

$5n = 5(4) = 20$

Check: Are 12 and 20 in the ratio 3:5?

$\frac{12}{20} = \frac{3 \times 4}{5 \times 4} = \frac{3}{5}$ ✓

Does 12 + 20 equal 32? ✓

Page 425

$\frac{2 \text{ pounds}}{8 \text{ ounces}} = \frac{32 \text{ ounces}}{8 \text{ ounces}}$ Change pounds to ounces. 1 pound = 16 ounces

$= \frac{4 \times \cancel{8 \text{ ounces}}}{1 \times \cancel{8 \text{ ounces}}}$ Cancel the greatest common factor and the common unit of measure.

$= \frac{4}{1}$ The ratio has no unit of measure.

Page 426

$\frac{250 \text{ words}}{5 \text{ minutes}} = \frac{250 \div 5 \text{ words}}{5 \div 5 \text{ minutes}} = \frac{50 \text{ words}}{1 \text{ minute}}$

Answer: Isabel's typing rate is 50 words per minute.

10.2 Proportion

Page 433

1. a. $\frac{7}{8} \stackrel{?}{=} \frac{14}{16}$

$8 \times 4 \stackrel{?}{=} 7 \times 16$

$112 \stackrel{?}{=} 112$ true

Answer: The sentence is a proportion because the product of the means is equal to the product of the extremes.

b. $\frac{5}{3} \stackrel{?}{=} \frac{25}{18}$

$3 \times 25 \stackrel{?}{=} 5 \times 18$

$75 \stackrel{?}{=} 90$ false

Answer: The sentence is not a proportion because the product of the means is not equal to the products of the extremes.

2. a. $\frac{x}{6} = \frac{20}{30}$ To form a proportion, the cross products must be equal.

$30x = 6 \cdot 20$

$30x = 120$

$\frac{30x}{30} = \frac{120}{30}$ Undo × by ÷.

$x = 4$

Check: Replace x by 4 in the original sentence. $\frac{x}{6} = \frac{20}{30}$

Then reduce each ratio to see if they are equal. $\frac{4}{6} \stackrel{?}{=} \frac{20}{30}$

$\frac{\cancel{2} \cdot 2}{\cancel{2} \cdot 3} \stackrel{?}{=} \frac{2 \cdot \cancel{10}}{3 \cdot \cancel{10}}$

$\frac{2}{3} = \frac{2}{3}$ ✓

Answer: $x = 4$

b. $\frac{7}{y} = \frac{21}{27}$ To form a proportion, the cross products must be equal.

$21y = 7 \cdot 27$

$21y = 189$

$\frac{21y}{21} = \frac{189}{21}$ Undo × by ÷.

$y = 9$

Check: Replace y by 9 in the original sentence. $\frac{7}{y} = \frac{21}{27}$

Then reduce each ratio to see if they are equal. $\frac{7}{9} \stackrel{?}{=} \frac{21}{27}$

$\frac{7}{9} \stackrel{?}{=} \frac{7 \cdot \not{3}}{9 \cdot \not{3}}$

$\frac{7}{9} = \frac{7}{9}$ ✓

Answer: $y = 9$

Page 434

1 A rate is given: $\frac{7 \text{ gallons}}{168 \text{ miles}}$

2 Write a second rate in the same order, where x represents the number of gallons of gas used for a trip of 840 miles. $\frac{x \text{ gallons}}{840 \text{ miles}}$

Write the proportion from the two rates.

$$\frac{7}{168} = \frac{x}{840}$$

3 Solve the proportion by making the cross products equal. $168x = 7 \cdot 840$

$168x = 5{,}880$

$\frac{168x}{168} = \frac{5{,}880}{168}$

$x = 35$

4 *Check:* If it takes 35 gallons to go 840 miles, is the rate of gas usage equal to 7 gallons for 168 miles?

$\frac{35}{840} \stackrel{?}{=} \frac{7}{168}$

$\frac{\not{5} \cdot 7}{\not{5} \cdot 168} \stackrel{?}{=} \frac{7}{168}$

$\frac{7}{168} = \frac{7}{168}$ ✓

Answer: At the given rate of gas usage, it will take 35 gallons to go 840 miles.

Page 436

First find the constant ratio from the given pair of values. $\frac{m}{n} = \frac{36}{9} = \frac{4}{1}$ or 4

Use the constant ratio to write a rule relating the variables. $\frac{m}{n} = \frac{4}{1}$ or $m = 4n$

Use the rule that relates the variables to find the value of m that corresponds to $n = 5$. $m = 4n$; $m = 4(5) = 20$

Check: If $m = 20$ when $n = 5$, are these numbers in the constant ratio 4 : 1?

$\frac{20}{5} \stackrel{?}{=} \frac{4}{1}$

$\frac{\not{5} \cdot 4}{\not{5} \cdot 1} \stackrel{?}{=} \frac{4}{1}$

$\frac{4}{1} = \frac{4}{1}$ ✓

Answer: $m = 20$ when $n = 5$

10.3 Percent

Page 443

1. a. $40\% = \frac{40}{100} = \frac{2 \cdot 20}{5 \cdot 20} = \frac{2}{5}$

b. $40\% = \frac{40}{100} = 0.40$

2. a. $\frac{40}{100} = \frac{x}{70}$

$100x = 2{,}800$

$\frac{100x}{100} = \frac{2{,}800}{100}$

$x = 28$

Answer: 28 is 40% of 70.

b. $\frac{x}{100} = \frac{28}{70}$

$70x = 2{,}800$

$\frac{70x}{70} = \frac{2{,}800}{70}$

$x = 40$

Answer: 28 is 40% of 70.

c. $\frac{40}{100} = \frac{28}{x}$

$40x = 2,800$

$\frac{40x}{40} = \frac{2,800}{40}$

$x = 70$

Answer: 28 is 40% of 70.

3. a. Basic question: What is 20% of 90?

$\frac{20}{100} = \frac{x}{90}$

$100x = 1,800$

$\frac{100x}{100} = \frac{1,800}{100}$

$x = 18$

Check: Is 20% of 90 equal to 18?

$0.20(90) \stackrel{?}{=} 19$

$18 = 18$ ✔

Answer: Jorge will save $18.

b. Basic question: What percent of 75 is 45?

$\frac{x}{100} = \frac{45}{75}$

$75x = 4,500$

$\frac{75x}{75} = \frac{4,500}{75}$

$x = 60$

Check: Is 60% of 75 equal to 45?

$0.60(75) \stackrel{?}{=} 45$

$45 = 45$ ✔

Answer: 60% of those interviewed liked breakfast orange juice.

c. Basic question: 275 is 75% of what number?

$\frac{75}{100} = \frac{274}{x}$

$75x = 27,500$

$\frac{75x}{75} = \frac{27,500}{75}$

$x = 366.6666\ldots$

$x \approx 366.67$

Check: Is 75% of 366.67 equal to 275?

$0.75(366.67) \stackrel{?}{=} 275$

$275.0025 \stackrel{?}{=} 275$ Round off

$275 = 275$ ✔

Answer: The regular price is $366.67.

Page 444

First find the amount of the decrease.

$250 \text{ lb.} - 200 \text{ lb.} = 50 \text{ lb.}$

Write the ratio: $\frac{\text{amount of decrease}}{\text{original amount}}$ or $\frac{50}{250}$

Reduce the fraction. $\frac{1 \cdot 50}{5 \cdot 50}$

Write the fraction as a percent. $\frac{1}{5} = 20\%$

Answer: Earl reduced his weight by 20%.

10.4 Similarity

Page 451

Write a proportion between the measures in the drawing and the actual measures, with x as the actual width of the chip.

$$\frac{\text{diagram measure}}{\text{actual measure}} = \frac{\text{diagram measure}}{\text{actual measure}}$$

$$\frac{1 \text{ in.}}{0.5 \text{ mm}} = \frac{2.0 \text{ in.}}{x \text{ mm}}$$

$$1 \cdot x = 0.5 \cdot 2.0$$

$$x = 1.0$$

Check: If the actual width is 1 mm, does

$$\frac{\text{diagram measure}}{\text{actual measure}} \stackrel{?}{=} \frac{1 \text{ in.}}{0.5 \text{ mm}}$$

$$\frac{2}{1} \stackrel{?}{=} \frac{1}{0.5}$$

By calculator: $2 = 2$ ✔

Answer: The actual width of the chip is 1 millimeter.

Page 451

a. $\angle P$ corresponds to $\angle M$

$\angle Q$ corresponds to $\angle N$

$\angle R$ corresponds to $\angle O$

b. $\overline{PQ}$ corresponds to $\overline{MN}$

$\overline{QR}$ corresponds to $\overline{NO}$

$\overline{PR}$ corresponds to $\overline{MO}$

SOLUTIONS FOR TRY THESE

Page 453

1. The sum of the measures of the angles of a triangle is always 180°.

   In $\triangle ABC$: $m\angle A + m\angle B + m\angle C = 180$

   $$60 + 40 + m\angle C = 180$$
   $$100 + m\angle C = 180$$
   $$m\angle C = 80°$$

   In $\triangle DEF$: $m\angle D + m\angle E + m\angle F = 180$

   $$60 + 40 + m\angle F = 180$$
   $$100 + m\angle F = 180$$
   $$m\angle F = 80°$$

   The third pair of angles must be equal in measure.

2. Since the triangles are similar, the measures of the corresponding sides are in proportion. Write a proportion that involves $\overline{DF}$.

   $$\frac{AC}{DF} = \frac{AB}{DE}$$
   $$\frac{13}{DF} = \frac{5}{10}$$
   $$5 \cdot DF = 13 \cdot 10$$
   $$5DF = 130$$
   $$DF = 26$$

   Answer: $DF = 26$ cm

   Check: If $DF = 26$ cm, are the ratios equal?

   $$\frac{13}{DF} \stackrel{?}{=} \frac{5}{10}$$
   $$\frac{13}{26} \stackrel{?}{=} \frac{5}{10}$$
   $$\frac{13 \cdot 1}{13 \cdot 2} \stackrel{?}{=} \frac{1 \cdot 5}{2 \cdot 5}$$
   $$\frac{1}{2} = \frac{1}{2} \checkmark$$

Page 455

Draw a diagram to illustrate the problem.

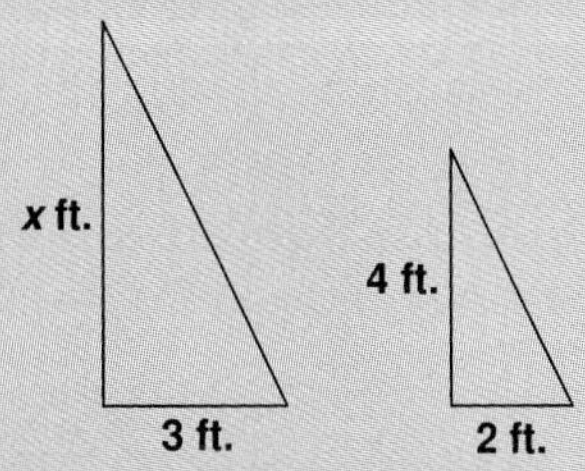

Mark the given lengths on the diagram. Note that the shadows are on the ground.

Apply the shadow rule to write a proportion, with x as the man's height.

$$\frac{\text{man's height}}{\text{man's shadow}} = \frac{\text{boy's height}}{\text{boy's shadow}}$$
$$\frac{x \text{ feet}}{3 \text{ feet}} = \frac{4 \text{ feet}}{2 \text{ feet}}$$
$$2x = 3 \times 4$$
$$2x = 12$$
$$x = 6$$

Answer: The man is 6 feet tall.

Check: If the man is 6 feet tall, are the measures in proportion?

$$\frac{\text{man's height}}{\text{man's shadow}} \stackrel{?}{=} \frac{\text{boy's height}}{\text{boy's shadow}}$$
$$\frac{6}{3} \stackrel{?}{=} \frac{4}{2}$$
$$\frac{2}{1} = \frac{2}{1} \checkmark$$

CUMULATIVE REVIEW • CHAPTERS 1-10

1. Is each of the following conclusions correct? If not, explain.
   a. Everyone on Flight 102 is going to Dallas. Sonya is on Flight 102.
   Conclusion: Sonya is going to Dallas.
   b. All the seniors in Ed's high school study a foreign language. Ed is studying Greek.
   Conclusion: Ed is a senior.

2. The graph shows John's temperature readings over a 24-hour period during his illness.

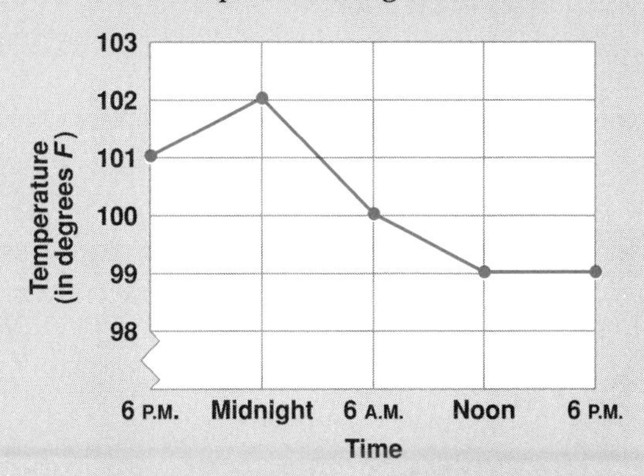

   a. During which 6-hour period was John's temperature rising?
   b. At what time was his temperature highest?
   c. What was his highest temperature?

3. Cara is buying a pencil. She has 13 pennies, 3 nickels, and a dime. In how many ways can she select a set of coins worth exactly 17¢? Make a table to show all the possible ways.

4. Draw a circle. If you draw 3 straight lines through the circle, what is the greatest number of sections into which you can divide the circle?

5. Evaluate:
   a. $5 + (-3)$ b. $|7| + |-7|$
   c. $12 \times \frac{1}{4}$ d. $-24 \div (-6)$
   e. 70% of 20 f. $1.35 - 0.8$
   g. $\frac{1}{3} + \frac{2}{5}$ h. $3.6 \div 0.3$

6. Use the order of operations to evaluate:
   a. $12 - 2(3)$ b. $8 + 18 \div 2$
   c. $3 + 4^2$ d. $10 - 4 - 2$

7. Arrange the numbers from least to greatest.
   a. $\frac{5}{6}, \frac{3}{8}, \frac{5}{12}$ b. $0, -1, -1.1, -1.01$

8. There are 500 videotapes available at the Stewart Avenue Library. The circle graph shows the percent of tapes in each category.

   a. In which category are there the fewest tapes?
   How many tapes are there of that kind?
   b. The number of Adventure tapes is equal to the combined total of which 2 other types?
   c. There are 10 more Comedy tapes than __?__ tapes.

9. Find the value of each expression if $a = 3$ and $b = -2$.
   a. $a + 2b$ b. $3a - b$
   c. $(a - b)^2$ d. $a^2 - 2ab + b^2$

10. Refer to the graph below to write the coordinates of each point.

a. A

b. B

c. C

d. D

e. A', the image when point A is reflected over the horizontal axis.

f. B', the image when point B is translated 2 units up.

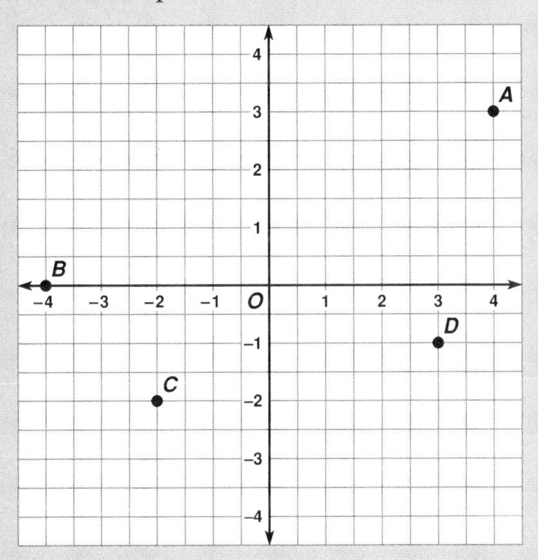

11. Evaluate:

a. $3a + 4b$ when $a = 5$ and $b = -2$

b. $5(x - y)$ when $x = 0.2$ and $y = 1.2$

c. $12pq$ when $p = \frac{1}{3}$ and $q = \frac{3}{4}$

12. For each verbal expression in *Column A*, find a matching symbolic expression in *Column B*.

Column A	*Column B*
a. The sum of a number and 2 is 8.	(1) $n - 8 = 2$
b. 8 more than the sum of a number and 2	(2) $n - 8 < 2$
c. 8 less than a number is 2.	(3) $n + 2 = 8$
d. A number decreased by 8 is less than 2.	(4) $2(n + 8)$
e. Twice the sum of a number and 8	(5) $n + 2 + 8$
f. A number decreased by the quotient of 8 and 2	(6) $2 < n < 8$
	(7) $8n - 2$
g. The product of a number and 8, decreased by 2	(8) $n - \frac{8}{2}$
h. A number is more than 2 and less than 8.	

13. Solve and check.

a. $x + 7 = 9$ **b.** $y - 2 = 6$

c. $3w = 12$ **d.** $\frac{1}{2}p = 6$

e. $3 - x = 7$ **f.** $2d - 12 = 36$

g. $3w + w + 4 = 16$ **h.** $8x = 2x + 12$

i. $3(x - 2) = 9$

14. Solve each inequality and graph the solution set on a number line.

a. $3x > 6$ **b.** $y - 4 \leq 2$

c. $2z - 5 \geq 3$ **d.** $\frac{1}{3}h < 6$

e. $-4x \geq 8$ **f.** $-3x + 1 < 16$

15. Use an equation to solve each problem. First guesstimate an answer.

a. In two days, the Bird-Watching Club spotted 13 different kinds of birds. If they saw 5 kinds the first day, how many different kinds of birds did they see the second day?

b. It took Dion 3 hours to ride his motorbike from home to Blue Lake, a distance of 111 miles. How many miles did he average per hour?

c. Stan had twice as many videotapes as Lonny. If altogether they had 57 videotapes, how many did Lonny have?

d. Carrie got a raise of $19 a month. At the new rate, her earnings for the year totaled $11,448. What were her monthly earnings before the raise?

---

16. Transversals s and t intersect parallel lines m and n, with $s \perp m$. Choose a description from *Column B* to make each of the sentences in *Column A* true.

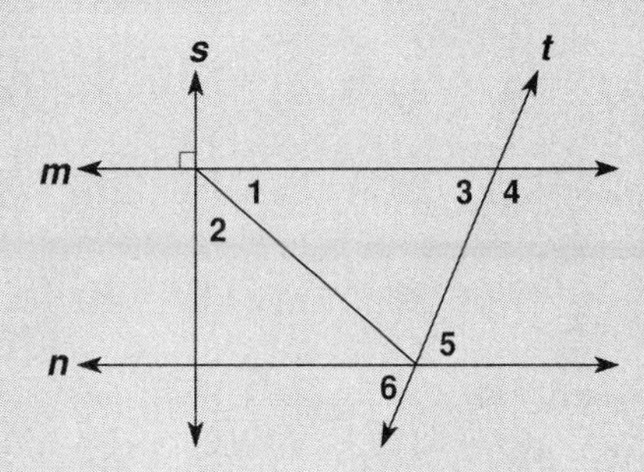

Column A	*Column B*
a. $\angle 1$ and $\angle 2$ are _?_ angles.	(1) corresponding
b. $\angle 5$ and $\angle 6$ are _?_ angles.	(2) alternate interior
c. $\angle 3$ and $\angle 6$ are _?_ angles.	(3) supplementary
d. $\angle 3$ and $\angle 5$ are _?_ angles.	(4) complementary
e. $\angle 3$ and $\angle 4$ are _?_ angles.	(5) vertical

17. Copy the following figures and draw the lines of symmetry.

a. rectangle

b. equilateral triangle

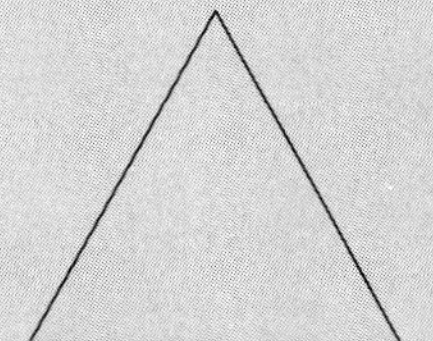

c. isosceles triangle

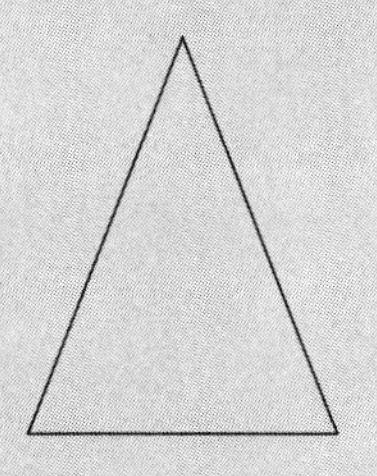

d. square

e. isosceles right triangle

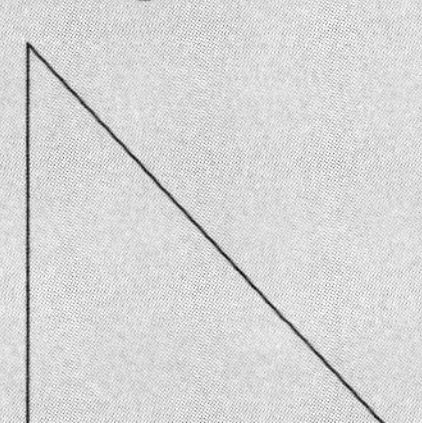

f. regular pentagon

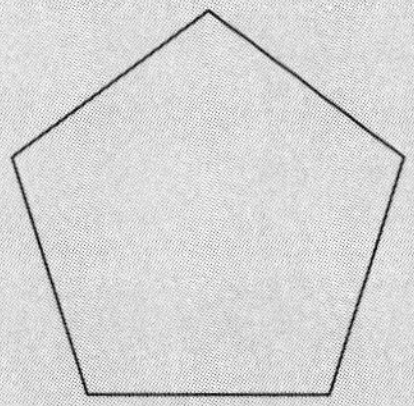

18. Draw a diagram for each problem, then use an equation to find the answer.

a. The measures of a pair of vertical angles are represented by $3a - 8$ and $a + 54$. Find the degree measure of each of the two angles.

b. In isosceles triangle ABC, the congruent sides are $AB = x + 5$ and $AC = 2x - 2$, and the perimeter is 32 cm. Find the length of each side of $\triangle ABC$.

c. In $\triangle RST$, $m\angle R = x - 10$, $m\angle S = 2x - 30$, and $m\angle T = 2x + 20$. Find the number of degrees in each angle.

d. In $\triangle DEF$, $\overline{DM}$ is the median to $\overline{EF}$. If $EM = 2x + 5$ and $FM = 4x - 1$, find the length of $\overline{EF}$.

19. Find each area.

a.

b.

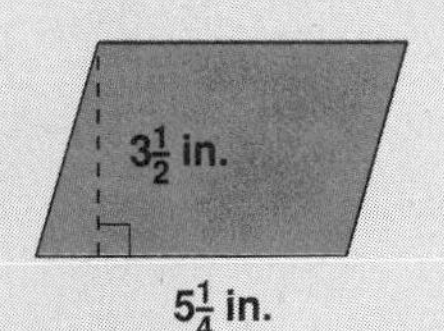

c.

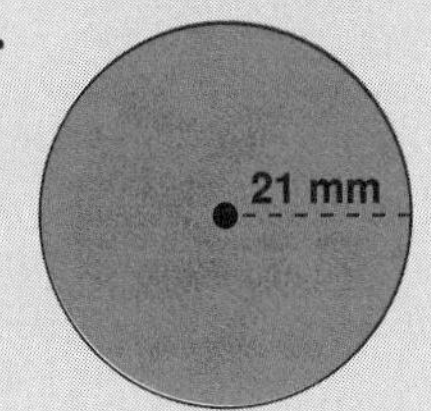

Use: $\pi = \frac{22}{7}$

20. Draw each figure and find its surface area.

a. a cube measuring 4 cm on each edge

b. a rectangular prism with length 6 in., width 4 in., and height 5 in.

21. Find the volume of each solid.

a.

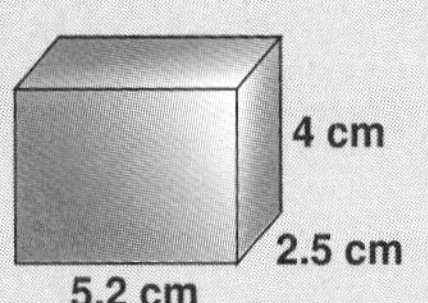

b.

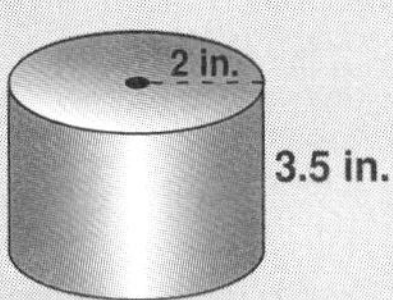

Use: $\pi = 3.14$

22. a. The scale on a map is 1 inch = 5 miles. If Richland and Belmont are 35 miles apart, how long is the segment on the map that represents the distance?

b. The blueprints for a house are drawn on a scale of 1 inch = 3 feet. What is the actual length of the kitchen if its length on the drawing is 4 inches?

23. a. The ratio of men to the total number of adults on a ski trip is 3 : 5. If there are 20 adults altogether, how many men are there?

b. The measures of a pair of supplementary angles are in the ratio of 2 : 3. Find the number of degrees in each angle.

24. Solve each proportion.

a. $\frac{x}{6} = \frac{10}{15}$ **b.** $\frac{3}{4} = \frac{x+1}{8}$

Probability 11

ACTIVITY

A bag contains 10 chips that are either black, white, or gray.

A chip is picked, its color noted, and it is then returned to the bag.

Looking at the results of 40 such picks, can you figure out how many of each color are in the bag?

11.1 From Experiment to Theory

Experiments That Count Outcomes

Try an experiment with a thumbtack to see how often it lands with the point up and how often it lands with the point down.

From a height of about 1 foot, drop a tack onto a desk 100 times.

Record your results in a table to show the number of times each possibility occurs (the ***frequency*** of each possibility).

For convenience in recording, use ***tally marks***, where 𝍸 means 5. Also for convenience, record the results in groups of 25.

Here is what a frequency table for this experiment might look like.

Trials	Point Up	Frequency	Point Down	Frequency
1st 25	𝍸 𝍸 ||||	14	𝍸 𝍸 |	11
2nd 25	𝍸 𝍸 𝍸 ||	17	𝍸 |||	8
3rd 25	𝍸 𝍸 𝍸	15	𝍸 𝍸	10
4th 25	𝍸 𝍸 𝍸 |	16	𝍸 ||||	9
100		62		38

Match your results against the results here.

Interpret your results. For example, the results here show these ratios:

$$\frac{\text{point up}}{\text{total trials}} \text{ is about } \frac{60}{100} \text{ or } \frac{3}{5} \qquad \frac{\text{point down}}{\text{total trials}} \text{ is about } \frac{40}{100} \text{ or } \frac{2}{5}$$

Based on these results, would you say that in one throw, you would have a better chance of getting point up or point down?

These results show that for every 5 throws, the chances are 3 will be point up and 2 will be point down. So, for 1 throw, you do have a better chance of getting point up.

Example 1 Try an experiment tossing a paper cup 50 times to see how often it lands:

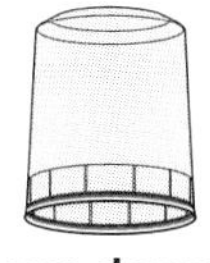

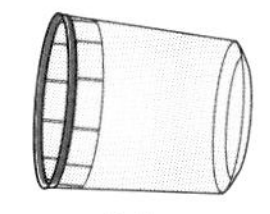

Record your results in a frequency table.

(Continues on next page)

Match your results against the results here.

Outcome	Tally	Frequency
Up	卌 卌	10
Down	卌 卌 ||||	14
Side	卌 卌 卌 卌 卌 |	26
	TOTAL	50

The results here can be written as 3 different ratios, comparing the number of times the cup lands up, down, or sideways to the total number of times the cup was tossed:

$\frac{\text{up}}{\text{total}} = \frac{10}{50}$ $\frac{\text{down}}{\text{total}} = \frac{14}{50}$ $\frac{\text{side}}{\text{total}} = \frac{26}{50}$

Based on these results, what outcome has the best chance of occurring should you toss the cup just once? In the 50 tosses, the sideways outcome occurred most frequently (more than the sum of the other two). So, in one toss, it is most likely that the outcome would be sideways.

In both the thumbtack and the cup experiments, one result turned out to be more likely than the others.

Example 2 Try an experiment tossing a coin 100 times to see how often *heads* turns up and how often *tails* turns up.

Record your results in a frequency table, keeping track of the results in groups of 25.

Match your results against the results here.

heads

tails

Trials	Heads	Frequency	Tails	Frequency
1st 25	卌 卌 ||||	14	卌 卌 |	11
2nd 25	卌 卌	10	卌 卌 卌	15
3rd 25	卌 ||	7	卌 卌 卌 |||	18
4th 25	卌 卌 卌 |	16	卌 ||||	9
100		47		53

Based on these results, which outcome (heads or tails) appears more likely?

These two outcomes over the 100 tosses are so close to each other (heads = 47, tails = 53) that we must conclude them to be *equally* likely.

Note that although the outcomes are close to equal over the 100 tosses, the results in the individual groups of 25 are less balanced.

What do you think would happen if the coin were tossed 1,000 times?

The greater the number of trials, the closer each ratio (heads : total or tails : total) would come to exactly 1 : 2.

Predict the chance of getting heads in one toss of the coin.

The chance of getting heads in one toss of a coin is 1 : 2.

Experiments That Count Outcomes

- In a situation that has a limited number of outcomes (for example, a thumbtack can only land point up or point down), you can carry out an experiment to see the likelihood of each outcome.
- The greater the number of trials in an experiment, the more reliable the results are.
- One outcome may be more likely than the others.
  (It is more likely that a paper cup will land on a side than on the top or bottom.)
- Outcomes may be equally likely.
  (It is just as likely that a *fair* coin will land heads up as tails up. For some coins that have been specially weighted, the outcomes are not equally likely.)
- The chance of getting a particular outcome is written as a ratio.
  (The chance of getting heads in one toss of a fair coin is 1 : 2.)

Try This *(For solution, see page 510.)*

Try an experiment with a key. Use the writing on the key to distinguish between the two sides.

As you would toss a coin, toss the key 100 times, and keep track of the two outcomes.

Record your results in a frequency table, in groups of 25.

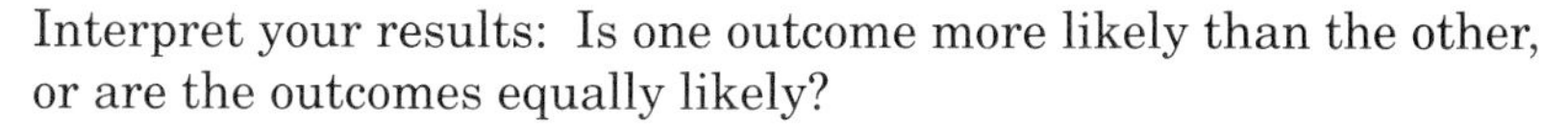

Interpret your results: Is one outcome more likely than the other, or are the outcomes equally likely?

Do the experiment again, this time with a very different size key. For example, if you used a house key the first time, now use a mailbox key.

Compare the results of the two experiments.

Writing a Probability Ratio

From an experiment, you can sometimes draw a conclusion about the likelihood of an outcome

Example 3 Try an experiment rolling a 6-sided number cube (a *die*) 30 times. Record your results in a frequency table.

Match your results against the results here.

Outcome	1	2	3	4	5	6	Total
Frequency	4	6	4	5	3	8	30

(Example continues on next page)

Can you draw a conclusion about the likelihood of the outcomes?

It appears that the outcomes may be equally likely. But the number of trials is really too small to be conclusive.

However, the shape of the cube suggests that the outcomes are equally likely. Try adding on the results of 30 more tosses.

Suppose you had the results of 600 tosses of a die and they showed that the outcomes were equally likely. What would the "perfect" frequency table look like?

Outcome	1	2	3	4	5	6	Total
Frequency	100	100	100	100	100	100	600

This perfect result, with all the frequencies equal may not happen in reality. But the more trials, the closer to this perfect idea (***theory***), you get.

From the results of this table, predict the chance (***probability***) of getting a "5" in one roll of the die.

To determine the probability of getting a "5", write a ratio.

$$\text{Probability of getting a 5} = \frac{\text{number of times 5 occurred}}{\text{total number of tosses}}$$

$$P(5) = \frac{100}{600} \text{ or } \frac{1}{6}$$

What is the probability of getting a "3" in one roll of the die?

To write this probability, you can use the theory without going back to the experiment.

$$P(3) = \frac{1}{6}$$

← There is only 1 three on the die.
← There are 6 numbers on the die.

Writing a Probability Ratio

- **The ratio that describes the chance that a particular outcome will occur is called a *probability.***
- **The numerator of the ratio is the number of ways in which the particular outcome could occur (the number of successes).**

  The denominator of the ratio is the number of ways in which all the different outcomes could occur (the total number of possibilities).

$$\textbf{Probability} = \frac{\textbf{number of successes}}{\textbf{total possibilities}}$$

Try This (For solution, see page 510.)

A soft drink company tested 3 new flavors on a group of 1,000 people. The table shows how many people preferred each flavor.

Flavor	1	2	3	Total
Frequency	109	513	378	1,000

a. What was the most preferred flavor? the least preferred?
b. Write a ratio to compare the number of people who preferred flavor 1 to the total number of people tested. Do the same for flavors 2 and 3.
c. Approximately what percent of the people tested preferred flavor 2? flavor 1? flavor 3?
d. If Sy is now asked to test the 3 flavors, what is the probability that he will choose flavor 2?

Example 4 This spinner is equally likely to stop in any of the six numbered sections. Stopping on a line does not count.

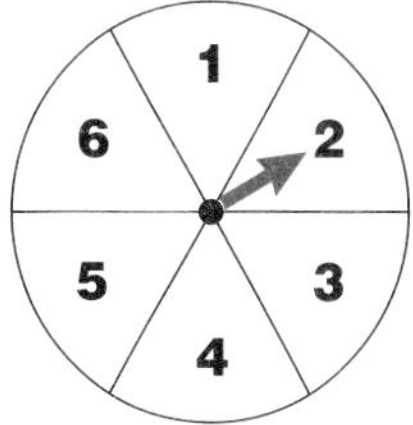

a. What is the probability that after one spin, the spinner will stop on "2"?

$P(2) = \frac{1}{6}$ ← There is only 1 section marked 2. ← There are 6 sections in all.

b. What is the probability that after one spin, the spinner will stop on an even number?

$P(\text{even}) = \frac{3}{6}$ ← There are 3 sections with even numbers. ← There are 6 sections in all.

$P(\text{even}) = \frac{1}{2}$ Reduce the ratio to simplest form.

c. Which is more likely to occur: that the spinner stops on an even number or that the spinner stops on "2"?

Compare the values of the two ratios.

Since $\frac{3}{6} > \frac{1}{6}$, it is more likely that the spinner will stop on an even number than it is that the spinner will stop on "2."

$P(\text{even}) = \frac{3}{6}$

$P(2) = \frac{1}{6}$

d. What is the probability that after one spin, the spinner will stop on a number less than 7?

$P(<7) = \frac{6}{6}$ ← All 6 sections have numbers less than 7. ← There are 6 sections in all.

$P(<7) = 1$ Reduce the ratio to simplest form.

e. What is the probability that after one spin, the spinner will stop on a number greater than 6?

$P(>6) = \frac{0}{6}$ ← None of the sections has a number greater than 6. ← There are 6 sections in all.

$P(>6) = 0$ Reduce the ratio to simplest form.

Values for a Probability Ratio

- **The values of probability ratios range from 0 through 1.**
- **A probability of 0 means that the outcome will never happen.**
- **A probability of 1 means that the outcome will surely happen.**
- **The closer a probability is to 1, the more likely the outcome.**
- **The closer a probability is to 0, the less likely the outcome.**

Try This *(For solution, see page 510.)*

Ten identical cards were marked, each with a different number from 0 through 9. The cards were turned face down and shuffled. Marla was asked to pick a card.

Find the probability that the card Marla picked had:

a. the number 4
b. an odd number
c. an even number
d. a composite number
e. a factor of 12
f. a positive multiple of 3
g. a number less than 4
h. a number greater than 6
i. a number less than 10
j. a number greater than 10

SELF-CHECK: SECTION 11.1

1. What can you learn from conducting an experiment that counts outcomes?

Answer: Such an experiment will give you an idea of the likelihood of the different outcomes.

2. What do we call a table that keeps track of the outcomes of an experiment?

Answer: a frequency table

3. What are tally marks used for?

Answer: to count

4. For what kinds of objects that can be tossed will the outcomes be equally likely?

Answer: Objects that have equal surfaces are well balanced, and their outcomes will be equally likely. For example, on a fair coin, heads or tails are equally likely.

5. Since the outcomes of tossing a fair coin (heads or tails) are equally likely, would you expect to get exactly 5 heads and 5 tails in 10 tosses? Explain.

Answer: Ten tosses are too few to see the behavior of the outcomes. But, the more tosses you do, the closer to equal outcomes you will get.

6. What do we call the ratio that describes the chance that a particular outcome will occur?

Answer: probability

7. Explain the meaning of a probability ratio of 3 : 5.

Answer: You have 3 out of 5 chances that this outcome will occur.

8. Which outcome is more likely: a probability of 1 : 5 or a probability of 3 : 5?

Answer: 3 : 5, because 3 : 5 has 3 chances out of 5 while 1 : 5 has only 1 chance out of 5.

9. What does a probability ratio of 1 mean?

Answer: The outcome will surely happen.

10. What does a probability ratio of 0 mean?

Answer: The outcome will never happen.

11. The values of probability ratios range from 0 through 1. What values show that an outcome is more likely? less likely?

Answer: The closer the probability is to 1, the more likely the outcome is; the closer a probability is to 0, the less likely the outcome.

EXERCISES: SECTION 11.1

1. Try each of the following experiments.

First predict whether one outcome is more likely than the others, or if the outcomes are equally likely.

As you carry out the experiment, keep track of your results in a frequency table.

Compare the results to your prediction.

a. Toss a cone-shaped paper cup 50 times.

Record the number of times it lands in one of these positions:

standing

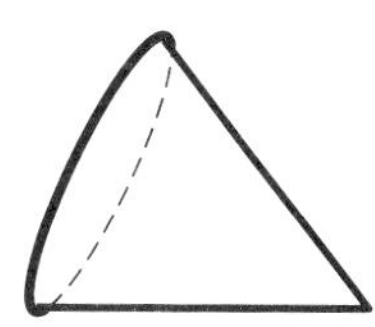

b. Fold a rectangular card in half.

Toss the folded card 50 times and record the number of times it lands in one of these positions:

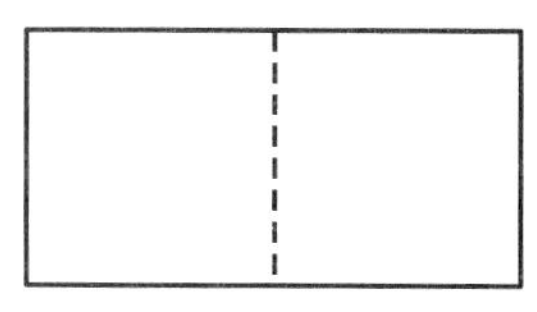

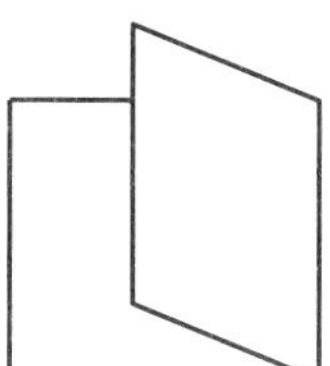

book

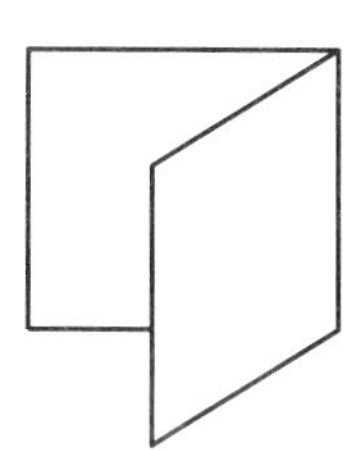

tent

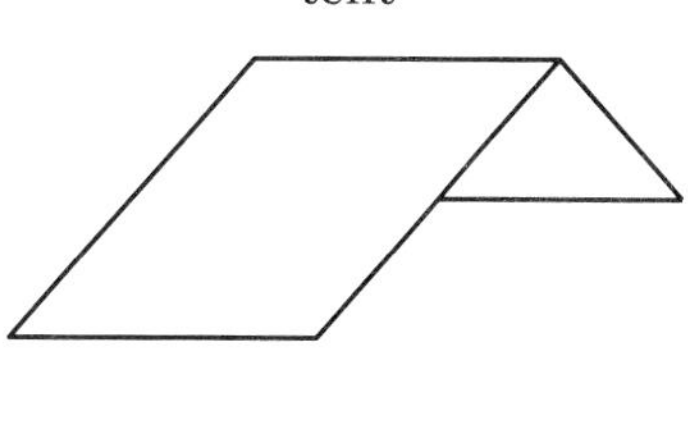

c. Mark one side of a dried bean to be able to distinguish between the two sides.

Toss the bean 50 times and record the number of times it lands marked side up or unmarked side up.

2. Choose the situation of each pair that would have equally likely outcomes. Explain your choice.

a. Each spinner may land in any of 3 numbered sections.

Spinner A

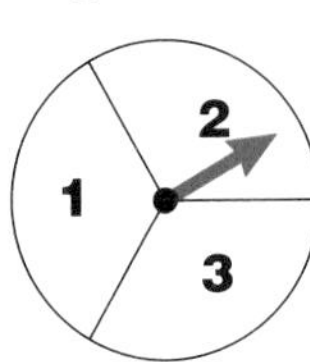

Spinner B

b. Without looking, a chip is tossed into a rectangular box whose bottom is divided into two sections. The chip lands in section 1 or section 2.

Box A

1	2

Box B

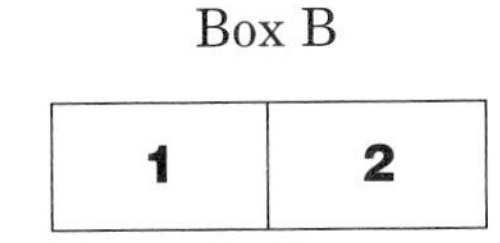

3. A consumer study asked 1,000 husbands their color preference when buying a sweater as a Valentine gift for their wives. The results of the survey are shown in the table.

Color	Gold	Green	Red	Total
Frequency	212	99	689	1,000

a. What was the most preferred color? the least preferred?

b. Write a ratio to compare the number of husbands who preferred the color red for the sweater gift to the total number of husbands surveyed.

Do the same for the colors gold and green.

c. As a result of this survey, do you agree or disagree with each of the following conclusions? Explain.

(1) Most husbands buy their wives sweaters as Valentine gifts.

(2) Seven out of ten husbands prefer red as the color for a sweater to be given as a Valentine gift to their wives.

(3) Men dislike the color green.

(4) About 40% of men who buy sweaters as Valentine gifts for their wives prefer the color gold.

d. Mr. Perkins is going to buy his wife a sweater as a Valentine gift. The probability that he will choose the color green is:

(1) 1 : 10 (2) 2 : 10 (3) 3 : 10 (4) 7 : 10

4. Draw a number line 6 inches long.

At the left, mark the value 0 to represent the probability of an impossible outcome.

At the right, mark the value 1 to represent the probability of an outcome that is certain.

Place the points A through E on your number line to represent where the probability of each of these situations would fall.

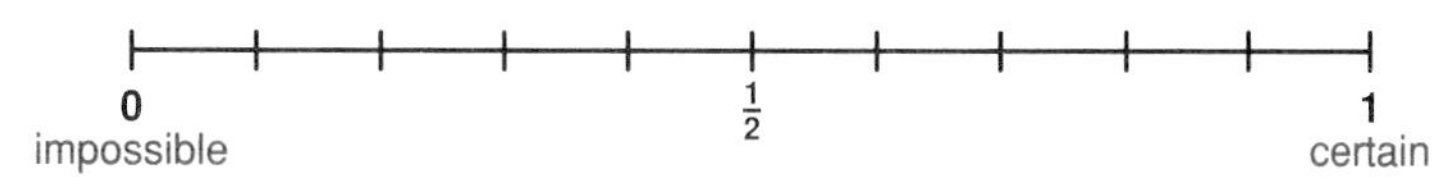

A. The sun will rise tomorrow.
B. The Statue of Liberty will collapse today.
C. You will pass your next math test.
D. If you toss a coin, you will get heads.
E. If you roll a die, you will get 5.

5. In each situation, the event described is done *at random* so that all choices are equally likely.

a. A coin is chosen from a jar that contains 50 coins, 7 of which are dimes.
Find the probability that the coin chosen is a dime.

b. A letter is chosen from the word PROBABILITY.
Find the probability that the letter chosen is a B.

c. A person is chosen from 15 brother-sister pairs.
Find the probability that the person chosen is female.

d. A book is chosen from 6 novels and 8 biographies.
Find the probability that the book chosen is a novel.

e. A color is chosen from the American flag.
Find the probability that the color chosen is green.

f. This textbook is opened to a right-hand page.
Find the probability that the page number is odd.

g.

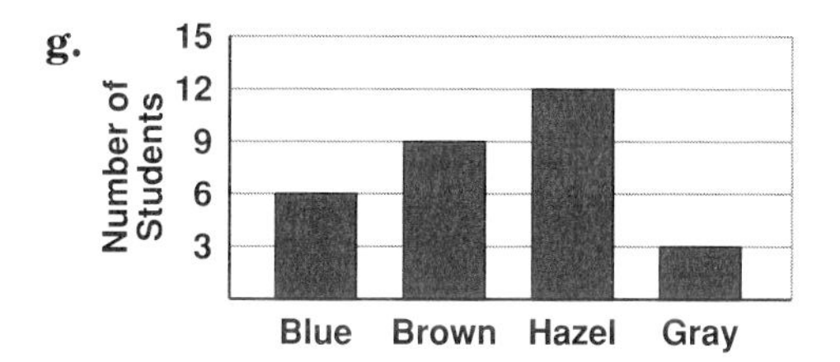

A student is chosen from a class in which the eye colors of all the students are shown in this bar graph.
Find the probability that the student chosen has brown eyes.

6. Each of the names of the days of the week was written on a slip of paper. The papers were shuffled and put in a box. Nancy was asked to pick a paper from the box without looking.

Find the probability that the paper Nancy picked had a name that:

a. began with the letter W
b. began with the letter S
c. ended with the letter Y
d. ended with the letter D
e. had 6 letters
f. had 8 letters
g. had 10 letters
h. had more than 6 letters
i. had fewer than 10 letters
j. had more than 5 letters

7. Each of the 26 letters of the English alphabet was written on a card. The cards were turned face down and shuffled. Sam was asked to pick a card.

Find the probability that the card Sam picked was:

a. the letter Z
b. the letter E
c. the letter Q
d. a vowel (A, E, I, O, U)
e. a consonant (not a vowel)
f. a letter from the first half of the alphabet
g. a letter that comes after the letter D
h. a letter that comes before the letter Z
i. the letter A or a letter that comes after the letter A

8. A figure is chosen at random from these figures.

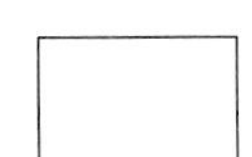  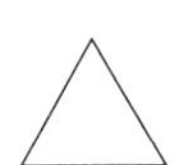 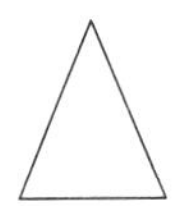

Find the probability that the figure chosen will have:

a. parallel sides
b. 180° as the sum of the measures of its angles
c. all its sides equal in measure
d. all its angles equal in measure
e. an obtuse angle
f. an acute angle
g. a right angle
h. an angle whose measure is 60°
i. the name *polygon*

9. A standard deck of playing cards has four suits in two colors: hearts and diamonds are red, clubs and spades are black.

There are 13 cards in each suit:
Ace, 2, 3, 4, 5, 6, 7, 8, 9, 10, Jack, Queen, King.

If one card is drawn at random from the deck, find the probability that the card drawn is:

a. red
b. black
c. not red
d. green
e. a King
f. a red King
g. a picture card
h. a Jack or a Queen
i. not a Jack

10. Without looking, Quint picked one of these 16 discs from a box.

Find the letter for which the probability that Quint picked it is equal to:

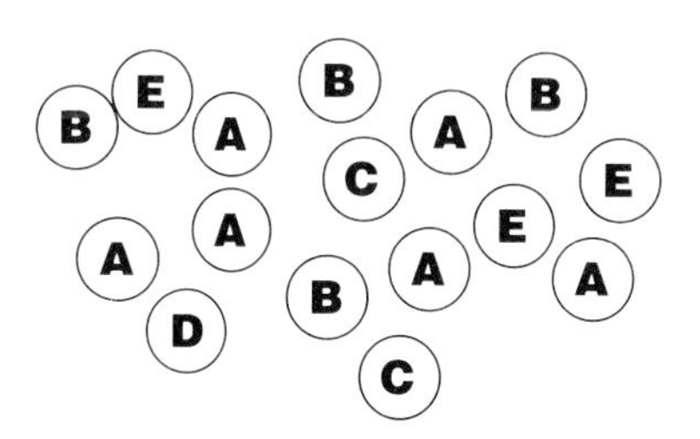

a. $\frac{1}{16}$ **b.** $\frac{3}{16}$ **c.** $\frac{1}{8}$ **d.** $\frac{3}{8}$ **e.** $\frac{1}{4}$

11. This spinner is equally likely to land in any of the 6 numbered sections.

Question: What is the probability that the spinner will land on 3?

Answer: $\frac{1}{6}$

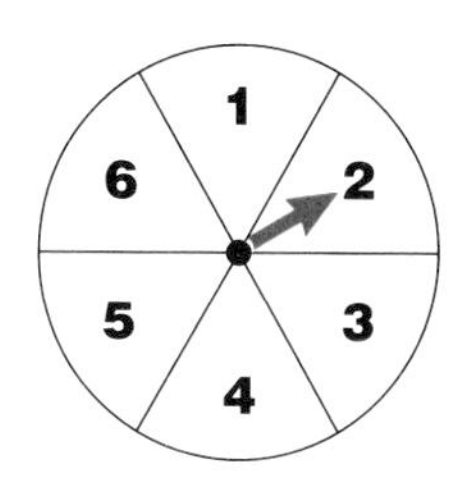

Write a probability question for each of these answers:

a. $\frac{1}{2}$ **b.** $\frac{5}{6}$ **c.** 1 **d.** 0 **e.** $\frac{1}{3}$

12. a. Do you agree or disagree with each of the following? Explain.

(1) Dan said if the probability of getting 3 in one roll of a die is 1 : 6, then the probability of not getting 3 is 5 : 6.

(2) Doris said that if the probability of rain tomorrow is 90%, then the probability of no rain tomorrow is 10%.

(3) Don said that if the probability that this spinner will land on C is 1 : 3, then the probability that the spinner will not land on C is also 1 : 3.

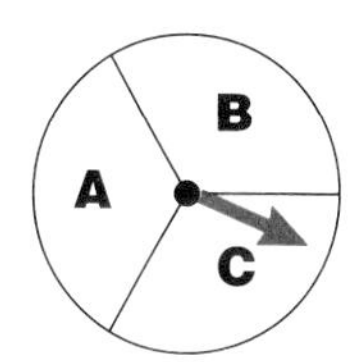

b. Suppose you know the value for the probability that an event will occur. Write a rule for finding the value that the event will not occur.

13. In two spins, which spinner is most likely to spin a sum greater than 7?

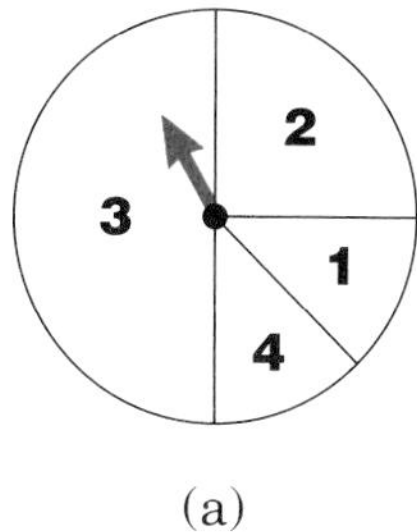

(a)

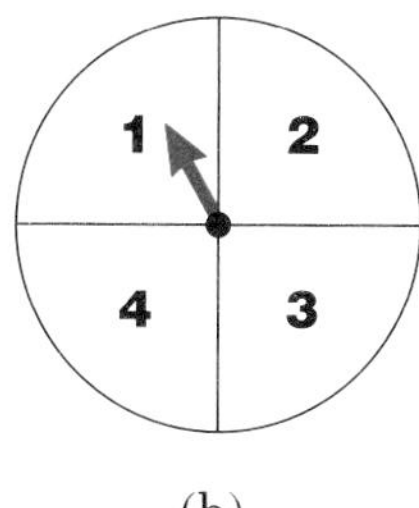

(b)

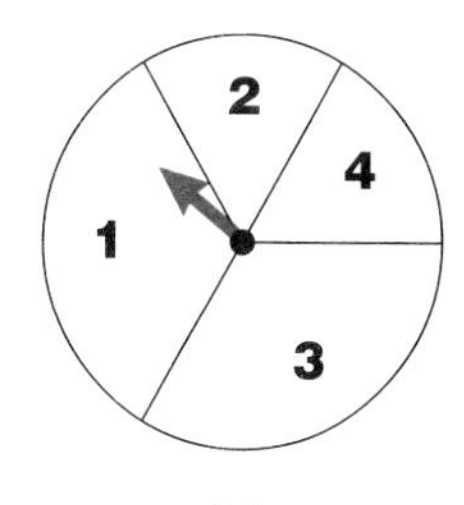

(c)

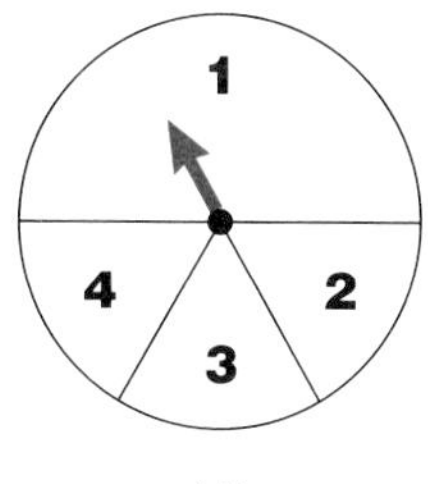

(d)

EXPLORATIONS

1. You have most likely had the experience of guessing at the answer to a test question. How well do you think you would do if you guessed at *all* the answers?

Try this guessing experiment with a partner.

Partner 1: Write the numbers 1-10. Next to each number choose one of the letters a, b, c, d.

This list will represent the answers to 10 questions, each of which had 4 choices given for the answer.

Partner 2: Write the numbers 1-10. For each number, guess which letter your partner has written as an answer.

After you have finished guessing, match your answers with your partner's answers.

Scoring 10 points for each match, what percent did you match?

Partners switch roles and do the experiment again.

Partners compare scores.

Answer these questions:

a. What is the probability of guessing correctly if there are 4 choices?

b. Describe some conditions that a person who is designing a multiple-choice test might have in mind. For example, the test maker would most likely not want all the answers to be the same.

Do you think you might be better at guessing now? Try the experiment again.

c. Do you have a better chance of guessing correctly on a true-false test? Explain.

Try this experiment again, beginning with a list of 10 true or false answers. First, predict your score. How close did you come to your prediction?

(Explorations continue)

2. This spinner has been spun 100 times.

a. Which of these frequency tables do you think would best describe a set of results for this experiment?

Explain.

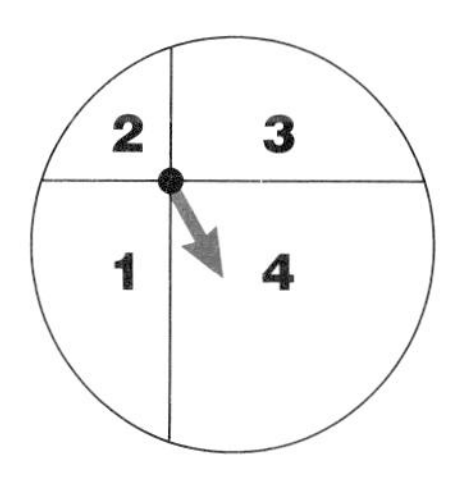

Table A

Outcome	1	2	3	4
Frequency	11	9	30	50

Table B

Outcome	1	2	3	4
Frequency	26	24	28	22

b. Now make your own spinner to carry out this experiment.

Be sure to copy the pattern on the spinner board as shown.

For the spinner, you can use a paper clip with the point of a pencil placed as shown. Hold the pencil steady with one hand and spin the opposite end of the clip with the other hand.

Keep track of your results in a frequency table. Do your results indicate that 4 is the most likely outcome or that all the outcomes are equally likely?

Are you surprised?

c. Try this to see what's happening.

Cut out a circle the size of the spinner board from the center of a piece of paper.

Place the paper with the circular hole over the spinner board and move the paper around until the fixed end of the spinner is at the center of the circular hole in the paper.

Do you think the areas of the four regions on the spinner board determine the outcome? If not, which of the following determines the outcomes?

(1) the length of the radius of the spinner board

(2) the angle measures at the fixed end of the spinner

(3) the lengths of the dividing lines

d. What can you conclude about the likelihood of the outcomes?

11.2 Sample Space

Listing All the Possibilities

If 2 coins—a penny and a dime—are tossed at the same time, what is the probability that they will both land heads up?

To answer this question, you need to be able to think of all the equally-likely possibilities.

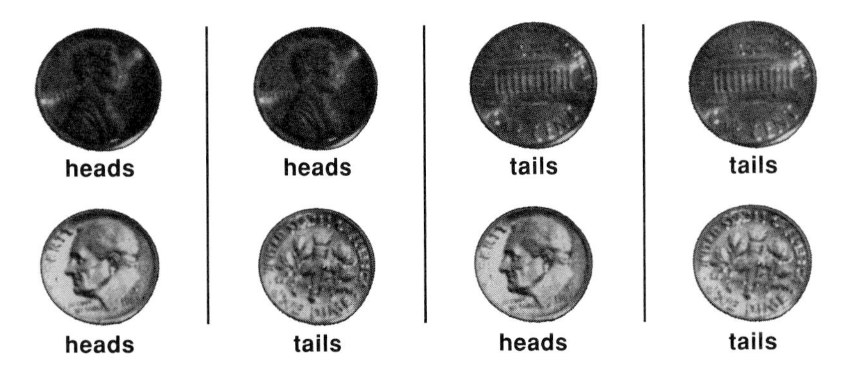

Of the 4 possible outcomes, only 1 shows both coins heads up. Thus, the probability of getting heads on both coins is 1 : 4.

A convenient way to see all the possibilities is to draw a ***tree diagram***.

From a starting point, list in a column the 2 possibilities for the penny: heads (H) or tails (T)

For each choice for the penny, list the 2 possibilities for the dime.

Read the diagram to write a list of all the possible outcomes.

Of the 4 possible outcomes, there is only 1 with both coins heads up: HH.

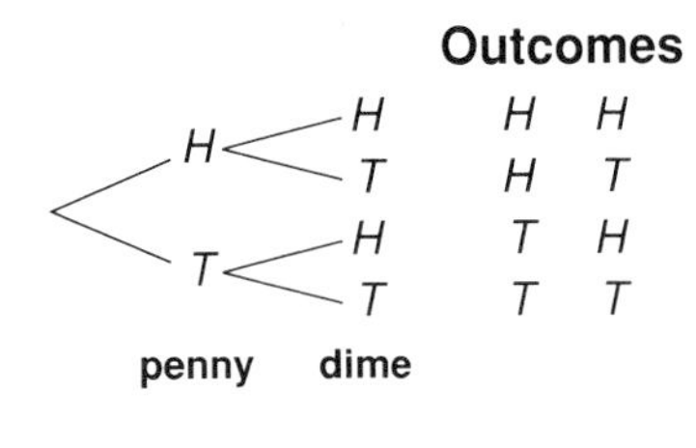

$$P(H\ H) = \frac{1}{4}$$

Example 1 Three coins—a penny, a nickel, and a dime—are tossed at the same time. What is the probability of getting heads on all 3 coins?

heads heads heads

Draw a tree diagram to help write the list of all possible outcomes.

From a starting point, list in a column the 2 possibilities for the penny.

For each possibility for the penny, list the 2 possibilities for the nickel.

For each possibility for the nickel, list the 2 possibilities for the dime.

Read the tree diagram to write the list of all possible outcomes.

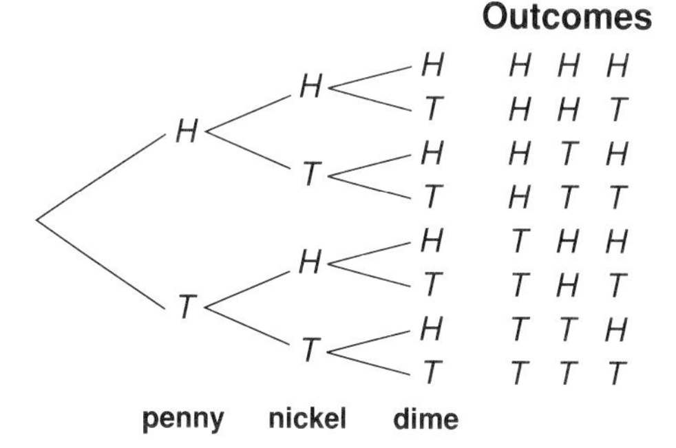

Of the 8 possible outcomes, only 1 shows heads on all the coins: HHH.

$$P(HHH) = \frac{1}{8}$$

Using the same list of possible outcomes (called the ***sample space***), we can find other probabilities. For example, what is the probability of getting two heads and one tail?

The list of 8 possible outcomes shows 3 times when there are two heads and one tail. HHT HTH THH

Thus the probability of getting two heads and one tail is 3 : 8. $P(2H, 1T) = \frac{3}{8}$

Name some other probabilities you can find using this sample space.

Example 2 Write the sample space (all the possible outcomes) for spinning these 2 spinners at the same time.

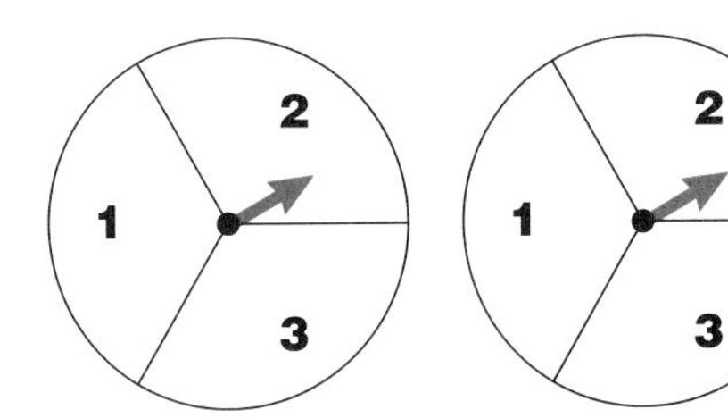

Draw a tree diagram to help write the sample space.

From a starting point, list in a column the 3 possibilities for the first spinner.

For each possibility for the first spinner, list the 3 possibilities for the second spinner.

Read the tree diagram to write the sample space.

There are 9 possible outcomes.

first spinner	second spinner	Outcomes
1	1	1 1
	2	1 2
	3	1 3
2	1	2 1
	2	2 2
	3	2 3
3	1	3 1
	2	3 2
	3	3 3

Use this sample space to find the probability that :

a. both spinners land on the same number

Of the 9 possible outcomes, there are 3 times when both spinners land on the same number. 1 1 2 2 3 3

Thus, the probability that both spinners land on the same number is 3 : 9 or 1 : 3.

$$P(\text{same number}) = \frac{3}{9} \text{ or } \frac{1}{3}$$

b. the sum of the numbers on which the two spinners land is five

Of the 9 possible outcomes, there are 2 times when the sum of the numbers is five. 2 3 3 2

Thus, the probability of getting a sum of five is 2 : 9.

$$P(\text{sum of } 5) = \frac{2}{9}$$

Try This *(For solution, see page 510.)*

a. Draw a tree diagram to help write the sample space for tossing a coin at the same time as rolling a die.

b. Using this sample space, find the probability of getting:

(1) heads on the coin and 3 on the die
(2) heads on the coin and an even number on the die
(3) tails on the coin and an odd number on the die
(4) tails on the coin and a number greater than 4 on the die
(5) heads on the coin and a number less than 3 on the die

The Counting Principle

A street vendor is selling cups of flavored ice.

There are 3 different size cups: small, medium, and large.

There are 4 different flavors: lemon, orange, cherry, and chocolate.

Steve wants to buy a cup of ice.
How many choices does he have?

To answer the question, you could draw a tree diagram that shows each of the 3 choices of cup size with each of the 4 choices of flavors.

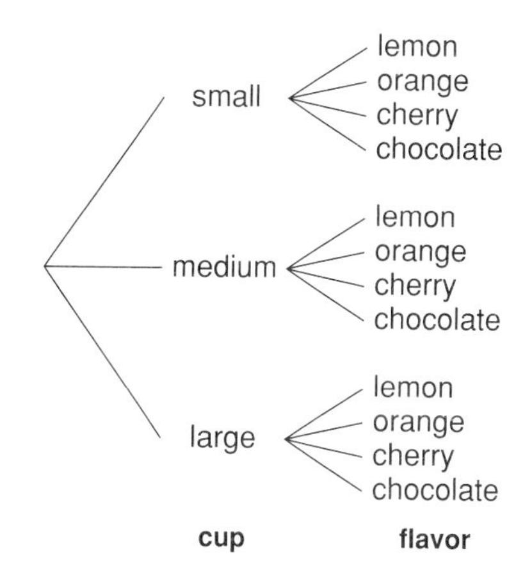

Since you are only interested in *how many* choices there are, just count the possibilities. There is no need to write out what all the possibilities are.

Do you see a way of *counting* the possibilities without drawing the tree diagram?

3	·	4	=	12
number of ways to choose cup		number of ways to choose flavor		total number of possibilities

The Counting Principle

You can find the total number of possible outcomes in a sample space by *multiplying* the number of choices for each stage.

Example 3 A deli offers 6 different kinds of sandwich meats, 5 different breads, and 3 different dressings.

How many different kinds of sandwiches can be made using a meat, a bread, and a dressing?

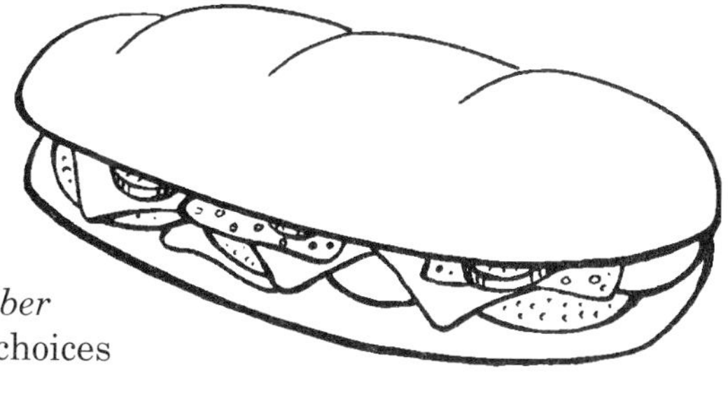

Since you are interested only in the *number* of different choices, not listing what the choices are, apply the Counting Principle.

6	·	5	·	3	= 90 ← total number of possibilities
bread		meat		dressing	

Try This *(For solution, see page 511.)*

Jerry has 4 shirts and 3 pairs of jeans. In how many ways can he select an outfit of a shirt and jeans?

SELF-CHECK: SECTION 11.2

1. What is another name for the set of all possible outcomes of an event?

Answer: the sample space.

2. What is a convenient way to get a list of the sample space?

Answer: Use a tree diagram.

3. If you just want to know how many outcomes there are in the sample space, do you need a tree diagram?

Answer: No, you can tell the number of outcomes by applying the Counting Principle.

4. What operation do you use in the Counting Principle?

Answer: multiplication

5. What numbers are you multiplying in the Counting Principle?

Answer: the number of choices for each stage of the event

EXERCISES: SECTION 11.2

1. Harry has 2 shirts, one blue and one black, and 2 pairs of jeans, one blue and one black, that can be mixed and matched.

a. Draw a tree diagram for choosing an outfit of a shirt and jeans.

b. From the tree diagram, list the sample space.
How many possible outcome are there?

c. Using this sample space, find the probability that:

(1) the shirt is blue

(2) the jeans are black

(3) the shirt is red

(4) both the shirt and the jeans are blue

(5) both the shirt and the jeans are the same color

(6) the shirt is blue and the jeans are not black

2. There is an equal likelihood of boys or girls being born in the Irwin family.
   a. Draw a tree diagram for the gender of the 3 children Mr. and Mrs. Irwin intend to have.
   b. From the tree diagram, list the sample space. How many possible outcomes are there?
   c. Using this sample space, find the probability that the children will be:
      (1) all boys
      (2) all of the same gender
      (3) a girl and 2 boys
      (4) a girl first, and then 2 boys
      (5) 2 girls and a boy
      (6) not a girl and 2 boys

3. A coin is tossed at the same time this spinner is spun.

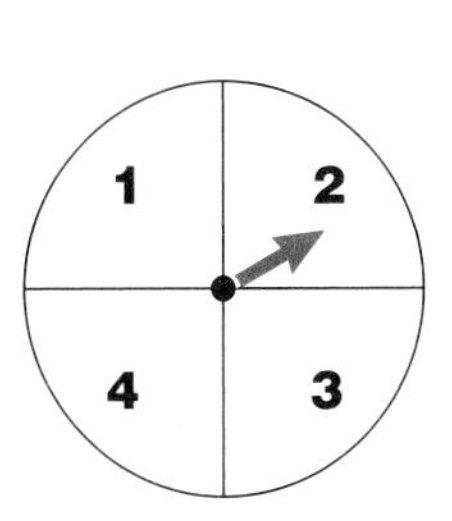

   a. Draw a tree diagram of the outcomes.
   b. From the tree diagram, list the sample space. How many possible outcomes are there?

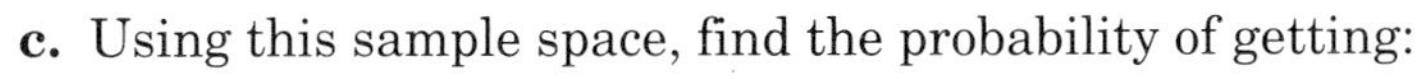

   c. Using this sample space, find the probability of getting:

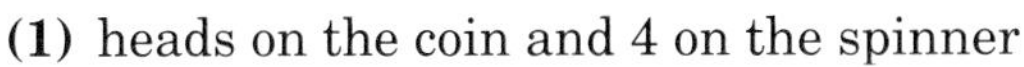

      (1) heads on the coin and 4 on the spinner

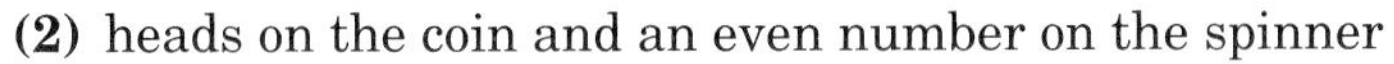

      (2) heads on the coin and an even number on the spinner
      (3) tails on the coin and an odd number on the spinner
      (4) tails on the coin and a number greater than 2 on the spinner
      (5) heads on the coin and a number less than 5 on the spinner
      (6) heads on the coin and not 3 on the spinner.

4. A card is selected from the four cards shown, and this spinner is spun.

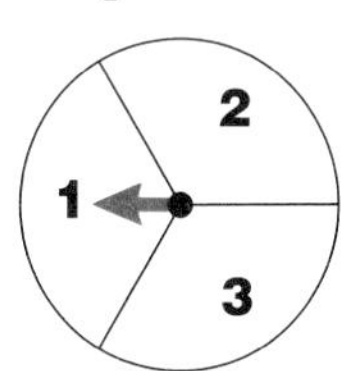

   a. Draw a tree diagram of the outcomes.
   b. From the tree diagram, list the sample space. How many possible outcomes are there?
   c. Using this sample space, find the probability of getting:
      (1) an Ace and 1 on the spinner
      (2) a red card and 1 on the spinner
      (3) a picture card and 2 on the spinner
      (4) a black card and an odd number on the spinner
      (5) not an Ace and 3 on the spinner
      (6) a blue card and 2 on the spinner

5. In a group of students, there are 3 girls (Amy, Asha, and Barb) and 3 boys (Len, Moe, and Max).

a. Draw a tree diagram for selecting 1 boy and 1 girl from the group.

b. From the tree diagram, list the sample space.
How many possible outcomes are there?

c. Using this sample space, find the probability that:

(1) the girl's name begins with A

(2) the boy's name begins with M

(3) the girl's name begins with C

(4) the boy's name does not begin with M

(5) the girl's name begins with B and the boy's name begins with M

(6) the girl's name begins with C and the boy's name begins with L

6. A black die and a white die are rolled at the same time. Consider the sum of the numbers shown on the two dice.

is not the same as

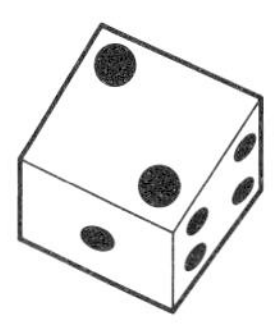

Why is it not convenient to draw a tree diagram to show the sample space?

Here's how to use a table to list the sums.

The possibilities for the black die are listed as the left heading of this table and the possibilities for the white die are listed as the top heading.

The first two rows of sums are shown.

Copy and complete the table of sums.

+	1	2	3	4	5	6
1	2	3	4	5	6	7
2	3	4	5	6	7	8
3	?	?	?	?	?	?
4	?	?	?	?	?	?
5	?	?	?	?	?	?
6	?	?	?	?	?	?

a. Study this table of sums, and describe some patterns that you see.

b. What is the lowest sum? the highest?

c. What sum appears most often?
Why do you think this is true?

d. Find the probability of each sum: $P(2)$, $P(3)$, . . ., $P(12)$.

e. What is the total when you add the probabilities of all the sums?
That is, $P(2) + P(3) + \ldots + P(12)$.

Why do you think this is true?

7. Use the Counting Principle to find the number of outcomes for each activity.

a. choosing a book and a video from 7 books and 3 videos

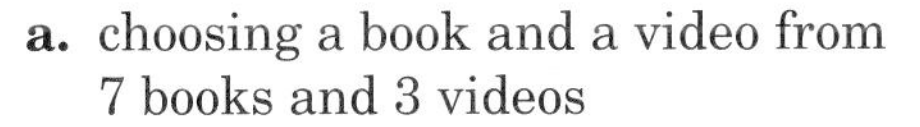

b. choosing an outfit from 3 pairs of jeans, 4 tops, and 2 pairs of shoes

c. traveling from West Pike to East Pike by using roads shown on this map

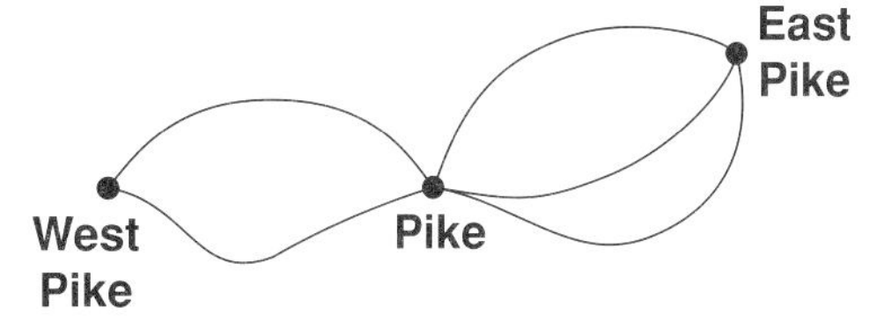

d. buying a car, given the choices shown in the table below

Model	Style	Color
Rocket Asteroid Shuttle	2-door 4-door	red blue brown black

e. tossing a coin three times

f. writing a code number consisting of a letter followed by two different digits

8. Morse Code uses dots and dashes to represent letters of the alphabet.

A •— B —••• C —•—• D ——• E •

a. How would you represent the word ACE in Morse Code?

b. Why do you think Samuel Morse decided to represent the letter E by a single dot?

c. There are 2 ways to represent a letter by a single dot or dash. Show 4 ways in which 2 symbols, (• and —) can be used.

d. How many ways are there to arrange:

(1) 3 symbols **(2)** 4 symbols

9. Use the Counting Principle to determine the number of outcomes in each sample space, and then find the probability.

a. Two dice are rolled. What is the probability of getting the same number on both dice?

b. An outfit consists of a jacket, a blouse, and a skirt. Each item has a choice of a red, a white, or a blue. Find the probability of selecting an outfit that is all red.

c. Find the probability of getting the result J2 by choosing a letter from the word NINJA and a digit from the number 2001.

EXPLORATIONS

1. Make a set of figures, each of which will have 3 characteristics: color, shape, and size.

   Color has 3 choices: red, blue, green.

   Shape has 4 choices: square, rectangle, triangle, circle

   Size has 2 choices: big, little

   a. With these choices for the 3 characteristics, how many different figures will you be able to make?

   b. Write a list, describing the 3 characteristics of each of the figures you will be able to make.

   Does your list have the same number of figures that you thought there would be?

   c. Examine all the figures and tell how many of them are:

   (1) red
   (2) not red
   (3) triangles
   (4) not triangles
   (5) big and green
   (6) big green circles
   (7) little or red
   (8) little red triangles
   (9) red and not little
   (10) red and not little circles
   (11) green or squares
   (12) green or not big squares

   Ask a partner to tell how many figures are red and square. Ask for other possibilities using the words *and*, *or*, *not*.

   d. Tell how many figures there would be if you added:

   (1) another color
   (2) another shape
   (3) another size
   (4) another color and another shape

2. Play these games with a partner to decide if the game is fair. Explain your conclusion.

   In all these games, one player, the "dealer", is Red and the other is Blue. Red and Blue have 50 chips each.

   a. Red pays Blue 1 chip, then tosses a coin. If Red gets heads, Blue pays Red 2 chips.

   b. Materials: 3 index cards marked as follows:

   one card: A on one side, B on the other side
   a second card: B on one side, C on the other side
   the third card: A on one side, C on the other side

   Red tosses the 3 cards. If the cards land so that each card shows a different letter, Blue pays Red 1 chip, otherwise Red pays Blue 1 chip.

   c. Materials: A hand of 5 cards each labeled 1, 2, 3, 4, 5.

   Red selects a card from his hand and places it face down.

   Blue pays Red 2 chips, then Blue selects a card from his own hand.

   If Blue's card matches Red's card, Red pays Blue 10 chips.

11.3 Independent and Dependent Events

Sample Spaces

Using the digits 4, 5, 6, how many different 2-digit numbers can you make, if you are allowed to repeat digits?

Draw a tree diagram to help you list this sample space.

From a starting point, list in a column the 3 possibilities for selecting the first digit.

Since repetitions are allowed, for each of the possibilities for the first digit, there are still 3 possibilities for the second digit.

Since repetitions are allowed, the selection of the first digit has no effect on the selection of the second digit. These two selections are ***independent events***.

1st	2nd	Outcomes
4	4	4 4
	5	4 5
	6	4 6
5	4	5 4
	5	5 5
	6	5 6
6	4	6 4
	5	6 5
	6	6 6

Read the tree diagram to list the sample space.

Allowing repetitions, there are 9 possible outcomes.

You can also determine the number of possible outcomes by applying the Counting Principle.

3	$\cdot$	3	=	9
number of ways to choose 1st digit		number of ways to choose 2nd digit		total number of possibilities

When a second event is *independent* of a first event, the number of possible outcomes for the second event is not affected by the first event.

Starting with the same digits 4, 5, 6, how many different 2-digit numbers can you make if you are *not* allowed to repeat digits?

Look back at the same tree diagram and eliminate the 3 outcomes that show repeats: 4 4 5 5 6 6

There are only 6 possible outcomes when no repetition of digits is allowed.

How would the tree diagram for no repeats look if you started that diagram "from scratch"?

From a starting point, list in a column the 3 possibilities for selecting the first digit.

Since repetitions are not allowed, for each of the 3 possibilities for the first digit, there are only 2 possibilities for the second digit.

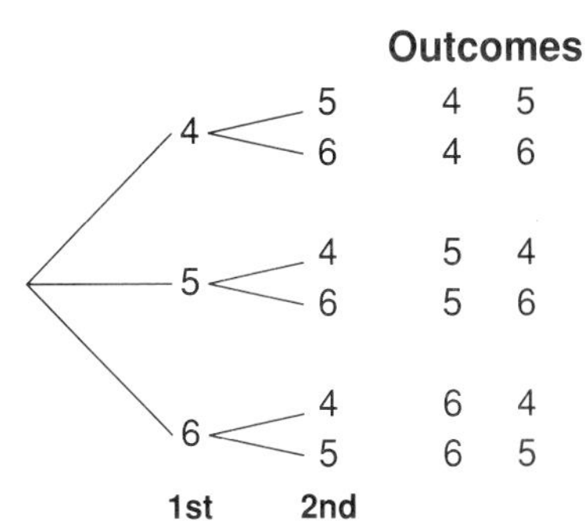

(Continues on next page)

Since repetitions are not allowed, the selection of the first digit *does* affect the selection of the second digit. These two selections are ***dependent events***.

You can also determine the number of possible outcomes by applying the Counting Principle.

3	·	2	=	6
number of ways to choose 1st digit		number of ways to choose 2nd digit		total number of possibilities

When a second event is *dependent* on a first event, the number of possible outcomes for the second event is affected by the first event.

Example 1

a. Use the Counting Principle to tell how many 3-letter arrangements are possible using the letters A, B, C:

(1) if repetitions are allowed

3	·	3	·	3	=	27
number of ways to choose the 1st letter		number of ways to choose the 2nd letter		number of ways to choose the 3rd letter		total number of possible outcomes

(2) if repetitions are not allowed

3	·	2	·	1	=	6
number of ways to choose the 1st letter		number of ways to choose the 2nd letter		number of ways to choose the 3rd letter		total number of possible outcomes

b. Draw a tree diagram to help write a list of the sample space for 3-letter arrangements of the letters A, B, C if no repetitions are allowed.

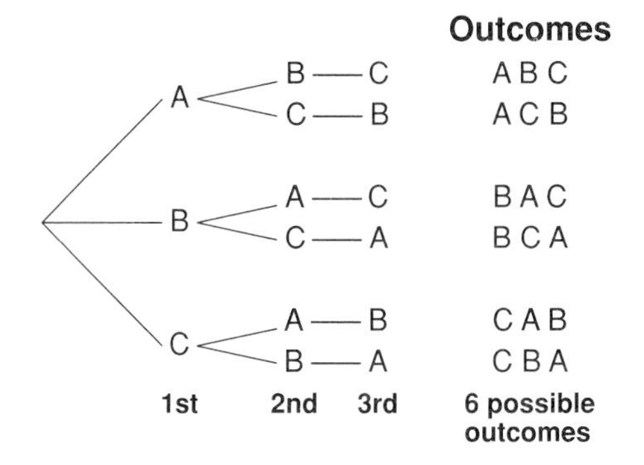

Example 2 Each possible 3-letter arrangement (no repetitions) of I, J, M is written on a card, and all the cards are placed in a box.

If JIM picks one card from the box, what is the probability that he picks his name?

1 ***Read the problem for information.***
You know that all possible 3-letter arrangements are to be made using 3 letters without repetition.

You have to find the probability of choosing 1 particular arrangement.

2 ***Make a plan.***
To write a probability ratio, you need to know the number of elements in the sample space.

Use the Counting Principle to find the number of elements in the sample space.

3 ***Carry out the plan.***
Apply the Counting Principle, with no repetitions allowed.

3	·	2	·	1	=	6
number of ways to choose the 1st letter		number of ways to choose the 2nd letter		number of ways to choose the 3rd letter		total number of possible outcomes

There are 6 possible ways of arranging the 3 letters.

Write the probability ratio

$$P(\text{J I M}) = \frac{1}{6}$$

← Only 1 arrangement spells JIM.
← There are only 6 possible arrangements.

4 ***Check.***
You could draw a tree diagram to verify that there are only 6 possible arrangements of the letters I, J, M, and that only 1 of these arrangements spells JIM.

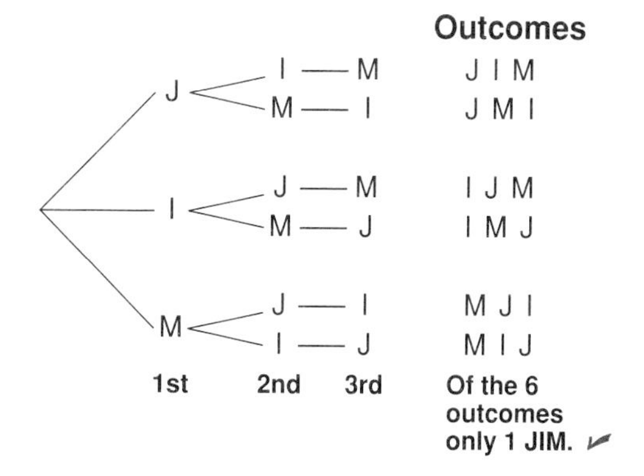

Answer: The probability that JIM will pick his name is 1 : 6.

Try These *(For solutions, see page 511.)*

1. Tell which of these pairs of events are independent and which are dependent. Explain.
   a. A die is rolled and a spinner is spun.
   b. Two cards are drawn from a deck of cards.
   The first card is replaced before the second is drawn.
   c. Two cards are drawn from a deck of cards.
   The first card is not replaced before the second is drawn.
2. Two-digit numbers are to be made by using the digits 7, 8, 9.
   Draw a tree diagram and list the sample space:
   a. if repetition of digits is not allowed
   b. if repetition of digits is allowed
3. Each possible 3-letter arrangement (no repetitions) of R, T, A is written on a card, and all the cards are placed in a box.

   If one card is selected at random from the box, what is the probability that the 3-letter arrangement chosen spells an English word?

Probability of a Combined Event in Which the Individual Events Are Independent

This bag contains exactly 2 black marbles and 1 white marble.

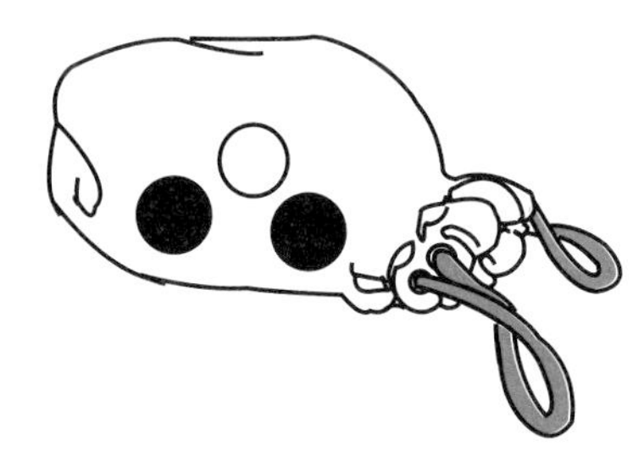

If, without looking, you pick a marble from the bag, what is the probability that it will be black?

$$P(\text{black}) = \frac{2}{3} \begin{matrix} \leftarrow \text{number of black} \\ \leftarrow \text{total number} \end{matrix}$$

Put the marble that you picked back into the bag. What is the probability of picking a black marble again?

The probability of picking a black marble again is still 2 : 3. Since the first marble has been returned to the bag, its selection has no effect on the selection of the second marble. These two selections are *independent* events.

Using the same bag of marbles, what would you say is the probability of first picking a black marble and then picking a white marble?

(Continues on next page)

Draw a tree diagram to verify your prediction about P(black, white).

Represent the 2 different black marbles as B_1, B_2.

There are 3 possibilities for selecting the first marble.

Since the first marble is put back, there are still 3 possibilities for the second marble.

Read the tree diagram to list the samples space.

There are 9 possible outcomes, 2 outcomes in which the first marble is black and the second marble is white.

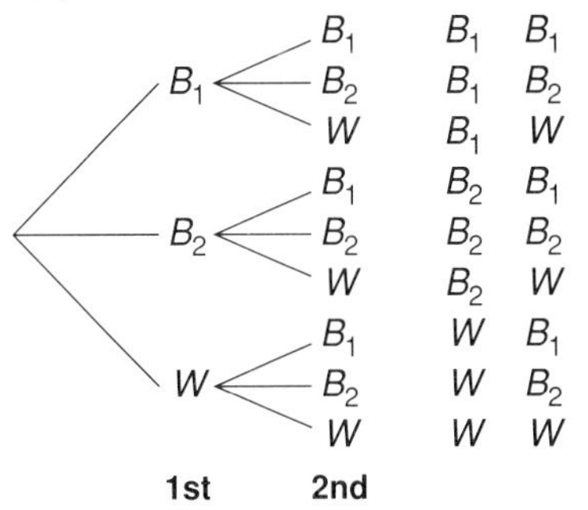

$B_1 W \quad B_2 W$

Thus, the probability of first getting a black marble and then a white marble is 2 : 9.

$$P(\text{black, white}) = \frac{2}{9}$$

How does this combined probability relate to the individual probabilities?

$$P(\text{black, white}) = P(\text{black}) \cdot P(\text{white})$$
$$= \frac{2}{3} \cdot \frac{1}{3} = \frac{2}{9}$$

The probability of a combined event in which the individual events are independent is the product of the probabilities of each event.

Example 3 If two standard dice are rolled, what is the probability of getting 3 on both?

This event is a combination of two independent events.

Find the probability of each event, and take their product.

$$P(3, 3) = P(3) \cdot P(3)$$
$$= \frac{1}{6} \cdot \frac{1}{6}$$
$$= \frac{1}{36}$$

To Calculate the Probability of a Combined Event in Which the Individual Events Are Independent

1. Calculate the probability of the first event.
2. Calculate the probability of the second event.
   This probability is not affected by the first event.
3. Multiply the two probabilities.

Example 4 From a standard deck of playing cards, a card is drawn and replaced. After replacing the first card, a second card is drawn.

What is the probability of getting a red card on the first draw and a black card on the second draw?

Since the first card is replaced before the second card is drawn, the two selections are independent events.

Find the probability of each event, and take their product.

$$P(\text{red, black}) = P(\text{red}) \cdot P(\text{black}) = \frac{26}{52} \cdot \frac{26}{52} = \frac{1}{2} \cdot \frac{1}{2} = \frac{1}{4}$$

Try This *(For solution, see page 512.)*

A coin is tossed and a die is rolled.

a. Are these events independent or dependent? Explain.

b. Draw a tree diagram and list the sample space for the combined event of tossing the coin and rolling the die.

c. Use this sample space to find the probability of getting heads on the coin and 4 on the die.

d. How does the probability of this combined event $P(\text{heads}, 4)$ relate to the individual probabilities $P(\text{heads})$ and $P(4)$?

Probability of a Combined Event in Which the Individual Events Are Dependent

Reconsider the bag with the 3 marbles, of which 2 are black and 1 is white.

This time after selecting a first marble, the marble is *not* returned to the bag.

What is the probability now of first picking a black marble and then a white marble?

Look back at the tree diagram for this sample space (page 499).

Note that of the 9 possible outcomes for making two selections, there are only 6 outcomes in which the first marble is black. (Eliminate the last 3 outcomes in which the first marble is white: $W B_1$ $W B_2$ $W W$)

Of the 6 possible outcomes in which the first marble is black, there are 2 times in which the second marble is white: $B_1 W$ $B_2 W$

Thus, the probability of first getting a black marble and then a white is 2 : 6 or 1 : 3.

(Continues on next page)

How does this combined probability relate to the individual probabilities?

$$P(\text{first black}) = \frac{2}{3} \begin{array}{l}\leftarrow \text{number of black marbles} \\ \leftarrow \text{total number of marbles}\end{array}$$

$$P(\text{then white}) = \frac{1}{2} \begin{array}{l}\leftarrow \text{number of white marbles} \\ \leftarrow \text{number of marbles left after the 1st marble is not replaced}\end{array}$$

$$P\begin{pmatrix}\text{first} & \text{then} \\ \text{black,} & \text{white}\end{pmatrix} = P\begin{pmatrix}\text{first} \\ \text{black}\end{pmatrix} \cdot P\begin{pmatrix}\text{then} \\ \text{white}\end{pmatrix} = \frac{2}{3} \cdot \frac{1}{2} = \frac{1}{3}$$

The sample space for a dependent event is reduced when the first selection is not replaced.

Example 5 From a standard deck of playing cards, a card is drawn. Without replacing the first card, a second card is drawn.

What is the probability of getting a red card on the first draw and a black card on the second draw?

Since the first card is not replaced, the second selection depends on the first.

① Calculate the probability for the first selection.

$$P(\text{first red}) = \frac{26}{52} \begin{array}{l}\leftarrow \text{number of red} \\ \leftarrow \text{total number}\end{array}$$

② Since the first card is not replaced, the sample space for the second draw is reduced.

$$52 - 1 = 51 \leftarrow \text{total number left after the draw}$$

Calculate the probability for the second selection.

$$P(\text{then black}) = \frac{26}{51} \begin{array}{l}\leftarrow \text{number of black} \\ \leftarrow \text{total left}\end{array}$$

③ Multiply the two probabilities.

$$P\begin{pmatrix}\text{first} & \text{then} \\ \text{red,} & \text{black}\end{pmatrix} = P\begin{pmatrix}\text{first} \\ \text{red}\end{pmatrix} \cdot P\begin{pmatrix}\text{then} \\ \text{black}\end{pmatrix}$$

$$= \frac{26}{52} \cdot \frac{26}{51}$$

$$= \frac{\cancel{26} \cdot 1}{\cancel{26} \cdot \cancel{2}} \cdot \frac{13 \cdot \cancel{2}}{51} \leftarrow \text{Factor to cancel.}$$

$$= \frac{13}{51}$$

To Calculate the Probability of a Combined Event in Which the Individual Events Are Dependent

① **Calculate the probability of the first event.**

② **Calculate the probability of the second event.**
The sample space for this probability is reduced because of the first event.

③ **Multiply the two probabilities.**

Try This *(For solution, see page 512.)*

Two-digit numbers are to be made from the digits 5, 6, 7.
No repetition of digits is allowed.

a. Is the selection of the second digit independent of the selection of the first digit or is the selection of the second digit dependent on the selection of the first digit? Explain.

b. Draw a tree diagram and list the sample space for forming the 2-digit numbers from the digits 5, 6, 7.

c. Use this sample space to find the probability of getting a 2-digit number in which the first digit is even and the second digit is odd.

d. How does the probability of this combined event P(even, odd) relate to the individual probabilities P(first even) and P(then odd)?

SELF-CHECK: SECTION 11.3

1. If you roll a die and then toss a coin, does the result on the die have an effect on the result of the coin? What do we call these events?

Answer: The roll of the die has no effect on the toss of the coin.
These are independent events.

2. If you roll 2 dice, does the result on the first die have an effect on the second die? What kind of events are these?

Answer: The first roll has no effect on the second roll.
These are independent events.

3. Two marbles are drawn from a bag.

a. When does the first draw not affect the second draw?

Answer: If the first marble is returned to the bag before the second marble is drawn, the first draw has no effect on the second draw. These are independent events.

b. When does the first draw have an effect on the second draw?

Answer: If the first marble is not returned to the bag before the second marble is drawn, the first draw does affect the second draw. These are dependent events.

c. If the first marble is not returned to the bag before the second marble is drawn, how is the sample space for the second marble affected?

Answer: The sample space is reduced. There is 1 less marble in the bag from which the second marble can be drawn.

4. Two-digit numbers are to be made from these digits: 2, 3, 4, 5.

a. How many choices are there for the first of the two digits?

Answer: There are 4 choices.

b. If repetition is allowed, how many choices are there for the second of the two digits?

Answer: There are still 4 choices.

c. Use the Counting Principle to tell how many 2-digit numbers can be formed from these 4 digits if repetition is allowed.

4	•	4	=	16
number of ways to choose 1st digit		number of ways to choose 2nd digit		total number of outcomes

Answer: If repetition is allowed, there are 16 possible ways to make 2-digit numbers from the 4 given digits.

d. If repetition is not allowed, how many choices are there for the second of the two digits?

Answer: There are only 3 choices for the second digit if you cannot repeat a digit already selected for the first digit.

e. Use the Counting Principle to tell how many 2-digit numbers can be formed from these 4 digits if repetition is not allowed.

4	•	3	=	12
number of ways to choose 1st digit		number of ways to choose 2nd digit		total number of outcomes

Answer: If repetition is not allowed, there are 12 possible ways to make 2-digit numbers from the 4 given digits.

EXERCISES: SECTION 11.3

1. Try this experiment.

   Materials: a box (or bag) with 4 marbles of different colors

   Test A: Without looking, take 1 marble from the box.
   Record the color of the marble.
   Do not replace the marble.
   Take another marble.
   Record the color of the second marble.

   Start over with 4 marbles.

   Do this 50 times, recording your results in a frequency table.

   Test B: Without looking, take 2 marbles from the box at the same time.
   Record their colors.

   Start over with the 4 marbles.

   Do this 50 times, recording your results in a frequency table.

   a. Are these dependent or independent events?

   b. Compare the results of Test *A* with those of Test *B*.

   c. Does it matter if the marbles are taken one at a time or both together?

2. Do you agree or disagree? Explain.

   a. Dan said that the tossing of two coins are independent events.

   b. Don said that tossing one coin twice is a pair of dependent events.

   c. Doris said that drawing two cards from a deck are independent events if you replace the first card before drawing the second.

   d. Dave said that choosing three letters from the alphabet are dependent events.

3. For each of the following situations, draw a tree diagram and list the sample space.

   a. With no repetition allowed, 2-digit numbers are to be made from 6, 7, 8, 9.

   b. With repetition allowed, 2-digit numbers are to be made from 3, 4, 5.

   c. With no repetition allowed, 3-letter arrangements are to be made from Q, R, S.

   d. Two jelly beans are to be eaten from a handful of 3 red and 2 yellow jelly beans.

4. A penny is tossed and this spinner is spun.

   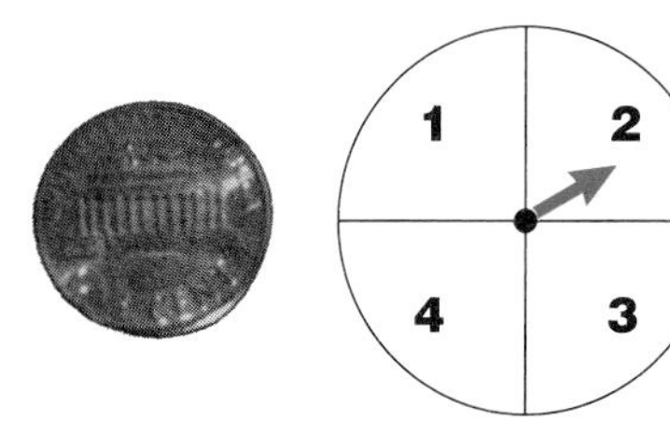

   a. Are these events independent or dependent? Explain.

   b. Draw a tree diagram and list the sample space for the combined event of tossing the coin and spinning the spinner.

   c. Use this sample space to find the probability of getting tails on the coin and 2 on the spinner.

   d. How does the probability of this combined event $P(\text{tails}, 2)$ relate to the individual probabilities $P(\text{tails})$ and $P(2)$?

5. With no repetition allowed, 2-letter arrangements are to be made form E, B, O, N.
   a. Are the selections of the letters independent or dependent events? Explain.
   b. Draw a tree diagram and list the sample space for forming the 2-letter arrangements.
   c. Use this sample space to find the probability of selecting a 2-letter arrangement in which the first letter is a vowel and the second letter is a consonant.
   d. How does the probability of this combined event P(vowel, consonant) relate to the individual probabilities P(first vowel) and P(then consonant)?
6. Find the probabilities of the individual events mentioned in each situation to find the probability of the combined event.
   a. Two dice are rolled.
      What is the probability of getting 4 on both?

   b. Two coins are tossed.
      What is the probability of getting tails on both?
   c. These two spinners are spun.
      What is the probability of getting an odd number on the first spinner and an even number on the second spinner?

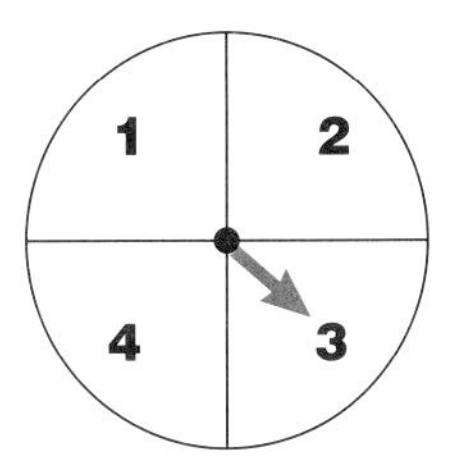

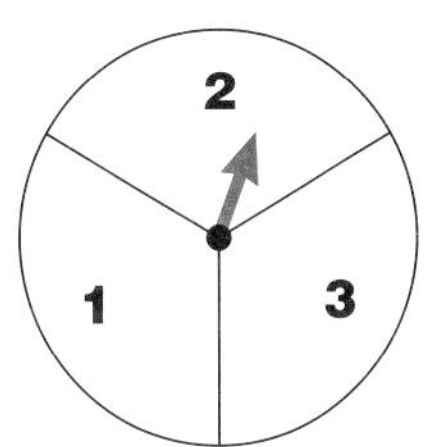

   d. Two marbles are selected from this bag that contains exactly 4 black and 2 white marbles.
      The first marble is not put back in the bag before the second marble is drawn.
      What is the probability that the first marble is white and the second is black?
   e. Two cards are drawn from a standard deck.
      The first card is not put back in the deck before the second card is drawn.
      What is the probability of getting first an Ace and then a King?

7. With no repetition allowed, two-digit numbers are to be formed from the digits 0, 1, 2, 3.

a. Draw a tree diagram and list the sample space.

b. If one two-digit number is selected at random, find the probability that it will be:

(1) odd **(2)** less than 30 **(3)** between 11 and 29

EXPLORATIONS

1. a. Use a tree diagram to answer this question.

This bag contains exactly 4 marbles, of which 3 are black and 1 is white.

A marble is picked and not returned to the bag.

After the first selection, a second marble is picked.

What is the probability that both marbles chosen are black?

b. Now see if you can use the individual probabilities *P*(first black) and *P*(then black) to get the probability of the combined event *P*(first black, then black).

Hint: The first black marble has an effect on both the numerator and the denominator of the probability ratio for the second black marble.

c. Find the probability of each combined event. If it is convenient, use a tree diagram as a check.

(1) Just before laundry day, Eric had only 2 brown socks and 2 black socks left in his sock drawer.

If Eric takes 2 socks from the drawer without looking, what is the probability that they match?

(2) In the Stone's freezer, there are 3 cheese pizzas, 2 vegetable pizzas, and 1 pepperoni pizza. Without looking, Mike takes a pizza and Tony takes a pizza.

What is the probability that the boys both have a cheese pizza?

(3) From a standard deck of playing cards, a card is drawn. Without replacement of the first card, a second card is drawn.

What is the probability that both cards are red?

2. Without replacement, three coins are drawn at random from a box that contains exactly 2 dimes, 5 nickels, and 3 pennies.

Find the probability that :

a. the three coins are all of different values

b. all three coins are nickels

c. two of the coins are dimes

d. none of the coins are dimes

e. the value of the coins is 11 cents

Things You Should Know After Studying This Chapter

KEY SKILLS

11.1 Use a frequency table to keep a count.
Write a probability ratio.

11.2 Draw a tree diagram to list all possible outcomes.
Apply the Counting Principle to find the number of possible outcomes.

11.3 Identify dependent and independent events.
Use a tree diagram or the Counting Principle to find the probability of a combined event.

KEY TERMS

11.1 frequency table • tally • probability • theory • at random

11.2 tree diagram • sample space • Counting Principle

11.3 independent/dependent events

CHAPTER REVIEW EXERCISES

1. Which of the following choices shows all possible values of the probability ratio P?

(a) $P \leq 1$ (b) $-1 \leq P \leq 1$

(c) $P \geq 0$ (d) $0 \leq P \leq 1$

2. For 300 spins of the spinner shown, which choice of frequencies best represents the expected outcomes?

	A	B	C
(a)	100	100	100
(b)	150	100	50
(c)	120	120	60
(d)	110	110	80

3. A marble is chosen at random from a jar that contains 2 red marbles, 2 blue marbles, and 4 yellow marbles. Match each outcome in *Column A* with its probability in *Column B*.

Column A	*Column B*
a. The marble is red.	(1) 0
b. The marble is not red.	(2) $\frac{1}{4}$
c. The marble is green.	(3) $\frac{1}{2}$
d. The marble is not green.	(4) $\frac{3}{4}$
e. The marble is yellow.	(5) 1

4. Find the probability of the event described.

a. An activity is chosen at random from this list that shows how many calories are burned in 20 minutes.

Activity	*Calories Burned*
Aerobic Dancing	105
Jogging	160
Ping Pong	58
Rowing Machine	206
Swimming	152

The activity chosen burns more than 150 calories.

b. A polygon is chosen at random from this set: {triangle, quadrilateral, pentagon}.

The polygon chosen has more than 3 sides.

c. A letter is chosen at random from the word STATISTICS.

The letter chosen is an S.

d. A book is chosen at random from 7 mysteries and 5 romances.

The book chosen is a mystery.

5. A standard 6-sided number cube is rolled once.

Find the probability of getting:

a. 3 **b.** an even number

c. 10 **d.** a whole number

6. The spinner on this board is spun once. Find:

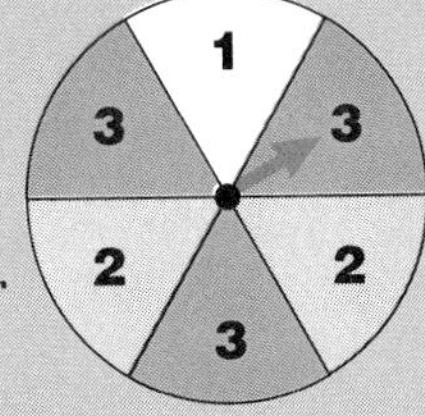

a. $P(1)$ $P(2)$

$P(3)$ $P(\text{not } 1)$

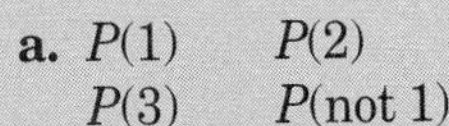

b. What do you expect for the value of $P(1) + P(2) + P(3)$?

Verify your prediction from your results in part **a**.

c. What do you expect for the value of $P(1) + P(\text{not } 1)$?

Verify your prediction from your results in part **a**.

7. Draw a spinner board that is divided into 5 equal parts. Label the sections A, B, or C so that:

$P(A) = \frac{1}{5}$ $P(B) = \frac{1}{5}$ $P(C) = \frac{3}{5}$

8. Tell which of these pairs of events are independent and which are dependent.

a. Two coins are tossed at the same time.

b. One coin is tossed and then another coin is tossed.

c. A marble is chosen from a bag. The marble is put back in the bag, and a second marble is chosen.

d. A card is chosen from a deck. The card is not put back in the deck, and a second card is chosen.

9. Draw a tree diagram to help list each sample space. Use the list to answer the question.

a. A coin is tossed at the same time that this spinner is spun.
List the possible outcomes.
What is the probability of getting heads on the coin and an even number on the spinner?

b. A coin is tossed and a card is selected from the set shown.

List the possible outcomes.
What is the probability of getting tails on the coin and a red card?

c. Three skirts (red, white, blue) may be mixed and matched with three blouses (red, white, blue) to make an outfit.
List the possible outcomes for an outfit of a skirt and blouse.
What is the probability that an outfit chosen at random will have a red skirt and a blouse that is not blue?

10. Use the Counting Principle to tell the number of possible outcomes.

a. There are 15 flavors of ice cream, 6 syrups, and 4 toppings.
How many different kinds of sundaes consisting of an ice cream, a syrup, and a topping can be made?

b. Two-digit numbers are to be formed from the digits 1–9.
How many different 2-digit numbers can be formed if repetition is allowed?
if repetition is not allowed?

SOLUTIONS FOR TRY THESE

11.1 From Experiment to Theory

Page 475

See the coin experiment of Example 2 for arrangement of table and possible results.

The outcomes of the key experiment are equally likely.

Page 477

a. Flavor 2 was most preferred and flavor 1 was least preferred.

b.

Flavor 1	Flavor 2	Flavor 3
$\frac{109}{1{,}000}$	$\frac{513}{1{,}000}$	$\frac{378}{1{,}000}$

c. Flavor 2: $\frac{513}{1{,}000}$ is about $\frac{500}{1{,}000}$ or $\frac{50}{100}$ or 50%

Flavor 1: $\frac{109}{1{,}000}$ is about $\frac{100}{1{,}000}$ or $\frac{10}{100}$ or 10%

Flavor 3: $\frac{378}{1{,}000}$ is about $\frac{400}{1{,}000}$ or $\frac{40}{100}$ or 40%

d. The probability that Sy will choose flavor 2 is about 50%, or 1 : 2.

Page 478

a. $P(4) = \frac{1}{10}$ ← There is 1 card with the number 4. ← There are 10 cards in all.

b. $P(\text{odd}) = \frac{5}{10}$ ← 5 cards have odd numbers: 1, 3, 5, 7, 9 ← There are 10 cards in all.

$P(\text{odd}) = \frac{1}{2}$ ← Simplify the ratio.

c. $P(\text{even}) = \frac{5}{10}$ ← 5 cards have even numbers: 0, 2, 4, 6, 8 ← There are 10 cards in all.

$P(\text{even}) = \frac{1}{2}$

d. $P(\text{composite}) = \frac{4}{10}$ ← composite numbers: 4, 6, 8, 9 ← There are 10 cards in all.

$P(\text{composite}) = \frac{2}{5}$

e. $P(\text{factor of } 12) = \frac{5}{10}$ ← factors of 12: 1, 2, 3, 4, 6 ← There are 10 cards in all.

$P(\text{factor of } 12) = \frac{1}{2}$

f. $P(\text{positive multiple of } 3) = \frac{3}{10}$ ← multiples of 3: 3, 6, 9 ← There are 10 cards in all.

g. $P(< 4) = \frac{4}{10}$ ← numbers less than 4: 0, 1, 2, 3 ← There are 10 cards in all.

$P(< 4) = \frac{2}{5}$

h. $P(> 6) = \frac{3}{10}$ ← numbers greater than 6: 7, 8, 9 ← There are 10 cards in all.

i. $P(< 10) = \frac{10}{10}$ ← All 10 numbers are less than 10. ← There are 10 cards in all.

$P(< 10) = 1$

j. $P(> 10) = \frac{0}{10}$ ← None of the numbers is greater than 10. ← There are 10 cards in all.

$P(> 10) = 0$

11.2 Sample Space

Page 488

a.

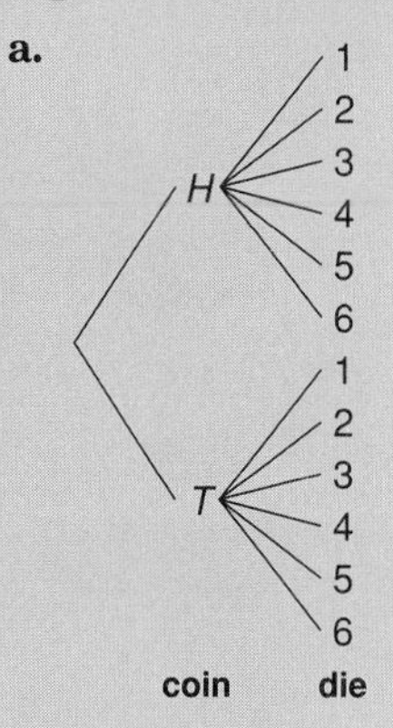

The outcomes in the sample space are:

$H1$	$H4$	$T1$	$T4$
$H2$	$H5$	$T2$	$T5$
$H3$	$H6$	$T3$	$T6$

There are 12 possible outcomes in all.

b. (1) Of the 12 possible outcomes, only one has heads on the coin and 3 on the die.

$$P(H, 3) = \frac{1}{12}$$

(2) Of the 12 possible outcomes, there are 3 that have heads on the coin and an even number on the die: $H2, H4, H6$.

$$P(H, \text{even}) = \frac{3}{12} = \frac{1}{4}$$

(3) Of the 12 possible outcomes, there are 3 that have tails on the coin and an odd number on the die: $T\,1, T\,3, T\,5$.

$$P(T, \text{odd}) = \frac{3}{12}$$
$$= \frac{1}{4}$$

(4) Of the 12 possible outcomes, there are 2 that have tails on the coin and a number greater than 4 on the die: $T\,5, T\,6$.

$$P(T, > 4) = \frac{2}{12}$$
$$= \frac{1}{6}$$

(5) Of the 12 possible outcomes, there are 2 that have heads on the coin and a number less than 3 on the die: $H\,1, H\,2$.

$$P(H, < 3) = \frac{2}{12}$$
$$= \frac{1}{6}$$

Page 490

$\underset{\text{shirt}}{4} \cdot \underset{\text{jeans}}{3} = 12$ ← total number of possibilities

Answer: Jerry can select an outfit of a shirt and jeans in 12 different ways.

11.3 Independent and Dependent Events

Page 498

1. In *independent* events, the second event is not affected by the result of the first.

In *dependent* events, the second event is affected by the result of the first.

a. Rolling a die and spinning a spinner are independent events.

b. If the first of two cards drawn from a deck is replaced before the second card is drawn, the second draw is independent of the first.

c. If the first of two cards drawn from a deck is not replaced before the second card is drawn, the second draw depends on the first.

2. a. repetition not allowed

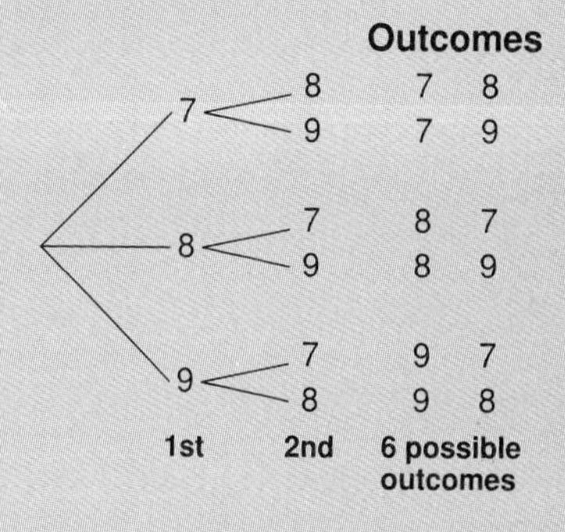

b. repetition allowed

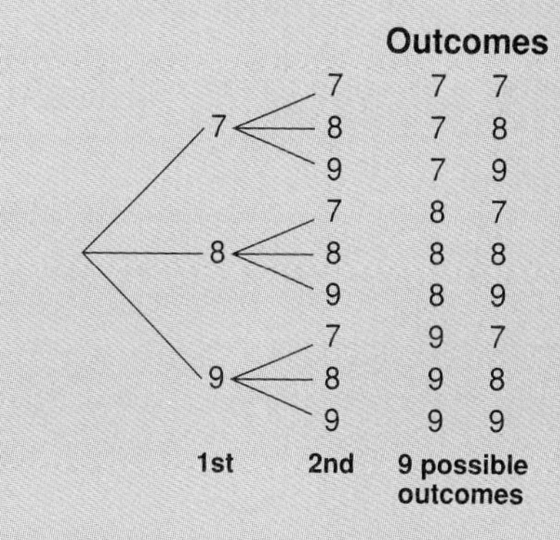

3.

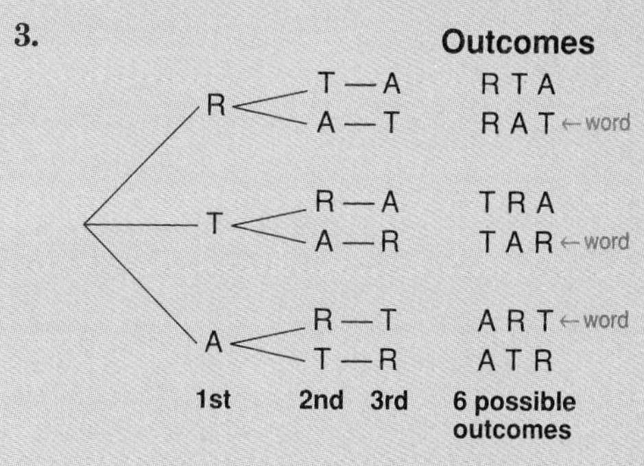

$$P(\text{word}) = \frac{3}{6} \begin{array}{l} \leftarrow \text{number of words: RAT, TAR, ART} \\ \leftarrow \text{total number of outcomes} \end{array}$$
$$= \frac{1}{2}$$

SOLUTIONS FOR TRY THESE

Page 500

a. Independent events.
The coin has no effect on the die.

b.

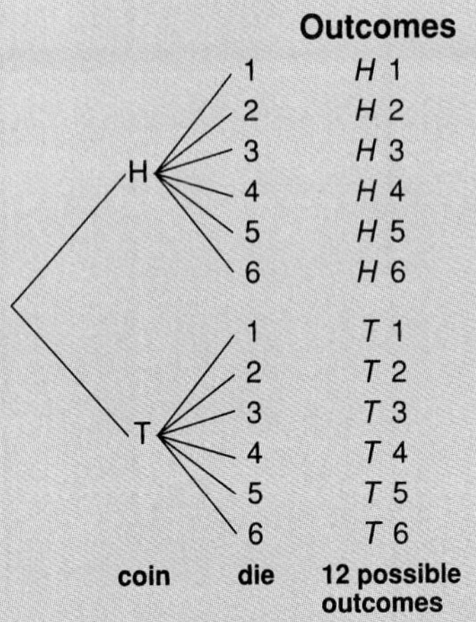

coin	die	Outcomes
H	1	H 1
	2	H 2
	3	H 3
	4	H 4
	5	H 5
	6	H 6
T	1	T 1
	2	T 2
	3	T 3
	4	T 4
	5	T 5
	6	T 6

12 possible outcomes

c. $P(\text{heads}, 4) = \frac{1}{12}$

d. $P(\text{heads}) = \frac{1}{2}$ $\qquad P(4) = \frac{1}{6}$

$$P(\text{heads}, 4) = P(\text{heads}) \cdot P(4) = \frac{1}{2} \cdot \frac{1}{6} = \frac{1}{12}$$

Page 502

a. Dependent events. When a selection is made for the first digit, it cannot be used again for the second digit.

b.

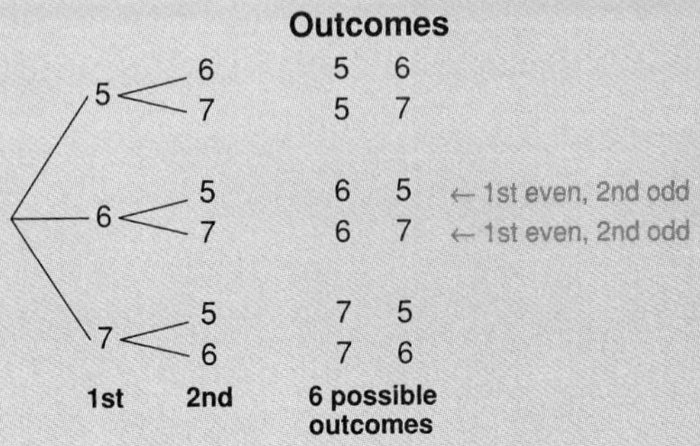

1st	2nd	Outcomes	
5	6	5 6	
	7	5 7	
6	5	6 5	← 1st even, 2nd odd
	7	6 7	← 1st even, 2nd odd
7	5	7 5	
	6	7 6	

6 possible outcomes

c. $P(\text{even, odd}) = \frac{2}{6}$ ← 2 ways to get even, odd: 65, 67 / ← total number of possible outcomes

$= \frac{1}{3}$

d. $P(\text{even}) = \frac{1}{3}$ ← only 1 even digit given: 6 / ← 3 digits given: 5, 6, 7

$P(\text{odd}) = \frac{2}{2}$ ← 2 odd digits given: 5, 7 / ← after 1st digit is chosen, only 2 left

$$P(\text{even, odd}) = P(\text{first even}) \cdot P(\text{then odd}) = \frac{1}{3} \cdot \frac{2}{2} = \frac{1}{3}$$

Statistics 12

Joe is a clerk in a stockroom. He has to report the total number of cartons that are stacked in the area pictured.

Aside from actually counting all the cartons, one by one, what are some ways that Joe can find out how many cartons there are?

12.1 Histograms and Stem-Leaf Plots

Frequency Tables

Here are the scores that the 30 students in Mr. Fried's math class got on their last test. The scores are given as percents.

66	75	75	90	90	85	81	78	61	73
78	88	83	82	91	92	67	66	71	88
93	99	76	63	71	88	88	87	87	86

If a score is chosen at random from this set of data, what is the probability that the score will be between 70% and 79%?

$P(70-79) = \frac{8}{30}$ ← There are 8 scores between 70% and 79%. ← There are 30 scores in all.

The probability of getting a score between 70% and 79% is 8:30.

Which is more likely: that a score chosen at random from this set of data is between 70% and 79%, or that a score chosen at random from this set of data is between 80% and 89%?

$$P(70-79) = \frac{8}{30} \qquad P(80-89) = \frac{11}{30}$$

Since $\frac{11}{30} > \frac{8}{30}$, a score between 80% and 89% is more likely than a score between 70% and 79%.

To analyze a set of data, it is convenient to have the data available in an organized fashion.

When you are studying different ways to organize and present data, you are working in the branch of mathematics called ***Statistics***.

It would be convenient to first organize the data about Mr. Fried's class in a frequency table that groups the scores in *intervals*.

Look at the scores to note the lowest value and the highest value. This will tell you the *range* of the scores, so that you will know where to begin and end the intervals.

Range = highest score – lowest score

In this case, the lowest score is 61% and the highest score is 99%.
The range is 99% – 61%, or 38%.

Grouping these scores in intervals of 10, the intervals can be 60–69, 70–79, 80–89, and 90–99.

Read through the scores, and count those in each interval. Keep track of your count by tally marks.

Interval	Tally	Frequency
60 – 69	𝍸	5
70 – 79	𝍸 III	8
80 – 89	𝍸 𝍸 I	11
90 – 99	𝍸 I	6
	TOTAL ▸	30

Example 1 Regroup the scores of the 30 students in Mr. Fried's class in a frequency table now using intervals of 5.

Interval	Tally	Frequency
60 – 64	||	2
65 – 69	|||	3
70 – 74	|||	3
75 – 79	~~||||~~	5
80 – 84	|||	3
85 – 89	~~||||~~ |||	8
90 – 94	~~||||~~	5
95 – 99	|	1
	TOTAL ▸	30

a. Use this display of the data to answer the following questions.

If a score is selected at random from this group of data:

(1) In which interval is the score most likely to fall?

Answer: 85–89, the interval with the highest frequency.

(2) In which interval is the score least likely to fall?

Answer: 95–99, the interval with the lowest frequency.

(3) Name three intervals of equal likelihood.

Answer: 65–69, 70–74, 80–84.

b. Compare the two frequency tables used for this set of data (the first in intervals of 10, the second in intervals of 5).

Both tables show such information as the location of the interval with the highest and lowest frequencies.

Which of these two frequency tables gives "better" information?

The frequency table with the smaller intervals gives more exact information. For example, using intervals of 5, you see that there are 3 intervals of equal likelihood. The table with the larger intervals of 10 did not reveal that information.

Try These *(For solutions, see page 562.)*

To determine the age appeal of a movie, the first 25 people entering a showing of the movie were asked their ages. Here are the results.

59	21	32	33	40
51	23	23	28	26
35	49	48	41	37
39	44	54	53	29
28	29	57	58	46

1. a. What is the lowest age?
b. What is the highest age?
c. What is the range of ages?

2. Arrange the data in a frequency table. Use intervals of 10, beginning with 20–29. (Keep this frequency table, to be used again.)

Histograms

The data in a frequency table that has been grouped in intervals can be displayed in a special type of bar graph called a ***histogram***.

Look at the histogram that can be drawn from each of the frequency tables prepared for the data of Mr. Fried's math class.

Using Intervals of 10

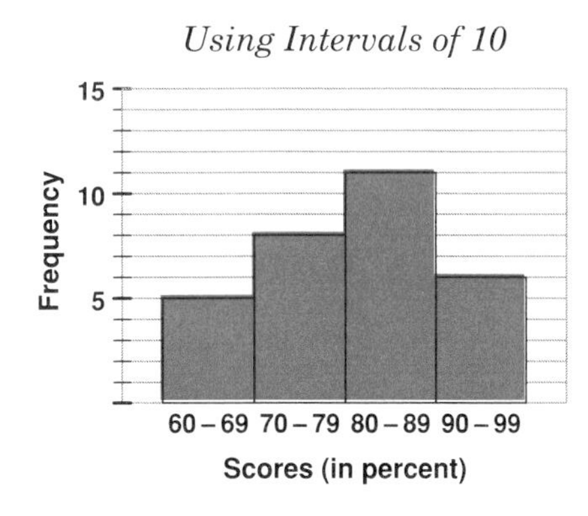

Using Intervals of 5

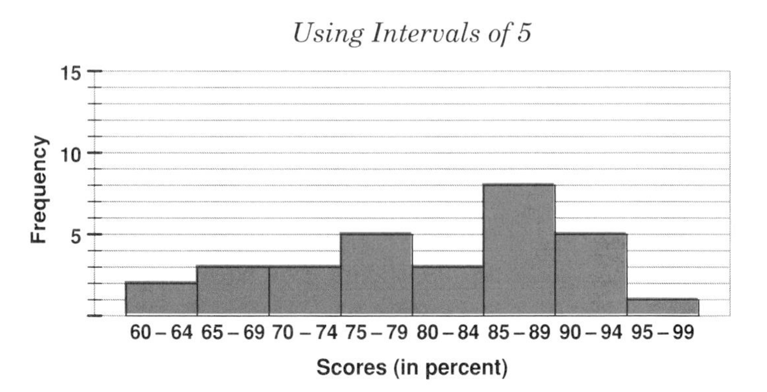

Like the frequency tables from which they come, the histograms that can be drawn for this set of data have advantages and disadvantages. For example, using intervals of 10 gives a more compact graph. Using intervals of 5 allows a more detailed analysis of the data.

How can the histogram that uses intervals of 5 be improved in appearance?
Since all the frequencies are under 10, the scale for frequency should be adjusted.
Use a bigger distance for the unit 1.

Making a Histogram for a Set of Data

1. **Organize the data in a frequency table that groups the data in equal intervals. Be sure the intervals cover the range of the data.**
2. **The vertical scale of the graph is for the frequency. This scale begins at 0.**
3. **The horizontal scale shows the intervals for the data. Arrange these equal intervals in order, from lowest to highest. Start the first interval away from the vertical axis.**
4. **For each interval, draw a bar whose height equals the frequency. The bars are connected to each other.**

Example 2 The Weather Bureau in Baltimore recorded these daily high temperatures in degrees Fahrenheit for the month of November:

46	31	33	42	25	29	37	44	45	36
35	40	52	48	35	39	40	42	51	29
45	30	26	52	44	54	46	43	45	42

Organize this data in a frequency table, using intervals of 5.

From the frequency table, draw a histogram to display the data.

Find the lowest temperature, 25°, and the highest, 54°. The range of the temperatures is 54°–25°, or 29°.

To include the lowest value, begin with 25–29 and continue with intervals of 5 through 50–54.

Interval	Tally	Frequency
25 – 29	||||	4
30 – 34	|||	3
35 – 39	~~||||~~	5
40 – 44	~~||||~~ |||	8
45 – 49	~~||||~~ |	6
50 – 54	||||	4
	TOTAL ➧	30

The vertical scale of the graph is for the frequency. Since the highest frequency value is 8, the scale 0–10 is convenient.

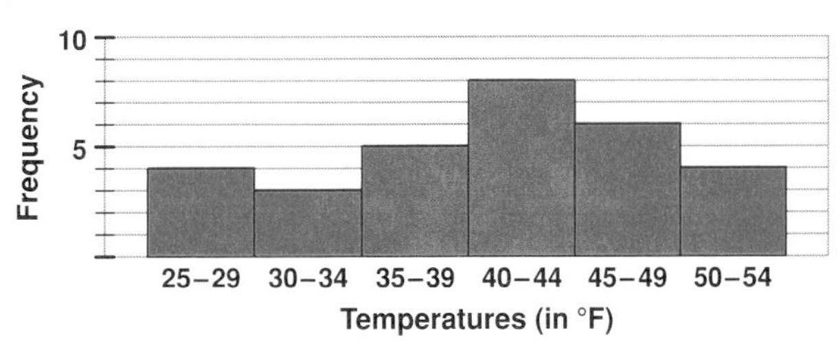

Use the histogram to answer these questions about the data.

If a temperature is chosen at random from this set:

a. What is the probability that the temperature chosen is in the interval 45–49?

Answer: $P(45-49) = \frac{6}{30}$, or $\frac{1}{5}$

b. What is the most likely interval for the chosen temperature?

Answer: 40–44, the interval with the highest frequency

Try These *(For solutions, see page 562.)*

Using the data collected from the age survey at the movie:

59	21	32	33	40
51	23	23	28	26
35	49	48	41	37
39	44	54	53	29
28	29	57	58	46

1. Use the frequency table previously made to draw a histogram for the data.

2. Use the histogram to answer these questions. If one person is chosen at random from the 25 people surveyed:

a. In which interval is the person's age most likely to fall?
b. In which interval is the person's age least likely to fall?
c. What is the probability that the person's age will be in the interval 40 – 49?
d. Which 2 intervals are of equal likelihood?

Stem-Leaf Plots

Let's have another look at the data from Mr. Fried's math class to study a different way of displaying the data.

66	75	75	90	90	85	81	78	61	73
78	88	83	82	91	92	67	66	71	88
93	99	76	63	71	88	88	87	87	86

Arrange these scores in 4 groups:

the 60's, the 70's, the 80's, and the 90's.

Within each group, arrange the scores in order from lowest to highest.

61	63	66	66	67						
71	71	73	75	75	76	78	78			
81	82	83	85	86	87	87	88	88	88	88
90	90	91	92	93	99					

Rotate this horizontal arrangement of numbers by a quarter turn counterclockwise so that the display shows vertical columns.

What do these columns of numbers resemble?

If you outlined these columns of numbers, you could make this display into a histogram with intervals of 10.

		88	
		88	
		88	
	78	88	
	78	87	
	76	87	99
67	75	86	93
66	75	85	92
66	73	83	91
63	71	82	90
61	71	81	90

Grouping by intervals, data can be arranged horizontally.

The horizontal arrangement can be written in a compact form by eliminating repetition. For example, here is how to shorten the row of numbers in the 60's:

61 63 66 66 67 can be written 6 | 1 3 6 6 7

(6 — All the numbers start with 6.) (1 3 6 6 7 — Just write the second digit for each number.)

Do the same for each row of numbers. That is, separate the beginning part of the number, called the ***stem***, from the end part of the number, called the ***leaf***.

61	63	66	66	67						
71	71	73	75	75	76	78	78			
81	82	83	85	86	87	87	88	88	88	88
90	90	91	92	93	99					

Stem	Leaf
6	1 3 6 6 7
7	1 1 3 5 5 6 8 8
8	1 2 3 5 6 7 7 8 8 8 8
9	0 0 1 2 3 9

Making a Stem-Leaf Plot for a Set of Data

1. **Decide on the stems, and list them in a column, in order from least (at the top) to greatest.**
2. **Put the leaves on the stems, in the order in which they appear in the data.**
3. **Rewrite the arrangement, this time listing the leaves in order.**

Example 3 Use the temperature data from the Baltimore Weather Bureau to make a stem-leaf plot.

46	31	33	42	25	29	37	44	45	36
35	40	52	48	35	39	40	42	51	29
45	30	26	52	44	54	46	43	45	42

To determine the stems, note that the temperatures can be arranged in 4 groups: the 20's, the 30's, the 40's, the 50's.

Stem	Leaf
2	
3	
4	
5	

Read through the data to put on the leaves.
First use the order that appears in the data.
Then rewrite the display with the leaves in order from lowest to highest.

Stem	Leaf
2	5 9 9 6
3	1 3 7 6 5 5 9 0
4	6 2 4 5 0 8 0 2 5 4 6 3 5 2
5	2 1 2 4

Stem	Leaf
2	5 6 9 9
3	0 1 3 5 5 6 7 9
4	0 0 2 2 2 3 4 4 5 5 5 6 6 8
5	1 2 2 4

Temperature (in °F)

Try This *(For solution, see page 562.)*

Using the data collected from the age survey at the movie:

59	21	32	33	40
51	23	23	28	26
35	49	48	41	37
39	44	54	53	29
28	29	57	58	46

1. Make a stem-leaf plot for the data.
2. Use the stem-leaf plot to answer these questions. Of the 25 people surveyed:
   a. How many were younger than 25?
   b. How many were younger than 35?
   c. How many were older than 45?

SELF-CHECK: SECTION 12.1

1. How do you choose the intervals for organizing data?

Answer: Find the highest and lowest values of the data so that you know where to begin and end the intervals.

The intervals must be of equal length. Often, we use intervals of 5 or 10.

The number of intervals requires judgment. You do not want too few or too many.

2. How do you find the *range* for a set of data?

Answer: Range = highest value – lowest value

3. How is a histogram different from a bar graph?

Answer: A histogram is a graph for data that has been grouped in intervals. The bars of a histogram are connected.

4. a. To arrange a set of data in a stem-leaf plot, how do you decide what the stems are?

Answer: The stem is the group of 10's into which the numbers fall.

b. What would you use as the stem for these numbers? 51 59 58 50 53

Answer: 5, since all these 2-digit numbers are in the 50's.

5. Name a way in which a stem-leaf plot gives more information about the data set than does a histogram.

Answer: From a stem-leaf plot you can read the actual data values in each interval.

A histogram tells only *how many* data values there are in each interval.

EXERCISES: SECTION 12.1

1. Make a frequency table to count the elements in each data set. Use the intervals given.

   a. *data set*: 24 grades on a test, given in percent

73	68	85	78	93	89	62	97
87	83	92	63	77	91	82	90
76	86	75	80	67	88	95	71

   Use intervals of 10 beginning with 60–69.

   b. *data set*: the 52 letters in this sentence

   I pledge allegiance to the flag of the United States of America.

   Use intervals of 5 letters of the alphabet: a-e, f-j, etc. Omit z.

   c. *data set*: 18 interest rates, given in percent

3.6	4.0	2.8	4.3	5.5	6.7
4.1	2.3	3.0	5.6	4.1	6.2
3.7	7.0	4.9	2.4	2.7	5.9

   Use intervals of 1%, beginning with 2.0–2.9

2. For each of the following sets of data:
   - Find the range.
   - Make a frequency table.

   When choosing intervals, keep in mind that the intervals must be of equal length, and that you do not want too few or too many intervals.

   a.

32	40	28	47	26	37
44	36	30	25	36	43
48	31	44	26	33	37

   b.

18	21	16	17	18	19
21	17	23	20	19	18
27	14	16	23	20	18

3. 

Although the overall cost of living continues to rise, technological advances considerably reduce some kinds of costs.

Here are the average rates, in cents, per minute of telephone use.

1985–86	37¢
1987–88	25¢
1989–90	19¢
1991–92	15¢

Make a histogram for the data.

4. Copy and complete each frequency table. Use the frequency table to make a histogram

a.

<table>
<tr><th>Interval</th><th>Tally</th><th>Frequency</th></tr>
<tr><td>21 – 30</td><td>|||</td><td>?</td></tr>
<tr><td>31 – 40</td><td>卌 ||</td><td>?</td></tr>
<tr><td>?</td><td>卌</td><td>?</td></tr>
<tr><td>?</td><td>|</td><td>?</td></tr>
</table>

b.

<table>
<tr><th>Interval</th><th>Tally</th><th>Frequency</th></tr>
<tr><td>0 – 4</td><td>|||</td><td>?</td></tr>
<tr><td>5 – 9</td><td>卌 卌 |</td><td>?</td></tr>
<tr><td>10 – 14</td><td>卌 卌 卌 卌 ||</td><td>?</td></tr>
<tr><td>15 – 19</td><td>卌 |||</td><td>?</td></tr>
<tr><td>?</td><td>卌 |</td><td>?</td></tr>
</table>

c.

<table>
<tr><th>Interval</th><th>Tally</th><th>Frequency</th></tr>
<tr><td>12.1 – 14.0</td><td>卌</td><td>?</td></tr>
<tr><td>14.1 – 16.0</td><td>|||</td><td>?</td></tr>
<tr><td>?</td><td>卌 ||</td><td>?</td></tr>
<tr><td>?</td><td>卌 ||||</td><td>?</td></tr>
<tr><td>?</td><td>卌 卌 ||</td><td>?</td></tr>
</table>

5. This histogram shows the results of a survey that asked students how many hours of television they had watched on Monday.

Use the histogram to answer these questions about the data.

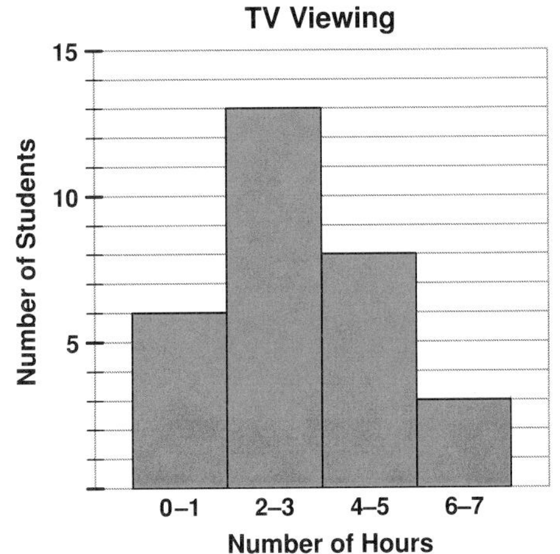

a. How many students watched for 2–3 hours?

b. How many students watched more than 5 hours?

c. How many students were in the survey?

d. What percent of the students watched fewer than 6 hours?

e. If one response is chosen at random:

(1) In which interval is it most likely to fall?

(2) What is the probability that it will fall in the interval 6–7?

6. For each set of data:

- Prepare a frequency table.
- Use the frequency table to make a histogram.
- Use the histogram to answer the questions about the data.

a. Here are the daily round-trip mileages for a group of commuters.

4	12	15	20	26	33	40	29
42	40	6	9	10	38	36	31
44	44	16	10	17	39	48	11

Use 5 intervals, beginning with 0–9.

(1) How many people drive less than 20 miles?

(2) For what percent of the people is the commute 40–49 miles?

(3) If one response is chosen at random, in which interval is it most likely to fall?

b. Here is the season record for the number of points scored in a game by a basketball player, Dave Starbuck.

22	17	31	25	30	26	47	31
19	43	16	27	23	34	28	32
29	36	31	40	35	19	24	31

Use 4 intervals, beginning with 10–19.

(1) In how many games did Starbuck score between 20 and 39 points?

(2) In what percent of the games did Starbuck score less than 30 points?

(3) Another player, Tiger, had a record of scoring at least 30 points in 1 out of 4 games. How does Starbuck's record compare to Tiger's?

c. Here is a 25-year record of average January temperatures, in degrees Fahrenheit, for Seattle, Washington.

42.5	40.5	40.1	40.5	44.9
37.1	43.2	45.1	39.0	44.4
34.8	37.8	44.4	39.4	41.8
38.8	38.7	38.7	37.0	39.7
41.2	33.1	40.9	42.4	41.1

Use 5 intervals, beginning with 33.0–35.9

(1) How many of the temperatures were less than 39°?

(2) What percent of the temperatures were between 29° and 45°?

(3) Millie is spending one January day sightseeing in Seattle between flights to Singapore. What is the most likely temperature interval for Millie's sightseeing?

7. A student studying conservation counted the number of rolls of paper towels bought by 40 customers at a supermarket.

1	4	6	0	0	2	6	1	1	1
4	0	0	6	4	1	1	1	2	0
1	0	0	0	1	1	6	6	4	4
2	0	0	2	6	6	0	0	1	1

a. What intervals would you use to group the data in 4 intervals? in 3 intervals?

b. Group the data in a frequency table of 4 intervals and make a histogram.

Answer the following questions about the data. Tell:

- if you could use either the table or the histogram
- if you must go back to the original data
- if you cannot draw a conclusion based on this survey

Explain.

(1) How many people bought at most 1 roll?

(2) How many people bought 2 rolls?

(3) How many people bought more than 1 roll?

(4) What is the range?

(5) How many people bought several rolls of paper towels because they were on sale?

(6) How many people prefer Brand X over Brand Y?

8. Write a list, row by row, of the data in each stem-leaf plot.

a. number of toll-free 800 calls received, per hour, by Mona's Mail Order House

Stem	Leaf
1	5 7
2	0 1 3 9
3	2 2 6
4	0 3

b. number of voters in a primary, per election district of Cumberland County

Stem	Leaf
21	7
22	1 4 5
23	0 6 6
24	2 3 3 9

9. Make a stem-leaf plot for each set of data.

a. heights, in centimeters, of tulip varieties

31	42	28	27
35	38	49	19
25	37	43	36
40	41	39	20

b. hours of jury deliberation in Dale County Courts

17	28	6	16
9	12	19	20
33	23	1	19
5	14	14	10

c. number of words per page in Elton's senior report

214	208	197	182
233	227	190	184
227	188	191	203
204	199	226	182

d. Harlan High School's best math scores on a College Entrance Test

732	677	685	743
668	702	680	663
688	672	698	734
727	706	661	660

e. maximum recorded life spans, in years

Species	*Years Lived*
Cat	34
Elephant	70
Horse	46
Hummingbird	8
Orangutan	54
Ostrich	50
Parakeet	25
Rhinoceros	40
Seal	46
Sheep	20
Spider	4
Squirrel	15
Tortoise	116
Whale	87

10. This stem-leaf plot shows the daily commuting time, in minutes, of the employees at Acme Electronics.

Use the stem-leaf plot to answer these questions about the data set.

Stem	Leaves
4	2 2 5 8
5	2 4 6 8 8
6	0 0 5 8 8
7	0 3 4 5 5 6
8	
9	0 2 2 2 5 6 7
10	1 2 5 5 7 8

a. What is the range of the data?

b. For how many employees is the commuting time:

(1) less than 1 hour

(2) at most 1 hour

(3) at least 1 hour and 45 minutes

c. If one employee is chosen at random, what is the probability that the commuting time will be more than an hour and a half?

(Exercises continue)

11. Al Greene demonstrated the broad jump for a group of elementary school students.

He then recorded the distances, in inches, that the students jumped after a few practice jumps.

The data that Al collected are displayed in a histogram and as a stem-leaf plot.

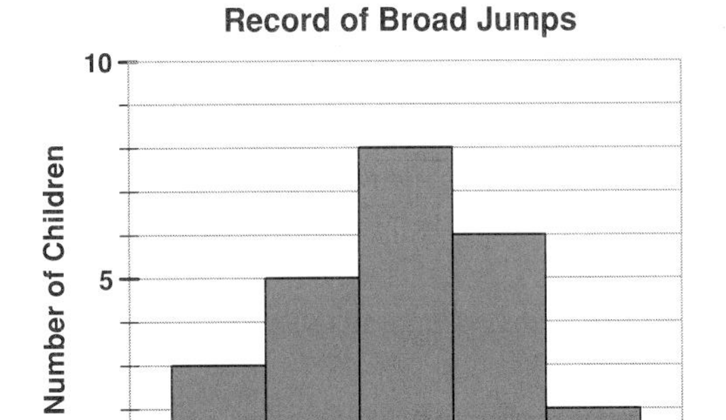

Stem	Leaves
1	7 8 9
2	3 5 6 6 7
3	2 4 5 6 8 8 8 9
4	0 1 3 4 6 9
5	0 2

Answer the following questions about the data set. Tell if you:

- can use either graph
- can use only the histogram
- can use only the stem-leaf plot
- cannot determine the answer from either graph

Explain. If either graph can be used, tell if one graph gives the answer more easily.

a. How many jumps were recorded?

b. What is the shortest distance recorded?

c. Which is the interval with the lowest frequency?

d. What was the greatest distance jumped?

e. What distance appears most often?

f. What is the range of the data?

g. What is the probability that Annie, a student picked at random, jumped at least 50 inches?

h. What is the probability that Annie's distance was less than 40 inches?

i. How many children jumped at least 3 feet?

j. How many children jumped no more than 2 feet?

EXPLORATIONS

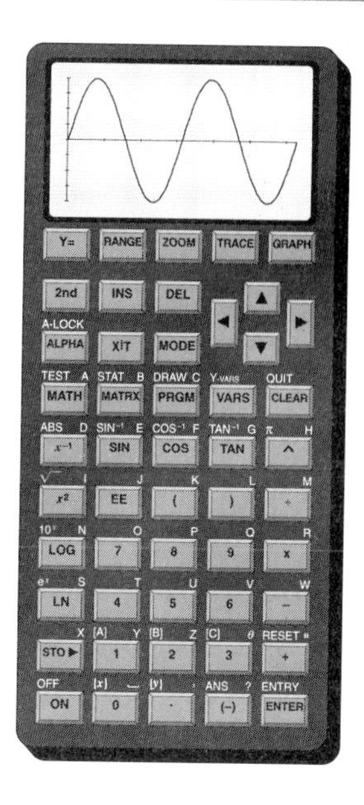

Advanced calculators do work with statistics. If your teacher has graphing calculators available, you can learn how to use the calculator to draw a histogram for a set of data.

Some basic keys on the calculator are:

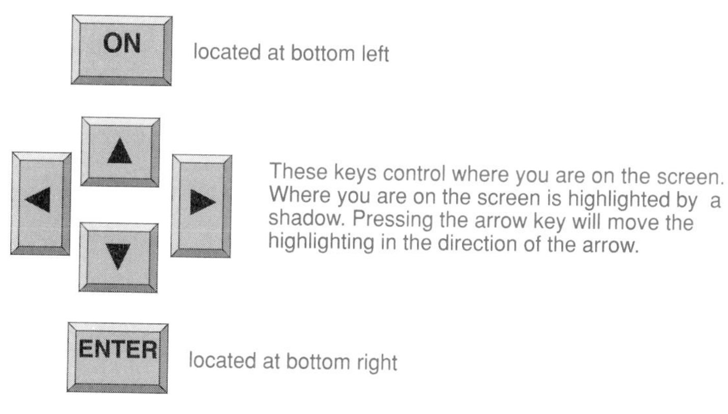

Here is a familiar data set for which you can use the calculator to draw a histogram. When you are finished, compare the result on the calculator with the histogram in the text after Example 1 (page 515).

data set: the scores that the 30 students in Mr. Fried's math class got on their last test

66	75	75	90	90	85	81	78	61	73
78	88	83	82	91	92	67	66	71	88
93	99	76	63	71	88	88	87	87	86

For convenience, work with the frequency table that groups the data into 4 intervals.

The calculator will draw a histogram that has 4 bars.

Interval	Frequency
60-69	5
70-79	8
80-89	11
90-99	6

(Exploration continues)

Step 1a To set the calculator so that it will be ready to operate with statistics, press the keys shown.

2nd — blue key, upper left corner   MATRX (STAT B) — column 2, row 3

The calculator will display at the top of the screen what it can do in statistics, called the *menu* for statistics.

CALC DRAW DATA

You want it to work with DATA.

Step 1b First you will clear any statistical data that may be in the calculator.

You want option 2, ClrStat, of DATA.

Press: ▶ ▶ 2 ENTER

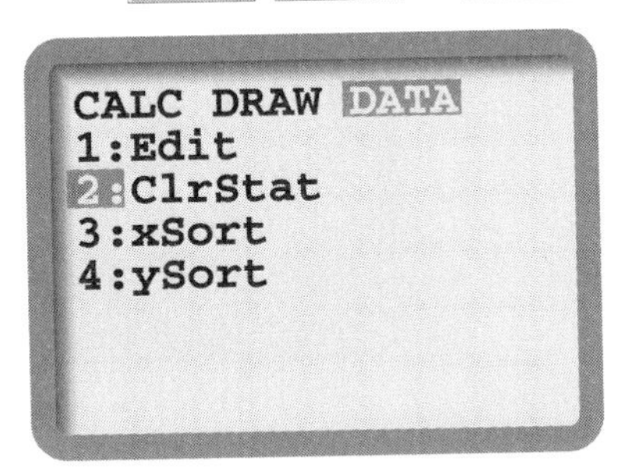

After data has been cleared, the calculator shows:

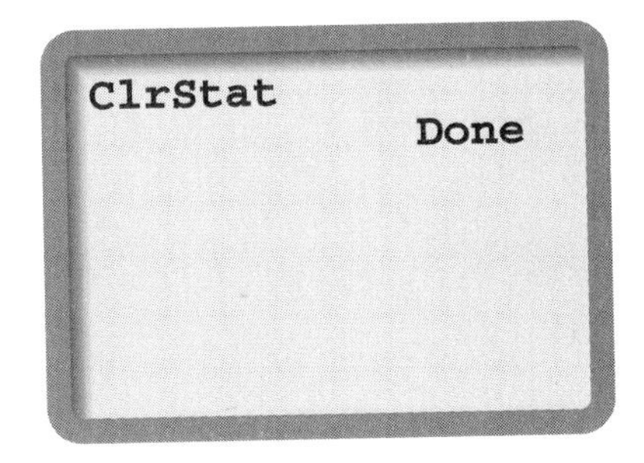

Step 1c Return to the calculator's menu for statistics.

Press: 2nd MATRX (STAT B)

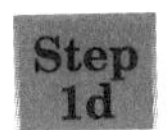

Now you want to be able to enter data. For this, you need option 1, Edit, of DATA.

Press: ▶ ▶ ENTER

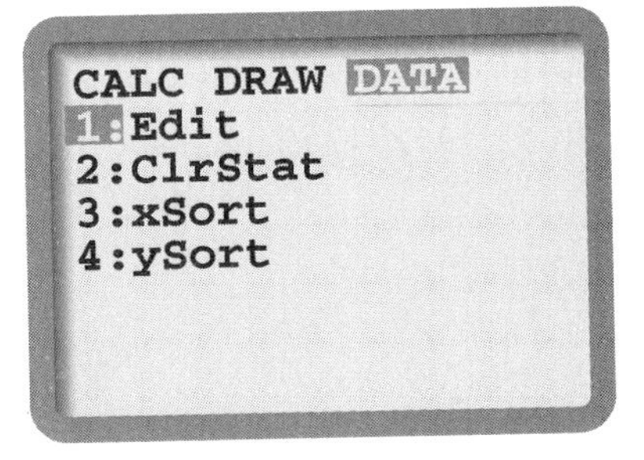

Step 2 The calculator uses x to represent the data value and y to represent the frequency.

The frequency on the calculator has been preset to 1.

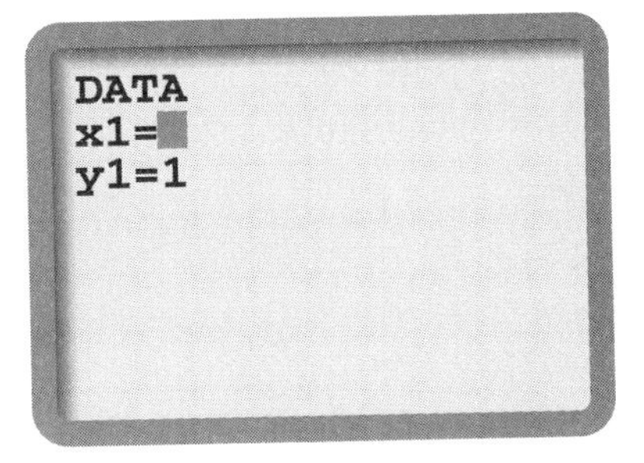

For each of the four intervals of the frequency table, you will enter the data value at the beginning of the interval and the frequency for the interval. The first interval begins at 60 and has a frequency of 5.

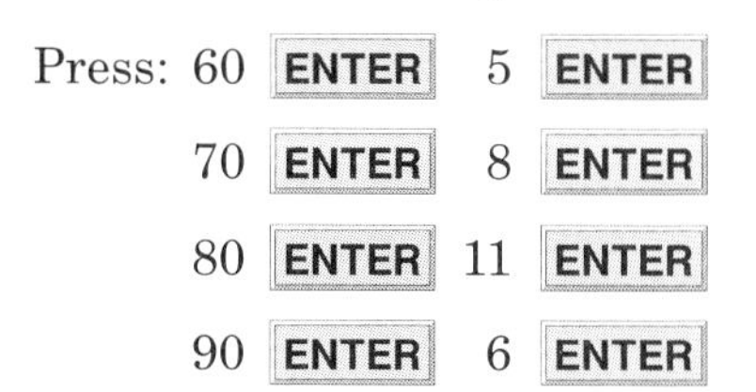

Before asking the calculator to draw the histogram for this data, you must set the range. Press: RANGE along the top second key

You will reset the lowest and highest values of x and y on the calculator.

The x-values come from the beginning of the first interval, 60, and the end of the last interval, 99, with interval widths of 10.

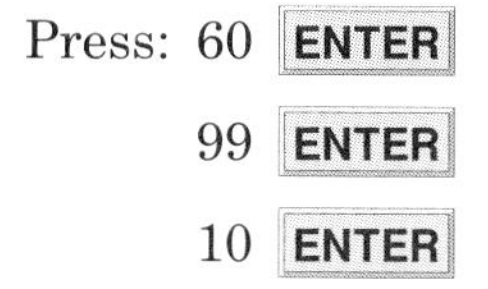

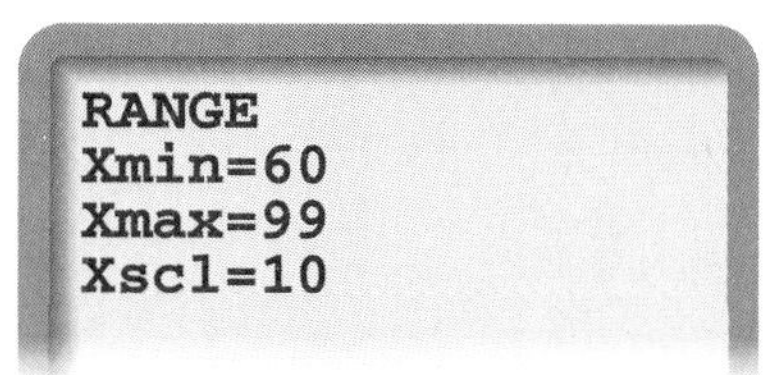

The y-values are the frequencies. You want the frequency scale to start at 0, and go higher than 8, say to 10. You want the scale to be marked off in units of 1.

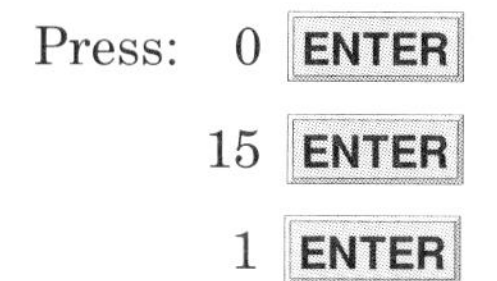

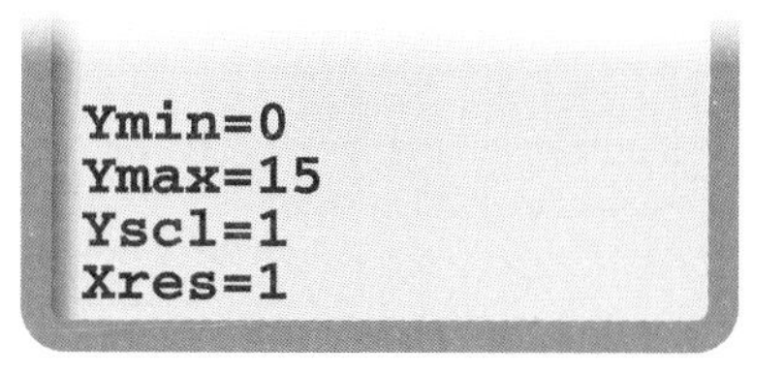

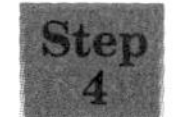

Now the calculator knows the data set and the range. Let's tell the calculator to draw the histogram.

Return to the statistics menu. Press: 2nd MATRX (STAT B)

You want to highlight DRAW and HIST. Press: ▶ ENTER

You want the calculator to draw the histogram. Press: ENTER

Well, there's the histogram. Would you like it to fit a little better on the screen? How do you think you can fix this?

Return to the range and set Xscl less than 10; try 8.

Then tell the calculator to draw the histogram again. Press:

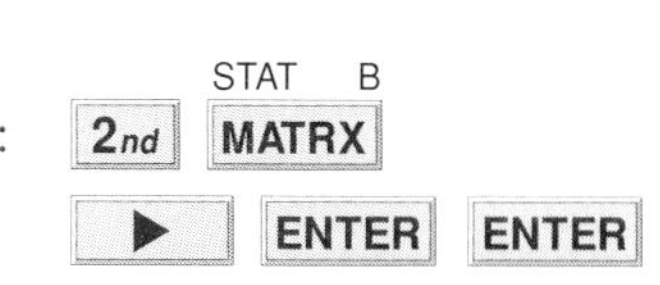

Try another, using the data and frequency table of Example 2 (page 517).

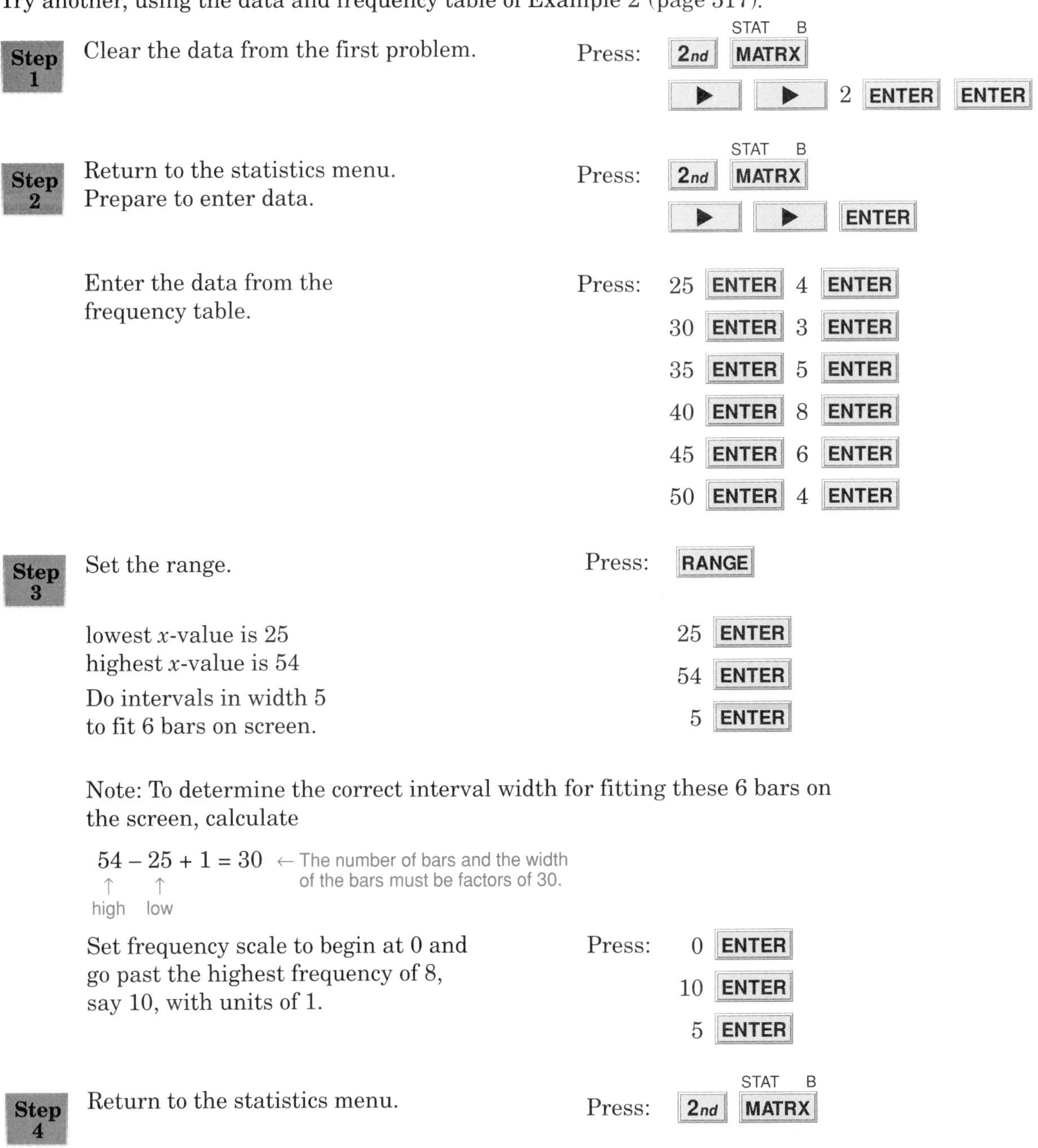

Step 1 Clear the data from the first problem. Press: 2nd MATRX (STAT B) ▶ ▶ 2 ENTER ENTER

Step 2 Return to the statistics menu. Prepare to enter data. Press: 2nd MATRX (STAT B) ▶ ▶ ENTER

Enter the data from the frequency table. Press:
25 ENTER 4 ENTER
30 ENTER 3 ENTER
35 ENTER 5 ENTER
40 ENTER 8 ENTER
45 ENTER 6 ENTER
50 ENTER 4 ENTER

Step 3 Set the range. Press: RANGE

lowest x-value is 25 — 25 ENTER
highest x-value is 54 — 54 ENTER
Do intervals in width 5 to fit 6 bars on screen. — 5 ENTER

Note: To determine the correct interval width for fitting these 6 bars on the screen, calculate

$54 - 25 + 1 = 30$ ← The number of bars and the width of the bars must be factors of 30.
(54 = high, 25 = low)

Set frequency scale to begin at 0 and go past the highest frequency of 8, say 10, with units of 1. Press: 0 ENTER 10 ENTER 5 ENTER

Step 4 Return to the statistics menu. Press: 2nd MATRX (STAT B)

Tell the calculator to DRAW the HIST. ▶ ENTER ENTER

Match the results on the calculator with the histogram in the text.

Draw other histograms on the calculator using data sets from the exercises. Match the results on the calculator with those histograms you had already drawn.

12.2 Statistical Averages

Different Types of Averages

Carlos Mendez owns a large citrus farm from which he has a monthly income of $8,000.

During the orange picking season, Mr. Mendez paid the four Ruiz brothers the following amounts for part-time work over a month:

$600 $700 $600 $800

Which one of these numbers would be a fair representation of the 5 numbers that represent monthly earnings on the Mendez farm?

8,000 600 700 600 800

If you do the familiar computation for average — add the 5 numbers and divide by 5 — do you think that the result, $2,140 is a fair representation of the 5 numbers? Explain.

The familiar computation for finding an average is useful when the numbers are pretty close to each other, that is, when the range of the numbers is relatively small.

Example 1 Find the average earnings of the Ruiz brothers for their month's work on the Mendez farm.

Since the earnings of the 4 brothers are pretty close to each other, it is fair to find an average by adding the 4 numbers and dividing by 4.

$$\frac{600 + 600 + 700 + 800}{4} = \frac{2,700}{4} = 675$$

Answer: For the month, the average of the earnings of the Ruiz brothers was $675.

This kind of average is called a *mean*.

The *mean* of a set of data is the sum of the data divided by the number of pieces of data.

To find one number to act as the representative of the earnings of Mr. Mendez and the Ruiz brothers, it would be appropriate to find a *middle* value.

This kind of average is called a *median*.

Example 2 Using the median, find the average monthly earnings from the farm of Mr. Mendez and the Ruiz brothers.

Arrange the 5 numbers in order and choose the middle number.

600 600 700 800 8,000
↑ median (700)

Answer: The median monthly earnings of the 5 people is $700.

The *median* of a set of data is the middle number, after the data has been put in order.

Suppose Mr. Mendez hired 6 more part-time workers to pick the oranges. Consider the monthly earnings of the Ruiz brothers and the 6 other part-time workers.

600 600 700 800 600 600 600 650 700 1,000

What one number would you choose to represent this group of 10 numbers?

For this group of 10 numbers that are close in value, it would be fair to use 600, the number that occurs most frequently, as the representative of the group.

This kind of average is called a *mode*.

The *mode* of a set of data is the number with the highest frequency.

Statistical Averages Used to Represent a Group of Data

The Mean

Find the mean of *n* pieces of data by adding the *n* pieces and then dividing the sum by *n*.

The Median

Find the median of a set of data by putting the data in order and then choosing the middle value.

The Mode

Find the mode of a set of data by selecting the number that appears most frequently.

Try This *(For solution, see page 563.)*

In the last 7 games that Jose bowled for his league, his scores were:

155 172 164 138 172 117 172

a. Find his mean score (to the nearest integer) for these games.
b. What effect does the low score of 117 have on the mean?
c. Find the median score for the 7 games.
d. What effect does the low score of 117 have on the median?
e. Find the mode of these 7 scores.

More About the Median

How would you find the middle for an even number of pieces of data?

Example 3 Find the median value for this data set:

20 16 7 14 15 10 12 18

Arrange the data in order. 7 10 12 14 15 16 18 20

There is no middle number in the data set of 8 numbers. ↑ median = 14.5

The median is halfway between 14 and 15.

Answer: The median is 14.5.

The *median* for an even number of data pieces may or may not be a member of the set.

More About the Mode

The 10 people who are in Steve's book club arranged themselves for a photo.

Here are their heights, in inches:

72 70 70 70 68

68 64 64 64 62

What would you say is the mode for this set of numbers?

Since 70 and 64 both have the same high frequency, this data set has *two* modes, one for the average height of the boys and one for the average height of the girls.

Since you see that a data set can have more than one mode, you might ask if a data set could have *no* mode.

Yes, a data set in which each piece of data appears only once has no mode.

For example: 60 80 70 85 90 65 75 95

A data set may have more than one mode.
A data set may not have any mode.

Example 4 Find the mean, median, and mode for this data set:

3.5 3.5 5 2 3.5

For the mean: find the sum of the 5 values and divide by 5

$$\frac{3.5 + 3.5 + 5 + 2 + 3.5}{5} = \frac{17.5}{5} = 3.5$$

For the median: arrange the data in order and choose the middle value

2 3.5 3.5 3.5 5

↑ median

For the mode: select the number with the highest frequency

The mode is 3.5 since it appears most often.

Answer: For this data set, the mean, median, and mode are the same value: 3.5

Try These *(For solutions, see page 563.)*

1. Find the median for this set of data:

66 73 65 81 71 62 92 85

2. Examine each data set for a mode. Explain your results.

a. test scores: 81 75 90 63 59 70 96

b. batting averages: .245 .230 .270 .250 .275 .300 .245 .310 .245 .290

c. heights: 5'4" 5'9" 5'4" 5'4" 5'1" 5'9" 5'7" 5'9" 5'0" 5'10"

More About the Mean

An equation can be useful in solving some problems that involve a mean.

Example 5 On 5 math tests, Laura achieved these grades:

94 86 92 80 78

What grade must Laura get on her next test so that her average for the 6 tests will be 85%?

1 *Read the problem for information.*

You know 5 of the 6 grades.
You want to find the 6th grade so that the average of the 6 grades will be 85%.

2 *Make a plan*

Before writing an equation to model the problem, guesstimate an answer.
Let x = the grade on the 6th test.

$$\frac{94 + 86 + 92 + 80 + 78 + x}{6} = 85$$ The average is the sum of the 6 grades divided by 6.

3 *Solve the equation.*

$$\frac{94 + 86 + 92 + 80 + 78 + x}{6} = 85$$

$$\frac{430 + x}{6} = 85$$ Combine like terms.

$$\frac{430 + x}{6} = \frac{85}{1}$$ Rewrite 85.

$$1(430 + x) = 6(85)$$ Cross multiply.

$$430 + x = 510$$

$$430 - 430 + x = 510 - 430$$ Subtract 430 from both sides.

$$x = 80$$

4 *Check your result.*

If 80 is the 6th grade, is the average equal to 85%?

$$\frac{94 + 86 + 92 + 80 + 78 + 80}{6} \stackrel{?}{=} 85$$

$$\frac{510}{6} \stackrel{?}{=} 85$$

$$85 = 85 \checkmark$$

Answer: For an 85 average, Laura needs to get 80% on the 6th test.

Try This *(For solution, see page 563.)*

On 4 Spanish tests, Tom achieved these grades: 82 89 78 86

What grade must Tom get on his next test so that his average for the 5 tests will be 85%?

Reading Averages From a Histogram

Which of the statistical averages are easy to read from a histogram?

Consider this histogram that displays the data for a group of 45 test scores.

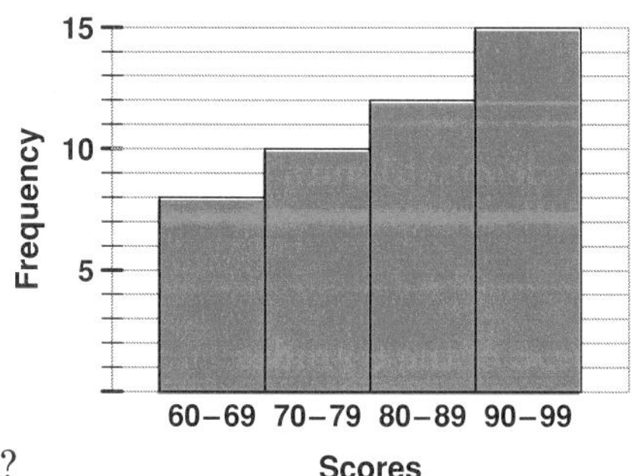

What can you conclude from the tallest bar?

The tallest bar tells you the interval that has the greatest frequency, which is called the ***modal interval***.

For this histogram, the modal interval is 90-99.

How can you find the interval that contains the median?

Since there are 45 scores, the median is the 23rd score.

median ↓

1 2 3 … 22 23rd 24 25 … 45

| 22 scores | | 22 scores |

$$\frac{45+1}{2} = \frac{46}{2} = 23$$

Add the frequencies of the intervals until you see where the 23rd score will fall.

$8 + 10 = 18$ ← not yet 23

$18 + 12$ ← passed 23

The 23rd score, or the median, lies in the interval 80-89.

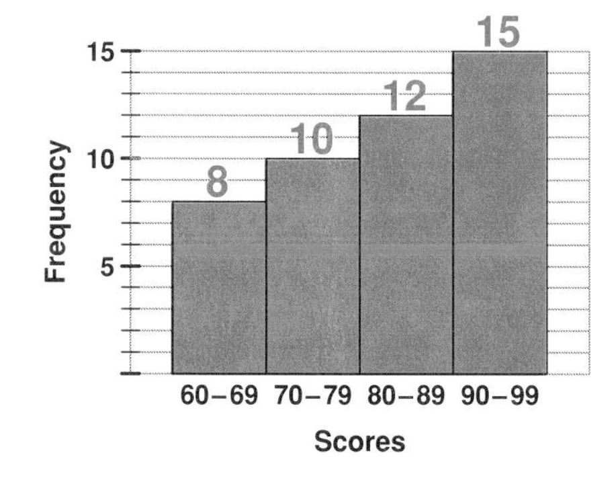

Reading Averages From a Histogram

- **The bar with the highest frequency shows the modal interval.**
- **To find the interval that contains the median value:**

Determine which value is the median value.

$$\text{median term} = \frac{\text{number of values} + 1}{2}$$

Beginning with the first bar, add the frequencies shown by the bars until you see the interval that contains the median value.

Example 6 From this histogram that displays the data for a group of 20 temperatures:

a. Find the modal interval.

The tallest bar tells you the interval with the highest frequency, which is the modal interval.

The modal interval is 40-44.

b. Find the interval that contains the median.

Determine which value is the median value.

$$\text{median term} = \frac{\text{number of values} + 1}{2}$$

$$= \frac{20 + 1}{2} = 10.5 \leftarrow \text{The median is halfway between the 10th and 11th values.}$$

Add the frequencies of the intervals until you see where the 10th and 11th values will fall.

$4 + 2 = 6$ ← not yet 10

$6 + 5 = 11$ ← the 11th value

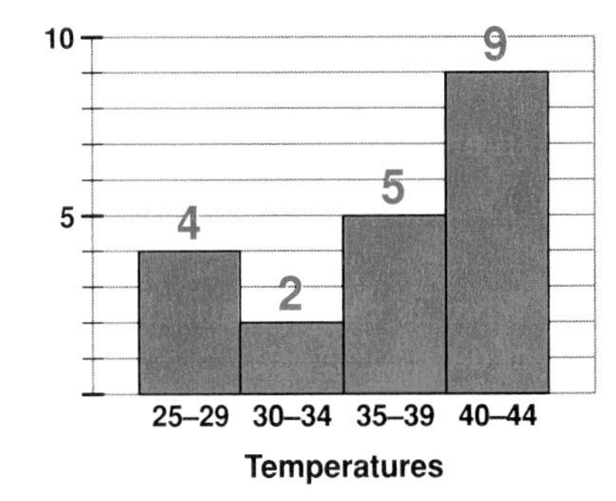

Answer: The median, or the value that is halfway between the 10th and 11th values, is in the interval 35-39.

Try This *(For solution, see page 563.)*

From this histogram that displays the data for a group of 25 ages, find:

a. the modal interval

b. the interval that contains the median

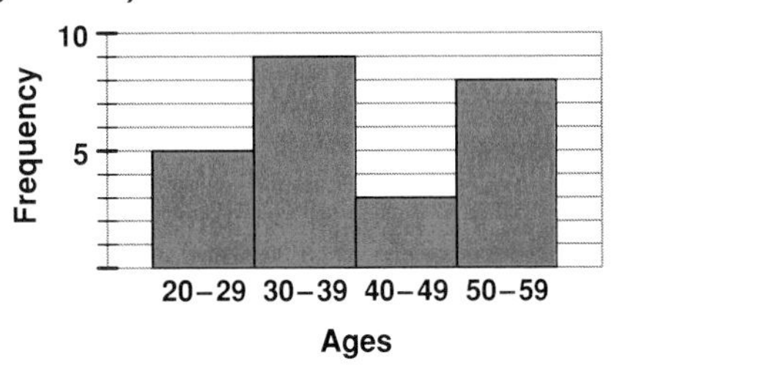

SELF-CHECK: SECTION 12.2

1. What is the statistical name for the familiar average in which you add the data values and divide the sum by the number of values?

   Answer: the mean

2. **a.** Make an observation about these 5 data values.

   10   12   11   17   9,000

   Answer: 4 of the 5 numbers are relatively close in size, and the 5th number is very much bigger.

   b. Explain why the mean is not a good representative for the average of these 5 numbers.

   Answer: 9,000 will overly influence the computation of the mean.

   c. What statistical average is a good representative of these 5 numbers?

   Answer: The median, the number that is exactly in the middle of the 5 numbers.

3. In order to find the median of a data set, what must you do first?

   Answer: Arrange the values in order.

4. What is the statistical name for the data value that occurs most often in a set of data?

   Answer: the mode

5. **a.** What does it mean to say that the mode for a data set is 12?

   Answer: 12 is the value that occurs most frequently in the data set.

   b. What does it mean to say that both 12 and 27 are the modes for a data set?

   Answer: 12 and 27 both have the same high frequency.

   c. What does it mean to say that a data set has no mode?

   Answer: All the data values occur with the same frequency.

6. **a.** Can you tell the mode of a data set by reading a histogram?

   Answer: No, the histogram will not tell you which single data score occurs most frequently.

   The histogram will tell you in which *interval* the data values occur most frequently.

   b. How do you read a histogram to find the modal interval?

   Answer: Look for the tallest bar.

EXERCISES: SECTION 12.2

1. **a.** Lay out polychips in these 3 groups:

   Rearrange the chips so that there are the same number of chips in the 3 groups. How many chips are in each group?

   b. Using the chips, what statistical average have you found for the numbers 3, 7, 8?

   c. In the same way, use polychips to find the mean of these data sets:

   (1) 5 8 14 **(2)** 6 9 12 13

2. Find the mean for each data set.

   a. time worked daily, in hours:

   6 8 8 7 8 5

   b. weekly earnings, in dollars:

   310 290 302 296 342 326

   c. test scores, as percents:

   75 87 82 90 81

   d. heights of basketball stars:

   5'10" 6' 6'3" 6'5" 5'11"

   e. weights of newborn babies:

   8lb. 2oz. 7lb. 12oz. 7lb. 10oz.
   6lb. 14oz. 6lb. 4oz. 5lb. 12oz.

3. **a.** Find the mean value of the 9 products in the body of this 3-by-3 multiplication table.

×	1	2	3
1	1	2	3
2	2	4	6
3	3	6	9

   b. Where does the mean value of the 9 products appear in the multiplication table?

   Study the position of the mean in the table. What patterns do you see that relate to the mean?

   c. Prepare a 5-by-5 multiplication table.

   From the table, predict the mean of the 25 products in the body of the table.

   Find the sum of the 25 products, and divide that sum by 25.
   Was your prediction correct?

4. Study each data set, and tell if you think the mean would be a fair representation of the average of the set. Explain your decision.

   If you think the mean is not a fair representation for the data set, tell which statistical average would be a better representation.

   a. 9 3 1 8 106

   b. 100 103 105 107 13

   c. 86 92 85 88 90

   d. The manager of Casual Footwear wants to reorder a model that has sold the following sizes today:

   $8\frac{1}{2}$ $7\frac{1}{2}$ 9 $7\frac{1}{2}$ 7 7 7

5. Find the median value for each data set.

   a. absences during a semester:

   7 2 3 1 6

   b. career home runs:

   5 8 10 11 10 23

   c. weights, in kilograms:

   2.9 1.8 2.0 2.6 1.7 2.4 1.5

   d. points won and lost:

   3 −4 5 −3 −4 6 0

   e. test scores:

   89 87 85 88 90 78

   f. temperatures, in °F:

   24 43 39 37 38 30 29 30

   g. thickness, in millimeters:

   1.04 1.21 1.10 1.01

6. For each data set, the given value of m represents one or more of the statistical averages (mean, median, mode).

   Tell which measure(s) m represents.

   a. 25 27 29 31 37 $m = 29$

   b. 50 55 56 62 65 67 $m = 59$

   c. 2 8 8 8 10 $m = 8$

   d. 17 19 21 23 80 82 $m = 22$

   e. 3 3 3 6 6 6 $m = 4.5$

7. Make up a set of:

a. 7 different numbers whose median is 20

b. 12 numbers whose median is 8

c. 5 numbers with a median of 6 and a mean of 5

8. Here are 3 data sets.

A:	4	5	10	4	12	5	
B:	7	10	−4	5	−2	8	4
C:	3	18	10	15	10	4	

a. Which sets have the same median value?

b. Which set has no mode?

c. For which set are the mean, median, and mode all equal?

9. From each set of data, one value is missing. Use the given mean to find the missing value.

a. The mean of 6 lengths is 58 cm.
The sum of 5 of the lengths is 300 cm.

b. The mean of 5 grades is 88%.
Four of the grades are 82, 94, 92, and 87.

10. Althea's teacher gave her the chance to raise her average by replacing her lowest grade with a project grade.

So far, Althea's grades are 82, 86, 78, 88, and 84.

a. What would her average be if she did not do a project?

b. What would her average be if her project grade is 85?

c. What grade must she get on her project to raise her average to 86?

d. What is the highest average Althea can get?

11. The histogram shows the distribution of grades in Hallie's Spanish class.

a. Which is the modal interval?

b. Which interval contains the median?

c. What is the probability that a student selected at random will have a grade in:

(1) the modal interval

(2) the interval that contains the median

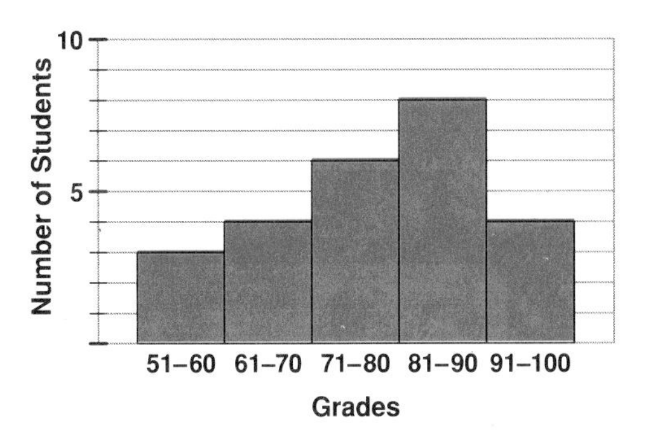

EXPLORATIONS

Here are the ages of 25 senior citizens who came to Family Service today for their flu shots.

Describe how you would find the mean of these 25 values.

65	67	68	65	65
67	73	73	73	67
68	68	70	70	68
70	70	70	67	68
67	73	68	68	67

Instead of adding 25 numbers, you can add fewer numbers if you consider the repetitions. For example, the value 65 appears 3 times, the value 67 appears 6 times, and so on.

Use a frequency table to summarize how many times each value appears.

Now describe how you would find the mean of the 25 values.

Find the mean of the 25 values.

Age	Frequency
65	3
67	6
68	7
70	5
73	4

Check your calculation against this calculation

$$\text{mean} = \frac{(65 \times 3) + (67 \times 6) + (68 \times 7) + (70 \times 5) + (73 \times 4)}{25}$$

$$= \frac{195 + 402 + 476 + 350 + 292}{25} = \frac{1{,}715}{25} = 68.6$$

Here's how to use an advanced calculator to compute the mean from a frequency table.

Step 1a Call up the statistics menu of the calculator.

Press: 2nd MATRX (STAT B)

Step 1b You want to clear any statistical data that may be in the calculator.

You want option 2, ClrStat, of DATA.

Press: 

```
CALC DRAW DATA
1:Edit
2:ClrStat
3:xSort
4:ySort
```

After data has been cleared, the calculator shows:

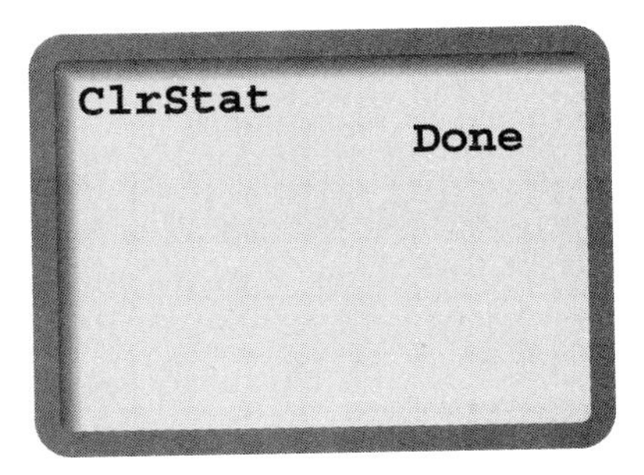

Step 1c Return to the statistics menu.

Press: 2nd MATRX (STAT B)

Step 1d Now you want to be able to enter data.

For this, you need option 1, Edit, of DATA

Press: ▶ ▶ ENTER

```
CALC DRAW DATA
1:Edit
2:ClrStat
3:xSort
4:ySort
```

Step 2 The calculator uses x to represent the data value and y to represent the frequency.

The frequency on the calculator has been preset to 1.

Refer to the frequency table. Enter each age and its corresponding frequency.

Press: 65 ENTER 3 ENTER
67 ENTER 6 ENTER
68 ENTER 7 ENTER
70 ENTER 5 ENTER
73 ENTER 4 ENTER

```
DATA
x1=
y1=1
```

```
DATA
x1=65
y1=3
x2=67
y2=6
x3=68
y3↓7
```

Step 3 Return to the statistics menu.

Press: 2nd MATRX (STAT B)

You want to ask for the calculation of the mean.

You want option 1 of CALC.

Press: ENTER ENTER

```
CALC DRAW DATA
1:1-Var
2:LinReg
3:LnReg
4:ExpReg
5:PwrReg
```

The calculator is displaying a whole array of statistical information about this data set.

Look for the mean value, which you already know is 68.6.

Note that the calculator displays the mean value first, calling it $\bar{x}$.

Look down the list of numbers to recognize other numbers from the calculation you did before: 1,715 and 25.

```
1-Var
 x̄=68.6
 Σx=1715
 Σx²=117793
 Sx=2.449489743
 σx=2.4
 n=25
```

Find the Mean From a Frequency Table

- **By Hand, Using a Simple Calculator to Assist**
  Multiply each data value by its frequency.
  Add these products.
  Divide the sum of the products by the total frequency.
- **On an Advanced Calculator**
  Call up the statistics menu.
  Clear previous data.
  Enter each data value and its corresponding frequency.
  Take option 1 of CALC.
  The mean is displayed as $\bar{x}$.

Find the mean value of each data set that is given in the frequency tables that follow.

First find the mean by hand, using a simple calculator to assist.

Then use an advanced calculator to find the mean.

a. amounts of breakfast calories for 10 dieters

Calories	Frequency
175	2
190	4
200	3
215	1

b. heights, in inches, of 30 men in a basketball gym class

Height	Frequency
68	5
69	7
70	3
71	10
72	5

12.3 Quartiles and Box-Whisker Plots

Quartiles

At the annual county-wide gathering of firefighters, there were 50 Dalmatians, the firehouse mascots.

The median age of the dogs was 7 years.

How many of the 50 dogs were 7 years old or younger?

The median age tells you the exact middle value. That is, 25 dogs were at the median age or below, and 25 dogs were at the median age or above.

median age of the 50 dogs
is 7 years
↓

25 dogs
7 years or younger

25 dogs
7 years or older

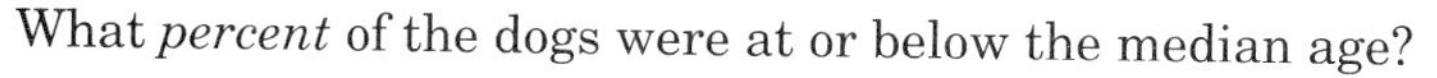

What *percent* of the dogs were at or below the median age?

Exactly half, or 50%, of the dogs were at or below the median age.

The median breaks a data group into 2 equal parts.
50% of the data is at or below the median value.
50% of the data is above the median value.

Further analysis of the ages of the dogs breaks the data into 4 equal parts, called *quartiles*.

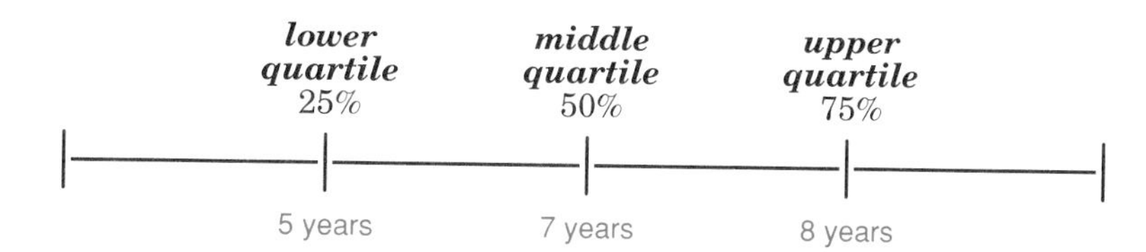

What information about the ages of the dogs can you read from the breakdown into quartiles?

The lower quartile shows that 25% of the dogs were 5 years old or younger.

The middle quartile, or median, shows that 50% of the dogs were 7 years old or younger.

The upper quartile shows that 75% of the dogs were 8 years old or younger.

The quartiles break a data group into 4 equal parts.

The lower quartile shows the value at or below which 25% of the values fall.

The middle quartile, or median, shows the value at or below which 50% of the values fall.

The upper quartile shows the value at or below which 75% of the values fall.

Example 1 Ann Reid employs 11 full-time workers in her interior design business.

From the list of annual salaries, calculate the quartile values.

To do the calculations for quartiles, the data values must be arranged in order.

Employee	*Annual Salary*	
Susan A.	$32,500	
Sato I.	27,380	
Carl P.	25,450	← upper quartile
Anna M.	25,450	
Carmen R.	21,950	
Lynn T.	19,210	← median value
Fred W.	17,560	
Heidi S.	16,225	
Juan M.	14,890	← lower quartile
Mike C.	13,670	
Clyde J.	12,400	

median term $= \frac{\text{number of values} + 1}{2} = \frac{11 + 1}{2} = \frac{12}{2} = 6$

The median is the 6th value in the ordered list: $19,210

lower quartile: Find the middle, or median, of the lower 5 values.

middle of lower half $= \frac{\text{number of values} + 1}{2} = \frac{5 + 1}{2} = \frac{6}{2} = 3$

The lower quartile is the 3rd value from the bottom in the ordered list: $14,890

upper quartile: Find the middle, or median, of the upper 5 values.

The upper quartile is the 3rd value from the top in the ordered list: $25,450

Example 2 Refer back to the list of Ann Reid's employees to answer these questions about the salaries.

a. What is the median salary and who earned it?

Answer: The median salary is $19,210, earned by Lynn T.

b. If Ms. Reid decides to raise the median salary by 5% next year, what would the median salary become?

raise = $19,210 x 0.05 = $960.50

new median salary = $19,210 + $960.50 = $20,170.50

c. How does Mike C.'s salary rank in the list?

Answer: Mike C.'s salary is below the lower quartile.

d. How much more than the lower quartile is the upper quartile?

difference = upper quartile – lower quartile

= $25,450 – $14,890

= $10,560

e. What is the range of salaries?

range = highest value – lowest value

= $32,500 – $12,400

= $20,100

f. What is the mean salary, rounded to the nearest dollar?

$$\text{mean} = \frac{\text{sum of the 11 salaries}}{11} = \$20{,}607$$

g. Whose salary makes the mean higher than the median?

Answer: Susan A.'s high salary raises the mean value.

To Find the Quartile Values of a Data Set

1. Arrange the data in order.

2. Find the median value for the entire data set.
The median divides the data set into 2 equal parts, a lower half and an upper half.

3. Find the middle value, or median, for each half of the data.
The lower quartile is the middle value of the lower half.
The upper quartile is the middle value of the upper half.

Example 3 In 1991, eighth-grade students from 14 countries across the world took a Math Proficiency Test. The highest possible score was 500.

From the list of average scores, calculate the quartile values.

Arrange the data in order to do the calculations.

You can number the ordered list, called ***ranking***. Countries with the same score have the same rank. Where is the United States on this list? Ouch!

Rank	*Country*	*Score*	
1	1. Taiwan	285	
2	2. Korea	283	
3	3. Switzerland	279	
3	4. Soviet Union	279	← upper quartile
4	5. Hungary	277	
5	6. France	273	
6	7. Israel	272	
			← median value
7	8. Canada	270	
8	9. Ireland	269	
8	10. Scotland	269	
9	11. Slovenia	266	← lower quartile
10	12. Spain	263	
11	13. United States	262	
12	14. Jordan	246	

The median is halfway between the 7th and 8th values: 271

The median divides the set of 14 values into two halves, each with 7 values.

The upper quartile is the middle value of the upper 7 values: 279

The lower quartile is the middle value of the lower 7 values: 266

Try This *(For solution, see page 564.)*

Spectacular as it is, Niagara Falls, at 182 feet high, is tiny compared to other falls across the world.

Here is an alphabetic list of the 8 highest waterfalls in the world.

Falls	*Height (ft.)*
Angel	3,212
Cuquenán	2,000
Great	1,600
Mardalsfoss	2,149
Ribbon	1,612
Sutherland	1,904
Tugela	2,014
Yosemite	2,425

a. Arrange the data in order, and rank them.

b. **(1)** Calculate the median. **(2)** How does the median divide the data set?
(3) Calculate the lower quartile. **(4)** Calculate the upper quartile.

c. What is the lowest data value? the highest data value?

(Keep this quartile information, to be used again.)

Box-Whisker Plots

Here is a graph of the data about the salaries of Ann Reid's employees. (See Example 1.)

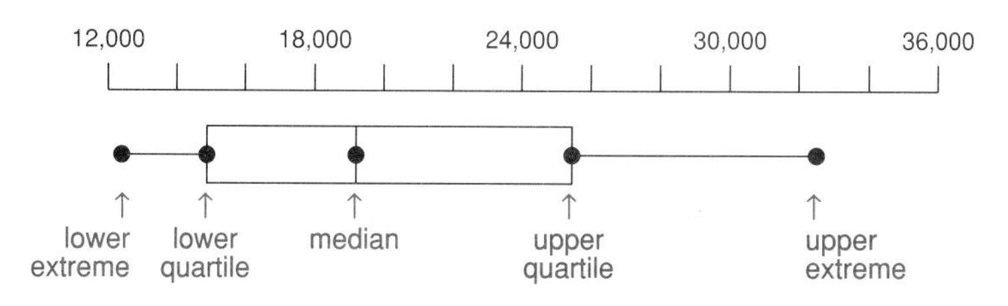

This kind of graph shows 5 important values from the data set:

1. the lowest value, called the *lower extreme*
2. the lower quartile
3. the median
4. the upper quartile
5. the highest value, called the *upper extreme*

The three quartile values are connected to make a *box*.
The two extreme values are connected to the box by lines, called *whiskers*.
This kind of graph is called a ***box-whisker plot***.

Example 4 Make a box-whisker plot for the Math Proficiency data for the 14 countries that participated. (See Example 3.)

1. The scores have already been put in order.
2. The quartile values have already been calculated.

   median = 271   lower quartile = 266   upper quartile = 279

3. Choose an appropriate scale, to include the extreme values.

   240 245 250 255 260 265 270 275 280 285 290

4. Use dots to mark 5 values: the 2 extremes and the 3 quartiles

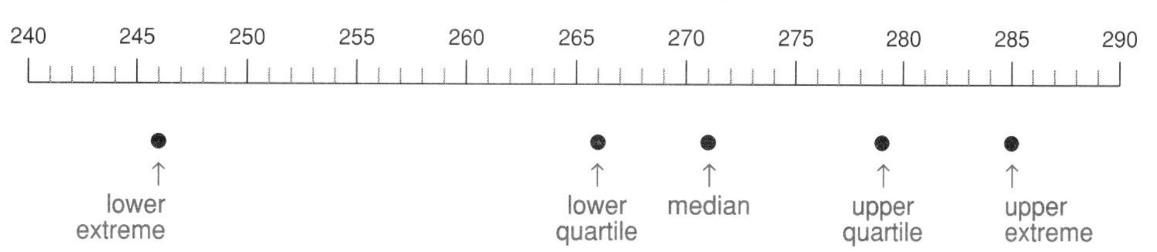

5. Connect the quartile values to make a box.

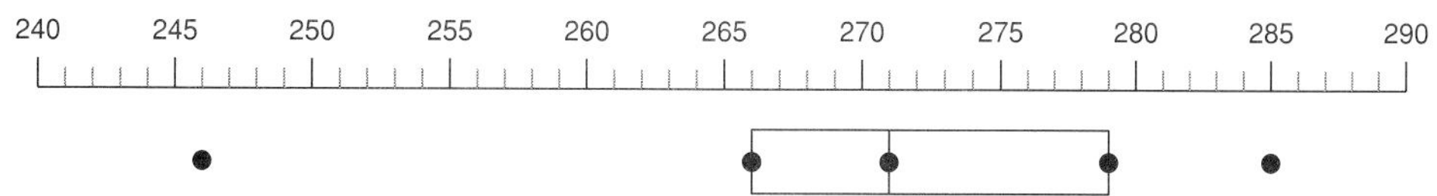

6. Use line segments to connect the extreme values to the box.

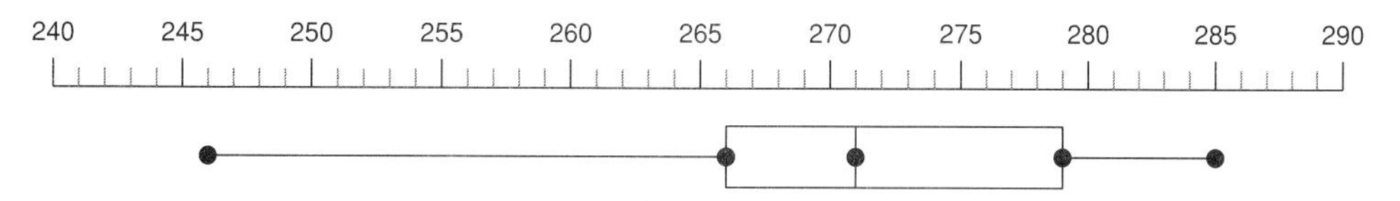

Locate where the score for the United States would fall on this graph.

Making a Box-Whisker Plot for a Data Set

1. **Arrange the values in order.**
2. **Calculate the median, the lower quartile, the upper quartile.**
3. **Choose an appropriate scale.**
4. **Use dots to mark 5 values: the 2 extremes and the 3 quartiles**
5. **Connect the quartile values to make a box.**
6. **Use line segments to connect the extreme values to the box.**

Example 5 Make a box-whisker plot for these 13 test scores.

73 62 93 87 83 94 77 80 79 95 68 85 81

Arrange the scores in order, and calculate the 3 quartiles

62 68 73 77 79 80 81 83 85 87 93 94 95

↑ lower quartile = 75 ↑ median ↑ upper quartile = 90

Using an appropriate scale, mark the 2 extremes and the 3 quartiles.

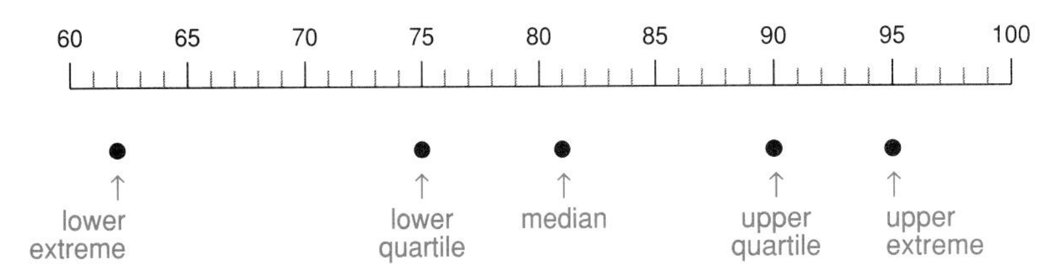

Connect the quartile values to make a box.

Use line segments to connect the extreme values to the box.

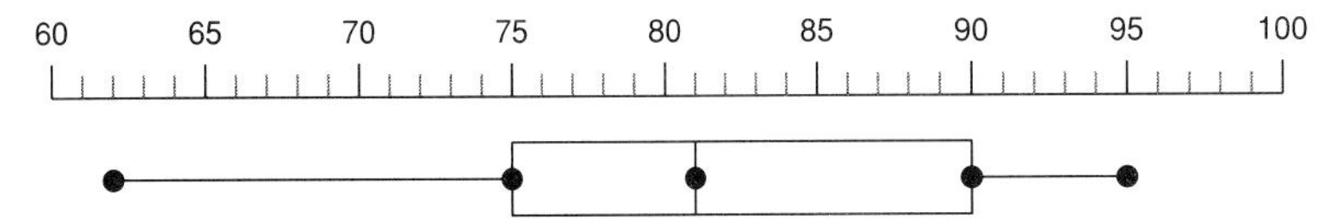

Example 6 Here are box-whisker plots of the average yearly rainfall in Seattle, Washington at the start of three decades: 1970, 1980, 1990. The data is given by month, in inches.

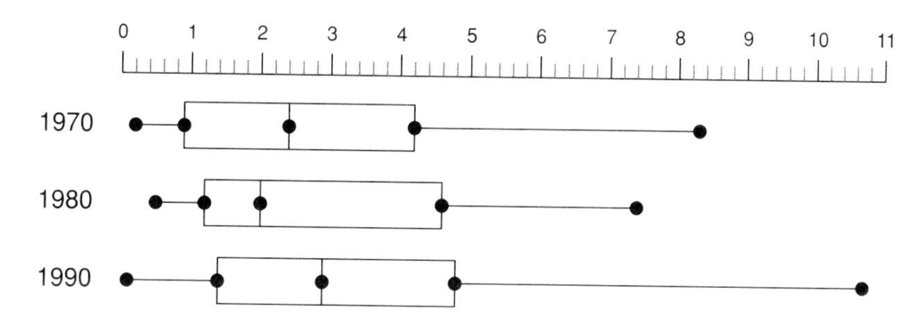

Compare the 3 plots.

a. In which year was the rainfall greatest? least?

As shown by the positions of the boxes, the rainfall was the greatest in 1990 and the least in 1970.

b. In which year was the highest monthly rainfall the greatest?

As shown by the extreme values, in 1990, the highest monthly rainfall was 10.7 inches, compared to a monthly high of 8.3 inches in 1970 and 7.4 inches in 1980.

c. In which year was the median lowest?

In 1980, the median was at 2.0, compared to 2.4 in 1970 and 2.9 in 1990.

Try This *(For solution, see page 564.)*

Use the quartile information you found for the waterfall data (page 548).

To make a box-whisker plot for the data:

(1) Choose an appropriate scale, to include the extreme values.

(2) Use dots to plot

the quartile data
- the lower quartile
- the median
- the upper quartile

the extreme values
- the lowest value
- the highest value

(3) Connect the quartile data to make a box.

(4) Use line segments to connect the extreme values to the box.

SELF-CHECK: SECTION 12.3

1. **a.** How does the median divide a data set?

   Answer: into 2 equal parts, a lower half and an upper half

   b. What is the middle value of the lower half called?

   Answer: the lower quartile

   c. What is the upper quartile?

   Answer: the middle value of the upper half

   d. Into how many parts do the quartiles divide a data set?

   Answer: 4 equal parts

   e. What are the names of the 3 quartiles?

   Answer: lower quartile, median, upper quartile

2. What does the rank of a score show?

   Answer: the location of the score in an ordered list

3. **a.** What is the lowest value of a data set called?

   Answer: the lower extreme

   b. What is the upper extreme of a data set?

   Answer: the highest value of the set

4. **a.** Which data values are connected to make the box of a box-whisker plot?

   Answer: the 3 quartile values

   b. What are the whiskers?

   Answer: the line segments that connect the extreme values to the box

EXERCISES: SECTION 12.3

1. Arrange each data set in order, and calculate:
   - the median
   - the lower quartile
   - the upper quartile

   a. ratings on a taste test

   3 8 5 6 4 6 7

   b. heights, in inches, of basketball players

   68 74 69 70 73 72 71 70 71 71 72

   c. ages of cousins at a family picnic

   9 17 3 $14\frac{1}{2}$ 26 5 $12\frac{1}{2}$ 21 16

d. top medal winners in the 25th Summer Olympics

Team	*Number of Medals*
Australia	27
China	54
Cuba	31
France	29
Germany	82
Great Britain	20
Hungary	30
Korea	29
Japan	22
Spain	22
U.S.A.	108
United Team	112

e. the most populous U.S. cities, with the populations rounded to the nearest tenth of a million

City	*Population*	*City*	*Population*
Chicago	2.8	New York	7.2
Dallas	1.0	Philadelphia	1.6
Detroit	1.0	Phoenix	1.0
Houston	1.6	San Antonio	0.8
Los Angeles	3.5	San Diego	1.1

f. National Basketball Association individual scoring leaders, 1992-93 season

Player	*Team*	*Score*
Barkley	Phoenix	25.6
Dumars	Detroit	23.6
Ewing	New York	24.2
Jordan	Chicago	32.6
Malone	Utah	27.0
Manning	Los Angeles	22.8
Olajuwon	Houston	26.1
O'Neal	Orlando	23.4
Robinson	San Antonio	23.4
Wilkens	Atlanta	29.9

2. Here is an alphabetic list of the 10 most populous U.S. states, with the populations rounded to the nearest tenth of a million.

State	*Population*	*State*	*Population*
California	30.4	New Jersey	7.8
Florida	13.3	New York	18.1
Illinois	11.5	Ohio	11.0
Michigan	9.4	Pennsylvania	12.0
North Carolina	6.7	Texas	17.1

a. Arrange the populations in order, and rank the states.

b. Which state has the largest population?

c. What is the population of the state that ranks 10th?

d. What is the range of these populations?

e. What is the median population? Which state populations are closest to the median?

f. How much more than the lower quartile is the upper quartile?

g. What is the mean population of these states?

3. Make a box-whisker plot for each of the following data sets, by:
- arranging the data in order and calculating the quartile values
- choosing an appropriate scale, to include the extreme values
- using dots to plot the quartile data and the extreme values
- connecting the quartile data to make a box
- using line segments to connect the extreme values to the box

a. this week's rentals at Vox Video

38 40 40 41 47 58 24

b. registers of kindergarten classes in North Plains

16 18 18 19 20 23 24 26

c. ages of a group entering an amusement park

13 18 20 10 7 3 4 13 17 27 10 11 26 4

4. Make a box-whisker plot for each of the following sets of data.

a. 1992 top earnings for chief executive officers of U.S. corporations, to the nearest million dollars

Use a scale from 20 to 140, in intervals of 20.

Corporation	*Earnings of CEO*
Hospital Corp of America	127
Primerica Corp	68
Toys R Us	64
U. S. Surgical	62
Mirage Resorts, Inc.	38
Medco Containment Services	30
General Dynamic	29
Torchmark	27
U. S. T., Inc.	25

b. altitudes of the highest U.S. cities, to the nearest ten feet

Use a scale from 4,500 to 7,000 in intervals of 500.

City	*Height*
Albuquerque, NM	4,950
Butte, MN	5,770
Carson City, NV	4,680
Cheyenne, WY	6,100
Colorado Springs, CO	5,980
Denver, CO	5,280
Gallup, NM	6,540
Flagstaff, AZ	6,900
Pueblo, CO	4,690
Santa Fe, NM	6,950

c. the windiest U.S. cities, average wind speeds in miles per hour

Use a scale from 10.5 to 13.5, in intervals of 0.5

City	*Wind Speed*
Boston, MA	12.9
Buffalo, NY	12.4
Cheyenne, WY	12.8
Cleveland, OH	10.9
Dallas, TX	10.9
Des Moines, IA	11.2
Great Falls, MT	13.1
Milwaukee, WI	11.8
Oklahoma City, OK	13.0
Omaha, NE	10.9
Providence, RI	10.9
Wichita, KS	12.7

Note that Chicago, nicknamed the "windy city," is not among the top 12.

Chicagos's average wind speed is 10.4 miles per hour.

5. This box-whisker plot is a summary of data obtained by asking 21 children to rate the taste of a new brand of soda on a scale from 1 (terrible) through 10 (terrific).

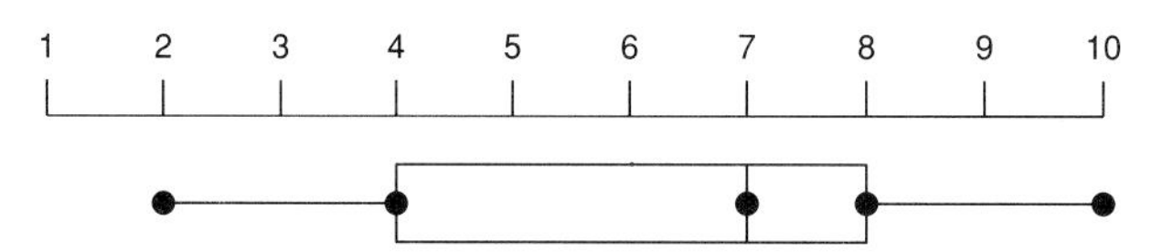

Interpret the plot to answer these questions:

a. What is the range? **b.** What is the median?

c. What is the difference between the upper and lower quartiles?

d. For a child picked at random, between which two consecutive values was the rating most likely to fall?

6. Here are box-whisker plots of the number of home runs hit by teams in the American League for 3 seasons: 1990, 1991, 1992.

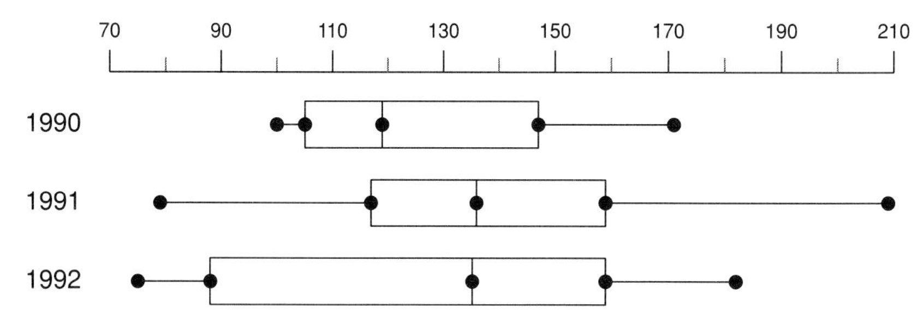

Compare the 3 plots.

a. In which year did the league have the greatest number of home runs?

b. In which year was the lowest number of home runs best?

c. In which 2 years was the upper quartile the same?

d. In which year was the lower quartile the worst?

EXPLORATIONS

Dan was told that his math score on the SAT was at the 80th ***percentile***. This means that 80% of all the scores for the test that Dan took were at or below Dan's score.

a. Some statistical measures that you have already used correspond to percentiles.

What percentile is:

the median?
the lower quartile?
the upper quartile?

b. Here are the heights, in inches, of 25 seventh-grade students.

Find:

(1) the 50th percentile (the median)

(2) the 60th percentile

(3) the 90th percentile

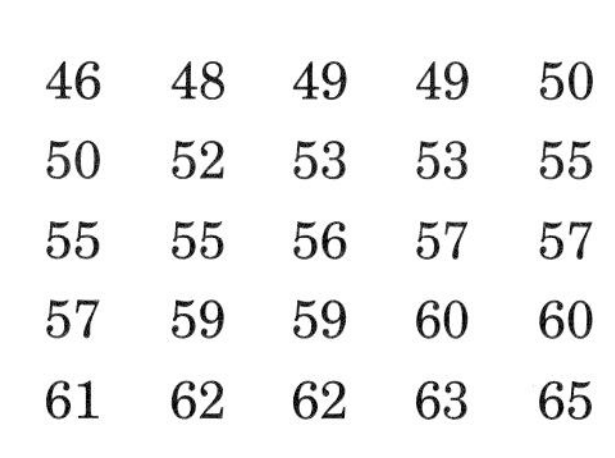

46	48	49	49	50
50	52	53	53	55
55	55	56	57	57
57	59	59	60	60
61	62	62	63	65

c. This histogram is a summary of the weights, in pounds, of 50 athletes.

Tell which interval contains:

(1) the lower quartile

(2) the 40th percentile

(3) the median

(4) the 95th percentile

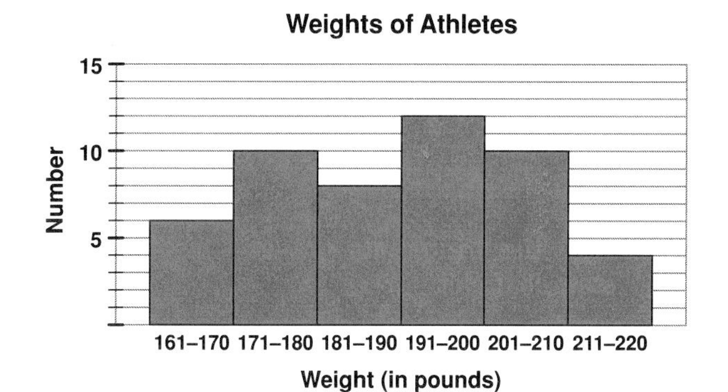

Things You Should Know After Studying This Chapter

KEY SKILLS

12.1 Find the range of a data set.
Make a frequency table to group a data set in intervals.
Use a frequency table to make a histogram.
Make a stem-leaf plot for a data set.
Read a histogram or a stem-leaf plot to answer questions about a data set.

12.2 Calculate the mean and median for a data set.
Identify a mode for a data set.
Identify the modal interval of a histogram.
On a histogram, find the interval that contains the median.

12.3 Calculate the 3 quartiles for a data set.
Make a box-whisker plot for a data set.
Read a box-whisker plot to answer questions about a data set.

KEY TERMS

12.1 statistics • intervals • range • histogram • stem-leaf plot

12.2 mean • median • mode • modal interval

12.3 quartile • extreme values • ranking • box-whisker plot

CHAPTER REVIEW EXERCISES

1. Carlita's bowling scores in 24 games are listed.

138	142	136	144	129	148	138	143
152	123	168	156	160	151	146	137
129	147	150	131	125	157	150	151

a. Make a frequency table, tallying the data in intervals of 10, starting with 121-130.
b. Which interval has the greatest frequency?
c. What is the range of the data?

2. The percent of family income spent for child care varies considerably among income groups, as shown in this table.

Monthly Income	less than $1,250	$1,250 – 2,499	$2,500 – 3,749	$3,750 or more
Percent for Child Care	21%	9%	7%	5%

Data rounded to the nearest percent.

a. Make a histogram for the data.
b. How does the percent spent on child care vary with respect to the income? Explain.
c. If the Hernandez monthly income is $3,300, about how much should they plan on spending for child care?

3. For each set of data:
- make a frequency table
- use the frequency table to make a histogram

a. daily temperature highs, in degrees Fahrenheit, for the last two weeks of March

Use intervals 55-59, 60-64, and so on.

63 72 58 60 67 73 78
68 63 54 63 70 75 68

b. boxes of cookies sold by 20 Girl Scouts

24 27 32 18 33 38 26 19 28 36
40 20 31 37 26 25 30 35 40 38

4. This stem-leaf plot shows the number of newspapers sold at 10 local newsstands. Make a list of the 10 pieces of data.

Stem	Leaf
6	8
7	0 5 9
8	7
9	3 6 8
10	5 5

5. Here is a list of activities that burn fewer than 100 calories in 20 minutes.

Activity	*Calories*
Baseball	81
Calisthenics	75
Dancing	74
Driving	45
Golf	85

Activity	*Calories*
Housework	63
Ice Skating	54
Office Work	45
Sailing	51
Sleeping	18

Activity	*Calories*
Softball	65
Standing	23
Walking (2 mph)	83
Watching TV	18

Make a stem-leaf plot for the data.

CHAPTER REVIEW EXERCISES

6. Match each data set in *Column A* with a description in *Column B*.

	Column A		*Column B*
a.	1 3 5 5 7 9	**(1)**	median = 2(mode)
b.	1 2 3 4 5 7 11 12 12 13	**(2)**	mean = median = mode
c.	1 2 3 3 5 6 7 8 9 10 12	**(3)**	median < mean
d.	4 6 6 7 8 12 13	**(4)**	mean > median > mode

7. The National Safety Council compiles statistics on fatal automobile accidents, in which one or more people died.

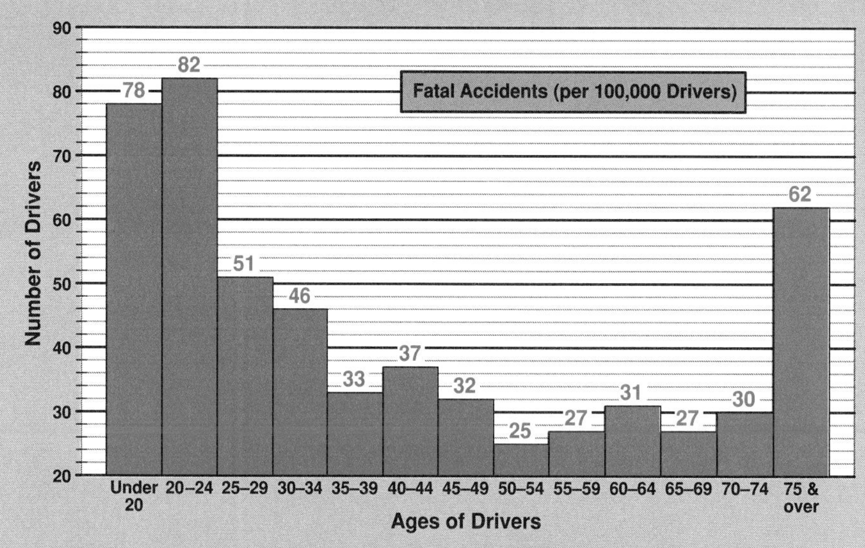

a. In which 3 age groups were the safest drivers?

b. How would you explain the relationship between unsafe driving and the age of the driver?

c. Which is the modal interval?

d. For the 561 accidents recorded, which interval contains the median?

8. A radio commercial offered a free brochure about health care in the United States.

The responses for one day, counted by state, are shown in the stem-leaf plot.

a. Find:

(1) the range **(2)** the mode **(3)** the median

b. What percent of the states had fewer than 100 responses?

c. What is the probability that a state chosen at random had over 150 responses?

Stem	Leaves
9	3 6
10	2 4 7 9
11	6 8
12	5 5 7 8 8
13	2 3 5 6 7 7
14	0 0 2 3 5
15	7 9 9 9
16	7 8 9
17	1 1 2 2 6 7 9
18	0 1 1 3
19	2 4 5 6 6
20	3 4 5

CHAPTER REVIEW EXERCISES

9. For each of the following data sets:

- Calculate the median, and the lower and upper quartile.
- Make a box-whisker plot for the data.

a. highest recorded numbers of families, to the nearest tenth of a million, watching a Super bowl on TV

Use a scale from 35 to 42, in intervals of 1.

Superbowl	*Year*	*Watching*	*Superbowl*	*Year*	*Watching*
XXVII	1993	42.0	XIX	1985	39.4
XX	1986	41.5	XXIII	1989	39.3
XVII	1983	40.5	XVIII	1984	38.8
XXI	1987	40.0	XIV	1980	35.3
XVI	1982	40.0	XIII	1979	35.1

b. the lengths, to the nearest 10 feet, of rivers of the world

Use a scale from 2,500 to 4,500, in intervals of 500.

River	*Length*	*River*	*Length*
Amazon	4,000	Mekong	2,600
Amur	2,740	Mississippi	3,700
Chang Jiang	3,960	Ob-Irtysh	3,360
Congo	2,720	Niger	2,590
Huang	2,900	Nile	4,140
Lena	2,730	Yenisey	2,540

10. Here are box-whisker plots of the average number of days per month of precipitation (rain, snow, sleet) for 3 cities.

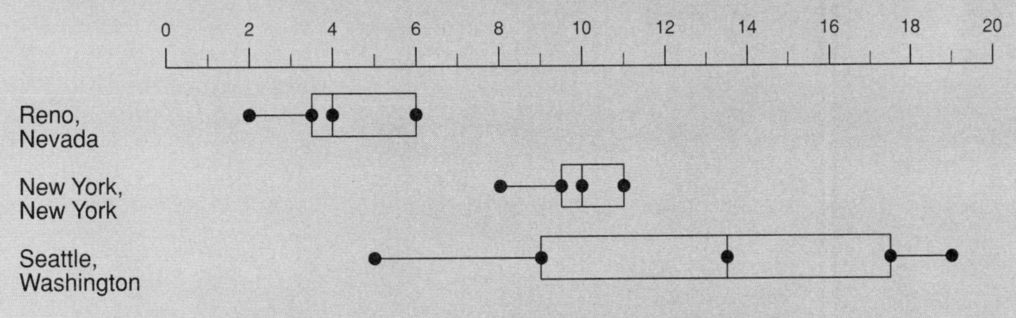

a. The plots for Reno and for New York do not have right-hand whiskers. What does this mean?

b. For which two cities is the range of data about the same?

c. Compare the upper extreme for Reno to a value for Seattle. How does the geography of these cities contribute to these values?

SOLUTIONS FOR TRY THESE

12.1 Histograms and Stem-Leaf Plots

Page 515

1. **a.** The lowest age is 21.
   b. The highest age is 59.
   c. The range is 59 - 21, or 38 years.

2.

Interval	Tally	Frequency
20 – 29	𝍸 III	8
30 – 39	𝍸	5
40 – 49	𝍸 I	6
50 – 59	𝍸 I	6
	TOTAL ▸	25

Page 518

1.

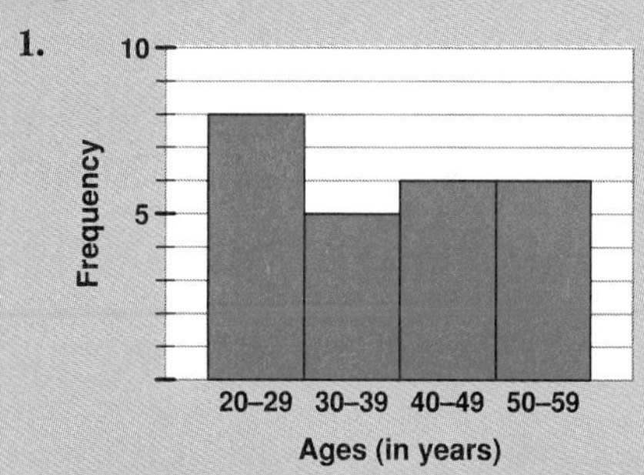

2. **a.** 20-29, the interval with the highest frequency
   b. 30-39, the interval with the lowest frequency
   c. $P(40\text{-}49) = \frac{6}{25}$
   d. 40-49 and 50-59

Page 520

1. To determine the stems, note that the data can be arranged in 4 groups: the 20's, the 30's, the 40's, the 50's

```
2 |
3 |
4 |
5 |
```

Read through the data to put on the leaves. First use the order that appears in the data.

```
2 | 1 3 3 8 6 9 8 9
3 | 2 3 5 7 9
4 | 0 9 8 1 4 6
5 | 9 1 4 3 7 8
```

Then rewrite the display with the leaves in order.

```
2 | 1 3 3 6 8 8 9 9
3 | 2 3 5 7 9
4 | 0 1 4 6 8 9
5 | 1 3 4 7 8 9
```

Ages (in Years)

2. **a.** Look in the 20's row.
   Answer: There are 3 ages less than 25.
   b. In the 30's row, there are 2 ages less than 35.
   And, all the ages in the 20's row are less than 35.
   There are 8 ages in the 20's row.
   Answer: There are 2 + 8, or 10, ages less then 35.
   c. In the 40's row, there are 3 ages greater than 45.
   And, all the ages in the 50's row are greater than 45.
   There are 6 ages in the 50's row.
   Answer: There are 3 + 6, or 9, ages greater then 45.

(Solutions continue)

SOLUTIONS FOR TRY THESE

12.2 *Statistical Averages*

Page 532

a. To find the mean: add the 7 scores and divide the sum by 7

$$\frac{155 + 172 + 164 + 138 + 172 + 117 + 172}{7} =$$

$$\frac{1{,}090}{7} \approx 156$$

Answer: To the nearest integer, the mean score is 156.

b. The low score brings down the mean value.

c. To find the median: arrange the 7 scores in order and choose the middle score

117 138 155 164 172 172 172

↑ median (164)

Answer: The median score is 164.

d. The low score does not bring down the median. There are 3 scores below the median and 3 scores above.

e. The score with the highest frequency is the mode.

Answer: The mode is 172.

Page 534

1. Arrange the data in order.

62 65 66 71 73 81 85 92

↑ median (between 71 and 73)

Since there are 8 numbers, there is no *middle* number.

The median is halfway between 71 and 73

Answer: The median score is 72.

2. **a.** No mode, each score appears only once.

b. 0.245 is the mode, the batting average with the highest frequency.

c. Both 5"4' and 5"9' are modes, since they appear with the same high frequency.

Page 535

Write an equation to model the problem.

Let x = the grade on the 5th test.

$$\frac{82 + 89 + 78 + 86 + x}{5} = 85$$

$$\frac{335 + x}{5} = \frac{85}{1}$$ Combine like terms. Rewrite 85.

$$1(335 + x) = 5(85)$$ Cross multiply.

$$335 + x = 425$$

$$335 - 335 + x = 425 - 335$$ Subtract 335 from both sides.

$$x = 90$$

Check:

$$\frac{82 + 89 + 78 + 86 + 90}{5} \stackrel{?}{=} 85$$

$$\frac{425}{5} \stackrel{?}{=} 85$$

$$85 = 85$$ ✓

Answer: Tom must get 90 on the next test.

Page 537

a. 30-39 is the modal interval, the interval with the highest frequency.

b. Determine which value is the median value.

$$\text{median term} = \frac{\text{number of values} + 1}{2}$$

$$= \frac{25 + 1}{2} = \frac{26}{2} = 13$$

The median value is the 13th value.

Add the frequencies of the intervals until you see where the 13th value will fall.

$5 + 9 = 14$ ← includes the 13th value

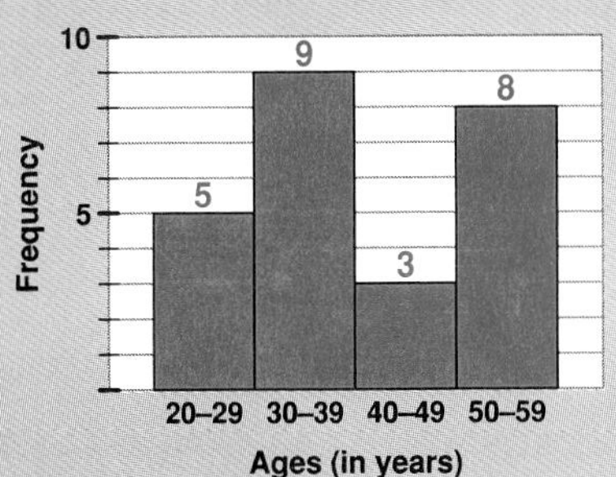

Answer: The interval 30-39 contains the median.

SOLUTIONS FOR TRY THESE

12.3 Quartiles and Box-Whisker Plots

Page 548

a.

Rank	*Falls*	*Height*	*Rank*	*Falls*	*Height*
1	Angel	3,212	5	Cuquenán	2,000
2	Yosemite	2,425	6	Sutherland	1,904
3	Mardalsfoss	2,149	7	Ribbon	1,612
4	Tugela	2,014	8	Great	1,600

b. **(1)** The median of the 8 heights is halfway between the 4th and 5th heights.

$$\frac{2{,}014 + 2{,}000}{2} = 2{,}007$$

(2) The median divides the data set into 2 equal parts, each with 4 data values.

(3) The lower quartile is the middle value of the lower 4 values.

That is, the lower quartile is halfway between the 2nd and 3rd values of the lower half.

$$\frac{1{,}612 + 1{,}904}{2} = 1{,}758$$

(4) The upper quartile is the middle value of the upper 4 values.

That is, the upper quartile is halfway between the 2nd and 3rd values of the upper half.

$$\frac{2{,}425 + 2{,}149}{2} = 2{,}287$$

Rank	*Falls*	*Height*	
1	Angel	3,212	
2	Yosemite	2,425	← upper quartile = 2,287
3	Mardalsfoss	2,149	
4	Tugela	2,014	← median = 2,007
5	Cuquenán	2,000	
6	Sutherland	1,904	← lower quartile = 1,758
7	Ribbon	1,612	
8	Great	1,600	

c. The lowest data value is 1,600 feet.

The highest data value is 3,212 feet.

Page 551

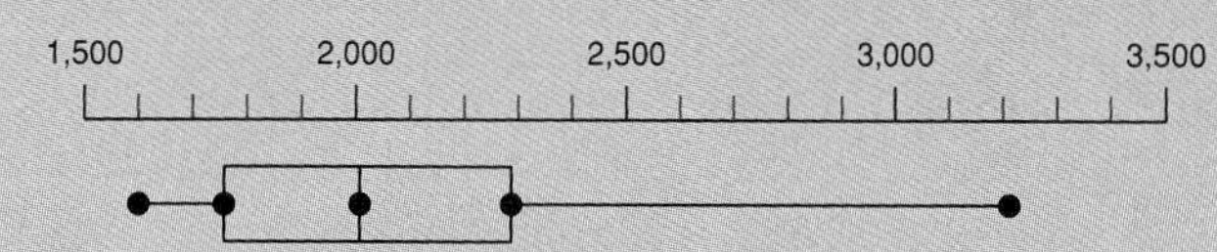

Polynomials

Use algebra chips to represent:

- the set of terms $3n^2, 4n, 5$
- the set of terms $2n^2, 2n, 3$
- the sum of the two sets
- the set of terms $2n^2, -4n, 2$
- the set of terms $4n^2, 2n, -3$
- the sum of the two sets

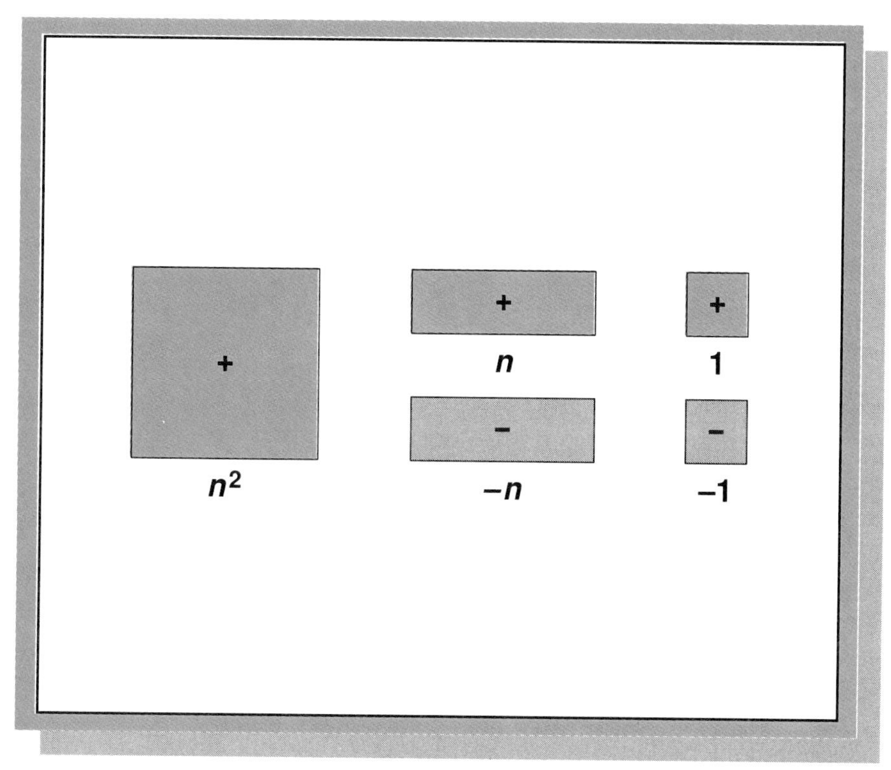

13.1 Addition

Adding Like Terms

Max placed a bookcase that is 4 feet 4 inches high on top of a cabinet that is 2 feet 7 inches high. How high is the piece of stacked furniture?

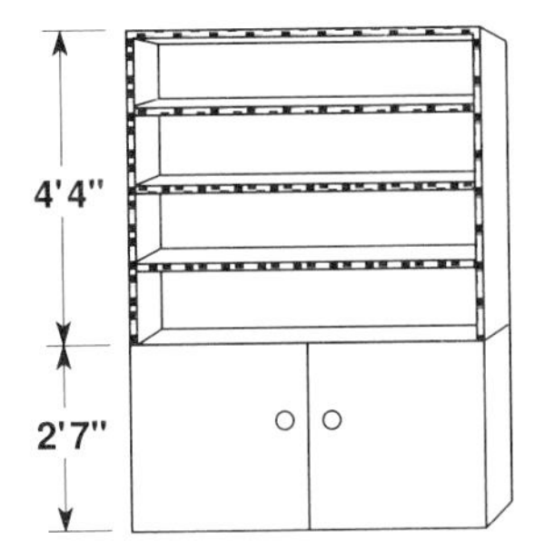

To get the height of the stack, add the like units:

	4 feet	4 inches
+	2 feet	7 inches
	6 feet	11 inches

In Arithmetic, only like units can be added.

How long is line segment AB?
The total length of $\overline{AB}$ is
$2n + 4n$, or $6n$ units

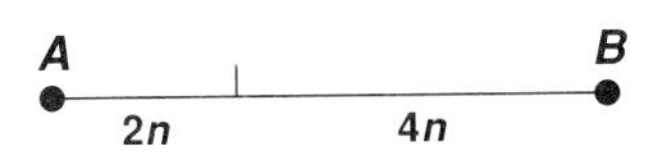

In Algebra, only like terms can be added.

Recall that a term has 3 parts.

exponent
$7n^2$
coefficient base

Like terms have the same base and the same exponent.

Like Terms		*Unlike Terms*	
$2n$ and $4n$	$2n^2$ and $4n^2$	$2n$ and $4x$	$2n^2$ and $2n^3$

To add like terms,
add the coefficients.

$$2n + 4n$$
$$= (2 + 4)n = 6n$$

Example 1 Using algebra chips, find the sum of $4n^2$ and $-2n^2$.

Lay out chips to represent each term.

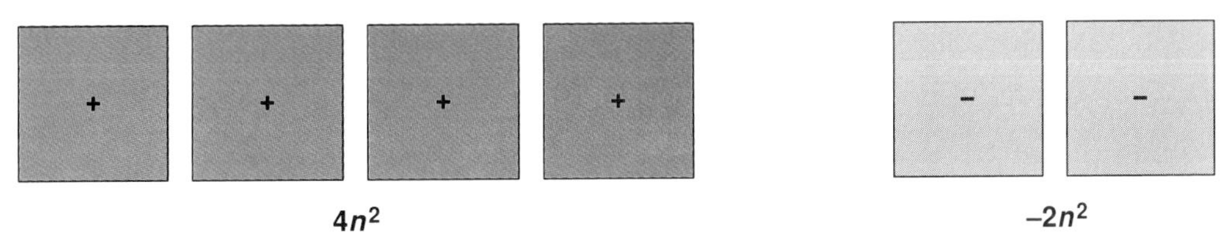

Remove zero pairs, and count the remaining chips.

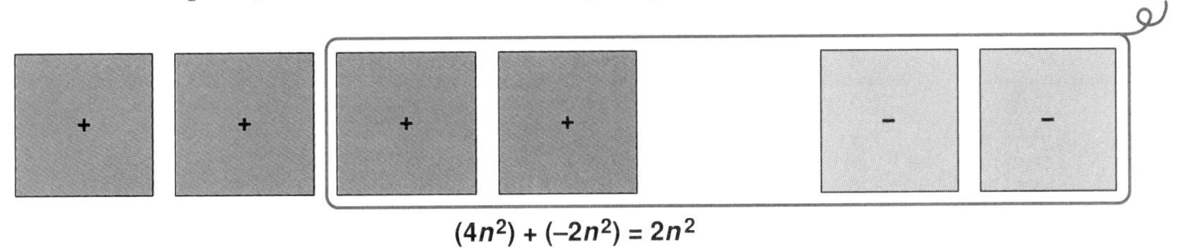

To find the sum of like terms, add the coefficients.
The base and exponent remain the same.

Example 2 Add like terms.

To add the coefficients, remember the rules of signs.

a. $\begin{array}{r} 7m^3 \\ +\ \underline{2m^3} \\ 9m^3 \end{array}$ **b.** $\begin{array}{r} -5xy \\ +\ \underline{-2xy} \\ -7xy \end{array}$ **c.** $\begin{array}{r} 9z^3 \\ +\ \underline{-2z^3} \\ 7z^3 \end{array}$ **d.** $\begin{array}{r} -8k \\ +\ \underline{3k} \\ -5k \end{array}$

When the signs are alike, add the absolute values and give the result the same sign as the two numbers.

When the signs are different, subtract the absolute values and give the result the sign of the number with the greater absolute value.

Try These *(For solutions, see page 606.)*

1. Using the basic algebra chips below, find the sum of each set of like terms.

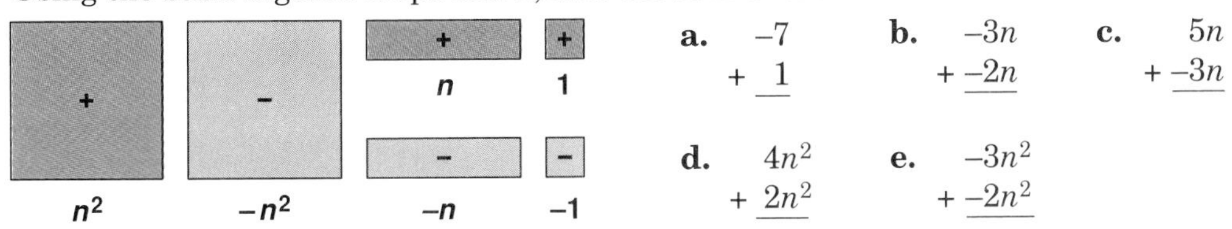

a. $\begin{array}{r} -7 \\ +\ \underline{1} \end{array}$ **b.** $\begin{array}{r} -3n \\ +\ \underline{-2n} \end{array}$ **c.** $\begin{array}{r} 5n \\ +\ \underline{-3n} \end{array}$

d. $\begin{array}{r} 4n^2 \\ +\ \underline{2n^2} \end{array}$ **e.** $\begin{array}{r} -3n^2 \\ +\ \underline{-2n^2} \end{array}$

2. Find the sum of each set of like terms by adding the coefficients.

a. $\begin{array}{r} 3m \\ +\ \underline{7m} \end{array}$ **b.** $\begin{array}{r} 10p^2 \\ +\ \underline{2p^2} \end{array}$ **c.** $\begin{array}{r} -8k \\ +\ \underline{-2k} \end{array}$ **d.** $\begin{array}{r} 7r^2 \\ +\ \underline{-2r^2} \end{array}$ **e.** $\begin{array}{r} -9b \\ +\ \underline{2b} \end{array}$

Adding Polynomials

A single term, such as $3x^2$, is called a ***monomial***.

A ***binomial*** has 2 terms.

$3x^2 + 4x$

A ***trinomial*** has 3 terms.

$3x^2 + 4x + 9$

In general, an algebraic number with 1, 2, 3, or more terms is called a ***polynomial***.

You can use algebra chips to represent a polynomial. Here are the basic chips.

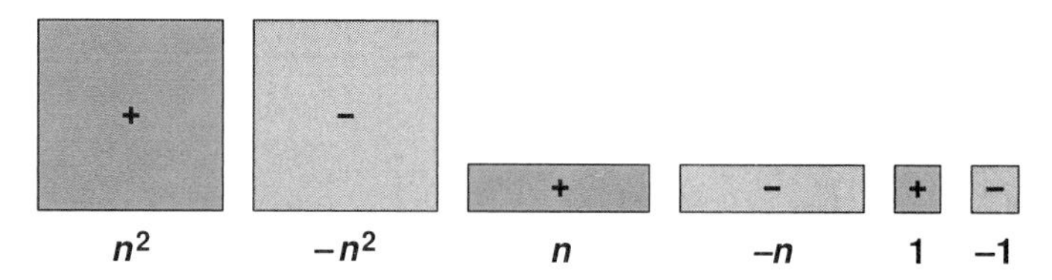

What polynomials are represented by each of the following arrangements?

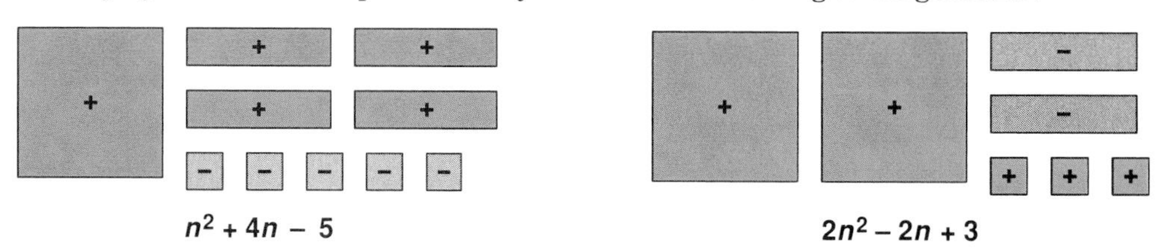

Use algebra chips to model the sum of the two polynomials. Here are the two sets of chips combined.

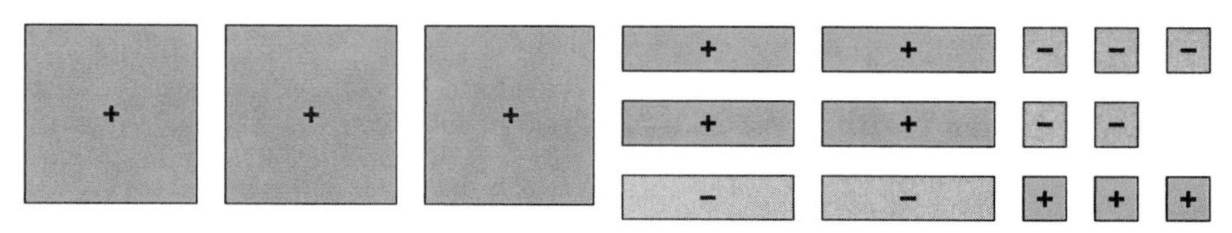

From the total number of chips, remove the equivalent of zero pairs.

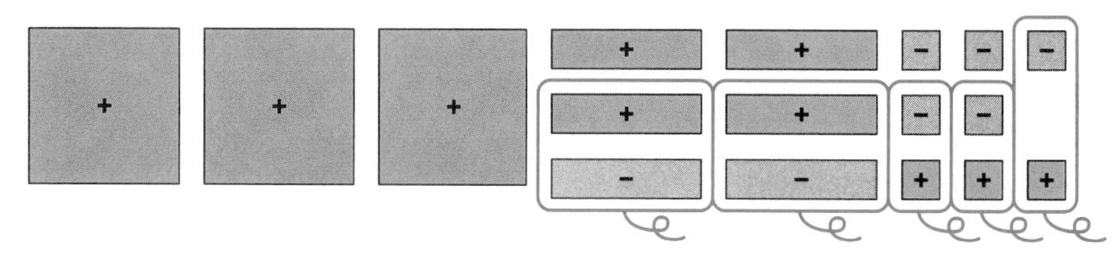

Count the remaining chips to tell the polynomial that is the sum: $3n^2 + 2n - 2$

Summarize the result of the model.

$$\begin{array}{r} n^2 + 4n - 5 \\ +\ \underline{2n^2 - 2n + 3} \\ 3n^2 + 2n - 2 \end{array}$$

To add polynomials, add like terms.

Example 3 Find the sum of the polynomials by adding like terms.

a. $\begin{array}{r} x^2 + 2x - 3 \\ + \underline{3x^2 + 5x - 6} \\ 4x^2 + 7x - 9 \end{array}$

b. $\begin{array}{r} 5y^2 - 3y + 7 \\ + \underline{-2y^2 - 5y + 2} \\ 3y^2 - 8y + 9 \end{array}$

c. $\begin{array}{r} 3a - 2ab + 5b \\ + \underline{-a + 7ab - 2b} \\ 2a + 5ab + 3b \end{array}$

Example 4 On her 3-day car trip, Judy covered $(2n + 3)$ miles on the first day, $(7n - 1)$ miles on the second day, and $(n + 2)$ miles on the third day. How many miles did Judy travel?

To find the sum of the 3 polynomials, add like terms.

$$\begin{array}{r} 2n + 3 \\ 7n - 1 \\ \underline{n + 2} \\ 10n + 4 \end{array}$$

Answer: Judy traveled a total of $(10n + 4)$ miles.

Try These *(For solutions, see page 607.)*

1. Use algebra chips to represent each polynomial.
   a. $2n^2 + 3n + 5$ b. $n^2 - 2n - 3$ c. $-2n^2 - 4n + 4$
2. Use algebra chips to find the sum of each pair of polynomials. Summarize the result of each model.
   a. $\begin{array}{r} n^2 + 2n + 5 \\ + \underline{n^2 + 4n + 3} \end{array}$ b. $\begin{array}{r} 2n^2 - 3n - 2 \\ + \underline{n^2 - 2n + 3} \end{array}$ c. $\begin{array}{r} -n^2 - 2n + 4 \\ + \underline{2n^2 - n - 2} \end{array}$
3. Find the sum of the polynomials by adding like terms.
   a. $\begin{array}{r} 2x^2 + 3x + 7 \\ + \underline{x^2 + 2x + 2} \end{array}$ b. $\begin{array}{r} 8z^2 - 2z - 4 \\ + \underline{z^2 - 3z + 3} \end{array}$ c. $\begin{array}{r} 6q^2 - 8q + 7 \\ + \underline{-3q^2 + q - 5} \end{array}$
4. In his home library, Tim has $(7n + 3)$ mystery books, $(2n + 4)$ novels, and $(6n - 9)$ science fiction books. How many books does Tim have in all?

SELF-CHECK: SECTION 13.1

1. Name the parts of the term $8y^3$.
   Answer: 8 is the coefficient, y is the base, 3 is the exponent.
2. Can there be a base that is not a variable?
   Answer: Yes, in 3^2, the base is 3.
3. Which of these terms are like? $2x$, $3x^2$, $4xy$, $2x^2$ Explain.
   Answer: $3x^2$ and $2x^2$ are like because they have the same base and the same exponent.

4. How would you add the like terms $3x^2$ and $2x^2$?

Answer: Add the coefficients: $3x^2 + 2x^2 = 5x^2$

5. Consider the terms $3x$ and 4.

a. Are they like or unlike? Explain.

Answer: Unlike, they do not have the same base.

b. How would you show the sum of the 2 terms?

Answer: $3x + 4$

c. Is this the same as $7x$? Explain.

Answer: No, $7x$ is the sum of $3x$ and $4x$.

6. Consider the polynomial $2x^2 + 5$.

a. How many terms are in the polynomial?

Answer: 2 terms: $2x^2$ and 5

b. What do we call a polynomial of 2 terms?

Answer: a binomial

7. Which polynomial is represented by these algebra chips?

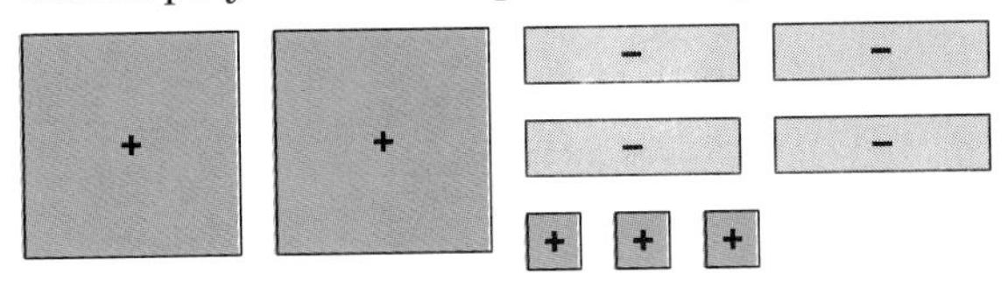

Answer: $2n^2 - 4n + 3$

EXERCISES: SECTION 13.1

1. Identify each set of terms as *like terms* or *unlike terms*. Give a reason for your answer.

a. $3t^2, 5t^2$ **b.** $3x, 3x^2$ **c.** $4a, 5ab$ **d.** 7, 16

2. Use algebra chips to find the sum of each set of like terms. Summarize the result of each model.

a. $\begin{array}{r} 5 \\ + \underline{-3} \end{array}$ **b.** $\begin{array}{r} 4n \\ + \underline{2n} \end{array}$ **c.** $\begin{array}{r} -3n \\ + \underline{-2n} \end{array}$ **d.** $\begin{array}{r} -6n \\ + \underline{3n} \end{array}$ **e.** $\begin{array}{r} 5n \\ + \underline{-6n} \end{array}$

f. $\begin{array}{r} 4n^2 \\ + \underline{3n^2} \end{array}$ **g.** $\begin{array}{r} -3n^2 \\ + \underline{-4n^2} \end{array}$ **h.** $\begin{array}{r} -n^2 \\ + \underline{5n^2} \end{array}$ **i.** $\begin{array}{r} 5n^2 \\ + \underline{-3n^2} \end{array}$ **j.** $\begin{array}{r} 4n^2 \\ + \underline{-4n^2} \end{array}$

3. Do you agree or disagree? Explain.

a. Jim said that the sum of $3x$ and $5x$ is $8x$.

b. Joe said that the sum of $2y^2$ and $3y^2$ is $5y^4$.

c. Janet said that the sum of $-2z$ and $-4z$ is $6z$.

d. Jerry said you could show that $7y^2$ is not the sum of $2y$ and $5y$ by letting y have a value, say 3.

4. Add like terms.

a. $\begin{array}{r} 2x \\ +\ \underline{5x} \end{array}$	**b.** $\begin{array}{r} -7y^2 \\ +\ \underline{-2y^2} \end{array}$	**c.** $\begin{array}{r} -z^2 \\ +\ \underline{5z^2} \end{array}$	**d.** $\begin{array}{r} 8ab \\ +\ \underline{-2ab} \end{array}$	**e.** $\begin{array}{r} xyz \\ +\ \underline{-xyz} \end{array}$
f. $\begin{array}{r} -3m^2 \\ +\ \underline{5m^2} \end{array}$	**g.** $\begin{array}{r} -6pq \\ +\ \underline{-4pq} \end{array}$	**h.** $\begin{array}{r} 7r^3 \\ +\ \underline{r^3} \end{array}$	**i.** $\begin{array}{r} -t^2 \\ +\ \underline{3t^2} \end{array}$	**j.** $\begin{array}{r} -2w \\ +\ \underline{5w} \end{array}$

5. Express each sum in simplest form.

   a. Trixie's brother saved d dollars. Trixie saved twice as much as her brother. Find the sum of their savings.

   b. Ian walked at the rate of m miles per hour. Jay walked 3 times as fast as Ian. Find the sum of their walking rates.

   c. Sam shoveled snow at the rate of s^2 shovelfuls per minute. Sally shoveled twice as fast as Sam. Find the sum of their shoveling rates.

   d. Terry typed at the rate of w words per minute. Tim typed 1.5 times as fast as Terry. Find the sum of their typing rates.

   e. Cindy, Lindy, and Mindy jogged on Saturday morning. Cindy covered a distance of k kilometers. Lindy's distance was twice Cindy's, and Mindy jogged one-third as far as Lindy. Find the sum of the distances.

6. Represent the perimeter of each figure in simplest form.

   a. square *ABCD* **b.** equilateral $\triangle PQR$ **c.** rectangle *EFGH* **d.** right $\triangle JKL$

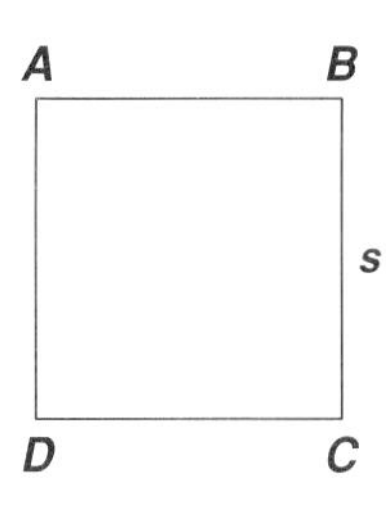

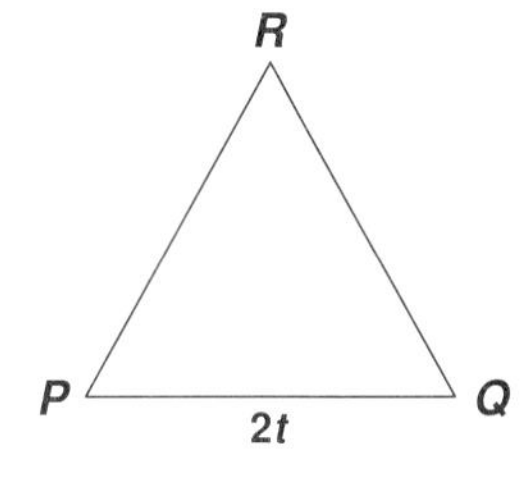

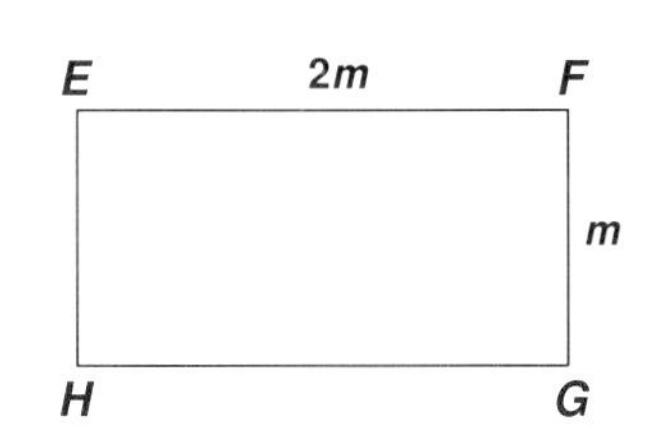

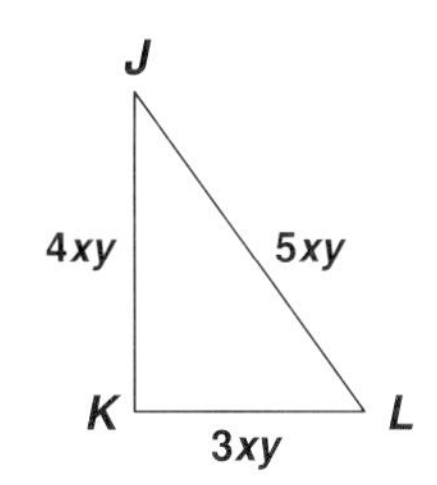

7. Use algebra chips to represent each polynomial.

   a. $2n^2 + 6n - 3$ **b.** $3n^2 - 2n + 5$ **c.** $-3n^2 - n + 2$

8. Use algebra chips to find the sum of each pair of polynomials. Summarize the result of each model.

a. $\begin{array}{r} 2n^2 + 4n + 3 \\ +\ \underline{3n^2 + 2n - 1} \end{array}$	**b.** $\begin{array}{r} n^2 - 2n + 5 \\ +\ \underline{2n^2 - n - 1} \end{array}$	**c.** $\begin{array}{r} 5n^2 + 4n - 4 \\ +\ \underline{-2n^2 - 3n + 2} \end{array}$
d. $\begin{array}{r} 6n^2 - 3n - 5 \\ +\ \underline{-n^2 + 2n - 2} \end{array}$	**e.** $\begin{array}{r} -3n^2 + 4n - 3 \\ +\ \underline{2n^2 - n + 3} \end{array}$	**f.** $\begin{array}{r} 3n^2 - 5n + 2 \\ +\ \underline{-2n^2 + 3n - 2} \end{array}$

9. Find the sum of each pair of polynomials by adding like terms.

a. $\begin{array}{r} 2c + 4 \\ + \underline{4c + 1} \end{array}$ **b.** $\begin{array}{r} 7q^2 + 3q \\ + \underline{2q^2 + 2q} \end{array}$ **c.** $\begin{array}{r} 8x^2 - 4x \\ + \underline{2x^2 - x} \end{array}$

d. $\begin{array}{r} 4p^2 + 3 \\ + \underline{3p^2 - 1} \end{array}$ **e.** $\begin{array}{r} -3d + 2 \\ + \underline{-2d - 5} \end{array}$ **f.** $\begin{array}{r} 4r^2 + 1.5r \\ + \underline{-2r^2 + 1.5r} \end{array}$

g. $\begin{array}{r} 2x^2 + 3x + 5 \\ + \underline{4x^2 + 2x + 3} \end{array}$ **h.** $\begin{array}{r} 3a^2 + a - 1 \\ + \underline{9a^2 + a + 5} \end{array}$ **i.** $\begin{array}{r} 6y^2 - 8y + 4 \\ + \underline{-y^2 + 3y + 7} \end{array}$

j. $\begin{array}{r} -2y^2 + 3y + 5 \\ + \underline{-3x^2 - 5y + 8} \end{array}$ **k.** $\begin{array}{r} -7z^2 - 4z - 3 \\ + \underline{3z^2 - 2z - 4} \end{array}$ **l.** $\begin{array}{r} -2r^2 + 6r + 5 \\ + \underline{5r^2 - 8r - 7} \end{array}$

m. $\begin{array}{r} 2 + 4w + 3w^2 \\ + \underline{3 - 2w + 4w^2} \end{array}$ **n.** $\begin{array}{r} -1 + 8x - x^2 \\ + \underline{-3 - 2x + 4x^2} \end{array}$ **o.** $\begin{array}{r} 9 - 2t + 2t^2 \\ + \underline{-2 + 5t - 5t^2} \end{array}$

10. Do you agree or disagree? Explain.

a. Latoya added $(2x + 3)$ and $(5 + 4x)$ to get $7x + 7x$.

b. Larry said that, to add $(2x + 3)$ and $(5 + 4x)$, it is a good idea to put the polynomials in the same order.

c. Leroy added $(2x + 3)$ and $(5 + 4x)$ by first rewriting the sum as $(2x + 3)$ and $(4x + 5)$. His result was $6x + 8$.

d. Laverne added $(2x + 3)$ and $(5 + 4x)$ by first rewriting the sum as $(3 + 2x)$ and $(5 + 4x)$. Her result was $8 + 6x^2$.

11. Arrange each pair of polynomials so that they are in the same order. Find each sum.

a. $(2x^2 + 3x + 5)$ and $(7 + 2x + 3x^2)$

b. $(3y^2 + 2y - 1)$ and $(4 - y - 2y^2)$

c. $(8 - 6r + r^2)$ and $(3r^2 + 5r - 4)$

d. $(5 + 4t - 3t^2)$ and $(t^2 - 2t - 6)$

12. Express each sum in simplest form.

a. The ages of the three Thomas children are $2y + 7$, $3y - 5$, and $y + 1$. Find the sum of the ages.

b. Yesterday, two shipments of books arrived at the library. One shipment contained $(4b^2 - 5b - 7)$ books and the other had $(2b^2 + 3b + 8)$ books. What was the total number of books?

c. This week at her job, Pam earned $(3d^2 + 4d + 2)$ dollars in regular wages and $(d^2 - 2d + 1)$ dollars in overtime. Find her earnings for the week.

d. Mrs. Searle bought a VCR, which was priced at $(2d^2 - 2d - 5)$ dollars.

The tax on her purchase came to $(d + 7)$ dollars.

How much did Mrs. Searle pay?

13. For each set of polynomials:

- Find the sum by adding like terms.
- Verify your result by letting $x = 2$, and evaluating each polynomial and the sum.

a. $(3x + 7) + (x - 3)$

b. $(2x^2 + 5x + 1) + (3x^2 + 2x - 1)$

c. $(5x^2 + 3x + 10) + (2x^2 - 3x + 8)$

d. $(5 + 2x - 3x^2) + (6 - 3x + 4x^2)$

e. $(1 - 2x + 4x^2) + (-2 + 4x - 2x^2)$

f. $(-2 - 3x + x^2) + (1 + 4x - x^2)$

EXPLORATIONS

1. To decode this message:

$\frac{?}{2x-3}\ \frac{?}{3x}\quad \frac{?}{-2x+1}\ \frac{?}{5}\ \frac{?}{x}\quad \frac{?}{x-1}\ \frac{?}{5}\ \frac{?}{-2x^2}$

$\frac{?}{x}\ \frac{?}{2x-3}\ \frac{?}{x+2}\quad \frac{?}{-2x+1}\ \frac{?}{-1}\ \frac{?}{-1}\quad \frac{?}{x+1}\ \frac{?}{x}$

$\frac{?}{2x-1}\ \frac{?}{3x}\ \frac{?}{-1}\quad \frac{?}{x-3}\ \frac{?}{5}\ \frac{?}{x^2-1}\ \frac{?}{-2x^2}\ \frac{?}{x-4}\ \frac{?}{-2x+1}$

$\frac{?}{-1}\ \frac{?}{5}\ \frac{?}{-x+1}\ \frac{?}{x^2}\ \frac{?}{x}\quad \frac{?}{x^2}\ \frac{?}{4x-1}\ \frac{?}{-x}\ \frac{?}{3x}\ \frac{?}{2x-1}$

Simplify each expression below. Locate your answer in the code. Each time the answer appears in the code, write the letter or symbol that is next to the expression in the blank above the answer. Use a separate sheet of paper onto which you have copied the code.

A	$6 + 8 + (-9)$	N	$(-x^2) + (-x^2)$
B	$(x + 5) + (-4)$	O	$(x + 4) + (x - 7)$
C	$(2x - 3) + (-x + 2)$	R	$(-2x - 4) + (x + 5)$
D	$(-x^2 - 3) + (x^2 + x)$	S	$(x - 2) + (-3x + 3)$
E	$(x - 1) + (-x)$	T	$x + (x - 1)$
G	$(3 - 2x) + (x - 3)$	U	$2(x + 1) + (-x)$
H	$(-x^2 + x) + (x^2 + 2x)$	W	$(x^2 + 7) + (-8)$
I	$(3x + 4) + (x - 5)$	Y	$(3x + 5) + (-2x - 5)$
L	$(3x^2 + 1) + (-2x^2 - 1)$	'	$(2x - 2) + (-x - 2)$

2. **a.** Which two of these fractions do you think are like terms? Explain.

$$\frac{x}{3} \qquad \frac{3}{x} \qquad \frac{2x}{3}$$

b. **(1)** What do you think is the sum of $\frac{x}{3}$ and $\frac{2x}{3}$?

(2) Verify your result by finding the value of each fraction and the value of the sum when x is a specific value, say $x = 2$.

13.2 Subtraction

Subtracting Like Terms

At birth, Rachel Hope weighed in at 5 pounds 8 ounces. After her first month, she weighed 6 pounds 11 ounces.

How much weight had she gained?

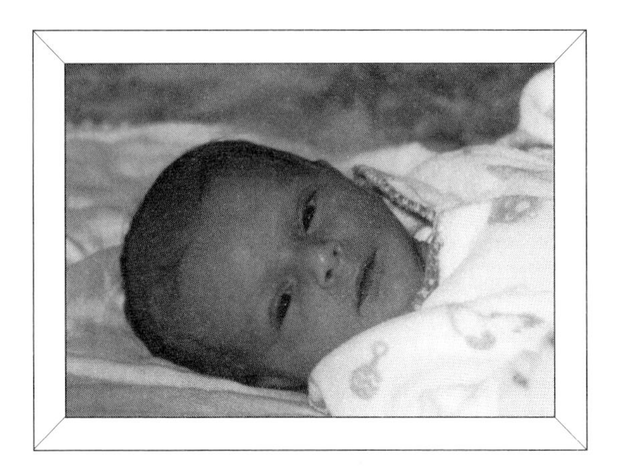

To find the weight gain, subtract the like units:

	6 lb.	11 oz.
−	5 lb.	8 oz.
	1 lb.	3 oz.

In Arithmetic, only like units can be subtracted.

How long is line segment AB?

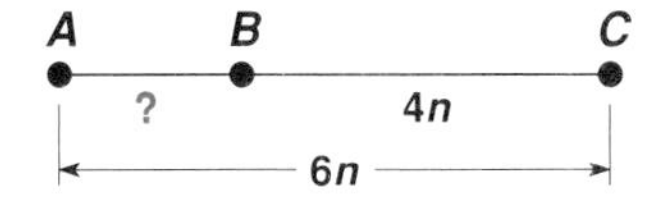

The total length of $\overline{AB}$ is $6n - 4n$, or $2n$ units.

In Algebra, only like terms can be subtracted.

Example 1 Using algebra chips, subtract $-n^2$ from $3n^2$.

Lay out chips to represent the minuend, the term from which you are to subtract.

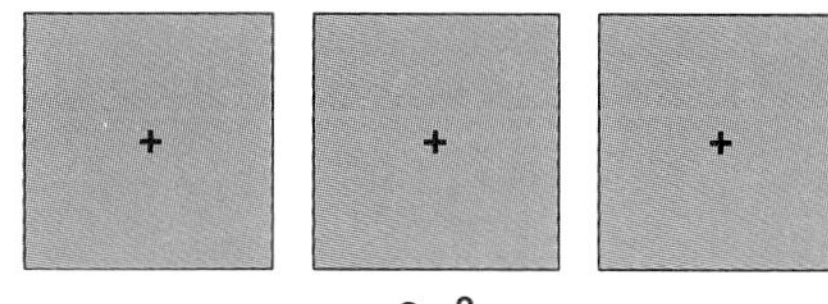

$3n^2$

Now you will want to remove one $-n^2$-chip, which represents the subtrahend, the term you are subtracting. But there are no $-n^2$-chips in the model. You must insert a zero pair. Since you will need one $-n^2$-chip, insert one zero pair of n^2-chips.

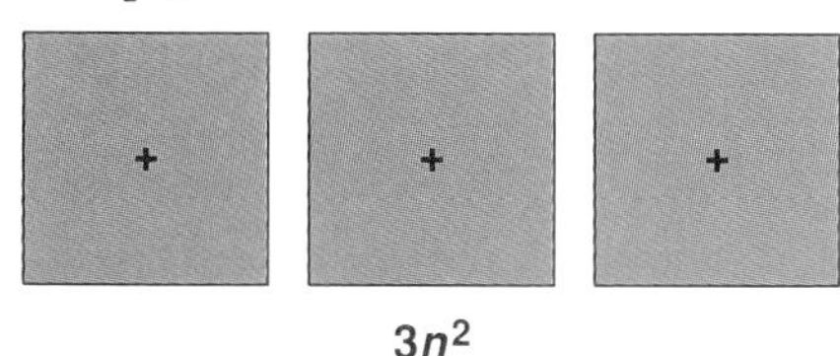

$3n^2$

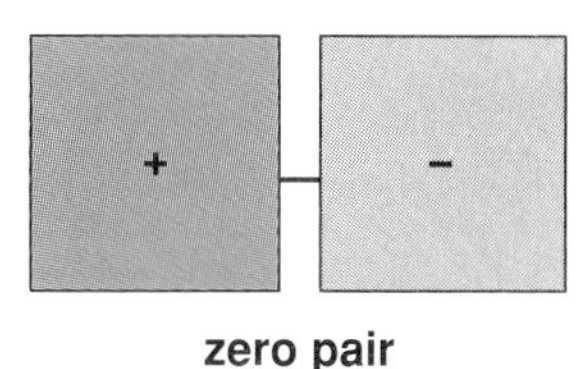

zero pair

Now you will be able to remove the chip that represents the subtrahend. Remove one $-n^2$-chip.

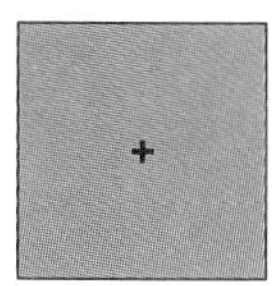

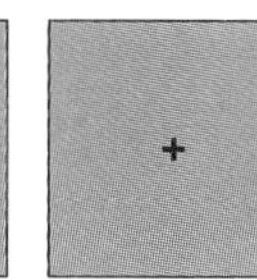

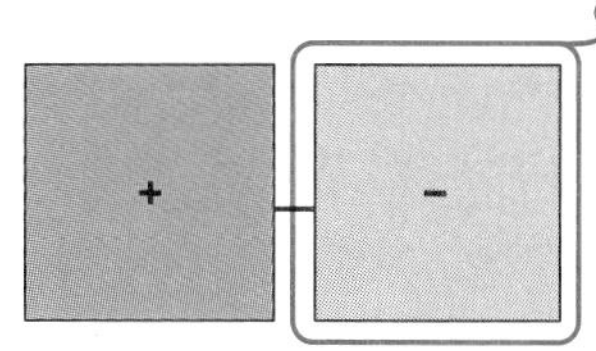

The remaining chips represent the difference: $4n^2$

Summarize the result of this model.

$$\begin{array}{r} 3n^2 \\ -\ \underline{-n^2} \\ 4n^2 \end{array}$$

The result is the same as adding the opposite of the subtrahend.

Subtraction changes to addition. → $\begin{array}{r} 3n^2 \\ \overset{+}{\ominus}\ \overset{+}{\ominus} n^2 \\ \hline 4n^2 \end{array}$ Change the sign of the subtrahend.

To find the difference of like terms, add the opposite of the subtrahend.

Example 2 Subtract like terms.

Change subtraction to addition of the opposite of the subtrahend.

a. $\begin{array}{r} 9z^2 \\ -\ \underline{-3z^2} \end{array}$ becomes $\begin{array}{r} 9z^2 \\ \overset{+}{\ominus}\ \overset{+}{\ominus} 3z^2 \\ \hline 12z^2 \end{array}$

b. $\begin{array}{r} -7m^2 \\ -\ \underline{-2m^2} \end{array}$ becomes $\begin{array}{r} -7m^2 \\ \overset{+}{\ominus}\ \overset{+}{\ominus} 2m^2 \\ \hline -5m^2 \end{array}$

c. $\begin{array}{r} -5x^2 \\ -\ \underline{3x^2} \end{array}$ becomes $\begin{array}{r} -5x^2 \\ \overset{+}{\ominus}\ \overset{-}{\oplus} 3x^2 \\ \hline -8x^2 \end{array}$

↑ When no sign appears, + is understood.

d. $\begin{array}{r} 4y^2 \\ -\ \underline{2y^2} \end{array}$ becomes $\begin{array}{r} 4y^2 \\ \overset{+}{\ominus}\ \overset{-}{\oplus} 2y^2 \\ \hline 2y^2 \end{array}$

Try These *(For solutions, see page 608.)*

1. Using algebra chips, find the difference of each set of like terms. Summarize the result of each model.

   a. $\begin{array}{r} 4n \\ -\ \ 2n \\ \hline \end{array}$ **b.** $\begin{array}{r} -3n \\ -\ -2n \\ \hline \end{array}$ **c.** $\begin{array}{r} -5 \\ -\ \ 3 \\ \hline \end{array}$ **d.** $\begin{array}{r} 6n \\ -\ -2n \\ \hline \end{array}$ **e.** $\begin{array}{r} -5n^2 \\ -\ \ 3n^2 \\ \hline \end{array}$

2. Find the difference of each set of like terms by adding the opposite of the subtrahend.

   a. $\begin{array}{r} 7m \\ -\ \ 3m \\ \hline \end{array}$ **b.** $\begin{array}{r} 3k \\ -\ \ 7k \\ \hline \end{array}$ **c.** $\begin{array}{r} -6y \\ -\ -4y \\ \hline \end{array}$ **d.** $\begin{array}{r} 8ab \\ -\ -2ab \\ \hline \end{array}$ **e.** $\begin{array}{r} -7x^2 \\ -\ \ 4x^2 \\ \hline \end{array}$

Subtracting Polynomials

To find the difference $(2n^2 + 3n + 5) - (n^2 + n + 4)$:

Lay out algebra chips to represent the first polynomial.

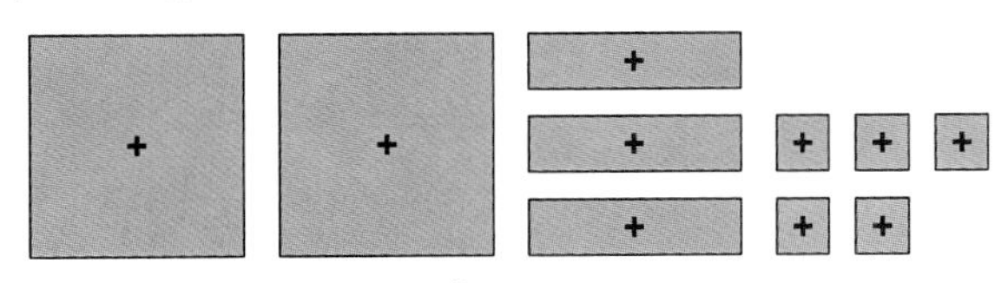

$2n^2 + 3n + 5$

Now remove the chips that represent the second polynomial. Remove one n^2-chip, one n-chip, and four 1-chips.

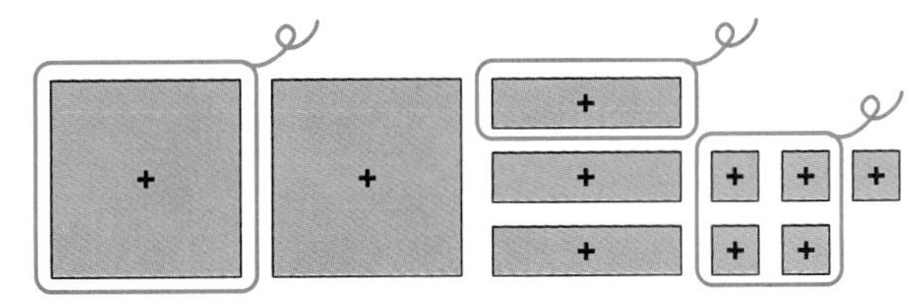

The remaining chips represent the difference: $n^2 + 2n + 1$

Summarize the result of this model.

$$\begin{array}{rr} & 2n^2 + 3n + 5 \\ - & n^2 + \ \ n + 4 \\ \hline & n^2 + 2n + 1 \end{array}$$

The result is the same as adding the opposite of the subtrahend.

$$\begin{array}{rr} & 2n^2 + 3n + 5 \\ \overset{+}{\ominus} & \overset{-}{\oplus}\, n^2 \overset{-}{\oplus}\ n \overset{-}{\oplus} 4 \\ \hline & n^2 + 2n + 1 \end{array}$$

← The sign of each term of the polynomial changes to the opposite sign.

↑ Subtraction changes to addition.

Example 3 Use algebra chips to find the difference $(3n^2 + 3n + 2) - (2n^2 - 2n + 1)$.

Lay out the chips to represent the first polynomial.

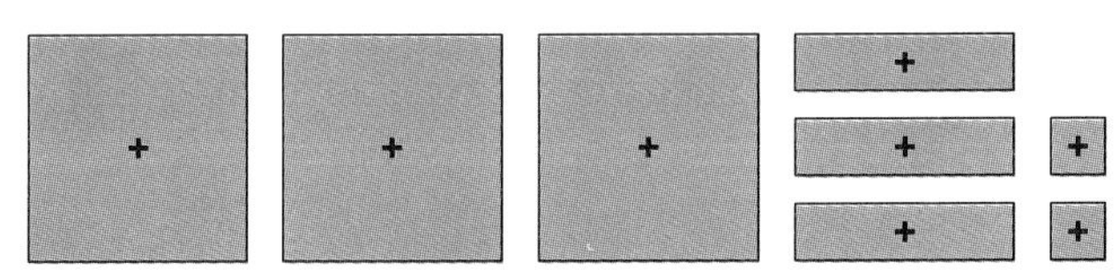

Now you want to remove the chips that represent the second polynomial. The second polynomial requires that you remove two $-n$-chips. But, there are no $-n$-chips in the model. You must insert zero pairs. Since you will need two $-n$-chips, insert two zero pairs of n-chips.

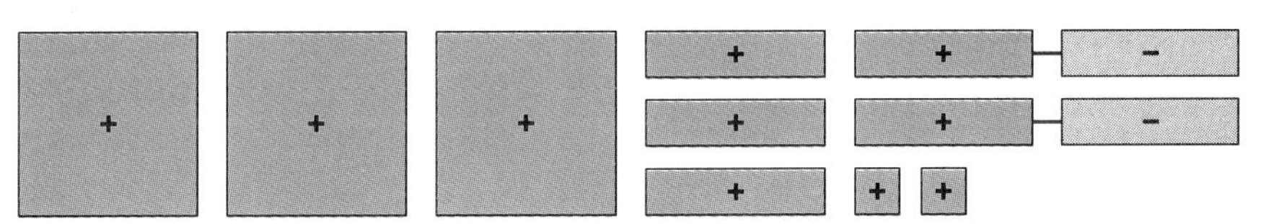

Now you will be able to remove the chips that represent the second polynomial. Remove two n^2-chips, two $-n$-chips, and one 1-chip.

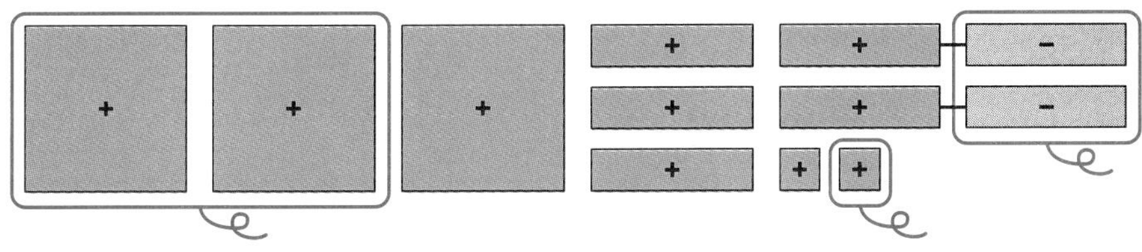

The remaining chips represent the difference: $n^2 + 5n + 1$

Summarize the result of this model.

$$\begin{array}{r} 3n^2 + 3n + 2 \\ -\ \underline{2n^2 - 2n + 1} \\ n^2 + 5n + 1 \end{array}$$

The result is the same as adding the opposite of the subtrahend.

$$\begin{array}{r} 3n^2 + 3n + 2 \\ \underline{\overset{+}{\ominus}\overset{-}{\oplus}2n^2\overset{+}{\ominus}2n\overset{-}{\oplus}1} \\ n^2 + 5n + 1 \end{array}$$

Example 4 Do this subtraction by adding the opposite polynomial.

$$\begin{array}{r} 6t^2 + 7t + 8 \\ -\ \underline{2t^2 - 4t + 3} \end{array}$$

Rewrite

$$\begin{array}{r} 6t^2 + 7t + 8 \\ -\ \underline{2t^2 + 4t - 3} \end{array} \quad \text{as} \quad \begin{array}{r} 6t^2 + 7t + 8 \\ \underline{\overset{+}{\ominus}\overset{-}{\oplus}2t^2\ \overset{-}{\oplus}4t\overset{+}{\ominus}3} \end{array}$$

← Change the sign of each term of the subtrahend (second polynomial).

Change subtraction to addition.

Then add by combining like terms.

$$\begin{array}{r} 6t^2 + 7t + 8 \\ +\ \underline{-2t^2 - 4t + 3} \\ 4t^2 + 3t + 11 \end{array}$$

To subtract polynomials, add the opposite of the subtrahend.

Example 5 For her clothing shop, Pasha ordered $(2n + 5)$ belts from Milly's Mail Order, and then placed an additional order for $(7n - 6)$ belts. After the season, $(n - 2)$ belts from Milly's remained in Pasha's store. Write an expression for the number of Milly's belts that Pasha had sold. Use the 4 key problem-solving steps.

1 ***Read the problem for information.***
You know the number of belts in each of 2 orders, and you know the number that remained after sales.

You want to find the number of belts sold.

2 ***Make a plan.***
Find the sum of the 2 orders of belts.
From this sum, subtract the number of belts remaining after sales.

3 ***Carry out your plan.***

Add to find the total.

$$\begin{array}{r} 2n + 5 \\ +\ \underline{7n - 6} \\ 9n - 1 \end{array}$$

From the sum, subtract the number left.

$$\begin{array}{r} 9n - 1 \\ -\ \underline{\ \ n - 2} \end{array} \qquad \begin{array}{r} 9n - 1 \\ \ominus \oplus n \ominus 2 \\ \hline 8n + 1 \end{array}$$

To subtract, add the opposite.

4 ***Check your answer.***
If Pasha ordered $(9n - 1)$ belts and sold $(8n + 1)$ belts, would $(n - 2)$ belts be left?

$$\begin{array}{r} 9n - 1 \\ -\ \underline{8n + 1} \\ ? \end{array} \qquad \begin{array}{r} 9n - 1 \\ \ominus \oplus 8n \oplus 1 \\ \hline n - 2 \end{array} \checkmark$$

Answer: Pasha sold $(8n + 1)$ belts from Milly's.

Try These *(For solutions, see page 610.)*

1. Use algebra chips to do each subtraction. Summarize the result of each model.
   a. $\begin{array}{r} 4n^2 + 5n + 3 \\ -\ \underline{2n^2 + 3n + 2} \end{array}$ **b.** $\begin{array}{r} 2n^2 + 2n - 4 \\ -\ \underline{n^2 - 3n - 1} \end{array}$ **c.** $\begin{array}{r} 3n^2 - 2n + 3 \\ -\ \underline{-n^2 + 2n + 3} \end{array}$
2. Do each subtraction by adding the opposite polynomial.
   a. $\begin{array}{r} 7x^2 + 8x + 6 \\ -\ \underline{3x^2 + 4x + 4} \end{array}$ **b.** $\begin{array}{r} 9y^2 + 4y - 7 \\ -\ \underline{3y^2 - y - 4} \end{array}$ **c.** $\begin{array}{r} 6z^2 - z + 8 \\ -\ \underline{-z^2 + 3z + 5} \end{array}$
3. Solve each problem.
   a. Kaitlin has $(12n + 5)$ cards in her baseball card collection. Timothy has $(6n - 1)$ cards in his baseball card collection. How many more cards than Timothy does Kaitlin have?
   b. Mr. Keenan ordered $(100x + 12)$ biology texts for his department, and then ordered an additional $(15x - 3)$ biology texts. Because of a budget cut, Mr. Keenan had to return $(22x + 4)$ of the texts. How many biology texts was he able to keep?

SELF-CHECK: SECTION 13.2

1. Name the parts of this subtraction: $9 - 2 = 7$

Answer: 9 is the minuend, 2 is the subtrahend, 7 is the difference.

2. How is subtraction related to addition?

Answer: Subtraction is the same as adding the opposite of the subtrahend.

3. What signs would you change to do this subtraction? $6 - (-2)$

Answer: Change the subtraction to addition, and write the opposite of the subtrahend. $6 + (+2)$

EXERCISES: SECTION 13.2

1. Use algebra chips to find the difference of each set of like terms. Summarize the result of each model.

a. $\begin{array}{r} 5n \\ -\ \underline{2n} \end{array}$ **b.** $\begin{array}{r} -2n \\ -\ \underline{-3n} \end{array}$ **c.** $\begin{array}{r} 4 \\ -\ \underline{-2} \end{array}$ **d.** $\begin{array}{r} -5n \\ -\ \underline{3n} \end{array}$ **e.** $\begin{array}{r} 3n \\ -\ \underline{-5n} \end{array}$

f. $\begin{array}{r} 3n^2 \\ -\ \underline{n^2} \end{array}$ **g.** $\begin{array}{r} 4n^2 \\ -\ \underline{-n^2} \end{array}$ **h.** $\begin{array}{r} -2n^2 \\ -\ \underline{-3n^2} \end{array}$ **i.** $\begin{array}{r} -5n^2 \\ -\ \underline{2n^2} \end{array}$ **j.** $\begin{array}{r} 3n^2 \\ -\ \underline{-3n^2} \end{array}$

2. Do you agree or disagree? Explain.

a. Rose said that $9x - 2x$ is 7.

b. Rita rewrote $(5n) - (-3n)$ as $(5n) + (+3n)$.

c. Roy used algebra chips to model $6n - 2n$. He first laid out six n-chips, then he inserted two zero pairs of n-chips.

d. Rob used algebra chips to model $(-4n) - (5n)$. He first laid out four $-n$-chips, then he inserted five zero pairs of n-chips.

3. Subtract like terms.

a. $\begin{array}{r} 8x \\ -\ \underline{5x} \end{array}$ **b.** $\begin{array}{r} 3r \\ -\ \underline{8r} \end{array}$ **c.** $\begin{array}{r} 7q \\ -\ \underline{-5q} \end{array}$ **d.** $\begin{array}{r} -5ab \\ -\ \underline{-2ab} \end{array}$ **e.** $\begin{array}{r} -9z \\ -\ \underline{5z} \end{array}$

f. $\begin{array}{r} 8x^2 \\ -\ \underline{x^2} \end{array}$ **g.** $\begin{array}{r} 2y^2 \\ -\ \underline{-y^2} \end{array}$ **h.** $\begin{array}{r} -6b^2 \\ -\ \underline{-8b^2} \end{array}$ **i.** $\begin{array}{r} -9k^2 \\ -\ \underline{2k^2} \end{array}$ **j.** $\begin{array}{r} 6t^2 \\ -\ \underline{6t^2} \end{array}$

4. Express each difference in simplest form.

   a. Ken baked $10c$ cookies for his birthday party. Before the party started, Ken ate c cookies. Find the number of cookies left for the party.

   b. Joe bought $6g^2$ gallons of paint to paint the outside of his house. When he was finished painting, g^2 gallons remained. Find the number of gallons Joe used.

   c. Pat earns d dollars per week. One-third of his weekly salary is withheld for taxes. Find the amount Pat takes home.

   d. David bought a jacket whose original price was d dollars. The jacket was on sale at a 20% discount. Find the sale price.

   e. For her book shop, Ms. Howard ordered $14b$ books from Pearl Publishers. She placed a second order at Pearl for $12b$ books. After three months, $8b$ books from Pearl remained in Ms. Howard's shop. Find the number of books from Pearl that Ms. Howard had sold.

5. Find each length in simplest form.

   a. The perimeter of this right triangle is $40x$ units. If the measures of the legs are $8x$ and $15x$ units, find the measure of the hypotenuse.

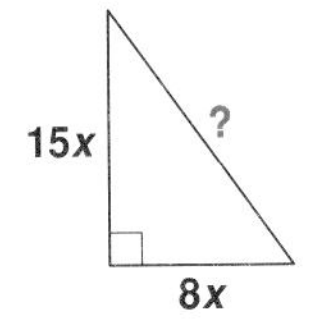

   b. The perimeter of this rectangle is $22k$. If the length is $7k$, find the width.

P = 22k
?
7k

   c. The perimeter of this isosceles triangle is $11t$. If the measure of each leg is $3t$, find the measure of the base.

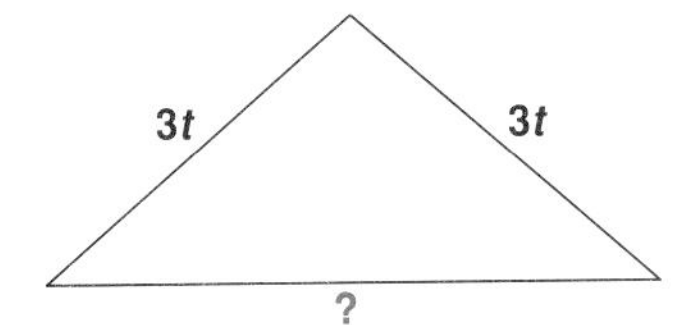

6. Use algebra chips to find the difference of each pair of polynomials. Summarize the result of each model.

   a. $\begin{array}{r} 5n + 4 \\ -\ \underline{2n + 3} \end{array}$

   b. $\begin{array}{r} -4n - 5 \\ -\ \underline{-3n - 2} \end{array}$

   c. $\begin{array}{r} 6n - 3 \\ -\ \underline{-2n + 1} \end{array}$

   d. $\begin{array}{r} 3n^2 - 4n + 5 \\ -\ \underline{2n^2 - 2n - 3} \end{array}$

   e. $\begin{array}{r} -4n^2 + 3n - 2 \\ -\ \underline{-2n^2 - 2n - 3} \end{array}$

   f. $\begin{array}{r} 5n^2 - 4n - 3 \\ -\ \underline{-n^2 + 3n - 3} \end{array}$

7. **a.** Use algebra chips to represent

   (1) the polynomial $2n^2 - 3n + 5$ **(2)** the polynomial $-2n^2 + 3n - 5$

   b. Answer these questions about the two polynomials in part a.

   (1) How are these polynomials related?

   (2) Without doing the addition, tell what you think the sum of these polynomials is. Verify your answer by adding the algebra chips.

8. Write the opposite of each polynomial.

   a. $2x$ **b.** $-5n$ **c.** $x + 3$ **d.** $a^2 - a$

   e. $-y + 1$ **f.** $b^2 - 3b + 2$ **g.** $-2y^2 + 4y - 3$ **h.** $1 - 5x + 3x^2$

9. Do each subtraction by adding the opposite polynomial.

   a. $\begin{array}{r} 4w + 3 \\ -\ \underline{3w + 1} \end{array}$ **b.** $\begin{array}{r} 3r + 5 \\ -\ \underline{r + 4} \end{array}$ **c.** $\begin{array}{r} 8x^2 + 5x \\ -\ \underline{6x^2 + 3x} \end{array}$

   d. $\begin{array}{r} 7k + 5 \\ -\ \underline{2k - 1} \end{array}$ **e.** $\begin{array}{r} 6y - 4 \\ -\ \underline{-2y + 2} \end{array}$ **f.** $\begin{array}{r} -3m^2 + 2m \\ -\ \underline{-2m^2 - 4m} \end{array}$

   g. $\begin{array}{r} 9p^2 + 4p + 2 \\ -\ \underline{2p^2 + 2p - 4} \end{array}$ **h.** $\begin{array}{r} 5y^2 + 2y - 3 \\ -\ \underline{y^2 - 3y + 1} \end{array}$ **i.** $\begin{array}{r} 3n^2 - 5n - 7 \\ -\ \underline{-n^2 + 3n + 3} \end{array}$

   j. $\begin{array}{r} -2w^2 + 3w + 1 \\ -\ \underline{-4w^2 - 2w - 5} \end{array}$ **k.** $\begin{array}{r} 4z^2 - 6z + 8 \\ -\ \underline{-z^2 - 3z - 2} \end{array}$ **l.** $\begin{array}{r} -t^2 - 5t - 3 \\ -\ \underline{-t^2 + 5t - 3} \end{array}$

   m. $\begin{array}{r} 2 + 4m + 3m^2 \\ -\ \underline{1 + 3m + 2m^2} \end{array}$ **n.** $\begin{array}{r} 2 - 2p - 6p^2 \\ -\ \underline{5 - 4p + 2p^2} \end{array}$ **o.** $\begin{array}{r} -5 + 5g - 5g^2 \\ -\ \underline{-3 + 3g - 3g^2} \end{array}$

10. Write each result in simplest form.

    a. From the sum of $9x + 3$ and $x + 5$, subtract $3x + 7$.

    b. Subtract $3a + 2$ from the sum of $5a + 3$ and $-6a - 5$.

    c. Add $2y - 10$ to the difference of $6y - 2$ and $5y + 3$.

    d. Subtract the sum of $2w - 3$ and $3w + 2$ from the sum of $4w + 7$ and $w - 6$.

11. Write each answer in simplest form.

    a. The route of a bike race covers $(m^2 + 5m + 1)$ miles. If Dana has biked $(m^2 - 2m + 3)$ miles, how far must she go?

    b. The Castleton School District has $(3 + 8d + 2d^2)$ days in the school year. If the schools have been in session for $(1 - 6d + d^2)$ days, how many days are left?

    c. Interest amounting to $(d - 3)$ dollars was added to the $(d^2 + 7d + 5)$ dollars in Andy's bank account. If he then withdrew $(10d - 1)$ dollars, what was the balance left in his account?

EXPLORATIONS

1. Copy the crisscross puzzle onto graph paper. Then simplify each expression and fill in the appropriate spaces. Use a separate space for each digit, variable, + or – sign.

ACROSS

1. $8 + (-5)$
2. $(a + 5) + (-6)$
4. $4r + (-3r)$
5. $(2bd - 1) + (1 - bd)$
7. $9 + (x - 10)$
9. $(3p + 5) - (4p + 5)$
10. $(-5)(-6)$
11. $(e + 3) + (e - 3)$
12. $(a + b^2) + (b - b^2)$
14. $(2xy + 9) - (xy + 9)$
16. $(p + wp + 3) - (p + 3)$
18. $(5c - 8) + (4 - 4c)$
19. $(g + 7) + (-g + 8)$
20. $(5r + 2) - (6r + 2)$
22. $(2b + 20) + (-2b + 1)$
23. $(12 + 2t) - (4 + t)$
25. $8bc + (-7bc)$
27. $(x^2 + 4x) + (-x^2 + 6x)$

1 ?		2 ?	3 ?	?		4 ?		
5 ?	6 ?		?		7 ?	?	8 ?	
	9 ?	?				10 ?	?	
11 ?	?		12 ?	?	13 ?		14 ?	15 ?
?			?		?			?
16 ?	17 ?		18 ?	?	?		19 ?	?
	20 ?	21 ?				22 ?	?	
	23 ?	?	?		24 ?		25 ?	26 ?
		?		27 ?	?	?		?

DOWN

1. $b + 2b$
3. $(x^2 + 1) - (x^2 + 10)$
4. $(-3r + 1) + (4r - 4)$
6. $(d + e) + (-2e)$
8. $(5x + 3) + (5x - 3)$
11. $(8 + 3w) - (6 + 2w)$
12. $(b + a) - (b - c)$
13. $(3b + 2) - (2b + 6)$
15. $(4y - 2) - (3y + 3)$
17. $(2p - 9) + (-p + 1)$
19. $(a + 4b) + (7b - a)$
21. $(7r + 2) - (6r - 4)$
24. $(x^2 + 12) + (8 - x^2)$
26. $-3cf + 4cf$

2. Consider the fractions $\frac{6}{x}$ and $\frac{2}{x}$.

a. Explain why they are like terms.

b. (1) Explain how you would subtract these like fractions.

(2) What is the difference $\frac{6}{x} - \frac{2}{x}$?

(3) Verify your result by finding the value of each fraction and the value of the difference when x is a specific value, say $x = 2$.

13.3 Multiplication

Powers

How many square units are in this square whose side measures 10 units?

The number of square units inside is:

10^2 or 10×10 or 100

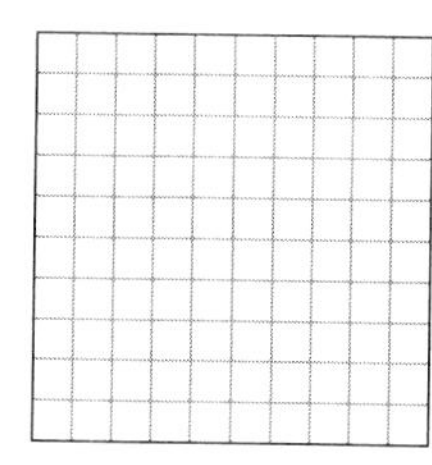

Recall that an exponent is a short way of writing factors that repeat.

$10^2 = 10 \times 10$
10 squared, or
10 to the second power

$5^3 = 5 \times 5 \times 5$
5 cubed, or
5 to the third power

$3^4 = 3 \times 3 \times 3 \times 3$
3 to the fourth power

How many factors of 7 do you think there are in $7^3 \times 7^2$?

$$7^3 \quad \times \quad 7^2$$
$$7 \times 7 \times 7 \quad \times \quad 7 \times 7$$

There are 5 factors of 7.

$$7^3 \times 7^2 = 7^5$$

How many factors of 4 do you think there are in $4^5 \times 4^3$?

$$4^5 \quad \times \quad 4^3$$
$$4 \times 4 \times 4 \times 4 \times 4 \quad \times \quad 4 \times 4 \times 4$$

There are 8 factors of 4.

$$4^5 \times 4^3 = 4^8$$

From these results, what rule can you follow to multiply powers of like bases?

To multiply powers of like bases, keep the base and add the exponents. $x^a \cdot x^b = x^{a+b}$

Example 1 Find each product.

a. $2^3 \cdot 2^4 = 2^{3+4} = 2^7$

b. $x^4 \cdot x^2 = x^{4+2} = x^6$

Whether the base is a number or a letter, when multiplying powers of like bases, keep the base and add the exponents.

c. $z \cdot z^3 = z^{1+3} = z^4$

z is the same as z^1.

d. $2y^3 \cdot 4y^6 = (2 \cdot 4)y^{3+6}$
$= 8y^9$

When there are coefficients, multiply them first.

Try These *(For solutions, see page 612.)*

1. Write the number 6^4 using factors instead of an exponent.
2. Write the number $3 \times 3 \times 3 \times 3 \times 3$ using an exponent instead of factors.
3. Larry said that $2^3 \times 2^4$ equals 4^7. Do you agree or disagree? Explain.
4. Find each product.

a. $5^4 \cdot 5^2$ **b.** $y^3 \cdot y^2$ **c.** $k \cdot k^2$ **d.** $k \cdot k^2$

Multiplying by a Monomial

For the familiar algebra chips, note the dimensions (shown on the sides of the chips pictured) and the areas (shown inside).

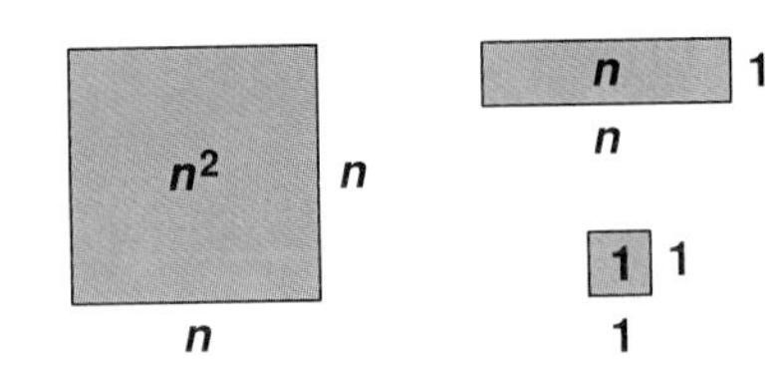

Lay out algebra chips in this pattern so that an n^2-chip and an n-chip together form a rectangle.

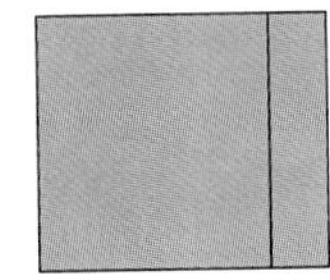

What are the dimensions of this rectangle?
What is the area of this rectangle?

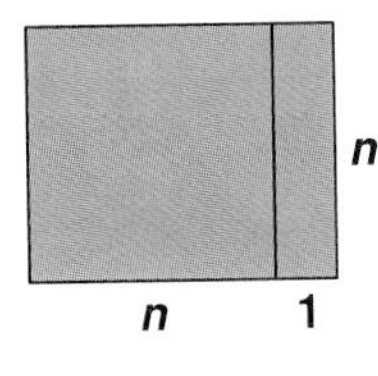

height $= n$
base $= n + 1$

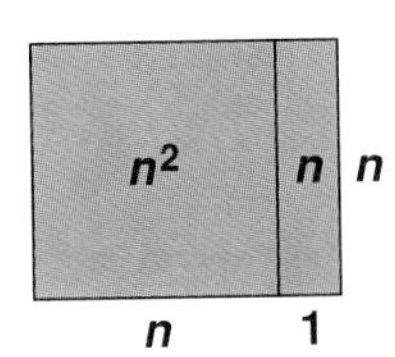

Area of Rectangle $=$ Area of n^2-chip $+$ Area of n-chip
$= n^2 + n$

Since base × height = area: $(n + 1) \times n = n^2 + n$

Example 2 Lay out 2 n^2-chips and 1 n-chip to form a rectangle. Find the dimensions and area of this rectangle.

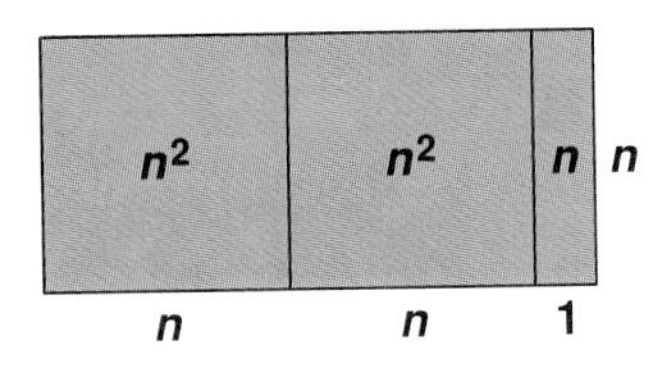

The dimensions of this rectangle are:

height $= n$ base $= 2n + 1$

Area of Rectangle $=$ Area of $2n^2$-chips $+$ Area of n-chip
$= 2n^2 + n$

Summarize the result of this model: $n(2n + 1) = 2n^2 + n$

When multiplying a polynomial by a monomial, each term of the polynomial is multiplied by the monomial.

Remember that coefficients are multiplied and exponents of like bases are added.

Example 3 Multiply $3x^2 + 4x$ by $2x$.

$2x\,(3x^2 + 4x)$	Multiply a polynomial by a monomial.
$2x \cdot 3x^2 + 2x \cdot 4x$	The monomial multiplies each term.
$2 \cdot 3x^{1+2} + 2 \cdot 4x^{1+1}$	Multiply coefficients. Add exponents of like bases.

Answer: $2x\,(3x^2 + 4x) = 6x^3 + 8x^2$

Try These (For solutions, see page 612.)

1. In each of the following models, n^2-chips and n-chips are laid out to form a rectangle. Find the dimensions and area of the rectangle.
Summarize the result of each model.

a.

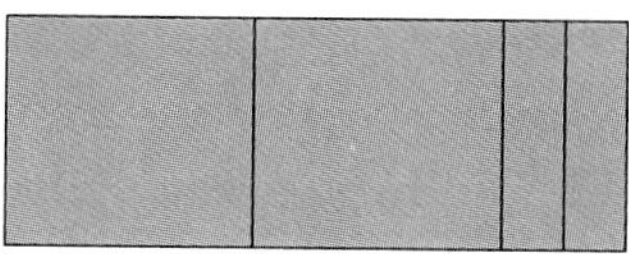

b.

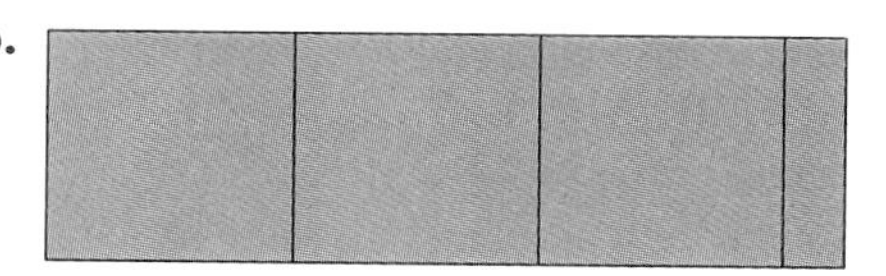

2. Multiply: **a.** $3x(2x + 1)$ **b.** $2q(3q^2 + 5)$ **c.** $4t(t^2 + 2t)$

Multiplying Two Binomials

Use algebra chips to build a rectangle whose dimensions are $(n + 3)$ by $(n + 2)$.

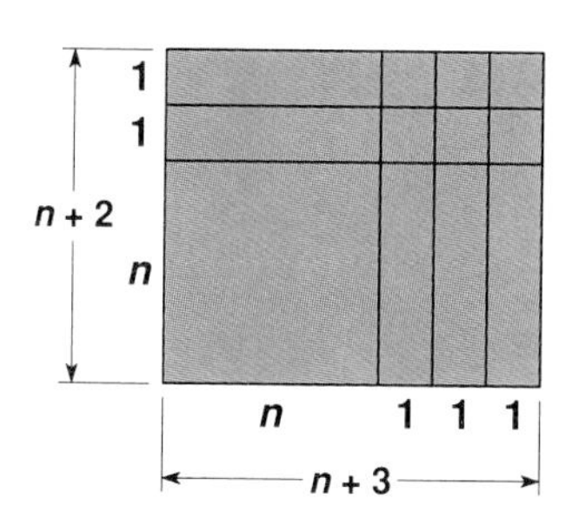

Find the area of this rectangle.

n	1	1	1
n	1	1	1
n^2	n	n	n

The rectangle contains one n^2-chip, five n-chips, six 1-chips.

Summarize the result: $(n + 3)(n + 2) = n^2 + 5n + 6$

Example 4 Use algebra chips to model the product $(n + 4)(n + 2)$.

Lay out chips to build a rectangle with dimensions $(n + 4)$ by $(n + 2)$.

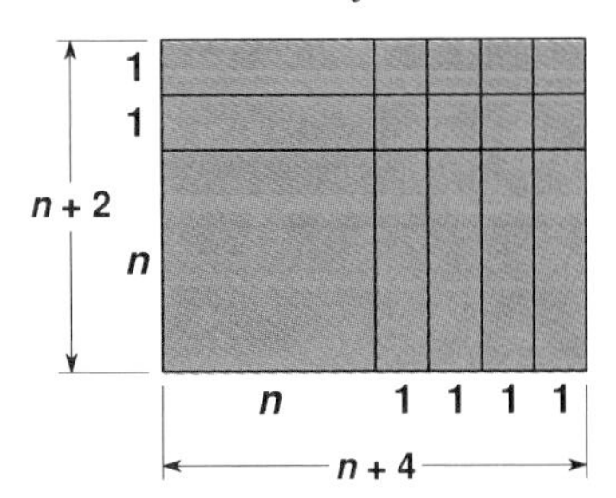

Count the chips that make up the rectangle.

n	1	1	1	1
n	1	1	1	1
n^2	n	n	n	n

The rectangle contains one n^2-chip, six n-chips, eight 1-chips.

Summary: $(n + 4)(n + 2) = n^2 + 6n + 8$

The results of these models show a way to multiply 2 binomials.

The first term and last term of the product come from the first and last terms of the binomials.

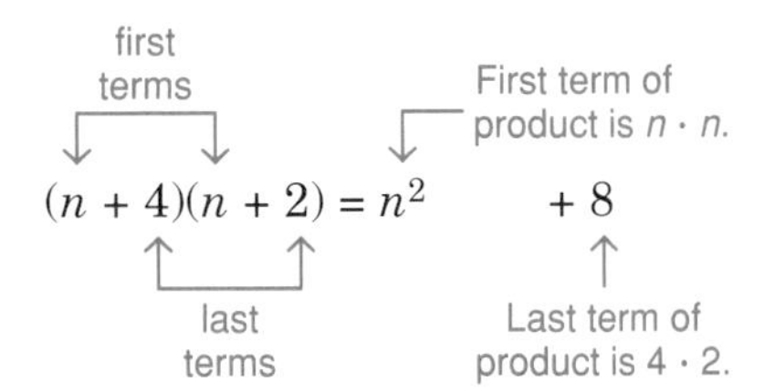

The middle term of the product comes from the outer and inner terms of the binomials.

outer terms; Middle term of product is $2n + 4n$.

$$(n + 4)(n + 2) = n^2 + 6n + 8$$

inner terms

Example 5 Find the product $(n + 1)(n + 3)$.

Multiply the first terms of the 2 binomials to get the first term of the product.

$$(n \quad)(n \quad) = n^2$$

Multiply the last terms of the 2 binomials to get the last term of the product.

$$(n + 1)(n + 3) = n^2 \quad + 3$$

Multiply the outer terms and multiply the inner terms. Add these results to get the middle term of the product.

$3n$ (outer); $3n + 1n$

$$(n + 1)(n + 3) = n^2 + 4n + 3$$

$1n$ (inner)

Answer: $(n + 1)(n + 3) = n^2 + 4n + 3$

FOIL Method for Multiplying Binomials

Multiply the First terms to get the first term of the product.

Multiply the Outer terms and multiply the Inner terms. Add these results to get the middle term of the product.

Multiply the Last terms to get the last term of the product.

Example 6 Find the product $(x + 4)(x - 2)$.

Multiply the first terms of the 2 binomials to get the first term of the product.

$$(x \quad)(x \quad) = x^2$$

Multiply the last terms of the 2 binomials to get the last term of the product.

$$(x + 4)(x - 2) = x^2 \quad - 8$$

Multiply the outer terms and multiply the inner terms. Add these results to get the middle term of the product.

$-2x$ (outer), $+4x$ (inner), $-2x + 4x$

$$(x + 4)(x - 2) = x^2 + 2x - 8$$

Answer: $(x + 4)(x - 2) = x^2 + 2x - 8$

Try These *(For solutions, see page 613.)*

1. **a.** Use algebra chips to build a rectangle whose dimensions are $(n + 5)$ by $(n + 2)$.
   b. Count chips to find the area of this rectangle.
   c. Summarize the result of the model.
2. Use algebra chips to model the product $(n + 4)(n + 3)$. Summarize the result of the model.
3. Use the FOIL method to find each product. (For parts a and b, compare your results to the results of the models in Try These 1 and 2.)
   a. $(n + 5)(n + 2)$ **b.** $(n + 3)(n + 4)$ **c.** $(x + 4)(x - 3)$

SELF-CHECK: SECTION 13.3

1. a. How do you read the number 4^3?

Answer: 4 cubed, or 4 to the third power

b. What does 4^3 mean?

Answer: 4^3 means 3 factors of 4, or $4 \times 4 \times 4$

2. a. How many factors of 2 are in the product $2^3 \cdot 2^4$?

Answer: There are 7 factors of 2 in the product.

$$\underset{2 \cdot 2 \cdot 2}{2^3} \cdot \underset{2 \cdot 2 \cdot 2 \cdot 2}{2^4}$$

b. How many factors of x are in the product $x^3 \cdot x^4$?

Answer: There are 7 factors of x in the product.

$$\underset{x \cdot x \cdot x}{x^3} \cdot \underset{x \cdot x \cdot x \cdot x}{x^4}$$

c. What rule do you apply when multiplying powers of like bases?

Answer: When multiplying powers of like bases, keep the base and add the exponents.

d. Does the rule apply to like bases that are numbers as well as to like bases that are letters?

Answer: Yes. $2^3 \cdot 2^4 = 2^{3+4} = 2^7$ and $x^3 \cdot x^4 = x^{3+4} = x^7$

3. For the product $2x(x + 4)$, which terms of the binomial $x + 4$ are to be multiplied by the monomial $2x$?

Answer: Both terms of $x + 4$ are to be multiplied by $2x$.
$2x(x + 4) = 2x^2 + 8x$

4. For the product $(n + 3)(n + 5) = n^2 + 8n + 15$:

a. Explain the relation among the first terms.

Answer: Multiply the first terms of the binomials to get the first term of the product. $(n \quad)(n \quad) = n^2$

b. Explain the relation among the last terms.

Answer: Multiply the last terms of the binomials to get the last term of the product. $(n + 3)(n + 5) = n^2 \qquad + 15$

c. Explain the relation among the middle terms.

Answer: Multiply the outer terms and multiply the inner terms. Add these results to get the middle term of the product.

5n (outer terms) 5n + 3n

$$(n + 3)(n + 5) = n^2 + 8n + 15$$

3n (inner terms)

EXERCISES: SECTION 13.3

1. Write each expression using factors instead of the exponent.

a. 5^4 **b.** 2^5 **c.** y^2 **d.** z^3 **e.** x^2y^3

2. Write each expression using an exponent instead of factors.

a. $4 \times 4 \times 4 \times 4 \times 4$ **b.** $y \cdot y \cdot y$ **c.** $x \cdot x \cdot x \cdot z \cdot z$

3. Explain why you agree or disagree with each statement.

a. Sarah said that: $3^2 \cdot 3^4 = 9^8$ **b.** Steve said that: $3^2 \cdot 3^4 = 9^6$

c. Shawn said that: $3^2 \cdot 3^4 = 3^8$ **d.** Sonya said that: $3^2 \cdot 3^4 = 3^6$

4. Find each product.

a. $3^3 \cdot 3^2$ **b.** $y^3 \cdot y^5$ **c.** $t^4 \cdot t$ **d.** $a^2 \cdot a^2$ **e.** $4^2 \cdot 4^5$

f. $3n(4n)$ **g.** $4x^2(5x^2)$ **h.** $-2t(3t^3)$ **i.** $n^2(4n^5)$ **j.** $-6c^2(-c)$

k. $5r^2(0.2r^3)$ **l.** $15x^4\left(\frac{1}{3}x^4\right)$ **m.** $3p\left(\frac{1}{3}p^3\right)$ **n.** $0.1m(1.4m)$ **o.** $10x(2y^2)$

p. $a(ab)$ **q.** $2t^3(3tz)$ **r.** $z(y^2z^2)$ **s.** $(xy)(xy)$ **t.** $ab(a^2b)$

5. **a.** Express the area of each plane figure in simplest form. (Leave π.)

(1)

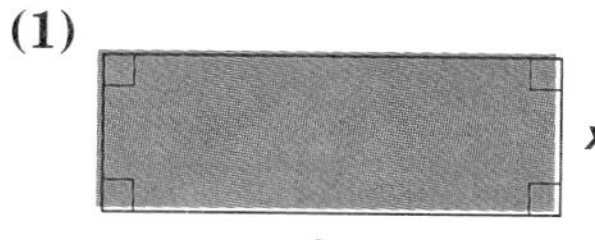

(2)

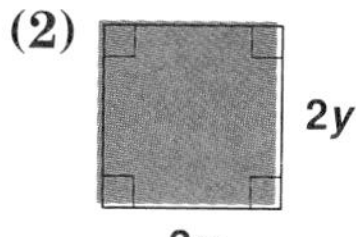

(3)

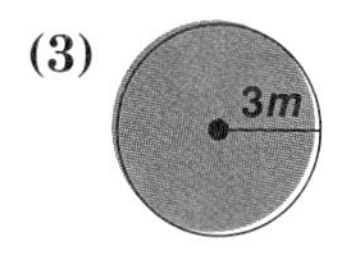

b. Express the volume of each solid figure in simplest form. (Leave π.)

(1)

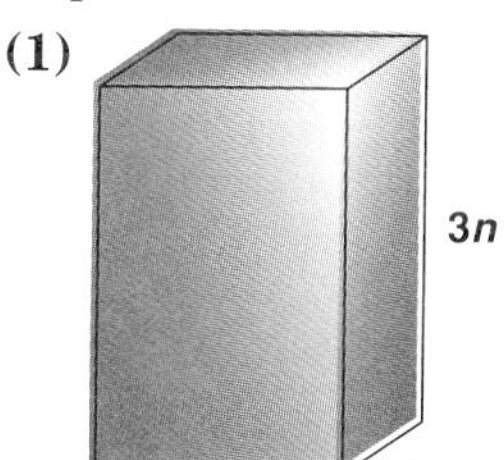

(2)

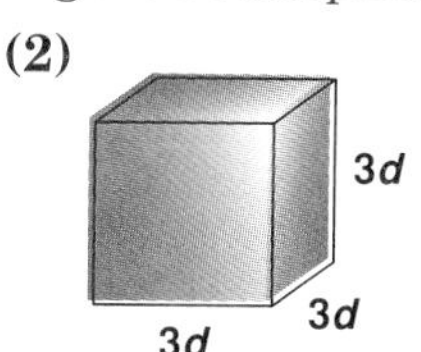

(3)

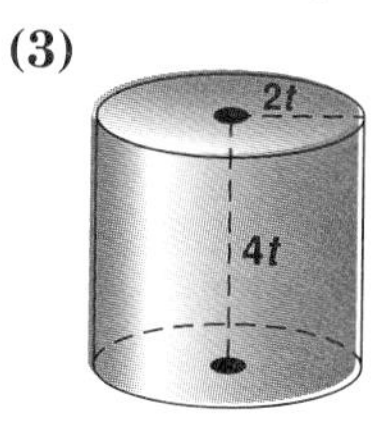

6. Express each result in simplest form.

a. Nola earns p dollars per hour. If she worked $4p$ hours, how much did she earn?

b. Jack traveled for $2q$ hours at the rate of $5q$ miles per hour. How many miles did he cover?

c. In an auditorium, there are $7r$ rows with $6r$ seats in each row. How many seats are in the auditorium?

d. For a survey, $3q$ questionnaires were prepared; each had 5 questions. If q^2 of the questions were repeats, how many different questions were prepared?

e. Ruth had $12b^2$ books to pack. After packing $2b$ books in each of $5b$ cartons, how many books had she left to pack?

7. **a.** Write the number $(2^3)^2$ using factors instead of exponents.
How many factors of 2 are there?

b. Write the expression $(x^2)^4$ using factors instead of exponents.
How many factors of x are there?

c. Without first writing out the factors, tell how many factors of y you think there are in the term $(y^4)^3$.

d. From your observations in parts a–c, write a rule that tells how to raise a power to a power. Use the symbols $(x^a)^b$ to write your rule.

e. Apply your rule to choose the equivalent of $(10^2)^3$ from the following:
(1) 10^5 (2) 10^6 (3) 100^6 (4) $1{,}000^5$

8. In each of the following models, n^2-chips and n-chips are laid out to form a rectangle.
Find the dimensions and area of the rectangles.
Summarize the result of each model.

a.

b.

c.
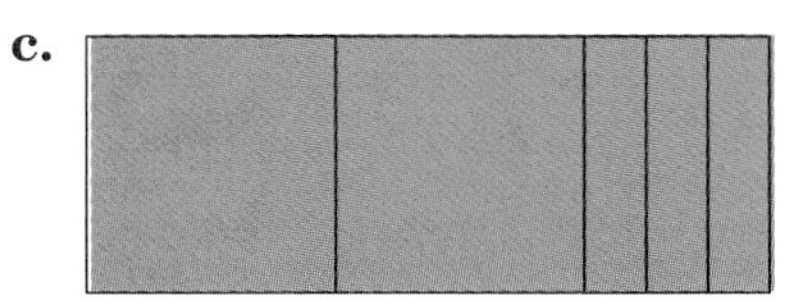

9. Multiply.

a. $3(x - 4)$ **b.** $-5(x + 3)$ **c.** $2a(a - 4)$

d. $-5(-2k + 1)$ **e.** $x(x + 2)$ **f.** $3t(t + 5)$

g. $m(m - 2)$ **h.** $5q(3q - 7)$ **i.** $2p(p^2 + 1)$

j. $3w(2w^2 - 3)$ **k.** $4y(y^2 + y)$ **l.** $-3k(4k^2 + k)$

m. $y(y^2 - 5y + 2)$ **n.** $x(x^2 + 5x + 1)$ **o.** $r(r^2 - r - 1)$

p. $3n(n^2 + 4n - 5)$ **q.** $2y(3y^2 - 2y + 1)$ **r.** $5a(2a^2 + 4a - 3)$

10. **a.** Express the area of each plane figure in simplest form.

(1)
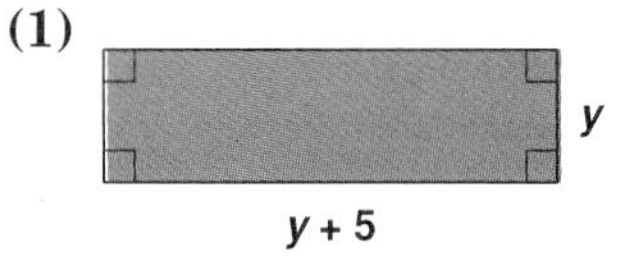

(2)
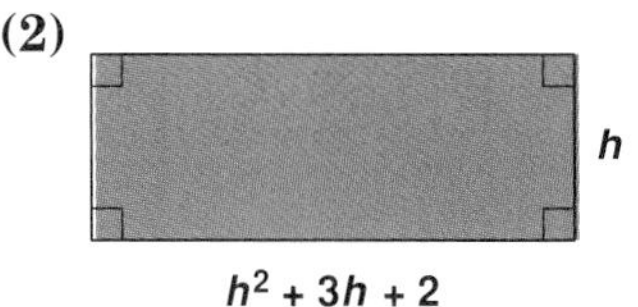

(3)
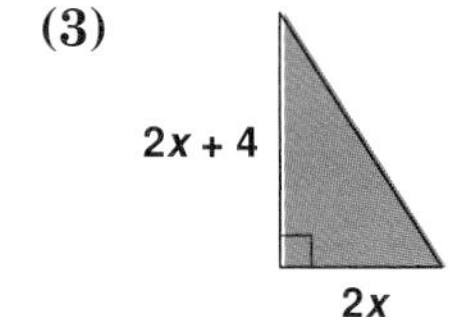

b. Express the volume of each solid figure in simplest form. (Leave π.)

(1) 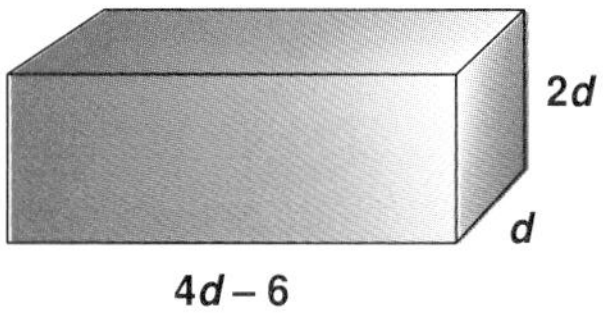
2d
d
4d – 6

(2)
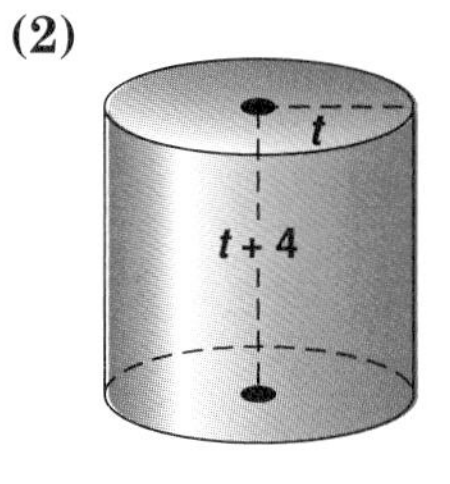

(3)
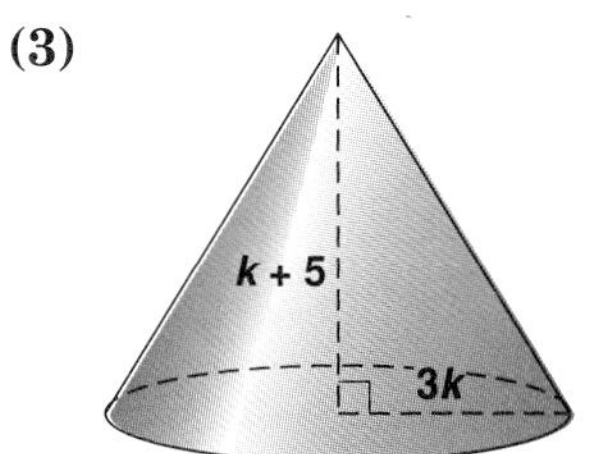

11. Express each result in simplest form.

a. Allie worked $3h$ hours a day for $(h + 8)$ days. How many hours did she work in all?

b. Jan did $2p$ pushups every morning for $(2p + 4)$ mornings. How many pushups did she do in all?

c. In Canyon county, there were $(e + 5)$ elementary school classes with an average of $4e$ children in each class. After an earthquake damaged the schools in Dale County, the Canyon elementary schools took in $(2e^2 - 3e + 5)$ more children. How many elementary school students were then in the Canyon schools?

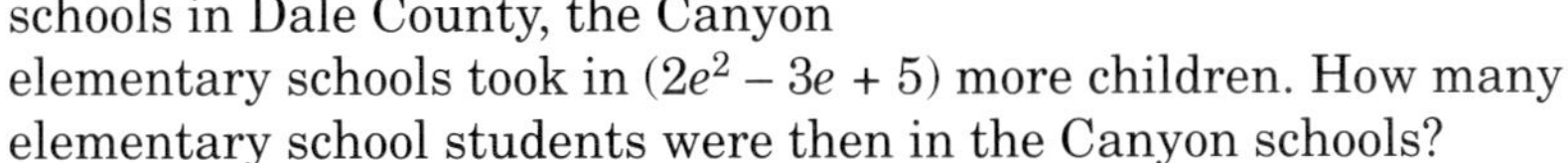

d. Ed's Electronics ordered $4r$ radios that cost Ed $(2r + 3)$ dollars each. If he sold all of the radios for a total of $(14r^2 + 16r)$ dollars, how much was Ed's profit?

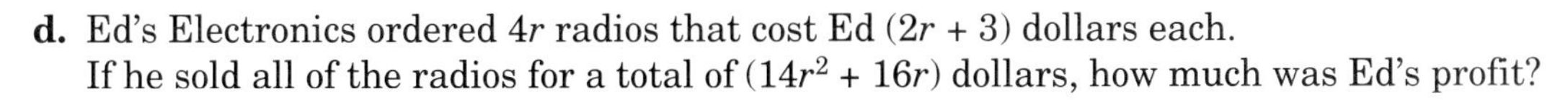

e. Marki had to drive a total of $(3m^2 + 4m + 1)$ miles to reach Culver City. Before lunch, she drove at the rate of $2m$ miles per hour for $(m + 3)$ hours. How many miles are left to drive?

12. Use algebra chips to build each rectangle with the given dimensions. Count the chips used to find the area of each rectangle. Summarize the result of each model.

a. $(n + 4)$ by $(n + 1)$ **b.** $(n + 5)$ by $(n + 2)$

c. $(n + 1)$ by $(n + 6)$ **d.** $(n + 3)$ by $(n + 5)$

e. $(n + 6)$ by $(n + 2)$ **f.** $(n + 4)$ by $(n + 4)$

13. Find the product of each pair of binomials.

a. $(y + 2)(y + 3)$ **b.** $(p + 8)(p + 2)$ **c.** $(k + 7)(k + 5)$

d. $(n + 6)(n - 1)$ **e.** $(x + 1)(x - 7)$ **f.** $(y + 2)(y - 4)$

g. $(z - 8)(z + 3)$ **h.** $(t - 10)(t + 1)$ **i.** $(n - 4)(n + 5)$

j. $(p - 3)(p - 5)$ **k.** $(t - 8)(t - 2)$ **l.** $(r - 6)(r - 3)$

m. $(w + 4)(w + 3)$ **n.** $(q + 2)(q - 5)$ **o.** $(k - 3)(k + 2)$

p. $(2b + 1)(b + 4)$ **q.** $(3h + 2)(h - 1)$ **r.** $(4p - 2)(p - 3)$

s. $(g + 2)(2g + 3)$ **t.** $(k - 1)(2k - 4)$ **u.** $(y + 1)(2y - 3)$

14. **a.** Express the area of each plane figure in simplest form.

(1)

$x + 2$

$x + 4$

(2)

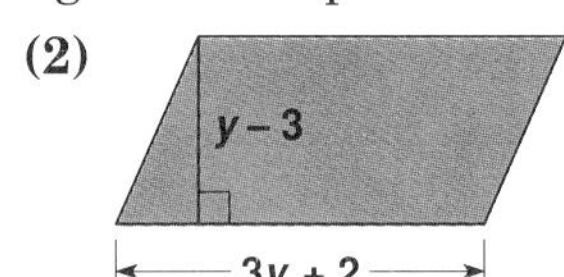

(3)

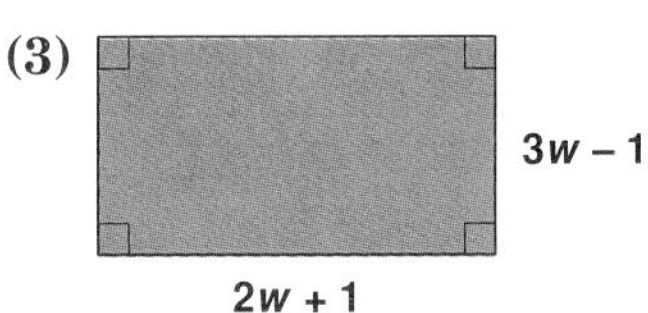

b. Express the volume of each solid figure in simplest form.

(1)

4

$x + 2$

$x + 5$

(2)

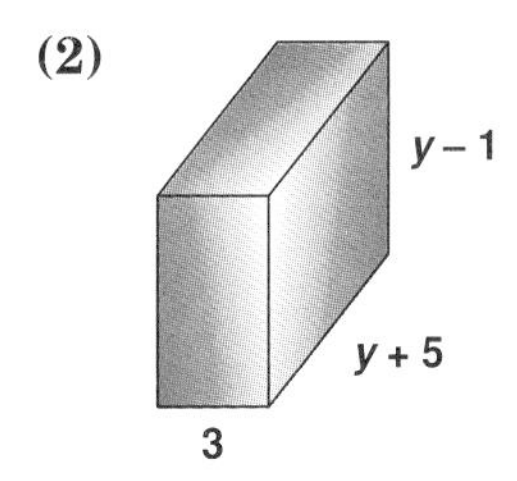

(3)

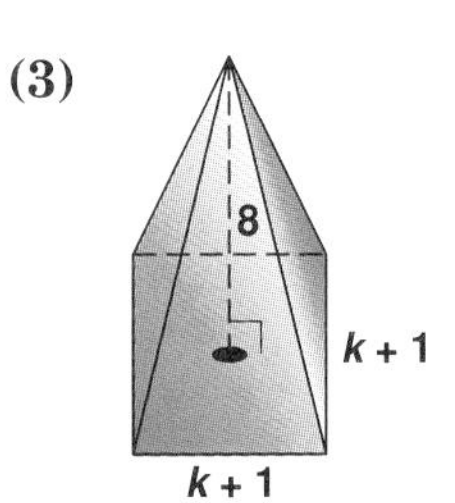

15. Express each result in simplest form.

a. Computan Manufacturing uses $(m + 1)$ microchips in each of its $(m + 4)$ Model HK computers. How many microchips are needed?

b. Lin's Music Shop stocked $(x + 1)$ copies of each of the $(3x + 2)$ top CDs. If $(x^2 + 2x + 1)$ CDs are already sold, how many are left?

c. Zach laid out a rectangular vegetable garden that was $(f + 2)$ feet by $(f - 1)$ feet. If he then increased each side by 2 feet, by how much had he increased the area of the garden?

d. Myra laid out a vegetable garden that was an $(f + 3)$ foot by $(f + 3)$ foot square. She then changed the shape of the garden to a rectangle by increasing the length by 4 feet. By how much had she increased the area?

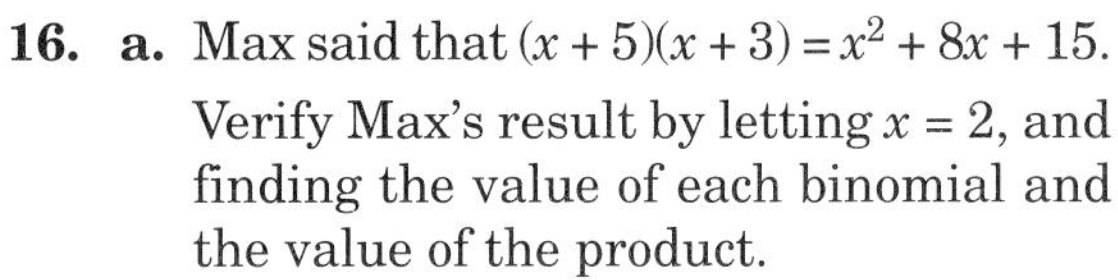

16. **a.** Max said that $(x + 5)(x + 3) = x^2 + 8x + 15$. Verify Max's result by letting $x = 2$, and finding the value of each binomial and the value of the product.

b. Determine if each product is correct or incorrect by letting $x = 2$. If a product is incorrect, give the correct answer.

(1) $(x + 9)(x + 2) = x^2 + 11x + 18$

(2) $(x - 7)(x + 3) = x^2 + 4x - 21$

(3) $(x - 2)(x - 5) = x^2 - 7x - 10$

(4) $(x + 2)(x - 8) = x^2 - 6x - 16$

EXPLORATIONS

1. a. (1) Together, these algebra chips form a square. What is the measure of each side of the square?

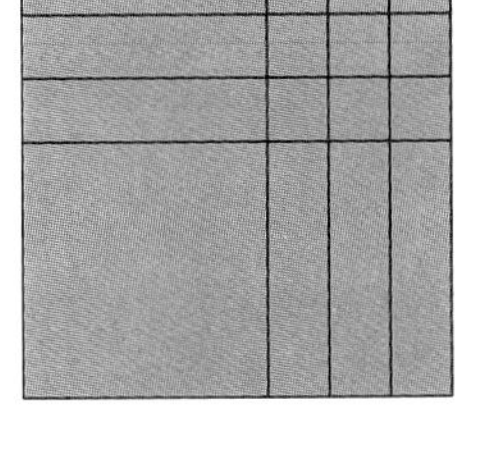

(2) Count chips to determine the area of the square.

(3) Use the FOIL method to find the product $(n + 3)(n + 3)$.

What do you notice about these two binomials?

What happens when you multiply the outer and inner terms of these two identical binomials?

Rewrite the multiplication $(n + 3)(n + 3)$ using an exponent instead of the factors.

b. (1) Using algebra chips, set up an $(n + 4)$ by $(n + 4)$ square.

(2) Count chips to determine the area of the square.

(3) Use the FOIL method to find the product $(n + 4)(n + 4)$, or $(n + 4)^2$.

c. (1) Study the results from parts a and b:

$(n + 3)^2 = n^2 + 6n + 9$ $\quad\quad$ $(n + 4)^2 = n^2 + 8n + 16$

Predict the product for $(n + 5)^2$.

Use the FOIL method to verify your prediction.

(2) Write a rule that tells how to square a binomial.

Use the symbols $(a + b)^2$ to write your rule.

(3) Apply your rule to choose the equivalent of $(x + 10)^2$ from the following:

(a) $x^2 + 100$ $\quad\quad$ (b) $x^2 + 10x + 100$ $\quad\quad$ (c) $x^2 + 20x + 100$

(4) Apply your rule to find each of these squares:

$(y + 7)^2$ $\quad\quad$ $(z + 9)^2$ $\quad\quad$ $(q + 11)^2$ $\quad\quad$ $(w + 20)^2$

d. The rule for squaring a binomial, $(a + b)^2 = a^2 + 2ab + b^2$, can be applied to finding the square of a number.

Example: Evaluate 23^2.

$23^2 = (20 + 3)^2$ — Rewrite 23 as the sum of two numbers whose squares you know.

$= 20^2 + 2(20)(3) + 3^2$ — Apply the rule for squaring a binomial.

$= 400 + 120 + 9$

$23^2 = 529$

Use this method to find these squares:

19^2 $\quad\quad$ 28^2 $\quad\quad$ 32^2 $\quad\quad$ 45^2

Check your results on a calculator.

2. **a.** **(1)** Use the FOIL method to find the product $(x + 3)(x - 3)$.

What do you notice about these two binomials?

What happens when you multiply the outer and inner terms of these binomials?

How many terms are in the product?

(2) Use the FOIL method to find the product $(x + 4)(x - 4)$.

b. **(1)** Study the results from part a:

$(x + 3)(x - 3) = x^2 - 9$ $\qquad$ $(x + 4)(x - 4) = x^2 - 16$

Predict the product for $(x + 5)(x - 5)$.

Use the FOIL method to verify your prediction.

(2) Write a rule that tells how to multiply binomials that differ only in the signs of the constant terms.

Use the symbols $(a + b)(a - b)$ to write your rule.

(3) Apply your rule to choose the equivalent of $(x + y)(x - y)$ from the following:

(a) $x^2 + y^2$ $\qquad$ (b) $x^2 - y^2$ $\qquad$ (c) $x^2 - 2xy - y^2$

(4) Apply your rule to find each of these products:

$(n + 9)(n - 9)$ $\qquad$ $(t + 10)(t - 10)$ $\qquad$ $(r + 20)(r - 20)$

c. The rule for multiplying two binomials that differ only in the signs of the constant terms, $(a + b)(a - b) = a^2 - b^2$, can be applied to simplify certain computations.

Example: Evaluate 25×15.

$25 \times 15 = (20 + 5)(20 - 5)$ Rewrite the given numbers as the sum and difference of two numbers whose squares you know.

$= 20^2 - 5^2$ Apply the rule for multiplying two binomials that differ only in the signs of the constant terms.

$= 400 - 25$

$25 \times 15 = 375$

Use this method to find these squares:

19×21 $\qquad$ 17×23 $\qquad$ 28×32 $\qquad$ 26×34

Check your results on a calculator.

13.4 Division

Powers

What is the width of this rectangle whose area is 32 square units and whose length is 8 units?

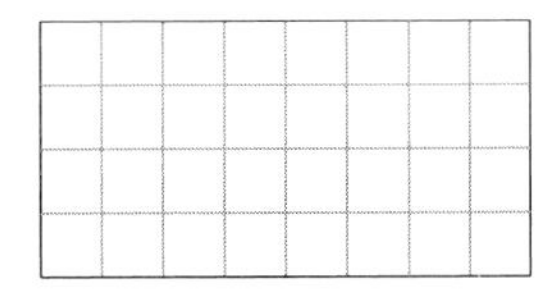

To find the width, divide the area by the length:

$$\text{width} = \frac{\text{area}}{\text{length}} = \frac{32 \text{ sq. units}}{8 \text{ units}} = 4 \text{ units}$$

Do this same arithmetic using powers of 2: $\frac{32}{8} = \frac{2^5}{2^3} = \frac{\cancel{2} \cdot \cancel{2} \cdot \cancel{2} \cdot 2 \cdot 2}{\cancel{2} \cdot \cancel{2} \cdot \cancel{2}} = 2^2$

How many factors of 3 do you think there are in $\frac{3^6}{3^2}$?

$$\frac{3^6}{3^2} = \frac{\cancel{3} \cdot \cancel{3} \cdot 3 \cdot 3 \cdot 3 \cdot 3}{\cancel{3} \cdot \cancel{3}} = 3^4$$

There are 4 factors of 3.

$$\frac{3^6}{3^2} = 3^4$$

How many factors of x do you think there are in $\frac{x^4}{x^3}$?

$$\frac{x^4}{x^3} = \frac{\cancel{x} \cdot \cancel{x} \cdot \cancel{x} \cdot x}{\cancel{x} \cdot \cancel{x} \cdot \cancel{x}} = x^1$$

There is 1 factor of x.

$$\frac{x^4}{x^3} = x^1 \text{ or } x$$

From these results, what rule can you follow to divide powers of like bases?

To divide powers of like bases, keep the base and subtract the exponents. $\frac{x^a}{x^b} = x^{a-b}$

Example 1 Find each quotient.

a. $\frac{7^9}{7^3} = 7^{9-3} = 7^6$ Whether the base is a number or a letter, when dividing powers of like bases, keep the base and subtract the exponents.

b. $\frac{x^5}{x^2} = x^{5-2} = x^3$

c. $\frac{y^3}{y} = y^{3-1} = y^2$ y is the same as y^1.

d. $\frac{12z^5}{4z^3} = \left(\frac{12}{4}\right)z^{5-3} = 3z^2$ When there are coefficients, divide them first.

Try These (For solutions, see page 614.)

1. Write the quotient $\frac{6^5}{6^2}$ using factors instead of exponents. Cancel common factors. How many factors of 6 are in the result?

2. Lenore said that $\frac{2^9}{2^3} = \frac{9}{3}$ or 3. Do you agree or disagree? Explain.

3. Find each quotient: **a.** $\frac{5^6}{5^2}$ **b.** $\frac{y^3}{y^2}$ **c.** $\frac{z^4}{z}$ **d.** $\frac{18m^5}{6m^3}$

Dividing by a Monomial

When picking up their stock of newspapers for delivery, Barb, Dana, and Tanya were asked by the dispatcher to equally share the papers that were in 3 stacks. The first stack had 18 papers, the second had 24 papers, and the third had 36 papers. How many papers did each of the 3 delivery girls get from these stacks?

From these 3 stacks, each girl took:

$$\frac{18 + 24 + 36}{3} \quad \text{or} \quad \frac{18}{3} + \frac{24}{3} + \frac{36}{3}$$

$$= \frac{78}{3} \qquad = 6 + 8 + 12$$

$$= 26 \text{ papers} \qquad = 26 \text{ papers}$$

In Arithmetic, the divisor of a sum of numbers divides *each* of the numbers.

Divide this set of algebra chips into 3 equal sets.

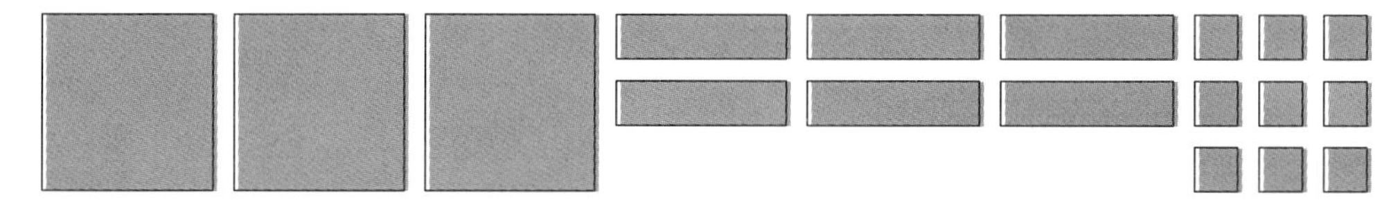

This set of chips represents the polynomial $3n^2 + 6n + 9$.

Here are the 3 equal sets of chips after the division.

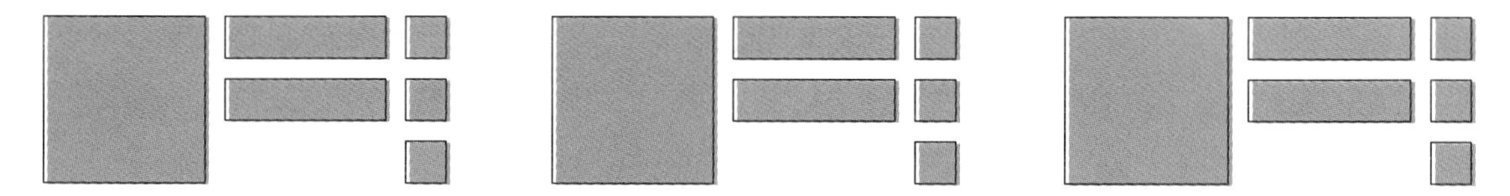

Each of these 3 equal sets contains one n^2-chip, two n-chips, three 1-chips.

Summarize the result of this model.

$$\frac{3n^2 + 6n + 9}{3} = \frac{3n^2}{3} + \frac{6n}{3} + \frac{9}{3} = n^2 + 2n + 3$$

To divide the polynomial by 3, divide each term by 3.

In Algebra, the monomial divisor of a polynomial divides *each* term.

Example 2 **a.** Divide $4x^2 + 12x - 16$ by 4.

$$\frac{4x^2 + 12x - 16}{4} = \frac{4x^2}{4} + \frac{12x}{4} - \frac{16}{4}$$ To divide the polynomial by 4, divide each term by 4.

$$= x^2 + 3x - 4$$

b. Verify your result by letting $x = 2$.

Find the value of the original polynomial when $x = 2$.

$4x^2 + 12x - 16$

$4(2)^2 + 12(2) - 16$

$4(4) + 24 - 16$

$16 + 24 - 16$

$40 - 16$

24

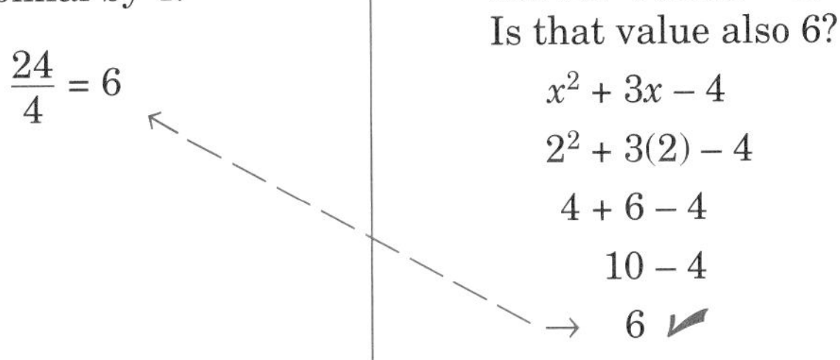

Divide the value of the polynomial by 4.

$\frac{24}{4} = 6$

Find the value of your answer when $x = 2$. Is that value also 6?

$x^2 + 3x - 4$

$2^2 + 3(2) - 4$

$4 + 6 - 4$

$10 - 4$

6 ✓

Lay out algebra chips to form a rectangle whose area is $4n^2 + 2n$. If the width of this rectangle is $2n$, what is the length?

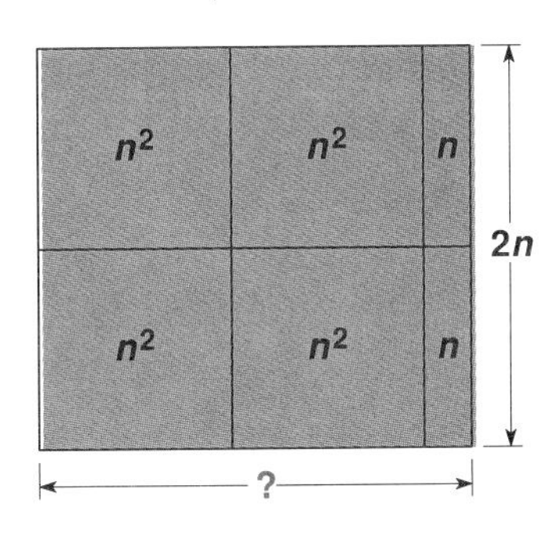

The length has the sides of two n^2-chips and the short side, 1 unit, of an n-chip. The length is $2n + 1$.

Summarize the result of this model.

$$\frac{\text{area}}{\text{width}} = \text{length} \quad \text{or} \quad \frac{4n^2 + 2n}{2n} = 2n + 1$$

Algebraically, each term of the polynomial that represents the area is divided by the monomial that represents the width.

$$\frac{4n^2 + 2n}{2n} = \frac{4n^2}{2n} + \frac{2n}{2n}$$

You can factor and cancel.

$$= \frac{\cancel{2} \cdot 2 \cdot \cancel{n} \cdot n}{\cancel{2} \cdot 1 \cdot \cancel{n}} + \frac{\cancel{2} \cdot 1 \cdot \cancel{n}}{\cancel{2} \cdot 1 \cdot \cancel{n}}$$

$$= 2\,n + 1$$

Note that when all factors of a term are cancelled, the result is 1, not 0.

Example 3 Divide $15x^3 + 6x$ by $3x$.

$$\frac{15x^3 + 6x}{3x}$$ Divide a polynomial by a monomial.

$$\frac{15x^3}{3x} + \frac{6x}{3x}$$ The monomial divides each term.

$$\frac{5 \cdot \cancel{3} \cdot \cancel{x} \cdot x \cdot x}{\cancel{3} \cdot 1 \cdot \cancel{x}} + \frac{\cancel{3} \cdot 2 \cdot \cancel{x}}{\cancel{3} \cdot 1 \cdot \cancel{x}}$$

$$5x^2 + 2$$

Answer: $\frac{15x^3 + 6x}{3x} = 5x^2 + 2$

Instead of writing out the factors to do the divisions, you can divide coefficients and subtract exponents of like bases.

Example 4 Divide $6x^4 + 12x^3$ by $3x$.

$\dfrac{6x^4 + 12x^3}{3x}$ Divide a polynomial by a monomial.

$\dfrac{6x^4}{3x^1} + \dfrac{12x^3}{3x^1}$ Divide each term by the monomial. $3x$ is the same as $3x^1$.

$\left(\dfrac{6}{3}\right)x^{4-1} + \left(\dfrac{12}{3}\right)x^{3-1}$ Divide coefficients. Subtract exponents of like bases.

$2x^3 + 4x^2$

Answer: $\dfrac{6x^4 + 12x^3}{3x} = 2x^3 + 4x^2$

Try These *(For solutions, see page 614.)*

1. Divide.

a. $\dfrac{3y + 6}{3}$ **b.** $\dfrac{10w^2 - 15w}{5}$ **c.** $\dfrac{12z^2 + 8z - 24}{4}$

2. Divide. To do the divisions in each term, factor and cancel.

a. $\dfrac{6x^2 + 3x}{3x}$ **b.** $\dfrac{21k^2 - 7k}{7k}$ **c.** $\dfrac{18m^2 + 6m}{2m}$

3. Divide. To do the divisions in each term, divide coefficients and subtract exponents of like bases.

a. $\dfrac{8t^4 + 4t^2}{2t^2}$ **b.** $\dfrac{15p^6 - 10p^4}{5p^2}$ **c.** $\dfrac{16x^4 + 4x^3 - 8x^2}{2x}$

SELF-CHECK: SECTION 13.4

1. a. How many factors of x are in the quotient $\dfrac{x^5}{x^2}$?

Answer: There are 3 factors of x in the quotient.

$$\frac{x^5}{x^2} = \frac{\cancel{x} \cdot \cancel{x} \cdot x \cdot x \cdot x}{\cancel{x} \cdot \cancel{x}} = x \cdot x \cdot x \text{ or } x^3$$

b. What rule do you apply when dividing powers of like bases?

Answer: When dividing powers of like bases, keep the base and subtract the exponents.

$$\frac{x^5}{x^2} = x^{5-2} = x^3$$

c. Does the rule apply to like bases that are numbers as well as to like bases that are letters?

Answer: Yes. $\frac{x^5}{x^2} = x^{5-2} = x^3$ and $\frac{4^5}{4^2} = 4^{5-2} = 4^3$

d. In division, how do you treat coefficients?

Answer: Divide coefficients. $\frac{8x^5}{2x^2} = \left(\frac{8}{2}\right)x^{5-2} = 4x^3$

2. For the quotient $\frac{8x+4}{4}$, which terms of the numerator are to be divided by 4?

Answer: All terms of the numerator are to be divided by 4.

$$\frac{8x+4}{4} = \frac{8x}{4} + \frac{4}{4} = 2x + 1$$

EXERCISES: SECTION 13.4

1. Rewrite each quotient using factors instead of the exponents. Write the answer using factors and then an exponent.

Example: $\frac{10^4}{10^2} = \frac{10 \cdot 10 \cdot 10 \cdot 10}{10 \cdot 10} = 10 \cdot 10 = 10^2$

a. $\frac{4^5}{4^2}$ **b.** $\frac{7^6}{7^4}$ **c.** $\frac{t^5}{t^3}$ **d.** $\frac{z^6}{z^2}$ **e.** $\frac{10m^4}{5m^2}$ **f.** $\frac{18q^3}{9q}$

2. Explain why you agree or disagree with each statement.

a. Erna said that $\frac{5^6}{5^2} = \frac{6}{2}$ or 3. **b.** Elie said that $\frac{5^6}{5^2} = 5^{6 \div 2}$ or 5^3.

c. Eddy said that $\frac{5^6}{5^2} = 5^{6-2}$ or 5^4.

3. Find each quotient.

a. $\frac{3^6}{3^4}$ **b.** $\frac{8^5}{8^2}$ **c.** $\frac{9^3}{9}$ **d.** $\frac{b^5}{b^3}$ **e.** $\frac{y^7}{y^4}$ **f.** $\frac{r^4}{r}$

g. $\frac{9n^4}{3n^2}$ **h.** $\frac{10w^6}{5w^4}$ **i.** $\frac{18p^5}{-2p^3}$ **j.** $\frac{-6z^7}{-3z}$ **k.** $\frac{-14b^3}{2b^2}$ **l.** $\frac{-8k^6}{-8k}$

4. Express each result in simplest form.

a. Marla drove a distance of $100m^3$ miles at the rate of $25m$ miles per hour. How many hours did she drive?

b. Carlos earned $50d^4$ dollars for $5d$ hours of work. What was his hourly rate?

c. The area of this rectangle is $32y^2$ square units. If the width is $8y$ units, what is the length?

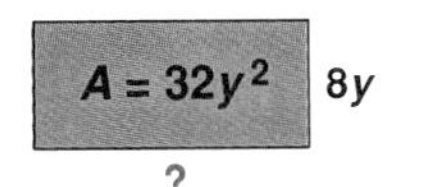

d.

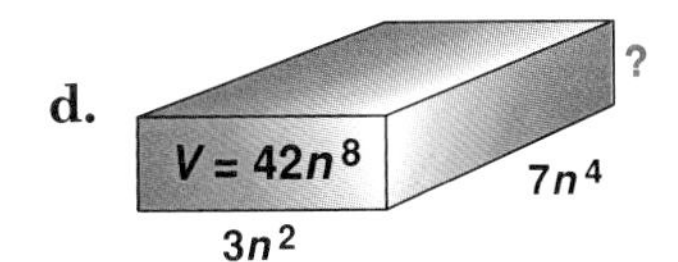

The volume of this prism is $42n^8$ cubic units. If the length is $3n^2$ units and the width is $7n^4$ units, what is the height?

e. $2m^2$ children equally shared $16m^3$ marbles. Cliff, one of the children, already had $7m$ marbles. How many marbles does Cliff have now?

5. Tell which polynomial is represented by the algebra chips in each set. Then tell which polynomial is represented after the required division.

a. Divide this set of algebra chips into 2 equal sets. Which polynomial is represented by each of the equal sets?

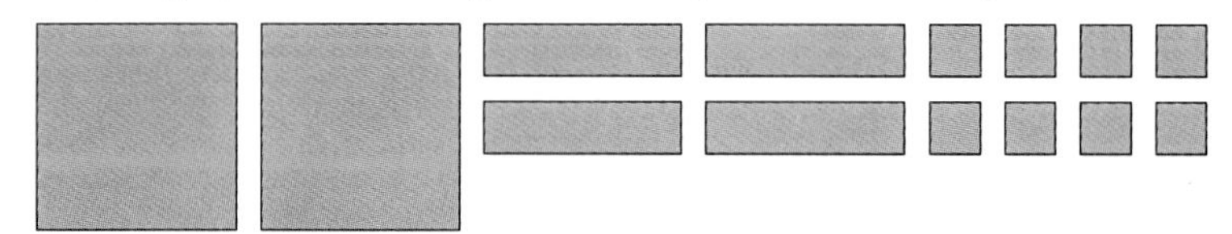

b. Divide this set of algebra chips into 3 equal sets. Which polynomial is represented by each of the equal sets?

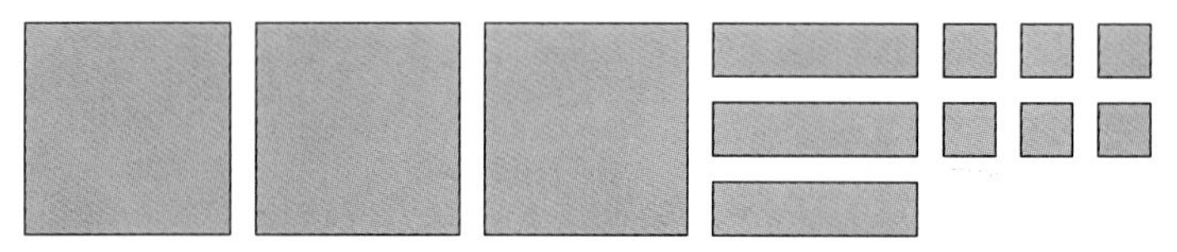

6. Divide. Verify your result by letting the variable = 2.

a. $\frac{10z + 15}{5}$ b. $\frac{16t^4 - 8t}{4}$ c. $\frac{20h^2 - 5h}{5}$

d. $\frac{12h + 9}{-3}$ e. $\frac{-64s^2 - 16}{-4}$ f. $\frac{28n^5 - 14n}{-7}$

g. $\frac{8d^2 + 16d + 8}{8}$ h. $\frac{21k^2 - 7k + 14}{7}$ i. $\frac{24y^2 - 12y - 6}{6}$

j. $\frac{9r^2 + 18r + 9}{-9}$ k. $\frac{36m^2 - 6m - 12}{-6}$ l. $\frac{48x^2 + 12x + 24}{-12}$

7. Divide. To do the divisions in each term, factor and cancel.

a. $\frac{t^3 + t^2}{t}$ b. $\frac{x^4 - x^2}{x^2}$ c. $\frac{w^3 - w^2}{w^2}$

d. $\frac{18m^2 + 6m}{6m}$ e. $\frac{25h^2 - 5h}{5h}$ f. $\frac{36x^2 + 18x}{9x}$

g. $\frac{50t^3 - 25t^2}{5t^2}$ h. $\frac{48y^2 + 12y}{-3y}$ i. $\frac{30t^3 - 10t^2}{-10t}$

j. $\frac{6x^3 + 4x^2 - 4x}{2x}$ k. $\frac{12t^3 + 6t^2 - 3t}{3t}$ l. $\frac{50q^3 - 25q^2 - 25q}{25q}$

8. Divide. To do the divisions in each term, divide coefficients and subtract exponents of like bases.

a. $\frac{6y^4 + 4y^3}{2y^2}$ b. $\frac{9z^5 - 6z^4}{3z^3}$ c. $\frac{8x^3 + 4x^2}{4x}$

d. $\frac{24m^5 + 16m^3}{-8m^2}$ e. $\frac{36p^4 - 24p^2}{-12p}$ f. $\frac{75k^4 - 50k^3}{-25k^2}$

g. $\frac{12t^4 + 6t^3 + 3t^2}{3t}$ h. $\frac{16y^5 - 8y^4 + 12y^3}{4y^2}$ i. $\frac{18y^6 + 27y^5 - 9y^4}{-9y^2}$

j. $\frac{25z^4 - 15z^3 - 20z^2}{-5z}$ k. $\frac{20t^5 - 40t^4 - 10t^3}{-10t^2}$ l. $\frac{-15k^6 + 45k^5 - 30k^3}{-15k}$

9. Explain why you agree or disagree with each statement.

a. Karla said that the quotient $\frac{x}{x}$ equals 0.

b. Kevin factored the quotient $\frac{x}{x}$ as $\frac{x \cdot 1}{x \cdot 1}$ and concluded that the result is 1.

c. Kathy concluded that any number divided by itself equals 1. She also said to except 0.

d. Kenny applied the rule of exponents to the quotient $\frac{x}{x}$ and concluded that the result is x^0.

e. Karl concluded that the value of any number raised to the zero power equals 1. He also said to except 0.

10. Divide each polynomial by the monomial shown. Express answers in simplest form.

a. $\frac{7x^5 + x^2}{x^2}$ **b.** $\frac{10m^2 + 5m}{5m}$ **c.** $\frac{5t^4 - 10t^3}{5t^3}$

d. $\frac{12w^3 + 6w^2 + 3w}{3w}$ **e.** $\frac{15y^3 - 10y^2 - 5y}{-5y}$ **f.** $\frac{14m - 42m^3}{7m}$

11. Express each result in simplest form.

a. The $(4n^2 + 12n)$ students on a school trip were seated in $2n$ buses. If the students were equally distributed among the buses, how many were in each bus?

b. Nick drove $(6m^2 - 3m)$ miles and used $3m$ gallons of gas. How many miles per gallon did he average?

c. In the first $2t$ years of a recycling program, the Town of Great Cove collected $(8t^3 - 6t^2 + 2t)$ tons of materials. What was the average number of tons per year?

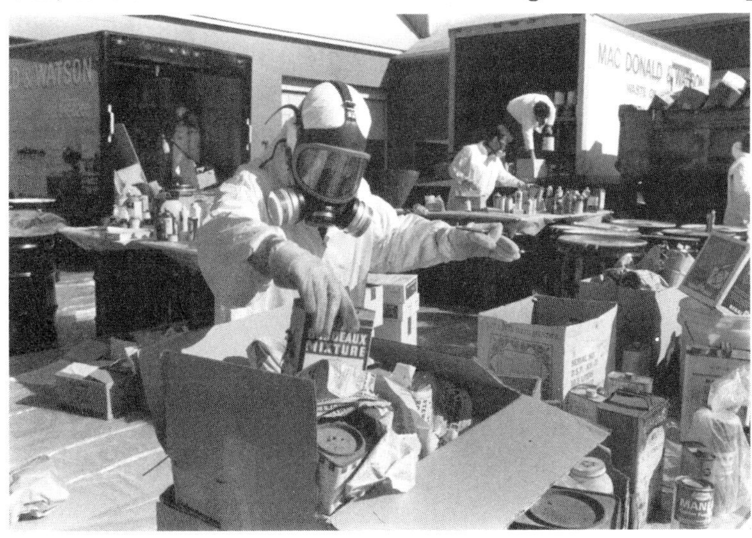

d. The Wyoming Trail Company puts $8g$ ounces of goose down into each of its down vests. How many vests can they make using $(16g^3 - 16g^2 - 24g)$ ounces of down?

e. The $3xy$ components being shipped by PQR Industries have a combined weight of $(12x^2y^2 + 15xy)$ pounds. What is the average weight of a component?

EXPLORATIONS

1. If $x^0 = 1$, evaluate each of the following expressions.

a. $2x^0$ **b.** $(2x)^0$ **c.** $2 + x^0$ **d.** $(2 + x)^0$

2. Do you agree or disagree? Explain.

a. Muriel said that the quotient $\frac{x^2}{x^3}$ equals x.

b. Miriam factored the quotient $\frac{x^2}{x^3}$ and concluded that the result is $\frac{1}{x}$.

c. Michael applied the rule of exponents to the quotient $\frac{x^2}{x^3}$ and concluded that the result is x^{-1}.

d. Mindy concluded that when the exponent of the numerator is less than the exponent of the denominator, the quotient has a negative exponent.

e. Matthew said that another way of writing x^{-1} is $\frac{1}{x}$.

3. Copy and complete the table for each number whose name is given.

Name	Standard Form	Exponential Form
ten million	10,000,000	10^7
one million	1,000,000	10^6
one hundred thousand	100,000	?
ten thousand	?	?
one thousand	?	?
one hundred	?	?
ten	?	?
one	?	?
one tenth	0.1 or $\frac{1}{10}$	?
one hundredth	?	?
one thousandth	?	?
one ten-thousandth	?	?

Things You Should Know After Studying This Chapter

KEY SKILLS

13.1 Add polynomials.

13.2 Subtract polynomials.

13.3 Multiply powers of like bases.
Multiply a polynomial by a monomial.
Multiply 2 binomials.

13.4 Divide powers of like bases.
Divide a polynomial by a monomial.

KEY TERMS

13.1 term • coefficient • base • exponent • like terms • unlike terms • polynomial • monomial • binomial • trinomial

13.2 minuend • subtrahend • difference

13.3 like bases • product

13.4 quotient

CHAPTER REVIEW EXERCISES

1. Which of the following terms is like $4x^2$?

(a) $4x^4$ (b) $4y^2$

(c) $2x^2$ (d) 42

After you have selected your answer, give a reason for rejecting each of the other choices.

2. Add.

a. $3x + 4x$ **b.** $2z^2 + 5z^2$

c. $8k^3 + (-4k^3)$ **d.** $\begin{array}{r} 2w + 5 \\ + \underline{3w + 2} \end{array}$

e. $\begin{array}{r} 5n^2 + 8n \\ + \underline{8n^2 - 5n} \end{array}$ **f.** $\begin{array}{r} -r - 7 \\ + \underline{2r + 3} \end{array}$

g. $\begin{array}{r} 3a^2 + 5a + 1 \\ + \underline{2a^2 - a + 6} \end{array}$ **h.** $\begin{array}{r} 3 - 5y + y^2 \\ + \underline{5 - 3y + y^2} \end{array}$

i. $\begin{array}{r} 2p^3 - 3p^2 + 2p \\ + \underline{2p^3 - 3p^2 - 2p} \end{array}$

3. Use algebra chips to do each subtraction.

a. $6n^2 - 2n^2$ **b.** $9n - (-3n)$

c. $-5n - (-2n)$ **d.** $\begin{array}{r} 3n^2 + 7n + 4 \\ - \underline{n^2 + 2n + 3} \end{array}$

e. $\begin{array}{r} 5n^2 - 3n - 5 \\ - \underline{n^2 + 2n - 1} \end{array}$ **f.** $\begin{array}{r} -3n^2 + 6n - 2 \\ - \underline{2n^2 - 4n - 2} \end{array}$

4. Subtract.

a. $7x^2 - 3x^2$ **b.** $9m^3 - (-2m^3)$

c. $10w - (-10w)$ **d.** $\begin{array}{r} 5p + 7 \\ - \underline{3p + 5} \end{array}$

e. $\begin{array}{r} 8b + 4 \\ - \underline{6b - 6} \end{array}$ **f.** $\begin{array}{r} 3k^2 - 7 \\ - \underline{-k^2 - 3} \end{array}$

g. $\begin{array}{r} 5x^2 - 3x + 2 \\ - \underline{3x^2 + 2x - 2} \end{array}$ **h.** $\begin{array}{r} 6r^2 - 6r - 6 \\ - \underline{5r^2 - 5r + 5} \end{array}$

i. $\begin{array}{r} -t^2 + t + 1 \\ - \underline{t^2 - t - 1} \end{array}$

5. Express each answer in simplest form.

a. Subtract $2a + 5$ from the sum of $18a + 3$ and $3a + 4$.

b. Add $2d^2 + 3d - 3$ to the difference of $10d^2 + 6d + 5$ and $7d^2 - d - 2$.

c. Subtract the sum of $8t + 2$ and $t - 4$ from the sum of $t^2 - 3t + 1$ and $t^2 - 6t + 3$.

6. Express each answer in simplest form.

a. Ashley and Frank went scuba diving. If Ashley dove down $7f$ feet and Frank dove $3f$ feet deeper, how far down did Frank dive?

b. The City Sports Complex covers a rectangular area $(5x - 3)$ meters long and $(2x + 4)$ meters wide.
What is the perimeter of the complex?

c. Before a raise to $(2x^2 + 2x - 1)$ dollars a week, Eldon's weekly salary was $(x^2 + 11x + 3)$ dollars.
What was the amount of the raise?

d. The Harris High Glee Club has $4g$ sopranos, half as many altos, $3g$ tenors, and 2 more baritones than altos.
What is the total number of singers?

e. In $\triangle ABC$, $AB = (2y - 8)$ cm and $BC = (y + 4)$ cm. If the perimeter of $\triangle ABC$ is $(5y + 2)$ cm, find AC.

7. a. Write 3^4 using factors instead of the exponent.

b. Write $5 \times 5 \times 5$ using an exponent instead of factors.

c. Do you agree or disagree with each statement? Explain.

(1) Cliff said that $2^4 = 8$.

(2) Carol said that $2^2 \times 2^4 = 4^8$.

(3) Carla said that $2^2 \times 2^4 = 2^8$.

(4) Claus said that $2^2 \times 2^4 = 2^6$.

(5) Ceila said that $(2^2)^4 = 2^8$.

d. Find each product.

(1) $5^2 \times 5^4$ **(2)** $x^3 \cdot x^4$

(3) $5t^2 \cdot 3t$ **(4)** $-3m(-2m^3)$

8. In this model, n^2-chips and n-chips are laid out to form a rectangle.

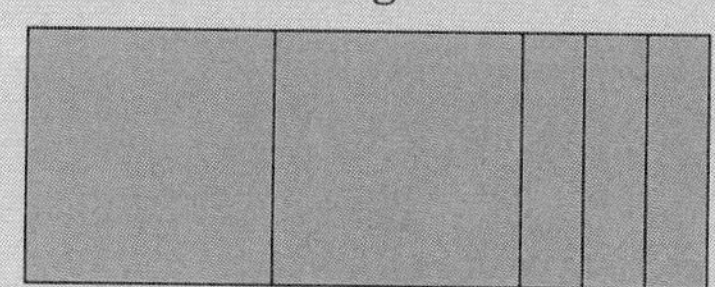

Find the dimensions and area of the rectangle.

9. Find each product.

a. $w(w + 4)$ **b.** $3z(z - 1)$
c. $-n(2n - 5)$ **d.** $5p(2p^2 - p + 3)$
e. $6m(3m^2 - 2m - 4)$ **f.** $-k(5k^2 - 3k + 2)$

10. Use algebra chips to build a rectangle that is $(n + 2)$ by $(n + 3)$. Count the chips used to find the area of the rectangle. Summarize the result of the model.

11. To multiply $(x + 4)(x + 3)$ by the FOIL method:

a. Explain how to get the first term of the product. What is the first term of the product?

b. Explain how to get the last term of the product. What is the last term of the product?

c. Explain how to get the middle term of the product. What is the middle term of the product?

12. Multiply.

a. $(x + 2)(x + 5)$ **b.** $(a - 3)(a - 7)$
c. $(d + 4)(d - 1)$ **d.** $(t - 5)(t + 2)$
e. $(3k + 2)(k + 3)$ **f.** $(2m + 5)(m - 2)$
g. $(r + 3)(r + 3)$ **h.** $(h - 4)(h - 4)$
i. $(m + 5)^2$ **j.** $(k + 4)(k - 4)$
k. $(y - 3)(y + 3)$ **l.** $(4 + h)(4 - h)$

13. a. Write the quotient $\frac{5^6}{5^2}$ using factors instead of exponents. Cancel common factors. How many factors of 5 are in the result?

b. Do you agree or disagree with each statement? Explain.

(1) David said that $\frac{5^6}{5^2} = \frac{6}{2}$.

(2) Donna said that $\frac{5^6}{5^2} = 5^3$.

(3) Danny said that $\frac{5^6}{5^2} = 5^4$.

c. Find each quotient in simplest form.

(1) $\frac{8^3}{8^2}$ **(2)** $\frac{t^5}{t^2}$ **(3)** $\frac{m^7}{m}$
(4) $\frac{16w^4}{2w}$ **(5)** $\frac{15x^3}{-5x}$ **(6)** $\frac{3r}{3r}$

14. Find each quotient in simplest form.

a. $\frac{24 + 12 + 36}{6}$ **b.** $\frac{3h - 12}{3}$
c. $\frac{8x^2 + 12x}{4x}$ **d.** $\frac{3u^3 - 3u^2}{3u^2}$
e. $\frac{10m^2 - 20m}{-10m}$ **f.** $\frac{30t^3 + 15t^2 + 45t}{15t}$

15. Express each answer in simplest form.

a. Wally's Windows is replacing all the windows in each of $3w$ buildings at Coral Cove. If each building has $(3w^2 + 12w)$ windows, how many windows must be replaced?

b. Rosa wants the area of her vegetable garden to be $(4y^2 + 8y)$ square feet. She has already marked off the width as $4y$ feet. How much should she mark off as the length?

c. How many square feet of sod are needed to cover a highway divider strip that is $(3x^2 + 2x)$ feet long and x feet wide?

d. Timothy drove $(16m^2 + 8m)$ miles on $4m$ gallons of gas. What was his average mileage per gallon?

e. If each pot of chili at Atlantic Antic can serve $(p + 3)$ people, how many can be served with $(p + 4)$ pots?

f. How many revolutions per minute are made by a fan that revolves $(4f^2 + 6f)$ times in $2f$ minutes?

SOLUTIONS FOR TRY THESE

13.1 Addition

Page 567

1. **a.** Lay out chips to represent each term.

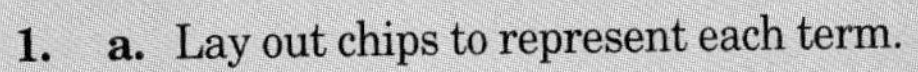

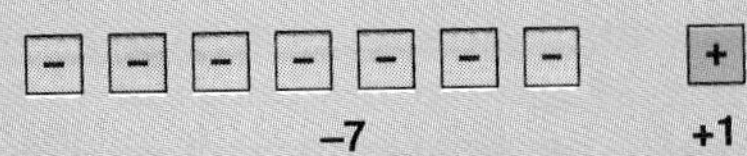

Remove zero pairs, and count the remaining chips.

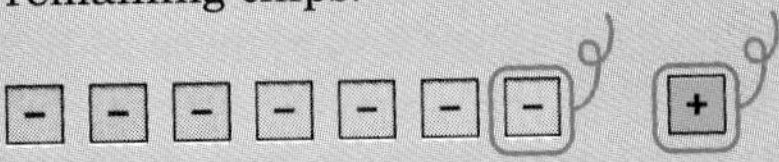

Answer: $(-7) + (1) = -6$

b.

Answer: $(-3) + (-2n) = -5n$

c.

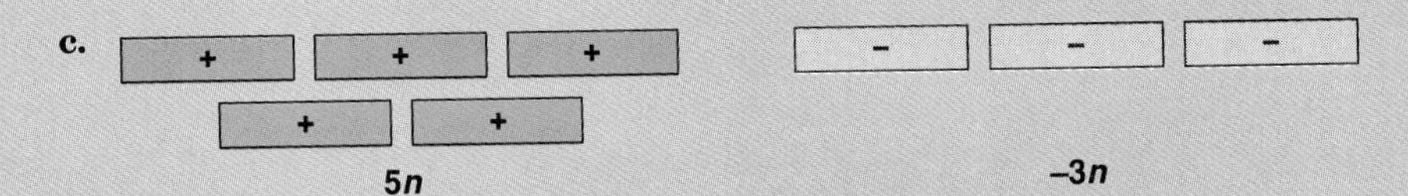

Remove zero pairs.

Answer: $(5n) + (-3n) = 2n$

d.

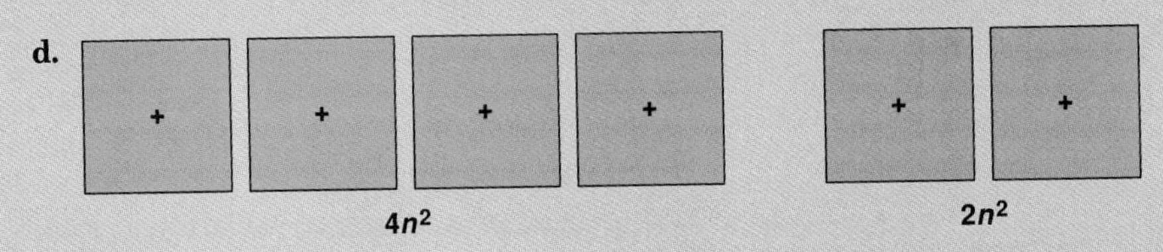

Answer: $(4n^2) + (2n^2) = 6n^2$

e.

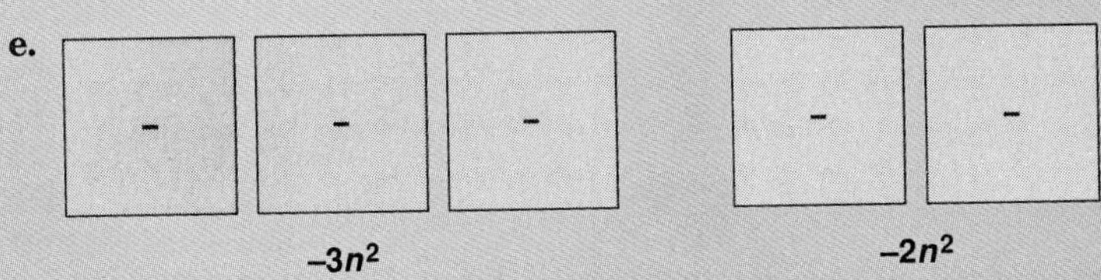

Answer: $(-3n^2) + (-2n^2) = -5n^2$

2. **a.** $\begin{array}{r} 3m \\ +\ 7m \\ \hline 10m \end{array}$ **b.** $\begin{array}{r} 10p^2 \\ +\ 2p^2 \\ \hline 12p^2 \end{array}$ **c.** $\begin{array}{r} -8k \\ +\ -2k \\ \hline -10k \end{array}$ **d.** $\begin{array}{r} 7r^2 \\ +\ -2r^2 \\ \hline 5r^2 \end{array}$ **e.** $\begin{array}{r} -9b \\ +\ 2b \\ \hline -7b \end{array}$

Page 569

1. **a.**

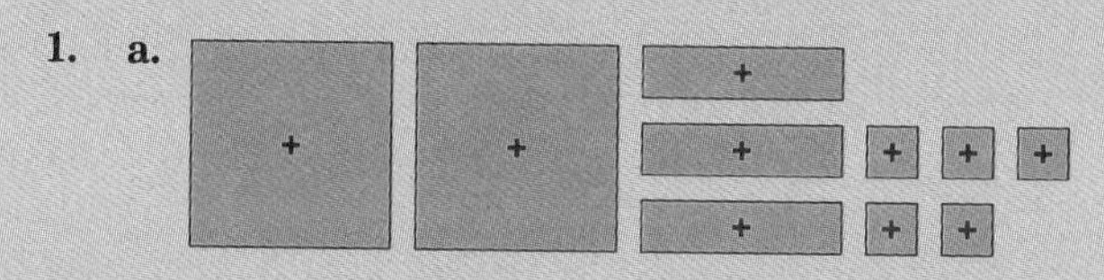

$2n^2 + 3n + 5$

b.

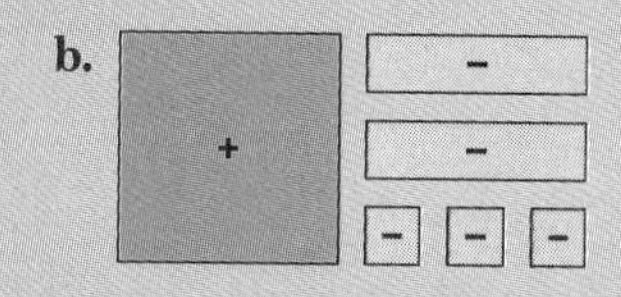

$n^2 - 2n - 3$

c.

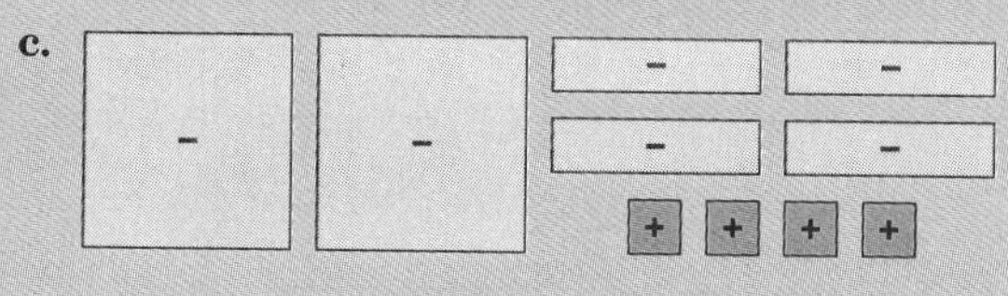

$-2n^2 - 4n + 4$

2. **a.**

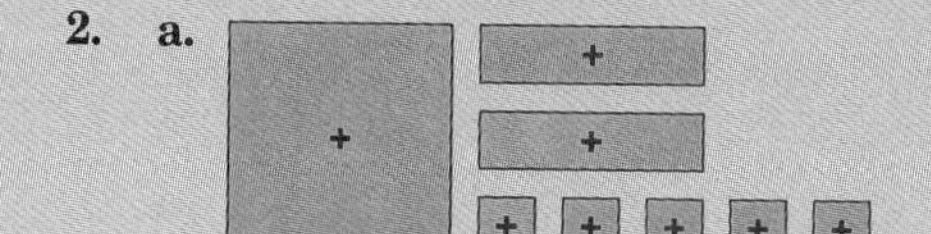

$n^2 + 2n + 5$

$n^2 + 4n + 3$

$(n^2 + 2n + 5) + (n^2 + 4n + 3) = 2n^2 + 6n + 8$

b.

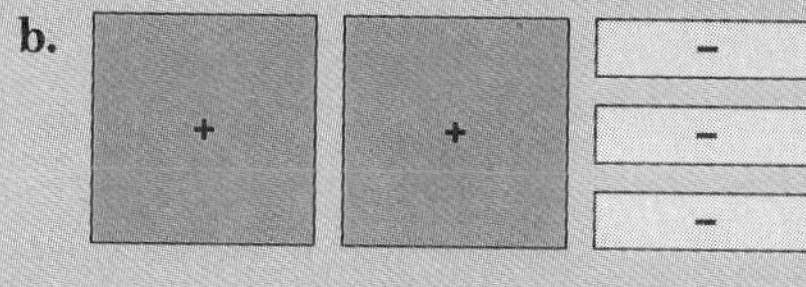

$2n^2 - 3n - 2$

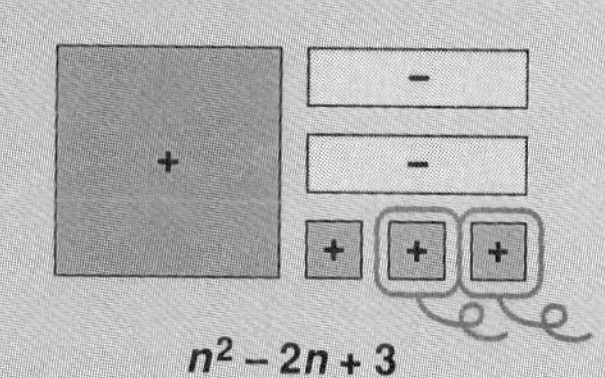

$n^2 - 2n + 3$

$(2n^2 - 3n - 2) + (n^2 - 2n + 3) = 3n^2 - 5n + 1$

c.

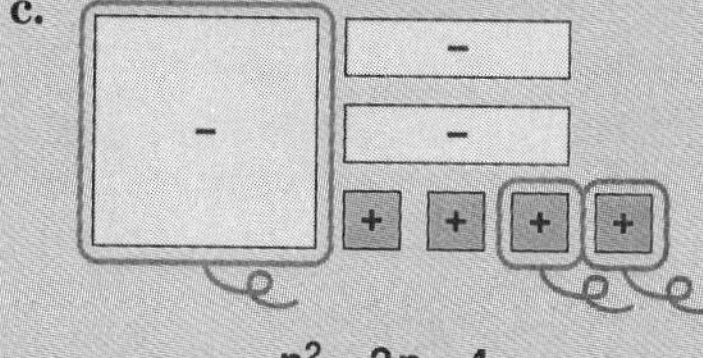

$-n^2 - 2n + 4$

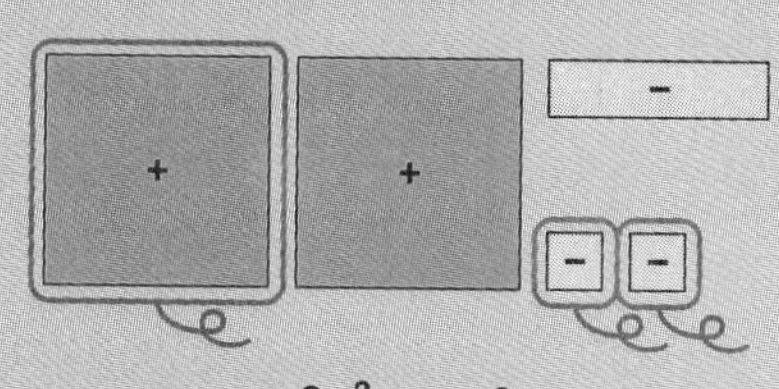

$2n^2 - n - 2$

$(-n^2 - 2n + 4) + (2n^2 - n - 2) = n^2 - 3n + 2$

SOLUTIONS FOR TRY THESE

3. **a.**
$$\begin{array}{r} 2x^2 + 3x + 7 \\ +\quad x^2 + 2x + 2 \\ \hline 3x^2 + 5x + 9 \end{array}$$

b.
$$\begin{array}{r} 8z^2 - 2z - 4 \\ +\quad z^2 - 3z + 3 \\ \hline 9z^2 - 5z - 1 \end{array}$$

c.
$$\begin{array}{r} 6q^2 - 8q + 7 \\ + -3q^2 + q - 5 \\ \hline 3q^2 - 7q + 2 \end{array}$$

4.
$$\begin{array}{r} 7n + 3 \\ 2n + 4 \\ 6n - 9 \\ \hline 15n - 2 \end{array}$$

13.2 Subtraction

Page 576

1. **a.** Lay out chips to represent the first term.

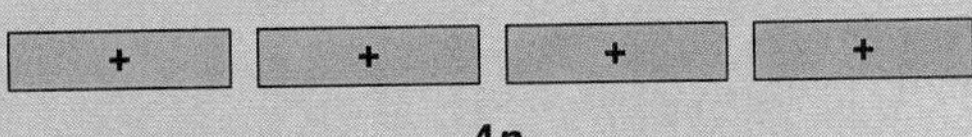

4n

Remove two n-chips.

The remaining chips represent the difference: $2n$

Summary: $(4n) - (2n) = 2n$

b. Lay out chips to represent the first term.

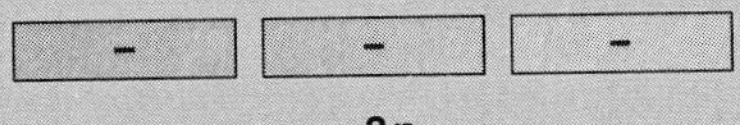

–3n

Remove two $-n$-chips.

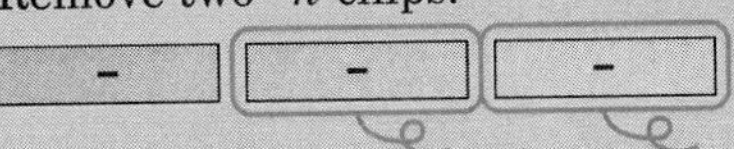

The remaining chips represent the difference: $-n$

Summary: $(-3n) - (-2n) = -n$

c. Lay out chips to represent the first term.

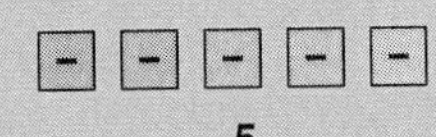

–5

Since you will need to remove three 1-chips, insert three zero pairs of 1-chips.

Remove three 1-chips.

The remaining chips represent the difference: -8

Summary: $(-5) - (3) = -8$

d. Lay out chips to represent the first term.

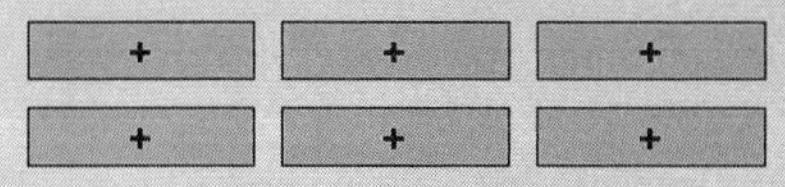

Since you will need to remove two n-chips, insert two zero pairs of n-chips.

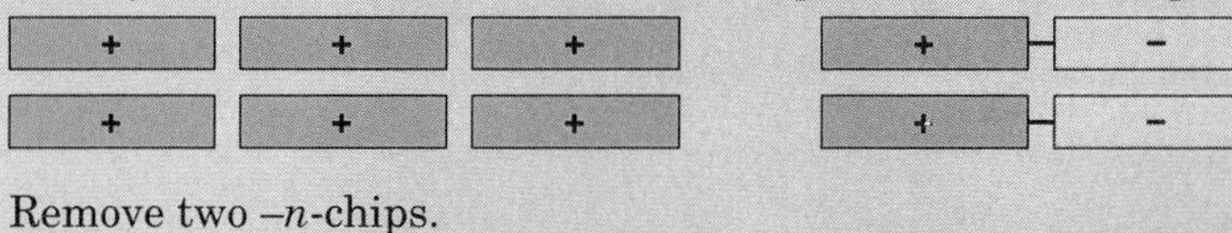

Remove two $-n$-chips.

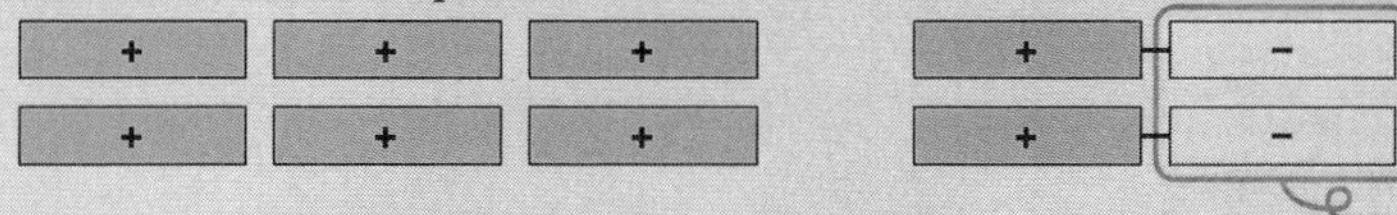

The remaining chips represent the difference: $8n$

Summary: $(6n) - (-2n) = 8n$

e. Lay out chips to represent the first term.

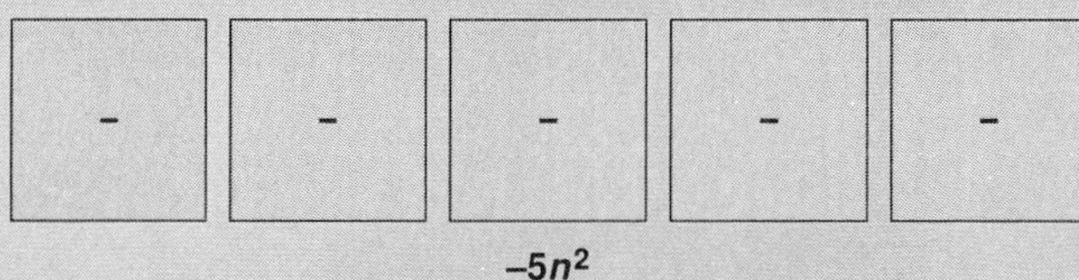

Since you will need to remove three n^2-chips, insert three zero pairs of n^2-chips.

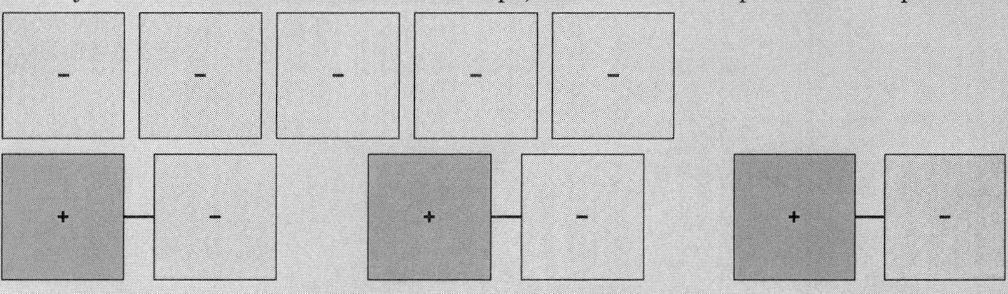

Remove three n^2-chips.

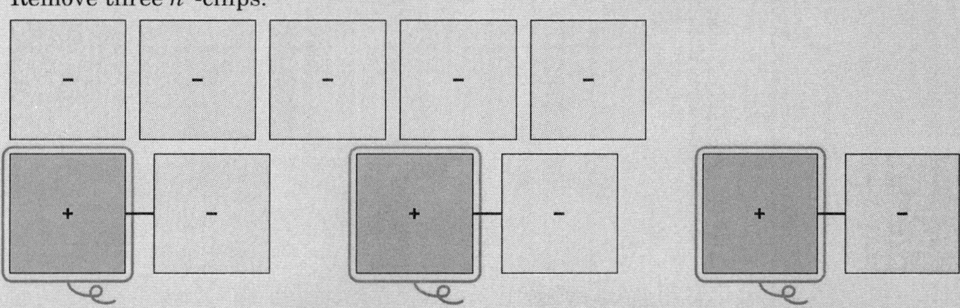

The remaining chips represent the difference: $-8n^2$

Summary: $(-5n^2) - (3n^2) = -8n^2$

2. Change subtraction to addition of the opposite of the subtrahend.

a. $\begin{array}{r} 7m \\ \overset{+}{\ominus}\overset{-}{\oplus}3m \\ \hline 4m \end{array}$ **b.** $\begin{array}{r} 3k \\ \overset{+}{\ominus}\overset{-}{\oplus}7k \\ \hline -4k \end{array}$ **c.** $\begin{array}{r} -6y \\ \overset{+}{\ominus}\overset{+}{\ominus}4y \\ \hline -2y \end{array}$ **d.** $\begin{array}{r} 8ab \\ \overset{+}{\ominus}\overset{+}{\ominus}2ab \\ \hline 10ab \end{array}$ **e.** $\begin{array}{r} -7x^2 \\ \overset{+}{\ominus}\overset{-}{\oplus}4x^2 \\ \hline -11x^2 \end{array}$

Page 578

1. **a.** Lay out chips to represent the first polynomial.

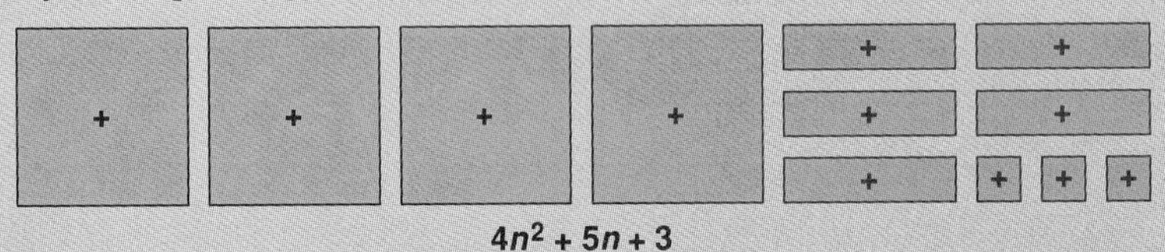

$4n^2 + 5n + 3$

Remove two n^2-chips, three n-chips, and two 1-chips.

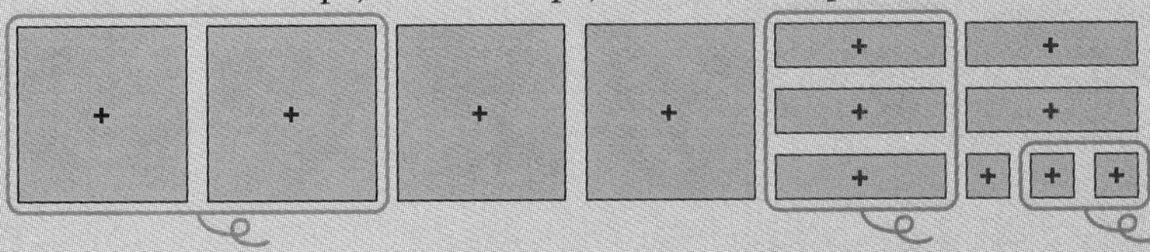

The remaining chips represent the difference: $2n^2 + 2n + 1$

Summary:

$(4n^2 + 5n + 3) - (2n^2 + 3n + 2) = 2n^2 + 2n + 1$

b. Lay out chips to represent the first polynomial.

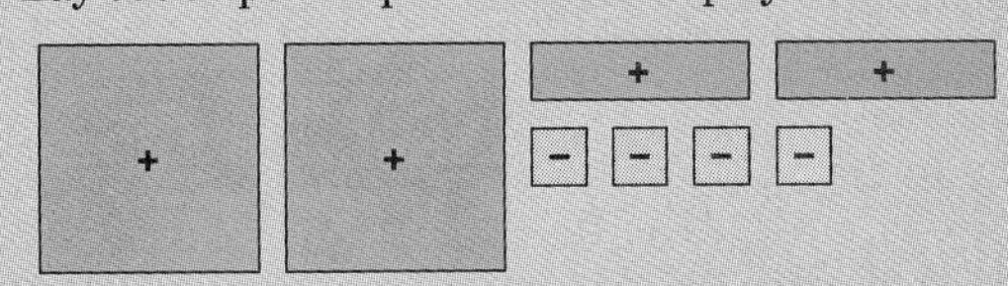

$2n^2 + 2n - 4$

Since you will need to remove three $-n$-chips, insert 3 zero pairs of n-chips.

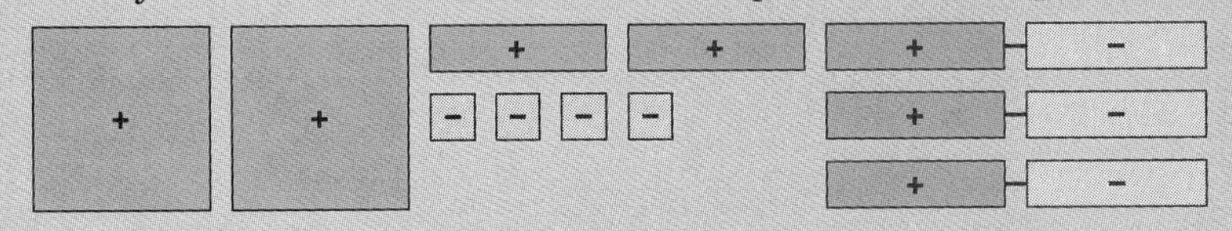

Now remove one n^2-chip, three $-n$-chips, and one -1-chip.

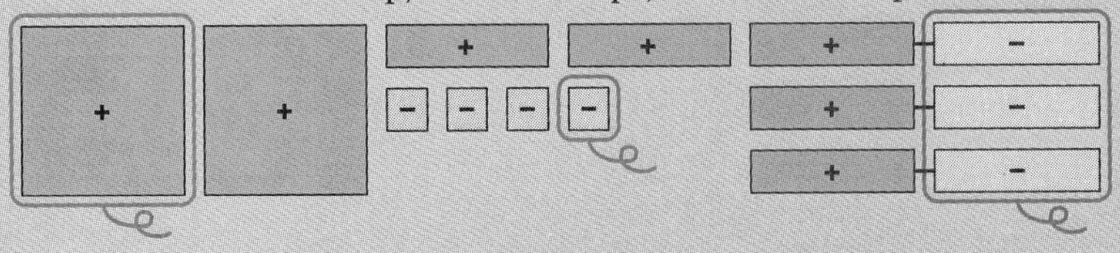

The remaining chips represent the difference: $n^2 + 5n - 3$

Summary: $(2n^2 + 2n - 4) - (n^2 - 3n - 1) = n^2 + 5n - 3$

SOLUTIONS FOR TRY THESE

c. Lay out chips to represent the first polynomial.

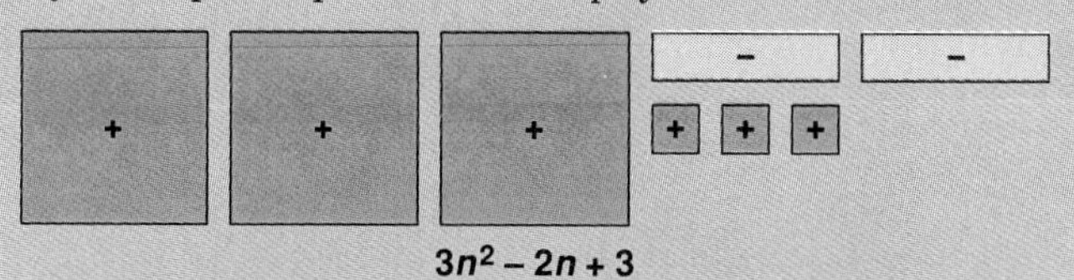

Since you will need to remove one $-n^2$-chip and two n-chips, insert one zero pair of n^2-chips and two zero pairs of n-chips.

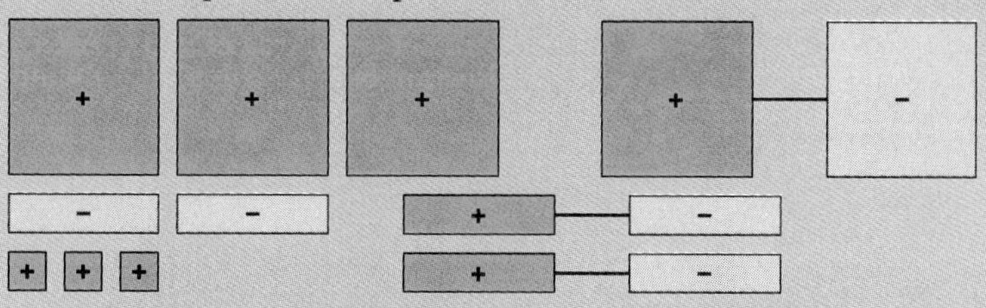

Now remove one $-n^2$-chip, two n-chips, and three 1-chips.

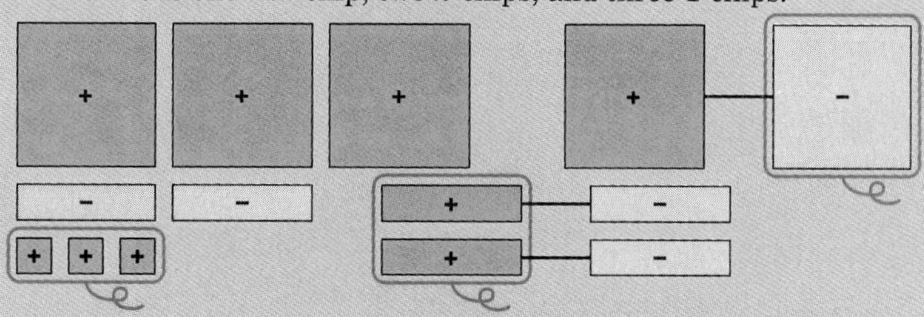

The remaining chips represent the difference: $4n^2 - 4n$

Summary: $(3n^2 - 2n + 3) - (-n^2 + 2n + 3) = 4n^2 - 4n$

2. **a.**
$$\begin{array}{r} 7x^2 + 8x + 6 \\ \underline{\ominus \oplus 3x^2 \oplus 4x \oplus 4} \\ 4x^2 + 4x + 2 \end{array}$$

b.
$$\begin{array}{r} 9y^2 + 4y - 7 \\ \underline{\ominus \oplus 2y^2 \ominus \ y \ominus 4} \\ 7y^2 + 5y - 3 \end{array}$$

c.
$$\begin{array}{r} 6z^2 - \ z + 8 \\ \underline{\ominus \ominus \ z^2 \oplus 3z \oplus 5} \\ 7z^2 - 4z + 3 \end{array}$$

(Solutions continue)

3. **a.** Find the difference.

$$\begin{array}{r} 12n + 5 \text{ Kaitlin} \\ -\ \ 6n - 1 \text{ Timothy} \\ \hline \end{array}$$

Subtraction is addition of an opposite.

$$\begin{array}{r} 12n + 5 \\ \ominus \oplus 6n \ominus 1 \\ \hline 6n + 6 \end{array}$$

Answer: Kaitlin has $(6n + 6)$ more cards than Timothy.

b. Add to find the total.

$$\begin{array}{r} 100x + 12 \\ -\ \ 15x - 3 \\ \hline \end{array}$$

Subtract the returns from the sum.

$$\begin{array}{r} 115x + 9 \\ -\ \ 22x + 4 \\ \hline \end{array}$$

Subtraction is addition of an opposite.

$$\begin{array}{r} 115x + 9 \\ \ominus \oplus 22x \oplus 4 \\ \hline 93x + 5 \end{array}$$

Check: From the total of $(115x + 9)$ books, if Mr. Keenan kept $(93x + 5)$ books, would he have returned $(22x + 4)$ books?

$$\begin{array}{r} 115x + 9 \\ -\ \ 93x + 5 \\ \hline ? \end{array} \qquad \begin{array}{r} 115x + 9 \\ \ominus \oplus 93x \oplus 5 \\ \hline 22x + 4 \end{array} \checkmark$$

Answer: Mr. Keenan was able to keep $(93x + 5)$ books.

13.3 Multiplication

Page 583

1. $6^4 = 6 \times 6 \times 6 \times 6$
2. $3 \times 3 \times 3 \times 3 \times 3 = 3^5$
3. Disagree, $2^3 \times 2^4 = 2^7$. When multiplying with like bases, the base remains the same and the exponents are added.
4. **a.** $5^4 \cdot 5^2 = 5^{4+2} = 5^6$
   b. $y^3 \cdot y^2 = y^{3+2} = y^5$
   c. $k \cdot k^2 = k^{1+2} = k^3$
   d. $2m^3 \cdot 3m^2 = (2 \cdot 3)m^{3+2} = 6m^5$

Page 585

1. **a.**

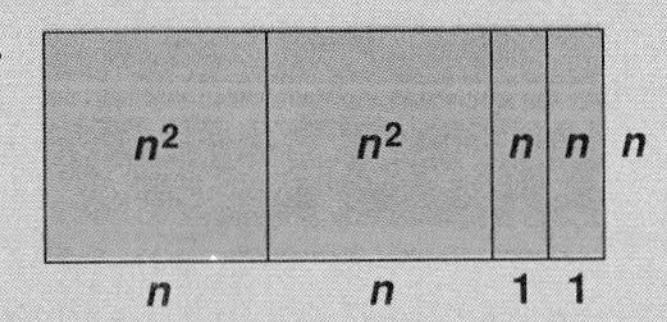

height = n
base = $2n + 2$

Area of Rectangle = Area of two n^2-chips + Area of two n-chips
$= 2n^2 + 2n$

Summary: $n(2n + 2) = 2n^2 + 2n$

b.

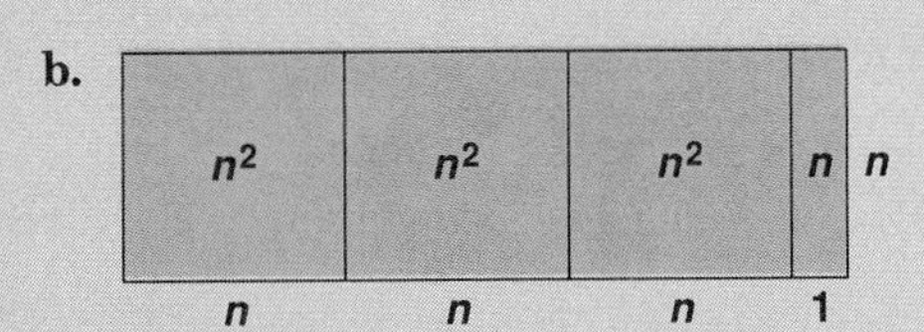

height = n
base = $3n + 1$

Area of Rectangle = Area of three n^2-chips + Area of n-chip
$= 3n^2 + n$

Summary: $n(3n + 1) = 3n^2 + n$

2. **a.** $3x(2x + 1)$ — Multiply a polynomial by a monomial.
$3x \cdot 2x + 3x \cdot 1$ — The monomial multiplies each term.
$3 \cdot 2x^{1+1} + 3 \cdot 1x$ — Multiply coefficients. Add exponents.
Answer: $3x(2x + 1) = 6x^2 + 3x$

b. $2q(3q^2 + 5)$ — Multiply a polynomial by a monomial.
$2q \cdot 3q^2 + 2q \cdot 5$ — The monomial multiplies each term.
$2 \cdot 3q^{1+2} + 2 \cdot 5q$ — Multiply coefficients. Add exponents.
Answer: $2q(3q^2 + 5) = 6q^3 + 10q$

c. $4t(t^2 + 2t)$ — Multiply a polynomial by a monomial.
$4t \cdot t^2 + 4t \cdot 2t$ — The monomial multiplies each term.
$4 \cdot 1t^{1+2} + 4 \cdot 2t^{1+1}$ — Multiply coefficients. Add exponents.
Answer: $4t(t^2 + 2t) = 4t^3 + 8t^2$

SOLUTIONS FOR TRY THESE

Page 587

1. a.

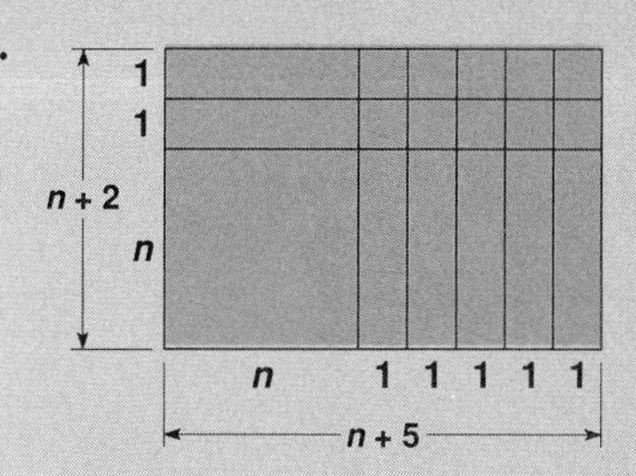

b.

n	1	1	1	1	1
n	1	1	1	1	1
n^2	n	n	n	n	n

The rectangle contains one n^2-chip, seven n-chips, ten 1-chips.

c. $(n + 5)(n + 2) = n^2 + 7n + 10$

2. Lay out chips to form a rectangle with dimensions $(n + 4)$ by $(n + 3)$.

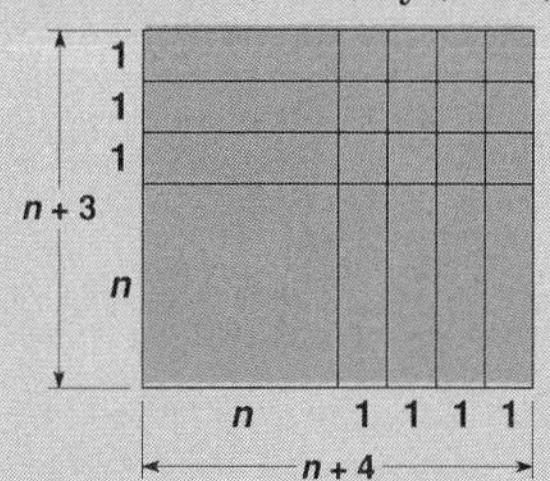

Count the chips to determine the area.

n	1	1	1	1
n	1	1	1	1
n	1	1	1	1
n^2	n	n	n	n

The rectangle contains one n^2-chip, seven n-chips, twelve 1-chips.

$(n + 4)(n + 3) = n^2 + 7n + 12$

---

3. a. Multiply the first terms of the 2 binomials to get the first term of the product.

$(n \quad)(n \quad) = n^2$

Multiply the last terms of the 2 binomials to get the last term of the product.

$(n + 5)(n + 2) = n^2 \qquad + 10$

Multiply the outer terms and multiply the inner terms. Add these results to get the middle term of the product.

Outer: $2n$; inner: $5n$; $2n + 5n$

$(n + 5)(n + 2) = n^2 + 7n + 10$

Answer: $(n + 5)(n + 2) = n^2 + 7n + 10$

b. Multiply the first terms of the 2 binomials to get the first term of the product.

$(n \quad)(n \quad) = n^2$

Multiply the last terms of the 2 binomials to get the last term of the product.

$(n + 3)(n + 4) = n^2 \qquad + 12$

Multiply the outer terms and multiply the inner terms. Add these results to get the middle term of the product.

Outer: $4n$; inner: $3n$; $4n + 3n$

$(n + 3)(n + 4) = n^2 + 7n + 12$

Answer: $(n + 3)(n + 4) = n^2 + 7n + 12$

c. Multiply the first terms of the 2 binomials to get the first term of the product.

$(x \quad)(x \quad) = x^2$

Multiply the last terms of the 2 binomials to get the last term of the product.

$(x + 4)(x - 3) = x^2 \qquad - 12$

Multiply the outer terms and multiply the inner terms. Add these results to get the middle term of the product.

Outer: $-3x$; inner: $+4x$; $-3x + 4x$

$(x + 4)(x - 3) = x^2 + 1x - 12$

Answer: $(x + 4)(x - 3) = x^2 + x - 12$

SOLUTIONS FOR TRY THESE

13.4 Division

Page 595

1. $\frac{6^5}{6^2} = \frac{6 \cdot 6 \cdot 6 \cdot 6 \cdot 6}{6 \cdot 6} = 6^3$

   There are 3 factors of 6 in the result.

2. Disagree, when dividing with like bases, the base does not cancel. To cancel, you must first write out all the factors.

   $\frac{2^9}{2^3} = \frac{2 \cdot 2 \cdot 2 \cdot 2 \cdot 2 \cdot 2 \cdot 2 \cdot 2 \cdot 2}{2 \cdot 2 \cdot 2} = 2^6$ or 64

3. **a.** $\frac{5^6}{5^2} = 5^{6-2} = 5^4$

   b. $\frac{y^3}{y^2} = y^{3-2} = y^1$ or y

   c. $\frac{z^4}{z} = z^{4-1} = z^3$

   d. $\frac{18m^5}{6m^3} = \left(\frac{18}{6}\right) m^{5-3} = 3m^2$

Page 598

1. **a.** $\frac{3y+6}{3} = \frac{3y}{3} + \frac{6}{3}$
   $= y + 2$

   b. $\frac{10w^2 - 15w}{5} = \frac{10w^2}{5} + \frac{15w}{5}$
   $= 2w^2 - 3w$

   c. $\frac{12z^2 + 8z - 24}{4} = \frac{12z^2}{4} + \frac{8z}{4} + \frac{24}{4}$
   $= 3z^2 + 2z - 6$

2. **a.** $\frac{6x^2 + 3x}{3x} = \frac{6x^2}{3x} + \frac{3x}{3x}$
   $= \frac{3 \cdot 2 \cdot x \cdot x}{3 \cdot 1 \cdot x} + \frac{3 \cdot 1 \cdot x}{3 \cdot 1 \cdot x}$
   $= 2x + 1$

   b. $\frac{21k^2 - 7k}{7k} = \frac{21k^2}{7k} + \frac{7k}{7k}$
   $= \frac{7 \cdot 2 \cdot k \cdot k}{7 \cdot 1 \cdot k} - \frac{7 \cdot 1 \cdot k}{7 \cdot 1 \cdot k}$
   $= 3k - 1$

   c. $\frac{18m^2 + 6m}{2m} = \frac{18m^2}{2m} + \frac{6m}{2m}$
   $= \frac{9 \cdot 2 \cdot m \cdot m}{2 \cdot 1 \cdot m} + \frac{3 \cdot 2 \cdot m}{2 \cdot 1 \cdot m}$
   $= 9m + 3$

3. **a.** $\frac{8t^4 + 4t^3}{2t^2} = \frac{8t^4}{2t^2} + \frac{4t^3}{2t^2}$
   $= \left(\frac{8}{2}\right) t^{4-2} + \left(\frac{4}{2}\right) t^{3-2}$
   $= 4t^2 + 2t^1$
   or $4t^2 + 2t$

   b. $\frac{15p^6 - 10p^4}{5p^2} = \frac{15p^6}{5p^2} + \frac{10p^4}{5p^2}$
   $= \left(\frac{15}{5}\right) p^{6-2} - \left(\frac{10}{5}\right) p^{4-2}$
   $= 3p^4 - 2p^2$

   c. $\frac{16x^4 + 4x^3 - 8x^2}{2x} = \frac{16x^4}{2x^1} + \frac{4x^3}{2x^1} + \frac{8x^2}{2x^1}$
   $= \left(\frac{16}{2}\right) x^{4-1} + \left(\frac{4}{2}\right) x^{3-1} - \left(\frac{8}{2}\right) x^{2-1}$
   $= 8x^3 + 2x^2 - 4x^1$
   or $8x^3 + 2x^2 - 4x$

The Straight Line 14

Using a 4-by-4 grid and only straight lines, graph the letters of the alphabet that spell your name.

Letting the point (0, 0) be the lower left corner of the grid, write the coordinates of each letter.

Change the location of (0, 0) to a different corner of the grid, and write new coordinates for each letter.

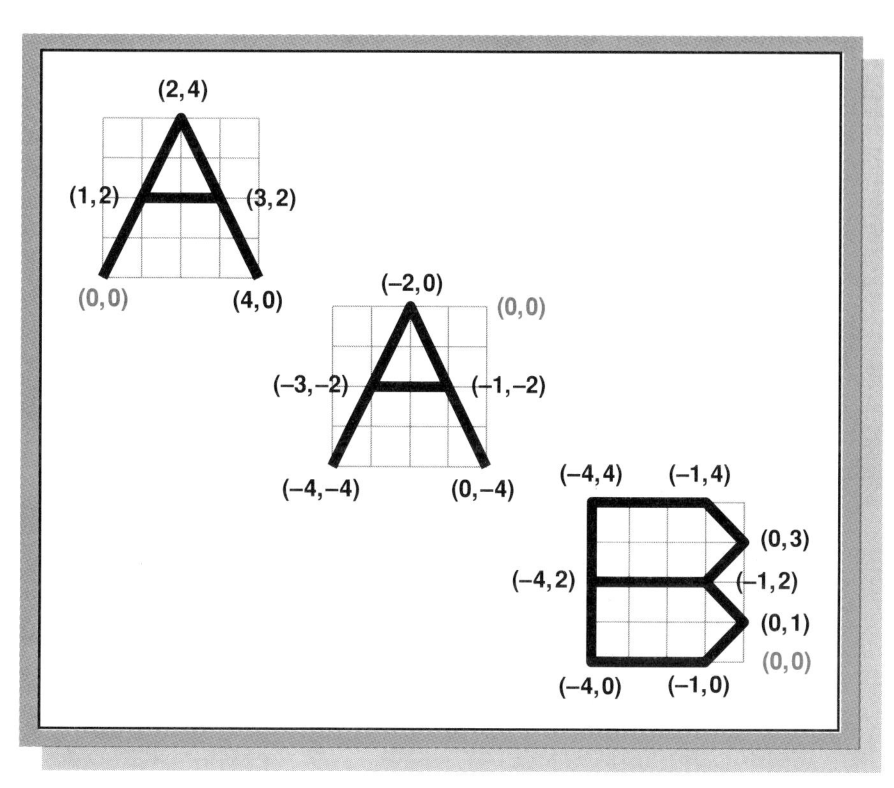

14.1 Graphing a Line

Writing an Equation, and Graphing

Name other points that you think could belong to this set:

$$\{(5, 5), (3, 3), (0, 0), (-2, -2)\}$$

Other points that could be in this set are (4, 4) and (–1, –1).

Any point for which the coordinates are the same number could be in this set.

Here is what these 6 points look like when they are graphed and connected.

The arrowheads show that this line extends in both directions to include all points that could belong to this set.

If (x, y) represents a point that belongs to this set, write a rule that tells the relationship between x and y.

For this set, $y = x$.

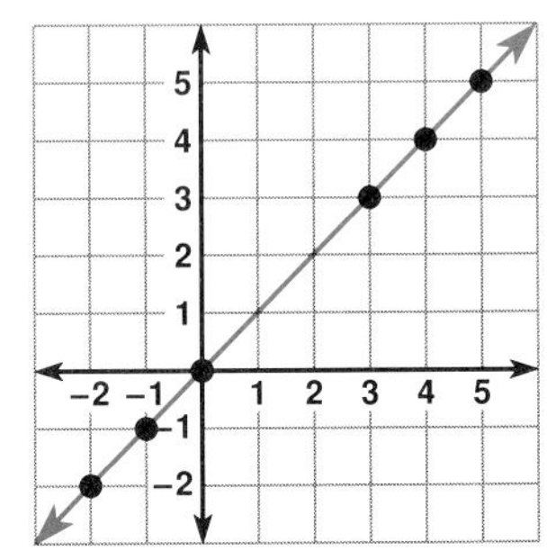

All points that are on this line follow the rule that their coordinates have the same number value.

The rule that tells the relationship between the coordinates of the point of a line acts as the name of the line. The rule is called the *equation of the line*.

Example 1 Consider this set: $\{(5, 6), (4, 5), (3, 4), (2, 3)\}$

a. Name another point that could be in this set.

Other points that could be in this set are (1, 2) and (0, 1). Any point for which the second coordinate is 1 more than the first coordinate could be in this set.

b. If (x, y) represents a point that belongs to this set, write a rule that tells the relationship between x and y.

For this set, $y = x + 1$.

c. Use the 6 points to graph the line whose equation is $y = x + 1$.

Plot the points in the familiar way, and connect them.

To show that the connecting line contains all the points (x, y) so that $y = x + 1$:

Use x to label the horizontal axis.
Use y to label the vertical axis.
Label the origin O.
Use arrowheads on the line.
Write the equation on the line.

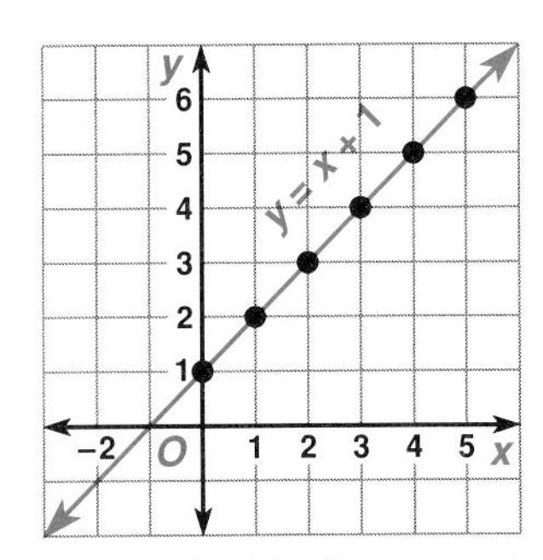

The graph of the line $y = x + 1$ is a diagonal line, also called an ***oblique*** line. This line intersects both axes, at 1 on the y-axis and at –1 on the x-axis.

Example 2 Consider this set: $\{(0, 3), (2, 1), (3, 0), (5, -2)\}$

a. Name another point that could be in this set.

Other points that could be in this set are $(1, 2)$ and $(-1, 4)$.
Any point for which the sum of the coordinates is 3 could be in this set.

b. If (x, y) represents a point that belongs to this set, write a rule that tells the relationship between x and y.

For this set, $x + y = 3$.

c. Use the 6 points to graph the line whose equation is $x + y = 3$.

Use x to label the horizontal axis and y for the vertical axis. Label the origin O.

Plot the 6 points, and use arrowheads on the connecting line.

Write the equation on the line.

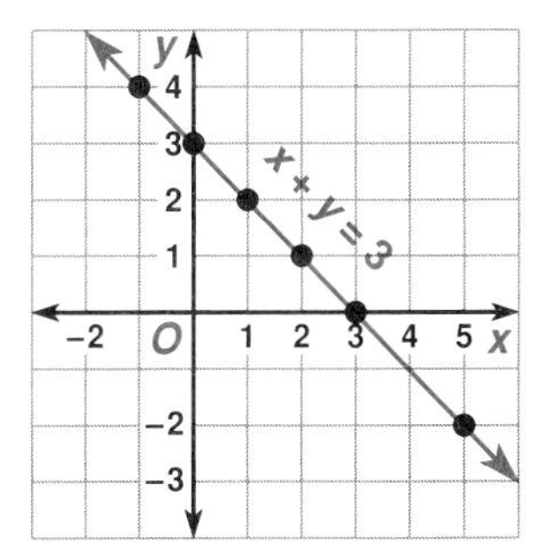

The graph of the line $x + y = 3$ is an oblique line. This line intersects both axes, at 3 on the y-axis and at 3 on the x-axis.

Examples 1 and 2 show lines for which the equation states a relationship between the two coordinates of each point on the line.

The names of some lines depend on the behavior of just one coordinate.

Example 3 Consider this set: $\{(-3, 2), (0, 2), (1, 2), (3, 2)\}$

a. Name another point that could be in this set.

Other points that could be in this set are $(4, 2)$ and $(-1, 2)$. Any point for which the second coordinate is 2 could be in this set. The first coordinate could be any number.

b. If (x, y) represents a point that belongs to this set, write a rule that tells the characteristic of a point in this set.

For this set, $y = 2$.

c. Use the 6 points to graph the line whose equation is $y = 2$.

Label the axes and the origin.

Use arrowheads on the line, and write the equation of the line.

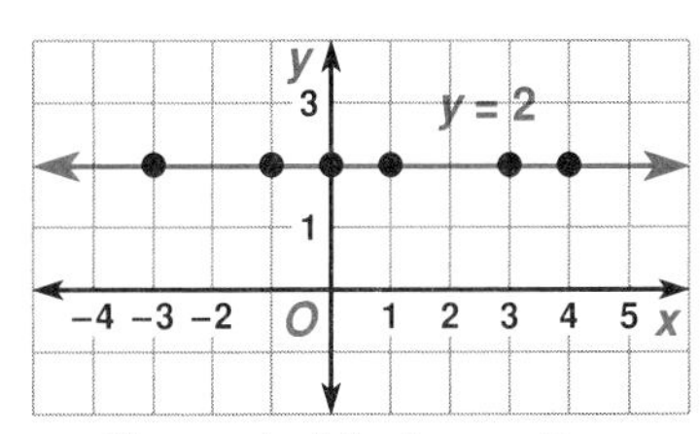

The graph of the line $y = 2$ is a horizontal line. This line intersects only the y-axis, at 2.

Try These *(For solutions, see page 669.)*

1. Consider this set of points: $\{(3, 6), (2, 4), (1, 2), (-1, -2)\}$
   a. Name 2 other points that could be in the set.
   b. Using words, write a rule that tells the relationship between the coordinates of a point in this set.
   c. If (x, y) represents a point in this set, write a rule that tells the relationship between x and y.
   d. Use 6 points to graph the line whose equation you have written.
   e. What type of straight line have you graphed?
   f. Which axes does the line intersect? Where?

2. Consider this set of points: $\{(2, 4), (2, 3), (2, 1), (2, -1)\}$
   a. Name 2 other points that could be in the set.
   b. Using words, write a rule that tells the special property of one of the coordinates of a point in this set.
   c. If (x, y) represents a point in this set, write a rule that tells the characteristic of a point in this set.
   d. Use 6 points to graph the line whose equation you have written.
   e. What type of straight line have you graphed?
   f. Which axes does the line intersect? Where?

Choosing Points to Graph a Line

Name a point for which the y-coordinate is twice the x-coordinate.

Name two other such points.

This table shows three points for which the y-coordinate is twice the x-coordinate.

x	1	2	−1
y	2	4	−2

Plot these points, and connect them to draw a graph of the line $y = 2x$.

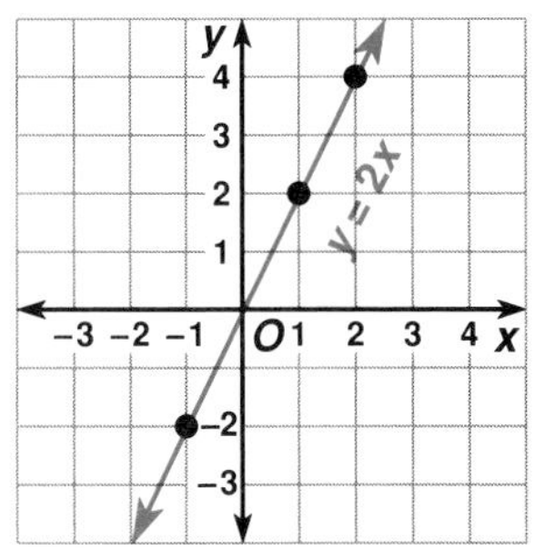

The graph of the line $y = 2x$ is an oblique line. This line intersects both axes, at the origin.

To draw the graph of a line whose equation you know, choose at least three points whose coordinates satisfy the equation.

Example 4 Draw the graph of the line whose equation is $y = x - 2$ by choosing at least 3 points for which the y-coordinate is 2 less than the x-coordinate.

First pick a value for x, say 5. Then the corresponding y-value is $5 - 2$, or 3.

A table is a convenient way of keeping track of the coordinates.

x	$x - 2$	y
5	$5 - 2$	3
3	$3 - 2$	1
0	$0 - 2$	-2

Points for which the y-coordinate is 2 less than the x-coordinate are (5, 3), (3, 1), and (0, –2).

Plot the points from the table to draw the graph of the line $y = x - 2$.

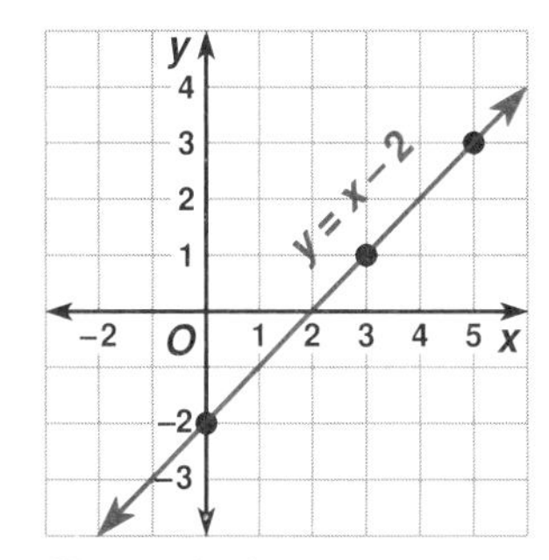

The graph of the line $y = x - 2$ is an oblique line. This line intersects both axes, at 2 on the x-axis and at –2 on the y-axis.

The table of Example 4 showed the calculation of the y-coordinate from the chosen x-coordinate.

Sometimes, you will not need to show calculations in the table.

Example 5 Draw the graph of the line whose equation is $x + y = 4$.

Choose at least three points for which the sum of the coordinates is 4.

x	y
1	3
3	1
0	4

Points for which the sum of the coordinates is 4 are (1, 3), (3, 1), and (0, 4).

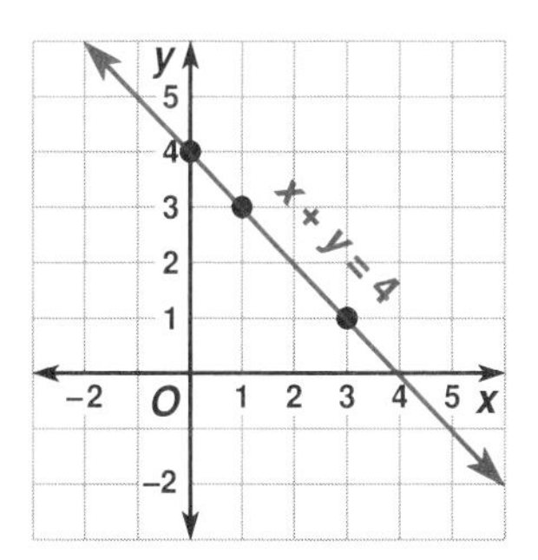

The graph of the line $x + y = 4$ is an oblique line. This line intersects both axes, at 4 on the y-axis and at 4 on the x-axis.

Example 6 Draw the graph of the line whose equation is $x = 3$.

Choose three points for which the x-coordinate is 3. The y-coordinate can be any number.

x	y
3	1
3	3
3	−2

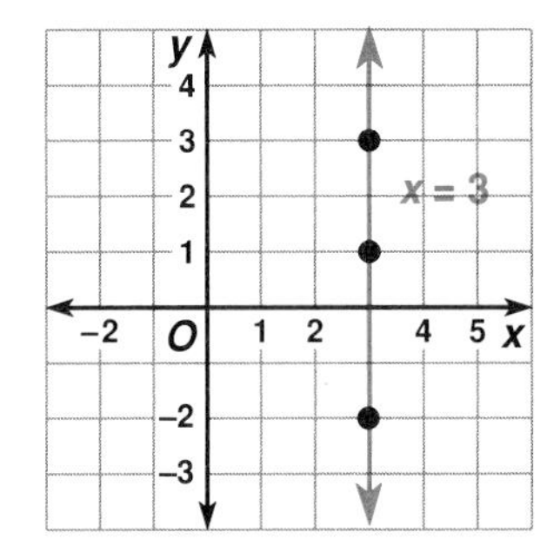

The graph of the line $x = 3$ is a vertical line. This line intersects only the x-axis, at 3.

Try These *(For solutions, see page 669.)*

1. **a.** What is the relationship between coordinates for points that are on the line $y = 3x$?
   b. Name three points that must be on the line $y = 3x$.
   c. Use the coordinates of the three points you named to draw the graph of the line $y = 3x$.
   d. What type of straight line is $y = 3x$?
   e. Which axes does the line $y = 3x$ intersect? Where?

2. **a.** What is the relationship between coordinates for points that are on the line $y = x + 4$?
   b. If a point on the line $y = x + 4$ has an x-coordinate of 2, what is the corresponding y-coordinate?
   c. Name two other points that must be on the line $y = x + 4$.
   d. Use the three points mentioned to draw the graph of the line $y = x + 4$.
   e. What type of straight line is $y = x + 4$?
   f. Which axes does the line $y = x + 4$ intersect? Where?

3. **a.** What is the relationship between coordinates for points that are on the line $y = x - 3$?
   b. Copy and complete this table to determine the coordinates of points that are on the line $y = x - 3$.

x	$x - 3$	y
6	6 − 3	?
4	?	?
3	?	?
1	?	?

   c. Use the coordinates of the points in the table to draw the graph of the line $y = x - 3$.
   d. What type of straight line is $y = x - 3$?
   e. Which axes does the line $y = x - 3$ intersect? Where?

SELF-CHECK: SECTION 14.1

1. Consider this set of points: $\{(1, -1), (2, -2), (3, -3), (4, -4)\}$

a. How are the coordinates of each point in this set related?

Answer: The second coordinate is the opposite of the first.

b. How can you write the rule for this set using symbols?

Answer: Let (x, y) represent a point in the set.
The rule for the set is $y = -x$.

c. To graph this set of points, what labels do you use on the axes?

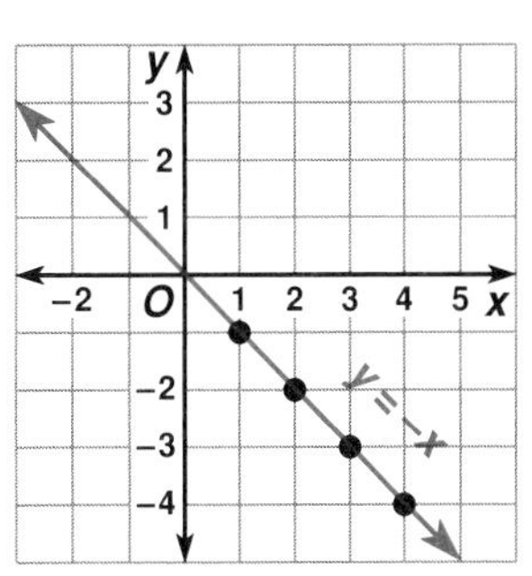

Answer: The horizontal axis is the x-axis and the vertical axis is the y-axis to correspond to the coordinates (x, y).

d. What type of straight line is $y = -x$?

Answer: oblique

e. Which axis does the line $y = -x$ intersect?

Answer: The oblique line $y = -x$ intersects both the y-axis and the x-axis, at the origin.

2. Consider the equation $y = x + 5$.

a. If a point is going to be on the graph of this equation, what does the equation tell you must be true about the coordinates of that point?

Answer: To be on the graph of this equation, the y-coordinate of the point must be 5 more than the x-coordinate.

b. If a point on the graph of this equation has an x-coordinate of 1, what is the y-coordinate?

Answer: If $x = 1$, then $y = 1 + 5$ or $y = 6$.
$(1, 6)$ is a point on the graph.

c. How can you find other points that will be on this graph?

Answer: Choose any value for x. Then the corresponding y-value must be 5 more. Choose another value for x. The corresponding y-value is 5 more.

d. Will the point $(3, 7)$ be on this graph? Explain.

Answer: No, the y-coordinate is not 5 more than the x-coordinate.

EXERCISES: SECTION 14.1

1. Explain why you agree or disagree with each of the following statements about this set of points whose elements all follow the same rule:

$$\{(5, 4), (3, 2), (1, 0), (0, -1), (-2, -3)\}$$

a. Jim said that (6, 6) could belong to this set.

b. Joan said that (–3, –4) could belong to this set.

c. Janet said that to be in this set, the y-coordinate of the point must be 1 less than the x-coordinate.

d. Jose said that if (x, y) represents a point in this set, then the rule for the set is $y = x - 1$.

e. Jack said that another way to express the rule for this set is to say that the x-coordinate is 1 more than the y-coordinate, $x = y + 1$.

2. For each of the following sets of points:

(1) Name 3 more points that could be in the set.

(2) Using words, write a rule that tells the relationship between the coordinates of a point in the set.

a. {(3, 3), (0, 0), (–2, –2)}

b. {(4, 8), (2, 4), (–3, –6)}

c. {(7, 6), (5, 4), (2, 1)}

d. {(3, 4), (0, 1), (–2, –1)}

e. {(1, 4), (2, 3), (5, 0)}

f. {(8, 2), (7, 1), (6, 0)}

3. For each of the following sets of points:

(1) Name another point that could be in the set.

(2) Using words, write a rule that tells the relationship between the coordinates of a point in the set.

(3) If (x, y) represents a point in the set, write a rule that tells the relationship between x and y.

(4) Using the given points of the set and the other point you have named, draw the line whose equation you have written.

(5) Tell where the line you have graphed intersects the axes.

a. {(–3, –3), (–1, –1), (0, 0)}

b. {(3, 9), (1, 3), (–2, –6)}

c. {(5, 3), (3, 1), (2, 0)}

d. {(3, 5), (0, 2), (–2, 0)}

e. {(0, 4), (1, 3), (2, 2)}

f. {(9, 4), (7, 2), (5, 0)}

4. For each of the following sets of points:

(1) Name another point that could be in the set.

(2) Using words, write a rule that tells the special property of the points in the set.

(3) If (x, y) represents a point in the set, write a rule that tells the special property of the points in the set.

(4) Using the given points of the set and the other point you have named, draw the line whose equation you have written.

(5) Identify the line you have drawn as *horizontal* or *vertical*.

(6) Tell which axis your line intersects, and where.

a. $\{(2, 1), (2, 4), (2, 7)\}$

b. $\{(4, 1), (2, 1), (0, 1)\}$

c. $\{(3, -2), (1, -2), (-3, -2)\}$

d. $\{(-2, 5), (-2, 3), (-2, 0)\}$

e. $\{(0, 6), (0, 2), (0, -1)\}$

f. $\{(-2, 0), (0, 0), (3, 0)\}$

5. a. Write a rule using x and y to describe each relationship between coordinates of points on a line.

(1) The y-coordinate is twice the x-coordinate.

(2) The y-coordinate is 3 more than the x-coordinate.

(3) The sum of the coordinates is 5.

(4) The y-coordinate is 1 less than the x-coordinate.

(5) The x-coordinate is 4 more than twice the y-coordinate.

(6) The x-coordinate is 2 less than three times the y-coordinate.

b. Another name for the x-coordinate of a point is ***abscissa,*** and another name for the y-coordinate is ***ordinate.***

(x, y) is also called (abscissa, ordinate)

Write a rule using x and y to describe each relationship between coordinates of points on a line.

(1) The ordinate is equal to the abscissa.

(2) The ordinate is 2 more than the abscissa.

(3) The ordinate is 1 less than twice the abscissa.

(4) The sum of the abscissa and ordinate is 12.

(5) The ordinate is 5 more than twice the abscissa.

(6) The abscissa is 3 less than four times the ordinate.

6. Explain why you agree or disagree with each statement.

a. Susan said that the point (1, 3) is not on the graph of the equation $y = x$.

b. Steve said that the point (1, 3) is on the graph of the equation $y = 3x$.

c. Sara said that the point (1, 3) is on the graph of the equation $x + y = 4$.

d. Saleem said that the point (1, 3) is on the graph of the equation $x = 3$.

7. Tell whether the given point is on the graph of the line with the given equation.

	Point	*On the Graph of This Line?*
a.	(2, 4)	$y = 2x$
b.	(3, 5)	$x + y = 2$
c.	(3, 0)	$y = 2x - 6$
d.	(10, 16)	$y - x = 6$
e.	(3, 1)	$2x + 5y = 11$
f.	(0,–4)	$y = 3x + 4$
g.	(–1, –4)	$y = 5x + 1$
h.	(0, –2)	$x - 2y = 4$

8. Explain why you agree or disagree with each statement.

a. Carrie said that to be on the graph of the line $y = 5x$, if the x-coordinate of a point is 2, then the corresponding y-coordinate is 10.

b. Carl said that to be on the graph of the line $x + y = 7$, if the x-coordinate of a point is 7, then the corresponding y-coordinate is also 7.

c. Carla said that to be on the graph of the line $y = x + 1$, if the abscissa of a point is 3, then the ordinate must be 4.

d. Claude said that to be on the graph of the line $y = 3x$, if the ordinate of a point is 6, then the abscissa must be 2.

e. Candie said that to be on the graph of the line $x = 4$, the x-coordinate must be 4, and the corresponding y-coordinate must also be 4.

f. Charlie said that to be on the graph of the line $y = 6x$, if the abscissa of a point is 12, then the corresponding ordinate is 2.

9. For each given point, find the value of the missing coordinate so that the point will lie on the graph of the given line.

	Point	*The Point Is On This Line*
a.	(2, ?)	$y = 5x$
b.	(0, ?)	$y = 3x + 1$
c.	(–3, ?)	$x + y = 10$
d.	(?, 8)	$y = 2x$
e.	(?, 5)	$x - y = 1$
f.	(?, 7)	$y = 2x - 1$
g.	(–4, ?)	$y - x = 7$
h.	(?, –3)	$x + y = 4$
i.	(0, ?)	$y = 5x - 2$
j.	(?, 0)	$y = 2x + 6$

10. For each line whose equation is given, copy and complete the table of points. Below the table, list the coordinates of the points. Then use these points to graph the line.

Identify the line you have graphed as *horizontal*, *vertical*, or *oblique*, and tell where the line intersects the axes.

a. $y = x + 7$

x	$x + 7$	y
0	0 + 7	7
1	?	?
3	?	?

Points on the line $y = x + 7$ are (0,7), (1, ?), and (3, ?).

b. $y = x - 3$

x	$x - 3$	y
5	5 − 3	2
3	?	?
0	?	?

Points on the line $y = x - 3$ are (5,2), (3, ?), and (0, ?).

c. $y = \frac{1}{2}x$

x	$\frac{1}{2}x$	y
8	$\frac{1}{2}(8)$	4
6	?	?
2	?	?

Points on the line $y = \frac{1}{2}x$ are (8, 4), (6, ?), and (2, ?).

d. $y = 2x + 1$

x	$2x + 1$	y
0	2(0) + 1	1
2	?	?
4	?	?

Points on the line $y = 2x + 1$ are (0, 1), (2, ?), and (4, ?).

e. $y = 3x - 2$

x	$3x - 2$	y
−1	3(−1) − 2	−5
0	?	?
2	?	?

Points on the line $y = 3x - 2$ are (−1, −5), (0, ?), and (2, ?).

f. $y = 4 - x$

x	$4 - x$	y
1	4 − 1	3
2	?	?
4	?	?

Points on the line $y = 4 - x$ are (1,3), (2, ?), and (4, ?).

g. $x + y = 8$

x	$x + y = 8$	y
0	0 + y = 8	8
1	?	?
3	?	?

Points on the line $x + y = 8$ are (0, 8), (1, ?), and (3, ?).

h. $y - x = 6$

x	$y - x = 6$	y
1	y − 1 = 6	7
0	?	?
2	?	?

Points on the line $y - x = 6$ are (1, 7), (0, ?), and (2, ?).

i. $x = 4$

x	y
4	5
4	?
?	?

Points on the line $x = 4$ are (4, 5), (4, ?), and (?, ?).

11. For each line whose equation is given, create a table of at least 3 points that are on the graph of the line. Then use your points to draw the graph.

Identify the line you have graphed as *horizontal*, *vertical*, or *oblique,* and tell where the line intersects the axes.

a. $y = -2x$
b. $y = x + 5$
c. $y = x - 6$
d. $y = 2x + 4$
e. $y = 2x - 4$
f. $y = 3x - 3$
g. $x + y = 9$
h. $y - x = 5$
i. $x - y = 6$
j. $x = 5$
k. $y = 2$
l. $x = -4$

12. **a.** Yolanda is 2 years older than Xavier, who is now 3 years old.

(1) Write an equation to express the relationship between Yolanda's age (y) and Xavier's age (x).

(2) Prepare a table of points that would be on the graph of this equation.

(3) Using the points from your table, draw a graph to represent this equation. Use a full sheet of graph paper.

(4) For this problem, can the graph extend in both directions? Explain.

(5) Mark on your graph the point that will tell you Yolanda's age when Xavier is 12 years old.

How old will Yolanda be when Xavier is 12?

b. For each situation involving Yolanda and Xavier, write an equation and draw the graph of the equation.

Each graph has a starting point because of the situation. Label this starting point S.

Use A to mark the graph with the point that answers the question, and answer the question.

(1) *Situation:* Yolanda's hourly wage is twice as much as Xavier's hourly wage.

Question: How much does Yolanda earn for an hour if Xavier earns $5.50?

(2) *Situation:* Xavier needs 10 more photocopies of each office memo than Yolanda needs.

Question: How many photocopies must Xavier make when Yolanda makes 9 copies?

(3) *Situation:* Together, Yolanda and Xavier can enter 20 cards in an exhibit of baseball cards.

Question: If Yolanda enters 12 cards, how many can Xavier enter?

13. **a.** The shaded figure is formed by the line $y - x = 3$, the y-axis, and the x-axis. Find the area.

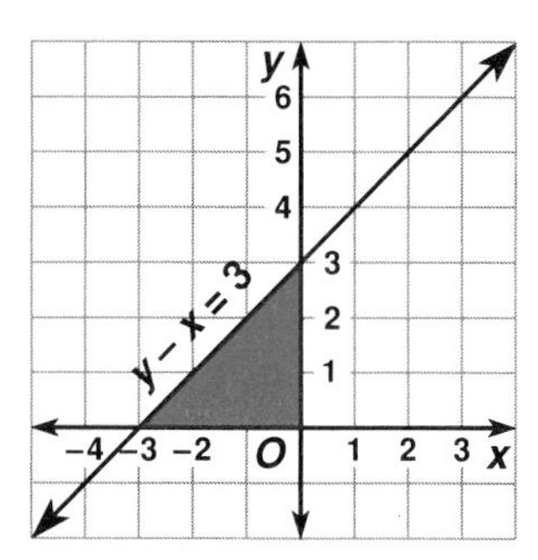

b. Find the area of the figure formed by:

(1) the line $x + y = 4$, the x-axis, and the y-axis

(2) the line $y - x = 5$, the x-axis, and the y-axis

(3) the line $x - y = 3$, the x-axis, and the y-axis

(4) the line $x + y = 6$, the line $y - x = 6$, and the x-axis

(5) the lines $x + y = 5$, $x - y = 5$, $y - x = 5$, and $x + y = -5$

EXPLORATION

Here's how to use an advanced calculator to draw the graph of an equation.

(1) Set the calculator to accept the equation you wish to graph.

Press: Y= — the first black key in the top row

(2) Enter the equation $y = x + 1$.

Press:   

X|T + 1 ENTER

↑ use this for the variable x located left of cursor keys

The calculator is ready to accept as many as 4 equations.

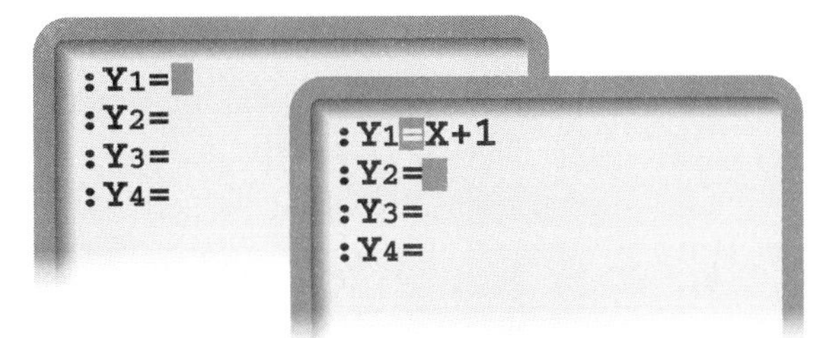

(3) Enter the range of the values to be used.

Press: RANGE — 2nd black key in the top row

The calculator is preset to use values of x and y from –10 to 10 in intervals of 1.

```
RANGE
Xmin=⁻10
Xmax=10
Xscl=1
Ymin=⁻10
Ymax=10
Yscl=1
Xres=1
```

Let's leave the x-values to range from –10 to 10, and change the y-values to range from –7 to 7.

Press: 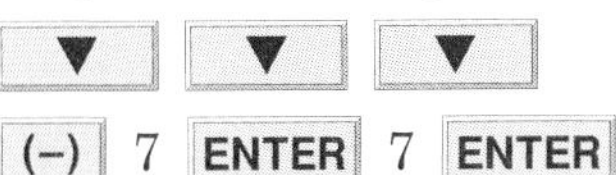

(–) 7 ENTER 7 ENTER

```
RANGE
Xmin=⁻10
Xmax=10
Xscl=1
Ymin=⁻7
Ymax=7
Yscl=1
Xres=1
```

(4) After the equation of the line and the range of values are entered, the calculator is ready to draw the graph.

Press: GRAPH — last black key in the top row

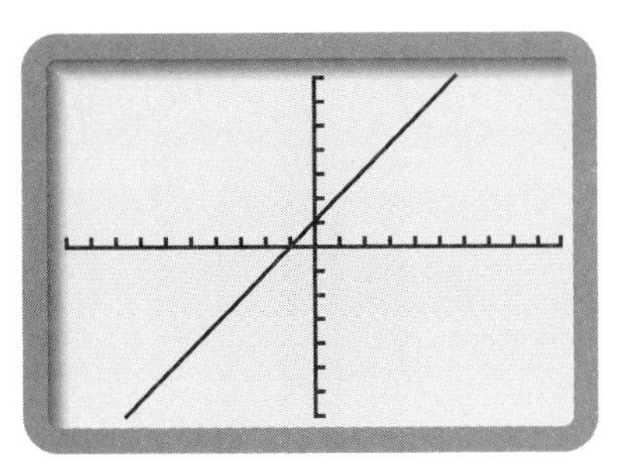

The graph is an oblique line that intersects the y-axis at 1 and the x-axis at –1.

Compare the calculator display with Example 1 (page 616).

Pressing TRACE will display a blinking cursor on the line. The coordinates of the point at which the cursor is located are shown at the bottom of the screen.

You can read the coordinates of other points on the line by using    to move the cursor.

Try Another

1. Clear the previous equation. — Press: **Y=** **CLEAR** ⟵ located under cursor keys

   Enter the new equation $y = 2x$. — Press: 2 **X|T** **ENTER**

2. Set the range; let's use x from –5 to 5 in intervals of 1, and y from –5 to 5 in intervals of 1. — Press: **RANGE** **(–)** 5 **ENTER** 5 **ENTER** 1 **ENTER** **(–)** 5 **ENTER** 5 **ENTER**

3. Draw the graph. — Press: **GRAPH**

   Compare the calculator display with the graph at the bottom of page 618.

What do you think you would have to do to an equation like $x + y = 4$ before you could enter it on the calculator?

Because the calculator will only accept an equation in the form $y =$, you must first get your equation with y alone.

$$x + y = 4$$
$$x - x + y = 4 - x \quad \text{Subtract } x \text{ from both sides.}$$
$$y = 4 - x$$

Now graph the equation $y = 4 - x$ on the calculator, and compare the result with Example 5.

14.2 The Slope of a Line

The Meaning of Slope

Mark and Mindy are each going to climb a hill.

From these diagrams of the hills, whose hill do think is harder to climb? Explain.

Mindy's hill is *steeper* than Mark's and, therefore, harder to climb.

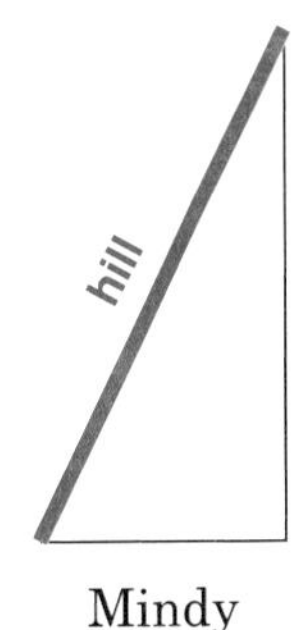

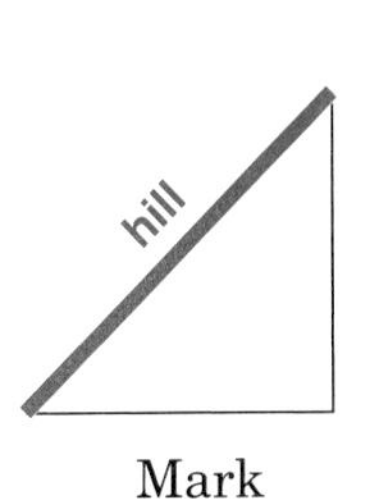

The steepness of a line is called the *slope*.

Here is a graph of Mindy's hill, which is the line $y = 2x + 1$.

Explain how you could travel along grid lines and get from point A to point B.

From B to C.

From A to C.

From P to A.

From P to C.

From A to B, go up 2 and right 1.

From B to C, go up 2 and right 1.

From A to C, go up 4 and right 2.

From P to A, go up 4 and right 2.

From P to C, go up 8 and right 4.

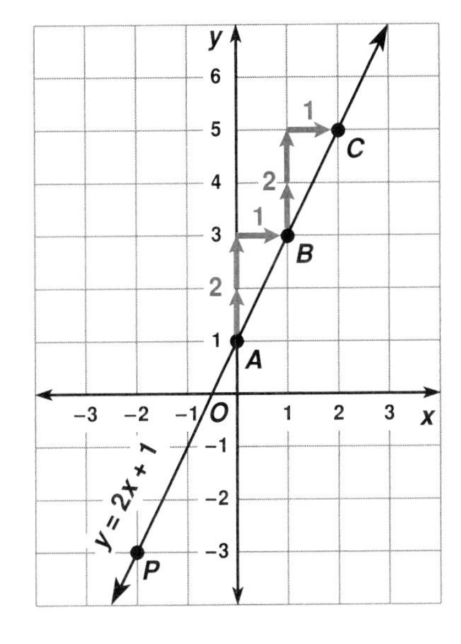

Notice that the direction ratios that travel along grid lines and get you from one point to another on the line $y = 2x + 1$ are equivalent:

$$\frac{\text{up}}{\text{right}} = \frac{2}{1} = \frac{4}{2} = \frac{8}{4}$$

The direction ratio, such as $\frac{\text{up}}{\text{right}}$, tells you the slope of the line.

Between any two points on a line, the slope is always the same ratio.

Example 1

a. Tell how to travel along grid lines to get between the points A, B, and C on the graph of the line $y = x + 1$.

From A to B, go up 2 and right 2.

From A to C, go up 5 and right 5.

From B to C, go up 3 and right 3.

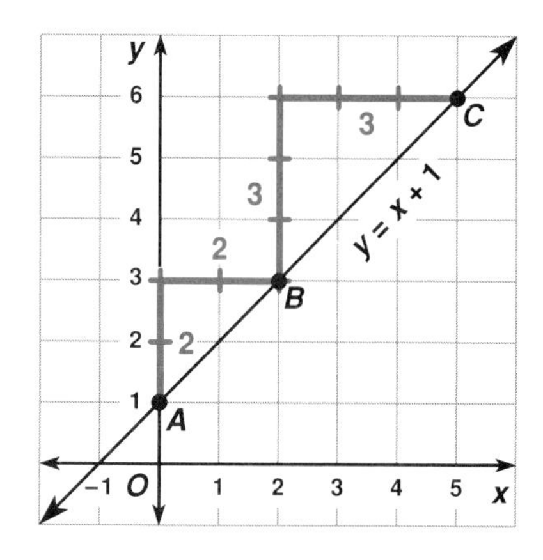

b. What is the slope of the line $y = x + 1$?

Read the slope of the line from the direction ratio $\frac{\text{up}}{\text{right}}$.

$$\frac{\text{up}}{\text{right}} = \frac{2}{2} = \frac{5}{5} = \frac{3}{3} = \frac{1}{1}, \text{ or } 1$$

The slope of the line $y = x + 1$ is 1.

To get from point R to point S along grid lines, you can go up 5 and left 3.

How would you get from S to R?

From S to R, go down 5 and right 3.

Notice that the two direction ratios give the same result, since the quotient of numbers with different signs is negative.

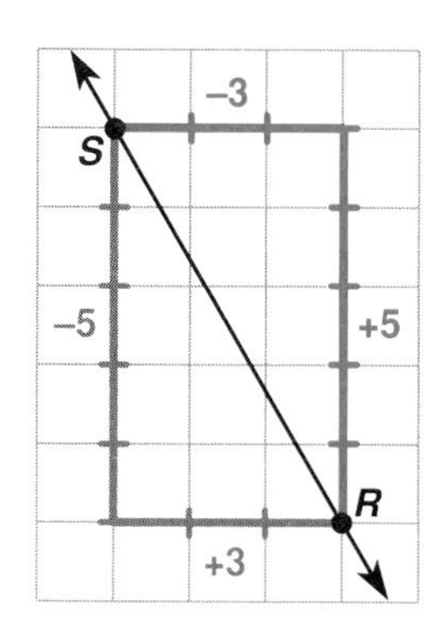

$$\frac{\text{up}}{\text{left}} = \frac{+5}{-3} = -\frac{5}{3} \qquad \frac{\text{down}}{\text{right}} = \frac{-5}{+3} = -\frac{5}{3}$$

The slope of a line is determined by a direction ratio, which gives the vertical direction as the numerator and the horizontal direction as the denominator.

$$\textbf{slope} = \frac{\textbf{up or down}}{\textbf{right or left}}$$

Example 2 Points $A(2, 1)$ and $B(6, 3)$ are on line AB, whose equation is $y = \frac{1}{2}x$.
Find the slope of this line.

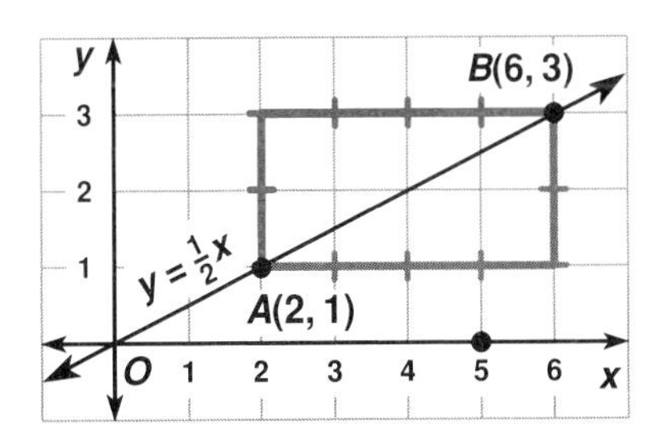

To find the slope of the line, you can use either the direction ratio from A to B, or the direction ratio from B to A.

From A to B: $\frac{\text{up}}{\text{right}} = \frac{2}{4}$, or $\frac{1}{2}$

From B to A: $\frac{\text{down}}{\text{left}} = \frac{-2}{-4}$, or $\frac{1}{2}$

Answer: The slope of the line $y = \frac{1}{2}x$ is $\frac{1}{2}$.

Try This *(For solution, see page 670.)*

a. For this line, whose equation is $y = 2x - 4$, tell how to travel along grid lines to get:

(1) from A to B **(2)** from B to A
(3) from B to C **(4)** from C to B
(5) from A to C **(6)** from C to A

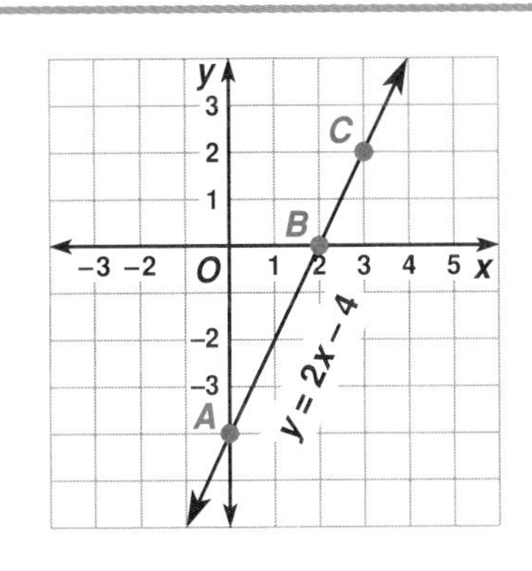

b. For this line, what is the value of:

(1) the direction ratio $\frac{\text{up}}{\text{right}}$

(2) the direction ratio $\frac{\text{down}}{\text{left}}$

c. What is the slope of the line $y = 2x - 4$?

Using Coordinates to Find Slope

Points $A(2, 1)$ and $B(4, 7)$ are on line AB, whose equation is $y = 3x - 5$.

Find the slope of this line.

From A to B, go up 6 and right 2.
The slope is the direction ratio:

$$\frac{\text{up}}{\text{right}} = \frac{6}{2}, \text{ or } 3$$

The slope of the line $y = 3x - 5$ is 3.

Instead of counting boxes along the grid to get from one point to another on the line, you can use the coordinates of the points, in this case: $A(2, 1)$ and $B(4, 7)$

From which coordinates of the points will you know whether you are going up or down, in the vertical direction?

The difference of the y-coordinates gives the vertical distance.

From A to B: $7 - 1 = +6$, meaning 6 units up

From which coordinates will you know whether you are going right or left, in the horizontal direction?

The difference of the x-coordinates gives the horizontal distance.
From A to B: $4 - 2 = +2$, meaning 2 units right

Thus, the slope of line AB, determined by taking the direction ratio from A to B is:

$$\frac{7-1}{4-2} = \frac{6}{2}, \text{ or } 3$$

The slope of a line is a ratio, determined from the coordinates of two points on the line.

$$\textbf{slope} = \frac{\textbf{difference in } y\textbf{-coordinates}}{\textbf{difference in } x\textbf{-coordinates}}$$

Example 3 Points $M(-2, 8)$ and $N(0, 4)$ are on line MN, whose equation is $y = -2x + 4$.

Use the coordinates of these points to find the slope of line MN.

To work with the differences in the coordinates, be sure you use the points in the same order for both differences.

Let's go from N to M, meaning that you subtract N values from M values.

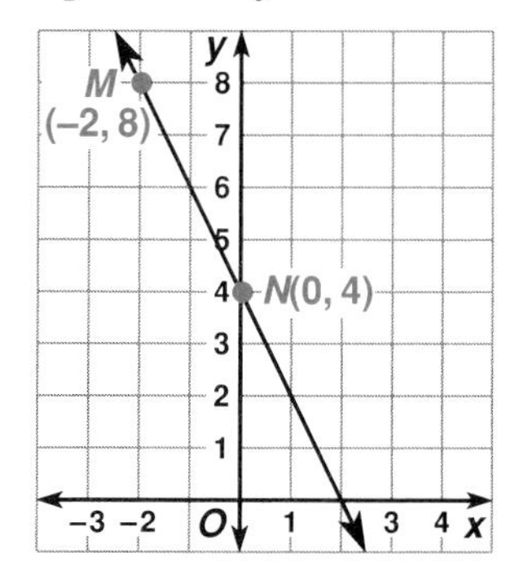

$$\text{slope} = \frac{\text{difference in } y\text{-coordinates}}{\text{difference in } x\text{-coordinates}}$$

$$\text{slope} = \frac{8-4}{-2-0} = \frac{4}{-2}$$ Up 4 left 2 takes you from N to M.

slope of $\overleftrightarrow{MN} = -2$

Answer: The slope of $y = -2x + 4$ is -2.

Using the Coordinates of Two Points to Find the Slope of a Line

1. Subtract *y*-coordinates.
2. In the same order, subtract *x*-coordinates.
3. $\text{slope} = \dfrac{\text{difference in } y\text{-coordinates}}{\text{difference in } x\text{-coordinates}}$

Example 4 Find the slope of the line PQ, on which are the points $P(-1, 2)$ and $Q(7, 8)$.

Subtract the coordinates of P from the coordinates of Q.

$$\text{slope} = \frac{\text{difference in } y\text{-coordinates}}{\text{difference in } x\text{-coordinates}}$$

$$\text{slope of } \overleftrightarrow{PQ} = \frac{8-2}{7-(-1)} = \frac{6}{8} = \frac{3}{4}$$

Answer: The slope of line $PQ = \frac{3}{4}$.

Try This *(For solution, see page 670.)*

Points $R(1, 3)$ and $S(2, 6)$ are on line RS, whose equation is $y = 3x$.

a. Use the coordinates of these points to find the slope of the line.

b. Verify your result by counting grid lines to get a direction ratio.

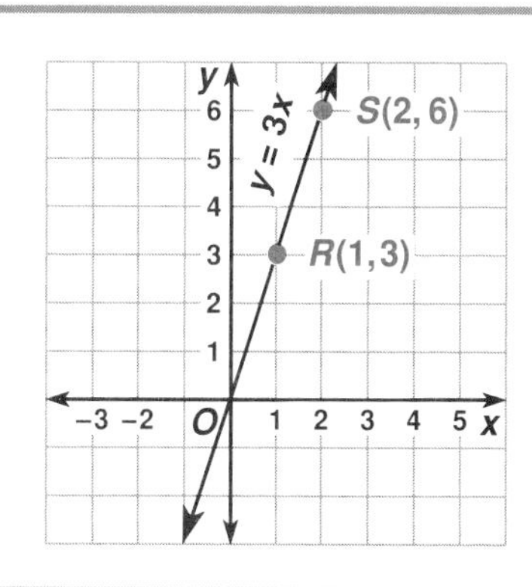

SELF-CHECK: SECTION 14.2

1. What is the steepness of a line called?

Answer: the slope

2. a. Write these directions for getting from one point to another on a line as a ratio: go up 1 and right 3

Answer: $\dfrac{\text{up}}{\text{right}} = \dfrac{1}{3}$

(Self-Check continues)

b. What does this direction ratio represent?

Answer: the slope of the line

c. What other direction ratio can you follow to get to points on this line?

Answer: go down 1 and left 3

$$\frac{\text{down}}{\text{left}} = \frac{-1}{-3} = \frac{1}{3}$$

3. When writing the direction ratios for slope, which directions go in the numerator of the ratio? in the denominator?

Answer: $\text{slope} = \dfrac{\text{up or down}}{\text{left or right}}$

4. Besides counting on the grid between two points to find the slope of a line, how else can you use the points?

Answer: To find the slope of a line from two points, use the coordinates of the points.

$$\text{slope} = \frac{\text{difference in } y\text{-coordinates}}{\text{difference in } x\text{-coordinates}}$$

EXERCISES: SECTION 14.2

1. For each of the following graphs, tell how to travel along grid lines to get:

(1) from A to B	**(2)** from B to A	**(3)** from B to C
(4) from C to B	**(5)** from A to C	**(6)** from C to A
(7) from P to A	**(8)** from B to P	**(9)** from P to C

a.

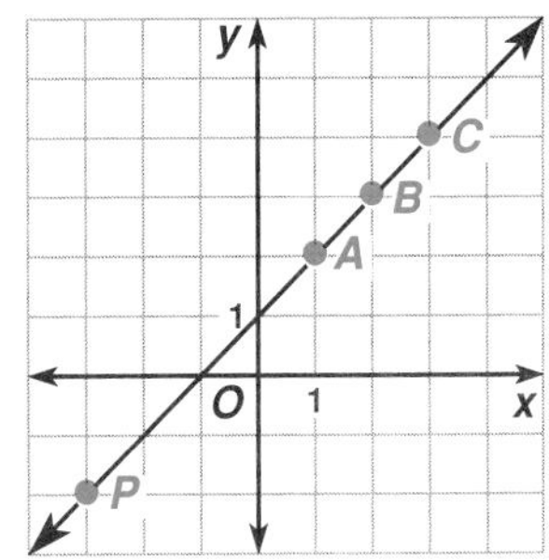

b.

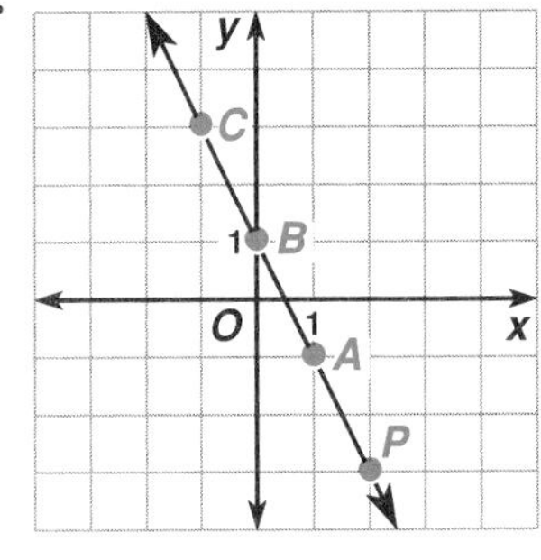

c.

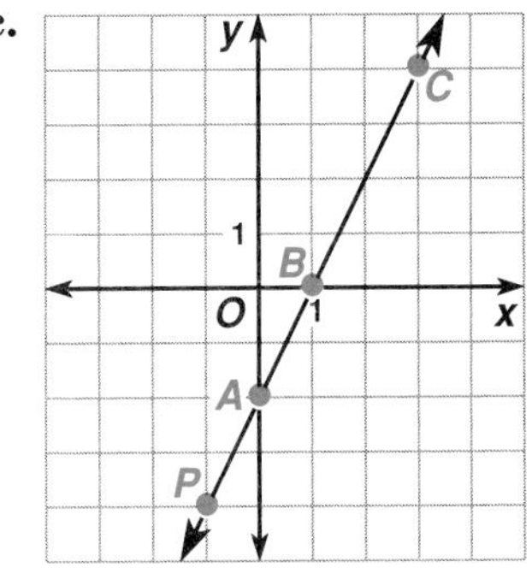

d.

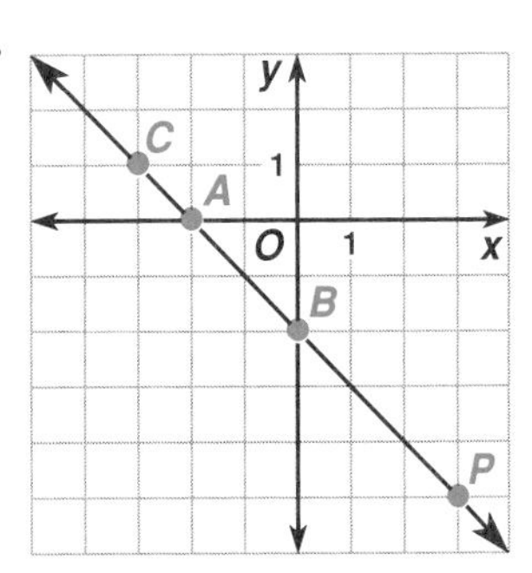

e.

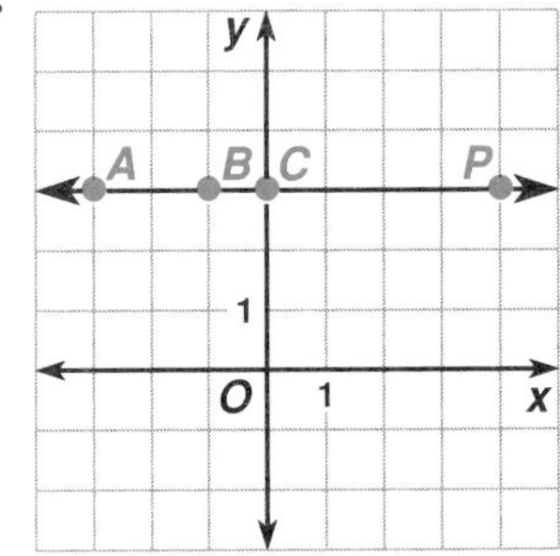

f.

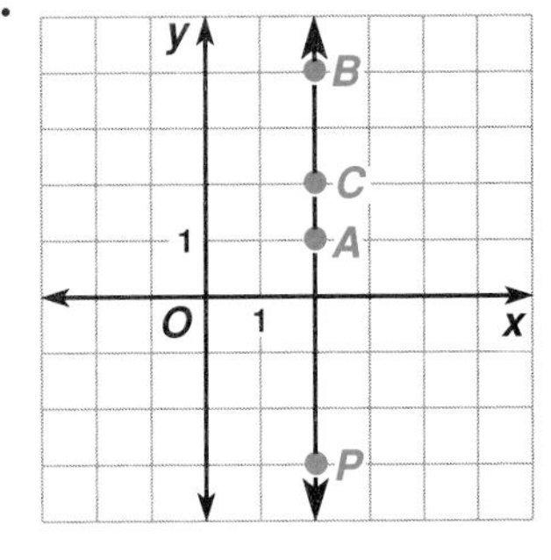

2. Explain why you agree or disagree with each statement.

a. Gina said that to get from $A(1, 2)$ to $B(3, 7)$, you go up 5 and right 2, giving the direction ratio $\frac{5}{2}$.

b. George said that to get from $P(-2, -1)$ to $Q(-3,-3)$, you go down 2 and left 1, giving the direction ratio $\frac{-2}{-1}$, or $\frac{2}{1}$.

c. Gail said that to get from $R(0, 2)$ to $S(4, 0)$, you go up 4 and left 2, giving the direction ratio $\frac{4}{-2}$, or $-\frac{2}{1}$.

d. Gino said that the direction ratios $\frac{6}{-2}$, $\frac{-6}{2}$, $-\frac{3}{1}$ are equivalent.

e. Greg said that the direction ratios $\frac{-4}{-2}$, $\frac{4}{2}$, $\frac{2}{1}$ are equivalent.

3. For each line, the coordinates of two points, A and B, are given. Use the direction ratio from A to B, $\frac{\text{up or down}}{\text{right or left}}$, to find the slope of the line. Express each ratio in simplest form.

a.

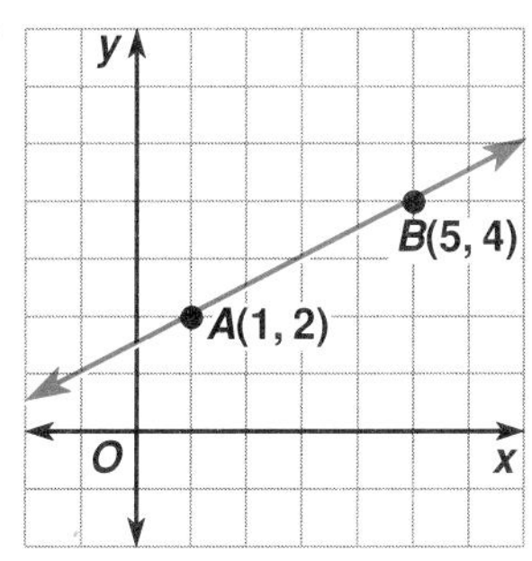

b.

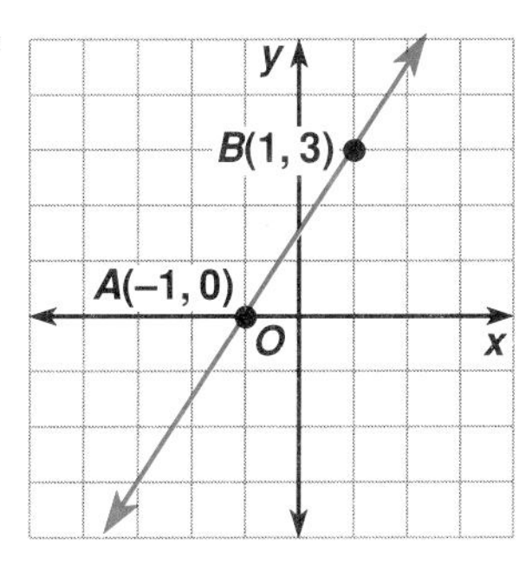

c.

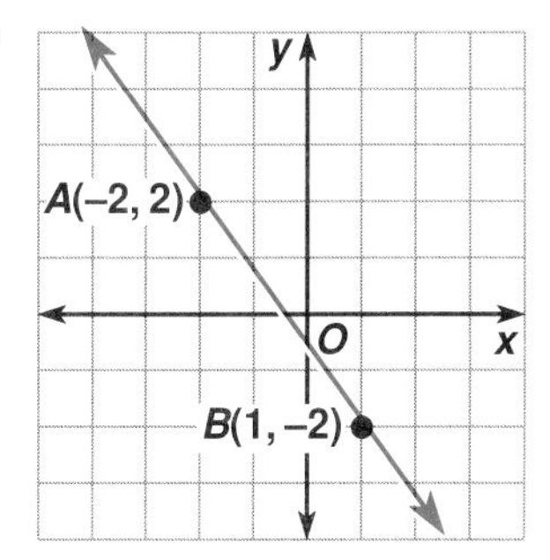

d.

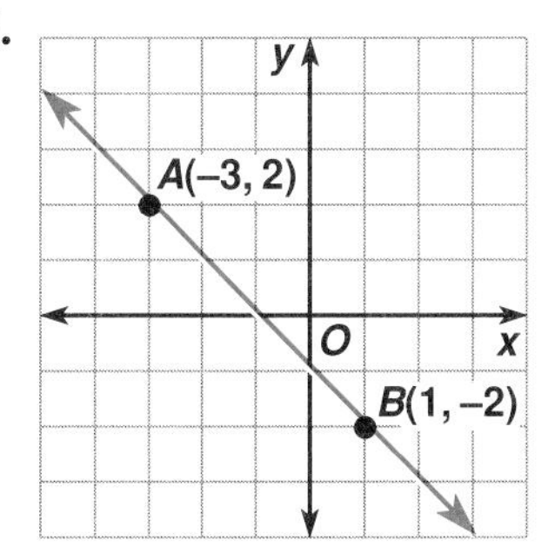

e.

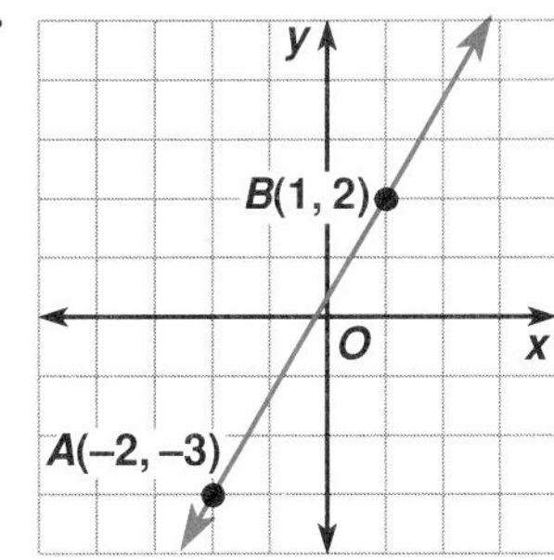

f.

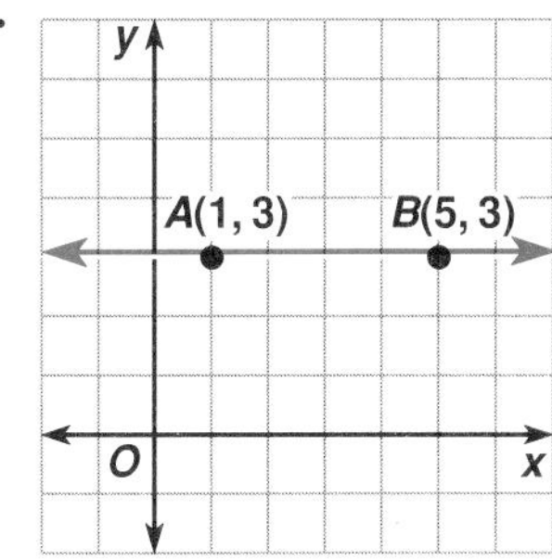

4. Use direction ratios to determine if each set of points lies on one line. Explain each decision.

a.

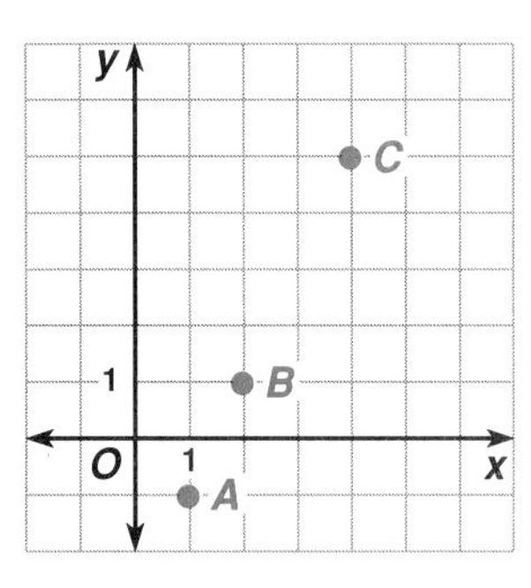

b.

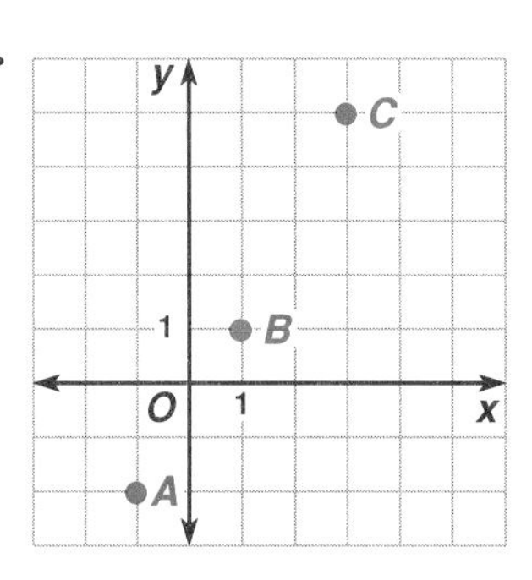

c.

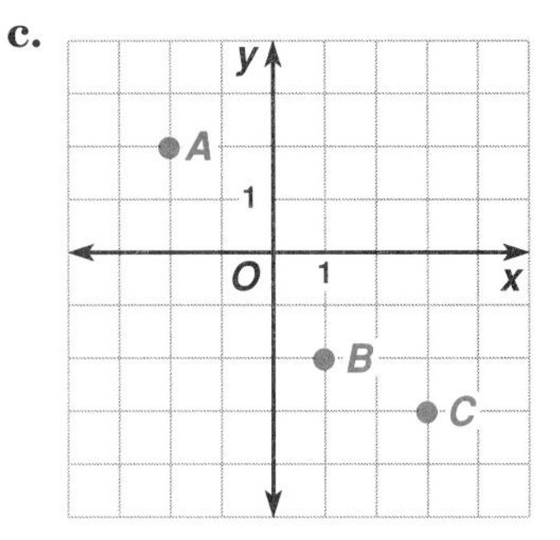

5. a. Lines AB, CD, and EF are shown in these diagrams.

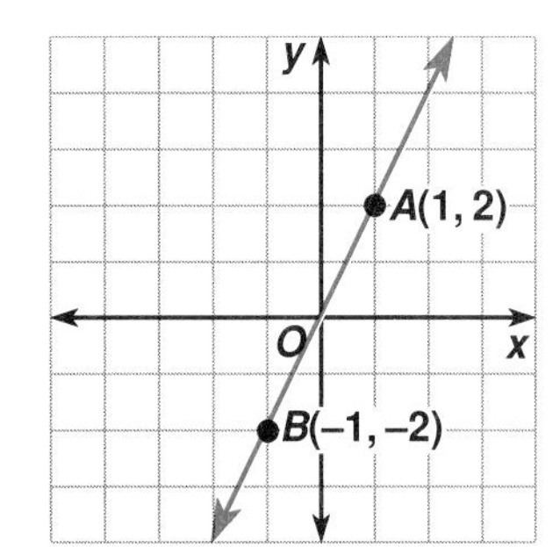

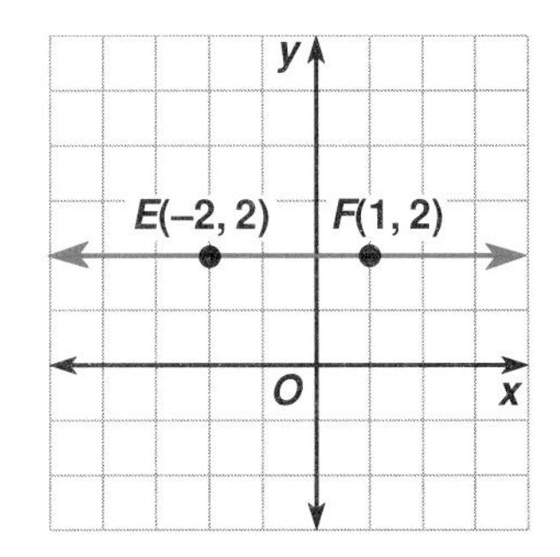

Use a direction ratio with the given points to determine which of the three lines has:

(1) a slope of 2 **(2)** a slope of −2 **(3)** a slope of 0

b. Based on your results in part a, tell which of the following lines, ℓ, m, or n, you would expect to have a positive slope, a negative slope, or a slope of 0.

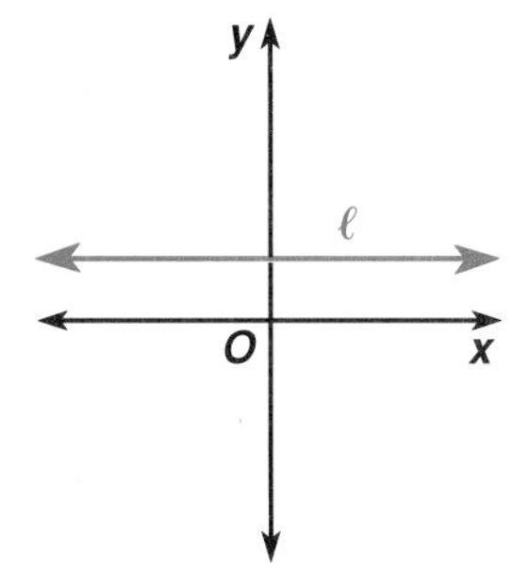

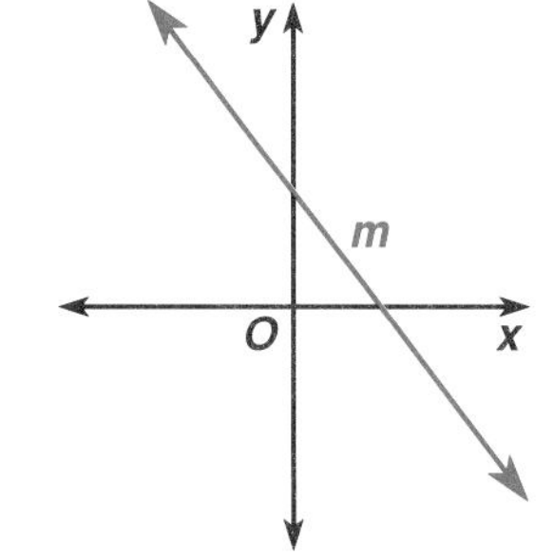

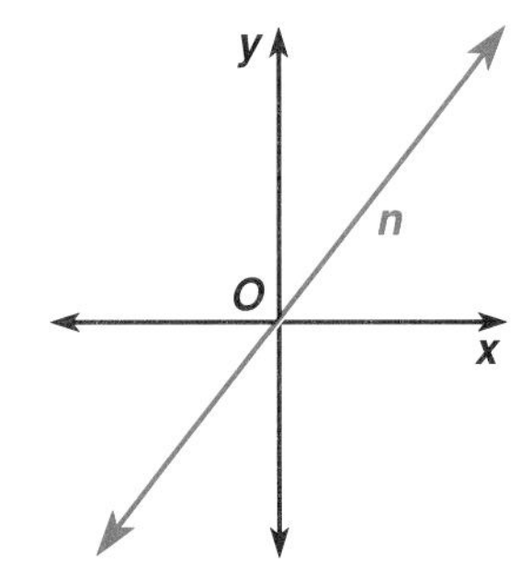

6. a. Use the coordinates of the two points shown on each line to determine the slope of the line.

(1)

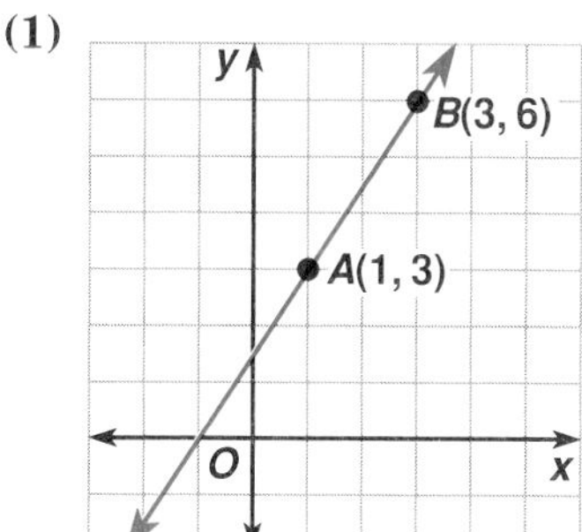

(2)

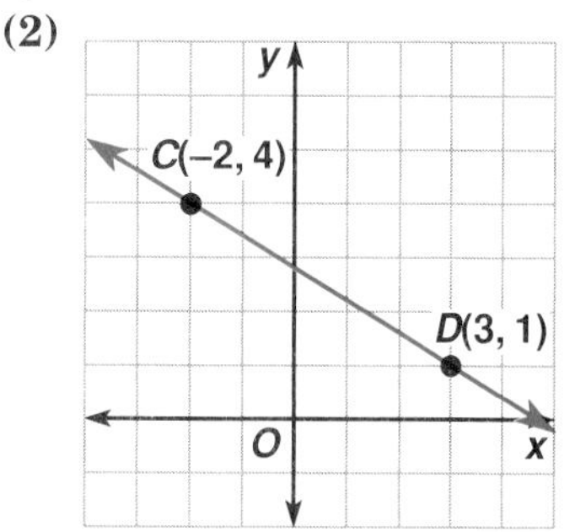

(3)

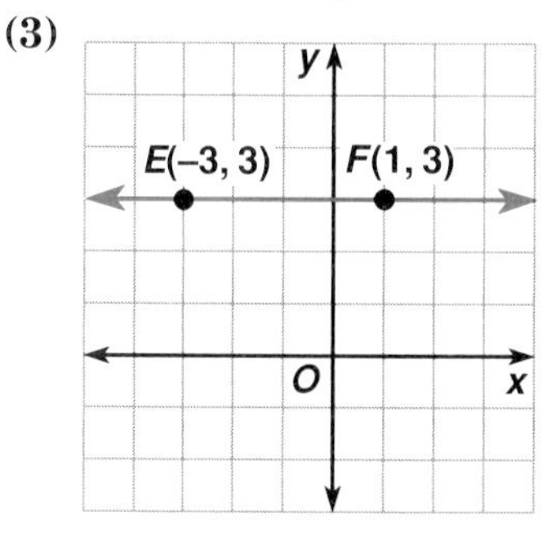

b. Find the slope of the line through each pair of points.

(1) (0, 0) and (2, 2) **(2)** (1, 3) and (5, 5)

(3) (2, 4) and (3, 2) **(4)** (−2, 1) and (1, 3)

(5) (−2, −3) and (4, −3) **(6)** (0, 2) and (2, 0)

(7) (−3, 5) and (5, −3) **(8)** (8, 0) and (6, 8)

(9) (−1, −2) and (−3, −4) **(10)** (−4, −3) and (−2, −7)

7. Each pair of points is on a line that has the given slope.
Use graph paper to model the situation. Find the missing coordinates.

a. $A(0, 5)$ and $B(x, 8)$ are on line AB, whose slope is $\frac{1}{2}$

b. $C(2,0)$ and $D(4, y)$ are on line CD, whose slope is 1

c. $E(3, y)$ and $F(5, 0)$ are on line EF, whose slope is 2

d. $G(4,7)$ and $H(2, y)$ are on line GH, whose slope is 1

e. $I(x, 2)$ and $J(3, 4)$ are on line IJ, whose slope is 2

f. $K(-2, -1)$ and $L(x, 2)$ are on line KL, whose slope is $\frac{3}{4}$

g. $M(x, 2)$ and $N(4, 1)$ are on line MN, whose slope is $-\frac{1}{3}$

EXPLORATIONS

Use one set of axes on graph paper to do all of the following.

a. (1) Plot $A(2, 10)$ and $B(-1, 4)$, to draw line AB.
Use the coordinates of A and B to determine the slope of line AB.

(2) Plot $C(-2, -2)$ and $D(0, 2)$, to draw line CD.
Use the coordinates of C and D to determine the slope of line CD.

(3) Plot $E(-1, -3)$ and $F(4, 7)$, to draw line EF.
Use the coordinates of E and F to determine the slope of line EF.

b. What is true about the slopes of these three lines?

c. Make an observation about the way the lines have fallen on the graph.

d. Draw a conclusion from your results.

14.3 The Equation y = mx + b

Reading the Equation of a Line

Here is a list of the equations of some of the lines you have worked with, and the values that were found as slopes.

Make an observation as to where in the equation of a line you can read the value of the slope.

Equation of Line	*Slope*
$y = 2x + 1$	2
$y = \frac{1}{2}x$	$\frac{1}{2}$
$y = 3x - 5$	3
$y = -2x + 4$	-2

When written in the form $y =$, the equation of a line shows the value of the slope of the line as the coefficient of x.

Example 1 The equation of this line is $y = 5x + 1$.

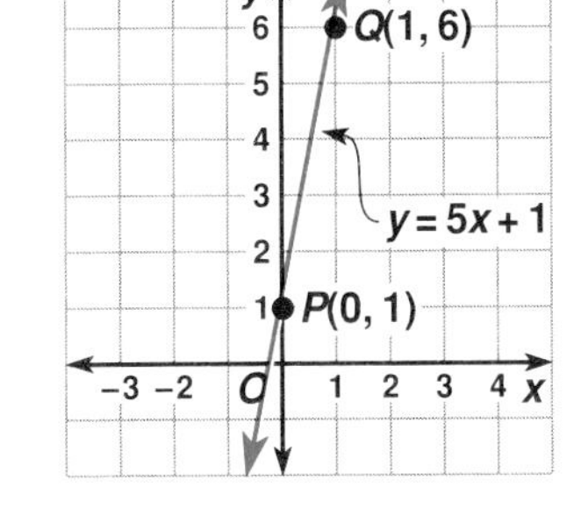

a. From the equation, write the slope of the line.

Since this equation is in the form $y =$, the slope is the coefficient of x.

Answer: The slope of the line $y = 5x + 1$ is 5.

b. Verify your result by writing the coordinates of two points on the line, and calculating the slope from the coordinates.

From the graph, you see that $P(0, 1)$ and $Q(1, 6)$ are points on the line. Is the slope 5?

$$\text{slope} = \frac{\text{difference in } y\text{-coordinates}}{\text{difference in } x\text{-coordinates}} = \frac{6-1}{1-0} = \frac{5}{1}, \text{ or } 5 \checkmark$$

Let's examine these three graphs of lines to see what other information about a line can be read from its equation.

First, make an observation about the slopes of the three lines.

From the equations, you see that the coefficients of x are the same value, 2. Thus, the slopes of all the lines are equal.

$y = 2x + 1 \qquad y = 2x + 3 \qquad y = 2x + 6$

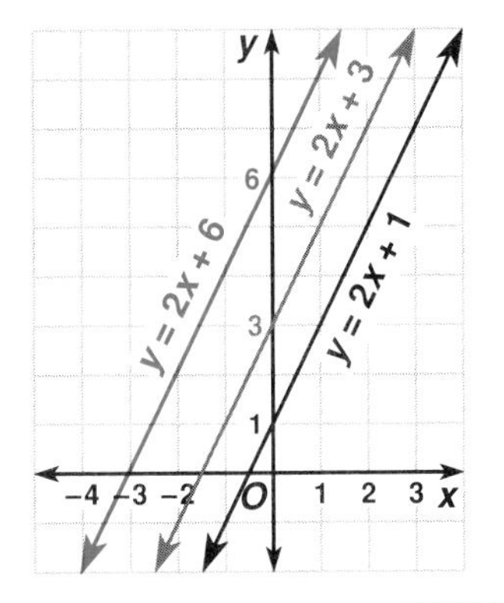

What special property do these lines with equal slopes have?

Lines with equal slopes are parallel.

Now study the constant term in each equation, and make an observation of how that number is related to the graph of the line.

The line $y = 2x + \mathbf{1}$ intersects the y-axis at 1.

The line $y = 2x + \mathbf{3}$ intersects the y-axis at 3.

The line $y = 2x + \mathbf{6}$ intersects the y-axis at 6.

When written in the form $y =$, the constant term of the equation is the number at which the line intersects the y-axis. This number is called the *y-intercept* of the line.

Example 2 Consider the equation $y = 3x + 2$.

a. From the equation, read the slope and y-intercept of the line.

Since the equation is in the form of $y =$, the coefficient of x is the slope and the constant term is the y-intercept.

$$\underset{\text{slope}}{y = \underset{\uparrow}{3}x + \underset{\uparrow}{2}}\quad (3 = \text{slope},\ 2 = y\text{-intercept})$$

Answer: For the line $y = 3x + 2$, the slope is 3 and the y-intercept is 2.

b. Verify your result by graphing the line.

To graph the line, create a table of points. Choose three values of x, and calculate the corresponding y-values.

x	$3x + 2$	y
1	$3(1) + 2$ $3 + 2$	5
2	$3(2) + 2$ $6 + 2$	8
−1	$3(-1) + 2$ $-3 + 2$	−1

Plot the three points (1, 5), (2, 8), and (−1, −1).

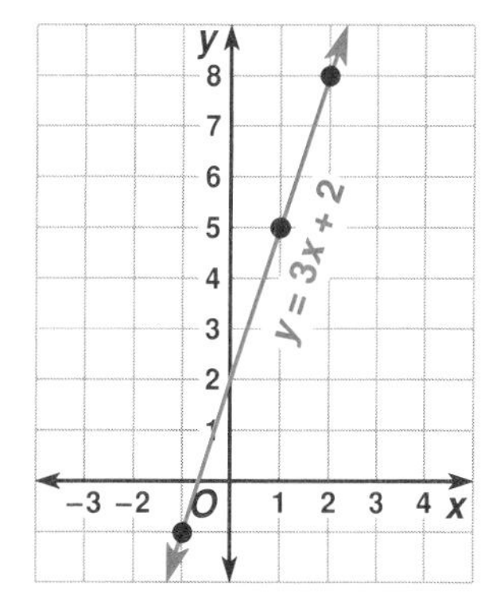

Does the graph intersect the y-axis at 2? ✓

Is the slope of the line equal to 3?

To answer this question, note the direction ratio between two points on the line. From (1, 5), you go up 3 and right 1 to reach (2, 8).

$$\text{slope} = \frac{\text{up}}{\text{right}} = \frac{3}{1}, \text{ or } 3\ ✓$$

Reading the Equation of a Line

When the equation is in the form $y =$, the coefficient of x is the slope and the constant term is the y-intercept.

$$y = \underset{\substack{\uparrow\\ \text{slope}}}{m}x + \underset{\substack{\uparrow\\ y\text{-intercept}}}{b}$$

Try These (For solutions, see page 670.)

1. Is the following statement true or false? Explain.
   $y = 2x + 4$ is an example of an equation of a line whose slope is 2.
2. Consider the equation $y = 3x + 4$.
   a. From the equation, read the slope and the y-intercept.
   b. Verify your result by graphing the line.
      To graph the line, copy and complete the table shown.

x	$3x + 4$	y
1	$3(1) + 4$	?
2	?	?
–1	?	?

3. Is the following statement true or false? Explain.
   The lines $y = 2x + 3$ and $y = 2x + 5$ are parallel.

Using Slope to Graph

$P(0, 1)$ is a point on a line, and the slope of the line is $\frac{1}{2}$. Describe how to use the slope to get from P on this line to other points on this line.

Use the direction ratio of the slope.

To get from P to another point Q on this line, go up 1 and right 2.

From Q, go up 1 and right 2 to R.

Use P, Q, and R to draw the line.

Point T is located at $(-2, 0)$ on this line. Describe how to get from P to T.

To get from P to T, go down 1 and left 2.

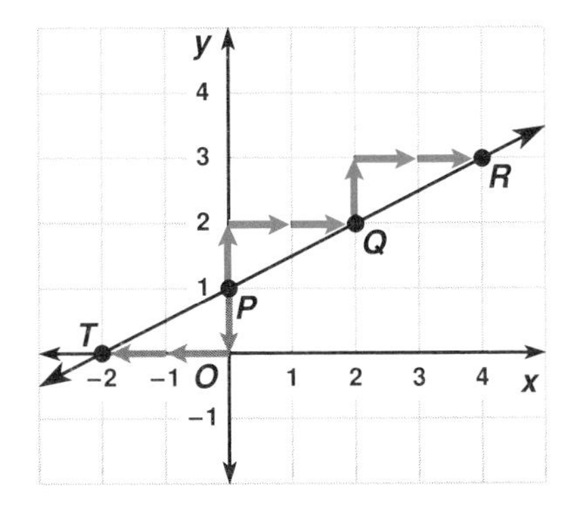

If you know the location of one point on a line, you can use the slope of the line to locate other points on the line and, thus, graph the line.

Example 3 Consider the line $y = \frac{1}{3}x + 2$.

a. From the equation, write the slope and the y-intercept.

slope ↓ y-intercept ↓

$y = \frac{1}{3}x + 2$ slope $= \frac{1}{3}$ y-intercept $= 2$

b. Locate the y-intercept, 2, on the y-axis at the point $P(0, 2)$.

From P, use the slope to locate other points on the line.

From P, go up 1 and right 3 to Q.

From Q, go up 1 and right 3 to R.

From P, go down 1 and left 3 to T.

From T, go down 1 and left 3 to W.

Draw the line through the 5 points you located.

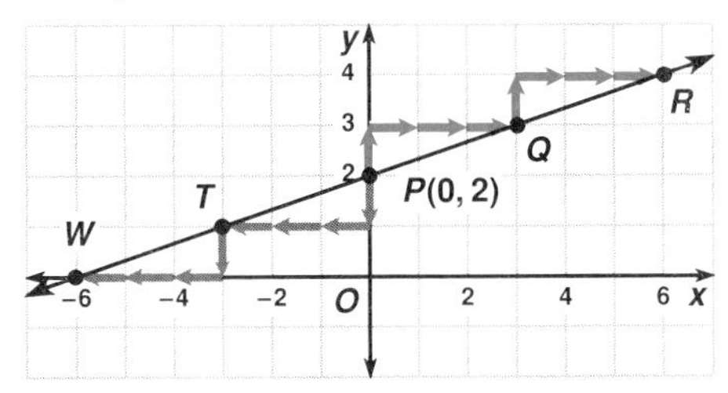

Example 4 Consider the line $y = 2x - 4$.

a. From the equation, write the slope and the y-intercept.

$$\underset{\text{slope}}{} \quad \underset{y\text{-intercept}}{}$$
$$y = 2x - 4$$

slope = 2, or $\frac{2}{1}$ — Rewrite the whole number as a fraction with a denominator of 1.

y-intercept = -4 — Be sure to include the sign with the y-intercept.

b. Locate the y-intercept, -4, on the y-axis at $P(0, -4)$.

From P, use the slope to locate other points on the line.

From P, go up 2 and right 1 to Q.

From Q, go up 2 and right 1 to R.

From P, go down 2 and left 1 to T.

From T, go down 2 and left 1 to W.

Draw the line through the points you located.

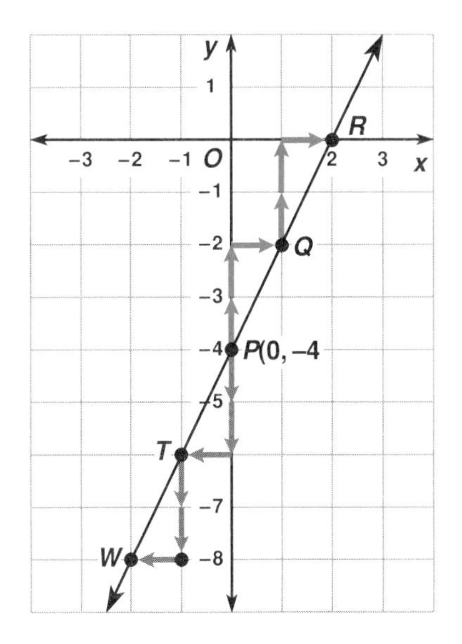

Try This *(For solution, see page 671.)*

Consider the equation $y = \frac{1}{3}x + 1$.

a. From the equation, write the slope and the y-intercept.

b. On graph paper, draw a set of axes and mark the y-intercept of the given line on the y-axis.

What are the coordinates of point P, the point at which the line will cross the y-axis?

c. **(1)** Use the slope of the line to locate three other points that are "above" point P on the line.

(2) Use the slope of the line to locate three other points that are "below" point P on the line.

(3) Draw the line.

SELF-CHECK: SECTION 14.3

1. Consider the equation of the line $y = 3x + 4$.

a. Which number in the equation represents the slope of the line?

Answer: 3, the coefficient of x

b. What does the number 4 represent?

Answer: The y-intercept, the number on the y-axis at which the line crosses the y-axis.

c. Since the y-intercept of the line is 4, what are the coordinates of the point at which the line crosses the y-axis?

Answer: (0, 4)

2. What special property do lines with equal slopes have?

Answer: Lines with equal slopes are parallel.

3. Consider the line $y = \frac{1}{2}x + 7$.

a. What is the y-intercept?

Answer: 7

b. From 7 on the y-axis, how can you get to another point on this line?

Answer: The slope $\frac{1}{2}$ tells you to go up 1 and right 2.

c. If you start at 7 on the y-axis and go up 1 and right 2 to another point on this line, you get to a point that is "above" 7 on the line.

How can you get to a point that is "below" 7 on the line?

Answer: There are many ways to get points "below" 7 on the line.

For example: go down 1 and left 2
go down 2 and left 4

d. Explain why going up 1 and right 2 gives the same slope as going down 1 and left 2.

Answer: Slope is a ratio.

$$\frac{\text{up}}{\text{right}} = \frac{1}{2} \qquad \frac{\text{down}}{\text{left}} = \frac{-1}{-2} = \frac{1}{2}$$

The quotient of two negative numbers is equivalent to the quotient of two positive numbers.

EXERCISES: SECTION 14.3

1. Explain why you agree or disagree with each statement.

a. Dan said the slope of the line $y = 2x + 5$ is 5, while Deva said that the slope of the line $y = 2x + 5$ is 2.

b. Dina said that the slope of the line $2y = 6x$ is 6 because you read the slope from the coefficient of x.

c. Dave said that to find the slope of the line $2y = 6x$, you must first put the equation in the form $y =$. After dividing both sides of the equation by 2, getting $y = 3x$, Dave said the slope of the line is 3.

2. For each of the following equations whose graph is given:

(1) Tell the slope of the line from the equation.

(2) Verify your result by using the coordinates of the two points shown to calculate the slope.

a.

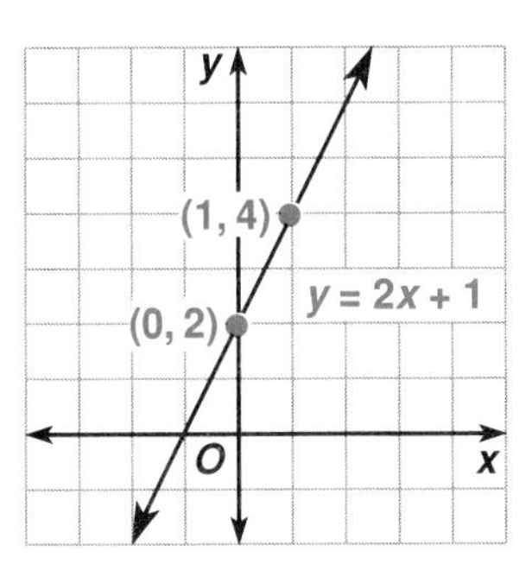

b.

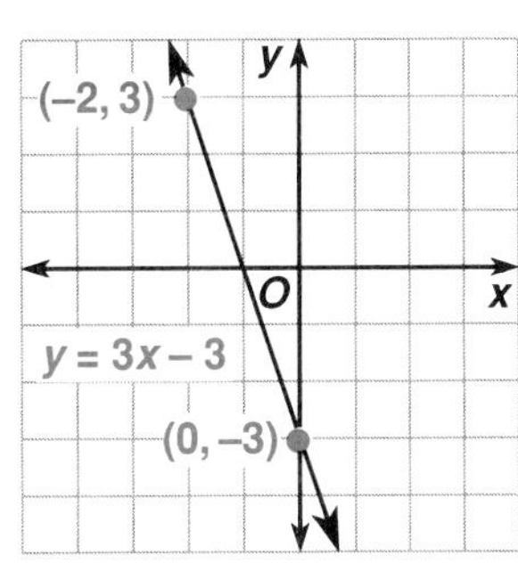

c.

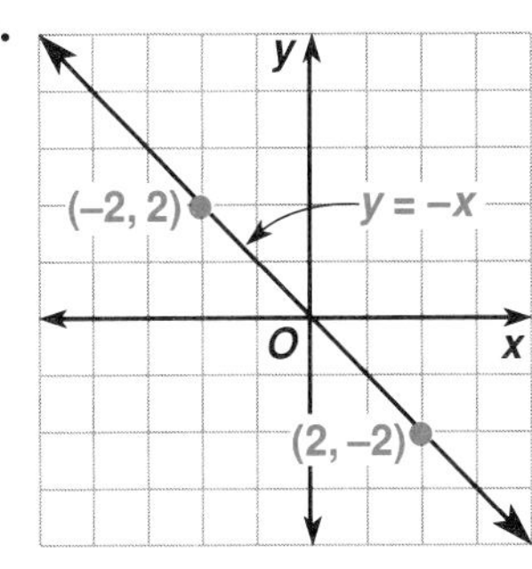

d.

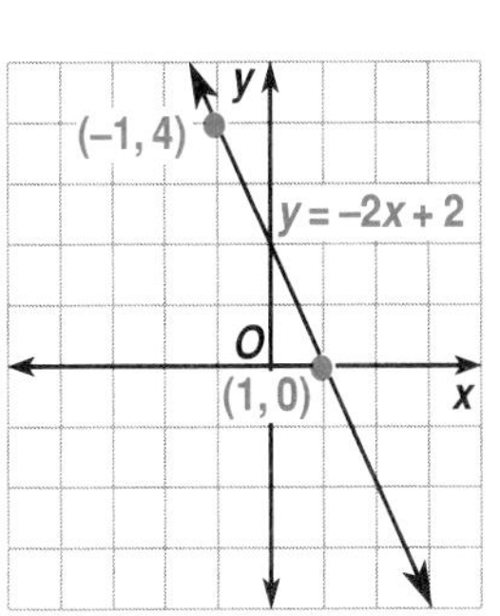

e.

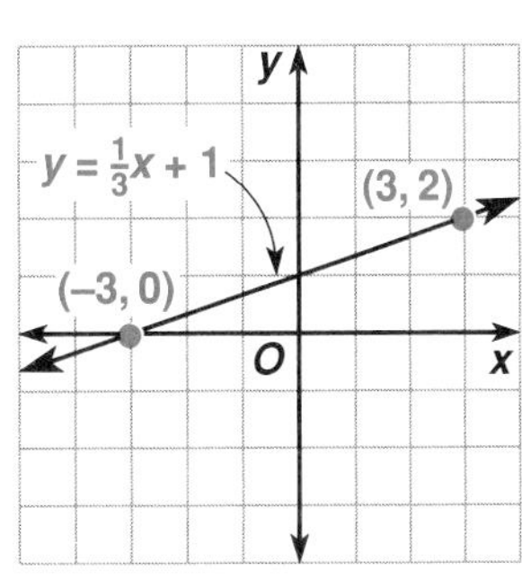

f.

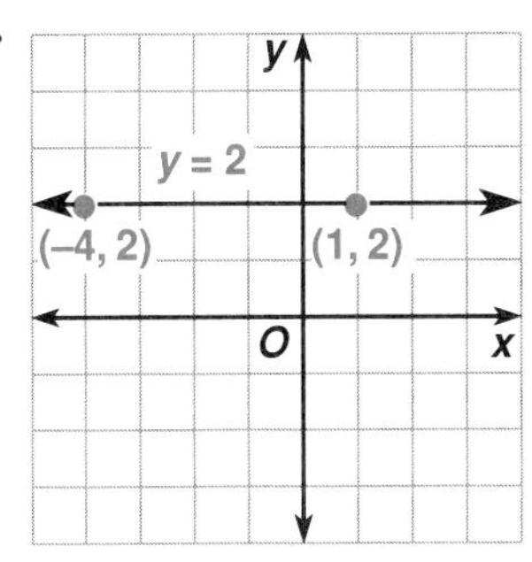

3. a. From the graph of each equation shown, tell the y-intercept.

(1)

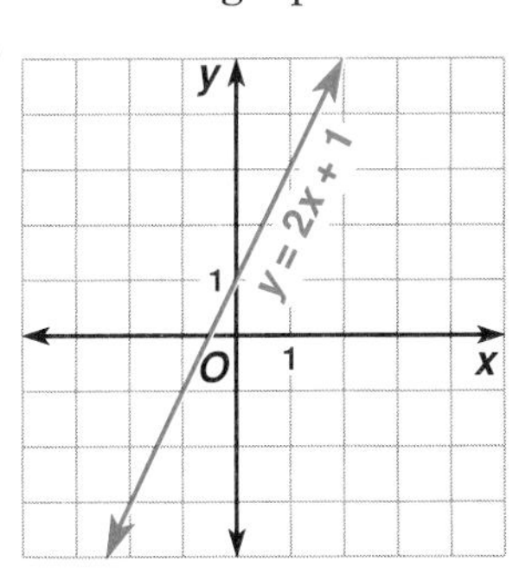

(2)

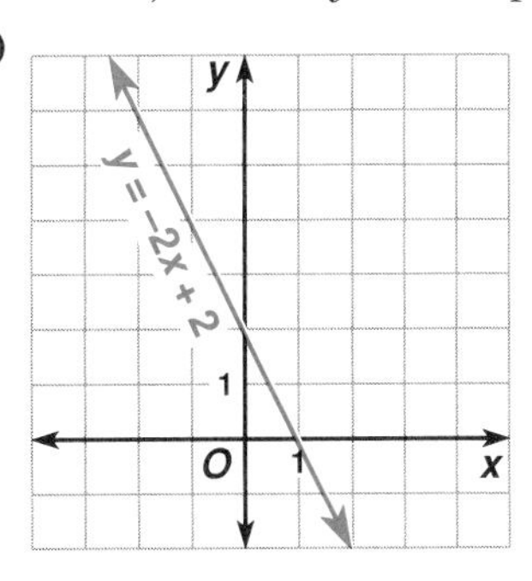

(3)

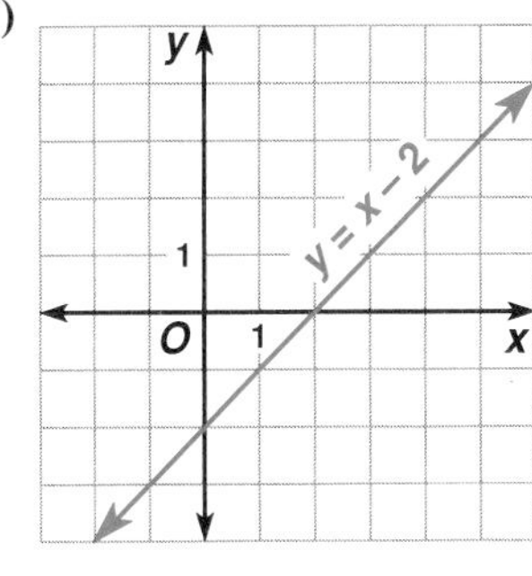

(4)

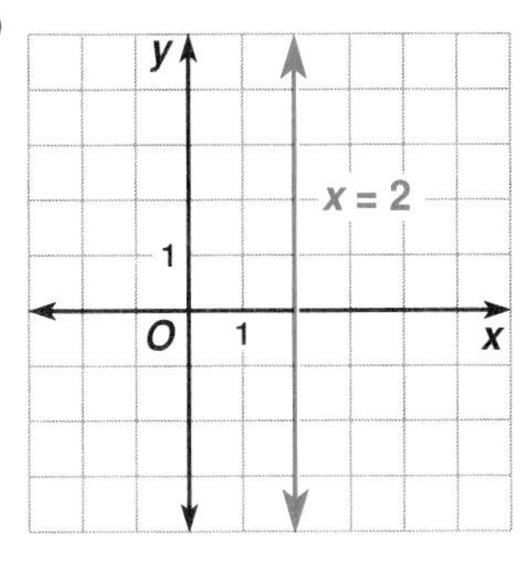

(5)

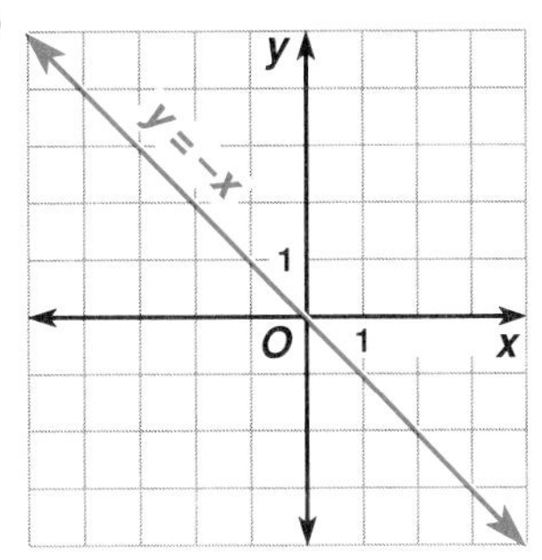

(6)

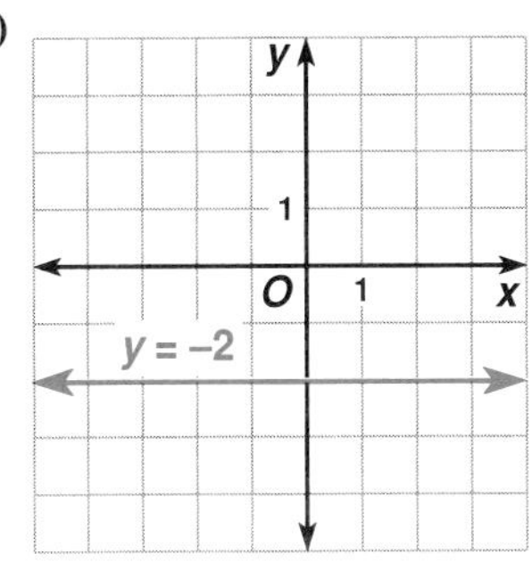

b. From each equation of a line, tell the slope and the y-intercept.

(1) $y = 2x + 1$	**(2)** $y = 3x + 4$	**(3)** $y = x + 5$
(4) $y = x$	**(5)** $y = 2x - 3$	**(6)** $y = -3x - 4$
(7) $y = \frac{1}{2}x - 2$	**(8)** $y = -\frac{2}{3}x + 1$	**(9)** $y = -x - \frac{1}{2}$
(10) $2y = 6x + 4$	**(11)** $3y = 9x - 3$	**(12)** $2y = 8x$
(13) $y = 3 + 2x$	**(14)** $y = 4 - 2x$	**(15)** $y = 3 - x$
(16) $x + y = 4$	**(17)** $2x + y = 8$	**(18)** $y - x = 5$

4. From each graph, find the slope and y-intercept of the line, and write an equation of the line.

a.

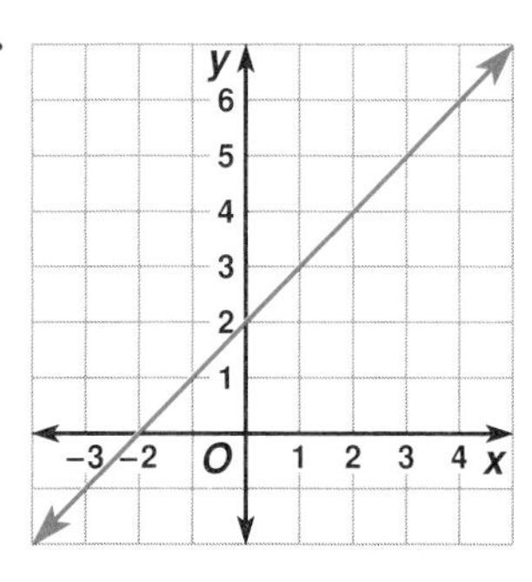

b.

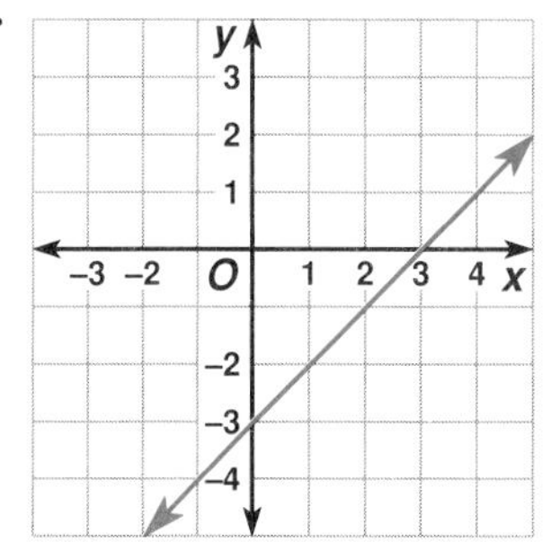

c.

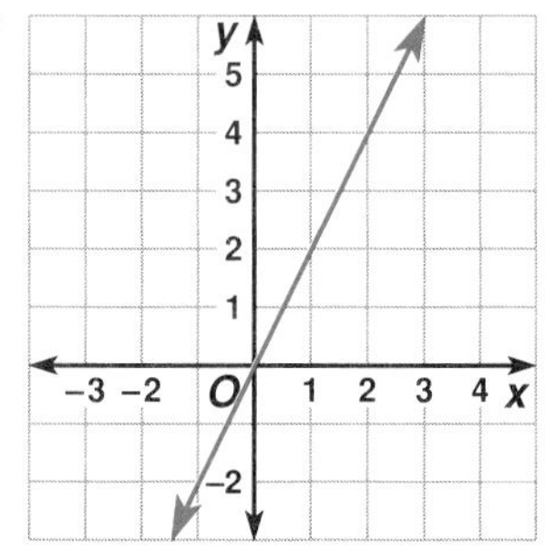

d.

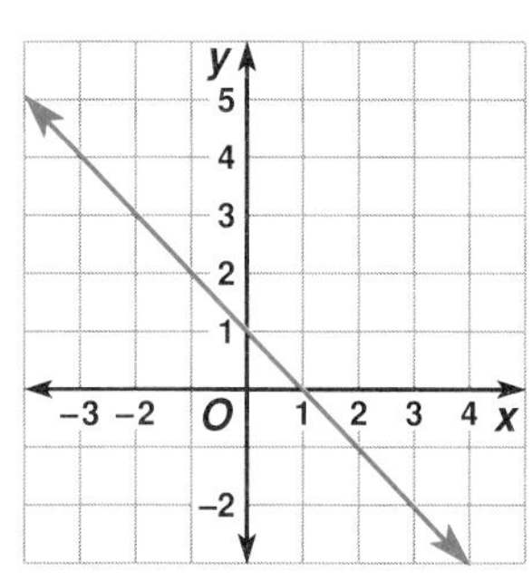

e.

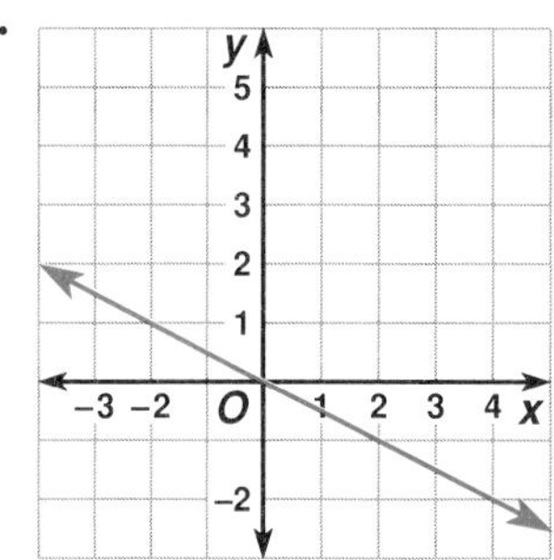

f.

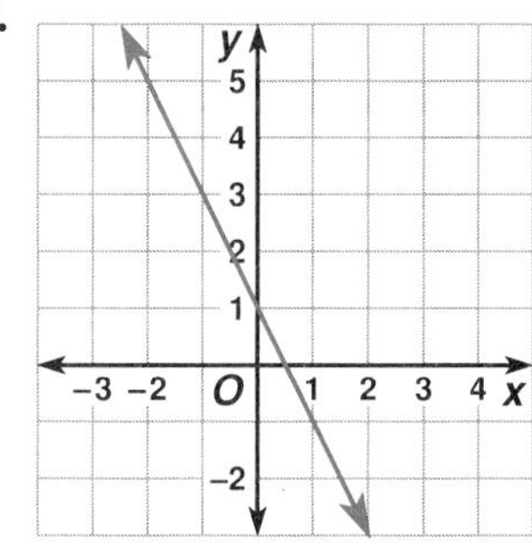

g.

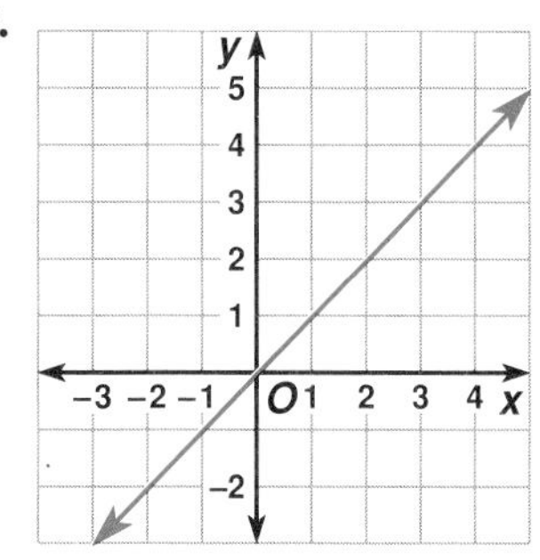

h.

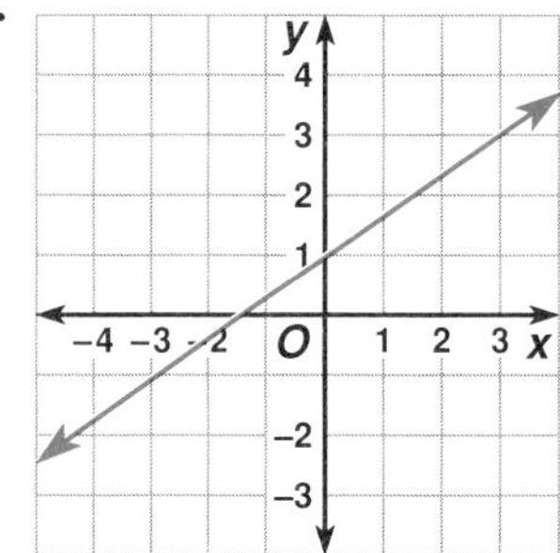

i.

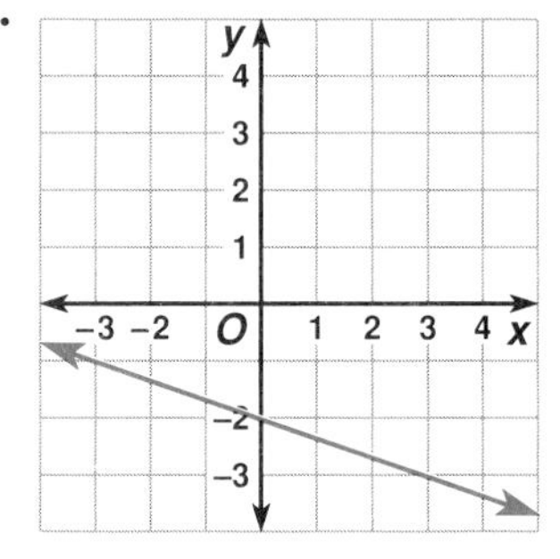

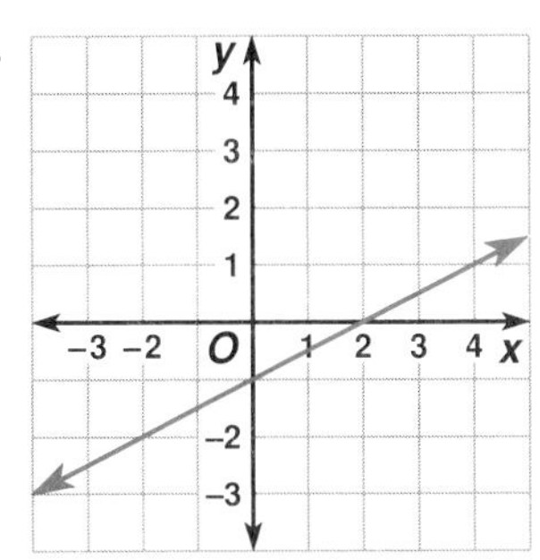

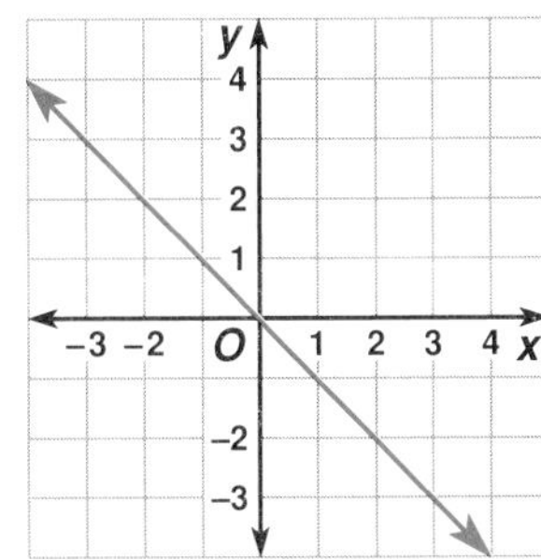

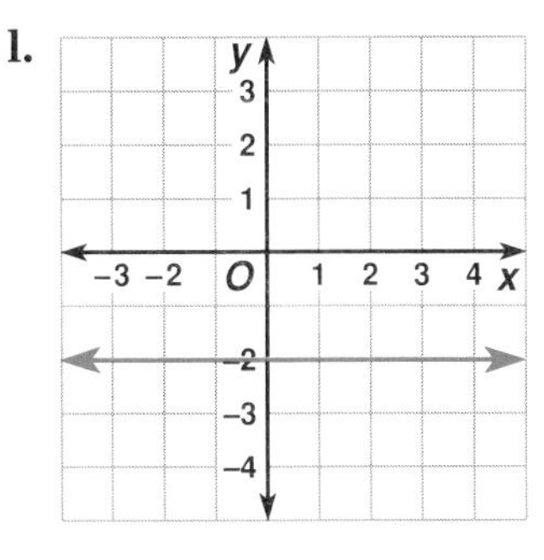

5. Explain why you agree or disagree with each statement.

a. Lisa said that when you graph the lines $y = 2x + 4$ and $y = 2x + 3$, the lines will be parallel.

b. Lazar said that parallel lines have the same y-intercept.

c. Lois said that all horizontal lines are parallel.

d. Leroy said that the equations $y = 3x + 4$ and $2y = 6x + 8$ represent the same line.

6. From each set of four equations, choose the two that represent parallel lines.

a. $y = 3x + 2$ $y = 2x + 3$ $y = 3x - 2$ $y = 23x$

b. $y = 4x - 2$ $y = 4 - 2x$ $y = -2x - 4$ $y = 2x + 4$

c. $y = 4x + 1$ $2y = 8x + 6$ $y = 1 - 4x$ $y + x = 4$

d. $2y = 6x + 4$ $y = 6x + 2$ $y = 2x + 6$ $3y = 9x + 3$

7. a. Line k is the result of translating the line $y = 2x + 1$ up by 3 units.
Write the equation of line k.

b. Line ℓ is the result of translating the line $y = 2x + 1$ down by 5 units.
Write the equation of line ℓ.

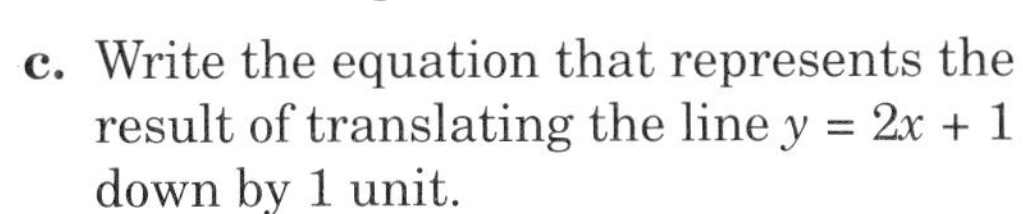

c. Write the equation that represents the result of translating the line $y = 2x + 1$ down by 1 unit.

d. Write the equation that represents the result of translating line k :

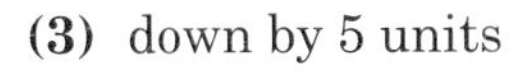

(1) up by 2 units   **(2)** down by 3 units   **(3)** down by 5 units

e. Write the equation that represents the result of translating line ℓ :

(1) up by 2 units   **(2)** down by 2 units   **(3)** up by 3 units

f. **(1)** How would line k have to be translated to produce line ℓ ?

(2) How would line ℓ have to be translated to produce line k?

g. Write the equation of the line that is parallel to line k, and:

(1) is 3 units above it   **(2)** is 6 units below it

(3) goes through the origin   **(4)** has a y-intercept of -2

8. Write the coordinates of the image P', the result of reflecting point P over the line shown. The line of reflection is the perpendicular bisector of the line segment PP'.

a. Reflect P over the y-axis.

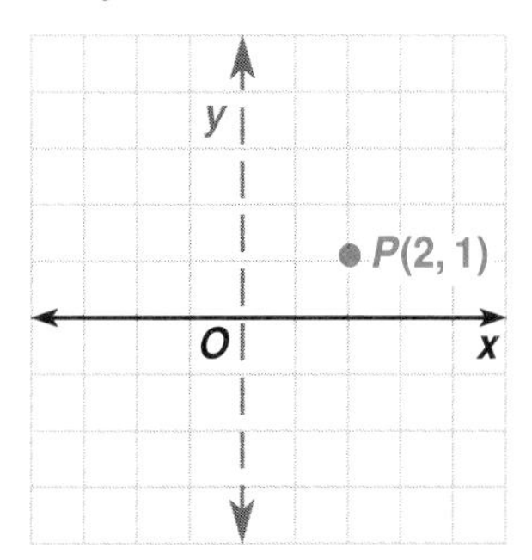

b. Reflect P over the x-axis.

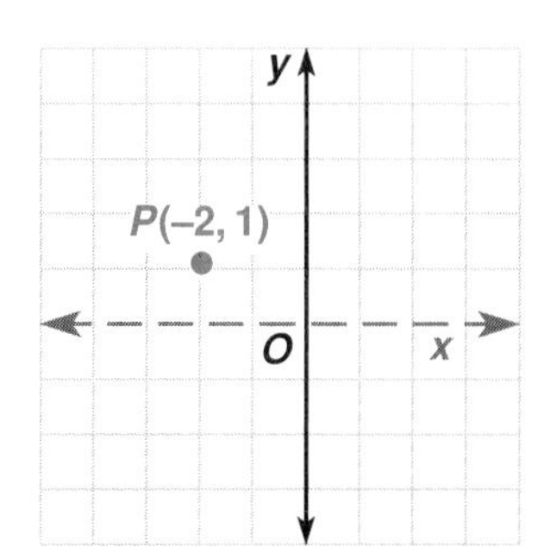

c. Reflect P over the line $y = x$.

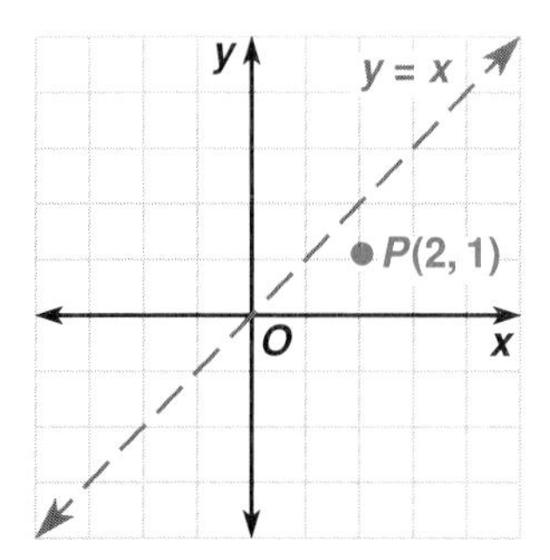

d. Reflect P over the line $y = -x$.

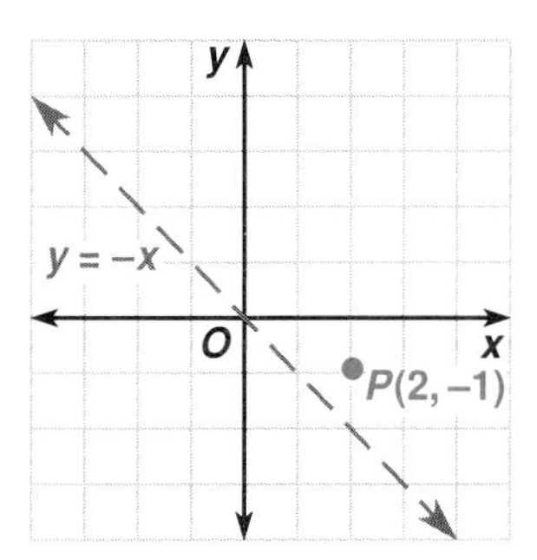

e. Reflect P over the line $y = 2$.

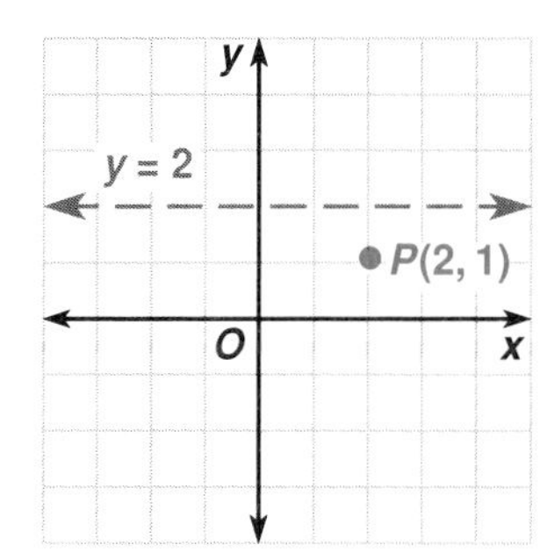

f. Reflect P over the line $x = 1$.

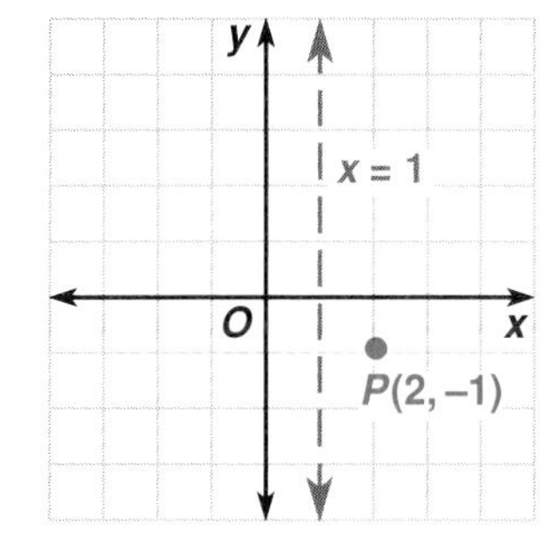

9. **a.** Graph the line whose slope is $\frac{1}{3}$ and whose y-intercept is 1 by following these steps:

(1) Use the y-intercept 1 to locate P, the point where the line crosses the y-axis.

Write the coordinates of P.

(2) Use the slope $\frac{1}{3}$ $\begin{matrix}\leftarrow \text{up 1} \\ \leftarrow \text{right 3}\end{matrix}$ to locate three points on the line that are "above" point P.

(3) Change signs to write another direction ratio that is equivalent to "up 1 and right 3."

$$\frac{1}{3} = \frac{-1}{-3} \quad \begin{matrix}\leftarrow \text{down 1} \\ \leftarrow \text{left 3}\end{matrix}$$

Use this direction ratio to locate three points on the line that are "below" point P.

(4) Draw the line, and write its equation.

b. Following these steps:

- From the equation, write the slope and the y-intercept.
- Plot the y-intercept. Call the point on the y-axis P.
- Using the direction ratio $\frac{\text{up}}{\text{right}}$ for the slope, locate three points on the line that are "above" P.
- Write the equivalent direction ratio $\frac{\text{down}}{\text{left}}$ for the slope, and use it to locate three points on the line that are "below" P.
- Draw the line and label it with its equation.

Graph each of these lines:

(1) $y = \frac{1}{4}x + 2$ **(2)** $y = \frac{2}{3}x + 2$ **(3)** $y = \frac{3}{4}x + 3$

(4) $y = \frac{1}{2}x - 4$ **(5)** $y = \frac{3}{4}x - 2$ **(6)** $y = \frac{2}{5}x - 1$

10. To result in a negative quotient, the signs of two numbers must be different.

Examples: $\frac{-6}{+2} = -3$ $\frac{+6}{-2} = -3$ $\frac{+1}{-3} = -\frac{1}{3}$ $\frac{-1}{+3} = -\frac{1}{3}$

Think backwards. If you have a fraction that is negative, then the numerator and denominator must have different signs.

Apply this thinking to slope.

The slope $-\frac{1}{3}$ is equivalent to different direction ratios:

$-\frac{1}{3} = \frac{-1}{+3}$ ← down 1 ← right 3 or $-\frac{1}{3} = \frac{+1}{-3}$ ← up 1 ← left 3

Write each of these slopes as two different direction ratios:

a. slope = $-\frac{3}{4}$ **b.** slope = $-\frac{1}{2}$ **c.** slope = $-\frac{2}{5}$ **d.** slope = $-\frac{2}{3}$

e. slope = $-\frac{3}{2}$ **f.** slope = $-\frac{5}{3}$ **g.** slope = $-\frac{4}{3}$ **h.** slope = $-\frac{6}{5}$

i. slope = -3 **j.** slope = -2 **k.** slope = -4 **l.** slope = -1

11. Following these steps:

- From the equation, write the slope and the y-intercept.
- Plot the y-intercept. Call the point on the y-axis P.
- From the slope, write two different direction ratios. Use these direction ratios to locate three points on the line "above" P and three points on the line "below" P.
- Draw the line and label it with its equation.

Graph each of these lines.

a. $y = -\frac{1}{4}x + 3$ **b.** $y = -\frac{3}{5}x + 1$ **c.** $y = -\frac{1}{3}x + 2$ **d.** $y = -\frac{1}{2}x - 2$

e. $y = -\frac{1}{4}x - 3$ **f.** $y = -\frac{2}{3}x - 4$ **g.** $y = -3x + 1$ **h.** $y = -4x + 2$

i. $y = -x + 3$ **j.** $y = -2x - 3$ **k.** $y = -x - 4$ **l.** $y = -3x$

EXPLORATIONS

1. a. Each of these graphs shows a pair of lines that intersect.

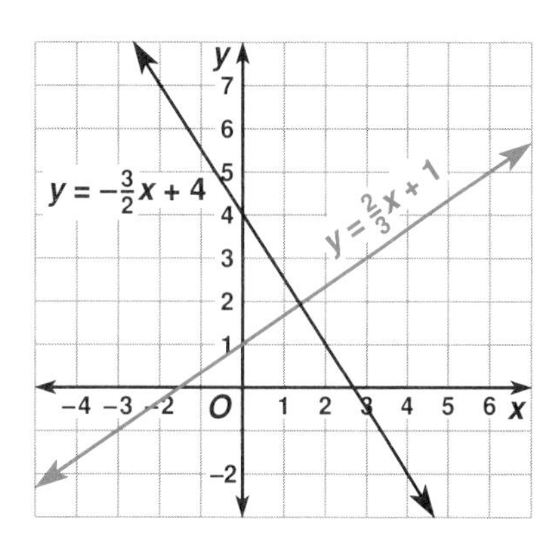

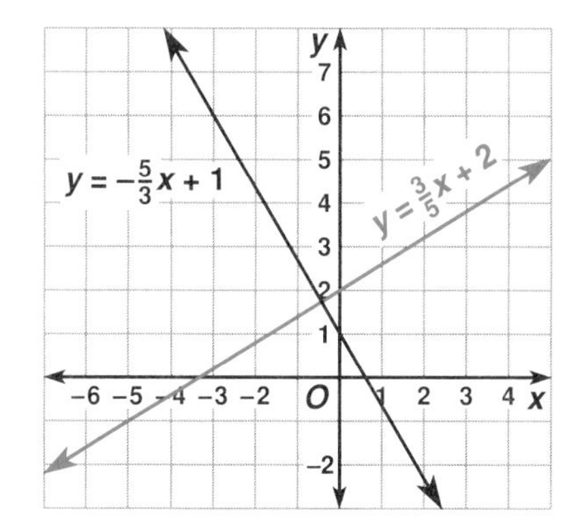

At what angle do the lines appear to intersect?
Use a protractor to verify your answer.

b. Examine the slopes of the pairs of lines in the graphs.

Pair of Lines	*Slope*	*Pair of Lines*	*Slope*
$y = \frac{2}{3}x + 1$	$\frac{2}{3}$	$y = \frac{3}{5}x + 2$	$\frac{3}{5}$
$y = -\frac{3}{2}x + 4$	$-\frac{3}{2}$	$y = -\frac{5}{3}x + 1$	$-\frac{5}{3}$

What pattern do you see for the slopes of the lines in each pair?
Draw a conclusion about the slopes of perpendicular lines.

c. Using your conclusion about the slopes of perpendicular lines, tell which of the following lines would be perpendicular to the line $y = 2x + 4$.

(1) $y = -2x + 1$ (2) $y = \frac{1}{2}x + 1$ (3) $y = -\frac{1}{2}x + 1$

(Explorations continue)

d. Let's see these graphs on a calculator.

To graph the first pair of lines:

① Press **Y=** to get the calculator to accept equations.

Press **CLEAR** to erase any previous equation.

To enter $y = \frac{2}{3}x + 1$, press: **(** 2 **÷** 3 **)** **X|T** **+** 1 **ENTER**

To enter $y = -\frac{3}{2}x + 4$, press: **(** **(−)** 3 **÷** 2 **)** **X|T** **+** 4 **ENTER**

② To set the range, press **RANGE** .

Because of the way the screen grid is divided, leaving the range at the preset numbers will not show you that these lines are really perpendicular.

Try it. Leave the range as set, and press **GRAPH** .

Now change the range to go from −4.8 to 4.8 for x in intervals of 1, and from −3.2 to 3.2 for y, in intervals of 1.

Press: **RANGE** **(−)** 4.8 **ENTER** 4.8 **ENTER** 1 **ENTER**

(−) 3.2 **ENTER** 3.2 **ENTER** 1 **ENTER**

③ Press: **GRAPH**

Following these 3 steps, graph the second pair of equations on the calculator.

Following the same steps, graph the pair of equations from part c that you said would be perpendicular. Are they?

(Explorations continue)

2. a. (1) Name 5 pairs of numbers x and y so that $x + y = 6$.

(2) Name 5 pairs of numbers x and y so that $x - y = 4$.

(3) Of the pairs of numbers you named, which pair satisfies both requirements, that $x + y = 6$ **and** $x - y = 4$?

b. (1) Use the 5 pairs of numbers (x, y) that you named in part a(1) to graph the equation $x + y = 6$.

On the same set of axes, use the 5 pairs of numbers (x, y) that you named in part a(2) to graph the equation $x - y = 4$.

(2) Describe your graph.

Tell how the pair of numbers you named as satisfying both equations $x + y = 6$ and $x - y = 4$ shows up on the graph.

Draw a conclusion about the coordinates of the point of intersection of a pair of lines on a graph.

c. (1) Create a table of 5 points that satisfy the equation $y = 2x + 1$.

Use the points in your table to graph $y = 2x + 1$.

(2) Create a table of 5 points that satisfy the equation $x + y = 4$.

Use the points in this table to graph $x + y = 4$ on the same set of axes on which you already graphed $y = 2x + 1$.

(3) From the graph, read the coordinates of the point of intersection of the lines $y = 2x + 1$ and $x + y = 4$.

d. Follow the 3 steps from Exploration 1(d) to graph the pair of lines $y = 2x + 1$ and $x + y = 4$ on the calculator.

How will you have to rewrite the equation $x + y = 4$ to enter it on the calculator?

Use the range –10 to 10 for x, in intervals of 1.

Use the same range for y.

To get an approximation of the coordinates of the point of intersection, press **TRACE** and move the arrow keys until you think the cursor is on the point of intersection.

The coordinates of the points you trace are shown at the bottom of the screen. How close can you come to (1, 3), the point that the graph you drew by hand showed as the point of intersection?

You can get a better look at the point of intersection on the calculator by adjusting the range. Let the range for both x and y go from –5 to 5, in intervals of 1.

14.4 Fitting a Line to Data

Mr. Taylor asked the 12 students in his tutoring class to report the number of minutes they had studied for their Math Skills Test. After Mr. Taylor received the grades for these students, he matched the grades against the time spent studying.

Would you expect there to be a relationship between the time spent studying and the grade on the test? Explain

Mr. Taylor wanted to show his students how important study time is. He showed the class a table of the information he had collected:

Study Time (minutes)	20	35	60	80	40	95	50	60	85	100	45	90
Grade (%)	64	60	75	92	73	97	76	80	91	95	67	88

Mr. Taylor asked Steve to prepare a graph to display this information.

Steve said to think of each pair of numbers as the coordinates of a point (time, grade).

Then the horizontal axis tracks the study time and the vertical axis tracks the grade.

Since all the numbers are positive, Steve positioned the axes to show only the first quadrant on his graph.

He used different scale numbers on the two axes.

Mr. Taylor wanted the class to study this graph of data points, which is called a ***scattergram***, to see if there was a relationship (***correlation***) between study time and grade.

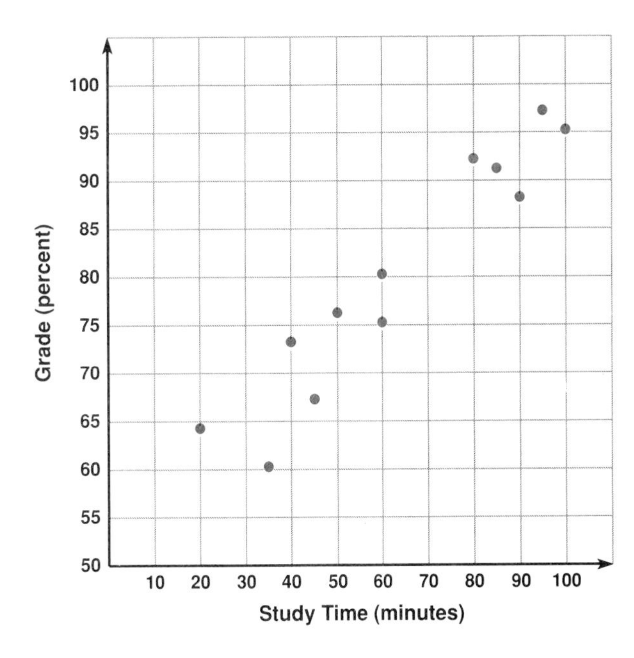

Steve said this scattergram shows that as the study time increases, the grades tend to get higher. But other students had difficulty understanding how the graph showed this.

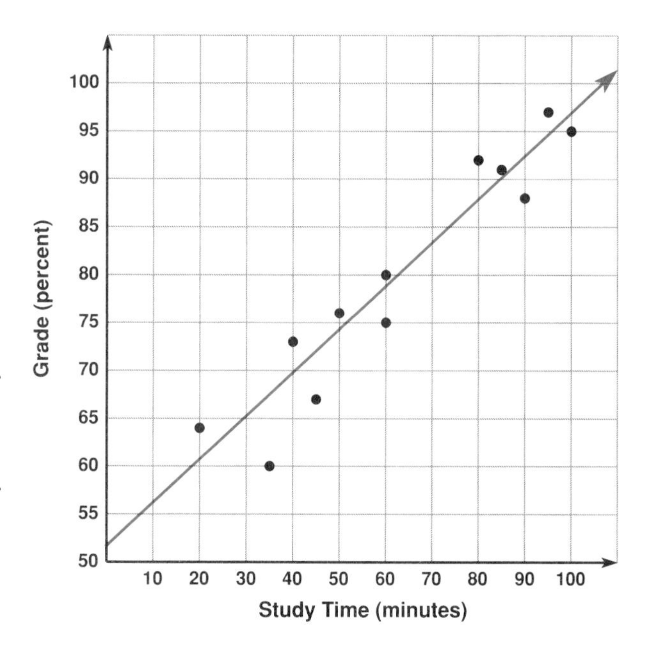

Mr. Taylor explained that a line could be fitted to these points so that the correlation would be easier to judge.

Clearly, the 12 points do not lie on one line. But you can fit a line between data points so that about half the points are just above the line and half the points are just below. One or more of the points might even be on the line.

Since the data points are clustered close to the line, the line acts as a model of the average of the points. It is called the ***trend line***.

In this case, the slope of the trend line is positive —that is, from left to right, the line goes up.

Since the slope of the trend line is positive, the data shows a positive correlation between study time and grade.

a. What does a positive correlation mean for this data?

As the study time increases, the grades get higher.

b. Does this mean that the student who studied longest got the highest grade? Or, that the student who studied the least got the lowest?

No, the student who studied longest (100 minutes) got the second highest grade (95%), and the student who studied least (20 minutes) got the second lowest grade (64%).

c. Did everyone who studied longer get higher grades than those who studied less?

No, one student who studied for 60 minutes got a lower grade (75%) than the student who studied 50 minutes and got 76%.

d. Did students who studied the same amount of time get the same grade?

No, two students studied 60 minutes each. One got 75% and the other got 80%.

e. Use the trend line to estimate the grade of a student who studied for 70 minutes.

Find 70 minutes on the horizontal axis and trace up to the point on the line. Read the corresponding grade from the vertical axis: just less than 85, about 83%.

f. Would you say that study time is the best predictor of test grades? Explain.

Although study time is important for achievement, it is not the only factor. Regular attendance, doing homework daily, and good study habits are also very important.

A scattergram is a plot of a set of data points that pair two different measures. Fitting a line to the points might show a correlation between the measures.

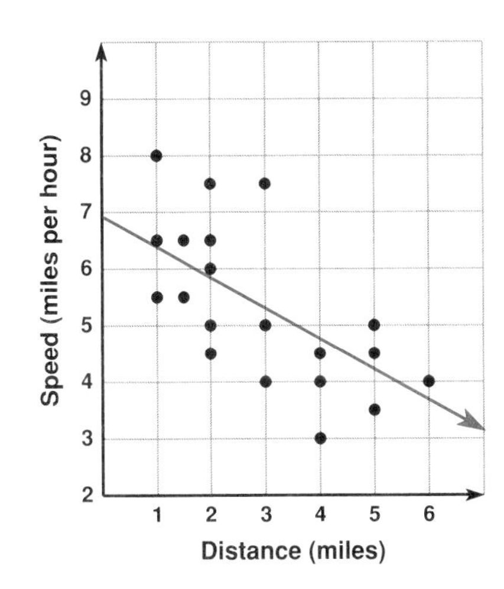

Example 1 Twenty students from Ms. Tano's gym class were asked to keep a record of their jogging. From the information, Ms. Tano matched the average distance a student covered with the average speed at which she ran.

This scattergram shows the results.

a. Describe how well the line shown fits the data.

Answer: Although the points are not clustered very close to the line, they are not that far away. This line is a reasonable fit for the data.

b. Describe the slope of the trend line.

Answer: From left to right, the trend line goes down, indicating a negative slope.

c. Is there a correlation for this data?

Answer: Yes, the slope of the trend line shows a negative correlation for the data.

d. What does a negative correlation mean for this data?

Answer: As the distance jogged increases, the speed of the jogger decreases.

e. At what speed did the jogger who covered 6 miles run?

Answer: 4 miles per hour (Locate the vertical coordinate for the point whose horizontal coordinate is 6.)

f. What distance did the fastest jogger cover?

Answer: 1 mile (The highest vertical coordinate is at 8. The corresponding horizontal coordinate is 1.)

g. Use the trend line to approximate the speed of a jogger who covers 2.5 miles.

Answer: 5.5 miles per hour (Locate 2.5 on the distance axis and trace up to the point on the line. Read the corresponding speed from the vertical axis.)

Scientists may spend years collecting data trying to determine correlations. Knowledge of such correlations is vital to us, such as links between cigarette smoking and lung cancer or heart disease.

The results of some studies may be contrary to what we expect.

Example 2 Leon collected data from classmates in his junior high school about the distance they live from school and the time it takes them to get to school in the morning.

Would you expect a correlation? Explain.

Here is a table Leon made of the data he collected from 21 classmates.

Distance (miles)	Time (min.)	Distance (miles)	Time (min.)	Distance (miles)	Time (min.)
0.5	5	2	10	4	10
1	5	2	15	4	20
1	10	2	40	4	40
1	20	2	50	4	60
1	30	2	60	5	20
1	55	3	30	6	10
2	5	3	50	6	40

a. Make a scattergram of this data showing distance on the horizontal axis, in intervals of 0.5 miles. Show time on the vertical axis in intervals of 5 minutes.

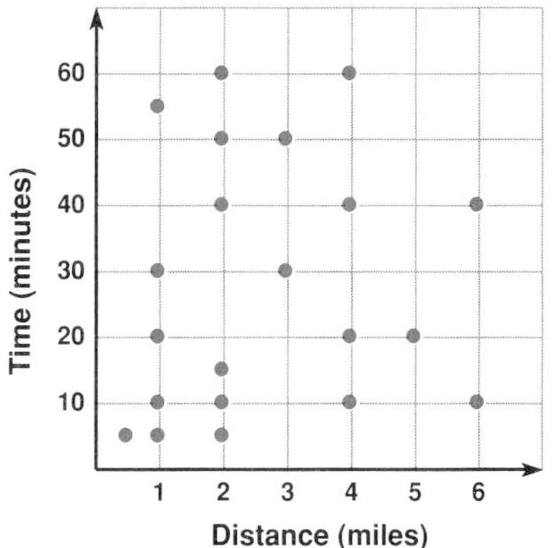

b. Does it look as though there is any trend for this set of data?

Answer: No, the points are scattered all over the graph.

c. What could account for the time differences for students who live the same distance from school?

Answer: Students come to school in different ways: walking, bicycling, school bus, parents drive them.

Some students do not come directly to school. They may have to drop off a younger sister or brother at day care; they may deliver morning newspapers; they may stop off at a relative's house for breakfast; they may meet friends and play some sport.

Try This *(For solution, see page 671.)*

The data in this scattergram is about domestic animals and animals in captivity.

It matches the gestation period (the average number of days it takes the animal to form, from conception to birth) against its life span (the average number of years it lives).

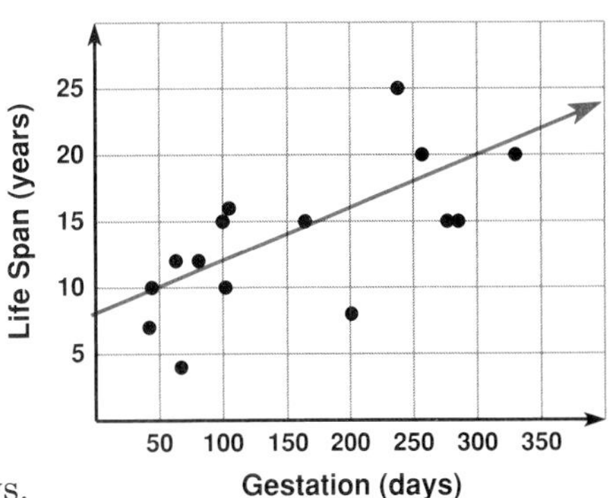

a. Is the slope of the trend line positive or negative? Explain.

b. What does a positive correlation mean for this data?

c. The lion has a gestation period of about 100 days. About how long is its life span?

d. Of the animals included in this data:

(1) The hippopotamus lives the longest. What is its life span?
(2) The guinea pig lives the shortest. What is its life span?

e. Use the trend line to estimate the life span of an animal whose gestation period is 250 days.

f. The gestation period for an Asian elephant is 645 days and its life span is 70 years.

(1) Is this point included in this scattergram? Explain.
(2) Is this data consistent with the trend of this scattergram? Explain.
(3) How does the data for an Asian elephant compare to the data for a human?

SELF-CHECK: SECTION 14.4

1. What do the points on a scattergram show?

Answer: a comparison between two measures, such as height and weight

2. If a line can be fitted to the data points of a scattergram, what does that show about, say, height and weight?

Answer: The trend line shows that there is a correlation between height and weight.

3. What does a positive correlation between height and weight mean?

Answer: As the height increases, the weight increases.

4. Give some reasons why it is useful to know about correlations between measures.

Answer: We can adopt good habits (studying for a test generally results in a higher grade).

We can prevent serious ailments (using sun block helps avoid certain skin cancers).

We can predict an outcome (the higher the level of education, the greater the income level).

EXERCISES: SECTION 14.4

1. For each scattergram:
   - Tell the two sets of measures that are being compared for the 20 people described.
   - Is there a correlation? If yes, what kind? Explain the meaning of the correlation.

   a. data from 20 students

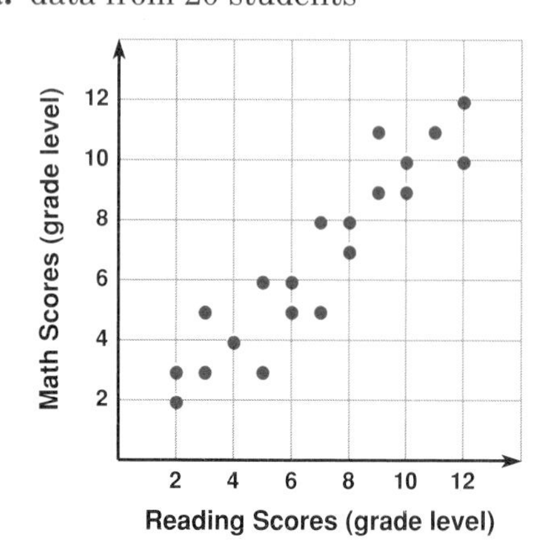

   b. data from 20 dieters

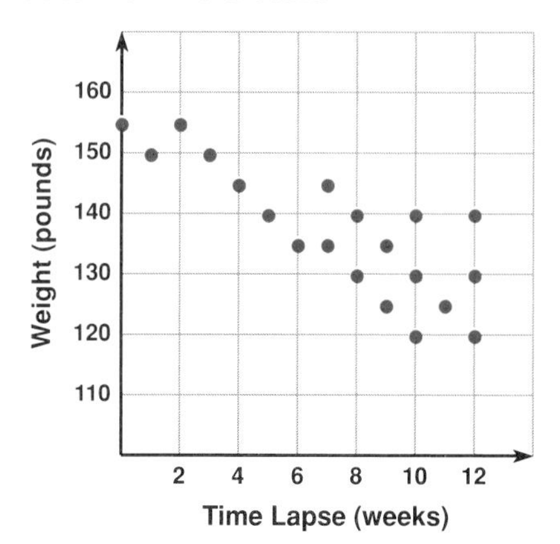

   c. data from 20 stockholders

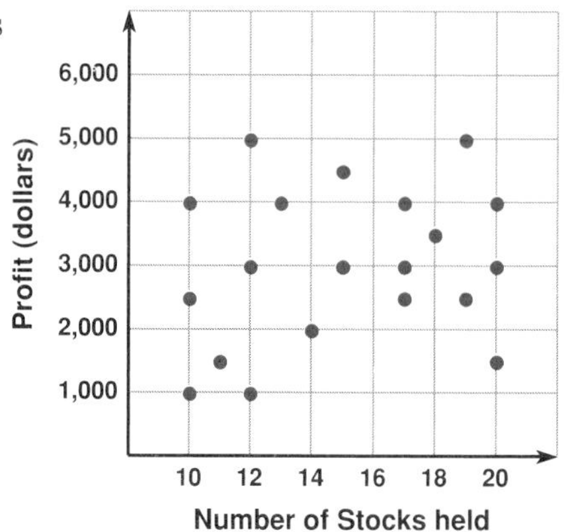

2. This scattergram records observations relating weekly rainfall to the growth of a tree seedling.

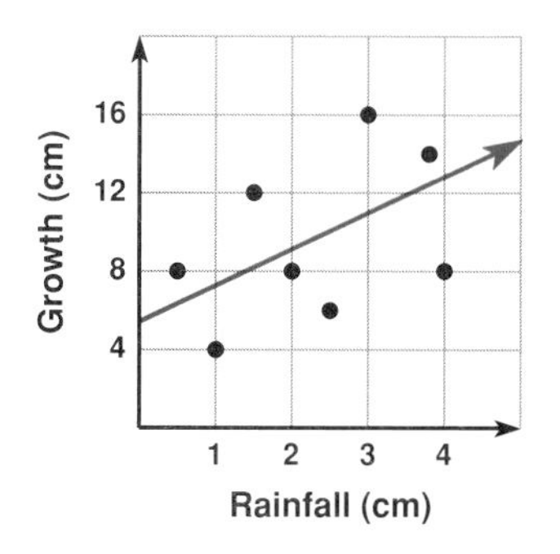

   a. How much did the seedling grow in a week that had 1 cm of rainfall?

   b. How much rainfall was there in a week in which there was the most growth?

   c. Does this data show a correlation between rainfall and growth? Explain.

   d. Use the trend line to estimate the growth for 3.5 cm of rainfall.

3. Saturated fats are said to raise blood cholesterol, which increases the risk of heart disease and cancer.

   Polyunsaturated fats tend to reduce blood cholesterol.

   For 22 types of oils and fats, this scattergram relates the percent of saturated fat to the percent of unsaturated fat.

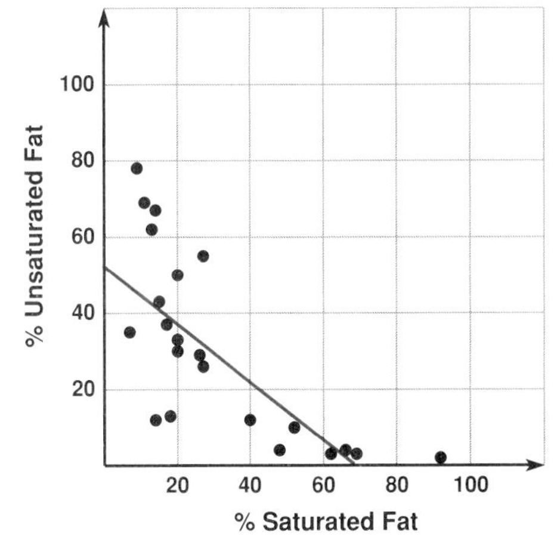

   a. For the products tested, as the percent of saturated fat increases, what happens to the percent of unsaturated fat? What kind of correlation is this?
   b. Coconut oil has the highest percent of saturated fat. About what percent is that?
   c. Canola oil has the lowest percent of saturated fat. About what percent of unsaturated fat is canola oil?
   d. Safflower oil is not quite as low as canola oil in saturated fat, but safflower oil is highest in unsaturated fat. About what percent unsaturated fat is safflower oil?
   e. Three of the products tested are each 20% saturated fat. Which is the most healthy?

4. This scattergram shows data for 29 buildings in New York City.

   The height of the building (rounded to the nearest 50 feet) is compared to the number of stories the building has.

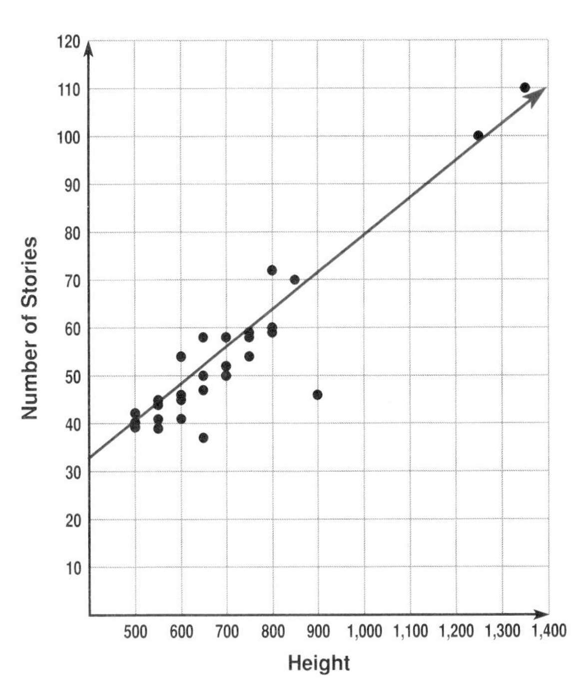

   a. For the buildings included, what general relationship is there between the height and the number of stories?
   b. Does Citicorp building, whose height is 900 feet, fit this pattern? Explain.
   c. The highest buildings in the city are the twin towers of the World Trade Center.

      What is the height of each tower? How many stories in each?
   d. What is the range of the number of stories for the four New York buildings that are each 650 feet high?
   e. Estimate the average height of one story of an average tall building. Explain your calculations

      How does this height compare to the height of a story in your home? in your school?

5. For each table of data:

- Use the horizontal axis for the numbers in the top row of the table, and the vertical axis for the numbers in the bottom row.
- Draw a scattergram.
- If there appears to be a correlation, sketch a trend line.

a. Compare the number of hours per week spent on exercise and on watching TV. Label both axes from 1 through 12.

Exercise	10	3	1	11	12	10	2	8	4	6
TV	4	8	12	5	1	2	10	5	8	7

b. Compare the results of rolling a die twice. Label both axes from 1 through 6.

1st Toss	2	4	1	5	3	5	3	2	6	1	4	4
2nd Toss	2	3	3	2	1	4	4	5	6	5	1	6

c. Compare math and science grades. Label both axes by tens from 60 through 100.

Math	88	94	73	99	82	70	87	95	85	95	90
Science	80	89	77	80	85	68	81	98	85	95	85

EXPLORATIONS

1. **a.** Here is a method for fitting a trend line to the points on a scattergram.

The points plotted on the scattergram that follows in Step 3 are taken from the coordinates in the table below, which pairs the scores 15 students got on a reading test with the scores they got on a math test.

To draw the trend line after the points are plotted on the scattergram, follow these steps.

① Separate the data into Sets I, II, and III of equal size.

	Set I					Set II					Set III				
Reading	65	69	73	75	76	78	80	81	83	87	87	90	91	94	96
Math	75	71	75	55	77	60	73	74	87	82	89	93	95	99	93

② Determine the "center" of each set of points by getting the median value for the reading scores and the median value for the math scores.

Recall that the median value is the middle value when the data are *arranged in order*.

Set I	Reading	65	69	73	75	76	← were already in order
	Math	55	71	75	75	77	← now in order

The median-median point for Set I is (73, 75).

Set II	Reading	78	80	81	83	87	← were already in order
	Math	60	73	74	82	87	← now in order

The median-median point for Set II is (81, 74).

Set III	Reading	87	90	91	94	96	← were already in order
	Math	89	93	93	95	99	← now in order

The median-median point for Set III is (91, 93).

③ Plot the 3 median-median points on the graph.

Using a ruler, draw a dotted line through the median-median points for Sets I and III.

Using a ruler, slide the dotted line down about one-third the vertical distance to the median-median point from Set II, and draw a solid line parallel to the dotted line.

This solid line is called the ***median-median line***, and it is a trend line for the data set.

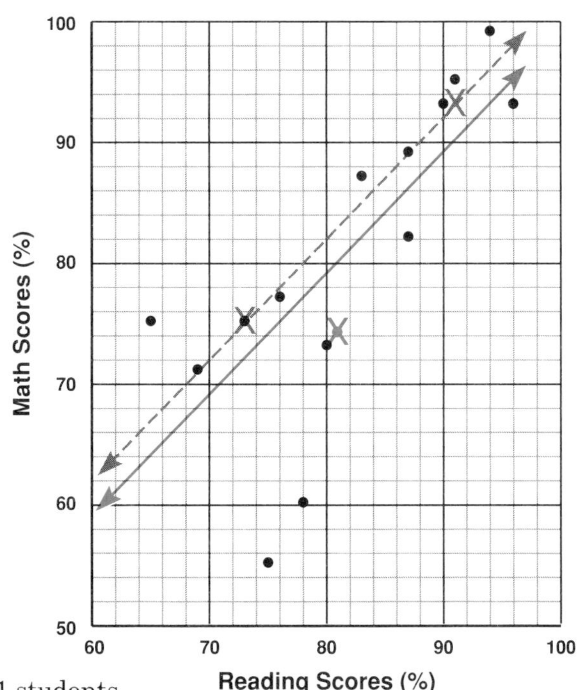

b. Here is a set of data (SAT scores) for 21 students.

Math	*Verbal*	*Math*	*Verbal*	*Math*	*Verbal*
340	520	410	490	490	420
350	550	420	560	490	510
360	540	430	540	500	440
360	520	440	430	520	380
380	500	450	450	520	390
390	480	460	390	530	460
400	440	470	450	540	470
←**Set I**→		←**Set II**→		←**Set III**→	

(1) Use a full sheet of graph paper to graph the data points, (Math, Verbal).

Horizontal axis, at bottom of page, used for Math Scores.
Scale from 330 to 550, in intervals of 10.

Vertical axis, at left of page, used for Verbal Scores.
Scale from 370 to 580, in intervals of 10.

(2) The data is already divided into 3 equal sets, I, II, III.

Find the median-median point for each set.

(3) Use X to plot the 3 median-median points on the graph.

Draw a dotted line between the median-median points for Sets I and III.

Use a ruler to raise or lower the dotted line about one-third the distance to the median-median point for Set II, and draw the median-median line as a solid line.

(4) Does the median-median line show a correlation for the data? Explain.

2. **a.** Here's how to draw a scattergram on a graphing calculator. Use the data from Mr. Taylor's tutoring class discussed on pages 651-652.

Study Time (minutes)	20	35	60	80	40	95	50	60	85	100	45	90
Grade (%)	64	60	75	92	73	97	76	80	91	95	67	88

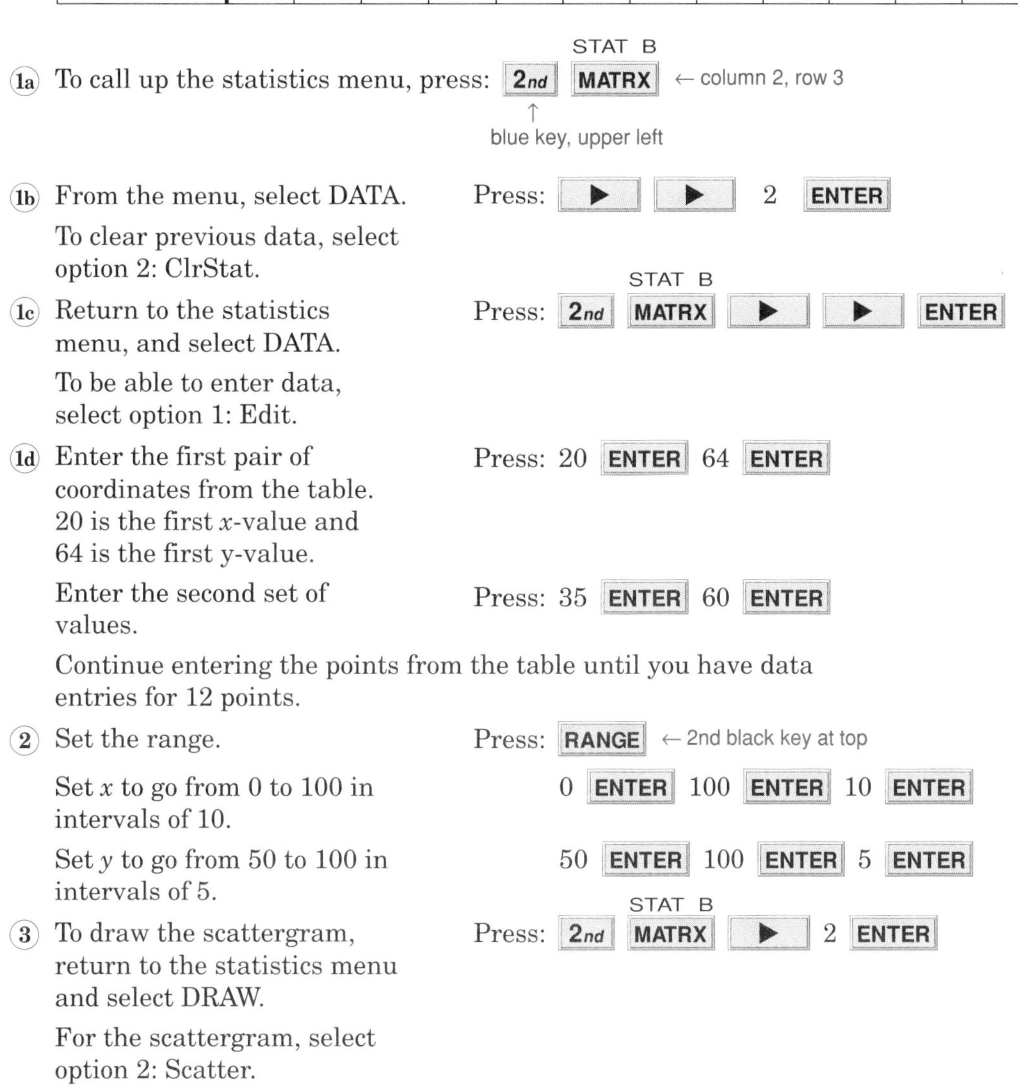

(1a) To call up the statistics menu, press: 2nd MATRX (STAT B) ← column 2, row 3
↑ blue key, upper left

(1b) From the menu, select DATA. Press: ▶ ▶ 2 ENTER
To clear previous data, select option 2: ClrStat.

(1c) Return to the statistics menu, and select DATA. Press: 2nd MATRX (STAT B) ▶ ▶ ENTER
To be able to enter data, select option 1: Edit.

(1d) Enter the first pair of coordinates from the table. 20 is the first x-value and 64 is the first y-value. Press: 20 ENTER 64 ENTER

Enter the second set of values. Press: 35 ENTER 60 ENTER

Continue entering the points from the table until you have data entries for 12 points.

(2) Set the range. Press: RANGE ← 2nd black key at top

Set x to go from 0 to 100 in intervals of 10. 0 ENTER 100 ENTER 10 ENTER

Set y to go from 50 to 100 in intervals of 5. 50 ENTER 100 ENTER 5 ENTER

(3) To draw the scattergram, return to the statistics menu and select DRAW. Press: 2nd MATRX (STAT B) ▶ 2 ENTER

For the scattergram, select option 2: Scatter.

Compare this result to the scattergram for Mr. Taylor's data shown on pages 651-652.

The calculator will even draw a trend line for the data set.

Note: The technical name for the trend line the calculator draws is *line of regression*. You will see the abbreviation LinReg.

And, the calculator will tell you the equation of the line, as well as how good the correlation is.

To get the calculator to graph the trend line on the scattergram, the equation of the trend line needs to be entered as one of the four equations the calculator can graph.

First, be sure previous equations are cleared.

(4) Return to the statistics menu, and select CALC.

Then select option 2: LinReg.

Press: 2nd MATRX (STAT B) 2 ENTER

Display:

LinReg

a = 50.7753304 ← *y*-intercept of trend line

b = .4588105727 ← slope of trend line

r = .9519572813

Using rounded values for slope and *y*-intercept, the equation of the trend line is: $y = 0.5x + 51$

(5) The calculator will write the equation, using its 10-place numbers for slope and *y*-intercept, and will enter the equation.

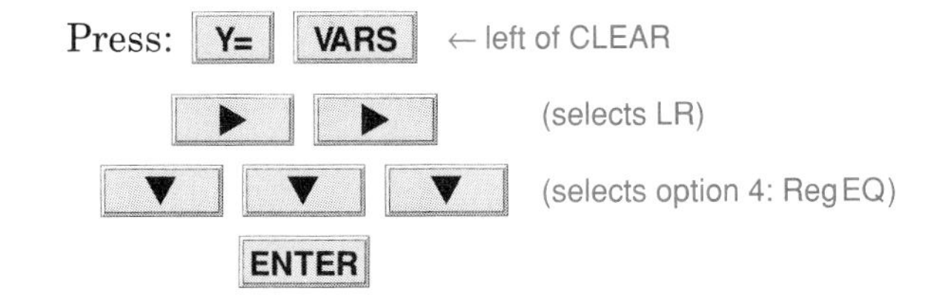

(6) Ready to graph.

Return to the statistics menu, and select DRAW.

Then select option 2: Scatter.

Compare this result to the scattergram with trend line for Mr. Taylor's data on page 651.

(Explorations continue)

b. Try another. First:

Clear previous data. Press: 2nd STAT B MATRX ▶ ▶ 2 ENTER

Clear previous equation. Press: Y= CLEAR

Use the data of Example 1, shown in this table:

Distance	Speed	Distance	Speed	Distance	Speed
1	5.5	2	6	4	4
1	6.5	2	6.5	4	4.5
1	8	2	7.5	5	3.5
1.5	5.5	3	4	5	4.5
1.5	6.5	3	5	5	5
2	4.5	3	7.5	6	4
2	5	4	3		

(1) Get ready to enter data. Press: 2nd STAT B MATRX ▶ ▶ ENTER

Enter data. Press: 1 ENTER 5.5 ENTER

Continue with the remaining 19 points from the data table.

(2) Set the range. Press: RANGE 0 ENTER 6 ENTER 1 ENTER

Do x from 0 to 6, in intervals of 1. 2 ENTER 9 ENTER 1 ENTER

Do y from 2 to 9, in intervals of 1.

(3) To draw the scattergram: from the statistics menu, select DRAW, Press: 2nd STAT B MATRX ▶ 2 ENTER

option 2: Scatter

(4) To include the trend line: from the statistics menu, select CALC, Press: 2nd STAT B MATRX 2 ENTER

option 2: LinReg

The display is information about the equation of the trend line. From the b-value, the slope is about –.5 . From the a-value, the y-intercept is about 7. The equation is: $y = -.5x + 7$

⑤ Have the calculator write the equation of the trend line in the spot for the first equation to graph.

Press: 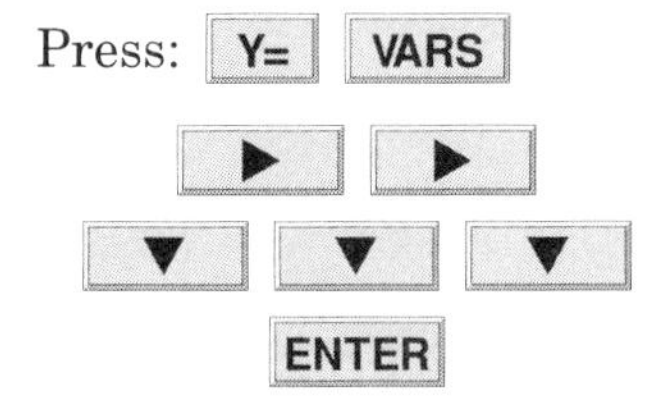

⑥ Redraw the scattergram, this time with the trend line.

Press: 2nd MATRX (STAT B) ▶ 2 ENTER

Compare this result with Example 1.

c. The calculator gives information about how good the correlation of the data is.

On the screen that tells you information about the trend line (called the line of regression on the calculator):

three values are listed:

a = the y-intercept of the trend line

b = the slope of the trend line

Note: These letters are different from the form of an equation $y = mx + b$, where m = slope and b = y-intercept.

The third value, r, tells how good the correlation is.

Values for r range from 0 (no correlation) to 1 (perfect correlation)
or from 0 (no correlation) to –1 (perfect correlation).

The closer r is to 1 (or to –1), the better the correlation.

Go back and look at the correlations the calculator gave:

for Mr. Taylor's problem, $r = .95$ (very close to 1, very high correlation)

for joggers of Example 1, $r = -.5$ (halfway between 0 and –1, pretty good correlation)

Things You Should Know After Studying This Chapter

KEY SKILLS

14.1 Graph a line from a set of points.
Choose points to graph a line.
Identify a line as horizontal, vertical, or oblique, and tell where the line intersects the axes.

14.2 Tell how to travel along grid lines to get from one point to another on the graph of a line.
Use a direction ratio to find the slope of a line.
Use the coordinates of two points on a line to find the slope of the line.

14.3 Read the slope and y-intercept from the equation of a line that is in the form $y = mx + b$.
Use the slope and y-intercept to graph a line.
Recognize that parallel lines have equal slopes.

14.4 Read a scattergram for information.
Identify a positive or negative correlation.
Recognize when there is no correlation.

KEY TERMS AND SYMBOLS

14.1 coordinates of a point: (x, y) or (abscissa, ordinate)
types of lines: horizontal, vertical, oblique

14.2 direction ratio = $\dfrac{\text{up or down}}{\text{right or left}}$

slope = $\dfrac{\text{difference in } y\text{-coordinates}}{\text{difference in } x\text{-coordinates}}$

14.3 $y = \underset{\substack{\uparrow \\ \text{slope}}}{m}x + \underset{\substack{\uparrow \\ y\text{-intercept}}}{b}$

14.4 scattergram • correlation • trend line

CHAPTER REVIEW EXERCISES

1. Consider this set of points: $\{(1,3), (2, 6), (3, 9)\}$

a. Which of the following points could belong to the set?

(1) $(4, 7)$ (2) $(1, -3)$

(3) $(9, 3)$ (4) $(-2, -6)$

b. Could $(0, 0)$ belong to this set? Explain.

c. Using words, write a rule that tells the relationship between the coordinates of a point in this set.

d. If (x, y) represents a point in the set, write a rule that tells the relationship between x and y.

e. Using the given points and the points you have found to belong to this set, draw a graph of the line whose equation you have written.

f. Does the point $(-1, -3)$ lie on this line? Explain.

g. Does the point $(3, 1)$ lie on this line? Explain.

2. Match each rule for the relationship between coordinates of points on a line in *Column A* with an equation for the line from *Column B*.

Column A	*Column B*
a. The y-coordinate is 1 more than the x-coordinate.	**(1)** $y = 1$ **(2)** $x = 1$ **(3)** $x = -1$
b. The x-coordinate is 1 less than the y-coordinate.	**(4)** $y = x + 1$ **(5)** $x = y - 1$
c. The abscissa is 1.	
d. The ordinate is 1.	

3. If the line $y = x + 4$ were graphed:

a. Would the point $(1, 5)$ be on the line? Explain.

b. What must be the value of y if the point $(2, y)$ is to be on the graph of the line?

c. What must be the value of x if the point $(x, 6)$ is to be on the graph of the line?

4. For each of the 3 graphs that follow:

a. Tell whether the line is horizontal, vertical, or oblique.

b. Tell which axes the line intersects and where.

c. Match these equations to the graphs:

Equation 1: $x = 2$

Equation 2: $y = 2$

Equation 3: $y = x$

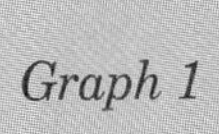

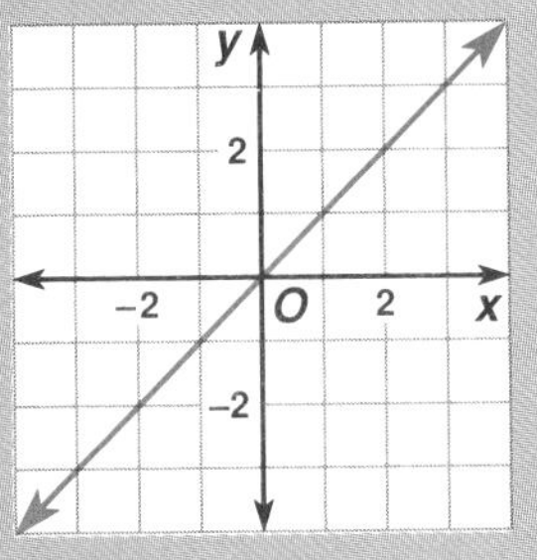

Graph 2

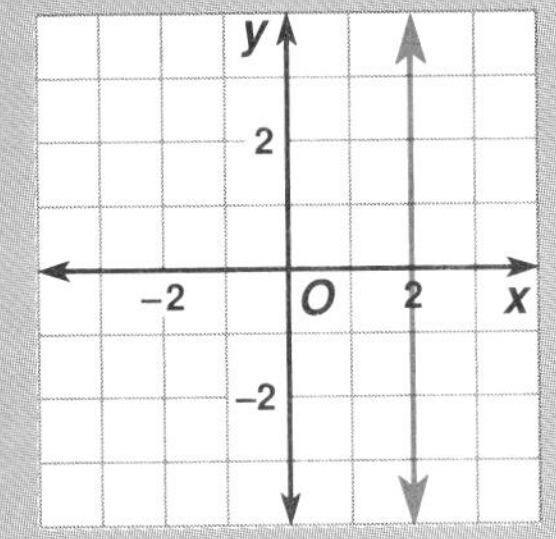

Graph 3

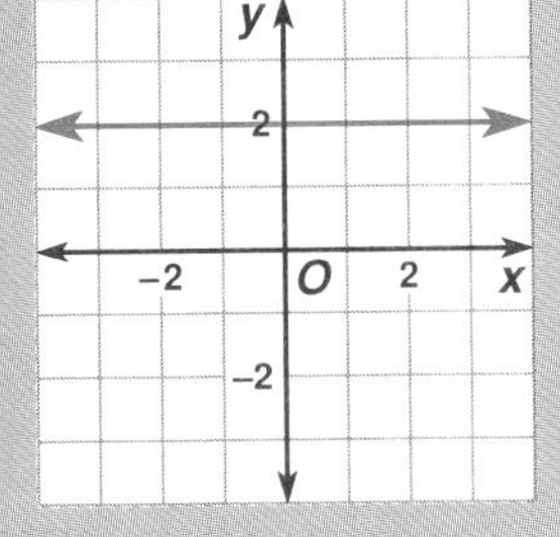

5. For each line whose equation is given, copy and complete the table of points. Below the table, list the coordinates of the points. Then use these points to graph the line.

a. $y = x + 3$

x	$x + 3$	y
0	0 + 3	3
2	?	?
–1	?	?

Points on the line $y = x + 3$ are (0,3), (2,?), and (–1,?).

b. $y = 2x - 1$

x	$2x - 1$	y
0	2(0) – 1	–1
2	?	?
–4	?	?

Points on the line $y = 2x - 1$ are (0, –1), (2, ?), and (4, ?).

c. $x = 3$

x	y
3	1
3	?
?	?

Points on the line $x = 3$ are (3, 1), (3, ?), and (?, ?).

6. For each line whose equation is given, create a table of at least 3 points that are on the graph of the line. Then use your points to draw the graph.

Identify the line you have graphed as horizontal, vertical, or oblique, and tell where the line intersects the axes.

a. $y = x + 4$

b. $y = 3x + 2$

c. $y = -4$

d. $x = 4$

7. a. Tell how to travel along the grid lines of this graph to get:

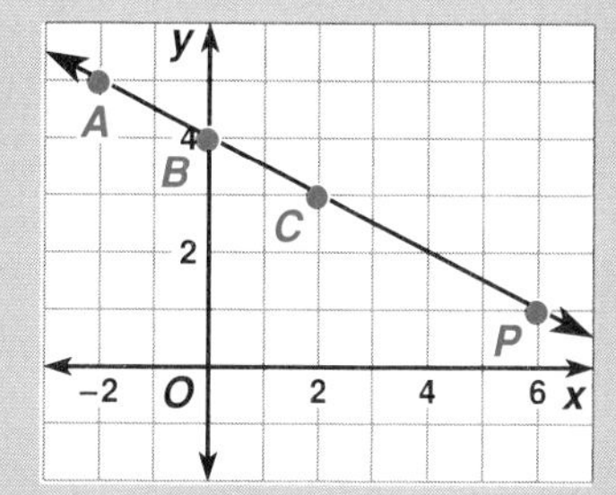

(1) from A to B (2) from B to A
(3) from B to C (4) from B to P
(5) from P to C (6) from P to A

b. What is the slope of this line that goes through the points A, B, C, and P?

c. Tell if each of the following is a correct or incorrect interpretation of the slope $\frac{1}{2}$?

(1) $\frac{\text{up } 1}{\text{right } 2}$ (2) $\frac{\text{down } 1}{\text{right } 2}$

(3) $\frac{\text{up } 1}{\text{left } 2}$ (4) $\frac{\text{down } 1}{\text{left } 2}$

8. Use the coordinates of each pair of points to find the slope of the line that can be drawn through these points.

Verify your result by plotting the pair of points on a graph, drawing the line, and giving the direction ratio that would take you from one point to the other.

a. $A(1, 1)$ $B(4, 5)$

b. $P(1, 1)$ $Q(0, 3)$

c. $G(1, 2)$ $H(5, 2)$

9. Tell the slope and y-intercept of each line.

a. $y = 2x + 4$

b. $y = 3x - 1$

c. $y = 5x$

d. $2y = 6x + 8$

e. $y + 1 = x$

f. $y + 2x = 3$

CHAPTER REVIEW EXERCISES

10. Following these steps:

- From the equation, write the slope and y-intercept.
- Plot the y-intercept. Call the point on the y-axis P.
- Using the direction ratio $\frac{\text{up}}{\text{right}}$ for the slope, locate three points on the line that are "above" P.
- Write the equivalent direction ratio $\frac{\text{down}}{\text{left}}$ for the slope, and use it to locate three points on the line that are "below" P.
- Draw the line and label it with its equation.

Graph each of these lines:

a. $y = \frac{2}{3}x + 1$ **b.** $y = \frac{3}{4}x - 2$

c. $y = 2x + 3$ **d.** $y = 3x$

11. Write each of these slopes as two different direction ratios:

a. slope $= -\frac{3}{5}$ **b.** slope $= -\frac{1}{4}$

c. slope $= -\frac{3}{7}$

12. Following these steps:

- From the equation, write the slope and y-intercept.
- Plot the y-intercept. Call the point on the y-axis P.
- From the slope, write two different direction ratios.

  Use these direction ratios to locate three points on the line "above" P and three points on the line "below" P.
- Draw the line and label it with its equation.

Graph each of these lines:

a. $y = -\frac{4}{3}x + 1$ **b.** $y = -\frac{1}{5}x - 3$

c. $y = -2x + 5$

13. This scattergram relates final exam grades and final report card grades for a science class.

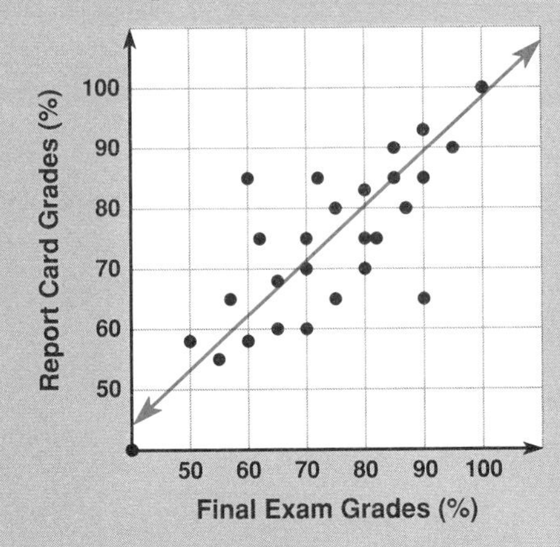

a. What kind of correlation does the trend line show for this data? Explain.

b. Which of these equations would be a good approximation for the equation of the trend line? Explain.

(1) $y = x$ (2) $y = -x$

(3) $y = 50$ (4) $x = 50$

c. Tom only achieved 60% on his final exam, yet his report card grade was 85%.

Does Tom's data fit the trend? Explain.

Give a reason that might explain Tom's situation.

d. Name the coordinates of another data point that does not fit the trend.

What might be a reason for this student's situation?

SOLUTIONS FOR TRY THESE

14.1 Graphing a Line

Page 618

1. $\{(3, 6), (2, 4), (1, 2), (-1, -2)\}$
   - **a.** Other points that could be in this set are $(-2, -4)$ and $(0, 0)$.
   - **b.** A point could be in this set if the second coordinate is twice the first coordinate.
   - **c.** $y = 2x$
   - **d.**

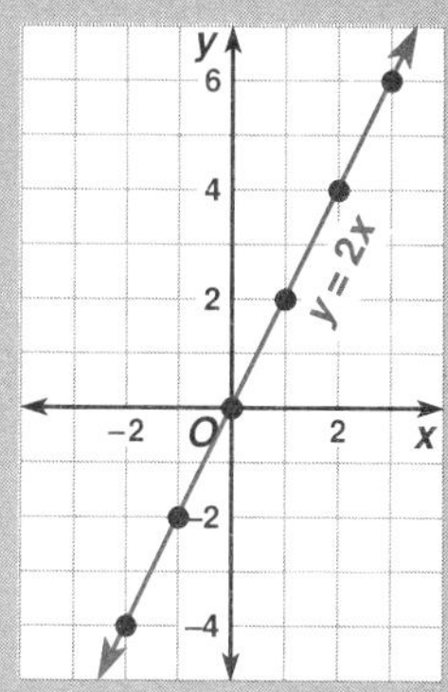

   - **e.** The line $y = 2x$ is an oblique line.
   - **f.** The oblique line $y = 2x$ intersects both the x-axis and the y-axis, at the origin.

2. $\{(2, 4), (2, 3), (2, 1), (2, -1)\}$
   - **a.** Other points that could be in this set are $(2, 0)$ and $(2, -4)$.
   - **b.** A point could be in this set if the first coordinate is 2. The second coordinate could be any number.
   - **c.** $x = 2$
   - **d.**

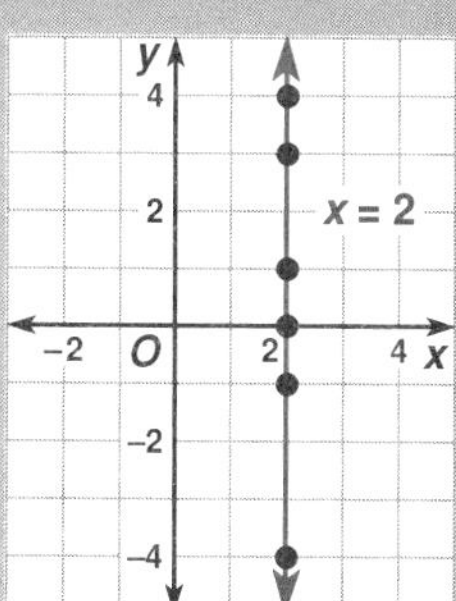

   - **e.** The line $x = 2$ is a vertical line.
   - **f.** The vertical line $x = 2$ intersects only the x-axis, at 2.

Page 620

1. - **a.** For points on the line $y = 3x$, the y-coordinate is 3 times the x-coordinate.
   - **b.** $(1, 3)$, $(2, 6)$, and $(3, 9)$ are points on the line $y = 3x$.
   - **c.**

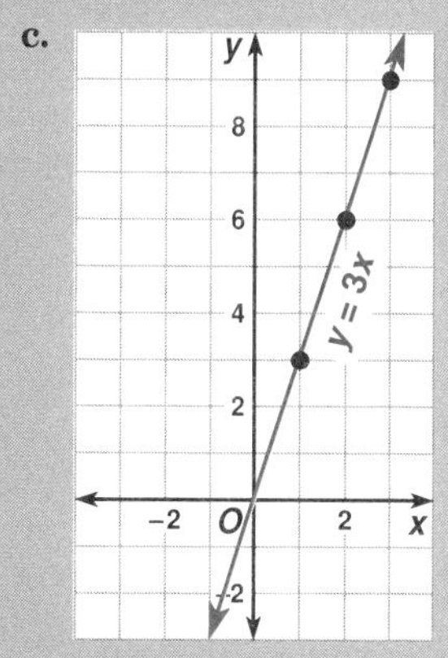

   - **d.** The line $y = 3x$ is an oblique line.
   - **e.** The oblique line $y = 3x$ intersects both the x-axis and the y-axis, at the origin.

2. - **a.** For points on the line $y = x + 4$, the y-coordinate is 4 more than the x-coordinate.
   - **b.** For a point on this line, if the x-coordinate is 2, then the y-coordinate is $2 + 4$, or 6.
   - **c.** $(1, 5)$ and $(-2, 2)$ are on the line $y = x + 4$.
   - **d.**

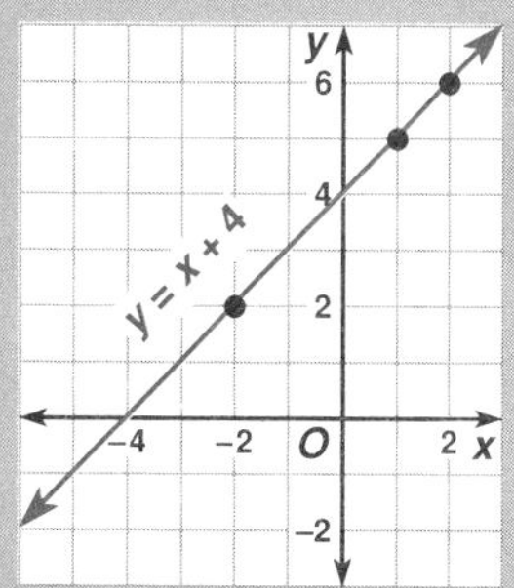

   - **e.** The line $y = x + 4$ is an oblique line.
   - **f.** The oblique line $y = x + 4$ intersects both axes, at 4 on the y-axis and at -4 on the x-axis.

3. a. For points on the line $y = x - 3$, the y-coordinate is 3 less than the x-coordinate.

b.

x	$x - 3$	y
6	6 – 3	3
4	4 – 3	1
3	3 – 3	0
1	1 – 3	–2

Points on the line $y = x - 3$ are (6, 3), (4, 1), (3, 0), and (1, –2).

c.

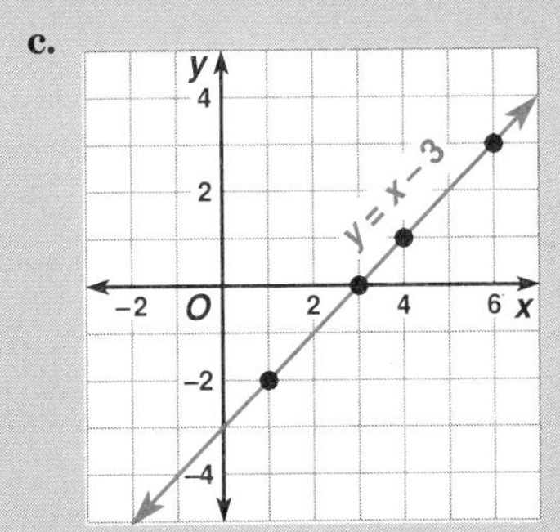

d. The line $y = x - 3$ is an oblique line.

e. The oblique line $y = x - 3$ intersects both axes, at 3 on the x-axis and at –3 on the y-axis.

14.2 The Slope of a Line

Page 631

a. **(1)** From A to B: go up 4 and right 2

(2) From B to A: go down 4 and left 2

(3) From B to C: go up 2 and right 1

(4) From C to B: go down 2 and left 1

(5) From A to C: go up 6 and right 3

(6) From C to A: go down 6 and left 3

b. **(1)** From A to B: $\frac{\text{up}}{\text{right}} = \frac{4}{2}$, or $\frac{2}{1}$

From B to C: $\frac{\text{up}}{\text{right}} = \frac{2}{1}$

From A to C: $\frac{\text{up}}{\text{right}} = \frac{6}{3}$, or $\frac{2}{1}$

For this line, the direction ratio $\frac{\text{up}}{\text{right}} = \frac{2}{1}$.

(2) From B to A: $\frac{\text{down}}{\text{left}} = \frac{-4}{-2}$, or $\frac{2}{1}$

From C to B: $\frac{\text{down}}{\text{left}} = \frac{-2}{-1}$, or $\frac{2}{1}$

From C to A: $\frac{\text{down}}{\text{left}} = \frac{-6}{-3}$, or $\frac{2}{1}$

For this line, the direction ratio $\frac{\text{down}}{\text{left}} = \frac{2}{1}$.

c. The slope of the line $y = 2x - 4$ is $\frac{2}{1}$, or 2.

Page 633

a. $R(1, 3)$ $S(2, 6)$

$$\text{slope} = \frac{\text{difference in } y\text{-coordinates}}{\text{difference in } x\text{-coordinates}} = \frac{6-3}{2-1} = \frac{3}{1}$$

Answer: The slope of the line $y = 3x$ is 3.

b. From R to S: go up 3 and right 1

$$\frac{\text{up}}{\text{right}} = \frac{3}{1}$$

14.3 The Equation y = mx + b

Page 640

1. True, $y = 2x + 4$ is an example of an equation of a line whose slope is 2.

When the equation of a line is in the form $y =$, the coefficient of x is the slope of the line.

2. a. $y = 3x + 4$

↑ slope ↑ y-intercept

The slope of the line $y = 3x + 4$ is 3 and the y-intercept is 4.

b.

x	$3x + 4$	y
1	3(1) + 4	
	3 + 4	7
2	3(2) + 4	
	6 + 4	10
−1	3(−1) + 4	
	−3 + 4	1

Plot the points
(1, 7), (2, 10), and (−1, 1).

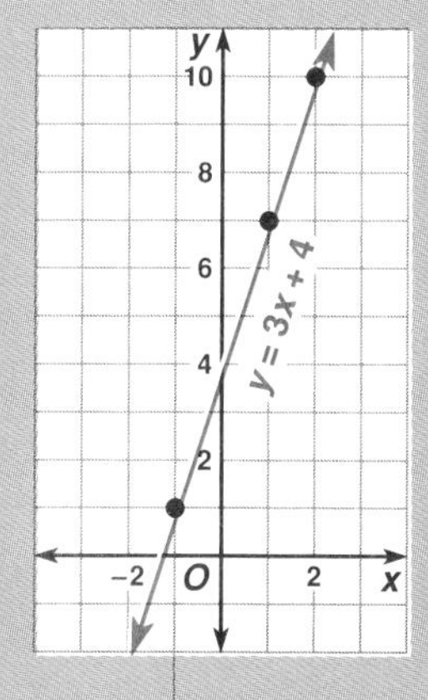

3. True, the lines $y = 2x + 3$ and $y = 2x + 5$ are parallel.
   These lines have equal slopes, 2.
   Lines that have equal slopes are parallel.

Page 641

a. $y = \frac{1}{3}x + 1$
   slope ↑ y-intercept ↑
   The slope of the line $y = \frac{1}{3}x + 1$
   is $\frac{1}{3}$ and the y-intercept is 1.

b. The line will cross the y-axis at 1.
   The coordinates of this y-intercept P are (0, 1).

c. To locate a point Q "above" P, go up 1 and right 3.
   From Q, go up 1 and right 3 to R.
   From R, go up 1 and right 3 to S.

   To locate a point T "below" P, go down 1 and left 3.
   From T, go down 1 and left 3 to W.
   From W, go down 1 and left 3 to Z.

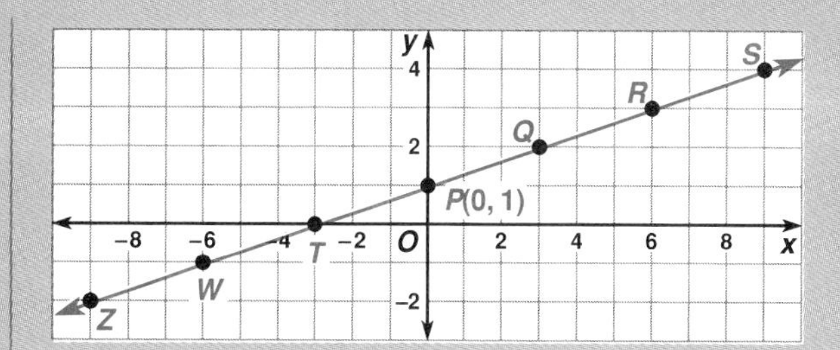

14.4 Fitting a Line to Data

Page 655

a. The slope of the trend line is positive.
   From left to right, the line goes up.

b. A positive correlation for this data means the longer the gestation period, the longer the life span.

c. Locate the vertical coordinates of the point whose horizontal coordinate is 100.
   Answer: 15 years.

d. (1) Locate the point with the highest vertical coordinate. Read the vertical coordinate from the vertical scale.
   Answer: The life span of the hippo is about 25 years.

   (2) Locate the point with the lowest vertical coordinate. Read the vertical coordinate from the vertical scale.
   Answer: The life span of the guinea pig is just under 5 years, about 4 years.

e. Locate 250 on the horizontal axis and trace up to the point on the line. Read the corresponding vertical coordinate.
   Answer: about 17 years

f. (1) No, the horizontal scale only goes up to 350.
   (2) Yes, the elephant has a long gestation period and a long life span.
   (3) The Asian elephant has the longest gestation period of any mammal, nearly 22 months.
   The gestation period for a human is about 9 months.
   The life span of a human is also about 70 years.
   Good news: The average life span for Americans is on the rise.

CUMULATIVE REVIEW • CHAPTERS 1-14

1. The table shows the number of hours worked last week by three employees of the Stock Supply Company.

Time Record					
Employee	M	Tu	W	Th	F
Elva Simpson	9	8	7	$8\frac{1}{2}$	$7\frac{1}{2}$
Lucy Starr	8	6	9	9	8
Jim Strand	8	7	9	10	8

   a. If the regular work week is 40 hours, which employee worked overtime?

   b. The hourly rate for these employees is $8.40 an hour for 40 hours, and $12.60 an hour for overtime. How much did each person earn?

2. For a health class project, Abdul and David kept count of their calorie intake for one week. The results are shown on the graph.

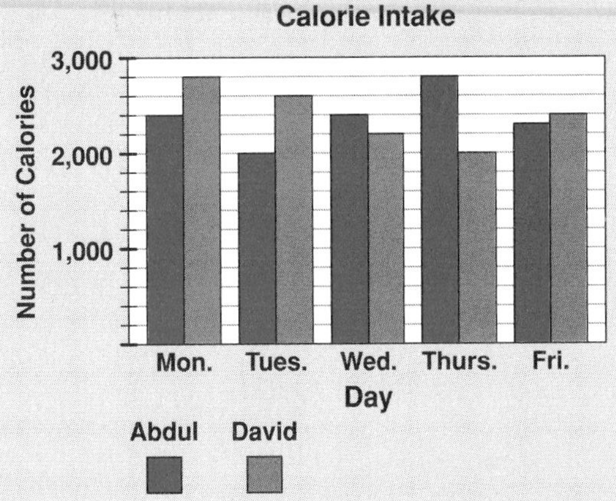

   a. On Monday, how many more calories did David consume than Abdul?

   b. On what day was there the greatest difference in their calorie consumption? Who ate more?

   c. Whose total intake for the week was greater? How much greater?

3. Find the missing numbers in each pattern.

   a. 1, 3, 6, 10, 15, ?, ?   b. 1, –2, 4, ?, ?, –32

4. For pothole repair of a 55-foot stretch of highway, one lane is being blocked off by placing traffic cones alongside. If there is to be a cone at each end of the work area, and the cones are to be no more than 10 feet apart, draw a diagram to determine how many cones are needed.

5. Write each group of numbers in order, from least to greatest.

   a. 5, –3, 0, 1, –4   b. $\frac{5}{6}, \frac{5}{8}, \frac{3}{4}, \frac{2}{3}$

   c. 0.2, 0.22, 0.202, 0.022, 0.0202

6. The 50 eighth-grade students at Kennedy School were surveyed to find how many people lived in each student's household. The circle graph shows the percent of households in each category.

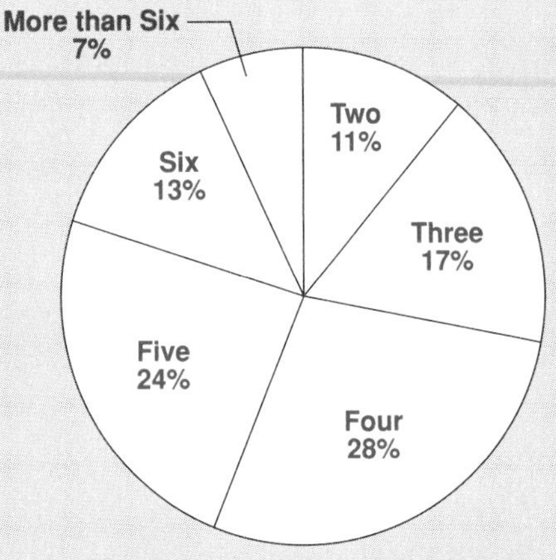

Number of People in Household

   a. In how many households are five people living?

   b. How many households have no more than four people?

   c. How many of the households might consist of a student living with one parent?

CUMULATIVE REVIEW • CHAPTERS 1-14

7. On graph paper, draw rectangle $ABCD$, with vertices $A(4, 3)$, $B(4, -1)$, $C(-2, -1)$, and $D(-2, 3)$. Find:

a. the perimeter **b.** the area

8. Use the order of operations to find the value of each expression when $a = 6$, $b = -2$, and $c = 2$.

a. $a + bc$ **b.** $(a - b)c$ **c.** $\frac{a + bc}{c}$ **d.** $ab - c^2$

9. Match each verbal expression in *Column A* with its symbolic equivalent from *Column B*.

Column A	*Column B*
a. Four less than twice x	(1) $x + 2 < 4$
b. Two more than x is less than four.	(2) $(x + 2) + 4$
c. The quotient of x and four is greater than two.	(3) $4x < x - 2$
d. Four more than the sum of x and two	(4) $2x - 4$
e. The product of x and four is less than the difference of x and two.	(5) $\frac{x}{4} > 2$

10. Solve each equation and check.

a. $y - 3 = 7$ **b.** $2n = -12$

c. $8 = \frac{w}{2}$ **d.** $3x + 5 = 11$

e. $2r + r + 6 = 18$ **f.** $4(d - 2) = 24$

11. Use an equation to solve each problem. First, guesstimate an answer.

a. After buying 2 new tee shirts at the mall, Shereel had 9 tee shirts altogether. How many did he have before the purchase?

b. Janet's average of 32 miles per hour when driving through town is 4 times her biking speed. What is her average speed for biking?

c. The 18 miles that Jay jogged last week was 4 miles less than twice the distance he jogged the week before. What distance did he jog the week before?

12. Solve each inequality and graph the solution set on a number line.

a. $p + 2 > 5$ **b.** $2d - 3 \leq 1$ **c.** $-3x < 12$

13. Draw a diagram to match each description.

a. $\overleftrightarrow{AB}$ and $\overleftrightarrow{CD}$ are perpendicular lines.

b. Lines m and n intersect, forming vertical angles 1 and 2.

c. $\overrightarrow{AB}$ and $\overrightarrow{AC}$ form acute angle BAC.

d. Transversal t intersects parallel lines p and q to form a pair of corresponding angles x and y.

e. $\triangle ABC$ is isosceles.

14. Use the algebraic expressions in each diagram to find the value of x and the degree measures of the angles.

a.

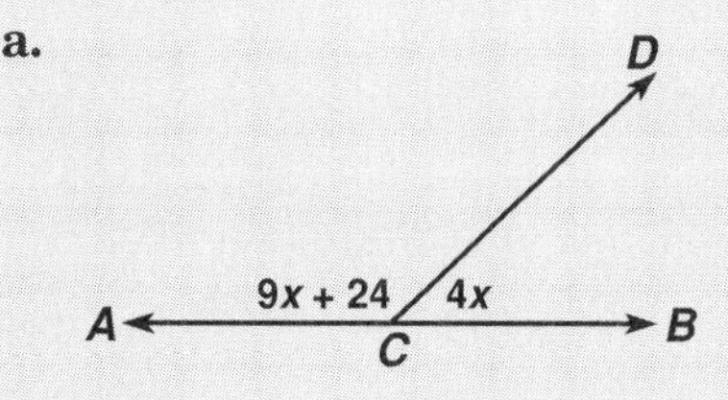

b.

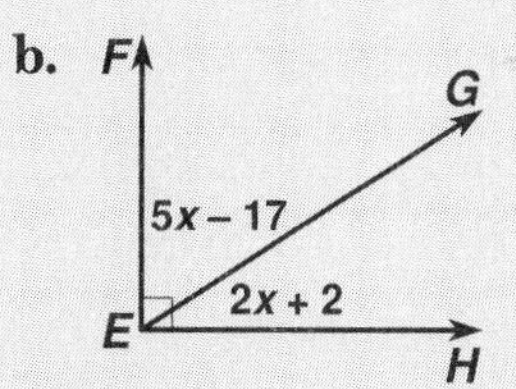

c.

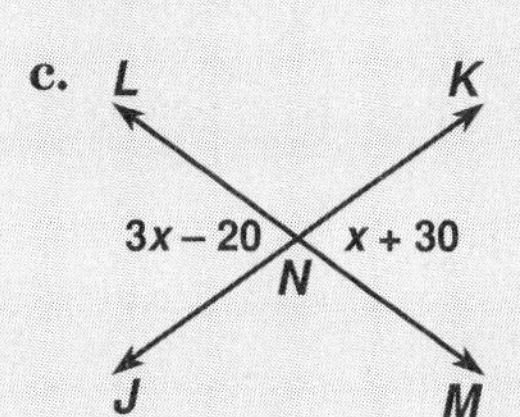

d.

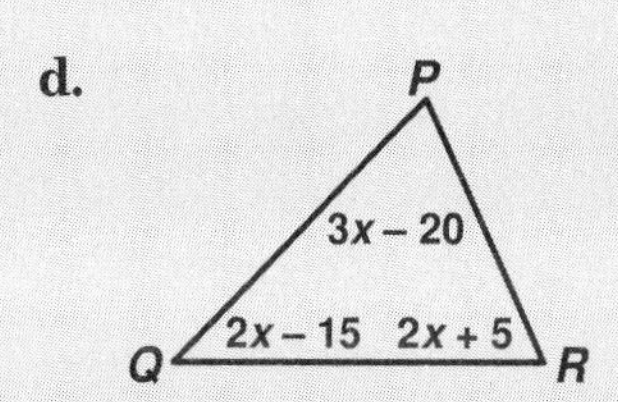

15. Find the area of:

a. a rectangle with base 9 ft. and height 7 ft.

b. a triangle with base $12\frac{1}{2}$ in. and height 8 in.

c. a parallelogram with base 12.2 cm and height 6.5 cm

d. a trapezoid with bases 3 m and 5 m and height 4 m

e. a circle with radius 14 mm (Use: $\pi = \frac{22}{7}$)

16. Find the surface area and volume of each solid.

a.

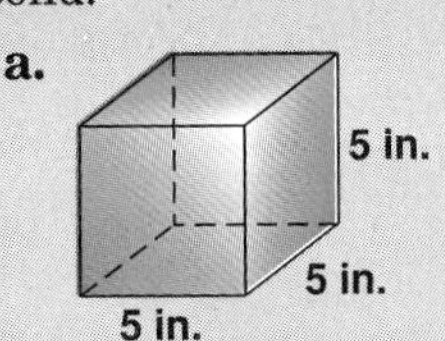

b.

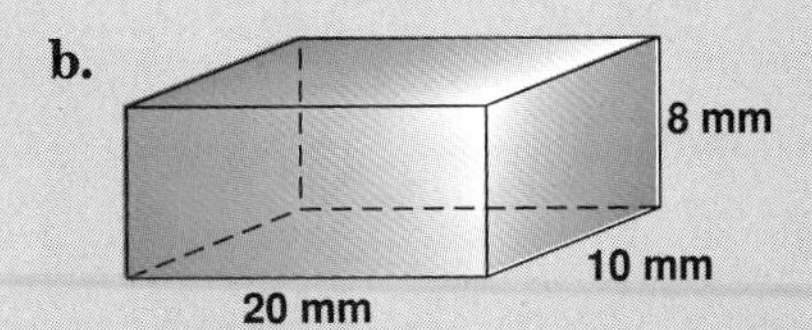

c.

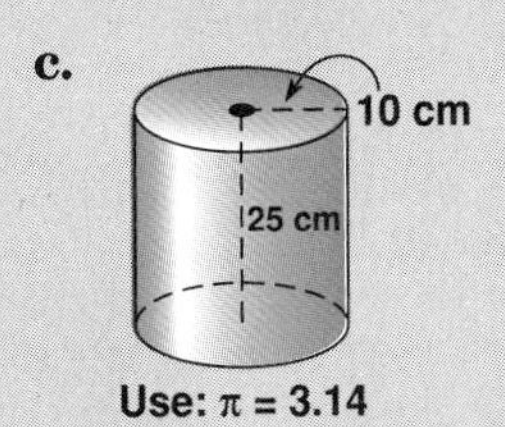

d.

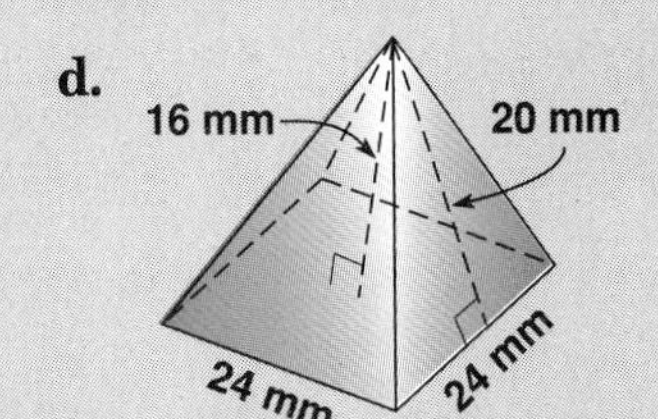

17. Solve each proportion and check.

a. $\frac{x}{12} = \frac{15}{20}$ **b.** $\frac{y}{9} = \frac{16}{24}$

18. Use an equation to solve each problem. First, guesstimate an answer.

a. The measures of a pair of complementary angles are in the ratio 11 to 7. Find the measure of each angle.

b. A ball team has a 5 : 3 ratio of wins to losses. If they played 40 games, how many games did they win?

19. Mrs. Stadler is expected to give birth to triplets. Draw a tree diagram and list the sample space to find the probability that all 3 babies will be girls.

20. Find the probability of each event, for one roll of the die and one spin of the spinner.

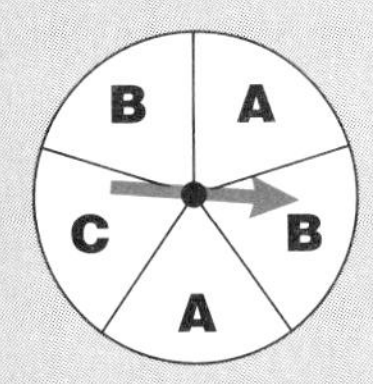

a. 6 on the die
b. C on the spinner
c. less than 3 on the die
d. A on the spinner
e. less than 7 on the die
f. Z on the spinner
g. 2 on the die and C on the spinner
h. an odd number on the die and B on the spinner
i. What is the probability that on three spins, the spinner will spell the word CAB?

21. From the menu shown, find how many different lunches can be ordered, if a lunch consists of a soup, a sandwich, and a drink.

Soup	Sandwich	Drink
Tomato	Egg Salad	Milk
Mushroom	Grilled Cheese	Coffee
	Tuna	Tea
	Bologna	

22. This histogram shows the heights of 24 students having a physical examination at Lincoln Rd. Elementary School.

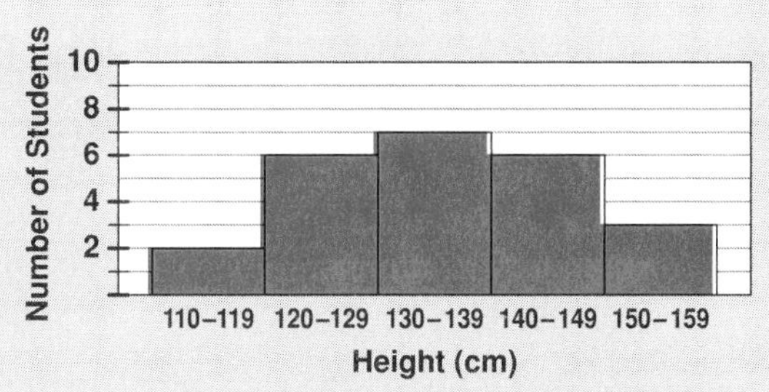

a. If a student is selected at random from the group of 24:

(1) In which interval is the height of the student most likely to fall?

(2) In which interval is the height of the student least likely to fall?

(3) Name two intervals of equal likelihood.

b. What is the range of the students' heights?

23. Using this temperature data collected from the Phoenix, Arizona Weather Bureau:

85	92	88	91	94	78	79
79	86	91	80	77	92	88
84	79	90	91	83	94	89

a. Make a stem-leaf plot for the data.

b. Use the stem-leaf plot to answer these questions. Of the 21 temperatures recorded:

(1) How many were below 83?

(2) How many were above 88?

24. a. For this data set,

9 12 12 13 18 19 29

find:

(1) the range **(2)** the mean

(3) the median **(4)** the mode

b. Which measure, mean, median, or mode, best represents the data? Explain.

25. The box-and-whisker plot represents Smithtown's average daily temperatures for April, in degrees Celsius.

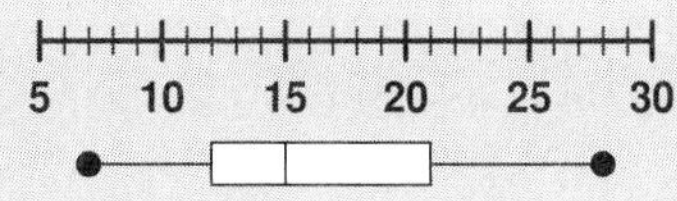

a. Find:

(1) the range of the data

(2) the median

(3) the lower quartile

(4) the upper quartile

b. If one day's temperature is chosen at random, it will most probably be between:

(1) 5° and 10° (2) 10° and 15°

(3) 15° and 20° (4) 20° and 30°

26. Perform the given operation.

a. Add:
$$\begin{array}{r} p^2 + 3p - 2 \\ + \; 2p^2 + p + 1 \\ \hline \end{array}$$

b. Subtract:
$$\begin{array}{r} 7x + 6 \\ - \; x - 3 \\ \hline \end{array}$$

c. Multiply: $4a(2a + 3)$

d. Multiply: $(3c^2)(2c^4)$

e. Multiply: $(y + 2)(y + 5)$

f. Divide: $\dfrac{15d^2 + 24d}{3d}$

27. Represent each answer in simplest form.

a. $\triangle ABC$ has a base measuring $(8x^2 + 6x)$ inches, and a height of $4x$ inches.
What is the area?

b. The length of rectangle $EFGH$ is $(2y^2 + 3y + 1)$ feet, and the width is $(y^2 + y - 1)$ feet.
Find the perimeter.

c. Flight 321 of Oriental Airways flew $(6m^3 + 8m^2 + 4m)$ miles in $2m$ hours.
What was the average speed?

28. **a.** Copy and complete this table of values, and use the points to graph the line $y = 2x - 1$.

x	$2x - 1$	y
0	$2(0) - 1$	-1
2	?	?
-1	?	?

b. Prepare a table of values to graph the line $y = x + 3$.

29. What is the slope and y-intercept of the line $y = 3x + 4$?

30. Which of these lines has a slope of $\frac{2}{3}$? Explain.

(a)

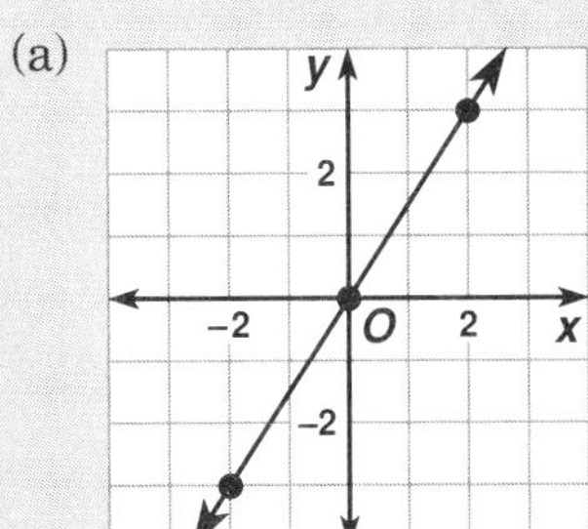

(b)

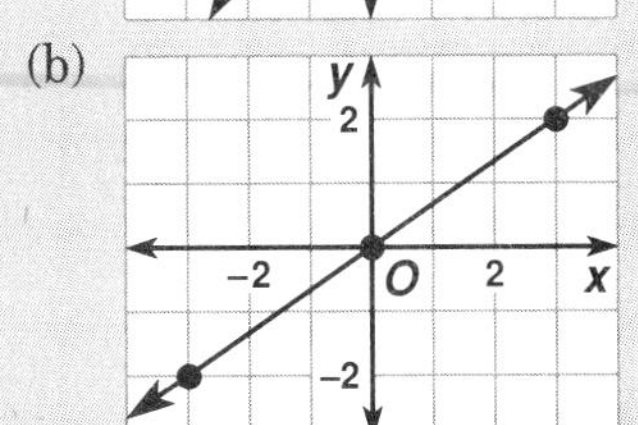

(c)

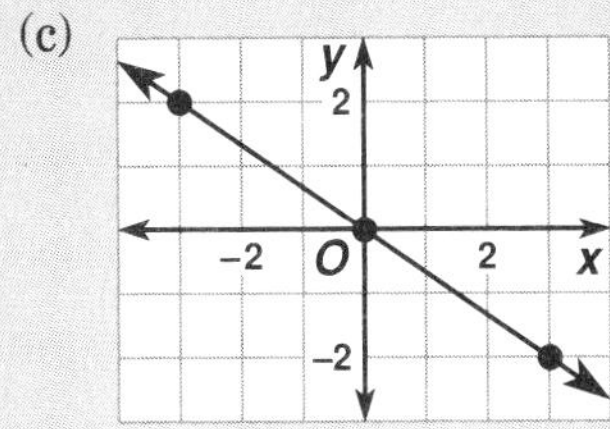

(d)

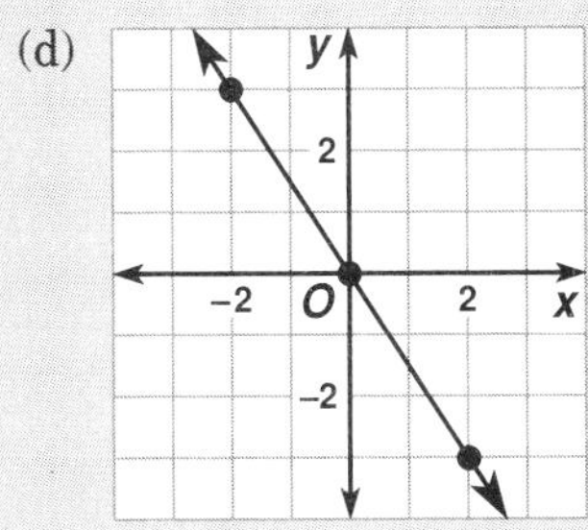

31. A line goes through the points (1, 3) and (3, 7).

a. Find the slope of the line.

b. If the line has a y-intercept of 1, write an equation of the line.

32. This scattergram relates population counts taken from the 1980 census to counts taken from the 1990 census for 15 places in New York State. The numbers, shown in thousands, have been rounded to the nearest hundred.

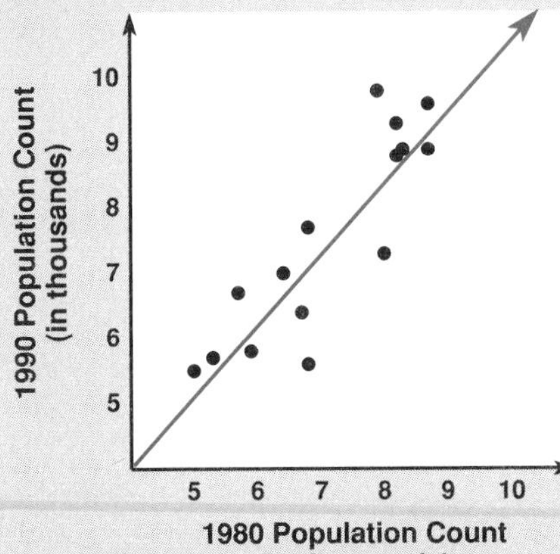

a. Does the trend line show a positive or negative correlation for the data? Explain.

b. Which of these equations would be a good approximation for the equation of the trend line? Explain.

(1) $y = x$ (2) $y = -x$

(3) $y = 5{,}000$ (4) $x = 5{,}000$

c. Does the trend show that these locations maintained their populations or significantly changed their populations? Explain.

d. The population of Brockport, NY rose from 7,900 in 1980 to 9,800 in 1990. Does Brockport's data fit the trend? Explain.

Appendix

Spiraled Review of Mathematical Skills

SKILL REVIEW 1

1. Write as a numeral: ten thousand thirty-two
2. Write in words: 154,201
3. Find the sum of 1,234 and 567.
4. Multiply: 731×25
5. Subtract 22 from 602.
6. Divide 126 by 3.
7. The graph shows the number of new members joining the Good Health Spa during a 4-month membership drive.

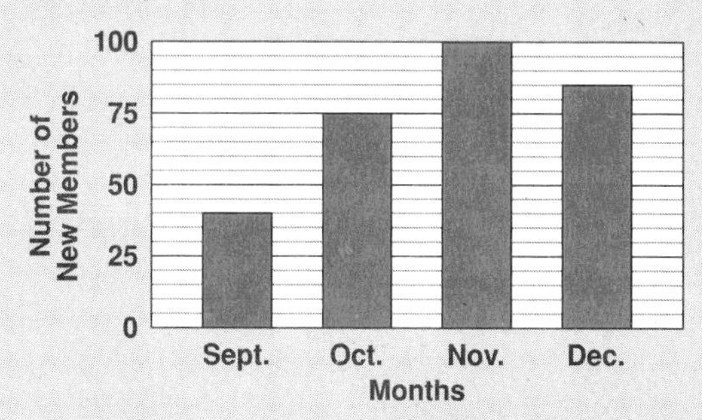

   In which month did about 40 new members join?
8. Round to the nearest hundred: 258,758
9. Juan divided 140 pencils into 7 equal bundles. How many were in each bundle?
10. The table shows some of the flights leaving Ingram Airport.

Flight Number	To	Departure Time (A.M.)	Gate
158	Altoona	6:52	2
63	Ithaca	8:14	4
144	Erie	9:03	1
86	Scranton	9:47	4
102	Easton	11:20	3

   What is the destination of the flight that leaves at 8:14 A.M.?
11. It takes Sam $4\frac{1}{2}$ hours to drive to his grandfather's house. If he leaves home at 8:30 A.M., at what time will he arrive?
12. Sue had $21.79 in her purse. After she lost a twenty dollar bill, how much did she have left?
13. Estimate which of the following numbers is closest to the sum:

$$\begin{array}{r} 5{,}437 \\ 694 \\ +\,7{,}889 \\ \hline \end{array}$$

   (a) 10,000 (b) 12,000 (c) 14,000 (d) 16,000
14. The graph shows the number of riders using the Intercity Bus Line one morning.

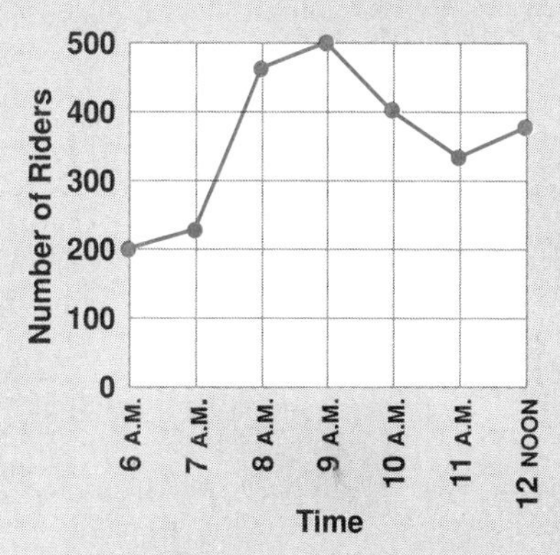

   Between which times was there the greatest increase in ridership?
   (a) 6 A.M.-7 A.M. (b) 8 A.M.-9 A.M.
   (c) 7 A.M.-8 A.M. (d) 9 A.M.-10 A.M.
15. Estimate the product: 58×304
   (a) 1,500 (b) 15,000 (c) 1,800 (d) 18,000
16. Which number is 5,999 rounded to the nearest ten?
   (a) 5,990 (b) 6,000 (c) 5,900 (d) 5,000
17. Choose the result when, from 1,000, seventy-nine is subtracted.
   (a) 921 (b) 931 (c) 1,021 (d) 831
18. Which number is 900 added to 1,111?
   (a) 1,120 (b) 1,201 (c) 2,011 (d) 1,011
19. What is 928 rounded to the nearest hundred?
   (a) 930 (b) 10,200 (c) 900 (d) 1,000
20. Subtract 9 from 92,010.
   (a) 92,001 (b) 2,010 (c) 91,991 (d) 81,991

SKILL REVIEW 2

1. Add:
$$\begin{array}{r} 232 \\ 47 \\ +\,3{,}125 \\ \hline \end{array}$$

2. Divide: $6{,}008 \div 4$

3. Subtract 1,924 from 2,000.

4. Multiply: 83×58

5. Find the sum of 1,009 and 9,001.

6. Write ten thousand ten as a numeral.

7. Round 123,456 to the nearest thousand.

8. What are the prime factors of 42?

9. Beth recycles 3 drink cans every school day, Monday through Friday. How many does she recycle for the week?

10. Find the cost of Tim's shopping trip if he bought a sweat shirt for $26, a backpack for $42, and a package of socks for $7.

11. Each ⌂ represents 20 homes that use solar heat.

Littleton
Troy
Brighton
Sayreville
Delmont

How many more homes use solar heat in Sayreville than in Troy?

(a) $3\frac{1}{2}$ (b) $4\frac{1}{2}$ (c) 60 (d) 70

12. Find the next number in the pattern: 1, 4, 7, 10, …

(a) 11 (b) 12 (c) 13 (d) 14

13. What is the least common multiple of 12 and 20?

(a) 4 (b) 24 (c) 240 (d) 60

14. The table shows the numbers of hours worked by part-time workers at Dan's Discount Store.

Number of Hours					
	Mon	Tue	Wed	Thu	Fri
Steve	5	3	6	4	7
Carla	3	3	5	6	6
Anne	4	8	3	4	8
Tonio	5	5	4	5	6

If the workers earn $8 an hour, how much more did Anne earn on Friday than Tonio did?

(a) $64 (b) $48 (c) $24 (d) $16

15. Mrs. Washington uses 4 eggs in each batch of her famous cookies. If she has $1\frac{1}{2}$ dozen eggs on hand, what is the greatest number of batches of cookies she can make?

(a) 2 (b) 4 (c) 6 (d) 8

16. What is the greatest common factor of 72 and 108?

(a) 216 (b) 72 (c) 36 (d) 6

17. What are the factors of 15?

(a) 2, 3, 5 (b) 1, 3, 15

(c) 1, 3, 5, 15 (d) 30, 150

18. What is the product of 234 and 101?

(a) 23,634 (b) 2,574

(c) 234,234 (d) 2,344

19. If the numbers are rounded to the nearest hundred, the estimate for 5,923 minus 5,111 is

(a) 0 (b) 800 (c) 900 (d) 1,000

20. Bus tokens cost $1.25 each. What is the maximum number of tokens Mrs. Chin can buy for $7?

(a) 4 (b) 5 (c) 6 (d) 7

SKILL REVIEW 3

1. Add: 2,173 + 38 + 117
2. Divide 3,648 by 12.
3. What is the remainder when 693 is divided by 5?
4. Write as a numeral: twelve thousand, three hundred four
5. Adding 1 to which digit will increase 32,784 by one hundred?
6. Divide: $11.7 \div 0.9$
7. How many times as large as 374 is 37,400?
8. Find the greatest common factor of 72 and 96.
9. Which letter is at $\frac{3}{5}$ on the number line?

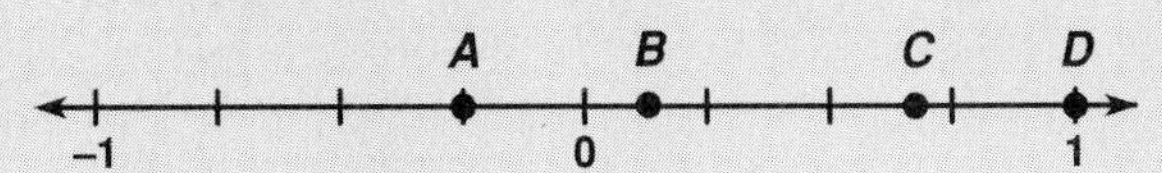

10. Evaluate: $2 \times 3 + 4 \times 5$
11. A photocopier charges 10¢ for the first copy and 7¢ for each additional copy. What is the cost for 18 copies?
12. Find the sum of $\frac{1}{2}$ and $\frac{1}{3}$.
13. The graph shows how much money Ray spent in one year on various items. What is the total amount Ray spent on discs and food?

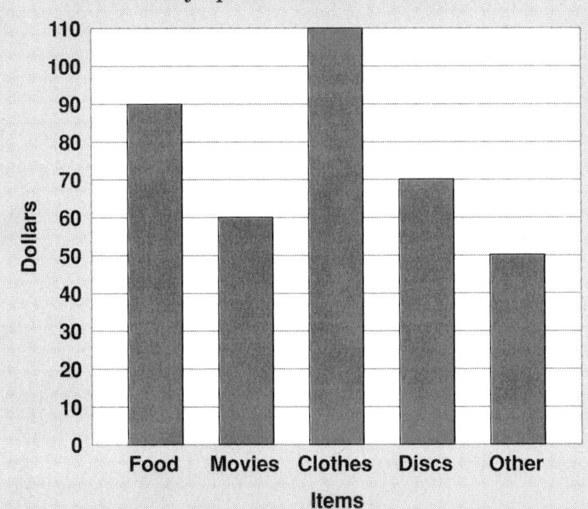

14. Each container represents 14 liters of oil. What is the total number of liters of oil represented?

15. Which number is prime?
    (a) 10 (b) 15 (c) 17 (d) 21
16. Which expression is false?
    (a) $12 \times 1 = 12$
    (b) $12 \times 1 = 12 \times 0$
    (c) $12 \times 0 = 0$
    (d) $12 \times 1 = 1 \times 12$
17. Four people ate pieces of a pizza. Which fractional part of the pizza was the largest?
    (a) $\frac{1}{3}$ (b) $\frac{1}{4}$ (c) $\frac{1}{6}$ (d) $\frac{1}{8}$
18. Which set of numbers is listed in order from least to greatest?
    (a) $\frac{1}{2}, \frac{2}{3}, \frac{3}{4}, \frac{3}{5}$
    (b) $\frac{3}{4}, \frac{3}{5}, \frac{1}{2}, \frac{2}{3}$
    (c) $\frac{1}{2}, \frac{3}{5}, \frac{2}{3}, \frac{3}{4}$
    (d) $\frac{1}{2}, \frac{2}{3}, \frac{3}{5}, \frac{3}{4}$
19. What is the least common denominator of $\frac{5}{6}$ and $\frac{3}{8}$?
    (a) 6 (b) 8 (c) 24 (d) 48
20. Which of the following shows the decimals listed in order, from least to greatest?
    (a) 1.1, 0.11, 0.011
    (b) 0.011, 0.11, 1.1
    (c) 1.1, 0.011, 0.11
    (d) 0.11, 0.011, 1.1

SKILL REVIEW 4

1. Write the numeral for five thousand sixty-three.

2. Add: 4,068
   17
   + 358

3. From 6,142, subtract 897.

4. Multiply: 762.34
   × 5.1

5. Divide: $23.2\overline{)95.12}$

6. Find 25% of 848.

7. What is $\frac{5}{6}$ of 24?

8. Divide: $12 \div \frac{1}{2}$

9. Zaluki worked from 8 A.M. until 2:20 P.M. How long did she work?

10. What is the largest prime number less than 25?

11. Find the value of: $3 + 7 \times 5$

12. What is the best estimate of 62×39?
    (a) 1,800 (b) 2,000
    (c) 2,200 (d) 2,400

13. When rounded to the nearest hundred, which number rounds to 400?
    (a) 347 (b) 351
    (c) 451 (d) 480

14. Which number is not a factor of 80?
    (a) 4 (b) 6
    (c) 16 (d) 20

15. Julia ran $\frac{2}{7}$ of a mile and then $\frac{6}{7}$ of a mile. How many miles did she run?
    (a) $\frac{8}{14}$ (b) $1\frac{1}{7}$
    (c) 1.7 (d) 1.17

16. In the pictograph, 1 [book] represents 10,000 books sold. What is the total number of books sold?

    (a) 3,500 (b) 35,000
    (c) 30,500 (d) 305,000

17. The sum of –2 and 9 is
    (a) –18 (b) –11
    (c) –7 (d) 7

18. The fees at a yoga class are:
    $25 for the first 3 classes
    $8 for each additional class
    What is the cost of taking 12 classes?
    (a) $171 (b) $100
    (c) $97 (d) $79

19. The circle graph shows a family's monthly budget. With an income of $3,600 per month, how much is spent on food?

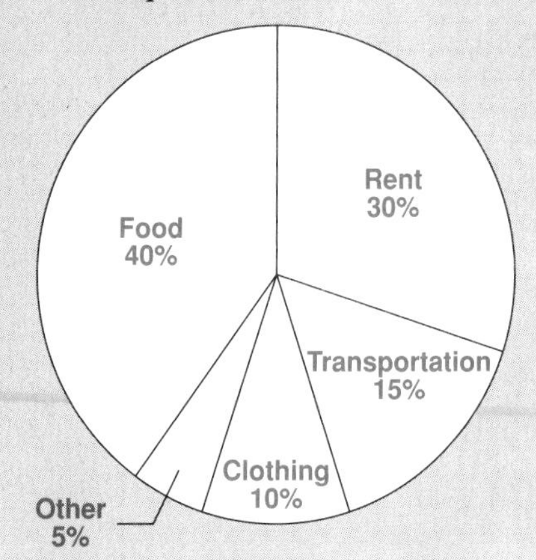

    (a) $900 (b) $1,440
    (c) $144 (d) $14.40

20. What are the coordinates of the point P shown?

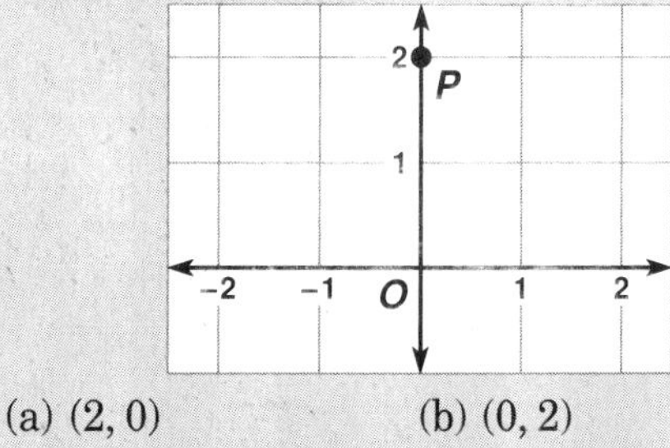

    (a) (2, 0) (b) (0, 2)
    (c) (–2, 0) (d) (0, –2)

SKILL REVIEW 5

1. Subtract 5,832 from 7,000.

2. Add: $43.9 + 0.17 + 8.2$

3. In the number 123.456, which digit is in the hundredths place?

4. Find the value of 5^3.

5. Add: $3\frac{1}{4} + 2\frac{2}{3}$

6. Reduce $\frac{16}{64}$ to lowest terms.

7. Multiply: 87.5×100

8. Divide: $0.5\overline{)3.75}$

9. Find the value of: $6 + 2 \times 3$

10. What is the least common denominator of $\frac{1}{3}$, $\frac{5}{6}$, and $\frac{3}{8}$?

11. Naomi bought two bunches of flowers that cost $7 per bunch. What change should she get from a $20 bill?

12. The bar graph shows, for 5 successive weeks, the weight of grass clippings Kayla collected to sell as mulch.

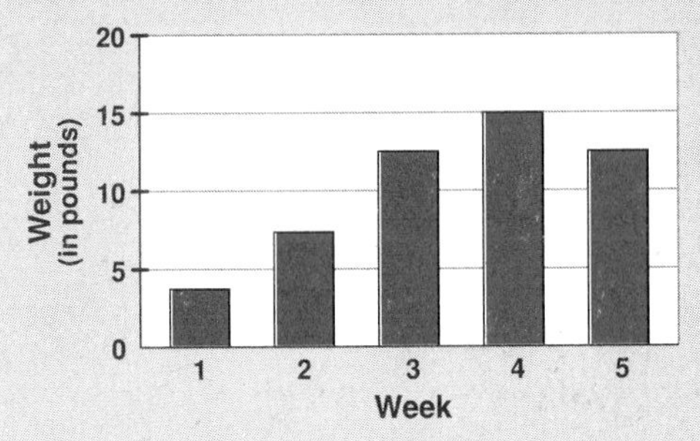

In which week was the amount collected half the amount in week 4?

13. Which group of fractions is arranged from smallest to largest?

(a) $\frac{1}{5}, \frac{1}{3}, \frac{1}{2}$ (b) $\frac{1}{2}, \frac{1}{5}, \frac{1}{3}$

(c) $\frac{1}{3}, \frac{1}{2}, \frac{1}{5}$ (d) $\frac{1}{2}, \frac{1}{3}, \frac{1}{5}$

14. Which number is a factor of 91?

(a) 3 (b) 7

(c) 9 (d) 11

15. Sy earns $500 per week. Here are the tax deductions from his paycheck:

Federal	$72.89
State	28.35
Social Security	38.25

What is the amount of Sy's take-home pay?

(a) $139.49 (b) $639.49

(c) $360.51 (d) $360

16. Which number is equivalent to $6\frac{3}{4}$?

(a) 6.34 (b) 0.675

(c) 6.75 (d) 6.8

17. Which fraction is equivalent to 80%?

(a) $\frac{1}{8}$ (b) $\frac{80}{100}$

(c) $\frac{8}{100}$ (d) $\frac{80}{1}$

18. Which unit can be used to measure the liquid in a container?

(a) gram (b) quart

(c) meter (d) inch

19. What number corresponds to the point P on the graph?

(a) (–3, 0) (b) (0, –3)

(c) (3, 0) (d) (0, 3)

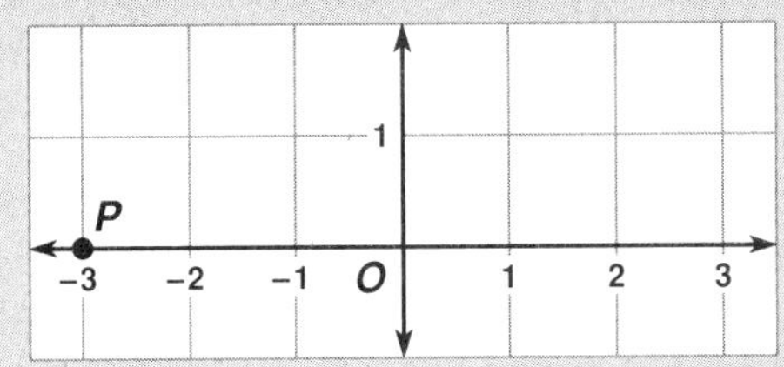

20. What is the value of $3x^2 + 4$ when $x = 4$?

(a) 148 (b) 28

(c) 52 (d) 60

SKILL REVIEW 6

1. Add:
$$\begin{array}{r} 3,784 \\ 672 \\ +\,4,205 \\ \hline \end{array}$$

2. Divide: $1,208 \div (-4)$

3. Simplify: $\frac{18}{24}$

4. Subtract 8,729 from 9,032.

5. Round 326,159 to the nearest thousand.

6. Multiply: 5.7×6.2

7.

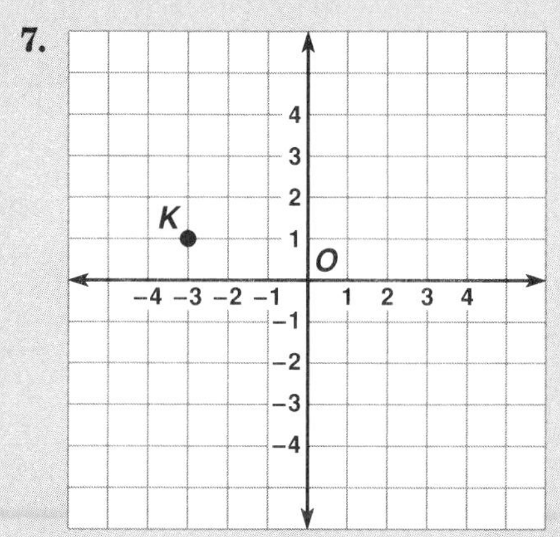

What are the coordinates of point K on the graph?

8. Find the value of $2a - 3b$ if $a = 3$ and $b = -2$.

9. Julie had 2 quarters and 6 dimes. What is the greatest number of 20-cent stamps she can buy?

10. What is the greatest common factor of 12, 20, and 24?

11. Find x if $5x - 3 = 17$.

12. Mrs. Castro paid \$47.53 for children's clothes. She returned two items that didn't fit, a shirt at \$13.47 and sandals costing \$24.50. What was the cost of the remaining clothes?

(a) \$9.56 (b) \$85.50 (c) \$10.53 (d) \$23.77

13. Which is a prime number?

(a) 15 (b) 27 (c) 29 (d) 35

14.

Which letter is at -0.5 on the number line?

(a) A (b) B (c) C (d) D

15. Lee used $\frac{3}{4}$ cup flour to make pancakes and $\frac{1}{2}$ cup flour in a cookie recipe. What was the total number of cups of flour used?

(a) 1 (b) $1\frac{1}{4}$ (c) $1\frac{3}{8}$ (d) $1\frac{1}{2}$

16. Which is the closest estimate of a 15% tip on a meal costing \$20.50?

(a) \$2 (b) \$2.50 (c) \$3 (d) \$4.50

17. Evaluate: $10 - 2 \times 3 + 1$

(a) 25 (b) 32 (c) 3 (d) 5

18. Which unit would be most appropriate for the contents of a package of cookies?

(a) grams (b) meters
(c) centimeters (d) liters

19. In a laboratory experiment, one test was performed at $-12°C$ and another test was performed at $23°C$. What was the difference in degrees between the two temperatures?

(a) 11° (b) −11° (c) 25° (d) 35°

20.

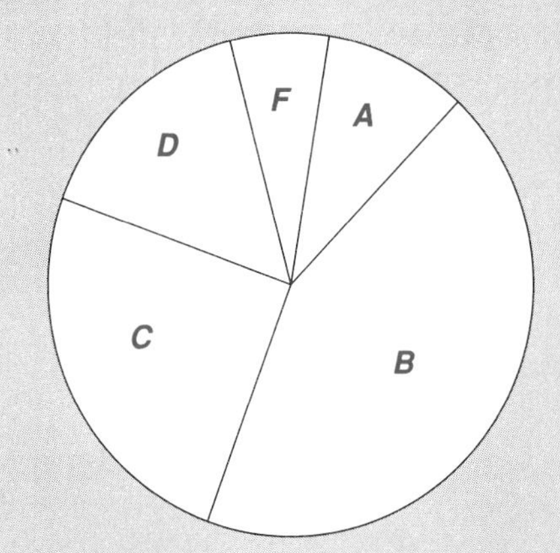

The circle graph shows the distribution of grades in Ms. Winkle's class. What grade was earned by about 25% of the students?

(a) A (b) B (c) C (d) D

SKILL REVIEW 7

1. Subtract: $\begin{array}{r} 503 \\ -392 \\ \hline \end{array}$

2. Add: $1\frac{1}{2} + 1\frac{1}{2}$

3. Multiply: $\frac{3}{7} \times \frac{7}{8}$

4. Reduce: $\frac{6}{20}$

5. Round to the nearest tenth: 561.521

6. Divide: $(-10x) \div (-5x)$

7. Divide: $\frac{5}{7} \div \frac{5}{14}$

8. What fractional part of the diagram is shaded?

9. Find the decimal equivalent of 92%.

10. Solve for x: $\frac{1}{2}x = 10$

11. Find the coordinates of point K:

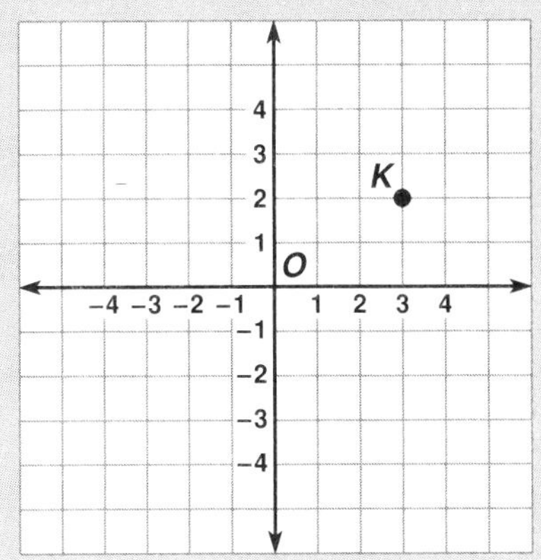

(a) (3, 2) (b) (3,4) (c) (2, 2) (d) (–3, –2)

12. Evalutate when $x = 2$: $2(x + 5)$

(a) 10 (b) 14 (c) 12 (d) 7

13. Which fraction is equivalent to $\frac{12}{15}$?

(a) $\frac{3}{5}$ (b) $\frac{4}{3}$ (c) $\frac{3}{4}$ (d) $\frac{4}{5}$

14. Which fraction is greater than $\frac{1}{2}$?

(a) $\frac{1}{3}$ (b) $\frac{3}{4}$ (c) $\frac{3}{10}$ (d) $\frac{3}{6}$

15. Out of a possible 50 points, 5 students received the homework scores shown. Who had the lowest score?

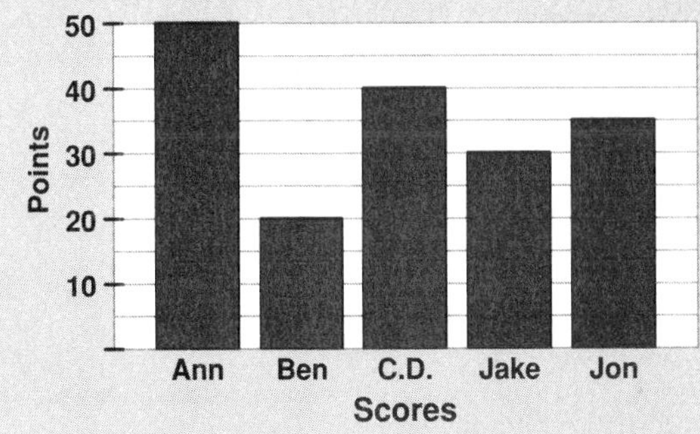

(a) Ann (b) Ben (c) Jake (d) Jon

16. Which number is a common factor of 12, 15, and 18?

(a) 2 (b) 5 (c) 6 (d) 3

17. Find 75% of 200.

(a) 150 (b) 75 (c) 100 (d) 15,000

18. Find a number smaller than –17.

(a) –16 (b) 0 (c) 20 (d) –20

19. Which graph represents $x < 2$?

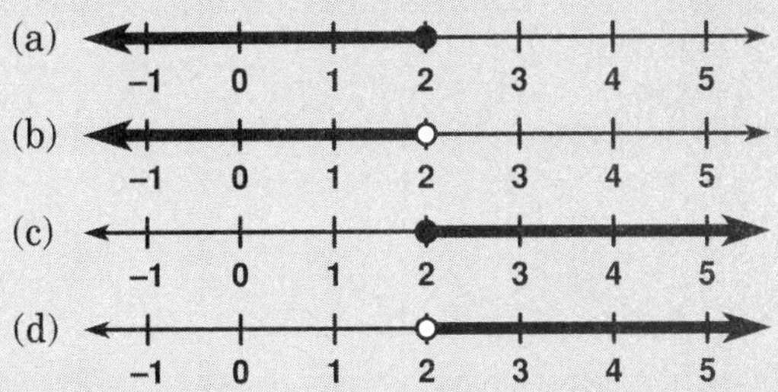

20. If three markers of equal value cost a total of \$2.97, which equation would be used to find the cost of one marker?

(a) $x + 3 = 2.97$ (b) $2.97x = 3$
(c) $3x = 2.97$ (d) $x = 3(2.97)$

SKILL REVIEW 8

1. Add: 3,521 + 677 + 45

$$\begin{array}{r} 3{,}521 \\ 677 \\ +\ \ 45 \\ \hline \end{array}$$

2. Subtract: $17.1 - 11.2$
3. Multiply: 302×70
4. Divide: $1.1\overline{)2.75}$
5. Reduce: $\frac{14}{21}$
6. Add: $\frac{2}{3} + \frac{1}{6}$
7. Find the perimeter of this rectangle:

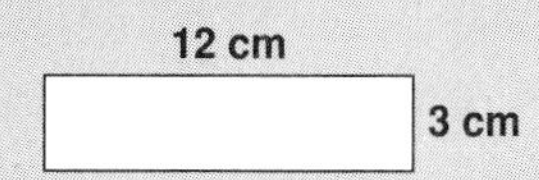

8. Write 2% as a decimal.
9. Evaluate: $3 + 5^2$
10. Solve for x: $8x + 1 = 9$
11. What is the distance between points A and B?

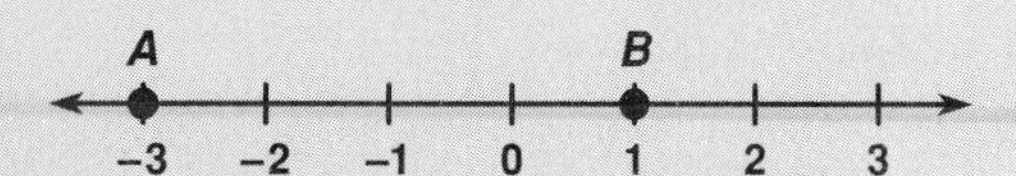

12. What is the product of –3 and 2?
13. What are the coordinates of point P?

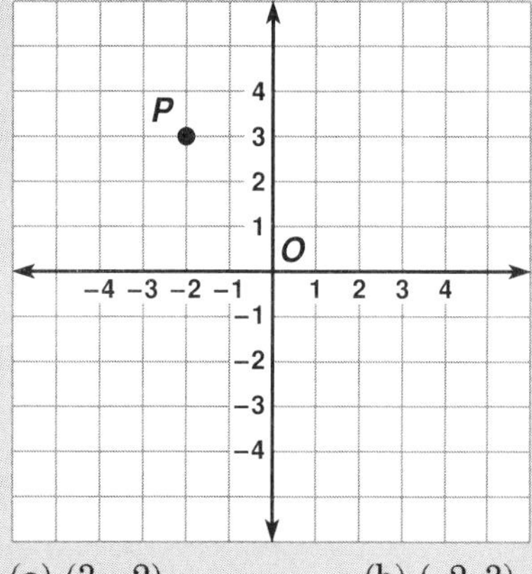

(a) (3, –2) (b) (–2, 3)
(c) (–3, 2) (d) (2, –3)

14. Which graph represents the inequality $x \leq 0$?

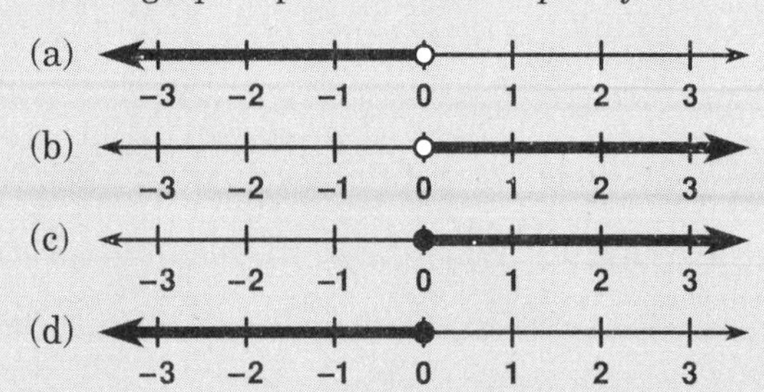

15. What value of x will make this open sentence true?

$$2x + 4 = -2$$

(a) $x = 1$ (b) $x = 3$
(c) $x = -1$ (d) $x = -3$

16. Evaluate: $(-6) - (-2)$

(a) –8 (b) –4
(c) 8 (d) 4

17. What value of x does *not* make this open sentence true?

$$x \geq -4$$

(a) –5 (b) –4
(c) –3 (d) 0

18. What is the measure of the complement of an angle of 40°?

(a) 140° (b) 40°
(c) 50° (d) 320°

19. If n represents a number, which expression represents 2 less than n?

(a) $2 - n$ (b) $2 \div n$
(c) $n - 2$ (d) $n \div 2$

20. Which expression has a value of 19?

(a) $5 + 4 \times 2 + 1$ (b) $7 - 1 \times 3 + 1$
(c) $10 - 2 \times 3 - 5$ (d) $5 + 7 \times 2$

SKILL REVIEW 9

1. Add: $50 + 17 + 219$

2. Round 24,115 to the nearest hundred.

3. Multiply: 507×23

4. The table shows the wins and losses of 6 soccer teams.

Team	Wins	Losses
Bulldogs	12	0
Terriers	10	2
Retrievers	6	6
Spaniels	5	7
Collies	2	10
Hounds	1	11

How many more games did the Terriers win than the Collies?

5. Add: $3.21 + 12.9$

6. Subtract: $\frac{4}{5} - \frac{1}{3}$

7. What is $\frac{2}{3}$ of 18?

8. Toni bought a pen for $1.80 plus 6% sales tax. What was the total cost?

9. Solve for x: $3x + 4 = 16$

10. Add: 12 and -9

11. What is the remainder when 1,232 is divided by 13?
    (a) 6 (b) 8 (c) 9 (d) 10

12. Juan correctly answered 17 questions out of 20. What percent of the questions did he answer correctly?
    (a) 17 (b) 85 (c) 15 (d) 3

13. Which number is between 2.4 and 2.5?
    (a) 2.502 (b) 2.452
    (c) 2.400 (d) 2.542

14. Ellen bought a coat that was reduced by 40%. If the original price was $120, what was the sale price?
    (a) $48 (b) $72 (c) $80 (d) $100

15. Which fraction has the greatest value?
    (a) $\frac{2}{3}$ (b) $\frac{5}{6}$ (c) $\frac{8}{9}$ (d) $\frac{11}{18}$

16. 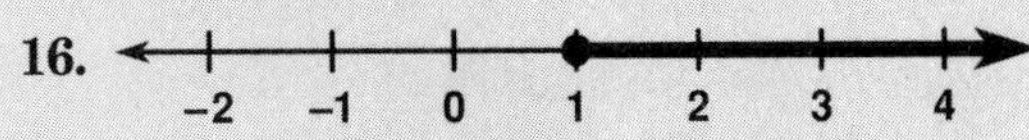

    Which open sentence is represented on the graph?
    (a) $x < 1$ (b) $x \leq 1$ (c) $x > 1$ (d) $x \geq 1$

17. If x represents a number, which expression represents 2 less than the number?
    (a) $2 - x$ (b) $x - 2$ (c) $2 < x$ (d) $2 > x$

18. Which line segment is parallel to $\overline{FC}$?

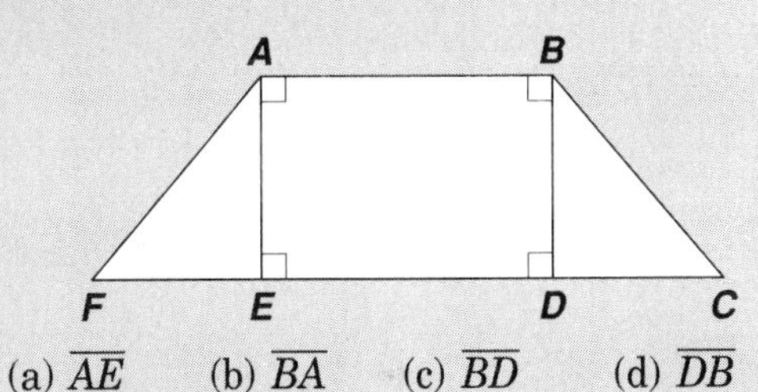

    (a) $\overline{AE}$ (b) $\overline{BA}$ (c) $\overline{BD}$ (d) $\overline{DB}$

19. What is the measure of the supplement of an angle of 30°?
    (a) 180° (b) 90° (c) 150° (d) 60°

20. Find the circumference of this circle.

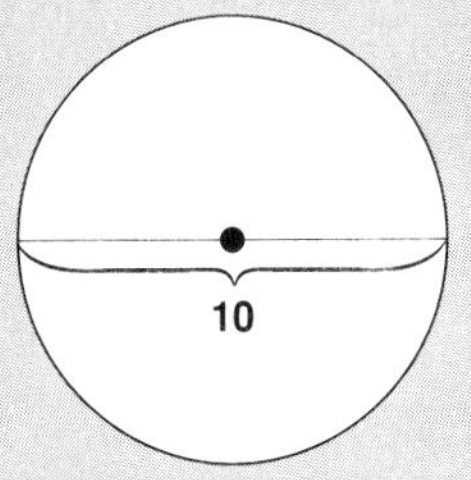

    (a) 20π (b) 100π (c) 25π (d) 10π

SKILL REVIEW 10

1. Write in numerals: two thousand forty

2. Subtract 21 from 5,000.

3. Find the product of 904 and 27.

4. Divide: $3.2\overline{)5.12}$

5. Reduce to lowest terms: $\frac{64}{128}$

6. Find the sum of –9 and 2.

7. Express in lowest terms: $\frac{7}{8} \div \frac{7}{4}$

8. On the ruler shown, which letter represents $1\frac{1}{4}$ inches?

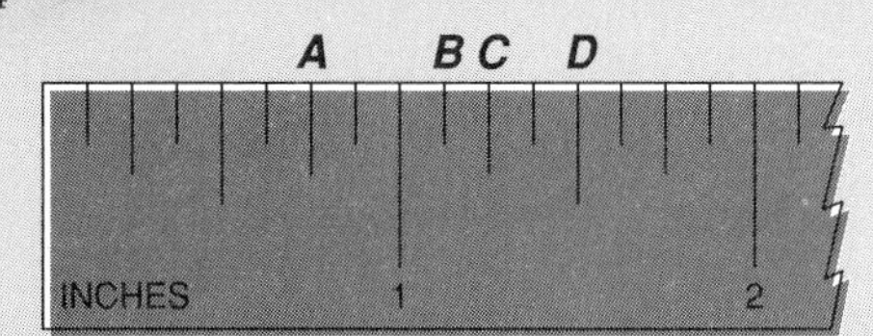

9. Combine like terms: $3 + 3x + 2x$

10. Solve for x: $2x - 7 = 5$

11. Solve for x: $\frac{x}{6} = \frac{4}{3}$

12. Which is a composite number?

(a) 2 (b) 12 (c) 13 (d) 17

13. Which number is equivalent to $7\frac{3}{4}$?

(a) 7.34 (b) 0.734 (c) 7.75 (d) 0.775

14. Which number is between –2.06 and –2.16?

(a) –2.05 (b) –2.17 (c) –2.0 (d) –2.1

15. Find the perimeter of this rectangle.

(a) 15 (b) 16 (c) 8 (d) 64

16. What number is 12% of 100?

(a) 0.12 (b) 1.2 (c) 12 (d) 120

17. On the graph, what is the number of units between points A and B?

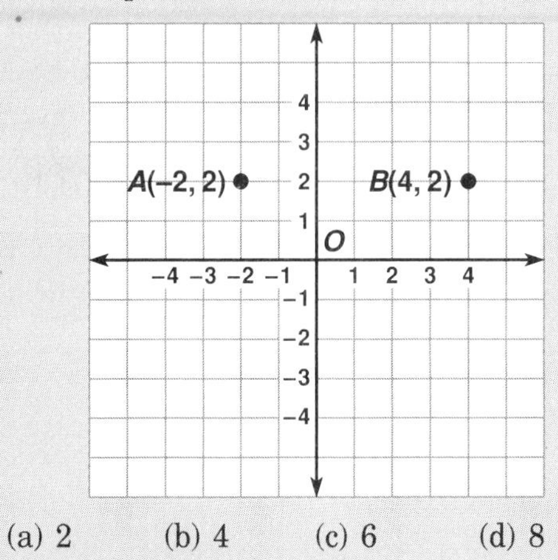

(a) 2 (b) 4 (c) 6 (d) 8

18. The value of $2 + 3^2$ is

(a) 25 (b) 11 (c) 10 (d) 8

19. Which graph represents $x \geq 2$?

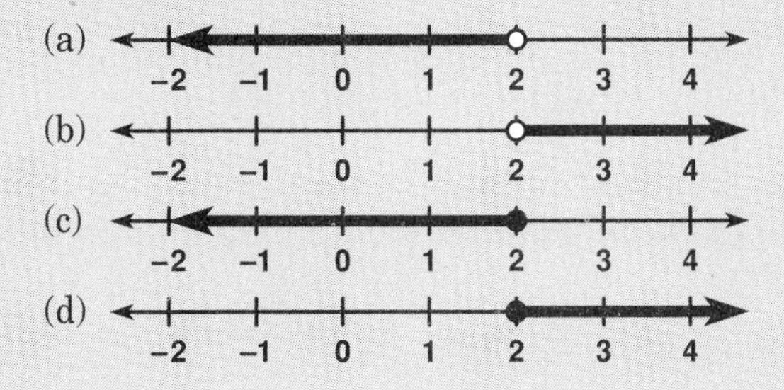

20. If 6 items of equal value cost a total of $2.16, which equation could you use to find the cost of one item?

(a) $x = 6(2.16)$

(b) $2.16x = 6$

(c) $6x = 2.16$

(d) $x + 6 = 2.16$

SKILL REVIEW 11

1. Add: 3,856 + 750

2. Subtract: 53.24 – 18.08

3. What is the value of 3^3?

4. Divide: $(-24) \div (-8)$

5. Evaluate: $20 - 10 - 2 \times 3$

6. Multiply: $\frac{1}{2} \times \frac{3}{4}$

7. What is the greatest number of 32-cent stamps Blair can buy with $2.00?

8. Solve for x: $4(x + 5) = 28$

9. Which point on the graph is located at (4, –2)?

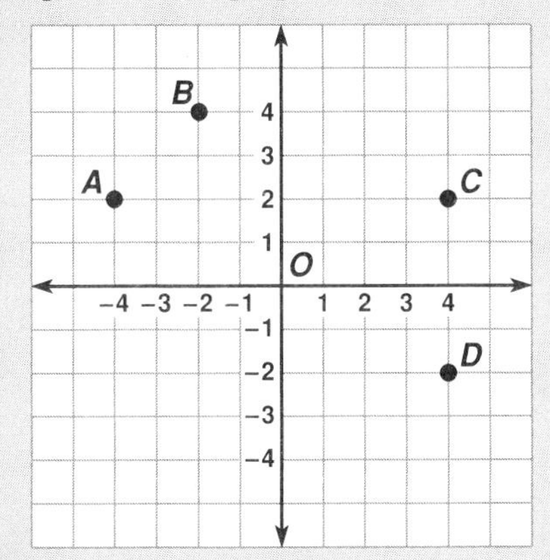

10. Find the area of a square whose perimeter is 12 feet.

11. Aaron's grades in his major subjects are 83, 78, 65, and 86. What is his mean (average) grade?

12. What is the greatest common factor of 36 and 90?
    (a) 3 (b) 9 (c) 18 (d) 3,240

13. Which fraction has the greatest value?
    (a) $\frac{2}{3}$ (b) $\frac{3}{5}$ (c) $\frac{3}{4}$ (d) $\frac{5}{8}$

14. A marble is taken at random from a jar containing 3 red marbles and 5 blue marbles. What is the probability that the marble taken is red?
    (a) $\frac{3}{8}$ (b) $\frac{3}{5}$ (c) $\frac{5}{8}$ (d) $\frac{1}{15}$

15. Each [bowl] in the following diagram represents 20 bowls of soup served in a diner.

How many bowls of soup were served?
    (a) $2\frac{1}{2}$ (b) 25
    (c) 40 (d) 50

16. A formula for changing Celsius (C) to Fahrenheit (F) temperature is:
$$F = \frac{9}{5}C + 32$$
What Fahrenheit temperature corresponds to 25°C?
    (a) 13F° (b) 25F°
    (c) 57F° (d) 77F°

17. In the proportion $\frac{6}{y} = \frac{9}{12}$, what is the value of y?
    (a) 7 (b) 8 (c) 9 (d) 10

18. If 16 out of 20 questions were answered correctly on a test, what percentage of the questions were answered correctly?
    (a) 16% (b) 4%
    (c) 64% (d) 80%

19. Which gives the best approximation of 10.8×12.4?
    (a) 10×12 (b) 11×13
    (c) 10×13 (d) 11×12

20. Which inequality is represented on the graph?

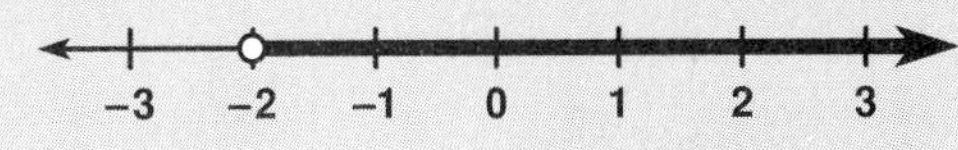

    (a) $x < -2$ (b) $x > -2$
    (c) $x \geq -2$ (d) $x > 2$

SKILL REVIEW 12

1. Multiply: $\begin{array}{r} 521 \\ \times\ 38 \\ \hline \end{array}$

2. Write the numeral for forty-four thousand, two hundred two.

3. Add: $17.1 + 0.22 + 5.8$

4. Divide: $0.5\overline{)15.05}$

5. Evaluate: 5^3

6. Subtract: $\frac{2}{5} - \frac{1}{3}$

7. Add: $(-3) + 2$

8. Write 15% as a fraction in lowest terms.

9. Find the perimeter of the triangle.

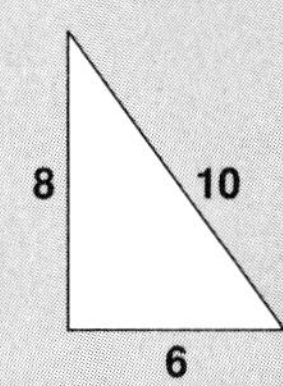

10. If one letter is chosen at random from the letters of the word MATHEMATICS, what is the probability that the letter will be an M?

11. What fractional part of the figure is shaded?

(a) $\frac{1}{4}$ (b) $\frac{3}{4}$
(c) $\frac{1}{2}$ (d) $\frac{1}{6}$

12. Which angle is acute?

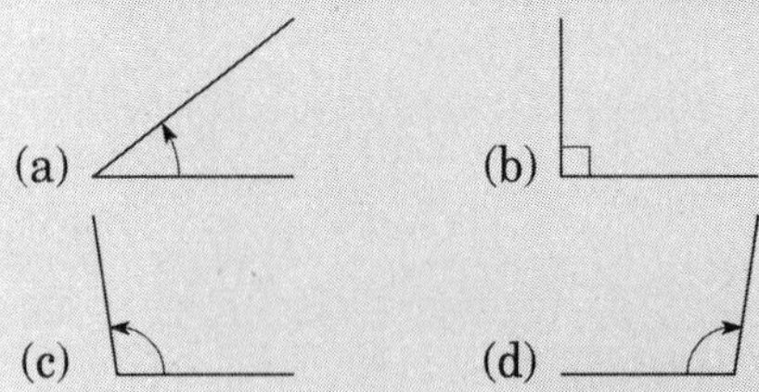

13. Solve for x: $10x + 4 = 44$
(a) 4.8 (b) 4.4
(c) 4.0 (d) 0.4

14. This weekend, Rick's Record Shop sold 250 discs and 150 tapes. What was the ratio of discs to tapes?
(a) 3 : 5 (b) 5 : 3
(c) 3 : 1 (d) 5 : 1

15. The prime factorization of 105 is
(a) 3×35 (b) 5×21
(c) 7×15 (d) $3 \times 5 \times 7$

16. Which of these sets of integers are in order from least to greatest?
(a) 0 –2 2 –4
(b) –2 –4 0 2
(c) –4 –2 0 2
(d) 2 0 –2 –4

17. On a map, 1 inch represents 150 miles. How many miles are represented by 2.5 inches?
(a) 250 (b) 350
(c) 375 (d) 475

18. Janell bought a sweater on sale at a 25% discount. If the original price was \$40, what was the sale price?
(a) \$10 (b) \$15
(c) \$30 (d) \$65

19. What is the mean of these 5 numbers?
1 4 5 7 7
(a) 7 (b) 5
(c) 4.8 (d) 4.0

20. Which inequality does this graph represent?

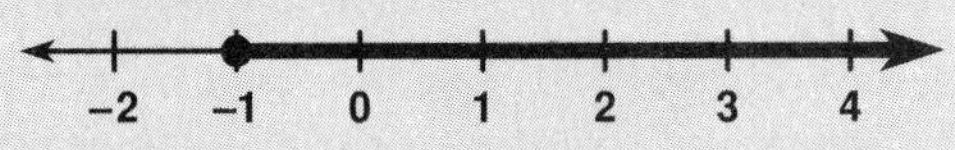

(a) $x > -1$ (b) $x \geq -1$
(c) $x < -1$ (d) $x \leq -1$

SKILL REVIEW 13

1. Subtract 47 from 1,043.

2. Divide: $24\overline{)2,568}$

3. Multiply 341 by 207.

4. Add: 325.008 + 32.58

5. Subtract: $7\frac{3}{8} - 2\frac{5}{8}$

6. 9 is 50% of what number?

7. This recipe gives the ingredients for 6 servings.

Rice Recipe
8-oz. envelope rice mix
2 cups water
$\frac{1}{2}$ teaspoon salt
3 tablespoons butter

How much water would be used for 12 servings?

8. Solve for x: $3x - 2 = 10$

9. Add: $-4 + (-2)$

10. When written as a decimal, 2.25% is equal to

(a) 2.25 (b) 0.225
(c) 0.0225 (d) 0.025

11. The total number of hours between 2 P.M. and 9 A.M. is

(a) 5 (b) 9 (c) 15 (d) 19

12. Marissa bought 5 book covers that were priced at 2 for $1. How much did she spend?

(a) $1.25 (b) $2.50
(c) $5.00 (d) $10.00

13.

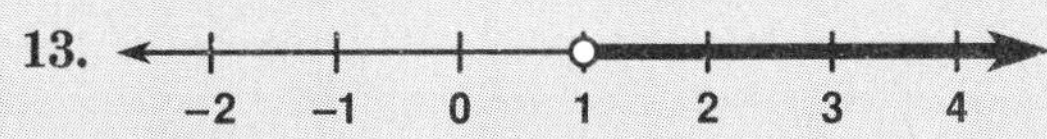

Which open sentence is represented by this graph?

(a) $x \leq 1$ (b) $x \geq 1$
(c) $x < 1$ (d) $x > 1$

14. Which expression is true?

(a) $0 > -4$ (b) $-4 > -1$
(c) $-4 > 0$ (d) $-1 > 0$

15. What is the value of x that makes this open sentence a proportion?

$$\frac{8}{12} = \frac{x}{9}$$

(a) 2 (b) 3
(c) 4 (d) 6

16. Evaluate: $17 - 2 \times 3$

(a) 45 (b) 11
(c) 14.7 (d) 1.1

17. How many square feet are in the area of the circle whose diameter measures 10 feet?

(a) 100π (b) 200π
(c) 25π (d) 50π

18. Data Set: 30 33 35 40 40

The median for this data set is

(a) 35 (b) 36
(c) 36.5 (d) 40

19. A committee has 3 boys and 5 girls. One member is chosen at random. What is the probability that a girl is chosen?

(a) 5 : 3 (b) 3 : 5
(c) 5 : 8 (d) 5 : 15

20. This triangle has the same number of square units of area as

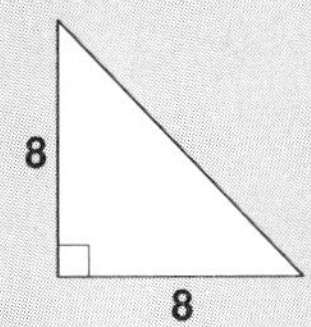

(a)
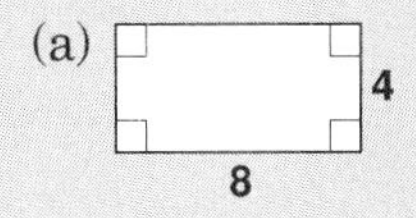

(b)
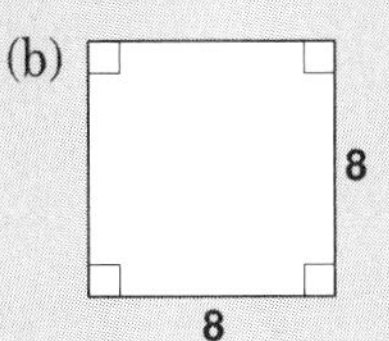

(c)
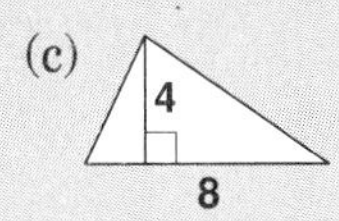

(d)
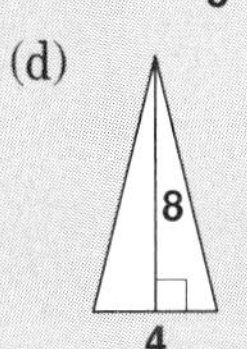

SKILL REVIEW 14

1. Add:

$$\begin{array}{r} 283 \\ 71 \\ +\ 8{,}609 \\ \hline \end{array}$$

2. Write the numeral for: thirty-two thousand forty

3. Bus tokens cost \$1.25 each. If Carlos has \$6.00, what is the greatest number of tokens he can buy?

4. Subtract 2,963 from 4,000.

5. Divide: $1{,}208 \div 4$

6. What is the median of these numbers?
   27, 19, 31, 22, 30

7. On five tests, Kay's grades were 88, 74, 83, 90, and 80. What is the mean (average) of her grades?

8. This graph shows the amounts of rainfall for the months of April, May, June, and July. How much more rain fell in April than in July?

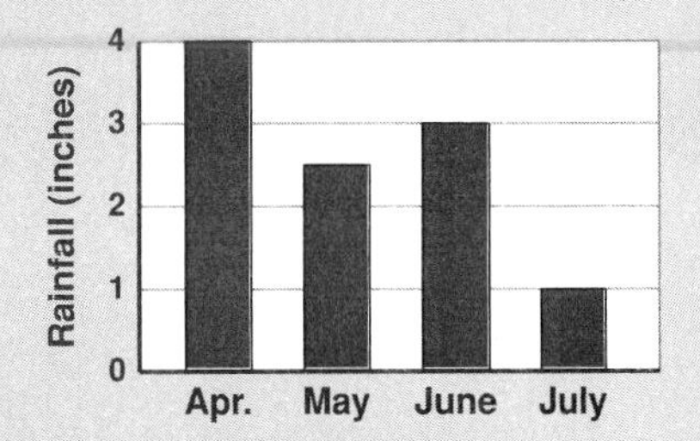

9. Reduce $\frac{72}{84}$ to lowest terms.

10. If $x = 2$ and $y = 3$, what is the value of $x + 2y$?

11. What is the total number of square centimeters in the area of this triangle?

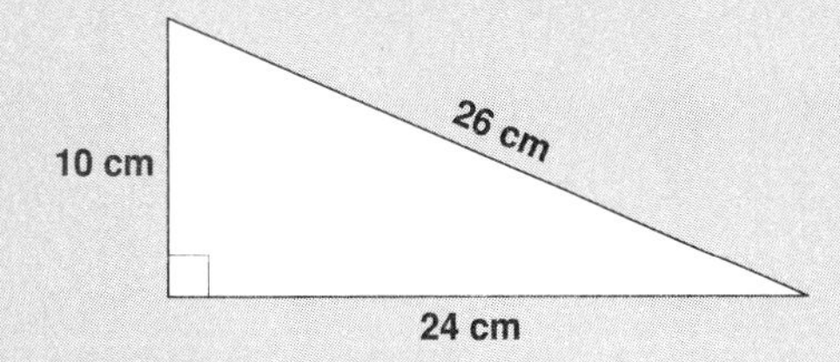

12. What is the sum of +16 and –7?

13. Solve for x: $2(x + 3) = 24$

14. Round 32.749 to the nearest tenth.

15. A bag contains 3 red chips, 4 white chips, and 6 blue chips. If Myra picks one chip at random from the bag, what is the probability she will choose a white chip?
   (a) $\frac{1}{4}$ (b) $\frac{3}{13}$
   (c) $\frac{4}{13}$ (d) $\frac{6}{13}$

16. Which number is equal to 3^2?
   (a) 5 (b) 6
   (c) 9 (d) 32

17. Narda is making a down payment of \$25 on her purchase of a \$275 TV set. If she pays the balance in monthly payments of \$25, how many months will it take her?
   (a) 12 (b) 11
   (c) 10 (d) 9

18. Which integer has the greatest value?
   (a) –17 (b) –15
   (c) –12 (d) –1

19. The prime factorization of 60 is
   (a) $2 \times 3 \times 10$ (b) $1 \times 3 \times 20$
   (c) $2 \times 3 \times 5$ (d) $2 \times 2 \times 3 \times 5$

20. Which of these points is not on the graph of the line below?
   (a) (–1, 1) (b) (0, –2)
   (c) (2, 4) (d) (–2, 0)

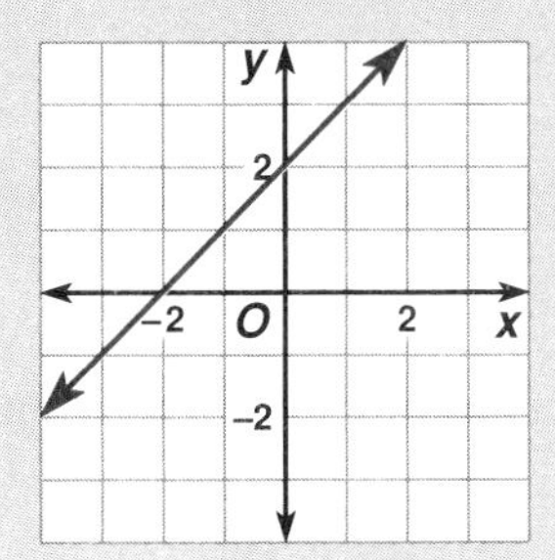

Index

A

B

C

D

E

F

G

H

I

N

O

P

Q

R

S

T

U

V

W

X

Y

Z

Photo Credits

Page	*Credit*
1	FPG
33	C. Mead / FPG
57	Mike Valeri / FPG
67	Chuk Muhlstock / FPG
72	Hurrgiimalla
81	Camera Hawaii / FPG
96	Mike Valeri / FPG
97	Michael S. Thompson / Comstock
121	John Elk / Stock Boston
124	Jonathan A. Meyers / FPG
126	Courtesy of Schindler Elevator Corp.
130	Fritz Photography /FPG
139	Courtesy of Concept II, Inc.
148	Courtesy of Empire State Building, managed by Helmsley-Spear, Inc.
154	Photo from European / FPG
163	The News World / FPG
168	Adele Hodge / FPG
174	FPG
196	Courtesy of World Book Educational Products
198	H. Armstrong Roberts
202	Ellis Herwig / Stock Boston
206	Spencer Grant / FPG
210	Mike Valeri / FPG
219	Georg Gerster / Comstock
219	Richard Laird / FPG
224	Gerard Fritz / FPG
227	Peter Menzel / Stock Boston
231	Courtesy of Human Resource Center Albertson, NY
236	Courtesy of WoodPlay, Inc.
258	Jeff Isaac Greenberg / Photo Researchers, Inc.
260	Spencer Grant / FPG
263	Hans Jesse / FPG
267	Mike Valeri / FPG
271	Jeffry W. Myers / FPG
271	Courtesy of KRANSCO
295	Gaetano Barone / FPG
298	Lester Tinker / FPG
319	Courtesy of Photocraft & Schwinn

Page	*Credit*
362	Kerwin B. Roche / FPG
362	Courtesy of Guide Dog Foundation for the Blind, Inc.
387	H. Armstrong Roberts
387	Ron Rovtar / FPG
388	Courtesy of Pella / Rolscreen Co.
407	Paul DePaola / FPG
408	George Bellerose / Stock Boston
416	Courtesy of Medieval Times Dinner & Tournament
425	Jeffry W. Myers / FPG
428	Kaya Hoffman
431	Jeffry W. Myers / FPG
438	Joe Crachiola / FPG
441	Grant Leduc / Stock Boston
443	Jam Photography / FPG
453	H. Armstrong Roberts
457	Courtesy of The Goodyear Tire & Rubber Co.
459	Len Rue, Jr. / FPG
521	H. Armstrong Roberts
525	Irene Vandermolen / FPG
526	Mike Valeri / FPG
531	Georg Gerster / Comstock
539	Court Mast / FPG
540	Jeffrey Sylvester / FPG
545	Barbara Sutro / FPG
546	Barbara Alper / Stock Boston
548	Maurice Fitzgerald / FPG
553	Will Burgess / Reuters / Bettmann
554	Elizabeth Hamlin / Stock Boston
555	Don Knight / FPG
569	Blumebilo / FPG
572	Dick Luria / FPG
574	Paul F. Occhiogrosso
580	Jeffrey Sylvester / FPG
581	Mike Valeri / FPG
591	H. Armstrong Roberts
592	Charles Schneider / FPG
601	Robert Houser / Comstock
651	Ron Rovtar / FPG